J	Joule
kD	Kilodalton
kJ	Kilojoule
K_M	Michaelis constant
LDL	Low-density lipoprotein
Mb	Myoglobin
NAD^+	Nicotinamide adenine dinucleotide (oxidized form)
NADH	Nicotinamide adenine dinucleotide (reduced form)
$NADP^+$	Nicotinamide adenine dinucleotide phosphate (oxidized form)
NADPH	Nicotinamide adenine dinucleotide phosphate (reduced form)
P_i	Phosphate ion
PAGE	Polyacrylamide gel electrophoresis
PCR	Polymerase chain reaction
PEP	Phosphoenolpyruvate
PKU	Phenylketonuria
PLP	Pyridoxal phosphate
PP_i	Pyrophosphate ion
PRPP	Phosphoribosylpyrophosphate
PS	Photosystem
RF	Release factor
RFLPs	Restriction-fragment-length polymorphisms
RNA	Ribonucleic acid
RNase	Ribonuclease
mRNA	Messenger RNA
rRNA	Ribosomal RNA
tRNA	Transfer RNA
snRNP	Small nuclear ribonucleoprotein
S	Svedberg unit
SDS	Sodium dodecylsulfate
SSB	Single-strand binding protein
SV40	Simian virus 40
T	Thymine
TDP	Thymidine diphosphate
TMP	Thymidine monophosphate
TTP	Thymidine triphosphate
U	Uracil
UDP	Uridine diphosphate
UMP	Uridine monophosphate
UTP	Uridine triphosphate
V_{max}	Maximal velocity
VLDL	Very low density lipoprotein

Concepts in Biochemistry

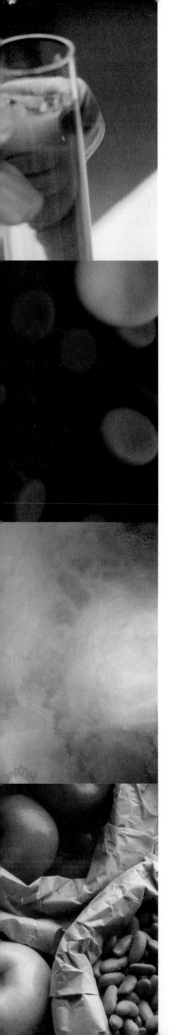

Concepts in Biochemistry

second edition

Rodney Boyer

Hope College

JOHN WILEY & SONS, INC.

New York • Chichester • Weinheim • Brisbane • Singapore • Toronto

Library of Congress Cataloging-in-Publication Data
Boyer, Rodney F.
    Concepts in biochemistry / Rodney Boyer—2nd ed.
        p.   cm.
    Includes bibliographical references and index.
    1. Biochemistry.   I. Title

QD415.B69 2001   572—dc21   2001044095
ISBN 0470-00379-0

Printed in Canada

10   9   8   7   6   5   4   3

Dedicated to Christel

P R E F A C E

To the Student

Biochemistry, the study of the molecules of life, was for much of its existence just a part of the biological and chemical sciences. Today, however, biochemistry has evolved and matured into a major discipline. This is quite evident by observing the large sizes and level of activities of Biochemistry Departments on every university campus and even at many smaller colleges. But, there are other, more important ways to measure the significance of biochemistry. Because of the creativity and determination of biochemists and other scientists, our understanding of the molecular nature of life is growing at an incredible rate. The tools of biochemistry have been used to explain biological processes such as the origin of life, cell development and differentiation, metabolism and energy, the origins and cures for diseases, and even human behavior. The principles of biochemistry are now reaching into chemistry, the health sciences, nutrition, agriculture, physiology, immunology, neuroscience, cell biology, bioengineering, agriculture, ecology, computer science, and psychology. Not only is biochemistry expanding; other disciplines are using the tools of biochemistry to solve their unique problems.

You are enrolled in this course probably because the academic preparation for your chosen career is among the fields listed above and requires knowledge in biochemistry. Most colleges and universities offer a one-semester, survey biochemistry course for students who require an introduction but have neither the time nor the more extensive background needed for an intermediate level, full-year course that is usually designed for biochemistry majors. This book was planned especially for the brief, survey course.

My goals in writing this book were:

- to offer students, not majoring in the field, a view of modern biochemistry by identifying the most important biochemical concepts and theories, and explaining them at a level that is appropriate to the students' backgrounds and career goals.
- to develop a lively, even conversational tone, but strive for an accurate and clear presentation.
- to present numerous biochemical applications that students can identify in their own career choices and in their everyday lives.

I believe that these goals are accomplished and enhanced in the second edition with the addition of the following educational tools:

- A CD-ROM entitled *Interactive Concepts in Biochemistry,* which contains 55 multimedia units divided in the categories Interactive Animations, Concept Review, Cutting-Edge Biochemistry, and Structure Tutorials, is available with each book. Studying these modules, which are coordinated with the chapters and referenced in the text with a CD icon in the margin (as shown here), will help you readily grasp even the most difficult biochemical concepts.

- 40 special-interest boxes, called Windows on Biochemistry, focus on health-related issues, laboratory techniques, biochemical applications, and other enrichment topics.
- Numerous end-of-chapter Study Exercises, divided into five categories—Understanding Terms, Reviewing Concepts, Working Problems, Writing Biochemistry, and Applying Biochemistry—give you extra opportunities not only to learn biochemical principles but also to apply them to the solution of real-life problems. The questions in the ⬤Applying Biochemistry category test your understanding of the material presented in the CD-ROM.

My hope is that when you complete this course, you agree that the above goals have been achieved and that using this book has enhanced your understanding and appreciation of biochemistry. In addition, I hope that your newly gained knowledge will encourage you to continue as a life-long student of biochemistry.

To the Instructor

The field of biochemistry has, in my opinion, undergone a major change of focus. Earlier biochemistry studies were based primarily on the structure and function of proteins with emphasis on the biological processes of metabolism and bioenergetics. Topics that were considered "other" (molecular biology, molecular genetics, cell biology, etc.) were hidden in the backs of biochemistry textbooks and seldom covered in classes. Experimental design in biochemical research now tends to place increased emphasis on the nucleic acids. Genetic approaches are routinely used to study all biological processes including the traditional core topics of biochemistry: structure, metabolism, and bioenergetics.

The modern blending of biochemistry and molecular biology has brought us to a very special time in the molecular life sciences. Biochemists and other life scientists have recently uncovered the molecular mechanisms of many fundamental biological processes especially in the areas of immunology, DNA structure and function, RNA structure and function, cell differentiation and development, and signal transduction. And in the year 2001, we have witnessed the unlocking of a new treasure chest of information that will impact biochemical endeavors into the foreseeable future and beyond. We now know the sequence of the human genome. Completion of this cooperative work (called the Human Genome Project), by Celera Genomics Corp. and the National Human Genome Research Institute, will not just influence the course of future biochemistry, it will lead the way. It will also expand research in other fields, especially the chemical sciences, computer science, physics, and the pharmaceutical sciences. This exciting breakthrough in biology formed the background and motivation for the second edition of *Concepts in Biochemistry*. The new edition has been updated throughout to describe the influence of the Human Genome Project on biochemistry.

A Balanced Approach with a Modern Perspective

Most colleges and universities offer a one-semester, introductory biochemistry course for which this book is intended. Brief, survey courses usually follow the same content order as the one-year, majors' level, biochemistry courses except that there is much less detail and theory, and more focus on applications. Students usually take the course during their sophomore or junior years after courses in introductory chemistry, organic chemistry, and introductory biology. This text presents an appropriate balance of theory and applications, a modern view of the new directions of biochemistry, and the proper mixture of chemistry and biology that should capture the interests of the diverse groups of students that usually make up these classes. The text is also appropriate for use in one-semester courses developed as a result of the new American Chemical Society requirements for 3-credit hours of biochemistry coursework for certified chemistry majors.

Integration of Theory, Concepts, and Applications

This book is divided into four parts:

  I. Molecules and Life
 II. Dynamic Function of Biomolecules
III. Storage and Transfer of Biological Information
IV. Metabolism and Energy

Part I sets the stage for future study by introducing organisms and cell structure and defining the environment in which biomolecules interact. Chapter 2 presents an early introduction to cell communication—the roles of DNA and RNA, and the processes of signal transduction.

Part II initiates a discussion of the conformational side of biochemistry and focuses on the structures and dynamic functions of proteins, carbohydrates, and lipids. This Part has been enhanced in the second edition with the presentation of proteomics, bioinformatics, and new material on protein structure and function.

Part III defines the informational side of biochemistry: the nucleic acids and their roles in the storage and transfer of biological information. These topics have traditionally appeared as the final chapters in biochemistry textbooks; however, in the first edition of *Concepts of Biochemistry,* they were brought forward to reinforce the idea that the latter topics of metabolism and bioenergetics are shaped by genetic information. Responses from students and instructors have been very positive about the effectiveness of this nontraditional order; hence, it is retained in the second edition. Because most students have some understanding and a natural fascination for molecular biology, molecular genetics, biotechnology, and related advances in health, their interests are sparked by presenting the topics in this new order. This modern, student-friendly organization sets the book apart from the competition because the early placement of nucleic acids enhances the traditional coverage of protein structure and function, and metabolism. As an example, it is now possible to present metabolism in a more contemporary manner, emphasizing gene regulation and integration.

Part IV focuses on bioenergetics and metabolism. The chemical reactions that result in degradation and synthesis are described along with details on regulation. A major stumbling block for students in the study of metabolism is the chemical understanding of the reactions. Early in Part IV, the general types of biochemical reactions are outlined and their chemistries are described in detail. The major change in Part IV of the new edition is the addition of Chapter 20, Integration, Coordination, and Specialization in Metabolism. In this chapter, the larger picture of how metabolic pathways in a whole organism are interlinked and coordinated is explored with emphasis on the biochemical responsibilities of each major organ. Chapter 20 concludes with a special section exploring the biochemistry of exercise.

The book does not change the traditional topics covered in current biochemistry courses, only their order and balance. These modifications, which are driven by new directions in biochemistry, will allow for valuable pedagogical changes in the use of class time. A syllabus outlining the scheduling of topics for a typical semester is available from the author.

New Pedagogical Aids in the Second Edition

Several improvements are introduced in the second edition to enhance student and instructor use:

- **Interactive Concepts in Biochemistry CD-ROM**—This interactive CD-ROM/Web site contains 55 multimedia units divided in the categories Interactive Animations, Concept Review, Cutting-Edge Biochemistry, and Structure Tutorials, that are coordinated with chapters and are referenced in the text. A *Multimedia Instructor Handbook* is available to accompany the CD-ROM.

 CD-ROM Reference

- **Windows on Biochemistry**—40 "boxes" are integrated into the text that present a special focus on topics such as biochemical applications, health-related issues, laboratory techniques, and other enrichment topics beyond the scope of the standard text material.

- **Study Exercises**—End-of-chapter problems for student study are increased to an average of 35 per chapter. These exercises are divided into five types—Understanding Terms, Reviewing Concepts, Working Problems, Writing Biochemistry, and Applying Biochemistry. Answers to most exercises are available in the Appendix. The questions in the ⊙Applying Biochemistry category require students to read and study the multimedia units in the CD-ROM, thus acquiring enhanced skills in solving practical problems.

Ancillaries for the Second Edition

- **Transparency Acetates**—More than 100 four-color figures, photographs, and tables from the text help you illustrate lectures.

- **Brooks/Cole Chemistry Resource Center Web Site**—This feature offers various student resources such as a Chemistry Time Line, a Molecule Guide, a link to Chemical Education Resources, and more.
- **Student Manual and Test Items**—This includes multiple-choice test items with answers. This is also available in electronic format.
- *ExamView* **Computerized Testing**—Electronic testing that allows for custom exams and tracking student progress.
- **Multimedia Presentation Manager**—This easy-to-use tool allows you to quickly assemble art and database files for insertion into your lectures.

Acknowledgments

Writing and publishing a textbook require the assistance of individuals who are creative, talented, and hard-working. All of these qualities were present in the individuals assembled to produce this book. The project was initiated by Jennifer Huber, Sponsoring Editor at Brooks/Cole. Other individuals at Brooks/Cole who assisted at various stages were Marcus Boggs, Sponsoring Editor; Tom Novack, Production Editor; Pat Quest, Photo Researcher; Tom Ziolkowski, Marketing; Laura Hubrich, Marketing; Joohee Lee, Permissions Editor; and Sue Ewing, Permissions Specialist. Emily Levitan, Editorial Assistant, was instrumental in coordinating the many manuscript drafts and organizing the reviews. I am especially indebted to Marcia Craig of Graphic World Publishing Services for her work as Production Service Editor. I thank Marcia especially for her continuous vigilance over the project and for always giving advance notice of the editing and proofreading schedules.

The preparation of the Second Edition of *Concepts in Biochemistry* was dependent on the many reviewers who, with full teaching and research schedules, still found time to critique yet another chapter of the manuscript. My reviewers, listed below, indeed are knowledgeable scientists and dedicated teachers:

Robert C. Bateman, Jr., University of Southern Mississippi
Donald C. Beitz, Iowa State University
William E. Boernke, Nebraska Wesleyan University
Bill Bordeaux, Huntington College
John Boyle, Mississippi State University
Todd A. Carlson, Grand Valley State University
Do Ren Chang, Averett College
Randolph A. Coleman, College of William & Mary
Shawn O. Farrell, Colorado State University
Anthony R. Jude, University of Arkansas at Little Rock
Reef (Philip D., II) Morse, Illinois State University
Marilee Benore Parsons, University of Michigan–Dearborn
Scott Pattison, Ball State University
Richard Posner, Northern Arizona University
Michael S. Stekoll, University of Alaska–Southeast
Dan M. Sullivan, University of Nebraska
Malcolm Watford, Rutgers University

I also thank my wife, Christel, for her encouragement and for her assistance in many of the tasks associated with writing. Finally, I wish to report that our blue-point Himalayan, Mausi, my constant writing companion, has now been napping under the desk lamp for 15 years.

I encourage all users of this text to send comments that will assist in the preparation of future editions.

Rodney Boyer
boyer@hope.edu
or
boyerbiochem@aol.com

About the Author

Rod Boyer is on the faculty at Hope College where he has taught, researched, and written biochemistry for 26 years. Born and reared in Nebraska, he earned his B.A. in chemistry and mathematics at Westmar College (Iowa) and his Ph.D. in physical organic chemistry at Colorado State University. After three years as an NIH Postdoctoral Fellow with M.J. Coon in the Department of Biological Chemistry at the University of Michigan Medical School, he joined the chemistry faculty at Hope. There he has directed the work of over 75 undergraduate students in research supported by the NIH, NSF, Dreyfus Foundation, Howard Hughes Medical Institute, and the Petroleum Research Fund (ACS). With his students, he has published numerous papers in the areas of ferritin iron storage and biochemical education. He spent a sabbatical year as an American Cancer Society Scholar in the lab of Nobel laureate Tom Cech at the University of Colorado, Boulder. Rod is also the author of *Modern Experimental Biochemistry* (Third Edition, 2000, Benjamin/Cummings) and serves on the Editorial Board for the journal *Biochemistry and Molecular Biology Education (BAMBEd)*. He is a member of the American Society for Biochemistry and Molecular Biology (ASBMB) and the Education and Professional Development Committee that recently designed the undergraduate biochemistry degree recommended by the ASBMB. In his spare time, Rod enjoys nature, classical music, and reading.

BRIEF CONTENTS

C O N T E N T S

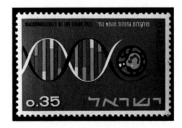

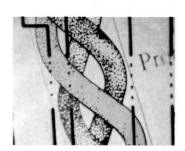

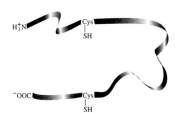

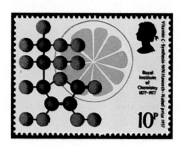
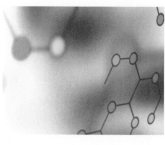

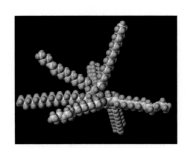

Part III
Storage and Transfer of Biological Information   241

Chapter 10 DNA and RNA: Structure and Function   243

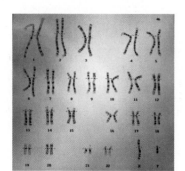

Chapter 11 DNA Replication and Transcription: Biosynthesis of DNA and RNA 271

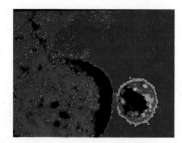

Chapter 12 Translation of RNA: The Genetic Code and Protein Metabolism 305

Part IV
Metabolism and Energy 363

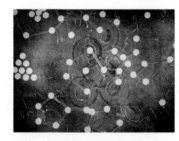

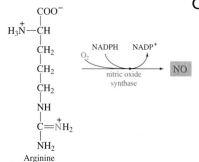

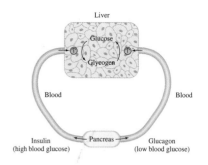

Part I

Molecules and Life

Chapter 1

Biochemistry:
Setting the Stage

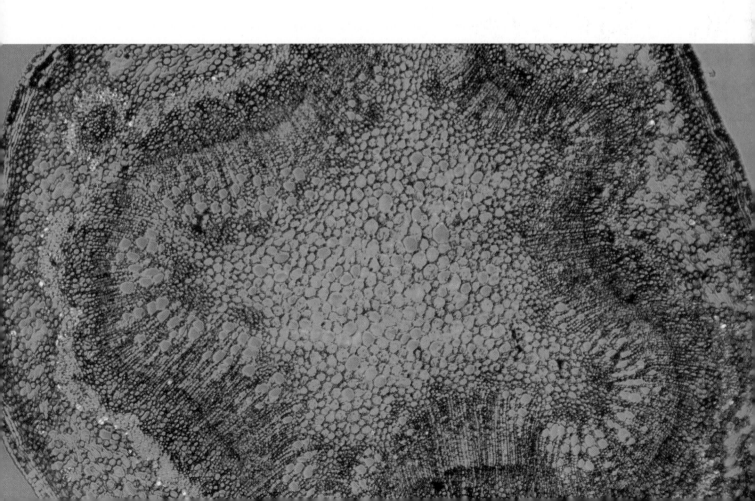

Welcome to biochemistry! This subject may not be entirely new because you have gained some biochemical understanding in previous experiences in biology and chemistry classes. Here you will encounter many more details of the complex, but always exciting, concepts in biochemistry. Even though the word *biochemistry* has become commonplace in our language, a precise definition is difficult. The simplest definition is "the chemistry of the living cell." As the word implies, biochemistry has components of biology and chemistry, and biochemists must be well-versed in both subjects. The overall goal of biochemistry is to describe life's processes using the language of molecules. Even the smallest living cell contains thousands of organic and inorganic chemicals, many of them large molecules called **macromolecules.** All biological processes—including vision, thinking, motion, digestion, immunity, and disease conditions—result from the actions of molecules large and small. Therefore, to describe biological processes, one must first have a knowledge of the chemical structures of the participating biomolecules. Second, one must have an understanding of the biological function or purpose of the cellular molecules. It is also important to understand the details of cell structure because many biological processes are compartmentalized; that is, they occur only in certain parts of the cell (often in vesicles or organelles surrounded by membranes). The respective roles of chemistry and biology in achieving the goals of biochemistry are readily apparent.

In addition to molecular structure and function and the relationships between these characteristics, biochemists also think about **bioenergetics**—the study of energy flow in living cells (Figure 1.1). Some molecular events of cells and organisms such as muscle contraction, synthesis of molecules, and membrane transport require the input of energy (are **endergonic**), whereas others, like metabolic degradation of food and absorption of light from the sun, release energy (are **exergonic**). How cells use biochemical reactions and biomolecules to transfer energy from exergonic to endergonic events will be pivotal in our understanding of the processes of life.

Other views of biochemistry result from dividing the discipline into two perspectives:

1. *Conformational:* Discovering the chemical structures and three-dimensional arrangements of biomolecules. This defines the chemical side of biochemistry.
2. *Informational:* Defining a language for communications inside cells and organisms. These processes represent the biological side of biochemistry.

All cells of a particular type have the same mixture of molecules. The unique molecular properties of identical cells are determined by the genes that are expressed by those cells. Genetic information in the form of DNA is decoded to yield proteins that are the major structural and functional molecules of cells. The flow of information from DNA will be important in our discussions of cellular communication and molecular recognition, and it describes how this book is organized.

Understanding biochemistry is not just important for those who will become biochemists and use the concepts daily in their work. Many aspects of everyday life are related directly to the subject matter of biochemistry. Here are some of the ways that biochemistry enriches our lives:

1. Biochemical studies lead us to a fundamental understanding of life. All of us have a natural curiosity about how our bodies work. How does a brain cell store mathematical or

Figure 1.1

The nature of biochemistry showing the important concepts and their relationships.

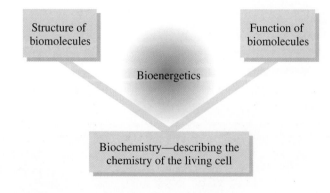

A running cheetah uses energy derived from metabolism of carbohydrates, fats, and proteins.

© *Bruce Coleman/Bruce Coleman Limited*

chemical formulas? What are the biochemical similarities and differences among the many forms of life? How do organisms store and transfer information necessary to reproduce themselves? What primary molecules and processes were involved in the origin of life? How is food digested to provide cellular energy? Research in biochemistry is providing answers to these questions.

2. Biochemistry has a profound influence on our understanding of medicine, health, nutrition, and the environment. Results from biochemical studies have already led to a molecular understanding of diseases such as diabetes, sickle cell anemia, phenylketonuria, cystic fibrosis, and hypercholesterolemia. Future targets for biochemical and biomedical studies include AIDS, cancer, Alzheimer's disease, and schizophrenia. Recombinant DNA technology and its ability to probe chromosomal regions for genetic mutations will play a major role in the diagnosis and treatment of these disease conditions. Recombinant DNA will also aid in the design of new plants for agricultural purposes, which should solve some of the world's food and nutrition problems. The study of enzymes (biological catalysts) and metabolism provides a foundation for the rational design of new drugs and detailed understanding of nutrition.

3. Biotechnology, the application of biological cells, cell components, and biological processes to technically useful operations, will also advance from biochemical discoveries. Already enzymes are used in the pharmaceutical industry to synthesize complex drugs. Various strains of microorganisms have been selected and altered for producing fuel alcohol from corn and other plant material, for cleaning up oil spills and toxic waste dumps, and for mining metals from natural ores.

💿 Cutting Edge: Hot careers

1.1 The Roots of Modern Biochemistry

Early History of Biochemistry

People of early civilizations in Egypt, China, India, Rome, and elsewhere did not understand the biochemical principles underlying the baking of leavened bread, the fermentation of fruit juices, or the treatment of maladies with plant and animal materials. However, the lack of knowledge did not prevent their enjoyment of the results of these biochemical processes. Early studies in biology, which concentrated on the treatment of illness and the attainment of good health, were firmly rooted in and combined with philosophy and religion.

The Chinese in the fourth century B.C. believed that humans contained five elements: water, fire, wood, metal, and earth. When all elements were present in proper balance, good

Biochemists study the molecular characteristics of living organisms on land and in water.

© NASA

health resulted. An imbalance in the elements caused illness. Chinese physicians discovered in the seventh century A.D. that night blindness could be treated with pig and sheep livers. Modern biochemists and physicians know that night blindness is caused by a deficiency of vitamin A, a biochemical abundant in liver.

The early Greeks, including Plato (428–348 B.C.), attempted to explain the body in terms of cosmological theories and stressed diet for treatment of disease. The Greek term for digestion, *pepsis,* a word indicating inner heat, is the origin of the word *pepsin,* a digestive enzyme. The Greek physician Galen (129–199 A.D.) campaigned for a pharmacological approach to good health using plant and animal products for disease treatment.

Arabic biology, which flourished after the founding of Baghdad in 762 A.D., was greatly influenced by early Greek scientific literature. However, the Arabs were not content with the abstract nature of Greek science so they advanced the use of Greek pharmaceutical recipes by determining and classifying the strength and chemical nature of natural drugs. The Greek and Arabic scientific literature did not arrive in western Europe until the 11th century A.D. During the next several centuries, medical schools, which followed the teachings of the Greeks, were established at Bologna, Paris, and Toledo. A key figure in European science, Paracelsus (1493–1541 A.D.), began a move away from the ancient medical doctrines of Aristotle, Galen, and the Arab scientist Avicenna (980–1037). Paracelsus studied medicine at several European universities, but it is doubtful whether he ever completed requirements for a medical degree. He spent his life writing and traveling from city to city expounding on his revolutionary ideas about medicine and biology. His concepts of biochemistry have been described by Pachter: "As a biochemist, he (Paracelsus) asserted that man is made out of the same material as the rest of creation, feeds on the substances that make up the universe, and is subject to the laws which govern their growth and decay. At the same time, each living being is unique, individually constituted and follows his own destiny."[1] Even now, over 450 years after Paracelsus' death, we are impressed with the correctness of his views.

Influenced by Paracelsus, biologists in the 17th and 18th centuries began in earnest a more molecular approach to the study of biological materials and processes. A favorite

[1]H. Pachter, *Paracelsus. Magic into Science* (New York, 1951), Henry Schuman.

Grape harvest and wine production as depicted in an early Egyptian wall painting. Fermentation of carbohydrates in grape juice to ethanol is carried out by yeast.

© George Holton/Photo Researchers

theme for study was the digestive process, for many scientists began to recognize that this could be explained by chemical principles. During the 19th century, any biological process that could not be understood in chemical terms was explained by the doctrine of **vitalism.** Vitalists argued that it was the presence of a vital force (life force or spirit) that distinguished the living organic world from the inanimate inorganic world. The experiment that destroyed the ideas of vitalism was the synthesis of urea, an organic chemical found in natural cells. In 1828, using only the inorganic and therefore "lifeless" chemicals ammonia and cyanic acid, the German chemist Friedrich Wöhler synthesized urea:

$$\underset{\text{Ammonia}}{NH_3} + \underset{\text{Cyanic acid}}{N\equiv C-OH} \longrightarrow \underset{\substack{\text{Ammonium} \\ \text{cyanate}}}{N\equiv C-O^-NH_4^+} \xrightarrow{\text{heat}} \underset{\text{Urea}}{H_2N-\overset{\displaystyle O}{\overset{\|}{C}}-NH_2}$$

It is difficult to pinpoint a specific time or event that marked the start of modern biochemistry. Many science historians usually select the above in vitro (without biological cells) synthesis of urea by Wöhler as the starting point. The significance of this event was commemorated by the issuance of a postage stamp on the 100th anniversary of Wöhler's death.

German stamp commemorating Wöhler and his synthesis of urea.

The Road to Modern Biochemistry

There is more than a single thread from these historical beginnings to present day biochemistry. Two separate and distinct avenues of scientific inquiry have led to our current state of biochemical knowledge (Figure 1.2). One avenue can be traced through the physical sciences and emphasizes structural characteristics of biomolecules. This approach has applied the basic laws of physics and chemistry to explain the processes of the living cell. For example, Linus Pauling in the 20th century used the tool of X-ray crystallography to study the structure of amides and peptides. The other avenue traveled by the biologists, especially microbiologists, cell biologists, physiologists, and geneticists, is characterized mainly by a study of cell organization and function. The first use of the term *biochemistry* is unclear; however, early scientists who considered themselves biochemists traveled on the physical sciences pathway. The two avenues of study converged in 1952 with the announcement by James Watson and Francis Crick of the double helix structure for DNA. Here the application

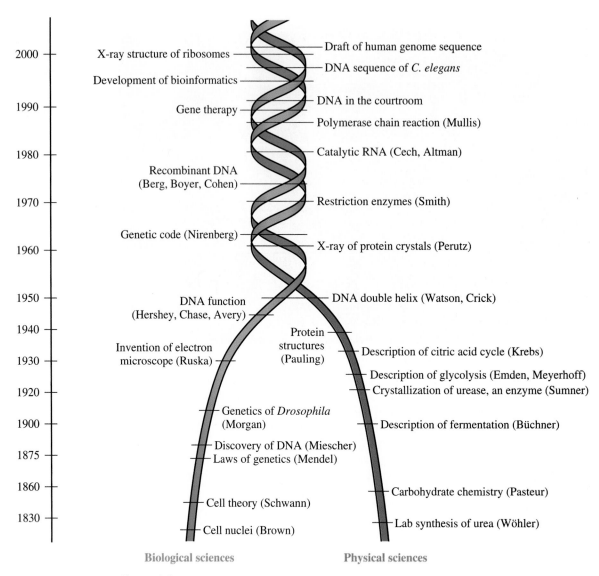

2000 — X-ray structure of ribosomes ——————— Draft of human genome sequence
                                                                        —— DNA sequence of *C. elegans*
        Development of bioinformatics ——————
1990 —   Gene therapy ——————— DNA in the courtroom
                                            ————— Polymerase chain reaction (Mullis)
1980 —   ————— Catalytic RNA (Cech, Altman)
        Recombinant DNA
        (Berg, Boyer, Cohen) ————————
1970 —   ————— Restriction enzymes (Smith)
        Genetic code (Nirenberg) ——————
1960 —   ————— X-ray of protein crystals (Perutz)

1950 —   DNA function   ————— DNA double helix (Watson, Crick)
        (Hershey, Chase, Avery) ——
1940 —   Protein
                                structures
        Invention of electron (Pauling)
1930 — microscope (Ruska) ——   ——— Description of citric acid cycle (Krebs)
                                            ——— Description of glycolysis (Emden, Meyerhoff)
1920 —   ——— Crystallization of urease, an enzyme (Sumner)
                Genetics of *Drosophila*
1900 —   (Morgan)   ——— Description of fermentation (Büchner)
                Discovery of DNA (Miescher)
1875 —   Laws of genetics (Mendel)

1860 —   ——— Carbohydrate chemistry (Pasteur)
                Cell theory (Schwann)
1830 —   Cell nuclei (Brown)   ——— Lab synthesis of urea (Wöhler)

            Biological sciences   **Physical sciences**

Figure 1.2

The origins of biochemistry from two perspectives, the physical sciences and the biological sciences. The dates of selected important events in the development of biochemistry are noted in the scale on the left.

of physics (crystallography), chemistry (structure and bonding), and biology (storage and transfer of genetic information) all came together to help solve what was the most exciting and complex biological problem at that time: the structure of the genetic material, DNA. The growth of knowledge in biochemistry since that time has been explosive. Some of the major events are noted in Figure 1.2.

The term **molecular biology** is often used to describe studies at the interface of chemistry and biology. The term was first coined in 1938 by officials at the Rockefeller Foundation to describe a new funding program that provided financial support for the application of the tools of the physical sciences to biology, biochemistry, cell biology, and genetics. Biochemistry and molecular biology have similar goals. However, their approaches to solving problems are different. Molecular biologists tend to emphasize the study of genetic material (RNA and DNA), especially its role in biological information transfer, and they use more biological experimental approaches involving organisms, recombinant DNA, and molecular genetics. As described earlier, biochemists focus on the structure and function of all biomolecules and energy relationships among them. Biochemists often use tools designed for chemical and physical measurements but are now increasingly using tech-

Cutting Edge:
Visionaries of science

James Watson (*left*) and Francis Crick discuss an early model of the DNA double helix.

niques of molecular biology. The boundaries between biochemistry and molecular biology are rapidly disappearing. It is now important for biochemists to know both the cycle of cell development and the citric acid cycle of metabolism. It is essential for molecular biologists to know and understand the chemical structures of biomolecules. Biochemistry and molecular biology are becoming indistinguishable because they seek answers to the same question: What is life?

1.2   Living Matter Contains C, H, O, N, P, and S

Chemical Elements in Biomolecules

Of the 100 plus chemical elements, only about 31 (28%) occur naturally in plants and animals. As shown in Figure 1.3 the elements present in biological material can be divided into three categories:

1. Elements *found in bulk form and essential for life:* Carbon, hydrogen, oxygen, nitrogen, phosphorus, and sulfur make up about 92% of the dry weight of living things.
2. Elements *in trace quantities in most organisms and very likely essential for life,* such as calcium, manganese, iron, and iodine.
3. Trace elements that *are present in some organisms and may be essential for life,* such as arsenic, bromine, and molybdenum.

We do not understand exactly how these elements were selected by primitive life forms during the early stages of evolutionary development. The elements found in biomolecules do not have similar properties and characteristics. Nearly all groups of the periodic table of elements are represented in biological materials. Both metals and nonmetals are present. One of two hypotheses may explain the selection: there was a deliberate choice because of an element's favorable characteristics or there was a random selection from the alphabet soup of elements present in the earth's crust, atmosphere, and universe. If the latter were true, then we would expect to find approximately the same ratios of elements in the universe as we find in biological organisms. A comparison of the elemental composition of the earth's crust and the universe with that of living matter shown in Figure 1.4 refutes the latter hypothesis.

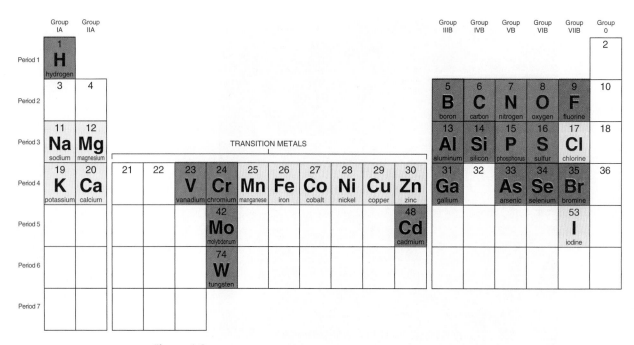

Figure 1.3

The biochemist's periodic table. Elements in red are present in bulk form in living cells and are essential for life. Those in yellow are trace elements that are very likely essential. Those elements in blue are present in some organisms and may be essential.

We must conclude that elements were selected according to their abilities to perform certain structural functions or to provide specific reactivities. For example, carbon forms multiple covalent bonds with other carbon atoms as well as with other elements such as nitrogen, hydrogen, oxygen, or sulfur. This feature allows the construction of long carbon chains and rings with the presence of reactive functional groups containing nitrogen, oxygen, and sulfur as in proteins, nucleic acids, lipids, and carbohydrates. Iron was selected by evolutionary forces because it is able to bind the oxygen molecule in a reversible fashion. Elements found in the earth and atmosphere may have been tested by trial and error in living organisms during millions of years. Those elements that most effectively performed the necessary tasks and, most importantly, allowed the plant or animal to thrive were retained.

Combining Elements into Compounds

The combination of chemical elements into biomolecules provides for great variety in chemical structure and reactivity. Compounds representing all three states of matter (gases, liquids and solids) are present in living cells. One of the most recent advances in biochemistry is the discovery of an enzyme that catalyzes the synthesis of the gas nitric oxide (NO) in the brain and other organs where it serves to regulate biological processes. Nature's molecules include examples of cations, anions, covalent compounds, ionic compounds, metal ions, coordination complexes, and polymers. Several well-known examples illustrate the diverse array of organic and organometallic chemicals that perform a variety of cellular roles. The **amino acids** are building blocks for protein structure. The **carbohydrates** are involved as nutrients in energy metabolism as well as playing roles in cell structure and molecular recognition. **Lipids** are a diverse collection of organic compounds that display low water solubility. The lipids have primary functions as energy molecules in metabolism, as components for the construction of cell membranes, and as hormones. The **vitamins,** a broad assortment of organic compounds, ensure proper growth and development by functioning in major biochemical processes. Prominent among the natural organometallic compounds are **heme** and **chlorophyll.** Both consist of a substituted porphyrin ring coordinated with a metal

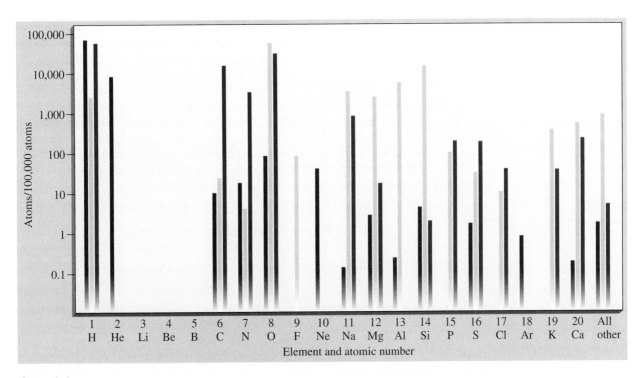

Figure 1.4
Elemental composition of the universe (*blue*), the earth's crust (*pink*), and the human body (*purple*).

ion. Heme, a porphyrin ring with iron (Figure 1.5a), is found in the oxygen transport proteins myoglobin and hemoglobin, in respiratory proteins such as cytochrome *c,* and in enzymes such as catalase. Chlorophyll, a magnesium–porphyrin complex (Figure 1.5b), is abundant in green plants and algae, where it functions as a receptor of light energy.

1.3 Biological Macromolecules

Many molecules present in biological cells are very large by the standards of inorganic and organic chemistry. Three major classes of natural macromolecules are found in biological cells: the **nucleic acids, proteins,** and **polysaccharides.** (Lipids are also considered a major class of biomolecules, but because they are usually not macromolecules, they are not

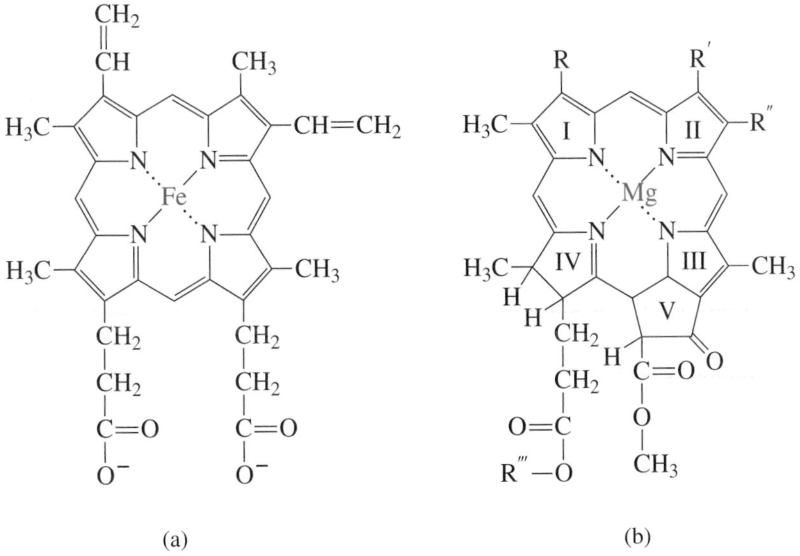

Figure 1.5

Two natural organometallic compounds: (a) heme, containing a porphyrin ring and iron; and (b) chlorophyll, containing a porphyrin ring and magnesium.

(a)

(b)

described in this section but in the previous paragraph.) These macromolecules participate in a complex array of biological processes, such as storing and transferring genetic information (nucleic acids), catalyzing biochemical reactions (proteins called enzymes), holding cells and organisms together (structural proteins and polysaccharides), transporting small molecules across cell membranes or from one location in an organism to another location (transport proteins), and protecting an organism against disease agents (protein antibodies). Although their structures and functions are quite different, all natural macromolecules have one common characteristic; they are polymers constructed by combining together hundreds, thousands, and sometimes millions of smaller, prefabricated molecules called monomers (Figure 1.6).

The reaction process that connects monomer units is called **condensation,** a chemical combining that results in the loss of a small molecule, usually water:

$$\underset{\substack{\text{Carboxylic}\\\text{acid}}}{R-\overset{\overset{\displaystyle O}{\|}}{C}-OH} \;+\; \underset{\text{Amine}}{R'-NH_2} \;\underset{H_2O}{\overset{H_2O}{\rightleftarrows}}\; \underset{\text{Amide}}{R-\overset{\overset{\displaystyle O}{\|}}{C}-NHR'}$$

In the example given here, a carboxylic acid is condensed with an amine to form a new compound with an amide bond. This represents the chemistry for combining amino acids to make proteins. During metabolic turnover in the living cell, the reverse process, called **cleavage** or **hydrolysis** (if water is used), also becomes important. The product formed by condensation or polymerization of thousands of glucose monomers is either **starch** or **cellulose,** depending on the type of chemical linkage between the glucose units. These macromolecules are called polysaccharides because they are composed of many saccharide (sugar) molecules. Because the monomeric units comprising these polysaccharides are chemically identical, they are, in general, termed **homopolymers.**

The proteins are the products of joining together amino acids by amide bonds. Because 20 different amino acids are available as monomeric building blocks, the resulting proteins are **heteropolymers.** By combining many different amino acids in proteins, a level of molecular complexity and structural diversity results that is not possible for starch or cellulose. Different proteins are formed by changing the order (sequence) and number of amino acid monomers. This allows the construction of a vast array of different protein molecules, each with its own physical, chemical, and biological characteristics. All molecules of a specific protein (for example, hemoglobin) within a species of organism, however, normally have an identical sequence of amino acids.

The nucleic acids are heteropolymers of monomeric units called **nucleotides. Deoxyribonucleic acid (DNA),** the chemical storage form of genetic information, is com-

Figure 1.6

Types of natural polymers: (a) Starch, a homopolymer formed by joining many identical glucose molecules. (b) Cellulose, a homopolymer formed by joining many identical glucose units. Note that different types of bonding are used in starch and in cellulose. (c) Protein, a heteropolymer formed by linking together amino acids. (d) Nucleic acid, a heteropolymer formed by combining different nucleotides, A, G, C, and U.

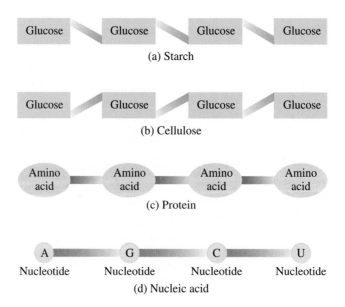

prised of the monomers deoxyadenosine 5′-monophosphate (dAMP), deoxyguanosine 5′-monophosphate (dGMP), deoxycytidine 5′-monophosphate (dCMP), and deoxythymidine 5′-monophosphate (dTMP). Each DNA molecule in the human chromosome contains millions of nucleotides. **Ribonucleic acid (RNA),** which is involved in the transfer of genetic information and in biological catalysis, is a heteropolymer of adenosine 5′-monophosphate (AMP), guanosine 5′-monophosphate (GMP), cytidine 5′-monophosphate (CMP), and uridine 5′-monophosphate (UMP). The genetic information present in nucleic acids is coded by the sequence of nucleotides.

In the last part of this book, we will focus on the important processes of **metabolism.** This refers to the thousands of biochemical reactions that define the synthesis (anabolism) and breakdown (catabolism) of all biomolecules including the macromolecules. The metabolic turnover of proteins, nucleic acids, polysaccharides, and lipids requires complex cellular machinery, strict control to assure reproducibility, the presence of reaction catalysts called enzymes (usually proteins), and complicated processes for exchange of chemical energy. Catabolism (exergonic) and anabolism (endergonic) are linked through the transfer of energy in the molecular form of **adenosine triphosphate (ATP).** All the information necessary to direct cellular tasks resides in the DNA of the organism, but the message in the form of nucleotide sequence must first be decoded and made into functional proteins before biochemical reactions can occur.

1.4 Organelles, Cells, and Organisms

The complexity of biomolecule synthesis and degradation is overshadowed by the next stage of organization—the self-assembly of macromolecules into higher levels of order. Our discussion about biomolecules has proceeded in a stepwise fashion beginning with the chemical elements and continuing with the precise combining of atoms to make small molecules and the joining together of these monomers to make functional polymeric biomolecules.

Let us continue in this hierarchical ascent from atoms to higher levels of organization. After the stage of biological macromolecules, we encounter **supramolecular assemblies,** organized clusters of macromolecules. Prominent examples are **cell membranes** (complexes of proteins and lipids), **chromatin** (complexes of DNA and proteins), **ribosomes** (complexes of RNA and proteins), and materials of a fibrous nature, such as the protein-containing **cytoskeleton.** The extensive presence and biological significance of supramolecular assemblies in cells and organisms illustrate the ability of biomolecules to recognize and interact with one another in a specific way. **Molecular recognition** is the result of an exact fit between the surfaces of two molecules. Molecules that are complementary diffuse together to form a complex that displays some biological activity. The molecules are held together by weak and reversible chemical forces. These chemical forces will be described further in later chapters.

Viruses (Figure 1.7) are another example of supramolecular assemblages. Biochemically, most viruses consist of a single DNA or RNA molecule wrapped in a protein package.

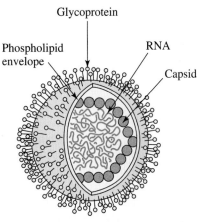

(a) Influenza virus (globular)

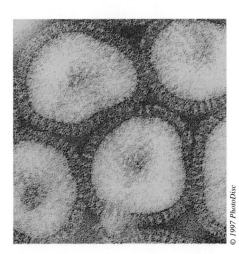

© 1997 PhotoDisc

Figure 1.7

Viruses are nonliving packages of nucleic acid and protein. On the left are schematic diagrams of four types of viruses and on the right an electron micrograph of each.

(continued)

Figure 1.7—*continued*

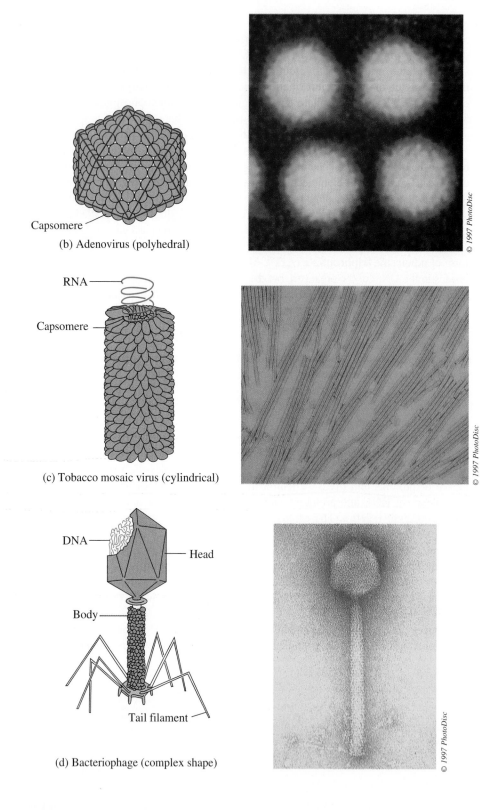

Capsomere

(b) Adenovirus (polyhedral)

© 1997 PhotoDisc

RNA

Capsomere

(c) Tobacco mosaic virus (cylindrical)

© 1997 PhotoDisc

DNA

Head

Body

Tail filament

(d) Bacteriophage (complex shape)

© 1997 PhotoDisc

Viruses cannot exist independently and are usually not considered a life-form. Instead, they are deemed parasites since they are unable to carry out metabolism or reproduction without the assistance of a host cell. When viruses infect a cell, they take control of the cell's metabolic machinery and force it to synthesize nucleic acids and proteins for new virus particles. Viruses are the cause of many plant and animal maladies and their presence in the world has resulted in much human suffering; however, an enormous amount of biochemistry has been learned from studies of their actions (see Chapter 10, Section 10.5).

After supramolecular assemblies, the next higher level of organization is the fundamental unit of life, the cell. Scientists have long recognized two basic classifications of organisms: (1) the **eukaryotes,** organisms, including plants and animals, whose cells have a distinct membrane-enclosed nucleus and well-defined internal compartmentation, and (2) the **prokaryotes,** simple, unicellular organisms, mainly bacteria and blue-green algae, with neither a distinct cell nucleus nor internal cellular compartmentation. This classification is achieved primarily by microscopic observation and, hence, is based on morphological cell structure and anatomy. If cells are classified by genetic analysis (DNA and RNA sequences), then three distinct types are recognized. In 1977, Carl Woese at the University of Illinois discovered through genetic analysis of ribosomal RNA that **archaebacteria** or **archae** (ancient bacteria) are different from prokaryotes and eukaryotes. The archaebacteria are also distinctive when their living conditions are examined; many are able to thrive in an environment of high acidity, high salt, high temperature, and absence of oxygen. They can be found in abundance in the hot springs of Yellowstone National Park or in volcanic areas on land and sea. The ability to grow at high temperatures (some bacteria grow at temperatures up to 110°C) is of special interest because nucleic acids and proteins normally become unraveled at temperatures above 60°C to 70°C. Chemists and biochemists are eager to find how archaebacteria stabilize their macromolecules of DNA and protein under these extreme conditions. (See Window on Biochemistry 1.1.) In our consideration of the biochemical characteristics of all organisms, we will observe the traditional division of cell types into prokaryotes and eukaryotes. The archaebacteria, when classified according to morphological observation, more closely resemble the prokaryotes.

Prokaryotic Cells

The prokaryotic organisms, although the least developed, are the most abundant and widespread of organisms. The structural features of this simple type of cell are illustrated in Figure 1.8 and biochemically characterized in Table 1.1. Several characteristics of prokaryotic cells can be generalized:

1. The size may range from 1 to 10 μm in diameter. Bacteria, an abundant prokaryotic organism, have three basic shapes: *spheroidal* (cocci), *rodlike* (bacilli), and *helically coiled* (spirilla).
2. The cellular components are encapsulated within a cell membrane and rigid cell wall. Occasionally the membrane may infold to form multilayered structures called mesosomes. The outside surface is often covered by flagella, which are appendages for locomotion, and pili, which are structural features responsible for the transfer of DNA during sexual conjugation and for attachment to surfaces.
3. The interior of the cell, called the **cytoplasm,** is a gel-like, heterogeneous suspension of biomolecules including small molecules, soluble enzymes, **ribosomes** (supramolecular particles of RNA and protein), and coiled DNA in the nucleoid region.
4. Each cell has one chromosome, a single copy of DNA (the **genome**). Several copies may be present in a rapidly growing cell that replicates by simple division.

© Jim Sternberg/Photo Researchers

Archaebacteria usually thrive in extreme environments like these Yellowstone hot springs.

Interactive Animation: Cell structure

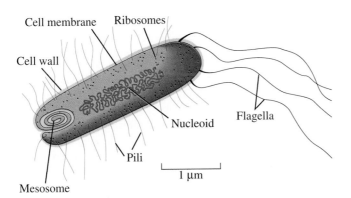

Cell membrane Ribosomes
Cell wall
Nucleoid
Flagella
Pili
Mesosome
1 μm

Figure 1.8
Schematic diagram of a typical prokaryotic cell.

Table 1.1
Molecular composition and biological function of prokaryotic cell components

Structural Feature	Molecular Composition	Biological Function
Cell wall, pili, and flagella	Polysaccharide chains cross-linked by proteins; coated with lipopolysaccharide; pili and flagella are extensions of the cell wall	Protection against mechanical and hypertonic stress; flagella assist in movement; pili assist in sexual conjugation
Cell membrane, mesosome	Bilayer of 40% lipid, 60% protein, perhaps some carbohydrate; mesosome is infolded membrane	Permeable boundary that allows for entry and exit of nutrients, waste; mesosome may play a role in DNA replication
Nucleoid region	Contains chromatin, a complex of chromosomal DNA and histone proteins	The genome; storage of genetic information; site of DNA replication
Ribosomes	Complexes of RNA (65%) and protein (35%)	The sites of protein synthesis
Vacuoles	Nutrients stored as small molecules or polymers	Storage of fuel molecules for energy metabolism
Cytoplasm	Small molecules, soluble proteins, enzymes, nutrients, inorganic salts; dissolved in aqueous solution	Region where many metabolic reactions occur

The prokaryotic organism that has been the object of most biochemical studies is the *Escherichia coli* (*E. coli*) bacterium. Indeed, we know more biochemistry about this organism that lives in our gut than any other, including humans. Because so much is understood about this bacterium, it has recently been possible to prepare pictures of the interior of a living cell. Dr. David Goodsell at the University of California, Los Angeles, has combined molecular composition data with structural information to provide the first true-to-life pictures of a living cell. These pictures show the relative distribution of molecules at the proper scale. A typical rodlike *E. coli* cell has dimensions of 2.95 μm by 0.64 μm. A cell weighs in at 2×10^{-12} g and has a volume of 0.88 μm^3.

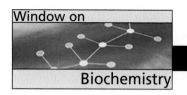

Window on

Biochemistry

1.1 Life at the Extremes—Some Like It Hot

As humans venture to regions of the earth that have extreme and inhospitable environments, we are discovering that other life forms arrived there first. Scientists exploring harsh regions (hot, rocky, subterranean earth; deep-sea hydrothermal vents; hot-water geysers; salt waters; alkali deserts; glacial ice; and permafrost) are finding evidence for the presence of living organisms. Extremophilic microbes include primarily archaebacteria (ancient bacteria) and eubacteria (true bacteria). Three stages of life forms are present in these environments: (1) active cells that manage to eke out a living, (2) dormant forms that thrive when cultured in the laboratory, and (3) remains of the dead. By studying the biochemistry and genetics of these microorganisms, it is possible to ponder questions about the origin of life and about how the molecules of life have changed over millions of years. In addition, these organisms provide a treasure chest of ancient enzymes, proteins, and nucleic acids that may have value in modern biotechnology and biomedicine. The discovery of life in harsh environments has also raised hopes for finding life on other planets and solar systems. Here we will survey forms of life found in unusual environmental conditions that once were thought to be incompatible with life.

- Heat: Viable cultures of *Pyrococcus furiosus* have been isolated from regions of Vulcano Island, Italy at temperatures of 112°C. *Thermus aquaticus,* found in the 110°C mineral hot springs of Yellowstone National Park, are the source of *Taq* DNA polymerase that is used in the polymerase chain reaction (PCR, page 349).
- Cold: Scientists at the Astrobiology Institute at NASA are studying microorganisms found in permafrost, polar ice caps, and glaciers of the Arctic and Antarctica.

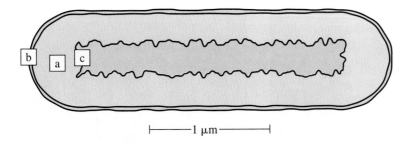

Figure 1.9

A schematic of an *E. coli* cell magnified 50,000 times. Region *a* is cytoplasm, region *b* is the cell wall, and region *c* is the nucleoid with DNA.

There are between 4000 and 6000 different kinds of molecules in a given cell, of which about 3000 are proteins (only about 1000 have been characterized). The information to make these proteins is stored in a single copy of DNA, which has a molecular mass of 2.5×10^9 kilodaltons (one dalton = one atomic mass unit).

An *E. coli* cell magnified to 50,000 times is shown in Figure 1.9. Three regions are indicated by small lettered squares: (a) the cytoplasm, (b) the cell wall, and (c) the nucleoid region. Each of these regions is magnified another 20 times in Figures 1.10a,b,c so that the actual drawings show a cell with a total magnification of 1 million times. To put these figures in perspective, it would require 600 of the cubes shown in Figure 1.10a to represent the entire volume of cytoplasm in a single cell. One almost gets the urge to walk through the cell and take a closer look. This may be more difficult than it seems. The pictures are snapshots of an instant in time; they do not show the actual movement of molecules in the gel-like cytoplasm. Dr. Goodsell has estimated that the average speed of a 160,000-kilodalton molecular mass protein is 500 cm/s. At this rate, the protein molecule would cover a distance of 10 nanometers (nm, its approximate size) in about 2 nanoseconds (ns). Even this small movement would result in much pushing and shoving because of the tight quarters inside the cell.

Cold-tolerant organisms include archaea, true bacteria, cyanobacteria, and even eukaryotic life such as yeast and moss. Many life forms found under these conditions [−10°C, 3.6 km (2.3 miles) deep] are cryopreserved, but resume metabolic activities when thawed in the lab. Cultures of microbes are also found growing on powdered rock formed by moving glaciers. The rock surfaces contain water, iron–sulfur minerals, and other essential nutrients. The extreme conditions existing at the polar regions of Earth may be similar to ice caps on Mars and to the icy moons of Saturn.

■ **Acidic:** Microorganisms, currently unclassified, have been found thriving on hydrated calcium sulfate in gypsum mines (pH = 0–1).

■ **Radioactive:** Microbes are found growing on seams of an oil-like form of carbon containing uranium and iron sulfides in South African gold mines at depths of 3.2 km and pressures of two atmospheres. The bacteria may be concentrating both gold and radioactive elements within their cells.

■ **Salty:** *Halobacterium halobium* thrives in sunny, salt lakes or brine ponds where the sodium chloride concentrations may reach 3 *M*. (See Window on Biochemistry 5.2, p. 116.)

■ **High pressure:** Scientists investigating ocean hydrothermal vents near the Juan de Fuca Ridge off the Oregon–Washington coast of the U.S. and in the Marianas Trench of the Pacific have found an abundance of *Methanogens*, strict anaerobes that use hydrogen gas to reduce carbon dioxide to methane. At 5–7 miles below the surface, pressures can reach several hundred atmospheres.

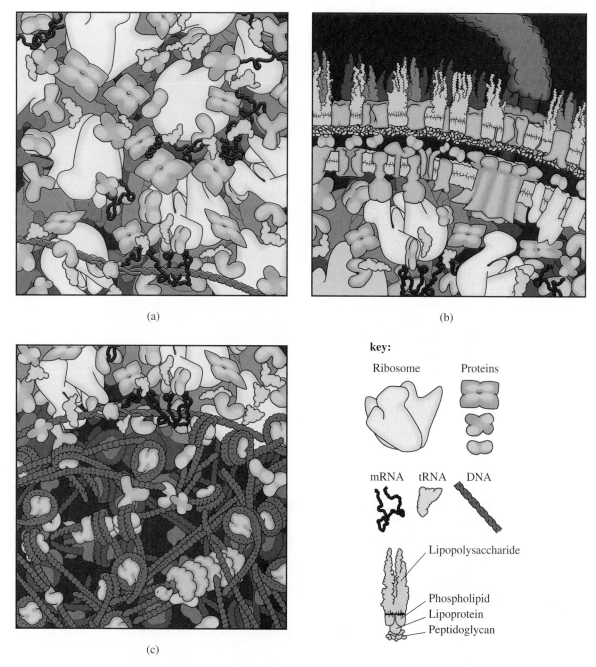

(a)

(b)

(c)

key:

Ribosome

Proteins

mRNA   tRNA   DNA

Lipopolysaccharide

Phospholipid
Lipoprotein
Peptidoglycan

Figure 1.10

The shapes of biomolecules are shown in the key. From Figure 1.9: (a) Region *a* is the cytoplasm magnified 20 times. (b) Region *b* is the cell wall magnified 20 times. (c) Region *c* is the nucleoid region magnified 20 times. *Source:* Adapted from Goodsell, 1991.

Eukaryotic Cells

The class of eukaryotes includes plants, animals, fungi, protozoans, yeasts, and some algae. The cells found in these organisms have little in common with the prokaryotes. The complex eukaryotic cells are much larger, with diameters ranging from 10 to 100 μm. They are surrounded by a plasma membrane made up of protein and lipid (see Figure 1.11). This is a chemical barrier through which all molecules that enter or exit the cell must pass. A unique feature of the eukaryotic cell is the compartmentation of cellular components and, therefore, the compartmentation of biological function. These compartments called **organelles** are actually membrane-enclosed packages of organized macromolecules that perform a specialized function for the cell. A listing of organelles along with their constituent biomolecules and biological function is found in Table 1.2. The organelles found in all eukaryotic cells are the **nucleus;** the **endoplasmic reticulum,** which contains **ribosomes;** the Golgi apparatus; and the **mitochondria.** In addition to these, animal cells have specialized organelles, the **lysosomes** and **peroxisomes,** whereas plant cells have **chloroplasts** and **glyoxysomes, a** modification of the peroxisomes. Plant cells are surrounded by a plasma membrane and a

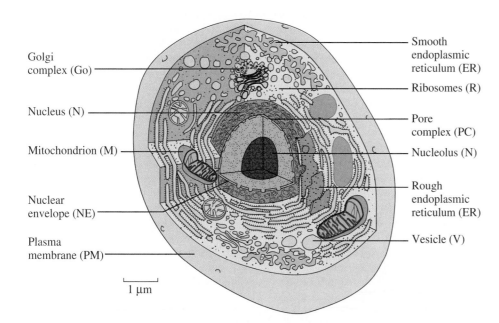

Golgi complex (Go)

Nucleus (N)

Mitochondrion (M)

Nuclear envelope (NE)

Plasma membrane (PM)

Smooth endoplasmic reticulum (ER)

Ribosomes (R)

Pore complex (PC)

Nucleolus (N)

Rough endoplasmic reticulum (ER)

Vesicle (V)

1 μm

Figure 1.11

Typical eukaryotic cells showing a schematic drawing (*above*) and an electron micrograph (*below*): (a) An animal cell. (b) A plant cell. *Source:* Based on Wolfe, 1993.

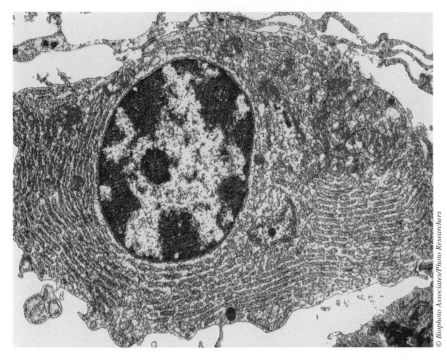

© *Biophoto Associates/Photo Researchers*

(a)

(continued)

Figure 1.11—*continued*

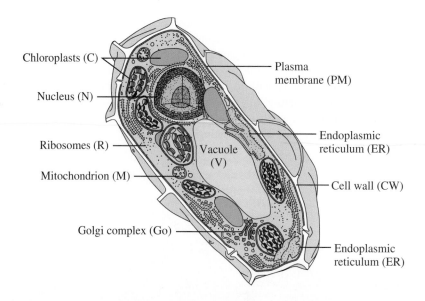

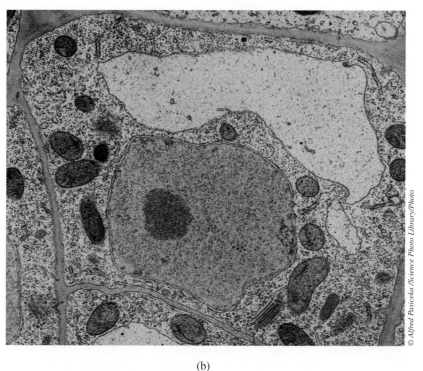

(b)

rigid cell wall composed primarily of the polysaccharide cellulose. Both plant and animal cells contain **vacuoles** for storage of nutrients and wastes, although these are more prominent in plant cells.

The organelles of a eukaryotic cell are not floating freely in a cytoplasmic sea. Rather, their movement and location are limited by the cytoskeleton, the three-dimensional fibrous matrix extended throughout the inside of the cell (see Figure 1.12). The function of the cytoskeleton is to give the cell shape, to allow it to move, and to guide internal movement of organelles. The fibers are composed primarily of protein. The prominent fiber types are

Table 1.2
Eukaryotic organelles, their constituent biomolecules, and biological function

Structural Feature	Molecular Composition	Biological Function
Cell membrane	Bilayer of proteins (50%) and lipids (50%) and some carbohydrate	Selectively permeable boundary for entry and exit of nutrients and waste; some important enzyme activities; location of receptors for signaling
Nucleus	Contains genomic DNA, and histone proteins as chromatin; RNA	Storage of genetic information; site of DNA replication and transcription to RNA
Endoplasmic reticulum with ribosomes	Flat, single-membraned vesicles of lipid and protein; ribosomes consist of RNA and proteins	Surfaces on which ribosomes bind for protein synthesis
Golgi apparatus	Flattened vesicles of lipid, protein, and polysaccharide	Secretion of cell waste products; site of protein processing
Mitochondria	Double-membraned with protein and lipids; interior (matrix) contains soluble and insoluble enzymes, RNA, and DNA	Site of energy metabolism and synthesis of high-energy ATP
Lysosomes (animal)	Single-membraned vesicles containing enzymes for hydrolysis	Metabolism of materials ingested by endocytosis
Peroxisomes (animal) or glyoxysomes (plant)	Single-membraned vesicles containing catalase and other oxidative enzymes	Oxidative metabolism of nutrients using O_2 to generate H_2O_2
Chloroplasts (plant)	Double-membraned organelles containing protein, lipid, chlorophyll, RNA, DNA, and ribosomes	Sites of photosynthesis. Convert light energy into chemical energy (ATP)
Cytoplasm	Cytoskeleton made of proteins; small molecules, soluble proteins, enzymes, nutrients, salts in aqueous solution	Provides shape to cell; region where many metabolic reactions occur

(1) the microtubules (diameter = 22 nm) composed of the protein tubulin, (2) the microfilaments (diameter = 6 nm) composed of actin, and (3) the intermediate filaments (diameter = 7–11 nm) (Figure 1.13). The major constituent protein in the intermediate filaments varies from cell to cell; for example, skin cells have an extensive array of intermediate fibers made of keratin.

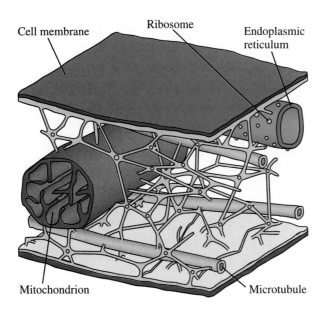

Figure 1.12
The interior of a cell showing the cytoskeleton. Microtubules and other filaments form an extended network in the cytoplasm.

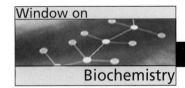

1.2   Relative Sizes of Objects Biochemists Study

The molecules and molecular particles that biochemists work with come in a wide range of sizes. We will concern ourselves here with two perspectives of size: (1) the **dimensions** of biochemical objects in terms of length and width or diameter and (2) the **molecular mass,** which is a measure of the quantity of material in an object. The dimensions for biochemical entities (molecules, particles, cells) are usually expressed in units of **nanometers** (nm, 1×10^{-9} meters or 10 angstroms, Å). A molecule of water has a diameter of about 0.3 nm; the amino acid alanine, a diameter of 0.5 nm; the oxygen-carrying protein hemoglobin, a diameter of about 5.5 nm; and a liver mitochondrion, a length of about 1500 nm (see graphic).

The standard, biochemical units for molecular mass are **daltons (D)** or **kilodaltons (kD,** 1000 daltons). One dalton is equal to the mass of a hydrogen atom. (You may also be familiar with the term *molecular weight* or *molar mass* of a substance, which is the mass of one mole of that substance, expressed in grams.) The mass of a water molecule is 18 D; alanine, 89 D; and hemoglobin, 64,500 D (64.5 kD). (Is it realistic to think of the mass of the mitochondrion in terms of daltons?)

Perhaps the smallest entity studied routinely by biochemists is the proton, H^+, which is approximately 0.1 nm in diameter. This tiny cation would be of interest to scientists measuring the pH (negative log of the $[H^+]$) of a solution and to someone studying the mechanism of acid-catalyzed reactions. A biochemist wishing to work with larger items of biological interest could use bacterial cells (2000–3000 nm long × 600 nm wide) or animals cells (at least 10–20 times larger than bacterial cells). For example, a liver cell has, on average, a length of 20,000 nm. Unfortunately, none of these objects from protons to animal cells is visible to the unaided eye which has a minimal viewing limit of about 0.2–0.3 mm. Therefore, these particles may be observed best with the help of molecular computer graphics programs linked to light microscopy, electron microscopy, nuclear magnetic resonance spectroscopy, and X-ray diffraction. Some animal cells are much larger; for example, vertebrate nervous system cells (like in the eel) are up to one meter long and are quite visible to the naked eye.

The size range we will be mostly concerned with in biochemistry has a lower limit of the proton and an upper limit of an average eukaryotic cell, which represents a total range of about one million. If the proton is represented in size by a shelled peanut, then a typical eukaryotic cell would have the size of volume enclosed by a sports stadium such as the Houston Astrodome. Reviewing the graphic reveals the enormous range of sizes and masses represented by biomolecules and other particles. Note that the scale is logarithmic so that differences are much greater than they appear.

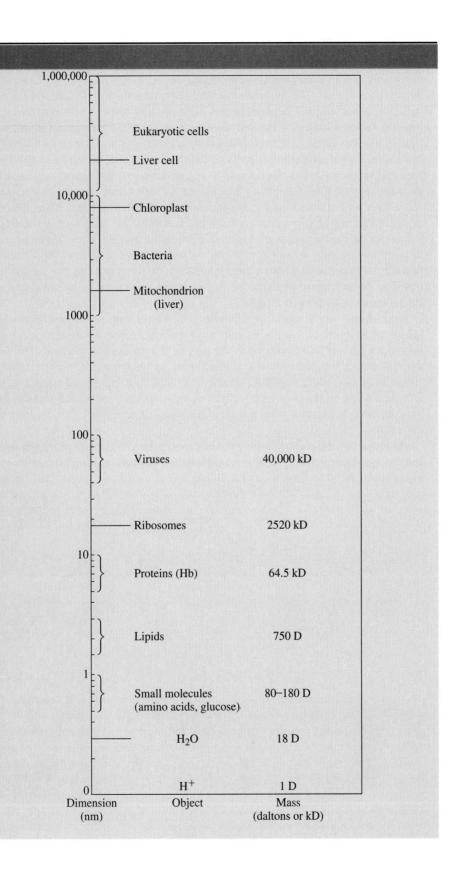

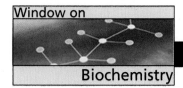

1.3   Capturing and Handling Cell Components

Biochemistry is an experimental science. Our understanding of biological processes at the molecular level comes from the laboratory observation of cells and cell constituents, especially cell organelles and biomolecules. Before one can do experimental biochemistry, the desired object of study must be separated from its natural surroundings. One tool useful for isolating cells, cell organelles, and biomolecules is the centrifuge. Using this instrument, scientists take advantage of the fact that biochemical entities have different physical characteristics: weights, sizes, densities, and shapes. Subjecting a biological sample to an extreme gravitational force, by spinning the sample at high speed, causes sedimentation of sample components at a rate that depends on their physical characteristics. Bacterial cells in a fermentation medium and erythrocytes (red blood cells) in blood plasma can be isolated by spinning a tube containing them at a rate of about 1000–2000 rpm (develops a gravitational field of about 1000g, 1000 times gravity). Whole cells sediment to the bottom of the tube leaving a liquid supernatant above them that can be decanted away from the cells.

Biochemists rarely work with whole cells, using **cell homogenates** instead. Whole prokaryotic or eukaryotic cells are homogenized in an aqueous, buffered solution, a process that breaks open the plasma membrane and cell wall (if present) and allows all the cell's internal components to be released into the solution. Homogenization is accomplished by one of several methods: grinding in a mortar with pestle and an abrasive substance, like sand; breaking the cells under high pressure; or using an electric homogenizer consisting of a rotating teflon pestle in a glass tube.

Now that the cell components are freed and suspended in solution, organelles may be isolated by the technique of **differential centrifugation.** The cell homogenate is subjected to successive centrifugations at increasing speeds (see figure).

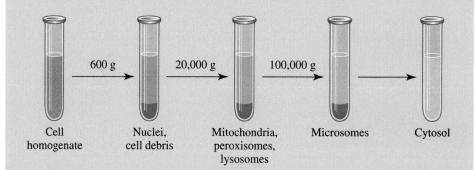

The supernatant from each run is decanted into another tube and centrifuged at a higher speed. Each type of cell organelle has a unique size, shape, and weight and can be isolated from the rest because it sediments at a different gravitational force. For example, mitochondria sediment in a tube spinning at a force of about 20,000g. After this spin, the biochemist can decant the supernatant leaving behind mitochondria for further study. The final centrifugation at 100,000g requires a high-speed ultracentrifuge and leaves a supernatant representing the cytoplasm and its soluble proteins and other molecules. Many of the enzymes of metabolism are present in the soluble cytoplasm. Now that the basic cell components are separated, the biomolecules present in each may be isolated on an individual basis (called "purification") using techniques of **chromatography** (see Window on Biochemistry 4.2, p. 88).

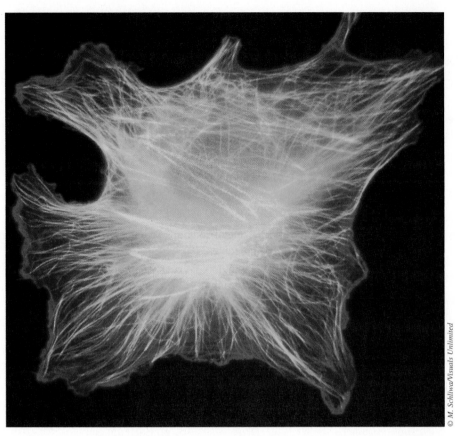

Figure 1.13
Micrograph of part of the cytoskeleton within a hamster kidney cell. Actin is shown in red, and tubulin is shown in green.

© M. Schliwa/Visuals Unlimited

1 μm

Summary

■ Biochemistry is the discipline of experimental science that studies the molecules of the living cell.

■ Biochemistry has many impacts on our lives: It attempts to explain the origin and cures for diseases; it improves nutrition; it develops methods to solve technological problems; it provides basic molecular understanding of life's processes.

■ The principles of biochemistry have been used by humans for thousands of years to ferment juices and bake bread. The

discipline originated by a combination of biology, chemistry, and physics.

■ The important classes of biomolecules include carbohydrates, lipids, proteins, and nucleic acids.

■ Scientists have recognized two main types of organisms: **eukaryotes,** organisms that have membrane-enclosed organelles, and **prokaryotes,** simple, unicellular organisms. The prokaryotes are sometimes divided into two groups, eubacteria and archaea.

Study Exercises

Understanding Terms

1.1 Define or explain the following important biochemical terms in 25 words or less.

- **a.** Prokaryotic cells
- **b.** Eukaryotic cells
- **c.** Cell plasma membrane
- **d.** Mitochondria
- **e.** Nucleus
- **f.** Bioenergetics
- **g.** Nucleic acids
- **h.** Viruses
- **i.** Differential centrifugation
- **j.** DNA
- **k.** Biomolecules
- **l.** Hemoglobin
- **m.** Cell organelle
- **n.** Biochemistry
- **o.** Macromolecules
- **p.** Vitalism
- **q.** Ribosomes
- **r.** Starch
- **s.** Metabolism
- **t.** Daltons
- **u.** Cell homogenate
- **v.** Condensation reactions
- **w.** Biotechnology
- **x.** Hydrolysis reactions
- **y.** Thermophilic bacteria
- **z.** Molecular biology

Reviewing Concepts

1.2 Describe the special structural features and some biological functions of the lipid class of biomolecules.

1.3 Draw stable structural molecules with the following molecular formulas.

a. $C_2H_5NO_2$ **d.** C_2H_4OS **g.** $C_2H_4O_3$
b. CH_4N_2O **e.** $C_3H_4N_2$ **h.** H_2O_2
c. $C_2H_4O_2$ **f.** $C_3H_7NO_2S$ **i.** H_3PO_4

Indicate the structures you drew that represent naturally occurring compounds.

HINT: The number of covalent bonds formed by each element is:

Carbon	4	Hydrogen	1
Oxygen	2	Sulfur	2
Nitrogen	3	Phosphorus	5

1.4 List five chemical elements present in biomolecules.

1.5 Name four gases that are essential for life.

1.6 List four metal ions present in living organisms.
HINT: Begin with Na^+.

1.7 List the major classes of biomolecules.

1.8 Without referring to figures in this book, draw the structure of a bacterial cell and an animal cell. Draw and label internal components of each type of cell.

1.9 Describe the differences between molecular biology and biochemistry.

1.10 Name at least one biological function for each of the following biomolecules.

a. DNA   **e.** Carbohydrates
b. RNA   **f.** Vitamins
c. Amino acids **g.** Proteins
d. Lipids   **h.** Water

1.11 Describe the major differences between prokaryotic and eukaryotic cells.

1.12 Which of the following are biopolymers (large-molecular-weight compounds made up of monomers)?

a. Glucose   **f.** Hemoglobin
b. Cellulose   **g.** Nucleotides
c. DNA   **h.** Proteins
d. Urea   **i.** Water
e. Amino acids **j.** O_2

1.13 For each cell organelle listed below, tell whether it is present in plant cells, animal cells, or both.

a. Nucleus   **e.** Chloroplasts
b. Ribosomes   **f.** Mitochondria
c. Lysosomes   **g.** Glyoxysomes
d. Peroxisomes

1.14 Biopolymers are composed of smaller monomeric units. Match the monomers used to synthesize each of the naturally occurring polymers.

Polymers	Monomers
_____ **1.** Nucleic acids	**a.** Glucose
_____ **2.** Polysaccharides	**b.** Amino acids
_____ **3.** Proteins	**c.** Nucleotides
_____ **4.** Cellulose	**d.** Monosaccharides
_____ **5.** Starch	

1.15 Describe how you would centrifuge a homogenate of heart muscle cells in order to isolate mitochondria.
HINT: See Window on Biochemistry 1.3.

1.16 What metal ions are present in each of the following biomolecules?

a. Hemoglobin   **c.** Cytochrome c
b. Chlorophyll **d.** Catalase

1.17 Briefly describe how you would prepare a cell extract of bean plant leaves.

1.18 Put the following biochemical entities in order according to size. Begin with the largest and proceed to the smallest.

a. Mitochondria   **e.** Glucose
b. Red blood cell **f.** O_2
c. DNA   **g.** Bacterial cell
d. Hemoglobin   **h.** Ethanol

1.19 Describe two ways in which DNA and RNA differ in terms of chemical structure.

Solving Problems

1.20 Predict in what cell organelle or region of the cell the following biochemical reactions or processes occur.

a. $2\ H_2O_2 \longrightarrow 2\ H_2O + O_2$ (animal cells)
b. DNA replication (animal and bacterial cells)
c. Hydrolysis of RNA ingested by endocytosis
d. Protein biosynthesis
e. $CO_2 + H_2O \xrightarrow{h\nu} (CH_2O) + O_2$
(synthesis of carbohydrates by photosynthesis)
f. $NADH + H^+ + {}^1\!/_2\ O_2 + ADP + P_i \longrightarrow$
$NAD^+ + H_2O + ATP$ (P_i = phosphate)

1.21 The enzyme hexokinase is required for the metabolism of the carbohydrate glucose. It is a soluble enzyme found in the cytoplasm. Describe how you could use centrifugation to prepare a cell fraction containing the enzyme separated from cell organelles.

1.22 The protein cytochrome c is necessary for converting energy from electron transport to the form, chemical bond energy in ATP. In which organelle would you expect to find the most cytochrome c?

1.23 Why was Wöhler's synthesis of urea from ammonium cyanate an important achievement in biochemistry?

1.24 Describe the contributions of chemists, biologists, and physicists to the origin of biochemistry.

1.25 What centrifuge fraction would you use to study the molecules involved in DNA function?

1.26 Speculate on why so many of the molecules in nature are based on the element carbon.

1.27 It is often said that in order for life to reproduce, develop, and thrive, there are three basic requirements: (1) a blueprint (directions), (2) materials, and (3) energy. What specific biomolecules or biological processes provide each of these needs?

1.28 Some scientists have argued that biochemical data obtained from cellular extracts (prepared by breaking open cells and thus killing them) are artifactual and that biochemistry should only be studied using living cells. Comment on the pros and cons of this statement.

1.29 Although you have not yet studied the biochemistry of metabolism, describe in your own words how food is converted to energy for muscle contraction.

1.30 Describe in general terms how energy from the sun is used by plants to make carbohydrates.

1.31 Use chemical structures to write condensation reactions of acetic acid, CH_3COOH, with each of the following molecules.

  a. CH_3CH_2OH, ethyl alcohol

  b. NH_3, ammonia

  c. CH_3COOH, acetic acid

  d. CH_3SH, methylthiol

  e. $CH_2-CH-CH_2$, glycerol
       $\ \ |\ \ \ \ \ |\ \ \ \ \ |$
       OH OH OH

1.32 DNA is more stable to hydrolysis reactions than RNA or protein. Is this consistent with the role of DNA in cells? Explain.

1.33 Consider a typical condensation/hydrolysis reaction:

$$\underset{\ }{R-\overset{\displaystyle O}{\overset{\|}{C}}-OH} + R'-NH_2 \rightleftharpoons R-\overset{\displaystyle O}{\overset{\|}{C}}-NHR' + H_2O$$

In a living cell, would you expect the equilibrium to lie toward the condensation products or the hydrolysis products? Explain.

1.34 Speculate on how proteins and other biomolecules in thermophilic organisms may be made to be more stable than the same molecules in organisms living under temperate conditions.

Writing Biochemistry

1.35 Find an article in a current newspaper or newsmagazine that is based on biochemistry or biomedicine. The topic should be related to concepts present in this chapter. Write a 50-word summary of the article emphasizing the importance of biomolecules or biological processes. Discuss ethical implications if appropriate.

 Applying Biochemistry

1.36 The unit on the history of biochemistry outlines the qualities that made Louis Pasteur, Francis Crick, and James Watson successful scientists and allowed them to make revolutionary discoveries. Make a list of these qualities and remember them because they will make you a successful student in biochemistry.

1.37 You will recall that Francis Crick in his youth was worried that all important scientific discoveries had already been made. A recently published book is focused on "the end of science" and is based on the idea that most major discoveries have been made and our scientific understanding of nature will not greatly increase in the future. Do you believe this idea and does this prediction change your mind about wanting to become a scientist?

1.38 In cell structure, the concept of "compartmentation" is an important one. This characteristic of cells is first obvious when we note that all living cells have some kind of boundary in the form of cell walls and/or membranes. The idea of compartmentation continues inside the cell in the form of vesicles and organelles. What special advantages does compartmentation provide for a living cell?

Further Reading

Atkins, P., 1987. *Molecules.* New York: W. H. Freeman.

Boyer, R., 2000. Centrifugation in biochemical research. In *Modern Experimental Biochemistry,* 3rd ed., pp. 189–208. San Francisco: Benjamin/Cummings.

Bullock, C., 2000. The Archaea—A biochemical perspective. *Biochem. Mol. Biol. Educ.* 28:186–191.

Cohen, S., 1984. The biochemical origin of molecular biology. *Trends Biochem. Sci.* 9:334–336.

Doolittle, W., 2000. Uprooting the tree of life. *Sci. Am.* 282(2):90–95.

de Duve, C., 1975. Exploring cells with a centrifuge. *Science* 189:186–194.

de Duve, C., 1991. *Blueprint for a Cell. The Nature and Origin of Life.* Burlington, NC: Carolina Biological Supply Co.

Goodsell, D., 1991. Inside a living cell. *Trends Biochem. Sci.* 16:203–206.

Goodsell, D., 1996. *Our Molecular Nature.* New York: Copernicus (Springer).

Graz, C., 1998. Teaching the origin of the first living systems. *Biochem. Educ.* 26(4):286–289.

Holliday, R., 2001. Aging and the biochemistry of life. *Trends Biochem Sci.* 26:68–71.

Ingber, D., 1998. The architecture of life. *Sci. Am.* 278(1):48–57.

Olson, A., and Goodsell, D., 1992. Visualizing biological molecules. *Sci. Am.* 267(5):76–81.

Orgel, L., 1998. The origin of life—a review of facts and speculations. *Trends Biochem. Sci.* 23:491–495.

Rasmussen, N., 1996. Cell fractionation biochemistry and the origins of "cell biology." *Trends Biochem. Sci.* 21:319–321.

Srere, P., 2000. Macromolecular interactions: Tracing the roots. *Trends Biochem. Sci.* 25:150–153.

Tribe, H., 2000. The making of a model of *Escherichia coli,* magnified two million times—The "Millennium Bug." *Biochem. Educ.* 28:120–123.

Chapter 2

The Flow of
Biological Information:
Cell Communication

The female gypsy moth secretes chemicals called pheromones to attract a male.

B iochemistry is much more than just a study of molecules and chemical reactions in the cell. Biochemistry has several dominant themes as discussed in Chapter 1: (1) the flow of biological information (molecular recognition and cellular communication), (2) the flow of energy and matter (bioenergetics and metabolism), and (3) the structure and function of biomolecules. The fundamentals of information flow are introduced in Chapter 2.

When flow of information is considered, one usually thinks about computer networks, radio waves, and wireless telephones. However, biochemists and molecular biologists have for many years been interested in learning how biological information is transferred from one generation to another, how cells and organisms respond to changes in their environment, how cells "speak" to each other via hormones and neurotransmitters, and how organisms communicate using chemicals called pheromones. What has been discovered is that the transfer of biological communication or the flow of information can be described using the basic laws of chemistry. DNA, RNA, proteins, and even some carbohydrates are information-rich molecules that carry instructions for cellular processes. "Reading" that information depends on specific, noncovalent interactions between molecules. Consider the storage and transfer of electronic information with a computer: Information is stored on a memory disk, retrieved for modification or processing, and transmitted to a printer or other computer. Genetic information is stored in a macromolecule, DNA. That information is passed on to the next generation by duplicating the DNA molecule. The information in DNA may also be processed into cellular RNA, a chemically and functionally different nucleic acid form. DNA contains the blueprint for construction of RNA, which then carries the message for synthesis of proteins. It is the protein molecules that are responsible for building cellular components and for maintaining the proper functioning of the cell. The process of information transfer from DNA is discussed in this chapter.

Other important levels of cellular communication control biochemical action. In order to coordinate the many activities of cells in a multicellular organism, it is essential for cells to communicate. The system that developed is based on chemical signaling. Most cells produce and secrete molecules that function as carriers of information. These chemical media-

Figure 2.1

The storage and replication of biological information in DNA and its transfer via RNA to synthesize proteins that direct cellular structure and function.

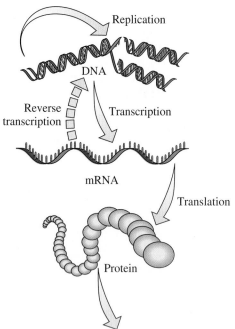

Cell structure and function
• Energy metabolism
• Synthesis and breakdown of biomolecules
• Storage and transport of biomolecules
• Muscle contraction
• Cellular communication (signal transduction)

tors interact with target cells where they evoke their desired biological effect. Biochemists have uncovered a number of complex strategies for transmission of the message to the inside of the target cell. A common mechanism for transmembrane communication is **signal transduction,** a process whereby the presence of a molecule exterior to the cell relays a command to an interior cell component. The components and mechanisms of signal transduction pathways are discussed later in this chapter. A schematic outline of cellular communication is shown in Figure 2.1. Note that DNA is the original source of all information. This chapter briefly describes the role of DNA as the repository of cellular information. Chapters 10 to 13 give details of the molecular events that occur in the use of DNA information.

Organisms at all levels of evolutionary development, single cellular or multicellular, have the ability to communicate with other members of the same species. Organisms produce and release chemicals in their environment in order to transmit messages to like species and to enemies. Some well-known examples are the pheromones, molecular signals released by insects, and perhaps even humans and other animals, that relay messages of a sexual nature between members of the opposite gender. Molecular details of this process appear in Chapter 9.

2.1 Storage of Biological Information in DNA

Biological Information

DNA, RNA, proteins, and some carbohydrates are informational molecules in that they carry directions for the control of biological processes. As we learned in Chapter 1, these groups of biopolymers are composed of monomeric units held together by covalent bonds, which are stable enough to store important data for relatively long periods of time. The informational content of these molecules is utilized by "reading" the sequence of the monomeric units. (Recall that such polysaccharides as starch and cellulose contain only one kind of monomer so they carry limited information). Reading the messages in the macromolecules depends on the formation of weak, noncovalent bonds between biomolecules. There are four types of noncovalent bonds that are of importance: van der Waals forces, ionic bonds, hydrogen bonds, and hydrophobic interactions (see Section 3.1).

The total genetic informational content for each cell, referred to as the **genome,** resides in the long, coiled macromolecule of DNA. Thus, DNA is the molecular repository for all genetic information. The informational message is expressed or processed in two important ways: (1) exact duplication of the DNA so it can be transferred during cell division to a daughter cell and (2) expression of stored information to first manufacture RNA and then manufacture the proteins that are the molecular tools that carry out the activities of the cell. In this indirect way, DNA exerts its primary effects by storing the messages for synthesis of proteins that direct the thousands of chemical reactions that occur in a cell (see Figure 2.1).

The DNA Molecule

The chemistry of the DNA molecule seems rather simple in view of the enormous amount of information stored and its fundamental role in the cell. As we learned in Chapter 1, a DNA chain is a long, unbranched heteropolymer, constructed from just four types of monomeric nucleotide subunits. Each monomer unit consists of three parts: an organic base containing nitrogen, a carbohydrate, and a phosphate. In 1952, Watson and Crick, using X-ray diffraction data and models, discovered that the DNA molecule is constructed of two strands interwoven into a three-dimensional helical structure (double helix; see Figure 2.2). The structural backbone of each strand, which makes up the outside of the molecule, is formed by covalent phosphodiester bonds, using the carbohydrate and phosphate groups of the nucleotide subunits. This arrangement brings the organic bases to the inside of the double helix. Neighboring bases on the same strand are stacked on top of each other (arranged like steps on a spiral staircase), which allows the formation of noncovalent interactions. Bases on opposite strands are also close neighbors, allowing the formation of **complementary base pairs** by specific hydrogen bonding. The optimal arrangement is adenine (A) in combination with thymine (T) and guanine (G) with cytosine (C). The language used for

This 1964 Israeli stamp, titled Macromolecules of the Living Cell, shows the DNA double helix and base pairing

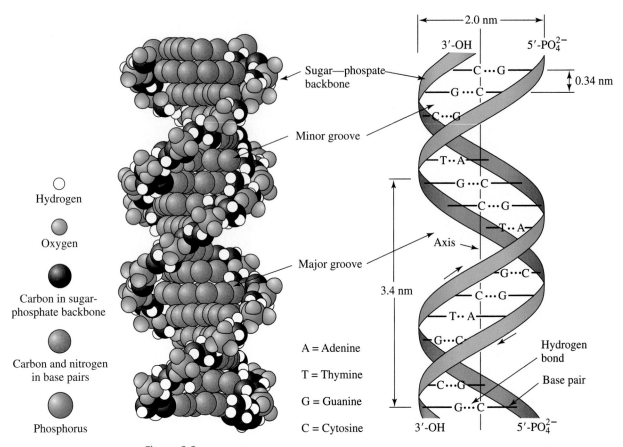

Figure 2.2

The Watson and Crick double helix model for DNA showing the stacking of nucleotide bases on the same strand and the hydrogen bonds between complementary nucleotide bases on opposite strands.

Cutting Edge:
Visionaries of science

Structure Tutorials: DNA

information storage in DNA consists of a four-letter alphabet: A, T, G, and C. The format for storage is a linear sequence of the nucleotides in DNA that vary in size and sequence with each species. The entire human genome, for example, is about 1 m long and contains an estimated 3 billion nucleotide base pairs. A DNA molecule from *E. coli* is about 2 μm long and contains about 4 million nucleotide pairs. We will return to details of DNA structure and function in Chapters 10 through 13.

The goal of the **Human Genome Project (HGP)** is to map and sequence the estimated 3 billion nucleotide base pairs of the human genome by the year 2005. The collaborative effort is sponsored by five governments (United States, United Kingdom, Japan, Germany, and France) and by the privately funded corporation, Celera Genomics. An initial stage of the project, which is now completed, was to sequence the genomes of other species including *Bacillus subtilis* (yeast) and *Caenorhabditis elegans* (nematode worm). A "rough draft" of the human genome sequence was announced in June 2000, five years ahead of the deadline. Results from the HGP are already having an impact on medicine and science. We now know the chromosomal positions of genes that are defective in many diseases including cystic fibrosis, muscular dystrophy, some forms of breast cancer, colon cancer, and Huntington's disease. The next step is to apply techniques of gene therapy to treat these diseases (see Chapter 13). New branches of science are being created to deal with the flood of data coming from the HGP. It is estimated that there are between 30,000 and 40,000 genes in the human genome that are expected to code for an equivalent number of proteins (the proteome). **Proteomics** is the name given to the broad field investigating the thousands of protein products made from the genome. Scientists trained in this area will attempt to classify and characterize the proteins, study their interactions with other proteins, and identify their functional roles. Those specializing in the subfield of **bioinformatics** will apply computers to organizing the mass of nucleic acid sequence data and studying relationships between protein sequence and structure.

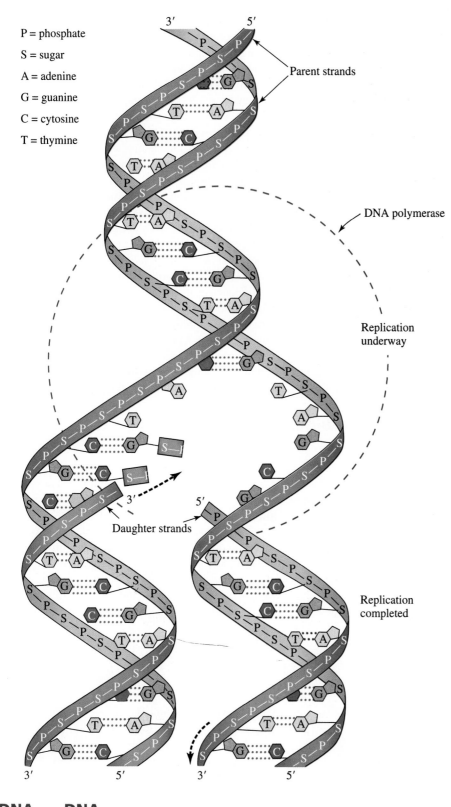

P = phosphate
S = sugar
A = adenine
G = guanine
C = cytosine
T = thymine

Parent strands

DNA polymerase

Replication
underway

Daughter strands

Replication
completed

Figure 2.3
The process of DNA replication
begins with strand unwinding.
Each strand serves as a template
for synthesis of a new daughter
DNA strand. The replication
process, catalyzed by the enzyme
DNA polymerase, occurs in a
semiconservative fashion with the
newly synthesized daughter molecules consisting of one new DNA
strand and one strand from the
parent DNA.

DNA → DNA

The duplication of DNA is a self-directed process. The DNA in concert with many accessory proteins dictates and directs the construction of identical DNA for progeny cells. The
process of DNA copying, called **replication,** begins with the unwinding of a short segment
of the two complementary strands (see Figure 2.3). Each strand is then used as a **template**
(pattern) for production of a new complementary partner strand. A new nucleotide subunit
brought into the process must first be held in position by hydrogen bonds and van der Waals

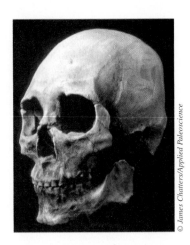

The skull of the 9400-year-old Kennewick man found along the Columbia River in Washington State, USA, in 1996.

forces to a complementary base on the template. Then it is covalently linked to the growing DNA chain by an enzyme called **DNA polymerase.** The chemical details of DNA replication are discussed in Chapter 11. The entire DNA molecule is duplicated, resulting in two identical molecules, one remaining in the parent cell and one for the daughter cell. The duplication process is called **semiconservative replication**—each duplex DNA molecule is composed of one original strand and one newly synthesized strand.

The question may be asked: why was DNA chosen for this most important role in the cell? The DNA molecule has been found to be especially stable under intra- and extracellular conditions. The covalent bonds linking the individual nucleotide subunits are chemically stable and not especially susceptible to hydrolytic cleavage in the aqueous environment of the cell. This results in a secure and durable storage form of genetic information that must remain undamaged and unchanged from generation to generation. Biochemists have recently discovered that DNA can be extracted from museum specimens and archaeological finds up to several million years old. Samples of DNA have been detected and analyzed from museum animal skins (140 years old), an Egyptian mummy (5000 years old), an 8000-year-old human brain, a 9400-year-old skeleton found in Washington State, USA, a 45,000-year-old plant, 65-million-year-old dinosaur bones, and a 120-million-year-old amber-preserved weevil. The DNA molecules extracted from ancient items are somewhat degraded in size and chemically modified by natural oxidation processes. However, by using a new technique in molecular biology called the **polymerase chain reaction,** it is possible to reconstruct the DNA closely to its original form and amplify the production of identical molecules (Chapter 13). Copies of the "antique DNA" are suitable for sequencing and can be compared to "modern DNA." These new developments now open up the possibility of studying directly the process of evolution. Indeed, a new field of "molecular archaeology" is emerging.

2.2 Transfer of Biological Information to RNA

DNA → RNA

In the previous section, we discovered how parental DNA is duplicated for genetic transfer to progeny cells. In this section we study the transformation of the message of DNA into the form of RNA. The word **transcription** is used to describe this process. DNA consists of a coded thread of information. During replication, the entire DNA molecule (from end to end) is duplicated. In contrast, transcription of DNA follows a somewhat different pathway. A significant fraction, but not all, of the message in DNA is expressed into RNA. The early hereditary studies of Mendel and others, showing transfer of characteristic traits to offspring, are best explained by dividing the genome into specific coding regions or units called **genes.** In prokaryotic cells, the sequence of bases in a gene is "read" in a continuous

Figure 2.4

Transcription of DNA to produce mRNA.

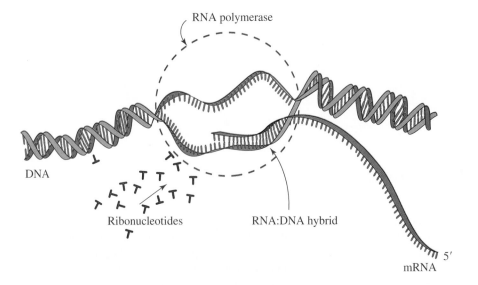

fashion with no gaps or interruptions. A gene in prokaryotic DNA can be defined as a region of DNA that codes for a specific RNA or protein product.

The process of transcription to produce RNA is similar to DNA replication except for the following changes as shown in Figure 2.4. (Details of the transcription process are discussed in Chapter 11.)

1. Ribonucleotides rather than deoxyribonucleotides are the monomeric building blocks.
2. The base thymine, which forms complementary base pairs with adenine, is replaced with uracil, which also pairs with adenine.
3. The RNA:DNA hybrid duplex product eventually unravels, releasing single-strand RNA and allowing the DNA template strands to rewind into a double helix.
4. The enzyme linking the nucleotides in the new RNA is **RNA polymerase.**

Many viruses, including those causing polio, influenza, and AIDS, and retroviruses that cause tumors, have a genome that consists of single-stranded RNA rather than DNA. Two different strategies are used by these viruses to assure multiplication. The retroviruses rely on a special enzyme, the structure of which is coded in their RNA and produced by the synthetic machinery of the infected cell. This enzyme, reverse transcriptase, converts the RNA genome of the virus into the DNA form that is incorporated into the host cell genome. Viral genetic information in this form can persist in the host cell in a latent and noninfectious state for years until stressful environmental conditions induce infection. Other RNA viruses affect multiplication by dictating the production of replicase, an enzyme that catalyzes the duplication of viral genomic RNA in a process similar to DNA replication.

Three Kinds of RNA

The transcription of cellular DNA leads to a heterogeneous mixture of three different kinds of RNA: ribosomal, transfer, and messenger. As shown in Table 2.1, the three types have some common characteristics: (1) They are all products of DNA transcription that is carried out by RNA polymerases (except in RNA viruses); (2) they are usually single stranded, except for small regions where a molecule may fold back onto itself; and (3) they all have functional roles in protein synthesis. The most abundant RNA is **ribosomal RNA (rRNA).** This is found in combination with proteins in the ribonucleoprotein complexes called ribosomes. In Chapter 1 ribosomes were defined as the subcellular sites for protein synthesis. Of the three major RNA forms, **transfer RNA (tRNA)** is the smallest with a size range between 73 and 93 nucleotides. This form of RNA combines with an amino acid molecule and incorporates it into a growing protein chain. There is at least one kind of tRNA for each of the 20 amino acids used in protein synthesis. **Messenger RNA (mRNA)** comes in a variety of sizes. Each type of mRNA carries the message found in a single gene or group of genes. The sequence of nucleotide bases in the mRNA is complementary to the sequence of bases in the template DNA. Messenger RNA is an unstable, short-lived product in the cell, so its message for protein synthesis must be immediately decoded and is done so several times by the ribosomes to make several copies of the protein for each copy of mRNA (Figure 2.5). Details of RNA structure and function are presented in Chapters 10 through 12.

**Interactive Animation:
Central dogma of biochemistry**

Table 2.1

Properties of the three kinds of RNA

Type of RNA	Relative Size	Base Pairing by Hydrogen Bonding	Biological Function
Transfer	Small	Yes, high level	Activates and carries amino acids for protein synthesis
Ribosomal	Three sizes, most are large	Yes, high level	Present with proteins in ribosomes, the cellular sites of protein synthesis
Messenger	Variable	No	Carries direct message for synthesis of proteins

Figure 2.5
Schematic diagram of the synthesis of proteins on ribosomes. Each copy of mRNA may have several ribosomes moving along its length, each synthesizing a molecule of the protein. Each ribosome starts near the 5′ end of an mRNA molecule and moves toward the 3′ end.

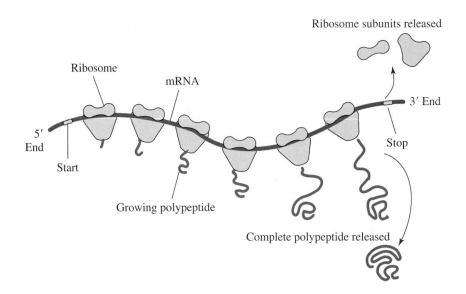

2.3   Protein Synthesis

mRNA → Proteins

The ultimate products of DNA expression in the cell are proteins. The full array of proteins made from the genomic DNA of an organism is called the **proteome.** Information residing in DNA is used to make single-stranded mRNA (transcription), which then relays the message to the cellular machinery designed for protein synthesis. Thus, mRNA serves as an intermediate carrier of the information in DNA. The message of DNA is in the form of a linear sequence of nucleotide bases (A, T, G, C); the message in mRNA is a slightly different set of nucleotide bases (A, U, G, C). Protein molecules, however, are linear sequences of structurally different molecules: amino acids. Two different "languages" are involved in the transformation from DNA and RNA to proteins; therefore, a **translation** process is required.

The Genetic Code

By studying the nucleotide base sequence of hundreds of genes and correlating them with the linear arrangement of amino acids in protein products of those genes, biochemists have noted a direct relationship. The two sequences are found to be collinear; that is, the sequence of bases in a gene is arranged in an order corresponding to the order of amino acids in the product protein (see Figure 2.6). A set of coding rules, called the **genetic code,** has been deciphered by studying the protein products of many synthetic and natural genes:

1. The coding ratio is a set of three nucleotides per amino acid incorporated into the protein; therefore, a triplet code is in effect.
2. The code is nonoverlapping; the three nucleotides on DNA are adjacent, treated as a complete set, and used only once for each translation step.
3. The sets of three nucleotides are read sequentially without punctuation.
4. A single amino acid may have more than one triplet code; that is, the genetic code is degenerate.
5. The code is nearly universal.
6. The triplet code also contains signals for "stop" and "start." The genetic code is translated in Figure 2.7.

Interactive Animation:
Protein synthesis

The presence or absence of a protein in the cell is usually controlled at the level of DNA. Control mechanisms, as discussed in Chapter 12, direct the production of protein by regulating the transcription of DNA. The control may be triggered by signal molecules inside or outside the cell. Intracellular signal molecules, which are often proteins, function by binding to a discrete region on the DNA that switches transcription on or off. DNA expression and other

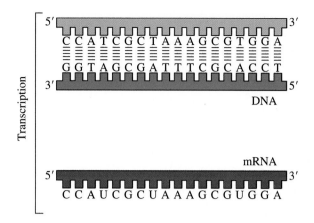

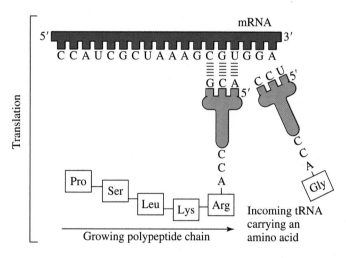

Figure 2.6
The collinear relationship between the nucleotide base sequence of a gene with the linear arrangement of amino acids in a protein. First, mRNA is formed as a complementary copy of one strand of the DNA. Sets of three nucleotides on the mRNA are then read by tRNA molecules. This involves the formation of hydrogen bonds between complementary bases. The amino acid attached to the tRNA is incorporated into the growing polypeptide.

metabolic activities may also be regulated by extracellular signal molecules, such as growth factors and hormones, through the intermediacy of **second messengers.** These processes of signal transduction are discussed later in this chapter. We consider the details of the translation process, the cell components required, and the regulation of protein synthesis in Chapter 12.

Exons and Introns

To summarize protein synthesis in prokaryotic cells, the sequence of DNA is "read" from a fixed starting point and the message is transcribed into the form of mRNA. The information in mRNA is translated by ribosomes into the language of amino acids, which are linked together by enzymes to form the protein product. It was always assumed that DNA expression in eukaryotic cells was similar if not identical to that in prokaryotic cells. To the surprise of many biochemists, it was discovered in 1977 that coding regions on eukaryotic DNA are often interrupted by noncoding regions; hence, it was said that such genes are split. The coding regions are called **exons**; the noncoding regions, intervening sequences or **introns.** The average size of an exon is between 120 and 150 nucleotide bases, or coding for 40 to 50 amino acids of a protein. Introns can be larger or smaller with a range of 50 to 20,000 bases in length. The reasons for the presence of introns in eukaryotic DNA are not completely understood. On one hand, some believe that introns contain "junk DNA" that serves no useful purpose and will eventually disappear as evolutionary forces continue to work. More likely, introns serve some purpose to more efficiently produce diverse arrays of proteins. The presence and extent of gene splitting depends on the evolutionary status of the organism. Introns are absent in prokaryotes, rare in lower eukaryotes (such as yeast), and rather common in vertebrates. This means that for eukaryotic cells we must change our idea that a gene is an isolated and fixed region carrying the message for a single protein.

Figure 2.7

The genetic code describing the relationship between a triplet of nucleotide bases on mRNA and the amino acid incorporated into a protein. The three nucleotide bases for each amino acid are read from the appropriate columns. For example, the codes for the amino acid phenylalanine are UUU and UUC.

	2nd position			
1st position (5′ end)	**U**	**C**	**A**	**G**
U	UUU UUC } Phe UUA UUG } Leu	UCU UCC UCA UCG } Ser	UAU UAC } Tyr UAA UAG STOP	UGU UGC } Cys UGA STOP UGG Trp
C	CUU CUC CUA CUG } Leu	CCU CCC CCA CCG } Pro	CAU CAC } His CAA CAG } Gln	CGU CGC CGA CGG } Arg
A	AUU AUC AUA } Ile AUG Met (START)	ACU ACC ACA ACG } Thr	AAU AAC } Asn AAA AAG } Lys	AGU AGC } Ser AGA AGG } Arg
G	GUU GUC GUA GUG } Val	GCU GCC GCA GCG } Ala	GAU GAC } Asp GAA GAG } Glu	GGU GGC GGA GGG } Gly

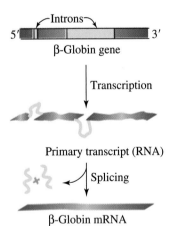

5′ ⌐Introns⌐ 3′

β-Globin gene

↓ Transcription

Primary transcript (RNA)

↙ Splicing

β-Globin mRNA

Figure 2.8

Processing of the split gene of the β chain of hemoglobin includes transcription of the entire gene to produce the primary RNA transcript and removal of the intervening sequences (introns) by splicing. Exons are shown in blue, introns in yellow.

The discovery of noncoding regions in DNA raised important questions. If some of the message in DNA is not used for protein synthesis, at what level is this information removed? Are both coding and noncoding regions transcribed into mRNA and translated into protein molecules that require shortening before they are functional, or is the mRNA modified before it is translated? It was discovered that newly synthesized mRNA is much longer than the form of mRNA actually translated by ribosomes. The final form of mRNA used for translation is the result of extensive and complicated chemical processing events, sometimes requiring several accessory enzymes. It is not uncommon for a gene (a region coding for a polypeptide or protein) to have two or more introns. For example, the gene for the β chain of hemoglobin (as shown in Figure 2.8) is split in two places by introns. These intervening sequences are spliced from the mRNA and the exons joined to produce the functional mRNA for the β chain.

As we will discover in later chapters, RNA processing in many organisms is carried out by **small nuclear ribonucleoproteins (snRNPs)** and specific enzymes. In some organisms, such as the ciliated protozoan *Tetrahymena thermophilia,* the RNA is processed by self-splicing without the assistance of enzymes or other proteins. The discovery of self-splicing or **catalytic RNA** raises some interesting evolutionary implications. This may point to RNA, a molecule that can function both as a replication and translation template and as an enzyme, as the first functional biomolecule in the origin of life. During evolutionary development, DNA became the replication template because it is chemically more stable than RNA, and proteins became catalysts (enzymes) because more efficient and diverse reactions were possible than with RNA. The role of catalytic RNA in processing RNA is discussed in Chapters 7 and 11.

2.4 Errors in DNA Processing

DNA Mutations

We have described DNA expression (replication, transcription, translation) as events that are carried out in a precise, accurate, and reproducible manner. All of the events are dependent on weak noncovalent interactions for recognition and binding. No mention was made regarding the possibility of errors in these processes or the result of chemical or physical changes to DNA. Throughout millions of years of development, organisms have evolved mechanisms that faithfully transfer genetic information; however, they have also developed repair processes for errors that may occur. The average error rate in replication is less than one wrong nucleotide inserted for every 10^9 nucleotides added. This is an error level that we can live with. In a rare event, a mistake may be made. The wrong nucleotide may be incorporated, a nucleotide may be deleted, or an extra one inserted. Changes of these kinds, called **mutations,** have consequences on that cell as well as on future generations since those changes will be continued. A mutation may be of two types: (1) If it is in an exon region of the DNA, it may alter the amino acid sequence of the protein and perhaps cause it to be functionally inactive; or (2) it may occur in a noncoding region (intron) and be without effect (a silent mutation). During DNA processing into new DNA, RNA, and proteins, several monitoring procedures are in effect to detect and correct expression errors before they become incorporated in proteins. These repair systems are discussed in Chapter 11.

Some individuals have inborn errors in their DNA that are transferred from generation to generation. These errors may be benign or may cause specific diseases. For example, **sickle cell anemia,** a disease condition found most often in people of African descent, is characterized by changes in the individual's hemoglobin (Figure 2.9). As we will discover in Chapter 5, these individual's hemoglobin is made with two amino acids out of 546 that are different from normal hemoglobin. This minor change, which results in dysfunctional hemoglobin molecules, causes much pain and suffering and even early death.

Changes made in DNA by undesired chemical or physical events are not as easy to deal with as natural errors in replication, transcription, and translation. Ultraviolet light, ionizing radiation (radioactivity), and some chemicals induce irreparable changes in DNA that result in nonfunctional proteins (Chapter 11). In fact, there is evidence that these agents cause some forms of cancer. A complete understanding of the molecular events that transform a normal cell into a cancerous one is still lacking.

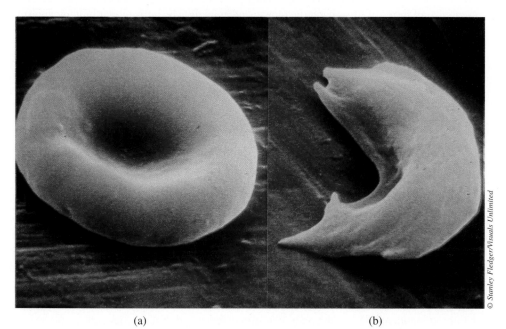

(a)   (b)

© Stanley Fledger/Visuals Unlimited

Figure 2.9

Red blood cells from a normal individual (a) and from an individual with sickle cell anemia (b). Normal hemoglobin is present in (a), whereas a variant hemoglobin HbS is present in (b), causing deformation.

2.5   Information Flow Through Cell Membranes

Information Transfer by Signal Transduction

The many biological activities occurring in a cell require exact coordination both from within and without. This is especially important for highly developed organisms that consist of various organs and types of tissues, each with a distinct type of cell and a specific biological role. Metabolic activities and other biological processes in cells are altered by the action of extracellular chemical signals, such as hormones, and growth factors. The molecular signals are secreted by many types of cells and are distributed throughout the organism. Each chemical messenger (the total number may be over 100 different kinds in a highly developed organism), whatever its site of origin, has a unique chemical structure and biological effect. Target cells using receptors on their surface recognize only those signal molecules intended for them. Some chemical mediators function in the local region of the secreting cell. Interesting examples of local mediators are the **prostaglandins,** molecules synthesized from fatty acids in a wide variety of cells. These molecular messengers, which have a diverse array of biological activities, including contraction of smooth muscle and blood platelet aggregation, are classified as lipids and are discussed in Chapter 9. Here we will focus on hormones, those signaling molecules synthesized and secreted by endocrine glands and transported to their site of action (target cell) via the bloodstream. These long-distance mediators have a wide range of structures that include amino acid derivatives, small peptides, proteins, and steroids. We are all familiar with the action of some hormones. Insulin and glucagon, for example, peptide products of the pancreas, regulate the rate of glucose metabolism. The steroid hormones, estrogens and androgens, produced by the gonads and adrenal cortex, respectively, regulate the development of secondary sex characteristics. Our understanding of how these and other hormones elicit their effects has increased greatly in recent years. We originally believed that all hormones interacted directly, inside the cell, with the proteins, enzymes, and nucleic acids whose activities they altered. Indeed, this is the pathway followed by some steroid hormones that, because of their nonpolar chemical nature, are able to diffuse readily through the membrane and proceed with direct delivery of their message. Other hormones are relatively polar and/or ionic and are unable to diffuse through the membrane of the target cell. Here a different strategy must be evoked. The current level of understanding is based on the concept of **signal transduction** (also called cell signaling), a process in which an extracellular chemical message is transmitted through the cell membrane to elicit an intracellular change. Signal transduction processes are found in essentially all forms of life—bacteria, plants, and animals. The intracellular biological result may involve metabolic regulation, cell development, muscle contraction, or response to environmental stimuli.

Interactive Animation: Signal transduction

Characteristics of Signal Transduction

The detailed step-by-step process of signal transduction will vary from organism to organism and from hormone to hormone, but a general chain of events involving several types of proteins can be outlined (Figure 2.10). The hormone proceeds from its source, via the bloodstream, to a target cell. Here it delivers its message by binding to specific receptors, usually protein molecules associated with the cell membrane. Most receptor molecules have three distinct regions: (1) a region (head) on the outside of the cell where the hormone binds, (2) a region that spans the cell membrane, and (3) a region (tail) that extends into the cytoplasm of the cell. Upon binding of the hormone to the receptor protein, the tail changes in shape and interacts with a **G protein,** a family of biomolecules associated with the inner membrane, so named because they bind guanine nucleotides (GDP, GTP). (Other signal transduction receptors we will encounter later include receptor tyrosine kinase and receptor-associated tyrosine kinase.) The G protein, thus activated, then passes the signal to an intracellular enzyme, usually **adenylate cyclase.** Depending on the type of G protein, adenylate cyclase is either stimulated or inhibited in its action. Adenylate cyclase is a ubiquitous enzyme that catalyzes the formation of **cyclic adenosine 3′,**

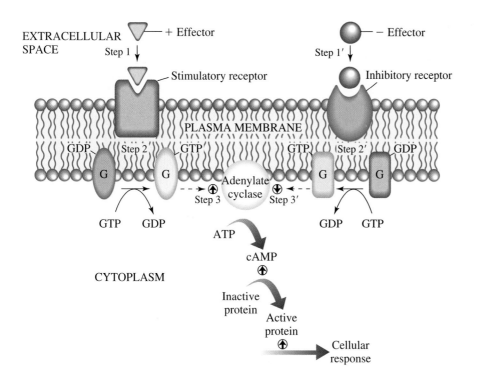

Figure 2.10

Details of the signal transduction process. Step 1: A hormone or other effector molecule binds to its receptor protein present in the membrane; an effector may be stimulatory (+, Step 1) or inhibitory (−, Step 1′). Step 2: The receptor stimulates interaction with a G protein, which activates the enzyme adenylate cyclase. Step 2′: An inhibitory process deactivates adenylate cyclase. Step 3: Adenylate cyclase produces cAMP, which causes a cascade of metabolic reactions including the conversion of an inactive protein (perhaps an enzyme) into an active protein. Step 3′: Cyclic AMP is not produced. In an example of epinephrine as the effector and a muscle cell as the target cell, the ultimate cellular response is an increase in intracellular cAMP and glucose concentration and an increase in metabolic energy. A stimulatory response is indicated by ⊕, and an inhibitory response is indicated by ⊖.

5′-monophosphate (cyclic AMP or cAMP) from ATP. An example of a **second messenger,** cAMP, is a short-lived, intracellular molecule that carries the command originally transmitted by the first messenger, the hormone. Once formed in the cell, cAMP or other second messenger acts by triggering a chain of reactions (a cascade event) regulated by protein kinase enzymes. Kinases catalyze the transfer of phosphoryl groups ($-OPO_3^{2-}$) from ATP to substrate molecules. Kinases act as molecular switches to turn metabolic processes on or off. To illustrate the flow of information by signal transduction, consider epinephrine (adrenaline), a product of the adrenal medulla, as an example of a first messenger (hormone) for muscle cells. In animals, strenuous muscular activity stimulates the release of epinephrine into the bloodstream. Being a relatively polar, water-soluble molecule, epinephrine is not able to diffuse through the nonpolar region of the muscle cell membrane; therefore, it must initiate its action by binding to a receptor on the outer surface of the membrane. Epinephrine binding to receptor proteins on the surfaces of muscle cells initiates the signal transduction process described above and in Figure 2.10. The ultimate intracellular response (after the complex cascade of reactions) is the stimulation of an enzyme that catalyzes the release of glucose from its storage form, glycogen. The increase of glucose concentration enhances the rate of metabolism, thus generating more cellular energy. Details on the actions of metabolic hormones, insulin, glucagon, and epinephrine are given in Chapter 20.

Recent detailed studies on signal transduction have uncovered other characteristics of the process. For example, receptors are now known to be physically connected to elements of the cytoplasm. Scaffolding and anchoring proteins hold groups of receptor proteins together to form networks for accurate transmission of information. Cell surfaces have many different types of receptors, and the scaffolding proteins are thought to organize and enhance the signal transduction process by holding all necessary extracellular and intracellular molecular components together in a single network. This new discovery corresponds well with studies showing that the cytoplasm is not an amorphous gel, but is made up of the fibrous matrix of the cytoskeleton and other proteins (see Section 1.4).

In related studies, other important second messengers have been discovered. Besides cAMP, we now know that cGMP, Ca^{2+}, and diacylglycerol transmit signals for intracellular regulation. Details of signal transduction systems related to specific roles will be covered in Part IV.

Two important characteristics of signal transduction should be considered at this time. *First, the chemical signal from hormone binding is amplified;* that is, it increases in magnitude at each of several steps of the cascade. Only one molecule of the hormone binding on the membrane is required to activate a molecule of adenylate cyclase; but this enzyme molecule can catalyze the formation of many molecules of cAMP. Each of these second messenger molecules can act to switch on protein kinase. Likewise, each activated protein kinase molecule can act on many target enzyme molecules. As a result of this reaction chain, the single binding event of a hormone molecule leads to an intracellular signal that may be amplified by several thousand times. The signal is enhanced even further because a target cell has many receptors for a specific type of hormone and each hormone–receptor interaction leads to the described amplification. *Second, hormones are usually released by the endocrine glands on demand.* Some change in the environment of the organism has made it necessary to change cellular conditions. For most hormones, it is desirable that its effect not be felt continuously by the target cell. The chemical signal delivered by the hormone, therefore, must be rapid and transient. A mechanism must be available for deactivation of the sig-

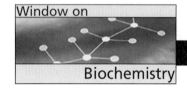
Window on

Biochemistry

2.1 Cell-Signaling Processes Are Altered by Drugs and Toxins

Biological processes that are dependent on cellular signaling may be turned on and off by the presence of foreign molecules and organisms. The mild stimulants caffeine (coffee), theophylline (tea), and theobromine (chocolate) act by inhibiting the enzyme cAMP phosphodiesterase, which catalyzes cAMP breakdown. Thus, the action of epinephrine and other hormones is prolonged in the presence of the stimulants. The newly approved abortifacient, RU486, acts by binding to progesterone receptors and inhibiting the normal function of the hormone in fertilization.

Many human maladies of the past and present result from bacterial and viral manipulation of signal transduction. For example, cholera, a disease that has been described in historical records for over 1000 years, is caused by the bacterium *Vibrio cholerae.* The cholera toxin, a protein, interferes with the normal action of cell-signaling G proteins, thereby causing the continuous activation of adenylate cyclase. Resulting high concentrations of cAMP in epithelial cells in the intestines cause uncontrolled release of water and Na^+, leading to diarrhea and dehydration. The nucleotide sequence of *V. cholerae* genome was announced in 2000, and its two chromosomes were found to contain about 4.1 million base pairs and about 3800 protein-encoded genes. *Yersinia pestis,* the bacterium that caused the "black death" plague in 14th-century Europe, acts by injecting into cells proteins that interfere with normal signaling processes. *Bordetella pertussis,* the bacterium that causes whooping cough, also produces a toxin that interferes with G proteins that inhibit adenylate cyclase. The human immunodeficiency virus (HIV) enters cells by binding to cell signaling receptors.

Maturity-onset (type 2) diabetes and cancer also may have causes linked to malfunctioning cell signaling. Patients with type 2 diabetes (insulin-insensitive) have a deficiency of insulin receptors. Cancer, which is usually described in terms of cell proliferation, is better understood when it is considered a disease of cellular communication. The message that a hormone or growth factor brings to the cell is often meant to regulate, by signal transduction, DNA replication, transcription, and other activities of cell development. A malfunction of this process may lead to uncontrolled cell growth. The link between signal transduction and cell proliferation has become stronger in recent years with the discovery that *ras* **proteins,** a family of proteins from the tumor-causing rat sarcoma virus, have amino acid sequences similar to G proteins. The *ras* proteins are the products of a virus-infected cell gene that has been mutated and transferred to the viral genome. The modified gene is called an **oncogene.** The *ras* protein products are shortened versions of G proteins and, therefore, cannot carry out the normal functions of

nal molecules. The process usually involves a reaction that leads to breakdown or chemical modification of the molecule. Second messengers, such as cAMP, are often very reactive, unstable molecules that are short-lived in the cell.

2.6 Diseases of Cellular Communication and Drug Design

Basic research in biochemistry, cell biology, microbiology, genetics, physiology, and other biosciences has moved us to greater understanding of the causes and possible cures for the many diseases that afflict humankind. Results from these studies point to a general premise: Many diseases are the result of genes that carry the wrong messages (mutations) or of a malfunction in the transmission of that message (cell signal transduction).

Perhaps the biggest breakthrough in medical research in recent years has been the development of biotechnology for the design of genetically engineered drugs (Chapter 13). Hundreds of new biotechnology firms have been formed, each with goals to develop drugs that could revolutionize medicine in the next decades. The approach taken by many new

G proteins in signal transduction. A consequence is that the infected cell may lose normal metabolic control and be transformed to a cancer cell. Not all oncogenes are of viral origin; some are mutated versions of normal cellular protein components. One chemotherapeutic drug that is known to act by altering signal transduction processes is tamoxifen, which works as an "anti-estrogen" to treat breast cancer. Estrogen is a female hormone that stimulates the growth of breast cancer cells. Tamoxifen interferes with estrogen action by binding to estrogen receptors without activating estrogen-sensitive genes, thus slowing or even stopping the proliferation of cancer cells.

Early French cartoon illustrating clothing that was used to protect against the bacterium, *Vibrio cholerae.*

Figure 2.11

New drugs are being targeted at various levels of cellular communication: (a) drugs that interfere with the process of protein synthesis, including DNA transcription to RNA and translation; (b) the design of drugs that bind to receptor sites to block the initiation of signal transduction by the natural effector; and (c) the design of drugs to disrupt signaling pathways.

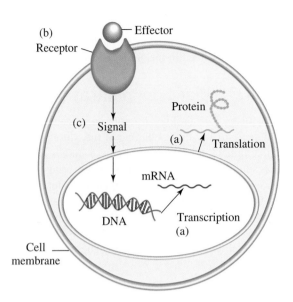

biotechnology laboratories is to develop drugs that block malfunctioning cellular processes. This can be done by halting the flow of misinformation. Biologically active molecules of all types—proteins, nucleic acids, carbohydrates, and lipids—are being prepared and tested for this purpose. The approaches taken by the pharmaceutical firms are numerous and different; however, there are similarities that allow classification according to the site of action for the newly discovered drugs. As shown in Figure 2.11, three levels of cellular communication are the targets for new drug products:

1. *Discovering drugs that function at the level of DNA.* Two approaches are possible. Many new drugs are designed to block the transcription of genes that lead to disease-causing proteins such as the cancer-causing *ras* proteins and those that cause inherited disorders such as cystic fibrosis. These drugs interfere with the transcription of DNA into mRNA. A second approach also interferes with protein synthesis, but at a different step— the translation of the message in mRNA into proteins. Drugs that target this level of DNA expression bind to mRNA, blocking the transmission of its message.

2. *Designing drugs that bind to receptor proteins on the outer surface of cells and, thus, block signal transduction processes.* These drugs can interfere with the action of viruses and inflammation-causing white blood cells, which act by binding to receptor sites. Drugs with this effect often have carbohydrate-like structures that are similar to biomolecules on the outer surfaces of viruses and white blood cells (Chapter 8).

3. *Designing small, nonpolar molecules that interfere with signaling pathways inside the cell.* These drugs have the potential of inhibiting cell proliferation (cancer) and inflammation or other immune disorders. Some of the targets for these drugs are the G proteins, enzymes that synthesize second messengers (for example, adenylate cyclase), and protein kinases.

Summary

- DNA, RNA, proteins, some carbohydrates, and other small molecules serve as informational molecules in that they carry instructions for the direction and control of biological processes. The chemical information in molecules is read by the formation of weak, noncovalent interactions.

- The genetic information in DNA flows through the sequence, DNA ⟶ RNA ⟶ proteins ⟶ cellular processes.

- Eukaryotic genes are discontinuous, consisting of coding regions called **exons** and noncoding regions called **introns.**

- Hormones and other internal messengers elicit their actions through **signal transduction,** a process using G proteins and an intracellular cascade of reactions to produce a **second messenger** inside the cell.

Study Exercises

Understanding Terms

2.1 Define the following terms used in this chapter.

- **a.** Chemical signaling
- **b.** Replication of DNA
- **c.** Template
- **d.** Hydrogen bonds
- **e.** Intron
- **f.** tRNA
- **g.** Second messenger
- **h.** Exon
- **i.** Signal transduction
- **j.** Mutation
- **k.** Oncogene
- **l.** Genetic code
- **m.** G protein
- **n.** Proteome
- **o.** Genome
- **p.** Human Genome Project
- **q.** Adenylate cyclase
- **r.** *Ras* proteins

Reviewing Concepts

2.2 What does it mean to say that "DNA, RNA, proteins, and some carbohydrates are information-rich molecules"?

2.3 Determine whether each of the following statements about DNA is true or false. Rewrite each false statement so it is true.

- **a.** DNA is a polymer composed of many nucleotide monomers.
- **b.** The Watson–Crick DNA helix consists of a single strand of polymeric DNA.
- **c.** In double-strand DNA, the base A on one strand is complementary to the base T on the other strand.
- **d.** The nucleotide bases in DNA include A, T, G, and U.

2.4 List three important differences between DNA and RNA.

2.5 Write the nucleotide sequence that is complementary to the single strands of DNA shown below.

- **a.** 5′ ATTTGACC
- **b.** 5′ CTAAGCCC

2.6 Write the nucleotide sequence that is complementary to the strands of RNA shown below:

- **a.** 5′ UACCG
- **b.** 5′ CCCUUU

2.7 Draw a complementary, double-strand polynucleotide that consists of one DNA strand and one RNA strand. Each strand should contain ten nucleotides and the two strands must be antiparallel, that is, one running from 5′ → 3′, the other from 3′ → 5′.

2.8 Assume that the short stretch of DNA drawn below is transcribed into RNA. Write out the base sequence of RNA product.

2.9 Draw the product synthesized by the action of reverse transcriptase on the following strand of nucleic acid.

2.10 What is the product of the following reaction?

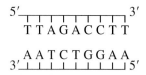

2.11 Draw the sequence of amino acids in a protein made from the following sequence of DNA.

2.12 Draw the products from the semiconservative replication of the DNA drawn below.

2.13 The peptide hormone oxytocin induces labor by stimulating the contraction of uterine smooth muscle. The human hormone has the following structure:

Cys-Tyr-Ile-Gln-Asn-Cys-Pro-Leu-Gly

- **a.** Write out a sequence of nucleotides in DNA that carries the message for this peptide structure.
- **b.** Write out the sequence of nucleotides in RNA that carries the message for this structure.

2.14 Describe the reaction catalyzed by each of the following enzymes.

- **a.** RNA polymerase
- **b.** Reverse transcriptase
- **c.** Adenylate cyclase

2.15 List three types of protein required for the signal transduction process.

2.16 Describe how the chemical signal from hormone binding to a receptor site can be amplified.

2.17 Name several diseases that may be caused by damage to the signal transduction processes in cells.

Solving Problems

2.18 Starch, glycogen, and cellulose are important carbohydrate biopolymers made by linkage of a single type of monomer, glucose. Can these molecules be considered "information rich"?

2.19 How can human DNA molecules, containing an estimated 3 billion nucleotide base pairs and being about 1 m long, fit into the cell's nucleus, which is only about 5 μm in diameter?

2.20 How many different trinucleotides can be formed by combining the nucleotide bases, ATG? Each trinucleotide must contain one of each base.

HINT: One combination is A-T-G.

2.21 What are some possible reasons for DNA being the storehouse of genetic information rather than RNA or proteins?

2.22 Why is the information transfer process mRNA → proteins called "translation"?

2.23 Explain why cancer can be described as a disease caused by a malfunction of cellular communication.

2.24 Why is it necessary for molecular recognition processes to be reversible?

2.25 What is the evolutionary significance behind the discovery that RNA can serve two roles—to transfer genetic information and to act as a catalyst for biochemical reactions?

2.26 The signal molecule cAMP is very reactive and breaks down within a few seconds of its synthesis. Why is this an advantage to the signal transduction process?

2.27 In signal transduction the binding of a first messenger to its receptor protein must be a reversible process. Why?

2.28 Use the concepts of signal transduction to explain how caffeine acts as a stimulant.

2.29 Describe the roles of scaffolding and anchoring proteins in the process of signal transduction.

2.30 Explain why a person with type 2 diabetes would show no physiological response to insulin injection.

Writing Biochemistry

2.31 "The Sixty-Second Paper." Immediately after reading this chapter, allow yourself just one minute to write down your answers to the following two questions: (1) What were the central concepts introduced in this chapter? and (2) What are the ideas introduced in this chapter that you found confusing or hard to understand?

2.32 Find a current article describing the Human Genome Project (HGP) in a popular science magazine like *Scientific American.* In 100 words, summarize the article by describing the HGP, its current status, and ethical implications of using the results for treatment of a disease.

 Applying Biochemistry

2.33 The CD-ROM unit on "AIDS Therapies" describes several approaches involving the use of drugs for treatment of AIDS. Medical research is moving closer to a possible vaccine against the disease. Find a recent article in your library or on the Internet that describes work toward the development of a vaccine. Write a 100-word summary of the article.

2.34 Your roommate looks over your shoulder as you review the CD-ROM unit on "Signal Transduction". He said that the process looks like magic to him. Try to write a 100-200 word explanation to convince your roommate, a philosophy major, that signal transduction processes are real.

2.35 Explain what is meant by "shotgun sequencing" DNA.

Further Reading

Borman, S., 2000. Proteomics: Taking over where genomics leaves off. *Chem. Eng. News* 78:31–37.

Bugg, C., Carson, W., and Montgomery, J., 1993. Drugs by design. *Sci. Am.* 272(6):92–97.

Cavenee, W., and White, R., 1995. The genetic basis of cancer. *Sci. Am.* 272(3):72–79.

Crick, F., 1958. On protein synthesis. *Soc. Exp. Biol.* 12:138–163.

Emmett, A., 2000. The human genome. *The Scientist* 14(15):1, 17–18.

Holliday, R., 2001. Aging and the biochemistry of life. *Trends Biochem. Sci.* 26:68–71.

Lander, E., 1996. The new genomics: Global views of biology. *Science* 274:536–539.

Langridge, W., 2000. Edible vaccines. *Sci. Am.* 283(3):66–71.

Lewis, R., 2000. TIGR introduces *Vibrio cholerae* genome. *The Scientist* 14(16):8.

Linder, M. and Gilman, A., 1992. G proteins. *Sci. Am.* 267(1):56–65.

Paabo, S., 1993. Ancient DNA. *Sci Am.* 269(5):86–92.

Palevitz, B., 1999. Ancient DNA—When is old too old? *The Scientist* 13(13):10.

Scott, J. and Pawson, T., 2000. Cell communication: The inside story. *Sci. Am.* 282(6):72–79.

Smith, C., 2000. The secret language of cells. *The Scientist* 14(2):24–26.

Thieffry, D., 1998. Forty years under the central dogma. *Trends Biochem. Sci.* 23:312–316.

Wilson, H., 1988. The double helix and all that. *Trends Biochem. Sci.* 13:275–278.

Chapter 3
Biomolecules in Water

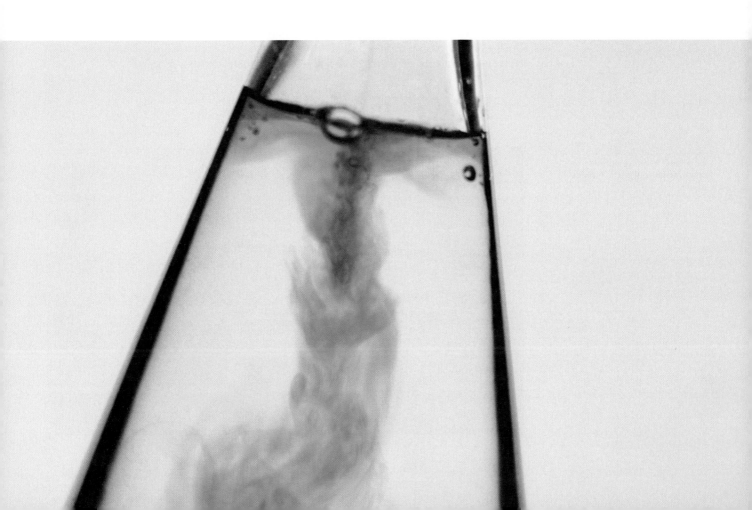

Table 3.1
Percent by weight of water in organs of the human body

Tissue or Organ	Percent by Weight of Water[a]
Skeletal muscle	79[b]
Heart	83[b]
Liver	71
Kidney	81
Spleen	79
Lung	79
Brain	77

[a]In adults.
[b]Fat-free tissue.

Water! Although a seemingly simple and abundant substance containing only the atoms of hydrogen and oxygen, it has extraordinary physical, chemical, and biological properties. Water is vital to all forms of life and comprises about 70% to 85% of the weight of a typical cell (Table 3.1). In addition, extracellular fluids such as blood, cerebrospinal fluid, saliva, urine, and tears are aqueous-based solutions. Many scientists believe that life began in an aqueous environment and, during the early stages of evolutionary development, all living organisms resided in water. Although plants probably evolved first, many forms of life developed lungs and were able to move to land. Some organisms (unicellular and multicellular) still require not just internal water but a constant extracellular environment that is aqueous. These organisms may live in rivers, lakes, and oceans or sheltered in the aqueous environment of another larger cell.

Water plays many roles in the cell and has great influence on the structure and behavior of all biomolecules. Water is important as a solvent and as a reactant molecule. A study of biomolecules is not complete without an understanding of the extraordinary properties of water as a solvent. Water provides a medium for metabolic reactions. It is literally the "matrix of life." Another crucial role played by water as a biological solvent system is to regulate proper cellular and extracellular conditions such as temperature and pH. With a high specific heat capacity, water is able to absorb large amounts of energy, in the form of heat, released from biochemical reactions. As a result, water as a solvent serves a role as temperature buffer. Water also acts as a solvent to dissolve substances that regulate hydrogen ion concentration (pH). Biological molecules will function properly only in an environment of constant pH. Buffering substances as simple as bicarbonate and as complex as proteins react with water to maintain a remarkably constant pH level in intracellular and extracellular fluids. Because most biological fluids are aqueous-based, water provides for the delivery to the cell of nutrients for growth and removal of wastes for general cleaning of cells. It is important to note that water, although a very effective solvent, is not a universal biological solvent as is sometimes declared. All biomolecules are not soluble in water and this, indeed, is fortunate. Because of this, organisms can develop compartmentation of cell structure and function by creating partitions (membranes) from molecules that are water-insoluble.

Water is not only a solvent, but also a participant in many important biochemical reactions. One of the most common biological reactions is cleavage of a chemical bond by water (hydrolysis) as observed in the initial steps of digestion of proteins, nucleic acids, and carbohydrates. Water is also a principal reactant in the photosynthesis process:

$$6\ CO_2 + 6\ H_2O \xrightarrow{h\nu} C_6H_{12}O_6 + 6\ O_2$$
$$\text{Carbohydrates}$$

Water, which plays roles as a biological solvent, reactant molecule, and temperature regulator, is essential to life.

© Douglas Faulkner/Photo Researchers

Here water acts as a reducing agent, a source of electrons to reduce carbon dioxide for the manufacture of glucose. The process of respiration, the final stage of energy metabolism in animals, generates water from O_2 by oxidation–reduction reactions (see Chapter 18).

Clearly, water is not just an inert bystander in the cell, but a selectively reactive molecule with unique properties that greatly affect biochemical molecules and biological processes. In this chapter, we shall examine some of those unusual properties of water and learn how they influence structure and reactivity of biomolecules.

3.1 Water, the Biological Solvent

Noncovalent Bonding in Biomolecules

Before we look at the physical and chemical properties of water, we need to consider bonding arrangements that are important in water molecules and in how water interacts with other molecules. In Chapter 1, we discovered that biomolecules large and small are made by the covalent combining of elements forming C—C, C—H, C—O, C—N, and other strong bonds. These bonds literally hold the atoms of a molecule together. Weaker interactions, called **noncovalent bonds,** bring together whole biomolecules for specific purposes. Four types of noncovalent bonds are important: **van der Waals forces, ionic bonds, hydrogen bonds,** and **hydrophobic interactions.** Properties and examples of these interactions are reviewed in Table 3.2. One of these types of stabilizing forces (hydrogen bonds) are at work when water dissolves polar molecules and when two strands of DNA are brought together in the double helix.

Properties of Noncovalent Bonds

Throughout our study of biochemistry we will encounter many examples where noncovalent molecular interactions bring together, in specific ways, two different molecules or different regions of the same molecule. Molecules have the ability to recognize and interact (bind) specifically with other molecules. We will use the term **molecular recognition** to describe this general phenomenon. The importance of these interactions in biology is that the combination of two molecules or the organized folding of a single molecule will lead to biological function not present in individual molecules or unfolded, randomly arranged molecules.

The interactions important in molecular recognition are often between a small molecule (called a ligand, L) and a macromolecule (M):

$$L + M \rightleftharpoons LM$$

LM represents a complex held together by noncovalent interactions with specialized biological function. The action of hormones is a good example. A hormone response is the consequence of weak, but specific, interactions between the hormone molecule and a receptor protein in the membrane of the target cell. Biochemical reactions also provide many examples of the importance of noncovalent interactions. Before a metabolic reaction can occur, a small substrate molecule must physically interact in a certain well-defined manner with a macromolecular catalyst, an enzyme. The biochemical action of a drug also depends on molecular interactions. The drug is first distributed throughout the body via the bloodstream. Drugs in the bloodstream are often bound to plasma proteins, which act as carriers. When the drug molecules are transported to their site of action, a second molecular interaction is likely to occur. The drug will likely bind to a receptor protein or other proteins. Many drugs elicit their effects by then interfering with biochemical processes. This may take the form of enzyme inhibition, where the drug molecule binds to a specific enzyme and prohibits binding of normal reactant and, therefore, inhibits catalytic action. Intramolecular noncovalent interactions (those within a molecule) also play significant roles in biomolecular processes—for example, stabilizing the folding of protein, DNA, and RNA molecules into regular, three-dimensional arrangements.

All molecular interactions that are the basis of molecular recognition have at least three common characteristics:

■ *First, the forces that are the basis of these interactions are noncovalent and relatively weak.* The strengths of these interactions are in the range of 1–30 kJ/mol compared to about

Featured in this Swiss stamp is a Greek god of water and waves

● Concept Review: Water, pH, and non-covalent bonding

● Cutting Edge: Molecular recognition

Table 3.2
Properties and examples of noncovalent interactions

Type	Brief Description and Example	Stabilization Energy (kJ/mol)	Length (nm)
Hydrogen bonds	Between a hydrogen atom bonded to an electronegative atom and a second electronegative atom	10–30	0.3
	Between neutral groups		
	Between peptide bonds		
van der Waals interactions	Between molecules with temporary dipoles induced by fluctuating electrons. This may occur between any two atoms in close proximity	1–5	0.1–0.2
Hydrophobic interactions	The presence of water forces nonpolar groups into ordered arrangements to avoid the water 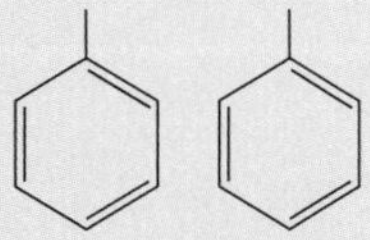 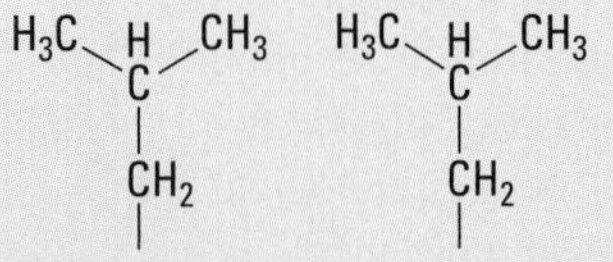	5–30	—
Ionic bonds	Interactions that occur between fully charged atoms or groups Na^+Cl^- $R{-}NH_3^+\ {}^-OOC{-}R$	20	0.25

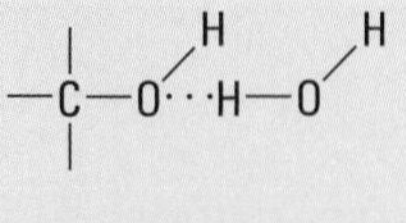

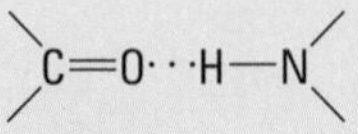

350 kJ/mol for a carbon–carbon single bond, a typical covalent bond. A single noncovalent bond is usually insufficient to hold two molecules together. DNA, RNA, and protein molecules have numerous functional groups that participate in noncovalent interactions. A collection of many of these interactions will lead to greatly stabilized complexes.

■ *Second, noncovalent interactions are reversible.* Noncovalent interactions are initiated when diffusing (wandering or moving) molecules or regions of a molecule come into close contact. Diffusion is brought about by thermal motions. An initial close encounter may not always result in the successful formation of a complex. A few weak bonds may form but may be disrupted by thermal motion, causing the molecules to dissociate. Therefore, bonds may constantly form and break until enough bonds have accumulated to result in an intermediate with a transient but significant existence. The complex can then initiate a specific biological process. An intermediate rarely lasts longer than a few seconds. Eventually, thermal motions cause the complex to dissociate to the individual molecules. Reversibility is an important characteristic of these interactions so that a static, gridlock situation does not occur. The biological process initiated by the complex LM must have a starting time and an ending.

■ *Third, the binding between molecules is specific.* Imagine that the interactions bring together two molecular surfaces. The two surfaces will be held together if noncovalent interactions occur. If on one surface there is a nonpolar molecular group (phenyl ring, hydrophobic alkyl chain, etc.), the adjacent region on the other surface must also be hydrophobic and

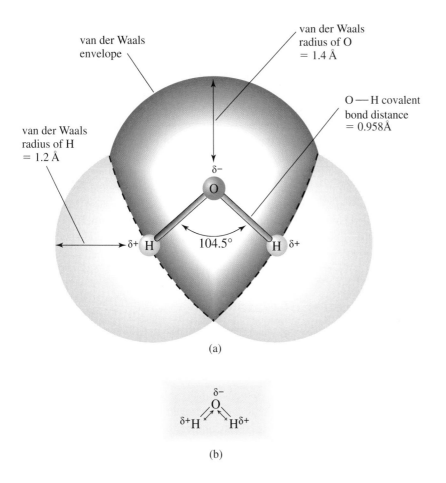

van der Waals envelope

van der Waals radius of O = 1.4 Å

O — H covalent bond distance = 0.958Å

van der Waals radius of H = 1.2 Å

$\delta-$

O

$\delta+$ H   104.5°   H $\delta+$

(a)

$\delta-$
O
$\delta+$H   H$\delta+$

(b)

Figure 3.1

(a) The structure of the water molecule showing the relative size of each atom by the van der Waals radius. Covalent bonds hold together oxygen and hydrogen atoms. The polar character, which is the result of electronegativity differences between oxygen and hydrogen, is indicated by the partial charges (δ^+ and δ^-) on atoms. (b) Water has a dipole moment because of its bent geometry. The arrows pointing to the more electronegative atom are used to show bond polarity.

nonpolar. If a positive charge exists on one surface, there may be a neutralizing negative charge on the other surface. A hydrogen bond donor on one surface can interact with a hydrogen bond acceptor on another. Simply stated, the two molecules must be compatible or complementary in a chemical sense so the development of stabilizing forces can hold molecules together. The concept of molecular recognition will take on many forms in our continuing studies of biochemistry.

The Structure of Water

The arrangement of hydrogen and oxygen atoms in the water molecule is nonlinear with an H—O—H bond angle of 104.5° (Figure 3.1a). The electronegativity value for the oxygen atom (3.5) is approximately one and one-half times that of hydrogen (2.1). Therefore, the electrons of the two covalent bonds are not shared equally; the oxygen atom has a stronger pull on the electrons and takes on a partial negative charge (δ^-). The hydrogen atoms are left with a partial positive charge (δ^+), because they do not have equal access to the bonding electrons. This gives rise to a molecule with a dipolar structure: a negative end sometimes called the "head" (oxygen) and positive ends sometimes called "tails" (hydrogens) (Figure 3.1b). The water molecule is electrically neutral (no net charge) but has a relatively large dipole moment because of its bent geometry. Water can be contrasted with another molecule of biochemical significance, CO_2, which also has polar bonds caused by electronegativity differences between the carbon and oxygen atoms but no dipole moment because it is linear (Figure 3.2).

The characteristics of water as described here have profound consequences for its structure and interactions with biomolecules. Water molecules can interact with each other by attraction of a positive tail (hydrogen) with a negative head (oxygen) as shown in Figure 3.3. This favorable interaction results in a hydrogen bond (see Table 3.2). This type of bond, which may be represented as X—H··A, is formed when an electronegative atom (A), such as oxygen or nitrogen, interacts with a hydrogen atom that is slightly positive or acidic as in

$$\delta-O=\overset{\delta+}{C}=O\delta-$$

Figure 3.2

The CO_2 molecule, although composed of polar bonds, has no dipole moment because it is linear. The electronegativity differences between C and O atoms are indicated by the partial charges (δ).

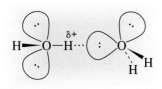

Figure 3.3

Hydrogen bond between two water molecules. The hydrogen atom (partially charged) of one water molecule interacts with a lone pair of electrons in an orbital of the oxygen atom of another water molecule.

🔘 Concept Review: Water, pH, and non-covalent bonding

Figure 3.4

Hydrogen bonds of biological importance: (a) between an alcohol and water or between alcohol molecules; (b) between a carbonyl group and water (X = H, R, OH, OR, or NH_2); (c) between two peptide chains, the carbonyl group of one peptide bonds to an N—H of another; (d) between complementary base pairs in DNA.

$X—H^{\delta+}$, where X may be nitrogen, oxygen, or sulfur. The hydrogen bond is strongest when the three atoms, $X—H \cdots A$, are in a straight line (180°) with the hydrogen atom interacting directly with a lone pair electron cloud of A. The hydrogen bond distance in water is about 0.18 nm (1.8 Å) and has a bond energy of about $20 \text{ kJ} \cdot \text{mol}^{-1}$ ($5 \text{ kcal} \cdot \text{mol}^{-1}$) compared to 0.096 nm (0.96 Å) and $460 \text{ kJ} \cdot \text{mol}^{-1}$ ($110 \text{ kcal} \cdot \text{mol}^{-1}$) for an O—H covalent bond.

Water structure has great significance in biochemistry because many biomolecules have atoms that can hydrogen bond with water, with themselves, and with other molecules. Some biochemical examples of hydrogen bonding are shown in Figure 3.4. Functional groups that participate in hydrogen bonding include (1) the hydroxyl groups in alcohols, organic acids, and carbohydrates; (2) carbonyl groups in aldehydes, ketones, acids, amides, and esters; and (3) N—H groups in amines and amides. The specific hydrogen bonding that occurs between complementary base pairs in DNA and RNA (Figure 3.4d) is discussed in Chapter 2. Although the strength of a single hydrogen bond may be small, the enormous number of potential hydrogen bonding groups in biomolecules more than makes up for their individual weakness.

3.2   Hydrogen Bonding and Solubility

Physical Properties of Water

The unusual physical properties of water are best illustrated by comparison to substances of similar structure and molecular weight (Table 3.3). Water has a higher boiling point, melting point, and viscosity than any other hydride of a nonmetallic element. These peculiar properties are the result of water's unusually high internal cohesiveness or the tendency of water molecules to "stick together," which is due to an extensive network of hydrogen bonds. Each water molecule theoretically can hydrogen bond with four neighboring water molecules (Figure 3.5a). In reality, the average number of hydrogen bonds to each molecule in liquid water at 10°C is about three. This number of hydrogen bonds decreases with increasing temperature. The theoretical number of four interacting neighbors for each water molecule is approached in crystalline ice (Figure 3.5b).

(a)

(b)

(c)

Thymine

Adenine

(d)

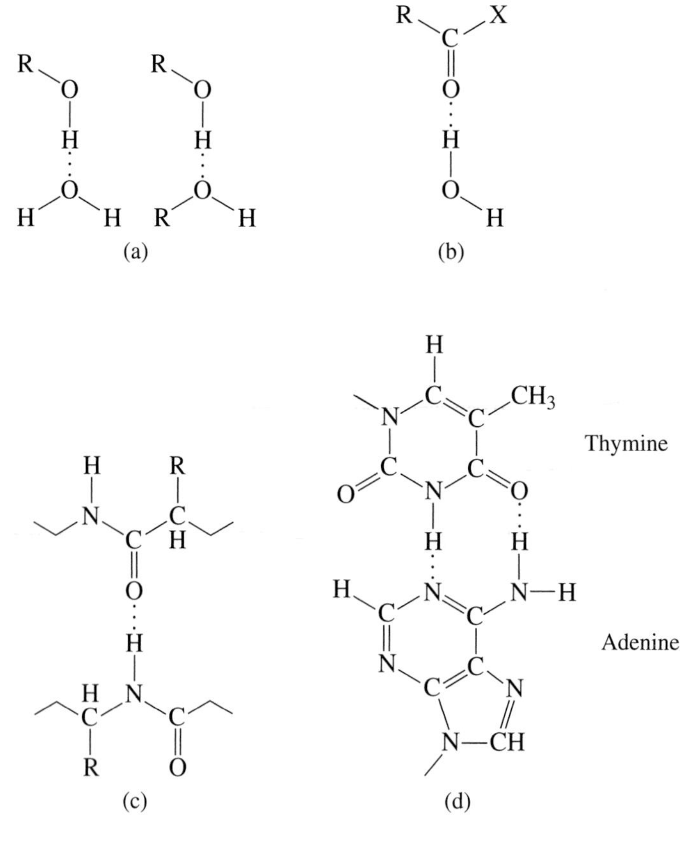

Property	H$_2$O	NH$_3$	CH$_4$	H$_2$S
Molecular weight	18	17	16	34
Boiling pt (°C)	100	−33	−161	−60.7
Freezing pt (°C)	0	−78	−183	−85.5
Viscosity^a	1.01	0.25	0.10	0.15

^{a}Units are centipoise.

Table 3.3
A comparison of some physical properties of water with hydrides of other nonmetallic elements: N, C, and S

A close examination of water structure by X-ray and neutron diffraction techniques reveals more detailed features. The network of hydrogen bonds as shown in Figure 3.5 is a snapshot representing an instant in time. The actual structure is dynamic with water molecules undergoing constant geometrical reorientations and forming new hydrogen bonds with other neighboring water molecules. This change happens for each molecule about once every 10^{-12} s. The term "flickering clusters" has sometimes been used to describe the constantly changing network of hydrogen bonds in liquid water. Using powerful X-rays generated at the European Synchrotron Radiation Facility in Grenoble, France, it recently has been shown that the hydrogen bonds that hold water molecules together in ice have substantial covalent character.

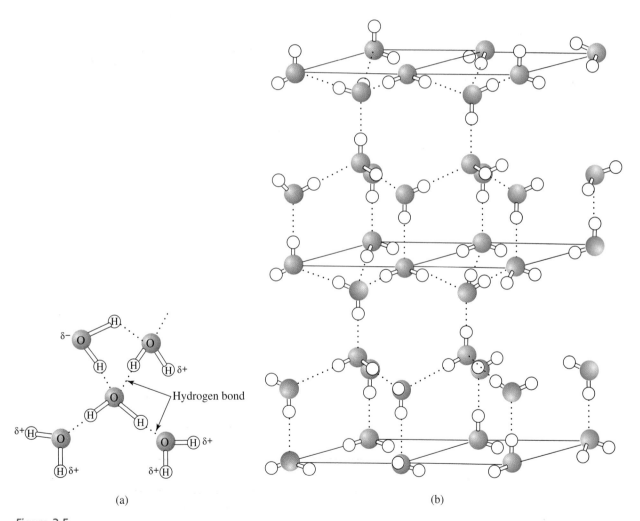

(a)　　　(b)

Figure 3.5
The network of potential hydrogen bonds in water. (a) The center water molecule may form hydrogen bonds with up to four neighboring molecules, but the average is about three. The network structure is constantly changing, with water molecules undergoing geometrical reorientations and forming new hydrogen bonds with other neighboring water molecules. (b) In ice, hydrogen bonding leads to the formation of a crystalline lattice.

Water as a Solvent

Water displays an exceptional capacity to dissolve many of the biomolecules found in living organisms. Biochemical substances that are ionic as well as those that are polar and uncharged are soluble in water. We examine here the interactions of water with some of these compounds.

Many uncharged biomolecules readily dissolve in water because they have polar functional groups that form favorable dipole–dipole interactions. A few examples of these compounds, including alcohols, amines, amides, and esters, are featured in Figure 3.4. Further examples are shown in Figure 3.6a. Extensive hydrogen bond networks are possible where the atoms of the polar functional groups combine with identical molecules, similar molecules, and/or water. Because of a favorable attraction to water molecules, ionic and polar compounds are said to be **hydrophilic,** a word of Greek origin translated as water (*hydro*) and loving (*philic*). Not all compounds containing polar functional groups are water soluble. Those with a relatively large hydrocarbon component (usually greater than four carbon atoms) are usually insoluble unless an ionic group or several polar groups are present. Cyclohexane is insoluble in water; but if one of the carbon–hydrogen groups is converted to an aldehyde or ketone and a hydroxyl group is substituted on each of the remaining five carbon atoms, the molecule becomes similar to a carbohydrate such as glucose and is water soluble.

Ionic compounds such as sodium acetate ($CH_3COO^-Na^+$) and monopotassium phosphate ($K^+H_2PO_4^-$) dissolve in water because their individual ions can be solvated (hydrated) by polar water molecules (Figure 3.6b). The negative dipole (oxygen atom) of water binds favorably with the Na^+ or K^+ forming a dipole–ion interaction. The acetate and phosphate anions are hydrated also by dipole–ion interactions. In this case, the partially positive hydrogen ends of water associate with the negative charges on the anion. The attraction between most ions and polar water molecules is strong enough to overcome the tendency of anions and cations to recombine. The great ion-solvating ability of water is illustrated by the Na^+, K^+, and Cl^- concentrations in human blood of 0.14 *M*, 0.004 *M*, and 0.10 *M*, respectively. Hundreds of other

Figure 3.6

Chemicals are made soluble in water by (a) dipole–dipole interactions. The partially charged positive atoms (hydrogen) of water and alcohol are attracted to oxygen atom dipoles of water and alcohol. The carbonyl group of an aldehyde, ketone, or acid can also be solvated by water. (b) Ion–dipole interactions. The positively charged sodium ion is surrounded by water molecules projecting their partially negative oxygen atoms (dipoles). The acetate ion interacts with the partially positive hydrogen atoms (dipoles) of water.

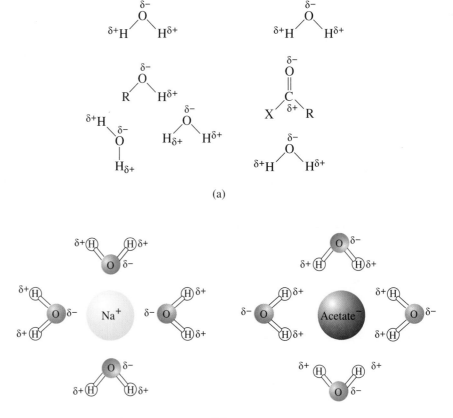

(a)

(b)

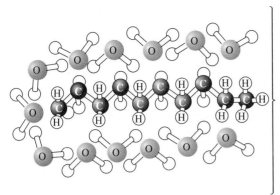

Water molecules in cage around hydrocarbon chain

Figure 3.7
Because hydrophobic molecules have no polar groups to interact with water, they have to be surrounded by a boundary of water molecules. The formation of this highly ordered cage of water requires much energy, which comes from hydrophobic interactions.

ionic substances are dissolved in blood. Even macromolecules, including proteins, nucleic acids, some lipids, and some carbohydrates under in vivo conditions exist as hydrated ions.

Nonpolar compounds are usually not water soluble because they contain neither ions nor polar functional groups that can interact favorably with water molecules. Hence, they are called **hydrophobic** (water fearing). Decane and benzene are examples of hydrophobic molecules (Figure 3.7). Some significant biochemicals have dual properties; they have both nonpolar and ionic characteristics. They are classified as **amphiphilic** (*amphi*, on both sides or ends, and *philic*, loving). This class of compounds is best illustrated by metal salts of long-chain carboxylic acids (Figure 3.8). A specific example, sodium stearate, has an ionic side or end (the carboxylate anion associated with the sodium cation) and a nonpolar hydrocarbon end. This molecule must be confused when placed in water solution. The observed result of this experiment is not the formation of a true solution but the self-assembly of the acid molecules into aggregates called **micelles**. The amphiphilic molecules avoid water contact in the hydrophobic region by pointing their hydrocarbon chains ("tails") to the water-free interior of the aggregate. The surface of each micelle is composed of the ionic "heads" stabilized by electrostatic interaction with metal cations and water. The favorable association of nonpolar hydrocarbon tails inside the micelle is defined as a **hydrophobic interaction** (see Table 3.2). In simplified terms, this situation is favorable because less energy is required to form the micelle than if the hydrocarbon chains were allowed to point out into the water, disrupting the network of hydrogen bonds. One important practical application of micellar solutions of sodium stearate and other similar compounds is their use as soaps to "solubilize" oil and grease in water. Owing to their dual chemical character, soaps are able to trap oil in the nonpolar region of the micelle and yet remain dispersed in aqueous solutions by hydration of the ionic region. Micelle formation is also the key to construction of biological membranes (see Chapter 9).

Much of this section has focused on how the structure of water is affected by the presence of solute molecules. It is also important to note changes in the structures of biomolecules brought about by the presence of water. Because of their dual character, long-chain carboxylic acid salts and other amphiphilic lipids take on a special arrangement in water. Proteins and nucleic acids also contain hydrophobic regions and ionic functional groups; therefore, these are amphiphilic substances. As we shall discover in later chapters, these biomolecules in water solution fold into conformations that bury hydrophobic regions in water-free areas and expose ionic and polar functional groups to water molecules. Important consequences result because it is often found that these complex and ordered arrangements of biomolecules are the only ones with biological activity.

3.3 Cellular Reactions of Water

Ionization of Water

It may be easy to view water as just a background material without much dynamic activity in the cell and organism; however, this is not a realistic picture. Although water is not usually considered a substance with robust chemical reactivity, it does display features of

**Concept Reviews:
Logarithms; Water, pH, and non-covalent bonding**

Formation of a micelle from the sodium salt of a long-chain carboxylic acid. The nonpolar hydrocarbon tails of the acid arrange themselves to avoid contact with water. The negatively charged carboxyl groups interact with water by forming ion–dipole interactions.

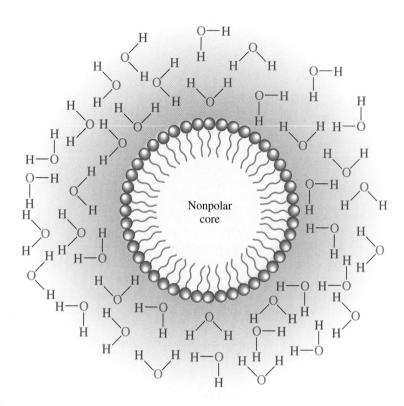

Sodium Stearate Micelle

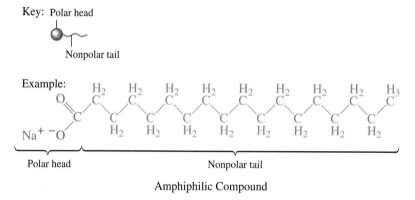

Key: Polar head

Nonpolar tail

Example:

Polar head

Nonpolar tail

Amphiphilic Compound

selective reactivity. Several examples of water as a participant in biochemical processes are presented in later chapters. Perhaps the most important reaction of water is its reversible self-dissociation or ionization to generate the hydronium ion (H_3O^+) and the hydroxide ion (OH^-):

$$H_2O + H_2O \rightleftharpoons H_3O^+ + OH^-$$

Although not as correct, because free H^+ does not exist in aqueous solution, the equation is often abbreviated as

$$H_2O \rightleftharpoons H^+ + OH^-$$

The extent to which this ionization reaction takes place is of special interest because it helps characterize the internal medium of cells. The more water molecules dissociated, the more ionic the medium. We can use the law of mass action to obtain a quantitative measure of the equilibrium point for the dissociation reaction:

$$K_{eq} = [H^+][OH^-]\backslash[H_2O]$$

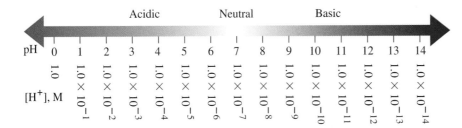

Figure 3.9
The pH scale for acids and bases.
The scale, which is logarithmic,
runs from 0 (very acidic) to 14
(very basic). A pH of 7 is neutral.

K_{eq} represents the **equilibrium constant** for the reaction, and brackets for each chemical entity indicate concentration units in moles per liter (M). If the K_{eq} for the ionization of pure water is determined from experimental measurements, it is possible to calculate a quantity for [H^+] and [OH^-] and, therefore, estimate the extent of self-dissociation. K_{eq} for pure water at 25°C has been determined to be $1.8 \times 10^{-16}\, M$. A value for [H_2O] can be estimated by dividing the weight of water in one liter (1000 g) by the molecular weight of water (18). This yields [H_2O] = 55.5 M. Therefore,

$$[H^+][OH^-] = K_{eq}[H_2O]$$

$$[H^+][OH^-] = (1.8 \times 10^{-16})(55.5)$$

$$[H^+][OH^-] = 1.0 \times 10^{-14}\, M$$

Because according to the chemical equation for dissociation H^+ and OH^- must have equal concentrations in pure water, then:

$$[H^+] = [OH^-] = \sqrt{1.0 \times 10^{-14}M} = 1.0 \times 10^{-7}\, M$$

Hydrogen ion concentrations in the exponential form ($1 \times 10^{-7}\, M$) are very small and, therefore, cumbersome for use in mathematical manipulation. So in 1909 Søren Sørensen introduced the term **pH** to more conveniently express [H^+]. He defined pH as the negative logarithm of the hydrogen ion concentration:

$$pH = -\log[H^+]$$

A [H^+] of $1 \times 10^{-7}\, M$ becomes a pH of 7. The entire pH scale from 0 to 14 is defined in Figure 3.9. The logarithmic feature of the pH scale is an important characteristic. Note that each digit increase or decrease of pH represents a tenfold change in [H^+]. A solution at pH 7 has a ten times greater [H^+] than a solution at pH 8.

The determination of pH is one of the most frequent measurements in the biochemistry laboratory. The structures of biomolecules and the efficiency of biochemical processes are dependent on [H^+]. A pH measurement is usually taken by immersing a glass combination electrode into a solution and reading the pH directly from a meter. The glass electrode is calibrated with solutions of known pH values.

As evident from the small size of [H^+], the extent of water self-dissociation is slight, but it does influence the ionic character of water-based matrices in cells and other aqueous-based biological fluids. The ionic environment and the presence of H^+ and OH^- promote the ionization of dissolved acidic and basic biomolecules. The pH values for several natural fluids are compared in Figure 3.10.

pH and pK

Calculations and measurements of pH as described in Section 3.3 can be carried out for all aqueous solutions. The pH of a solution will depend little on the hydrogen ions generated by the self-dissociation of water, but rather on the presence of other substances (acids or bases) that increase or decrease the H^+ concentration. Acids and bases are chemical substances that change the ionic properties of solutions. A useful definition of an acid is a substance that releases a proton (H^+) in water. A base is a substance that accepts a proton. An acid HA dissociates in aqueous solution accordingly:

$$\text{HA} \rightleftharpoons \text{H}^+ + \text{A}^-$$

Acid   Base

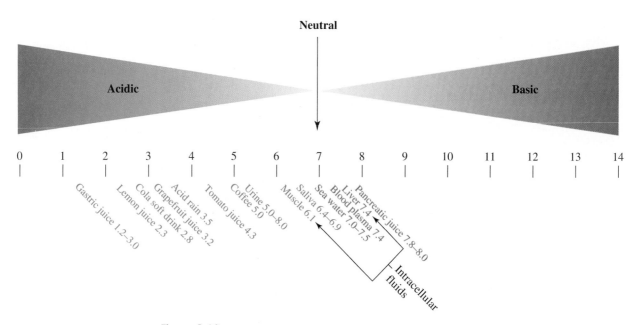

Figure 3.10
The pH values of some natural fluids. Note that many, but not all, are grouped around a neutral pH of 7.

This reaction describes the action of a whole range of acids from the strong mineral acids (HCl, H_2SO_4) to weak biological acids (acetic acid, lactic acid) to very weak acids (NH_4^+). In the equation, HA, as an acid, releases a H^+ to the solution. A^- is considered a base because it accepts a proton in the reverse reaction. HA and A^- have a special relationship; they are a **conjugate acid–conjugate base pair,** sometimes shortened to acid–base conjugate pair.

The strength of an acid, or the measure of its tendency to release a proton, can be indicated from its dissociation constant. For a general acid HA, the acid dissociation constant K_a is defined by

$$K_a = [H^+][A^-]\backslash[HA]$$

Acids common in biochemistry have a wide range of K_a values. Hydrochloric acid (HCl) is a strong acid with an immeasurably large K_a; acetic acid is much weaker with a K_a at 25°C of 1.75×10^{-5}; and NH_4^+, a very weak acid, has a K_a at 25°C of 5.62×10^{-10}. Note that the larger the K_a, the stronger the acid, and hence the greater the dissociation. Several acids of biochemical importance are listed with their dissociation constants in Table 3.4. As an exercise, write out the dissociation reaction for each acid (HCl, CH_3COOH, and NH_4^+) and identify the conjugate acid–conjugate base pairs.

Dissociation constants written as exponentials are not convenient for everyday use. These numbers are modified for easier use by the following definition:

$$pK_a = -\log K_a$$

The negative logarithm of the dissociation constant K_a is defined as pK_a (pH has a similar definition, the negative logarithm of $[H^+]$). Like K_a, the value pK_a is a quantitative measure of acid strength. The most common range of pK_a values for biochemical acids is from 2 to about 13 or 14. The smaller the value of pK_a, the stronger the acid. Note that this is the opposite of K_a, where a large value indicates a strong acid. Table 3.4 lists pK_a values along with K_a values for several biochemical acids.

Titration Curves

Values of pK (and dissociation constants) for acids can be determined experimentally by the procedure of titration. This consists of adding, with a buret, incremental amounts of a base to an acid sample dissolved in water and monitoring changes in the pH of

Acid	HA	K_a	pK_a
Formic acid	HCOOH	1.78×10^{-4}	3.75
Acetic acid	CH_3COOH	1.76×10^{-5}	4.75
Pyruvic acid	$CH_3COCOOH$	3.16×10^{-3}	2.50
Lactic acid	$CH_3CHOHCOOH$	1.38×10^{-4}	3.85
Malic acid	HOOC—CH_2—CHOH—COOH	(1) 3.98×10^{-4}	3.40
		(2) 5.5×10^{-6}	5.26
Citric acid	HOOC—CH_2—C(OH)—CH_2—COOH COOH	(1) 8.14×10^{-4}	3.09
		(2) 1.78×10^{-5}	4.75
		(3) 3.9×10^{-6}	5.41
Carbonic acid	H_2CO_3	(1) 4.3×10^{-7}	6.4
		(2) 5.6×10^{-11}	10.2
Phosphoric acid	H_3PO_4	(1) 7.25×10^{-3}	2.14
		(2) 6.31×10^{-8}	7.20
		(3) 3.98×10^{-13}	12.4
Ammonium ion	NH_4^+	5.6×10^{-10}	9.25

Table 3.4
Acids of biochemical importance

the solution with a pH meter. A graph is constructed by plotting the changes in pH on the vertical axis and the amount of base added on the horizontal axis. This activity yields a titration curve from which the pK_a can be determined (Figure 3.11). We will illustrate the construction of a titration curve using acetic acid, a weak biochemical acid, and sodium hydroxide, a strong base. The equation representing the chemistry of titration is

$$CH_3COOH + NaOH \rightleftharpoons CH_3COO^- Na^+ + H_2O$$
Acid   Base   Conjugate base

The data from the experiment are shown in Figure 3.11. The general shape of the curve, which is obtained for all weak acids, reveals useful information about the acid. The information is both structural and quantitative. The beginning pH of the acid solution, before addition of base, can be used to calculate the concentration of H^+ in the solution. Acetic acid is relatively weak (there is only slight dissociation), so most acetic acid molecules just prior to titration are in the form of CH_3COOH. The curve changes direction (inflects) at the

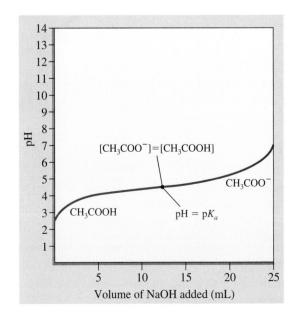

Figure 3.11

A titration curve. The experimental curve obtained from the titration of acetic acid with sodium hydroxide. The structure for acetic acid is shown at three pH values. Note that at the inflection point the two forms CH_3COOH and CH_3COO^- are present in equal molar concentrations. The pK_a (4.76) for the acid–base conjugate pair is measured here.

midpoint. At the inflection point, 0.5 mol of base has been added for each mole of acid present. Here, exactly one-half of the original acid has undergone dissociation so equal amounts of two forms of acetic acid are present: the undissociated form CH_3COOH (50%) and the conjugate base form CH_3COO^- (50%). The pH at this inflection point is equal to the pK_a of acetic acid. (We will show mathematical proof for this statement in the next section.) At the end point (or equivalence point, where 1 mol of base has been added), equal amounts of acetic acid and sodium hydroxide have reacted, so essentially all molecules of acetic acid are dissociated to the conjugate base $CH_3COO^-Na^+$. The titration experiment is valuable because it reveals the pK_a value as well as the ionic forms of acetic acid present at various pH values.

Our discussion to this point has concentrated on a monoprotic acid, which is an acid with only a single hydrogen atom that can dissociate per molecule. Many acids of biochemical importance have two or more acidic protons; that is, they are polyprotic. Some of these, listed in Table 3.4, include malic acid, citric acid, carbonic acid, and phosphoric acid. All of the acidic protons on a polyprotic acid do not dissociate at the same pK_a, but are released in sequence at different pK_a values. For example, phosphoric acid has three dissociable protons and, therefore, three pK_a values listed in Table 3.4, one value for each proton. The three-step ionization for phosphoric acid proceeds accordingly:

$$H_3PO_4 \xrightleftharpoons{pK_{a(1)} = 2.14} H_2PO_4^- \xrightleftharpoons{pK_{a(2)} = 7.20} HPO_4^{2-} \xrightleftharpoons{pK_{a(3)} = 12.4} PO_4^{3-}$$
$$+ \qquad\qquad + \qquad\qquad +$$
$$H^+ \qquad\qquad H^+ \qquad\qquad H^+$$

Note that the conjugate base for the first ionization reaction ($H_2PO_4^-$) becomes the proton donor (acid) for the second reaction, and so on. Titration curves for polyprotic acids become more complex than for monoprotic acids, but they are obtained in the same manner and they provide the same useful structural and quantitative information.

The Henderson–Hasselbalch Equation

In the previous section, we were able to use titration data to define the ionic forms of acetic acid present at three pH values: the beginning, midpoint, and the end of the titration. In addition, it was possible to estimate the concentration of each species at the three pH values. With the use of the **Henderson–Hasselbalch equation,** it is possible to calculate the concentration of acid and conjugate base at *all* points of the titration curve. The equation can be derived from the definition for dissociation constant:

$$pH = pK_a + \log [A^-]\backslash[HA]$$

If one thinks of this equation as containing four unknown but measurable quantities (pH, pK_a, $[A^-]$, and $[HA]$), then, if three are determined, the fourth can be calculated. The Henderson–Hasselbalch equation may be used for many purposes—for calculating $[A^-]$ and $[HA]$ separately or as a ratio $[A^-]/[HA]$, and for use in preparing laboratory buffer solutions. Note that when $[A^-] = [HA]$ (at the inflection point of the titration curve), the equation becomes

$$pH = pK_a + \log 1$$

$$\log 1 = 0$$

$$pH = pK_a$$

When a solution contains equal concentrations of acid (HA) and conjugate base (A^-), the pH of that solution is equal to the pK_a for the acid. Here are two study problems to illustrate the use of the Henderson–Hasselbalch equation.

Example 1
Lactic acid is present in our muscles after strenuous exercise. What is the value of the ratio $[A^-]/[HA]$ in a solution of lactic acid at a pH of 7.2?

Solution. The titration of lactic acid proceeds as follows:

$$CH_3CHCOOH + OH^- \rightleftharpoons CH_3CHCOO^- + H_2O$$

$$\underset{OH}{|} \qquad\qquad \underset{OH}{|}$$

Lactic acid (acid, HA) Lactate (conjugate base, A⁻)

We need to calculate the ratio [lactate]/[lactic acid]. From Table 3.4, the pK_a for lactic acid is 3.85. Therefore, the Henderson–Hasselbalch equation becomes

$$pH = pK_a + \log [lactate]\backslash[lactic acid]$$

Inserting known quantities:

$$7.2 = 3.85 + \log [lactate]\backslash[lactic acid]$$

$$\log [lactate]\backslash[lactic acid] = 7.2 - 3.85 = 3.35$$

$$[lactate]\backslash[lactic acid] = 10^{3.35} = 2239\backslash1$$

In a solution of lactic acid at pH 7.2, there are 2239 molecules (or moles) of the conjugate base lactate for each molecule (or mole) of lactic acid. Therefore, the solution at pH 7.2 consists of 99.9% lactate and 0.1% lactic acid.

Example 2

What is the pH of a lactic acid solution that contains 60% lactate form and 40% lactic acid undissociated form?

Solution. The ionization of lactic acid proceeds as shown in Example 1. The Henderson–Hasselbalch equation becomes

$$pH = pK_a + \log 0.60\backslash0.40$$

The pK_a for lactic acid is 3.85.

$$pH = 3.85 + \log 0.60\backslash0.40$$

$$pH = 3.85 + \log 1.5$$

$$pH = 3.85 + 0.18$$

$$pH = 4.03$$

Notice from these two examples how the concentration ratio of conjugate base (lactate) to acid (lactic) varies with pH.

These examples illustrate two of many applications for the Henderson–Hasselbalch equation. In the next section, we discover yet another practical use of the equation.

3.4 Buffer Systems

Maintaining a Constant pH

The hydrogen ion concentration of intracellular and extracellular fluids must be maintained within very narrow limits. A pH change in blood plasma of ±0.2 to ±0.4 may result in serious damage to an organism or even death. A constant pH ensures that acidic and basic biomolecules are in the correct ionic state for proper functioning. This is especially critical for enzymes that are sensitive to pH changes. Metabolic reactions generate high concentrations of organic acids that would change fluid pH values if buffering agents were not present. Consider the following experiment. If 1.0 mL of 10 M HCl is added to 1.0 L of 0.15 M NaCl solution at pH 7.0, the pH would tumble to pH 2.0. If 1.0 mL of 10 M HCl is added to 1 L of blood plasma, the pH would fall only from pH 7.4 to pH 7.2. There is nothing magical about blood. Its ability to maintain a constant pH is due to a heterogeneous mixture of biomolecules that can act to neutralize added acids and bases. Blood and other biological fluids contain **buffer systems:** reagents that resist changes in pH when H^+ or OH^- are added. Chemically, buffer systems contain acid–base conjugate pairs.

The titration curve for acetic acid (see Figure 3.11) displays a region in which the pH changes little with addition of OH^-. The center of this region is an inflection point that coincides with the pK_a of the titrated acid. The acetic acid solution at that point (0.5 mole OH^- per mole of acetic acid; pH = 4.76) shows the smallest change in pH and represents the most effective buffering range for the acetic acid–acetate conjugate pair. Remember that at this point there is an equal concentration of acetic acid (50%) and acetate (50%). The solution has a large amount of acetic acid to react with and neutralize added base *and* a large amount of acetate (base) to react with and neutralize added acid. Therefore, the solution is buffered or protected against pH changes caused by added acids or bases. In general, an

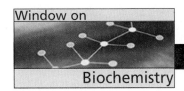

Window on

Biochemistry

3.1 Buffering Blood and Other Cellular Fluids

The major buffer system of blood and other extracellular fluids is the carbonic acid–bicarbonate conjugate pair:

$$H_2CO_3 \xrightleftharpoons[]{pK_a = 6.4} HCO_3^- + H^+$$

Carbonic acid   Bicarbonate

A base added to blood would be neutralized by the following reaction:

$$H_2CO_3 + OH^- \rightleftharpoons HCO_3^- + H_2O$$

The addition of an acidic substance to blood also results in neutralization:

$$HCO_3^- + H^+ \rightleftharpoons H_2CO_3$$

These reactions (see figure) illustrate how blood is protected against pH changes. The actual pH of blood (pH = 7.4) is at the upper limit of the buffering range of carbonic acid–bicarbonate (6.4 ± 1 = 5.4–7.4) and may not seem as efficient as desired. This inefficiency is remedied by a reserve supply of gaseous CO_2 in the lungs, which can replenish H_2CO_3 in the blood by the following series of equilibrium reactions:

$$CO_{2(g)} \rightleftharpoons CO_{2(aq)}$$

Lungs   Blood

$$CO_{2(aq)} + H_2O_{(l)} \rightleftharpoons H_2CO_{3(aq)}$$

$$H_2CO_{3(aq)} \rightleftharpoons HCO_{3(aq)}^- + H_{(aq)}^+$$

The reactions also work in reverse. A major product of metabolism, H^+ is removed from cells by the blood plasma. It is neutralized by reaction with HCO_3^- and leads to eventual release of $CO_{2(g)}$, which is exhaled from the lungs.

The carbonic acid–bicarbonate conjugate pair is the most important buffer system in biological fluids; however, it is not the only one. There is a diverse array of amino acids, peptides, and proteins with ionizable functional groups (—COOH and —NH$_3^+$) that assist in buffering. A major protein constituent of blood, hemoglobin also serves as a buffering agent. Details of the function of hemoglobin are discussed in Chapter 5.

Medical conditions caused by changes in blood pH are **acidosis** and **alkalosis.** An increase in the [H^+] of blood (acidosis) may have causes that are of metabolic

acid–base conjugate pair is most effective as a buffer system at a pH equal to its pK_a. The effective buffering range for an acid–base conjugate pair can be estimated:

$$\text{effective buffering range (pH)} = pK_a \pm 1$$

Acetic acid–acetate is an effective buffer in the pH range of 3.76–5.76. Phosphoric acid is an effective buffer in three pH ranges: (1) pH = 1.14–3.14, (2) pH = 6.20–8.20, and (3) pH = 11.4–13.4.

Biochemical processes that occur in cells and tissues depend on the strict regulation of the hydrogen ion concentration provided by natural buffers. When in vitro studies are done in the laboratory, artificial, buffered media are used to mimic the cell's natural environment. When cell components are isolated by using cell homogenates (see Window on Biochemistry 1.3), the constituent biomolecules are most stable when maintained in the natural pH range, which is usually 6–8. Sodium and potassium salts of phosphoric acid are among the most widely used buffers in biochemical research. The phosphate acid–base conjugate pair, $H_3PO_4^-$–HPO_4^{2-}, has a pK of about 7.2, which provides a useful pH range of 6.2–8.2 (see Table 3.4). Because phosphate is a normal constituent of the cell, its presence provides a natural environment. Other natural buffering agents that may be used in the laboratory include the acid–base conjugate pairs of acetic acid, citric acid, carbonic acid, and others listed in Table 3.4.

or respiratory origin. Metabolic acidosis occurs in individuals with untreated diabetes or those on starvation diets or high-protein, low-fat diets. All of these metabolic conditions lead to **ketosis,** the excessive generation of **ketone bodies,** which are acidic and increase the [H$^+$] of blood. Respiratory acidosis is caused by a change in [CO_2] that is often a symptom of pulmonary problems associated with emphysema or asthma. Untreated acidosis leads to coma and eventually death.

An increase in blood pH (alkalosis) also has metabolic or respiratory origins. Clinical administration of the salts of metabolic acids (sodium lactate or sodium bicarbonate) in excessive amounts or cases of severe vomiting cause metabolic alkalosis. Respiratory alkalosis is induced by hyperventilation (heavy breathing), which may result from hysteria, anxiety, or high-altitude sickness.

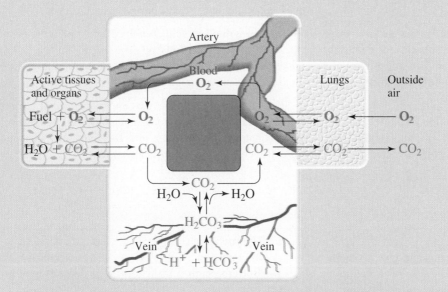

Control of blood pH by the carbonic acid–bicarbonate conjugate pair. There is some CO_2 in arterial blood and some O_2 in venous blood; not all of the CO_2 is exhaled as blood flows through the lungs. CO_2 in the lungs is in equilibrium with CO_2 in the blood. A high concentration of CO_2 in the lungs leads to respiratory acidosis while a low CO_2 concentration causes respiratory alkalosis.

Table 3.5
Some synthetic buffers including Good's buffers are listed with their abbreviations, structures, and ionization reactions. Each reaction shows the two predominant forms (acid and base) present in the useful pH range.

Name (Abbreviation)	Useful pH Range	Ionization Reaction
N-2-Acetamido-2-amino-ethanesulfonic acid (ACES)	6.4–7.4	$H_2NCOCH_2\overset{+}{N}H_2CH_2CH_2SO_3^- \rightleftharpoons$ $H_2NCOCH_2NHCH_2CH_2SO_3^- + H^+$
3-(Cyclohexylamino)-propanesulfonic acid (CHAPS)	10.0–11.0	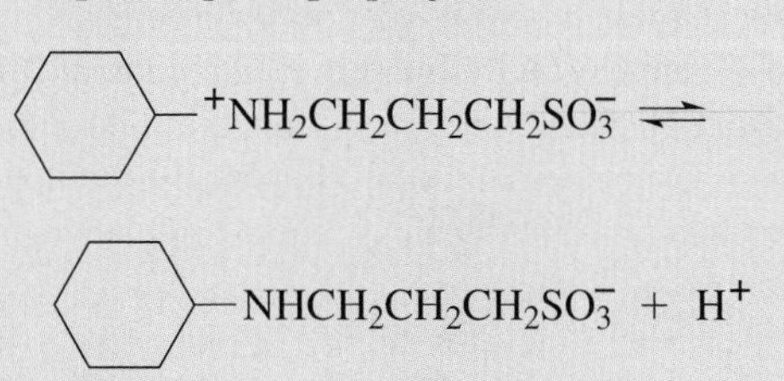
N-2-Hydroxyethyl piperazine-N'-2-ethanesulfonic acid (HEPES)	7.0–8.0	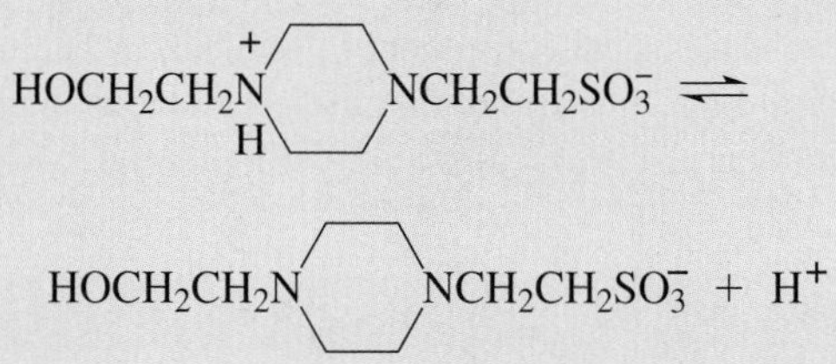
Tris(hydroxymethyl)-aminomethane (TRIS)	7.5–9.0	$(HOCH_2)_3\overset{+}{C}NH_3 \rightleftharpoons (HOCH_2)_3 CNH_2 + H^+$

Sometimes natural buffers like phosphate and citrate interfere with laboratory in vitro studies and it is necessary to use synthetic buffers. N. E. Good at Michigan State University synthesized and tested several acid salts that are now widely used as buffering agents (Table 3.5). The use of the synthetic buffer tris(hydroxymethyl)-aminomethane (TRIS) is now probably greater than that of phosphate salts.

Summary

■ Water, a nonlinear, polar molecule, serves at least three functions in the cell: It is an effective solvent, it is a reactant molecule, and it is a temperature buffer. As a solvent, water is able to dissolve biomolecules that are ionic and polar.

■ The most important reaction of water is its reversible ionization to generate the hydronium ion (H_3O^+ or the proton, H^+) and the hydroxide ion (OH^-). The extent of ionization is quantified by the pH scale ($pH = -\log [H^+]$).

■ The strength of an acid is defined by its pK_a, the negative log of its dissociation constant. The pK_a for an acid is equivalent to the pH at which there is an equal concentration of an acid and its conjugate base.

■ Blood and other cellular fluids are maintained at a constant pH by natural buffer systems including the carbonic acid–bicarbonate conjugate pair and the hydrogen phosphate–dihydrogen phosphate conjugate pair.

Study Exercises

Understanding Terms

3.1 Define each of the following terms and give a specific example, if appropriate.

a. An acid
b. A base
c. pH
d. pK
e. Henderson–Hasselbalch equation
f. Noncovalent interactions
g. A hydrophobic molecule
h. A hydrogen bond
i. Hydrophobic interactions
j. Buffer system
k. Acid–base conjugate pair
l. Metabolic acidosis
m. Micelle recognition
n. Good buffer
o. Amphiphilic molecules
p. Ionization of acids
q. Molecular recognition

Reviewing Concepts

3.2 Calculate the hydrogen ion concentration $[H^+]$ in each of the following solutions.

   a. Gastric juice, pH = 1.80
   b. Blood plasma, pH = 7.40
   c. Cow's milk, pH = 6.6
   d. Tomato juice, pH = 4.3
   e. Coffee, pH = 5.0
   f. Maple tree sap, pH = 7.1

3.3 The amino acid glycine, $H_3^+NCH_2COO^-$ has pK_a values of 2.4 and 9.8. Estimate the effective buffer range(s) for glycine.

3.4 Aspirin (acetosalicylic acid) has the following structure:

COOH

OCCH₃

O

   a. Draw the ionic structure for the predominant form of aspirin as it would exist in blood plasma.
   b. Draw the ionic structure for the predominant form of aspirin as it would exist in gastric juice.
   The pK_a for aspirin is 3.5.

3.5 Write acid dissociation reactions for each of the following biochemically important molecules. Show the ionization of all acidic protons.

   a. HCl
   b. CH_3COOH (acetic acid)
   c. NH_4^+ (ammonium ion)
   d. $CH_3(CH_2)_{13}CH_2COOH$ (palmitic acid)
   e. H_3^+N—CHCOOH   (an amino acid)
          |
          R
   f. H_3PO_4 (phosphoric acid)
   g. H_2O
   h. H_2CO_3 (carbonic acid)

3.6 Arrange the following natural solutions in decreasing order of acidity.

   Gastric juice   _____
   Blood   _____
   Acid rain   _____
   Cola   _____
   Coffee   _____

HINT: See Figure 3.10.

3.7 The nucleotide bases shown here form hydrogen bonds between the two strands of the double helix of DNA. Circle those atoms that may become involved in hydrogen bonding. Distinguish between acceptor atoms and donor atoms.

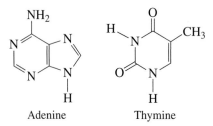

Adenine   Thymine

3.8 Determine whether each of the statements about noncovalent interactions is true or false. If false, change the statement so it is true.

   a. Ionic bonds are the result of electrostatic attraction between two ionized functional groups of opposite charge.
   b. Hydrogen bonds result from interaction of an anion with a hydrogen atom.
   c. Hydrophobic interactions are electrostatic attractions between nonpolar functional groups and water.
   d. H^+ and OH^- interact together by ionic bonding to form water.
   e. Hydrophobic interactions are important in the formation of micelles when the detergent sodium dodecanoate, $CH_3(CH_2)_{10}COO^-Na^+$, is added to water.

3.9 Identify the type of interaction that holds each of the following atoms or groups of atoms together. Choose from dipole–dipole, ion–ion, or dipole–ion.

   a. NaCl   **d.** R—O—H⋯O—R
   b. $Na^+(H_2O)_n$ |
   c. $CH_3COO^-Na^+$ H
                   e. $RNH_3^+Cl^-$

3.10 For each pair of molecules listed below, determine which one is more polar than the other.

   a. H_2O, CH_3OH
   b. H_2O, CH_3COOH
   c. $CH_3(CH_2)_3CH_3$, CH_3CH_2OH
   d. CH_2CH_2, CH_3CH_2OH
        | |
       HO OH
   e. $CH_3\overset{O}{\overset{\|}{C}}NH_2$, $CH_3\overset{O}{\overset{\|}{C}}H$
   f. CH_2CH_3, CH_3CH_3
       |
       NH_2
   g. $H_2N\overset{O}{\overset{\|}{C}}NH_2$, $CH_3\overset{O}{\overset{\|}{C}}NH_2$

3.11 What is the molar concentration of the HOH species in pure water?

3.12 Write the molecular formula for the conjugate acid of each of the following bases.
  a. OH^- d. CH_3COO^-
  b. HCO_3^- e. $H_2PO_4^-$
  c. $H_3^+NCH_2COO^-$

3.13 For each pair of molecules listed below, determine which one is the less polar (most nonpolar).

  a. CH_3CH_3, CH_2CH_3
                      |
                     OH

  b. CH_2CH_3, H_2O
     |
    OH

  c. $H_3^+NCHCOO^-$, $H_3^+NCHCOO^-$
          | |
         CH_3 CH
                          /   \
                        CH_3 CH_3

  d. CH_3CH_2COOH, $CH_3CHCOOH$
                            |
                           OH

  e. $HOOCCH_2CH_2COOH$, $CH_3CH_2CH_2COOH$

3.14 Write structures for the conjugate base of each of the following acids.
  a. H_2O
  b. $H_3^+NCH_2COOH$
  c. $CH_3(CH_2)_{10}COOH$
  d. HCO_3^-
  e. $CH_2(CH_2)_{10}CH_2$
        | |
       $^+NH_3$ $^+NH_3$

Solving Problems

3.15 Identify the type of bonding between each pair of atoms and molecules. In examples with more than one type, name the type of bonding indicated by the arrow.
  a. Na^+Cl^-
  b. $H—O—H$
  c. $H—O\text{---}H—O$
              | |
              H H

  d. $H_2N—C—NH—H\text{---}O$
          (i) ‖   (ii)   \
              O H

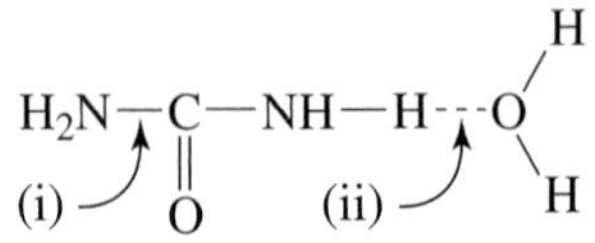

HINT: Select from covalent, hydrogen, and ionic bonding.

3.16 Determine whether each of the statements is true or false. If false, rewrite it so it is true.

  a. Noncovalent bonds are usually more easily broken then covalent bonds.

  b. The strength of a typical noncovalent bond is usually at least 300 kJ/mol.
  c. Noncovalent interactions often occur between only certain molecules.
  d. Noncovalent bonds reversibly break and reform at room temperature.

3.17 Define and contrast four types of noncovalent bonds that pair biomolecules for molecular recognition.

3.18 Describe the differences between hydrogen bonds and ionic bonds.

3.19 Predict which of the following compounds are soluble in water.
  a. $CH_3CH_2—OH$
  b. $CH_3(CH_2)_{10}CH_2—OH$
  c. CH_3CH_2COOH
  d.

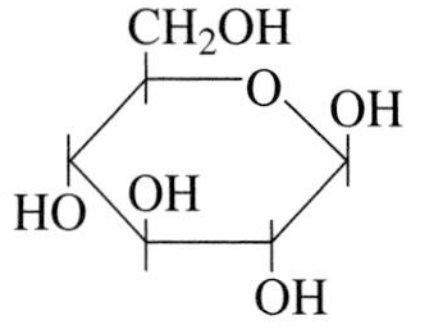

Glucose

  e. $H_3^+NCH_2COO^-$
     Glycine

  f. CH_2CHCH_2
      | | |
      OH OH OH
      Glycerol

  g.

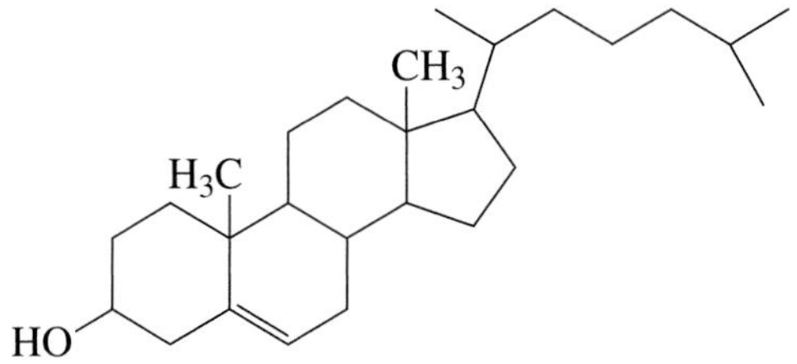

Cholesterol

3.20 Which of the following molecules will form micelles in water?
  a. $CH_3(CH_2)_{10}CH_2 - NH_3^+Cl^-$
  b. $CH_3(CH_2)_{10}CH_2COO^-Na^+$

  c.

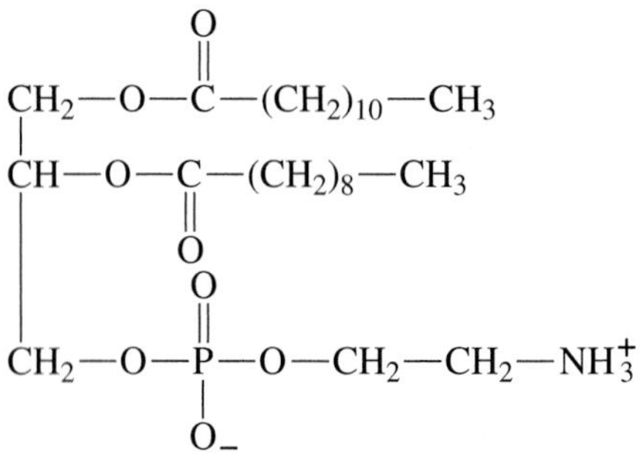

  d.

HINT: Look for polar head and nonpolar tail.

3.21 You need to prepare a buffer for use at pH 4.0. Which substance would be most effective?

   a. Lactic acid
   b. Acetic acid
   c. Phosphoric acid

HINT: Check Table 3.4 for pK values.

3.22 The acid–base conjugate pair H_2CO_3–HCO_3^- maintains the pH of blood plasma at 7.4. What is the ratio of bicarbonate to carbonic acid in blood?

3.23 Can you think of a possible emergency room treatment for a patient in a state of acidosis?

3.24 A buffer solution was prepared by mixing 0.05 mol of the sodium salt of the amino acid alanine with 0.1 mol of free alanine in water. The final volume is 1 L. The equilibrium reaction is

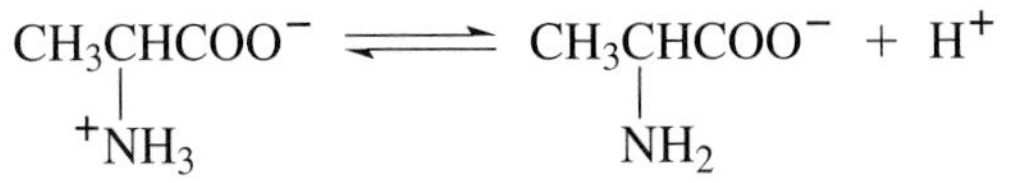

$$CH_3CHCOO^- \rightleftharpoons CH_3CHCOO^- + H^+$$

$^+NH_3$	NH_2
Alanine	Salt of alanine

pK_a = 9.9

What is the pH of the final solution?

HINT: Use the Henderson–Hasselbalch equation.

3.25 The pH of normal rainwater is approximately 5.6. The slight acidity compared to pure water is caused by dissolved CO_2. In some regions of the world, the acidity of rain has increased to a pH of approximately 3.5. This is caused by the presence of polluted air containing SO_2, SO_3, and NO_2. These oxides react with rainwater to form acids:

$$SO_2 + H_2O \rightarrow H_2SO_3$$
$$SO_3 + H_2O \rightarrow H_2SO_4$$
$$2\ NO_2 + H_2O \rightarrow HNO_3 + HNO_2$$

Calculate the ratio of bicarbonate to carbonic acid in normal rainwater and acid rain (pH = 3.5).

3.26 "Stomach acid," a result of our hectic, fast-paced lifestyles, is often treated with antacids. From your knowledge of acid–base chemistry, predict which of the following compounds would be an ingredient in over-the-counter antacids.

   a. $NaHCO_3$
   b. Ascorbic acid (vitamin C)
   c. $Mg(OH)_2$
   d. CH_3COOH (acetic acid)
   e. $NaAl(OH)_2CO_3$
   f. Aspirin
   g. Lemon juice
   h. $CaCO_3$

3.27 The aspirin product Bufferin contains magnesium carbonate, $MgCO_3$. What is the purpose of this ingredient?

3.28 Predict which of the compounds below can form hydrogen bonds with water. Show examples of hydrogen bonding for each one you select.

   a. CH_3CH_2OH
      Ethanol

   b. $CH_3CH_2CH_3$
      Propane

   c. 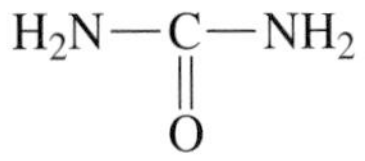 $H_2N—C—NH_2$
               $\|$
               O
      Urea

   d. 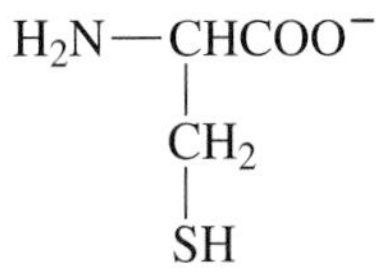 $H_2N—CHCOO^-$
                 $|$
                 CH_2
                 $|$
                 SH
      Cysteine

3.29 We will discover in the next chapter that amino acids are di- or triprotic acids. Classify each of the amino acids shown below as a diprotic or a triprotic acid and write all dissociation reactions showing removal of all acidic protons.

   a. 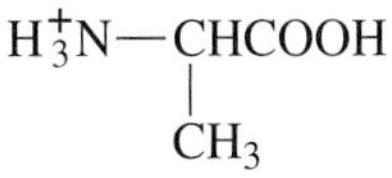 $H_3^+N—CHCOOH$
                  $|$
                  CH_3
      Alanine

   b. 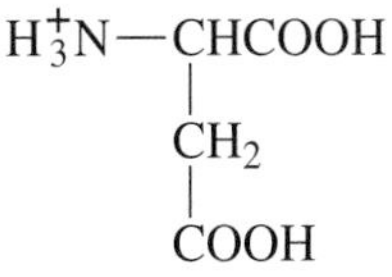 $H_3^+N—CHCOOH$
                  $|$
                  CH_2
                  $|$
                  $COOH$
      Aspartic acid

   c. 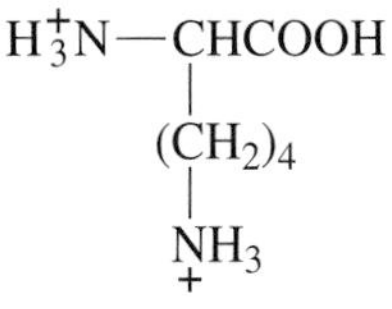 $H_3^+N—CHCOOH$
                  $|$
                  $(CH_2)_4$
                  $|$
                  NH_3
                  $+$
      Lysine

   d. 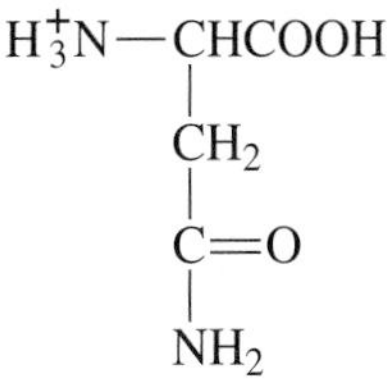 $H_3^+N—CHCOOH$
                  $|$
                  CH_2
                  $|$
                  $C=O$
                  $|$
                  NH_2
      Asparagine

3.30 Which of the compounds shown below would function as soaps or detergents?

   a. $CH_3(CH_2)_{12}CH_3$
   b. $CH_3(CH_2)_9CH_2COO^-K^+$

c. $CH_3(CH_2)_{10}CH_2OH$

d.

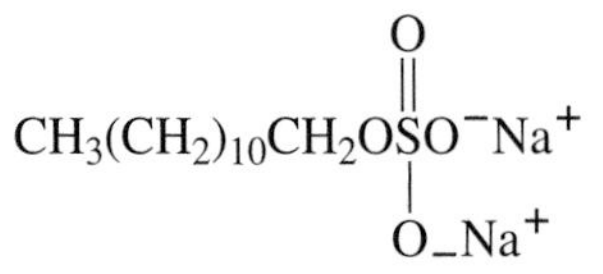

$$CH_3(CH_2)_{10}CH_2O\overset{\overset{\displaystyle O}{\|}}{\underset{\underset{\displaystyle O\_Na^+}{|}}{S}}O^-Na^+$$

3.31 Shown below is the structure of a dipeptide formed with the amino acids serine and cysteine. Circle all atoms that may become involved in hydrogen bonding with H_2O. Distinguish between those atoms that are hydrogen bond donors and hydrogen bond acceptors.

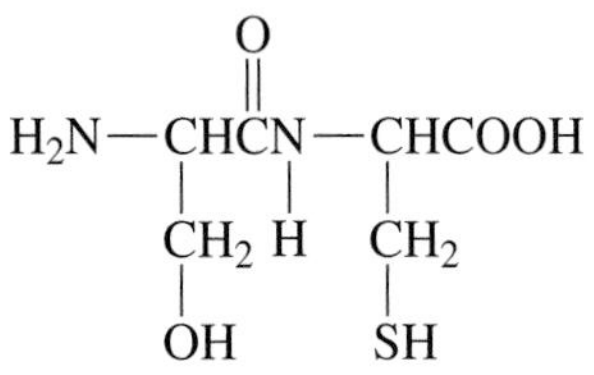

$$H_2N{-}\overset{\overset{\displaystyle O}{\|}}{\underset{\underset{\underset{\displaystyle OH}{|}}{\underset{\displaystyle CH_2}{|}}}{CH}}C\overset{}{\underset{\underset{\displaystyle H}{|}}{N}}{-}\overset{}{\underset{\underset{\underset{\displaystyle SH}{|}}{\underset{\displaystyle CH_2}{|}}}{CH}}COOH$$

Ser-Cys

3.32 What is the value of the ratio [lactate]/[lactic acid] in a solution of lactic acid at a pH of 5.0?

HINT: Use the Henderson–Hasselbalch equation:

$$pH = pK_a + \log [lactate]\backslash[lactic\ acid]$$

3.33 What is the pH of a lactic acid solution that contains 60% lactate form and 40% lactic acid?

3.34 Phosphate buffers are often used in biochemical research because they can be prepared in the physiological pH range of 7 and because phosphates are naturally occurring biomolecules. How many moles of monobasic sodium phosphate ($NaH_2PO_4 \cdot H_2O$) and dibasic sodium phosphate ($Na_2HPO_4 \cdot 7H_2O$) must be added to a liter of water to prepare a 0.5 M phosphate buffer of pH 7.0?

Writing Biochemistry

3.35 Your roommate does not have time to read this chapter before the next class exam. She has offered to pay you a handsome reward for writing a brief summary for her to review. Write a 100-word summary of Chapter 3 that contains a review of the most important concepts. Use the current Summary as a guide, but be creative in your approach and style.

3.36 "The Sixty-Second Paper." Your instructor will probably spend about one lecture on this chapter. Immediately after the lecture, use 1–2 minutes to write a paper answering the following questions: (1) What were the central points outlined in the lecture? and (2) What concepts were confusing?

 Applying Biochemistry

3.37 Although methane and water have similar molecular weights (16 and 18 daltons, respectively), they have quite different properties. Explain why methane is unable to form hydrogen bonds with itself and with other molecules.

3.38 Your roommate, an English major, makes heavy use of antacid tablets especially just before her creative writing papers are due in class. She has asked you to explain how the tablets work to neutralize stomach acid. Write the explanation in a 100-word paragraph.

Further Reading

Ball, P., 2000. *Life's Matrix: A Biography of Water.* New York: Farrar, Straus & Giroux.

Boyer, R., 2000. pH, Buffers, Electrodes, and Biosensors. In *Modern Experimental Biochemistry,* 3rd ed., pp. 29–41. San Francisco: Benjamin/ Cummings.

Colson, S. and Dunning, T. Jr., 1994. The structure of nature's solvent: water. *Science* 265:43–44.

Kegley, S., and Andrews, J., 1997. *The Chemistry of Water.* Sausalito, CA: University Science Books.

Klotz, I., 1995. Water, superwaters, and polywater. *The Chemical Intelligencer* Oct:34–41.

Norby, J., 2000. The origin and the meaning of the little p in pH. *Trends Biochem. Sci.* 25:36–37.

Rupley, J., Gratton, E., and Careri, G., 1983. Water and globular proteins. *Trends Biochem. Sci.* 8:18–22.

Wettlaufer, J., and Dash, J., 2000. Melting below zero. *Sci. Am.* 282(2): 50–53.

Zeuthen, T., 2001. How water molecules pass through aquaphorins. *Trends Biochem. Sci.* 26:77–79.

Chapter 4

Amino Acids, Peptides, and Proteins

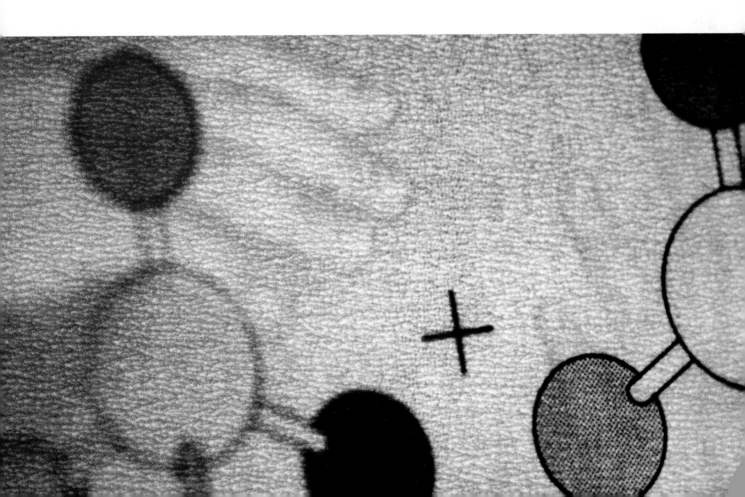

he term **protein** was first used by the Dutch chemist Gerardus Mulder in 1838 to name a specific group of substances abundant in all plants and animals. The importance of proteins was correctly predicted by Mulder, who derived the name from the Greek word *proteios,* meaning primary or first rank. Proteins, biopolymers constructed from amino acids, display a wide range of structures and functions. The amino acid building blocks are selected from a pool of 20 different molecules, each having a distinct chemical structure. Each type of protein has a specific structure that is defined by its genetically determined sequence of amino acids. The information necessary for the placement of amino acids in the proper proportion and order resides in the sequence of nucleotide bases in a gene (region of DNA). Now that the first reading of the human genome sequence has been completed (June 2000, see page 32), we know to an approximate degree of accuracy the order of nucleotides in DNA. The next challenge ahead is to understand the purpose of each and every protein product that is coded in the genome. In Chapter 2 we called this large group of proteins the proteome. Our scientific focus in the next few chapters will shift from genomics to **proteomics**—the broad study of all cellular proteins. More specifically, proteomics has the following goals: Classify and characterize all protein products coded by DNA, investigate regulation and expression levels (cellular concentrations and impact on health), determine the complete structure of each protein, study how proteins interact with each other and other biomolecules, and identify the functional role of each.

The unique amino acid composition and sequence in each type of protein allows it to fold into the precise three-dimensional arrangement necessary for carrying out its designated biochemical function. Proteins as a group of biomolecules display great functional diversity. Some proteins have roles in muscular contraction and structural rigidity; others serve in the transport and storage of small molecules. Antibodies (molecules for immune protection), enzymes (biological catalysts), and some hormones are proteins. The great variety in biological function is the direct result of the many variations in amino acid composition and sequence that are possible for a protein.

4.1   The Amino Acids in Proteins

Properties of Amino Acids

An amino acid is any organic molecule with at least one carboxyl group (organic acid) and at least one amino group (organic base). Using this broad definition, several hundreds of different amino acids are known to be present in plant and animal cells. In this chapter, how-

Table 4.1

Table of abbreviations of the 20 amino acids found in proteins

Name	One-Letter Abbreviation	Three-Letter Abbreviation
Glycine	G	Gly
Alanine	A	Ala
Valine	V	Val
Leucine	L	Leu
Isoleucine	I	Ile
Methionine	M	Met
Phenylalanine	F	Phe
Proline	P	Pro
Serine	S	Ser
Threonine	T	Thr
Cysteine	C	Cys
Asparagine	N	Asn
Glutamine	Q	Gln
Tyrosine	Y	Tyr
Tryptophan	W	Trp
Aspartate	D	Asp
Glutamate	E	Glu
Histidine	H	His
Lysine	K	Lys
Arginine	R	Arg

ever, we focus on a select group of amino acids: those 20 that are genetically coded for incorporation into proteins (Table 4.1). This set of amino acids shares a general structure comprised of a carbon center (the alpha carbon) surrounded by a hydrogen, a carboxyl group, an amino group, and a **side chain** represented by R (Figure 4.1). Because of this placement of functional groups around the α-carbon center, they are often called **α-amino acids.** The nature of the side chain, which can vary from a simple hydrogen atom to complex ring systems, determines the unique chemical and biological reactivity of each individual amino acid.

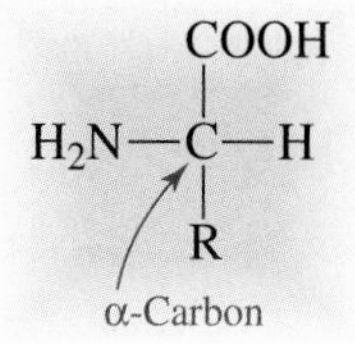

The tetrahedral alpha carbon in every amino acid, except glycine (R = H), has four different atoms or groups bonded to it, hence it is a chiral center. An important consequence of this arrangement is the existence of two nonsuperimposable stereoisomers (enantiomers), called D- and L-amino acids (Figure 4.2). The D and L isomers are mirror images of each other. Solutions of each pure stereoisomer show optical activity; that is, they rotate the angle of plane-polarized light. The angle of rotation is opposite for the two solutions. Although both D and L isomers of amino acids exist in nature, only the L isomers are used as building blocks for proteins. However, recent discoveries of cellular D-amino acids in free form and in proteins have intrigued biochemists (see Window on Biochemistry 4.1).

All of the 20 α-amino acids in pure form are white, crystalline, high-melting solids. They are soluble in water and insoluble in organic solvents such as acetone, chloroform, and ether. Aqueous solutions of amino acids conduct an electric current. The above properties are those expected of ionic compounds or salts, but such features are not present in the general structural formulas as shown in Figures 4.1 and 4.2. At physiological pH (about 7.4), amino acids exist as dipolar ionic species (Figure 4.3); that is, they have a positive and negative charge on the same molecule. These forms are sometimes called **zwitterions,** a German word meaning "inner salts."

As outlined in Figure 4.3 and in the reactions below, the carboxyl group with a pK_a of 2.3 will dissociate a proton and the NH_3^+ group with a pK_a of 9.7 will also dissociate a proton. Assuming that the R side chain has no charge, the net charge on the entire zwitterion form (structure B) of the amino acid (at neutral pH) is zero. Understanding the acid and base properties of amino acids is important because it allows one to predict the major ionic form of an amino acid present at any pH value. Predicting amino acid electrical charge becomes especially important when we discuss protein structure and function. The pK values for the 20 amino acids in proteins are listed in Table 4.2.

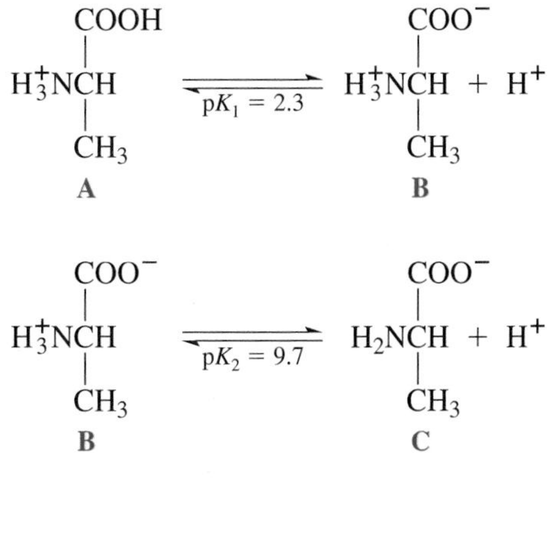

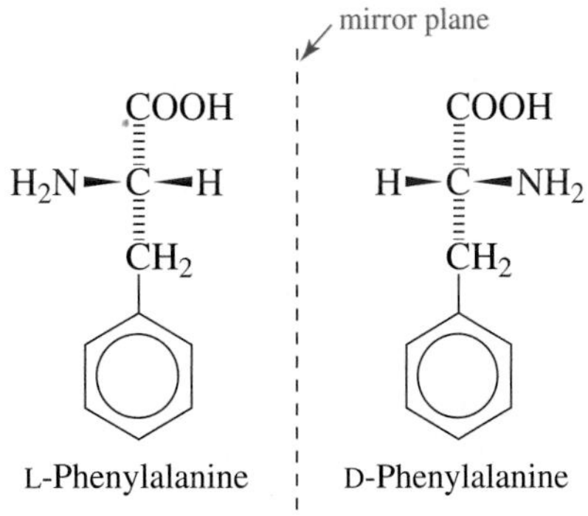

Figure 4.1

The general structure for the 20 amino acids in proteins. The four groups are linked covalently to the α carbon. All 20 amino acids have three of the groups including H, NH_2, and COOH. The fourth group, R, is different for each amino acid.

Figure 4.2

The D and L enantiomers for the amino acid phenylalanine. Note that the two structures are mirror images of each other. The dashed lines indicate bonds to groups behind the plane and ⌐ indicate bonds to groups in front of the plane of the α carbon.

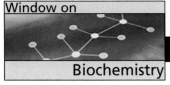

L-Alanine

Figure 4.3

The zwitterionic form for amino acids. Shown is L-alanine. Assume that the α carbon is in the plane of the page.

Because amino acids contain acidic and basic functional groups, they can be titrated as described for acetic acid in Chapter 3. The titration curve for the amino acid alanine (R = CH$_3$) is shown in Figure 4.4. When compared to that for acetic acid (see Figure 3.12), only minor differences are noted. Two inflection points are observed for alanine, which indicate two pK_a values, one to define proton dissociation from the carboxyl group (pK_1 = 2.3) and one to define the dissociation of a proton from the amino group (pK_2 = 9.7). The reactions above show the chemical changes for alanine (R = CH$_3$). Form A represents the predominant form of alanine at acidic pH. During titration, the more acidic proton (—COOH) reacts with added base to produce the zwitterion, B. When the pH of the titration solution is 2.3, about 50% of the alanine molecules are in form A and about 50% in form B. As more base is added, form B dissociates another proton (—NH$_3^+$), producing form C. When the pH of the titration solution is 9.7, about 50% of the alanine molecules are in form B and about 50% in form C. A titration curve of the same general shape as in Figure 4.4 would be obtained for all α-amino acids except those with an acid or base functional group on the side chain, R. Again, note how the structure of an amino acid varies with the pH of its environment.

Window on

Biochemistry

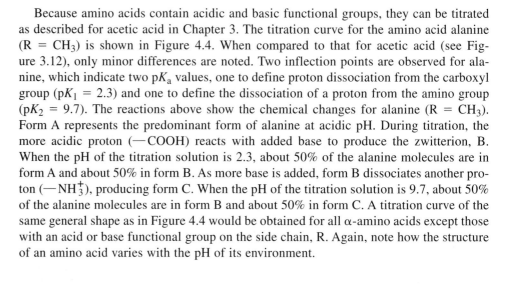

4.1 D-Amino Acids: Through the Other Side of the Mirror

The native proteins found in all organisms are constructed from L-amino acids; however, scientists are now finding the opposite enantiomers, D-amino acids, in many types of cells. The amino acids D-alanine and D-glutamate were first discovered in short peptides used as scaffolding in the cell walls of Gram-negative bacteria and archae. The presence of the unusual stereoisomers makes the cell walls less vulnerable to hydrolytic attack by enzymes. D-Valine is present in peptide antibiotics including valinomycin, actinomycin D, and gramicidin A. The D-amino acids found in bacterial cell walls and antibiotics are not incorporated by the normal protein synthesis processes found in the cell (DNA → mRNA → protein; as described in Sections 2.2 and 2.3). Rather, the peptides are synthesized by individual reactions that use free amino acids that are not linked to transfer RNAs.

Biochemists have recently found D-amino acids (especially aspartate) in proteins isolated from human teeth, eye lenses, erythrocytes, and some tumors. The proteins were those that had accumulated with age. Apparently, in aging protein, conversion of the natural L-aspartate to D-aspartate (racemization) occurs over time. The modified proteins usually show less biological effectiveness. Racemization of amino acids in proteins may be a process associated with aging.

Perhaps the most intriguing new discovery about D-amino acids is that D-serine is synthesized in mammalian brain where it functions as a neurotransmitter. During the past five years, researchers had found surprisingly high physiological concentrations of the rare enantiomer, but the source was not known. Biologists at the Johns Hopkins University School of Medicine have now shown that an enzyme in mouse brain catalyzes the conversion (racemization) of L-serine to its enantiomer, D-serine. By using antibody-labeling experiments, the enzyme, serine racemase, and D-serine were shown to be localized in glial cells surrounding neurons that contain the *N*-methyl-D-aspartate (NMDA) receptor. D-Serine and the amino acid glutamate bind to and thus activate the NMDA receptor. Activation of this receptor plays a role in memory and learning and may also be involved in nerve-cell death in strokes. The action of D-serine is unusual because other neurotransmitters like dopamine and serotonin are produced in neurons and secreted at nerve cell endings. Drugs that inhibit serine racemase may offer a treatment for stroke victims. Such drugs would slow the production of D-serine and thus lead to deactivation of NMDA receptors. During a stroke, the NMDA receptors are greatly stimulated, causing destruction of nerve cells.

Table 4.2
Table of pK values for the 20 amino acids found in proteins[a]

Name	pK_1	pK_2	pK_R
Glycine	2.4	9.8	
Alanine	2.3	9.9	
Valine	2.3	9.6	
Leucine	2.4	9.6	
Isoleucine	2.4	9.7	
Methionine	2.3	9.2	
Phenylalanine	1.8	9.1	
Proline	2.0	10.6	
Serine	2.1	9.2	
Threonine	2.6	10.4	
Cysteine	1.8	10.8	8.3
Asparagine	2.0	8.8	
Glutamine	2.2	9.1	
Tyrosine	2.2	9.1	10.9
Tryptophan	2.4	9.4	
Aspartate	2.0	10.0	3.9
Glutamate	2.2	9.7	4.3
Histidine	1.8	9.2	6.0
Lysine	2.2	9.2	10.8
Arginine	1.8	9.0	12.5

[a]pK_1 values are assigned to the α-carboxyl group, pK_2 values to the α-amino group, and pK_R to ionizable groups in the R side chain.

Classification of Amino Acids

Now that properties common to all 20 amino acids have been discussed, we focus on the feature that makes each amino acid chemically and biologically distinct: the R side chain. Structures for the side chains are shown in Figure 4.5. The R groups vary in size, polarity, charge, and chemical reactivity. Upon study of the R groups, it becomes evident that they

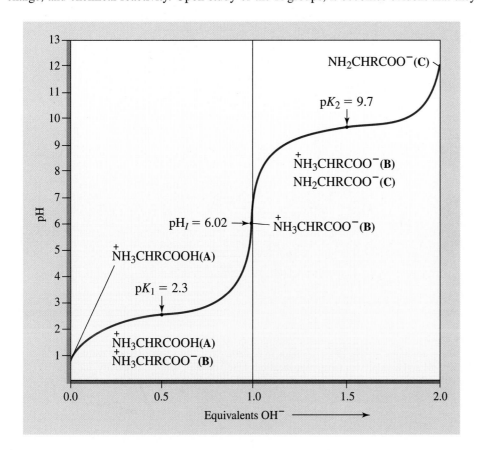

Figure 4.4
The acid–base titration curve for the amino acid alanine. The ionic species predominating at various pH values are indicated by letters A to C. There is only one primary species present at the beginning of the titration (A), at the end (C), and when the pH = pH$_I$ (B) (pH$_I$ is the pH at which the molecular structure has no net charge [isoelectric structure]). In between these points, two primary species are present in equal concentrations when pH values = pK values; (A) and (B) at pK_1 and (B) and (C) at pK_2.

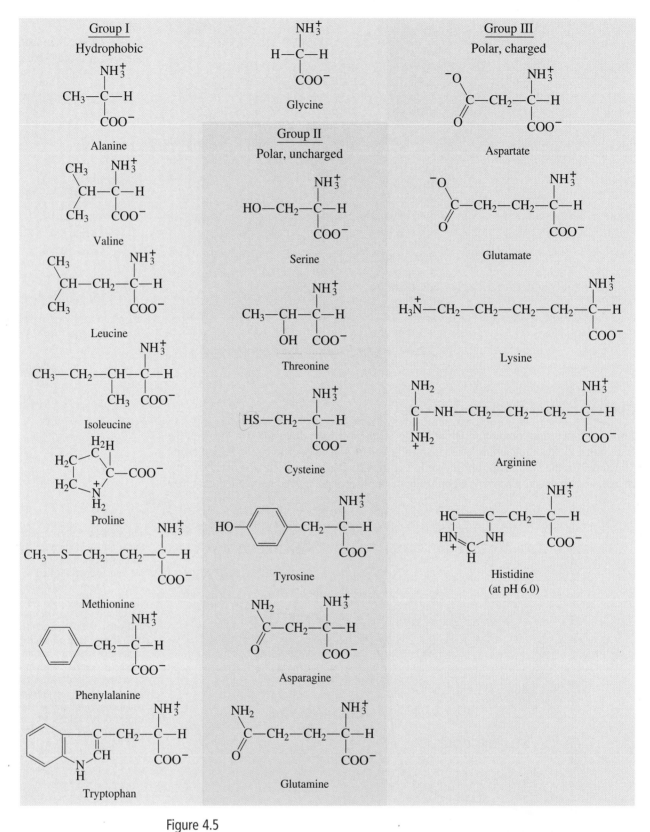

Figure 4.5

Complete structures for the 20 amino acids in proteins. They are shown as they would exist at physiological pH and divided into three classifications based on the chemical reactivity and polarity of the side chain R.

Group I, Hydrophobic	Group II, Polar, Uncharged	Group III, Polar, Charged
Gly	Ser	Asp
Ala	Thr	Glu
Val	Cys	Lys
Leu	Tyr	Arg
Ile	Asn	His
Pro	Gln	
Met		
Phe		
Trp		

Table 4.3
Classification of the amino acids based on side chain reactivity and polarity at pH 7.0

can be divided into classes based on polarity of the side chain (Table 4.3). Classification should assist in learning their structures and individual reactivity. In addition to names, structures, and group classification, students should also learn the one-letter or three-letter abbreviations assigned to each amino acid (see Table 4.1).

Group I: Amino Acids with Nonpolar Side Chains
Group I amino acids are glycine, alanine, valine, leucine, isoleucine, proline, phenylalanine, methionine, and tryptophan. All the side chains in group I amino acids have aliphatic or aromatic groups and therefore have hydrophobic character. Because the side chains are mainly hydrocarbons, there is little important chemical reactivity. Proteins dissolved in an aqueous solution will fold into a three-dimensional shape to bury hydrophobic group I amino acid residues into the interior.

Interactive Animation: Amino acids game

Group II: Amino Acids with Polar, Uncharged Side Chains at Physiological pH
The amino acids in group II are serine, cysteine, threonine, tyrosine, asparagine, and glutamine. Inspection of the side chains in this group reveals a wide array of functional groups but all have at least one heteroatom (N, O, or S) with electron pairs available for hydrogen bonding to water and other molecules. Perhaps the most interesting amino acid in this group is cysteine with an —SH (sulfhydryl group) on the side chain. The —SH group of a cysteine can react under oxidizing conditions with an —SH of another cysteine forming a **disulfide bond** (Figure 4.6). Reducing agents transform the disulfide bond back to sulfhydryl groups. This covalent bonding between cysteines becomes important in protein three-dimensional structures as we will see later in this and the next chapter.

Group III: Amino Acids with Polar, Charged Side Chains at Physiological pH
The amino acids in Group III are aspartate, glutamate, histidine, lysine, and arginine. Inspection of the side chains reveals functional groups that are either acidic or basic. At physiological pH, the side chains of aspartate and glutamate dissociate protons to form carboxylate

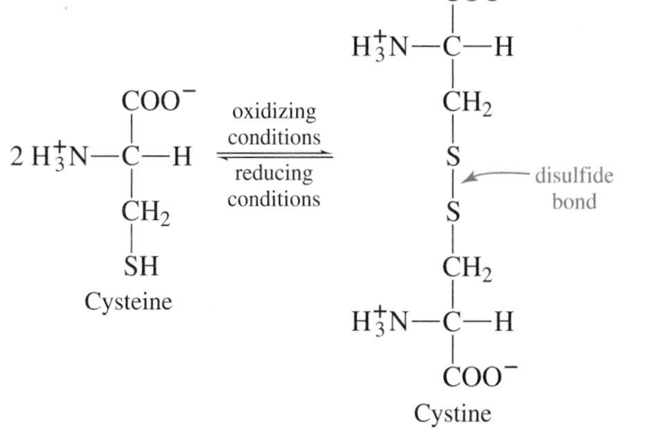

Figure 4.6
Two cysteine residues under oxidizing conditions form a disulfide bond. Reducing agents cause the cleavage of the disulfide bond and reverse the reaction.

Figure 4.7

Two ionic forms of histidine may be present in vivo. Which one would be predominant at a pH of 7.4?

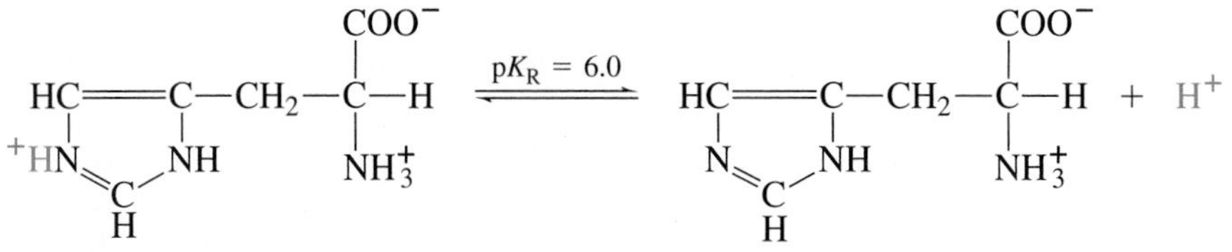

Histidine (His)

anions. The net charge on the molecules predominant at pH 7.4 is −1. Lysine, with an amino group on the side chain, and arginine, with a guanidino group, exist primarily in the +1 ionic state at physiological pH (see Figure 4.5). In histidine, the imidazole side chain has a pK_a value near physiological pH, so two major ionic forms are present at concentrations depending on the actual in vivo conditions (Fig. 4.7). At a physiological pH of 7.4, there will be a slight abundance of the zwitterionic form. Amino acids in Groups II and III are hydrophilic and tend to cluster on the exterior of a protein in an aqueous environment.

Reactivity and Analysis of Amino Acids

The chemical reactivity of α-amino acids is characteristic of their functional groups. We have already discussed the acid–base (titration) reactions of group I and II amino acids. When amino acids in group III are titrated, three inflection points (from which three pK_a values can be calculated) are observed. pK_1 is assigned to the α-carboxyl group, pK_2 to the α-amino group and pK_R to the ionizable group on the R side chain. (Can you identify the three functional groups in aspartic acid that are titrated?) Other reactions of amino acids are of importance. Carboxyl groups, under the proper reaction conditions, can be converted to esters or amides and amino groups converted to amides. Side chain hydroxyl groups, such as on serine, react with acids to form ester products. Some of these reactions occur under physiological conditions. As we will soon see, the carboxyl group of an amino acid can be linked to the amino group of another amino acid to form an amide or peptide bond.

Proteins occasionally contain unusual amino acids that are not in the set of 20. In most cases the different amino acids are derived by chemical reactions on standard amino acids after they have been incorporated into proteins. Some modified amino acids found in proteins are shown in Figure 4.8. The changed amino acids confer new biochemical properties on the protein. The cell often uses chemical modification of tyrosine to O-phosphotyrosine to regulate the activity of enzymes. The hydroxylated forms of proline and lysine are present in the structural protein collagen.

Because of the importance of amino acids in biological processes, biochemists have developed several methods for analysis of amino acid mixtures. The amino acid composition of a protein can be determined by (1) degrading the protein to its individual amino acids, (2) separating the amino acids, and (3) identifying and measuring the amino acids present. The technique of **chromatography** is widely used to separate, identify, and quantitatively measure

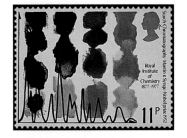

British stamp commemorating the development of chromatography.

Figure 4.8

Modified amino acid structures found in proteins. 4-Hydroxyproline and 5-hydroxylysine are found in collagen, a structural protein in animal tendons. Phosphoserine and phosphotyrosine change the reactivity of regulatory proteins.

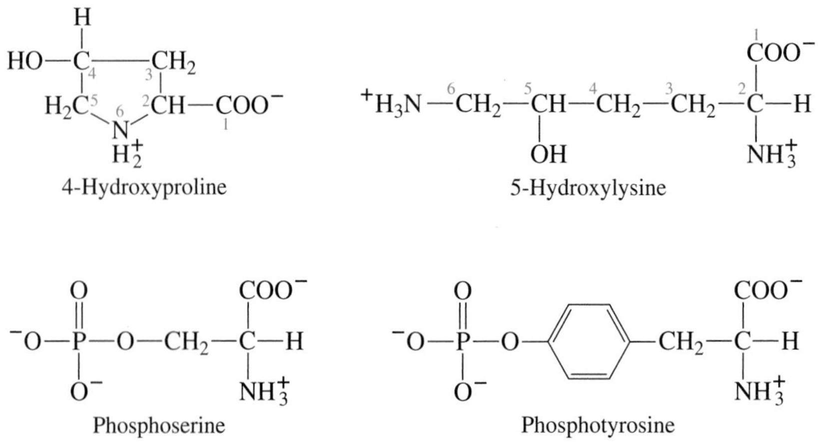

4-Hydroxyproline

5-Hydroxylysine

Phosphoserine

Phosphotyrosine

amino acids. Ion-exchange methods combined with **high performance liquid** chromatography (HPLC) are especially effective because they separate amino acids by acid–base properties (ionic charges) that are different for each amino acid. Two problems are often encountered during the analysis of amino acids: (1) Amino acids are usually present in extremely low concentrations so sensitive techniques are necessary, and (2) amino acids are not easily detected because they have few physical or chemical properties easily measured. Figure 4.9 shows

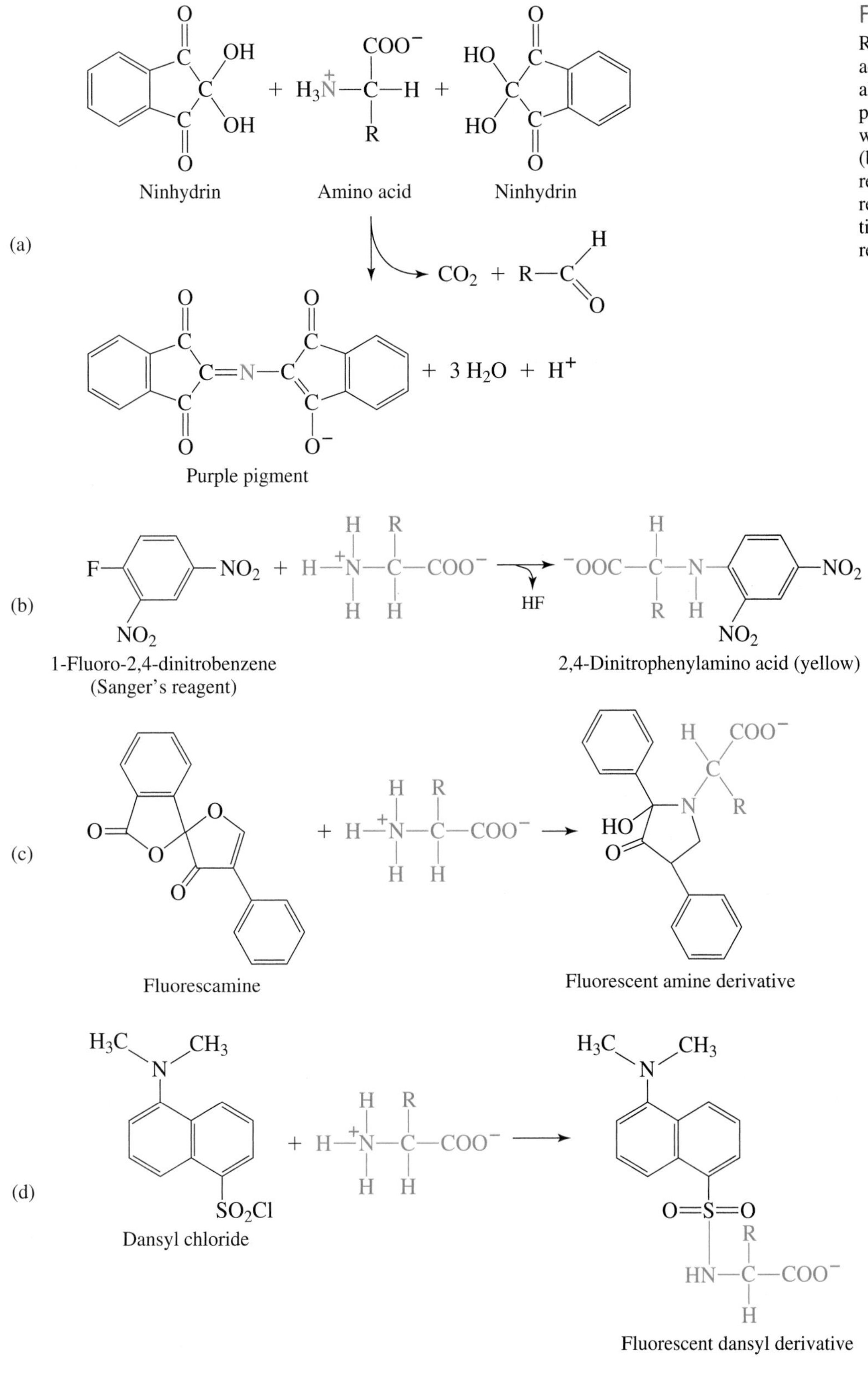

(a)

(b)

(c)

(d)

Figure 4.9

Reagents for the analysis of amino acids: (a) Ninhydrin reacts with all amino acids to give a blue-purple color except for proline, which gives a yellow color. (b–d) Sanger's reagent, fluorescamine, and dansyl chloride react with amino acids and peptides to produce colored or fluorescent products.

several chemical reagents now available that react with amino acids to form derivatives detectable by such sensitive techniques as ultraviolet–visible light (UV-VIS) absorption and fluorescence spectrophotometry. Ninhydrin reacts with amino acids to form a purple pigment (yellow with proline). Sanger's reagent (1-fluoro-2,4-dinitrobenzene) reacts with the amino group of α-amino acids to form yellow derivatives that are easily detected by spectrophotometry. The reagents fluorescamine and dansyl chloride lead to fluorescent derivatives of amino acids. These chemical reagents can be used to detect amino acids in the nanomole (10^{-9} mol) to picomole (10^{-12} mol) range.

4.2   Polypeptides and Proteins

Two α-amino acids can be linked together by formation of an amide or peptide bond:

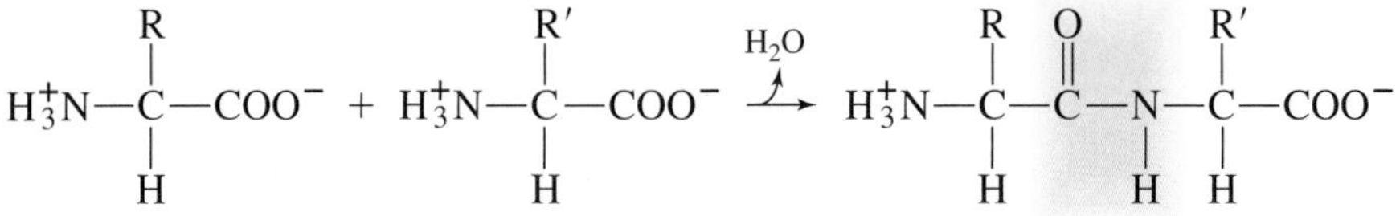

$$\underset{\substack{|\\H}}{\overset{\substack{R\\|}}{H_3^+N-C-COO^-}} + \underset{\substack{|\\H}}{\overset{\substack{R'\\|}}{H_3^+N-C-COO^-}} \xrightarrow{\ H_2O\ } \underset{\substack{|\ \ \ \ \ \ |\ \ |\\H\ \ \ \ \ H\ \ H}}{\overset{\substack{R\ \ \ O\ \ \ R'\\|\ \ \ ||\ \ |}}{H_3^+N-C-C-N-C-COO^-}}$$

The reaction, chemically defined as condensation (loss of a water molecule), brings together the carboxyl group of one amino acid with the amino group of the other to yield a dipeptide. The reaction can be repeated with additional amino acids joined onto the chain to form a tripeptide, tetrapeptide, pentapeptide, etc. Each amino acid, when incorporated into the polypeptide, is called a **residue.** Peptides with two to ten amino acid residues are usually named by the common chemistry prefix for numbers (di-, tri-, tetra-, penta-, hexa-, hepta-, octa-, nona-, and decapeptide). Products with 10 to 100 amino acids are called **polypeptides,** whereas those with more than 100 amino acids are called **proteins.** Most peptides and proteins isolated from cells and tissues contain between 2 and 2000 amino acids. Assuming the average molecular weight of all amino acids to be 110, the molecular weight of most peptide and protein chains ranges from 220 to 220,000 daltons, although much larger ones have been found.

No matter how many amino acids are joined by peptide bonds, there are almost always two distinct ends to the chain: an **amino terminus** (or N-terminus) and a **carboxyl terminus** (or C-terminus). In the dipeptide product shown in the reaction above, amino acid with side chain, R, is the N-terminus and amino acid with side chain, R', is the C-terminus. By convention, peptides are written and numbered from the amino end (on the left) to the carboxyl terminus (on the right) as in Figure 4.10. (Note that the two amino acids may be combined to form a dipeptide with amino acid R' as the amino terminus and R as the carboxyl terminus.) It is important to note that, although the α-amino and α-carboxyl ends of each amino acid (except those of the N and C ends) are changed in peptide formation, the R side

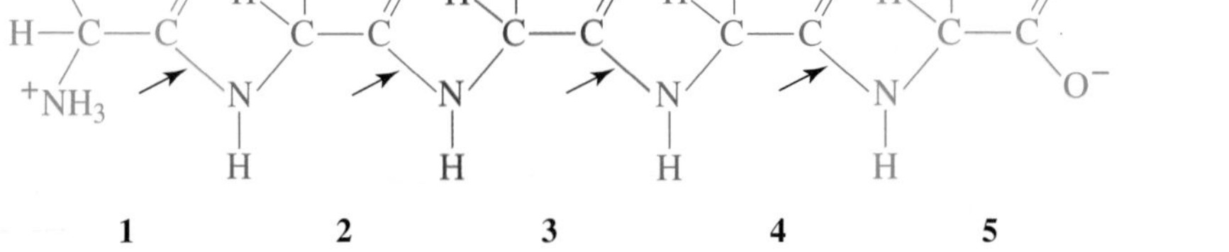

Figure 4.10

A short peptide showing correct numbering and direction. The amino terminus residue, always drawn to the left, begins with number 1. The structure of the pentapeptide is written Gly-Glu-Val-Ser-Lys. Arrows indicate peptide bonds.

chains of each amino acid residue remain unchanged. The chemical form and the electrical charge of the amino acid residue side chains will be of importance as we discuss the folding of proteins into their distinctive three-dimensional conformations.

The chemical reactions of peptides are characteristic of the functional groups present. The R side chains of amino acid residues display chemical reactivity similar to free amino acids. For example, hydroxyl groups in serine and tyrosine residues may be converted to esters with phosphoric acid (see Figure 4.8). Perhaps the chemical reaction of greatest importance is cleavage of peptide bonds by acid-, base-, or enzyme-catalyzed hydrolysis to yield free amino acids. The amino acid composition of proteins isolated from cells may be determined by hydrolysis in 6 M HCl. The digestion of protein in our food involves hydrolysis of peptide bonds by enzymes called **peptidases** or proteases in the stomach and intestines to yield free amino acids.

Peptide bond formation in vivo does not proceed as simply as implied in the above reaction. Peptide and protein synthesis, which is thermodynamically unfavorable, needs an energy source to activate the amino acids before incorporation into the growing polypeptide chain. In addition, instructions must be available for the placement of the right amino acids in the proper sequence. Proteins do not have a random sequence of amino acids but are ordered according to information in DNA as outlined in Chapter 2 and discussed in Chapter 12.

Proteins are greatly emphasized in biochemistry; but one should not think that only very long chains of amino acids are biologically functional. Glutathione, a tripeptide of glutamic acid, cysteine, and glycine, plays an important role in regulating oxidation–reduction reactions and destroying deleterious free radicals in the cell (Figure 4.11a). The pituitary gland hormones, oxytocin and vasopressin, are nonapeptides that are cyclized by formation of a disulfide linkage between two cysteine residues (Figure 4.11b,c). (Can you define an N-terminus and C-terminus for these peptides?) The hormone insulin, which regulates carbohydrate metabolism, contains 51 amino acids (Figure 4.11d). The enkephalins, peptides found in the brain and nervous system, play an important role in pain control (Figure 4.11e). A synthetic peptide (not produced naturally by organisms) that has become an important part of our lives is the sweetener, L-aspartyl-L-phenylalanine methyl ester (Aspartame or Nutrasweet®, Figure 4.11f). This dipeptide (about 200 times sweeter than common table sugar, sucrose) has many fewer calories than sucrose (16 calories for sucrose per teaspoon serving versus about 1 calorie for Aspartame) and is used in diet soft drinks and low-calorie food products.

4.3   Protein Function

Synthesis of proteins according to the messages stored in DNA leads to thousands of different molecules in a biological cell. Each type of protein, with its distinctive sequence of amino acid residues, has a characteristic shape, size, and biological function. Throughout the remainder of the book we will discuss, in detail, the diverse biological roles played by proteins. In preparation for this, we will introduce the major classes of proteins and describe general properties of protein structure (Table 4.4). We begin by classifying the proteins according to their biological roles.

Enzymes

All chemical reactions in an organism require a biological catalyst in order to bring biochemical reactions to completion under the rather mild conditions in the cell. For example, the digestion of carbohydrate in the diet begins in the mouth with the action of the enzyme, amylase. The replication of DNA during cell division requires the enzyme, **DNA polymerase.**

Structural Proteins

Many proteins provide mechanical support to cells and organisms. The strength of bone, skin, tendons, and cartilage is due primarily to the fibrous protein, collagen. The insoluble protein, keratin, is a major component of feathers, hair, and fingernails.

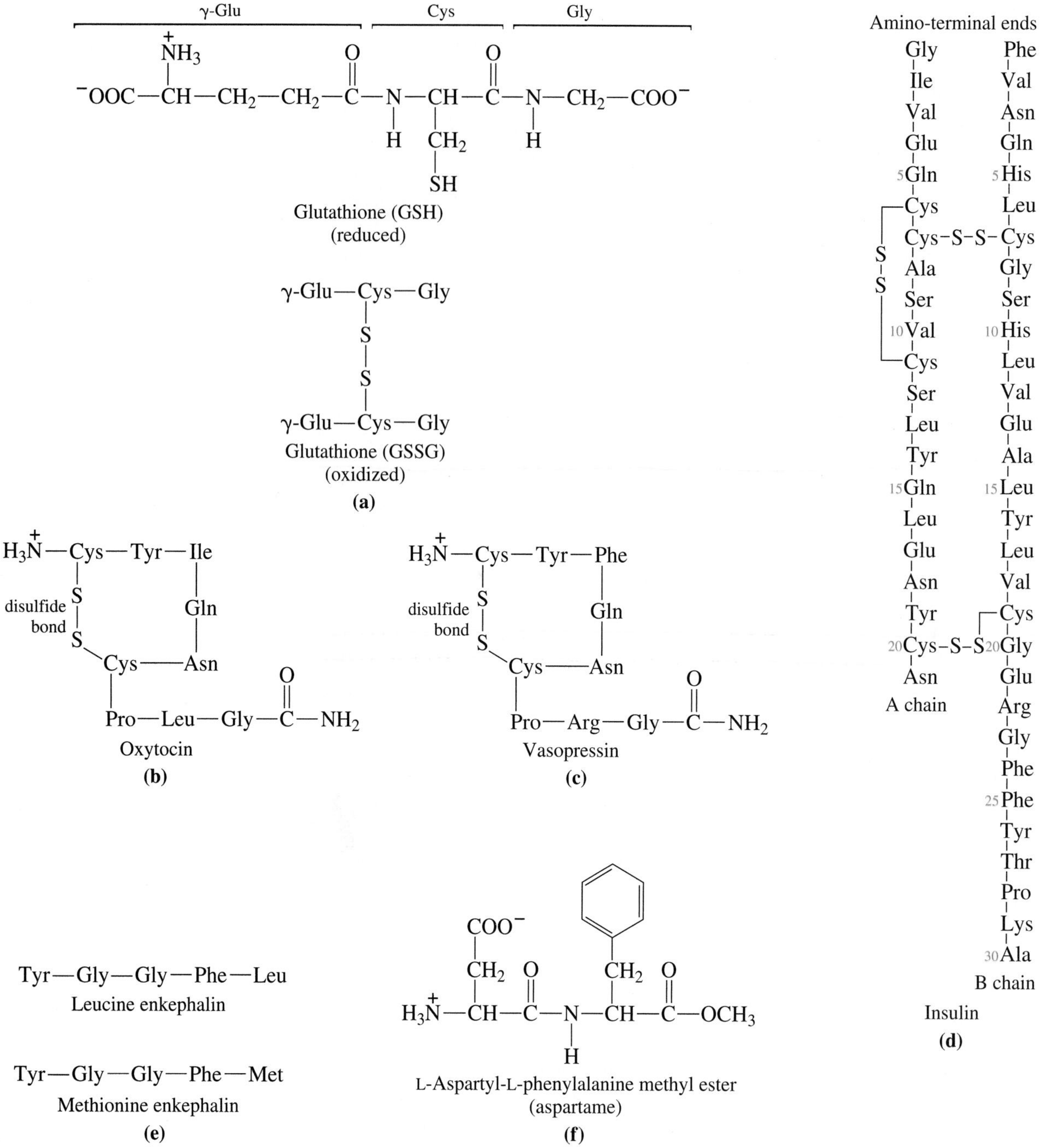

Figure 4.11

Structures of biologically important peptides: (a) glutathione in the reduced and oxidized form (a disulfide bond is present in the oxidized form); (b) oxytocin; (c) vasopressin; (d) insulin; (e) enkephalins; (f) aspartame.

Proteins of the intracellular cytoskeleton are used to assemble actin filaments, microtubules, and intermediate filaments (see Chapter 1, Section 1.4).

Immune Proteins

Structure Tutorial: Antibodies

Many plants and animals use proteins for defensive purposes. **Antibodies,** produced by higher organisms, are proteins that selectively bind and neutralize (destroy) foreign substances such as viruses and bacteria that may cause harm to the organism. Some snake venoms and the toxin of castor beans, ricin, are proteins.

Protein	Biological Function
	Enzymes
Phosphofructokinase	A glycolytic enzyme in carbohydrate metabolism that catalyzes phosphate group transfer from ATP to fructose 6-phosphate.
Citrate synthase	An enzyme in the citric acid cycle that combines acetyl-CoA and oxaloacetate to form citrate.
Trypsin	A digestive enzyme in vertebrates that catalyzes protein hydrolysis.
Ribonuclease	A hydrolytic enzyme produced by all organisms that catalyzes RNA hydrolysis.
RNA polymerase	An enzyme present in all organisms that catalyzes DNA-directed RNA synthesis.
Reverse transcriptase	An enzyme found in HIV (the virus that causes AIDS) that catalyzes RNA-directed DNA synthesis
	Structural Proteins
Collagens	Fibrous proteins found in all animals; form cable networks that serve as scaffolding for support of tissues and organs.
Elastins	Fibrous proteins found in connective tissue of the lungs and in large blood vessels such as the aorta, which have rubberlike properties that allow them to stretch to several times their normal length.
Keratins	Mechanically durable fibrous proteins present in vertebrates as major components of the outer epidermal layer and its appendages such as hair, nails, and feathers.
	Defense (Immune) Proteins
Antibodies	Globular proteins produced by the immune system of higher animals that participate in the destruction of biological invaders.
Interferons	Proteins, produced by higher animals, that interfere with viral replication.
	Transport and Storage Proteins
Hemoglobin	Globular heme–containing protein that carries oxygen from the lungs to other tissues of vertebrates.
Apolipoproteins	Components of lipoproteins such as low density lipoprotein (LDL) that participate in triacylglycerol and cholesterol transport.
Casein	Protein found in milk that stores amino acids.
Ferritin	Widely distributed protein that stores iron.
Myoglobin	A heme-containing protein found in vertebrates that binds oxygen.
	Regulatory and Receptor Proteins
Lac repressor	Genetic switch that turns off bacterial genes involved in lactose catabolism.
Insulin	A protein synthesized in the pancreas that acts as a signal for the fed state in higher animals.
Glucagon	A protein synthesized in the pancreas that acts as a signal for the starved state in higher animals.
	Muscle Contraction and Mobility Proteins
Actin	Component of skeletal muscle.
Myosin	Component of skeletal muscle.
Dynein	Protein that causes movement of sperm and protozoa by their flagella and cilia.

Table 4.4
Classes of proteins based on their biological functions

Transport and Storage Proteins

Many smaller biomolecules, including O_2, cholesterol, and fatty acids, must be transported throughout an organism to be used for energy metabolism and construction of cell components. Small molecules bind to proteins that carry them through the bloodstream and sometimes across cell membranes. Higher animals use blood **lipoproteins** to transport the very insoluble biomolecule, cholesterol. Many organisms use proteins to store nutrients for future use. Plants and animals store the micronutrient iron in the protein ferritin. The seeds of

Seeds like these of the Norway maple store proteins for germination.

plants store proteins, such as glutens, as an energy source for germination. Biomolecules, such as fatty acids, which are only slightly soluble in the blood of animals, are carried by albumins and globulins.

Regulatory and Receptor Proteins

The list of proteins that regulate cellular and physiological activity is long. Some well-known examples are the hormones insulin, prolactin, and thyrotropin. Other important examples are the **G proteins** that transmit hormonal signals inside cells (see Chapter 2, Section 2.6) and DNA binding proteins that assist in replication and regulation of protein synthesis (translation). The transmission of nerve impulses and hormone signals is mediated by **receptor proteins** located in cell membranes. When small signal molecules such as acetylcholine or epinephrine bind to receptor proteins on the surface of nerve cell membranes, a regulatory message is transmitted inside to the cellular machinery.

Muscle Contraction and Mobility Proteins

The proteins actin and myosin are functional components of the contractile system of skeletal muscle. The movement of sperm and protozoa by their flagella and cilia, respectively, is dependent on the protein dynein.

As is evident from these descriptions, there is great functional diversity among the proteins. However, there is a common thread that joins most of the proteins in their biological roles. During its action, a protein often undergoes selective binding or interaction with another molecule that is brought about by the formation of favorable, noncovalent interactions (see Chapter 3, Section 3.1). The other molecule may be small, for example, a reactant that binds to an enzyme, or a neurotransmitter molecule that binds to its receptor or the binding of a molecule for transport or storage. On the other hand, large molecules also interact with proteins. The formation of filaments or tubules in the cytoskeleton and the interaction of actin and myosin in muscle contraction involves the association of large molecules to produce supramolecular assemblies (see Chapter 1, Section 1.4). The action of antibodies depends on their binding to specific antigens, which may be viruses, bacteria, or foreign molecules. The synthesis rate of proteins is controlled by the binding of regulatory proteins to specific regions of DNA (Chapter 12, Section 12.4).

4.4 Protein Size, Composition, and Properties

Because each type of protein is coded by a gene, a unique region of DNA, all the molecules of that type will have identical amino acid composition and sequence. That distinct set of amino acid residues gives a protein its characteristic properties. One important property of a protein is its molecular weight. This constant may be calculated by summing the molecular weights of

Table 4.5
Molecular properties of some proteins

Protein	Molecular Weight[a]	Number of Amino Acid Residues	Number of Subunits
Insulin (bovine)	5,733	51	2
Cytochrome *c* (human)	13,000	104	1
Ribonuclease A (bovine pancreas)	13,700	124	1
Lysozyme (egg white)	13,930	129	1
Myoglobin (equine heart)	16,890	153	1
Chymotrypsin (bovine pancreas)	21,600	241	3
Hemoglobin (human)	64,500	574	4
Serum albumin (human)	68,500	550	1
Immunoglobulin G (human)	145,000	1320	4
RNA polymerase (*E. coli*)	450,000	4100	5
Ferritin (equine spleen)	450,000	4100	24
Glutamate dehydrogenase (bovine liver)	1,000,000	8300	40

[a]In daltons.

the amino acid residues. Because this is cumbersome, a molecular weight is often estimated by multiplying the number of amino acid residues by the average molecular weight of an amino acid residue, which is 110. Alternatively, the size of a protein can be defined in terms of mass, which has the units of **daltons.** Recall from Chapter 1 that one dalton is equivalent to one atomic mass unit. A protein with 250 amino acid residues has an approximate molecular weight of 27,500 daltons or 27.5 kilodaltons. The molecular sizes of several proteins are given in Table 4.5.

Some proteins are composed of a single polypeptide chain and are called **monomeric.** Cytochrome *c*, listed in Table 4.5, is an example. Others are **oligomeric**; that is, they consist of two or more polypeptide chains that are usually held together by noncovalent interactions. The term **multisubunit** is also used where each peptide is referred to as a **subunit.** The subunits of a multisubunit protein may be identical or different. Hemoglobin is tetrameric (four subunits) and comprised of two α-type protein chains and two β-protein chains. Iron storage protein, ferritin, contains 24 similar subunits.

Proteins may also be classified according to their composition. Proteins made up of only amino acid residues and no other biomolecules are **simple proteins.** The digestive enzymes trypsin and chymotrypsin are simple proteins. Proteins that contain some other chemical group such as a small organic molecule or metal atom are called **conjugated proteins.** The additional chemical group is called a **prosthetic group.** A hemoglobin molecule consists of four protein chains, each associated with a heme prosthetic group containing iron. The enzyme alcohol dehydrogenase from yeast is a tetrameric protein with a zinc atom associated with each subunit. The protein–prosthetic group as partners is usually more effective than the protein alone in its biological role.

The physical properties of a protein are also dependent on the nature of the amino acid residue R side chains. The electrical charge on a protein at physiological pH is determined by adding all the charges of the R side chains (at neutral pH). The positive charge at the N-terminus and the negative charge at the C-terminus neutralize each other. The charge on a protein becomes important when a protein is characterized by **electrophoresis** and solubility. Electrophoresis is a technique for purifying and measuring the molecular size of proteins (see Window on Biochemistry 4.2).

Protein solubility is best understood by dividing proteins into two categories. All proteins are not water soluble; in fact, insoluble proteins are essential for cell and organism structural integrity and rigidity (Table 4.6). **Globular proteins** are those that are water soluble and play a dynamic role in transport, immune protection, catalysis, and so on. These proteins are dissolved in biological fluids such as blood and cytoplasm and would be expected to have a relatively high content of amino acid residues with polar and charged R groups. In contrast, **fibrous proteins** are water insoluble and are represented by the structural proteins collagen and keratin. One would predict the fibrous proteins to contain a relatively high content of amino acid residues with nonpolar R groups. Silk, a fibrous protein, has an amino acid

Table 4.6
Examples of globular and fibrous proteins

Type of Protein	Function
Globular Protein	
Hemoglobin	Transport (oxygen transport)
Myoglobin	Storage (oxygen storage)
Ribonuclease	Enzyme (RNA hydrolysis)
Lysozyme	Enzyme (bacterial wall hydrolysis)
Cytochrome *c*	Electron transport
Immunoglobulin	Defense (antibody)
Actin	Movement (muscle protein)
Fibrous Protein	
Collagen	Structural protein
Keratin	Structural protein
Myosin	Movement (muscle protein)
Elastin	Elasticity

composition of 45% Gly, 30% Ala, 12% Ser, 5% Tyr, 2% Val, and 6% others. As we discover in the next chapter, fibrous proteins in an organism take on an ordered and somewhat rigid conformation. Fibrous proteins have a high tensile strength, whereas globular proteins, which are soluble in cytoplasm and other biological fluids, are more dynamic and flexible, but still ordered in their three-dimensional arrangements.

4.5   Four Levels of Protein Structure

The functional activity of a protein under cellular conditions depends on its folding into a distinct, three-dimensional structure called the **native conformation.** The folded arrangement of proteins can be defined in four levels as shown in Figure 4.12. The **primary structure** (Figure 4.12a) is defined as the sequence (order) of amino acid residues in a protein. The amino acid residues are held together by covalent peptide bonds. The amino acid sequence of a protein is of genetic significance because it is derived from the sequence of nucleotide bases in mRNA. The mRNA is transcribed from a sequence of nucleotides (gene) that resides on the DNA. Therefore, the primary structure of a protein is an inevitable result of the order of nucleotides in the DNA. The primary or covalent structure prepares the protein for the next three structural levels. Localized regions of the primary sequence fold into a regular, repeating **secondary structure,** such as an α-helix or a β-sheet shown in Figure 4.12b. Secondary structural elements can interact and pack into a compact globular unit called the **tertiary structure** (Figure 4.12c). The fourth level or **quaternary structure** (Figure 4.12d) defines the association of two or more polypeptide chains to form a multisubunit protein molecule.

Biochemists often try to predict the structural elements present in the biologically active shape of a protein before actual experimental measurements are made on the protein. In theory, one should be able to predict the native conformation of a protein by knowing just the sequence of amino acid residues. In spite of extensive studies of protein structures and research into the folding pathways for proteins with known sequences, we are still not able to forecast in detail all structural features. To date, the best way to achieve a complete and detailed picture of the secondary, tertiary, and quaternary structures of a protein is from X-ray crystallography.

4.6   Protein Primary Structure

Determination of Primary Structure

The primary structure of a protein is revealed by one of two experimental techniques: (1) isolation and purification of the protein followed by direct analysis of the amino acid sequence, or (2) determination of the nucleotide sequence of the gene that codes for the protein. The first option will be discussed here and the second option in Chapters 12 and 13.

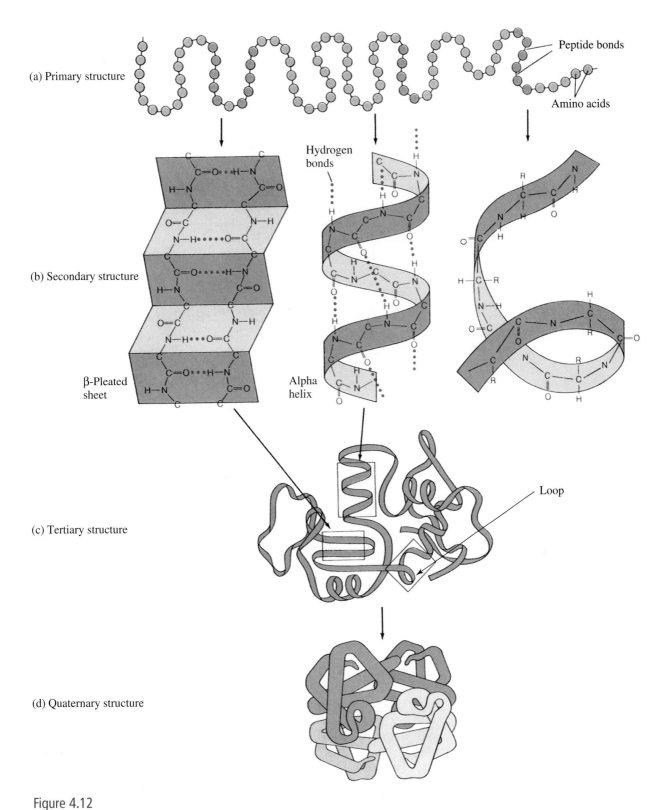

(a) Primary structure

Peptide bonds

Amino acids

(b) Secondary structure

Hydrogen bonds

β-Pleated sheet

Alpha helix

(c) Tertiary structure

Loop

(d) Quaternary structure

Figure 4.12

The relationships among the four levels of protein structure: (a) primary or covalent structure, (b) secondary structure, (c) tertiary structure, (d) quaternary structure.

The first step in protein analysis is to measure the composition of amino acids. The amino acid composition is best determined by complete hydrolysis of the protein. The least destruction to the amino acid products occurs under conditions of 6 *M* HCl and 110°C (Figure 4.13). The individual amino acid products are separated by ion-exchange chromatography and quantitative analysis by HPLC leads to the amino acid composition.

A protein may be characterized further by analysis of the N-terminal residue. As we discussed earlier in this chapter, most proteins have two ends: an amino end (N-terminus) and a carboxyl end (C-terminus). The amino terminal residue is identified by one of several available methods. Historically, the most important method is Sanger analysis. The protein is reacted with 1-fluoro-2,4-dinitrobenzene (Sanger's reagent). This chemical reacts with the end amino group to label the protein (see Figure 4.9). After hydrolysis of the tagged protein in 6 *M* HCl, the N-terminal amino acid is separated and identified as its 2,4-dinitrophenyl derivative. The first complete sequential analysis of amino acid residues in a polypeptide was reported by Frederick Sanger in 1953. After many years of work using the sequencing reagent 1-fluoro-2,4-dinitrobenzene, he deduced the sequence of the 51 amino acids in bovine insulin (see Figure 4.11d).

Edman Method of Protein Sequencing

If the rapid and accurate sequential analysis of a protein is desired, the method of choice is **Edman degradation** (Figure 4.14). This procedure, which uses the reagent phenylisothiocyanate, labels and releases just the N-terminal residue without destroying the remainder of the protein. The tagged amino acid from the N-terminus can be isolated and identified. The amino acid derivative is called a **phenylthiohydantoin.** A new N-terminal residue has now been uncovered on the protein (amino acid number 2 in the original protein). The protein product can be recycled many times through the Edman reaction, with loss of an amino acid from the N-terminus each time. An automatic analyzer, called a sequenator, is now available that can be programmed to control the proper mixing of reagents, the separation of products, and the identification of the phenylthiohydantoin derivatives of the amino acid residues. Approximately a microgram of polypeptide is required for complete sequential analysis by the Edman procedure.

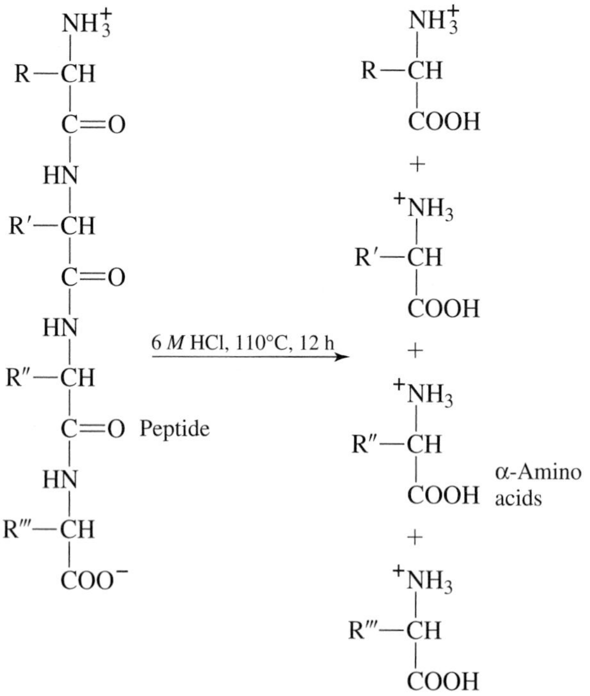

Figure 4.13

Hydrolysis of a protein by 6 *M* HCl. Each peptide bond is broken, resulting in a mixture of all the amino acids in the protein. The free amino acids are separated and identified by automated chromatography.

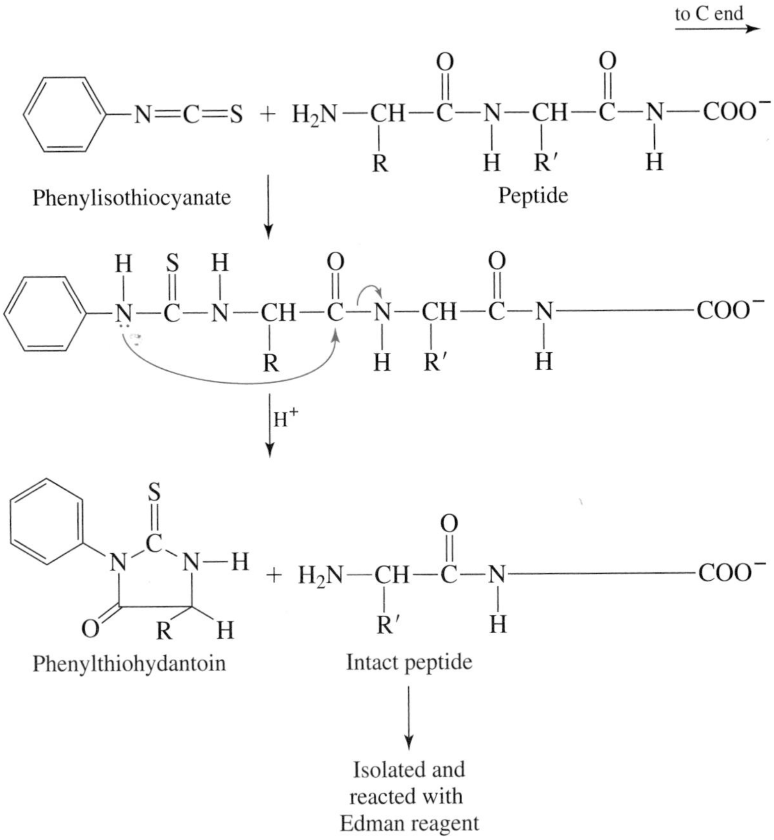

Figure 4.14
Reactions of the Edman degradation method to sequence a protein. Phenylisothiocyanate reacts with the amino group of the N-terminal residue. That modified amino acid can be released, isolated, and identified. Amino acids are removed and identified as phenylthiohydantoin derivatives. The intact peptide (minus the N-terminal residue) is isolated and recycled through another step of the Edman reaction.

The Edman method of sequencing is limited to an analysis of only the first 50 amino acids of a protein. Polypeptides and proteins with greater than 50 amino acid residues must be cut into smaller pieces for sequencing. Protein chains may be fragmented with chemical reagents and enzymes that split peptide bonds at known locations as shown in Figure 4.15. The reagent **cyanogen bromide** (CNBr), a very selective cutting agent, splits the peptide bond on the C side of every methionine residue in a protein. A protein with n methionine residues will result in $n + 1$ polypeptide fragments, unless two methionine residues are adjacent. (Can you predict the results from adjacent met residues?) Enzymes that may be used for specific cleavage are the digestive peptidases, trypsin and chymotrypsin. Trypsin is able to recognize the location of arginine and lysine residues in a protein by interacting with each unique side chain. The enzyme then catalyzes the hydrolysis of peptide bonds on the carboxyl side of the two amino acids. Chymotrypsin acts on the carboxyl side of amino acid residues phenylalanine, tryptophan, and tyrosine. The fragments, after separation by ion-exchange chromatography, may be sequenced by the Edman method.

Sequence analysis of proteins as described above is not always trouble-free. Difficulties may be encountered if disulfide bonds between cysteine residues are present. Such bonds must be broken with reducing agents such as mercaptoethanol and modified to prevent reformation. Sequencing very large proteins (those with 500 or more residues) is difficult because fragments made by reagents in Figure 4.15 are often much larger than 50 residues. Often two or more fragmentation methods must be used, which adds to the complexity of achieving complete sequential data.

Sequencing DNA

New techniques for protein sequence analysis have emerged from recombinant DNA research (Chapter 13). It is now possible to clone (make many identical copies of) long stretches of DNA. Genes carrying the message for selected proteins can be isolated in quantities sufficient for nucleotide sequencing. By using the genetic code, the structure of protein product from a gene can be deduced from the sequence of nucleotides in the DNA. Techniques for DNA sequencing are faster and more reliable than for protein sequencing. Determining protein sequences indirectly from DNA sequences will never completely replace direct analysis of the protein product. An amino acid sequence determined from DNA will be that of the initial protein product synthesized. Often that form of the protein (called the nascent form) is modified (converted to a biologically active form) by removal of short stretches of amino acid residues from the ends, formation of disulfide bonds, and chemical changes on amino acid residues. Chemical analysis of the active form of the protein will always be required so the two methods (DNA and protein sequencing) will continue to be complementary approaches to studying protein structure and function.

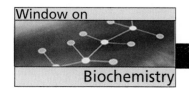

Window on

Biochemistry

4.2 Isolating and Purifying Proteins: Using the Tools of Biotechnology

Before a protein can be sequenced and characterized as described in Section 4.6, it must be isolated in a purified form; that is, the desired protein must be separated from all other proteins and other biomolecules. A typical experimental format used by today's biochemists to purify proteins is the following:

1. **Develop an assay to identify and quantify the protein.** A physical, chemical, or biological procedure is required to confirm the presence of the protein.
2. **Select a biological source for the protein and prepare a cell homogenate.** Proteins are often isolated from plant and animal tissue, biological fluids, and microbial cells. With the introduction of new techniques in DNA recombinant research, it is now possible to transfer, via a plasmid, the specific gene for a protein into another organism (usually *E. coli* cells) and allow that organism to synthesize the desired protein (see Chapter 13). The cell homogenate is prepared as described in Window on Biochemistry 1.3.
3. **Separate the cell homogenate into fractions by differential centrifugation** (see Window on Biochemistry 1.3). The cell extract after step 2 still contains thousands of different proteins and other biomolecules. The desired protein is separated from the others by taking advantage of properties that are unique to the protein: its polarity, charge, size (molecular weight), and ability to bind to other molecules.
4. **Perform chromatography: ion-exchange, gel filtration, and affinity.** In chromatography, the protein mixture to be examined is allowed to interact with two physically distinct entities: a mobile phase and a stationary phase. The mobile phase, usually an aqueous, buffered solution, moves the sample through a glass or plastic column containing the stationary phase, which is usually solid and has the ability to "bind" some types of molecules (see figure). In **ion-exchange chromatography,** the stationary phase, an ion-exchange resin, is charged and can be used to separate proteins with different charges. If the resin is positively charged, proteins with net negative charge will bind while neutral and positively charged proteins pass directly through the stationary phase. In **gel filtration chromatography,** the stationary phase consists of a cross-linked gel matrix that separates biomolecules on the basis of molecular size. Larger molecules pass more rapidly through the stationary phase than smaller molecules.

As discussed in Sections 3.1 and 4.3, many proteins bind selectively to other molecules. Because this binding is due to very specific interactions, it provides a highly selective means of separating molecules. The binding properties of proteins are exploited in the technique of **affinity chromatography.** Here, the stationary phase consists of a resin with attached biomolecules that can be recognized and bound by the desired protein. For example, the insulin receptor protein of some cells may be isolated using a column with a stationary phase containing bound insulin.

5. **Determine the protein's purity and molecular size by gel electrophoresis.** The final sample of protein is analyzed for purity by **electrophoresis,** a technique that separates molecules on the basis of charge and size. In this analytical method, proteins are allowed to

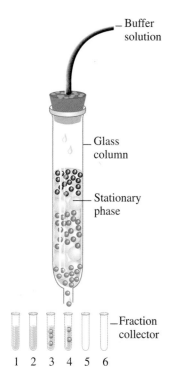

— Buffer solution

— Glass column

— Stationary phase

— Fraction collector

1   2   3   4   5   6

The principles of chromatography. The stationary phase is packed into a column. A solution containing the desired protein is applied to the top of the column. Aqueous buffer passes through the column to move proteins. Proteins shown in black, red, and green are separated according to their ability to "bind" to the stationary phase. Fractions containing different proteins are collected at the bottom of the column. Proteins that bind weakly to the column appear in earlier fractions.

Importance of Protein Sequence Data

The amino acid sequences of thousands of proteins are now available in computerized databases. By studying and comparing sequence information for different proteins, significant biochemical conclusions can be made:

1. Sequence data from numerous proteins can be searched for regions of identical or similar sequences. We are now discovering "families of proteins" that are related by common sequence features and therefore similar biological function. A comparison analysis may uncover proteins that contain large regions of **sequence homology** (identical sequence) or just one or two critical amino acid residues of the same identity at the same location in the protein chains. A protein of unknown function through this analysis may be added to one of the established protein families, and then insight into its biological

migrate in a polyacrylamide gel matrix subjected to an electric field. Under the experimental conditions for electrophoresis (basic pH), all proteins are negatively charged (but not necessarily the same net charge) and will move toward the anode (positive electrode). However, the gel matrix is cross-linked and acts like a molecular sieve allowing smaller protein molecules to move faster than larger molecules. Therefore, molecules are separated on the basis of both charge and size. After electrophoresis, the gel matrix is stained with a dye to detect the presence of proteins (see figure a). A highly purified protein sample will show just one band or perhaps one intense band and a few very light bands of protein impurities. A modification of polyacrylamide gel electrophoresis allows one to estimate the molecular weight of a protein. The protein, before electrophoresis analysis, is treated with **sodium dodecyl sulfate (SDS),** a negatively charged detergent that binds to most proteins in amounts proportional to the molecular weight of the protein. The SDS disrupts the native conformation of the protein molecules and places them all in a linear shape. Each protein now has approximately the same charge-to-mass ratio and the same shape, so separation by electrophoresis is based entirely on molecular size. By running a set of standard proteins of known molecular size on the same gel with the unknown, data are obtained to estimate the molecular weight of the unknown (figure b).

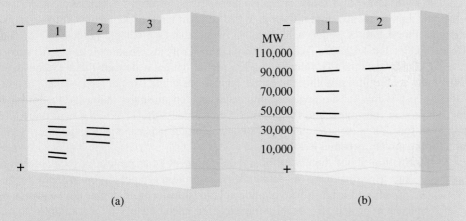

(a) The results of an electrophoresis experiment, using polyacrylamide gel, to analyze the purity of protein solutions. A crude extract was applied to position 1. In position 2 the crude extract was placed after some purification step, perhaps ion-exchange chromatography. Position 3 shows the purity of the protein after affinity chromotography. (b) An example of SDS electrophoresis to determine the molecular weight of a protein. A mixture of standard proteins of known molecular weight was electrophoresed in position 1. Position 2 contains an unknown, pure protein. Its molecular weight is estimated to be about 90,000.

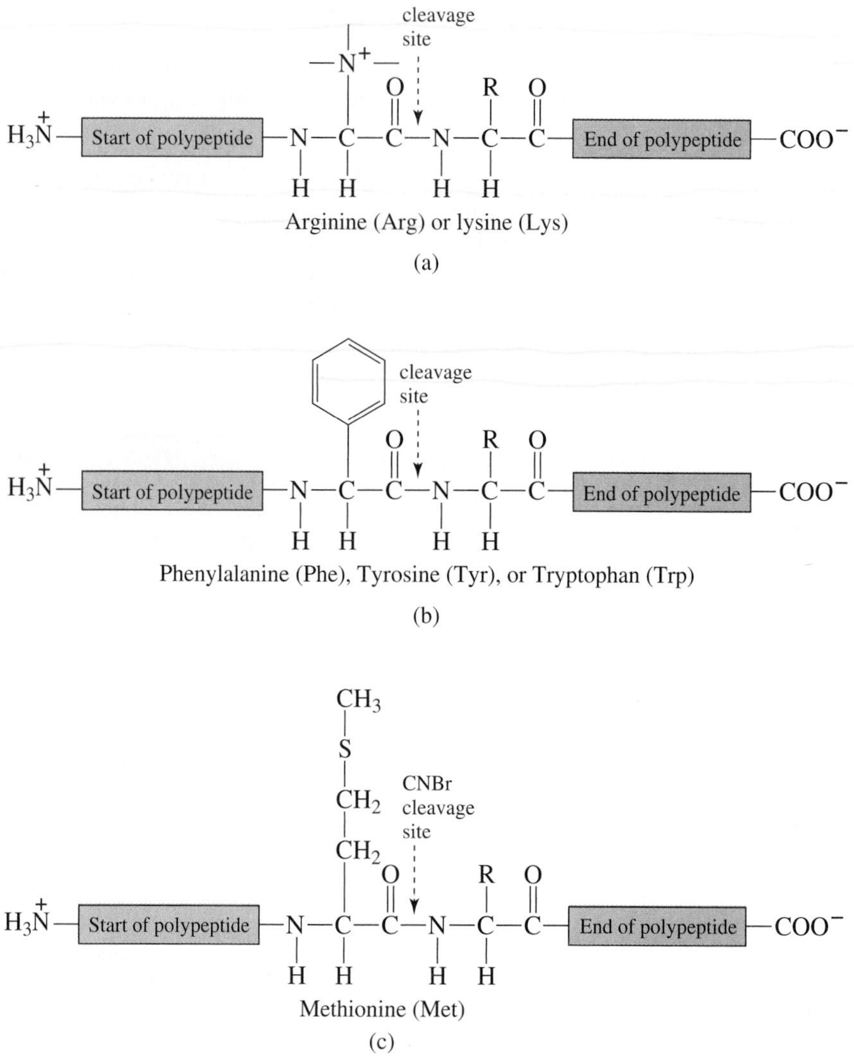

Figure 4.15

Reagents for the selective fragmentation of a protein. The reagents listed direct the cleavage of specific peptide bonds. (a) Treatment of a protein with trypsin results in the hydrolysis of peptide bonds on the C side of Arg and Lys residues. (b) Chymotrypsin catalyzes the cleavage of peptide bonds on the C side of Phe, Tyr, and Trp. (c) Cyanogen bromide cleaves the peptide bond on the C side of Met.

Arginine (Arg) or lysine (Lys)

(a)

Phenylalanine (Phe), Tyrosine (Tyr), or Tryptophan (Trp)

(b)

Methionine (Met)

(c)

function is discovered. A study of one important class of enzymes, the protein kinases, illustrates how sequence information is useful. Protein kinases are enzymes that catalyze the phosphorylation of amino acid residues in proteins (see Chapter 2, Section 2.5 and Chapter 4, Section 4.1). These reactions play a crucial role in cellular regulation. All known protein kinases contain a common sequence region (called a **domain**) that extends over 240 amino acid residues. Within this larger domain are found smaller regions that are individually responsible for binding reactants necessary for phosphorylation (ATP), for binding the protein substrates, and for providing a catalytic surface. When a new protein of unknown function is isolated and found to have a protein kinase domain, the likely function for the protein becomes evident.

2. Information about the evolutionary development of a protein is available from sequence data. By comparing the sequences of a protein type common to many organisms, it is possible to establish taxonomic relationships. For example, the sequence of cytochrome *c,* a protein for respiration (oxygen metabolism) in all aerobic organisms, has been determined in over 60 different species. Of the approximately 100 amino acid residues in cytochrome *c,* 27 residues are invariant in all organisms; that is, the same amino acid occupies the same numbered position in all 60 species. Approximately 55 residues are invariant in horse cytochrome *c* and yeast cytochrome *c.* The sequence of cytochrome *c* is identical in horse, pig, sheep, and cow. Indeed, sequence data may be used to make phylogenetic trees showing evolutionary relationships among organisms.

3. Amino acid sequences can be searched for changes that may lead to dysfunction of a protein. Normally, all of the molecules of a specific protein in a given species have an identical sequence. However, some individuals may produce a protein in which one or more amino acid residues is replaced by some other amino acid. The modified protein is often biologically inactive. A well-known example is the hemoglobin produced by individuals with the disease **sickle cell anemia** (see Chapter 5, Section 5.5). The sequence or composition change in the protein is usually the result of an inherited mutation in the gene. One or more nucleotides in the DNA have been replaced to change the message transmitted by mRNA. Research using recombinant DNA techniques developed from the Human Genome Project is now making it easier to detect such changes in chromosomal DNA.

4. Knowledge of the amino acid sequence of a protein may lead to conclusions about its three-dimensional structure (native conformation). As we have stated several times in this chapter, the manner in which a protein folds into its distinct biologically active form depends on its primary structure. We are not yet able to predict the detailed three-dimensional structure; however, in Chapter 5, we discuss some generalizations relating secondary, tertiary, and quaternary structure to the primary structure of a protein. Of the approximately 100,000 proteins of known primary structure, complete three-dimensional structures are known for only about 10,000, so much progress needs to be made in predicting protein structure. X-ray crystallography techniques that require protein crystals (very difficult to obtain) and extensive computer analyses are not able to keep up with the rapidly growing number of proteins isolated and sequenced. Protein structure analysis by nuclear magnetic resonance is advancing but only proteins with up to 250 amino acids can be studied.

Summary

- Proteins are biopolymers of varying size that are constructed from a set of 20 α-amino acids. These amino acids share a general structure composed of a carbon center surrounded by a hydrogen, a carboxyl group, an amino group, and a side chain (R) that differs for each amino acid. At physiological pH, amino acids exist as dipolar ions called **zwitterions.**

- The 20 α-amino acids are classified into three groups depending on the polarity of their side chains: (1) those with nonpolar side chains, (2) those with polar, uncharged side chains, and (3) those with polar, charged side chains.

- Peptides and proteins are formed by linking amino acids together using amide bonds. The α-carboxyl group of one amino acid combines with the α-amino group of another amino acid. Each type of protein has a well-defined amino acid composition and molecular weight.

- The molecular architecture of proteins may be organized into four levels of structure: **primary** defines the sequence of amino acids, **secondary** defines regular, repeating structural elements, **tertiary** defines the complete three-dimensional arrangement in space, and **quaternary** defines the association of two or more polypeptides.

- The primary structure of a protein may be determined either using the **Edman reagent,** which labels the amino acid at the amino terminus, or by cloning and sequencing the DNA region (gene) that encodes the protein.

- Proteins are isolated, purified, and characterized using techniques of chromatography and electrophoresis.

Study Exercises

Understanding Terms

4.1 Define the following terms. Use molecular structures where appropriate.

a. Protein
b. Amino acid enantiomers
c. Zwitterion
d. Prosthetic group
e. Sequence homology
f. Collagen
g. Multisubunit protein
h. Peptidase
i. Gel filtration
j. Glutathione
k. Affinity chromatography
l. Sodium dodecyl sulfate
m. Amino acid side chain
n. D-Serine
o. Proteomics
p. Amino terminus
q. Carboxyl terminus
r. Primary structure
s. α-Amino acids
t. Sanger's reagent
u. Electrophoresis
v. Immune proteins
w. Oligomeric
x. Protomers
y. Insulin
z. Chymotrypsin

Reviewing Concepts

4.2 Draw all the possible enantiomers of valine; of threonine; of glycine.

4.3 Draw the titration curve expected for glycine. Draw the structure(s) of glycine present at pH = 1.0, pH = 2.3, pH = 9.6, pH = 12.0. pK values are given in Table 4.2.

4.4 An ionization step for the amino acid serine is written below:

$$\underset{\textbf{A}}{\underset{\underset{\text{OH}}{|}}{\underset{\underset{\text{CH}_2}{|}}{\text{H}_3^+\text{N}-\overset{\overset{\text{COOH}}{|}}{\underset{|}{\text{C}}}-\text{H}}}} \rightleftharpoons \underset{\textbf{B}}{\underset{\underset{\text{OH}}{|}}{\underset{\underset{\text{CH}_2}{|}}{\text{H}_3^+\text{N}-\overset{\overset{\text{COO}^-}{|}}{\underset{|}{\text{C}}}-\text{H}}}} + \text{H}^+$$

Calculate the pH at which the following ratios of form A to form B (A:B) are present.

a. A:B = 0.1:1
b. A:B = 1:1
c. A:B = 1:2

HINT: Study Chapter 3 and the Henderson–Hasselbalch equation.

4.5 Consider the list of possible properties of amino acids below. Circle those properties that are true for all of the 20 amino acids in proteins.

a. Crystalline white solids at room temperature
b. High-boiling oils at room temperature
c. Soluble in organic solvents like acetone and chloroform
d. Water solutions conduct electric current
e. Insoluble in water
f. All are symmetric, nonchiral molecules
g. All contain the elements C, H, N, O, and S

4.6 In each pair of amino acids listed below, determine the one that has the more nonpolar side chain.

a. Gly:Ala
b. Ala:Ser
c. Val:Asp
d. Phe:Tyr
e. Pro:Lys

4.7 Listed below are several characteristics for the amino acids. List an amino acid that complies with each characteristic.

a. Alkyl group side chain
b. Aryl group side chain
c. Side chain that reacts with NaOH
d. A side chain that can form hydrogen bonds with water
e. A side chain that reacts with HCl
f. An amino acid that releases ammonia (NH_3) when heated with aqueous NaOH

4.8 Draw the molecular structure for the dipeptide Ala-Asp in its completely protonated form. Write ionization reactions showing the dissociation of three acidic protons.

4.9 Draw the ionic form of glutamic acid that would exist at pH = 12.

4.10 The following reagents are useful for characterizing proteins. Describe how each can be used and what information is obtained. Write all appropriate reactions.

a. Dansyl chloride   **e.** Trypsin
b. Sanger's reagent   **f.** SDS
c. 6 M HCl   **g.** Ninhydrin
d. Phenylisothiocyanate   **h.** Cyanogen bromide

4.11 Write out the structures for two dipeptides that could form by reacting the amino acids valine and serine under peptide forming conditions.

4.12 Describe differences between globular and fibrous proteins.

4.13 Estimate the number of amino acids in each protein below (MW = molecular weight).

a. Insulin, MW = 5733
b. Immunoglobulin G, MW = 145,000
c. Catalase, MW = 250,000

4.14 Estimate the molecular weight for each protein below.

a. Cytochrome c; 104 amino acid residues
b. Ferritin; 4100 amino acid residues

4.15 Name an amino acid for each of the properties listed below. Choose your answers from the 20 amino acids found in proteins.

a. A sulfhydryl side chain
b. Three ionizable groups
c. An amide side chain
d. Two chiral carbon centers
e. An imidazole side chain
f. No chiral carbon center

4.16 Draw the structure of histidine as it would exist at pH 1; at pH 5; at pH 11.

Solving Problems

4.17 Would you expect the peptides aspartame, glutathione, vasopressin, and insulin to be water soluble? Explain.

4.18 Complete the following reactions:

a. 2 Cys $\xrightarrow[\text{conditions}]{\text{oxidizing}}$

b. Gln + H_2O $\xrightarrow{\text{NaOH}}$

c. Gly-Ala + H_2O $\xrightarrow{\text{H}^+}$

4.19 Predict those amino acid residues among the standard 20 amino acids that would be buried inside a protein structure when the protein is dissolved in H_2O.

4.20 Explain how the primary structure of a protein can be obtained from the nucleotide sequence of DNA.

HINT: Review Chapter 2

4.21 Which of the following pairs of proteins could be separated by gel filtration?

a. Cytochrome *c* and hemoglobin

b. Ribonuclease A and lysozyme

c. Ferritin and myoglobin

HINT: See Table 4.5.

4.22 Each of the proteins listed below is treated with sodium dodecyl sulfate and separated by electrophoresis on a polyacrylamide gel. Each is run on an individual gel. Draw pictures of the final results.

a. Myoglobin

b. Hemoglobin (two α subunits, molecular weight = 15,500; two β subunits, molecular weight = 16,000)

4.23 You want to study the structure of the hormone insulin shown in Figure 4.11. The first step is to reduce the disulfide bonds with mercaptoethanol. This results in the formation of two polypeptides, A and B. Predict the results from treating A and B with each of the following reagents.

a. CNBr

b. Reaction with Sanger's reagent followed by 6 *M* HCl hydrolysis

c. Reaction of A and B with carboxypeptidase

d. Reaction of A and B with trypsin

4.24 What is the molecular basis of separation in each of the following chromatographic techniques?

a. Ion-exchange chromatography

b. Affinity chromatography

c. Gel filtration chromatography

4.25 Consider the dipeptide Asp-Phe and answer the following questions.

a. What amino acid is the N-terminus?

b. What amino acid is the C-terminus?

c. Circle all the ionizable groups in the dipeptide.

d. What is the net charge on the peptide at pH 7.0?

4.26 Assume that a protein has the following structure:

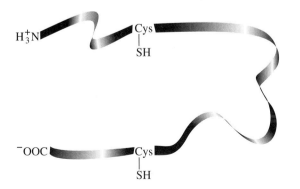

Draw the protein structure after treatment with an oxidizing agent.

4.27 From the list of amino acids below, select those with side chains capable of involvement in hydrogen bonding.

Ala	Phe	His
Ser	Thr	Tyr
Asn	Val	Cys

4.28 Your roommate, who is also in your biochemistry class, tells you that the Sanger reagent may be used for sequential analysis of a protein. How would you explain to your roommate that this is not completely correct?

4.29 Complete the following reaction:

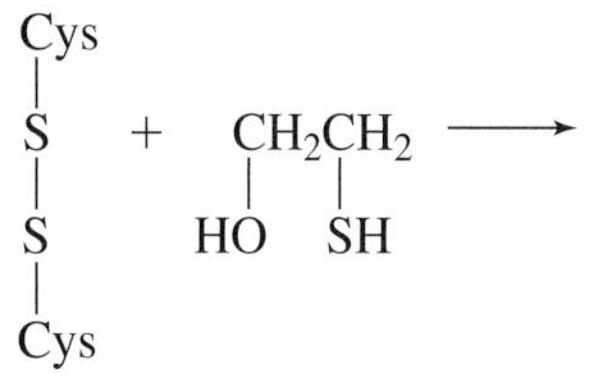

$$
\begin{array}{c}
\text{Cys} \\
| \\
\text{S} \\
| \\
\text{S} \\
| \\
\text{Cys}
\end{array}
\quad + \quad
\begin{array}{c}
\text{CH}_2\text{CH}_2 \\
|\quad\ | \\
\text{HO}\ \ \text{SH}
\end{array}
\quad \longrightarrow
$$

4.30 Assume that a protein has three cysteine residues. How many different arrangements are possible for intramolecular disulfide bonds?

4.31 Do you agree with the author's placement of glycine in group I (amino acids with nonpolar side chains) and histidine in group III (amino acids with polar, charged side chains)? What other classifications are possible for glycine and histidine? Explain.

4.32 What structural feature of a peptide does not allow for the identification of an N-terminus and a C-terminus?

4.33 Using molecular structures, write the reaction catalyzed by serine racemase. Circle the bond(s) broken during the reaction.

Writing Biochemistry

4.34 Select a reference from the Further Reading section below. Locate the article in your library and read it carefully. Write a 200-word summary of the article. You should emphasize concepts that were discussed in this chapter.

Applying Biochemistry

4.35 Life as we know it is dependent on 20 α-amino acids. These are used for protein synthesis as well as other functions. Do you think that life, as we understand it, would have been possible if the twenty amino acids were substituted by the 20 straight-chain hydrocarbons, CH_4 to $C_{20}H_{42}$?

4.36 Select your favorite amino acid and write a 100-word paper describing its biochemistry. You may need to use advanced biochemistry textbooks and the Internet.

Further Reading

Baker, W., and Panow, A., 1991. Disulfide groups in proteins. *Biochem. Educ.* 19:152–154.

Boyer, R., 2000. Isolation and purification of proteins. In *Modern Experimental Biochemistry,* 3rd ed., pp. 59–110, 111–140, 257–277. San Francisco: Benjamin/Cummings.

Branden, C., and Tooze, J., 1998. *Introduction to Protein Structure,* 2nd ed. New York: Garland Publishing.

Collins, F., et al., 2001. The human genome. *Nature* 409:745–748.

De Chadarevian, S., 1999. Protein sequencing and the making of molecular genetics. *Trends Biochem. Sci.* 24(5):203–206.

Goodsell, D., and Olson, A., 1993. Soluble proteins: Size, shape and function. *Trends Biochem. Sci.* 18:65–68.

Henry, C., 2000. Protein structure by mass spec. *Chem. Eng. News.* Nov. 27:22–26.

Lamzin, V., Dauter, Z., and Wilson, K., 1995. How nature deals with stereo-isomers. *Curr. Opin. Struct. Biol.* 5:830–835.

Lesk, A., 2000. *Introduction to Protein Architecture.* Oxford: Oxford University Press.

Rusting, R., 1992. Why do we age? *Sci. Am.* 267(6):130–141.

Saffran, M., 1998. Amino acid names. *Biochem. Educ.* 16(2):116–118.

Siegel, J., 2001. Single-handed cooperation. *Nature* 409:777–778.

Steinberg, D., 2001. Is a human proteome project next? *The Scientist* 15(7):1, 16–17.

Venter, J., 2001. The sequence of the human genome. *Science* 291:1304.

Wolosker, H., Blackshaw, S., and Snyder, S., 1999. Serine racemase: A glial enzyme synthesizing D-serine to regulate glutamate-*N*-methyl-D-aspartate neurotransmission. *Proc. Natl. Acad. Sci.* 96:13409–13414.

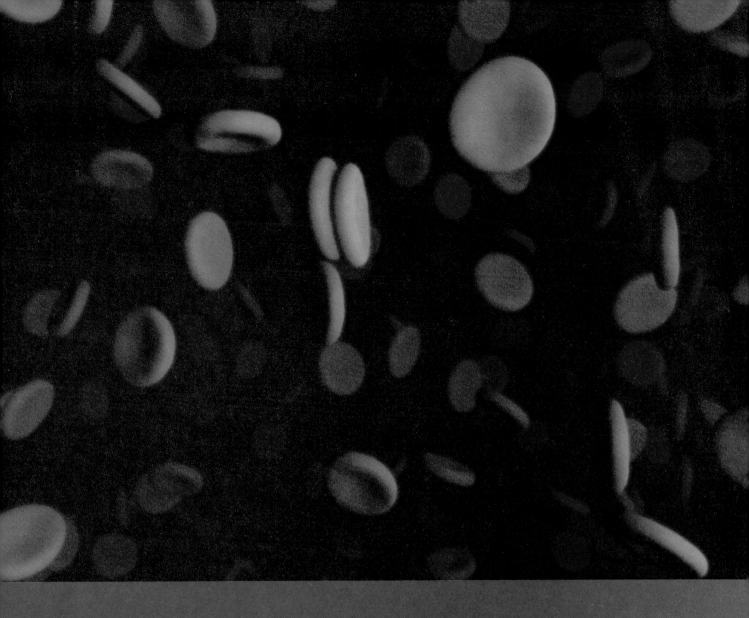

Part II

Dynamic Function
of Biomolecules

Chapter 5

Protein Architecture and Biological Function

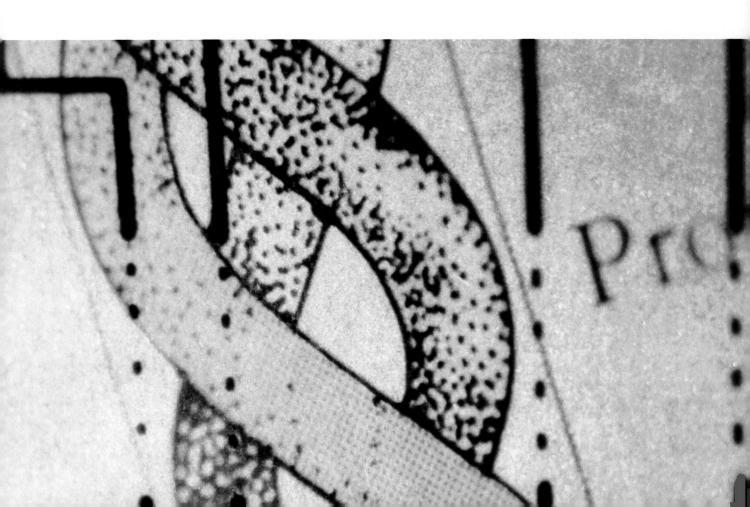

n Chapter 4, we discovered that protein architecture can be defined at four levels of structure. There, we focused on the first level, the concept of **primary** (or covalent) structure—the amino acid sequence that is determined by the sequence of nucleotides in the gene for each protein. Each type of protein molecule in its natural environment folds into a unique three-dimensional structure called the native conformation, which is defined by secondary, tertiary, and perhaps quaternary levels of organization. Most proteins are not biologically functional unless they are in their native conformation. All of the many factors of importance in the final folded arrangement are not completely understood; but we do know that the most important factors are the identity and sequence of the amino acids: the primary structure. In this chapter we study the elements of secondary, tertiary, and quaternary structure and learn general principles that govern the assembly of a protein into its highly organized, biologically active form. In conclusion, the structures of several proteins are explored and related to their biological function.

5.1 General Principles of Protein Design

Influence of Primary Structure

John Kendrew, working at Cambridge University, announced in 1958 the first three-dimensional model of a protein (sperm whale myoglobin), obtained by X-ray diffraction analysis as shown in Figure 5.1. Biochemists were surprised at the extreme complexity and apparent rigidity in the structure of this relatively small protein of only 153 amino acids. Since that first result from the late 1950s, the three-dimensional structures of nearly 10,000 proteins have been determined by X-ray and NMR analysis. With this large sample size and advanced computer techniques, we are now able to recognize some regularity among protein structures and, in fact, we can now define several general principles important in three-dimensional structures. *The most important principle emerging from such studies is the strong influence of the primary structure on the three-dimensional arrangement.* That is one reason it is so important to obtain an accurate amino acid sequence determination as a first step in protein analysis. Certain amino acids and sequences of amino acids favor the formation of specific secondary, tertiary, and quaternary structures. As we would predict, the main influence is the identity and sequence of amino acid side chains, because this is how proteins differ. (All proteins have the same primary, covalent backbone, a polymer of the

unit $-\text{NH}-\overset{}{\underset{\underset{R}{|}}{\text{CH}}}-\overset{\overset{O}{\|}}{\text{C}}-$; they differ in the branches comprised by the R groups.) Because the

primary structure of each type of protein is unique, we expect a unique or nearly unique three-dimensional structure for each protein. The influence of the side chains is noted especially in the natural tendency for globular proteins in an aqueous environment to fold by packing nonpolar amino acid residues (group I) into a hydrophobic core with a hydrophilic surface of polar amino acid residues (see Table 4.3). This arrangement buries nonpolar amino acid residues away from water and allows polar and charged amino acid residues (groups II and III) on the surface to interact with H_2O. The number of stabilizing interactions of protein with water and amino acid residues with other amino acid residues are at a maximum and the destabilizing contacts between hydrophobic amino acid residues and water are at a minimum. In fact, this stabilized arrangement is the main thermodynamic driving force in the folding of globular proteins. (Fibrous proteins are water insoluble and, because of their unique amino acid content and presence in nonpolar environments, such as membranes, they follow different folding rules.)

Noncovalent Bonds

A second principle of protein structure identifies weak noncovalent interactions as the most important stabilizing forces (review Section 3.1). The folding of a protein from a random coil (unfolded state) into a highly organized structure results in a large decrease in conformational entropy. From this observation, at first, we predict that the more ordered native

Iranian stamp celebrating the Fifth Biennial Biochemistry Symposium. The general structure of an amino acid (building blocks of proteins) is highlighted.

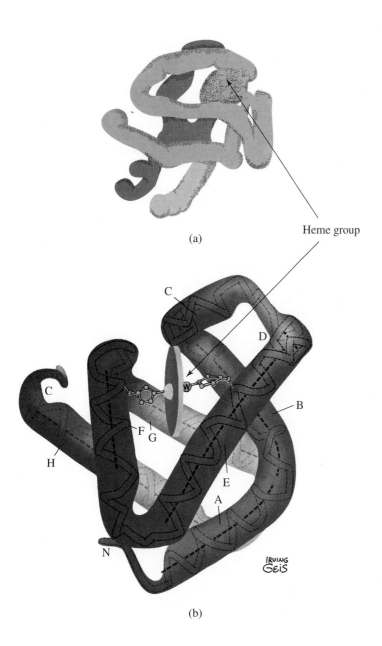

(a)

Heme group

C

D

C

B

F
G

H

W

E

A

N

IRVING
GEIS

(b)

Figure 5.1
The three-dimensional structure
of sperm whale myoglobin.
(a) An early model constructed
by J. Kendrew after analysis by
low-resolution X-ray diffraction.
(b) The structure of sperm whale
myoglobin deduced by high-
resolution X-ray diffraction. Myoglobin is comprised of eight
regions of α-helix indicated by A
through H. N and C indicate the
N-terminus and C-terminus,
respectively. The heme group is
held in position by coordination to
histidine side chains. The oxygen
binding site is at Ⓦ. This model
highlights secondary (*alpha* helix)
and tertiary structure, but does not
show the positions of individual
atoms. *Source:* © Irving Geis.
Geis Archives Trust. Rights
owned by Howard Hughes Medical Institute. Not to be used without permission.

conformation is thermodynamically less stable than the random coil and therefore not an important structure. However, other forces are at work that more than offset the loss in entropy and lead to protein structures that are stabilized. The difference in free energy between the unfolded state of a protein and its folded form is relatively small (on the order of 20–60 kJ/mol for most proteins) but large enough to expect the native conformation to have lower free energy and be more stable.

The chemical interactions that stabilize protein structures include hydrogen bonding between atoms of amino acid residues ($-N-H\cdots O=C$ for example) and atoms of amino acid residues with water ($C=O\cdots H-O_H$ for example). Figure 5.2 shows actual hydrogen bonding capabilities of some amino acid side chains. Interactions important in protein structure are shown in Figure 5.3. Hydrophobic interactions of nonpolar side chains clustered inside the core of the protein structure stabilize the structure by limiting contact with water. Ionic bonds and van der Waals forces also act to stabilize protein structure. The relative importance of various noncovalent forces changes from protein to protein since it depends on the amino acid composition and sequence.

Figure 5.2

Examples of amino acid side chains that hydrogen bond in protein structures. Atoms in red are hydrogen bond donor groups. Atoms in green are hydrogen bond acceptors.

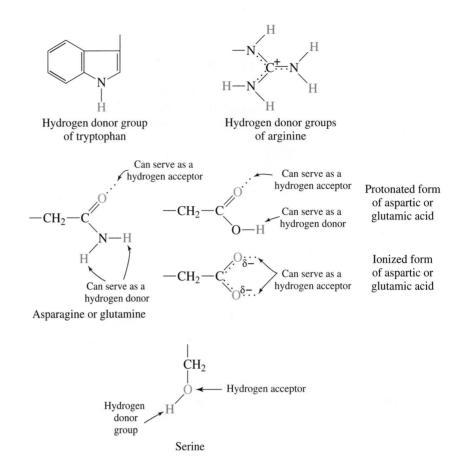

Hydrogen donor group
of tryptophan

Hydrogen donor groups
of arginine

Can serve as a
hydrogen acceptor

Can serve as a
hydrogen donor

Asparagine or glutamine

Can serve as a
hydrogen acceptor

Can serve as a
hydrogen donor

Protonated form
of aspartic or
glutamic acid

Can serve as a
hydrogen acceptor

Ionized form
of aspartic or
glutamic acid

Hydrogen acceptor

Hydrogen
donor
group

Serine

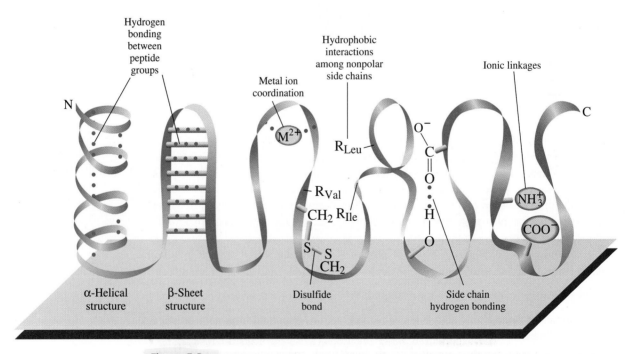

Hydrogen
bonding
between
peptide
groups

Hydrophobic
interactions
among nonpolar
side chains

Metal ion
coordination

Ionic linkages

N

C

M^{2+}

R_{Leu}

O^-

R_{Val}

CH_2 R_{Ile}

C

O

H

O

NH_3^+

S   S

CH_2

COO$^-$

α-Helical
structure

β-Sheet
structure

Disulfide
bond

Side chain
hydrogen bonding

Figure 5.3

Native conformation of a hypothetical protein showing the importance of noncovalent interactions in determining three-dimensional structure. Of the interactions shown the most important for 2° and 3° structure are hydrogen bonding, hydrophobic, and ionic. *Source*: Based on Campbell, 3rd ed., 1999.

Structure of Peptide Bonds

Another important finding from X-ray analysis of proteins is the rigid structure of the peptide unit. Each peptide group in a protein is in a locked "trans" configuration; that is, the hydrogen atom on the amide nitrogen is opposite to the oxygen atom of the carbonyl group (Figure 5.4a). The four atoms of the peptide group are in the same plane so the p orbitals can overlap to form partial double bonds between the nitrogen and carbon and the carbon and oxygen (Figure 5.4b). The most important consequence of this structural feature is the limited rotation about the amide carbon–nitrogen bond; hence, the peptide group acts as a single, planar unit. There is, however, complete rotational freedom about the amide unit and its two bonds to alpha carbons (Figure 5.4c).

Even though the three-dimensional structure for proteins appears unorganized and complicated, careful scrutiny of experimentally derived structures of many proteins reveals the occurrence of common areas containing regular patterns. This organization should not be a surprise because in Chapter 4 we noted the presence of common sequence regions called **domains.**

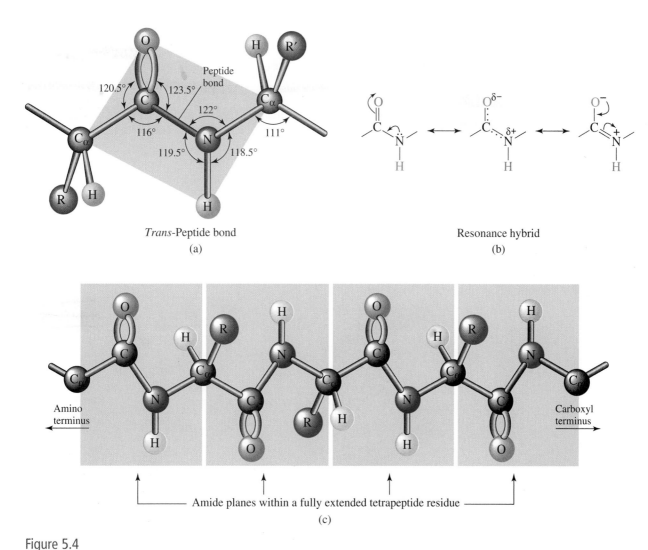

Trans-Peptide bond

(a)

Resonance hybrid

(b)

Amide planes within a fully extended tetrapeptide residue

(c)

Figure 5.4

(a) Each peptide bond in a protein structure is in a "trans" configuration and acts as a rigid and planar unit. Only limited rotation can occur around the C—N bond and free rotation is possible for Cα—N and Cα—C. (b) An abbreviated structure of the peptide unit showing resonance structures. Note, in the resonance hybrid, the partial double bond character of the C—N and C—O bonds. The peptide unit is locked in the "trans" arrangement because of partial double bond character of the carbon–nitrogen bond. (c) Planar amide units in a polypeptide. Native proteins are in the "trans" peptide arrangement. *Source:* Based on Lehninger et al, 1999.

(Domains can be defined as regions of perhaps 50 to 400 amino acid residues that act independently to play a structural role as well as a biologically functional role.) These regions of common amino acid sequence are expected to fold into similar, if not identical, conformations.

5.2   Elements of Secondary Structure

Structure Tutorial: Intro to chime with secondary structure

When a globular protein folds to bury hydrophobic amino acid residues in the interior, this organization also brings some of the polar polypeptide backbone into a nonpolar environment. The polar N—H and C═O peptide units in the interior of the protein chain are "neutralized" by hydrogen bond formation. Such hydrogen bonding is the foundation for secondary structure: regular, repeating interactions between amino acid residues that are "close together" in the linear sequence of the protein. The two most important types of secondary structure in globular proteins are the α-helix and the β-sheet. Helices and sheets are considered to be "regular" structures because they are comprised of repeating or periodic regions.

The α-Helix

The α-helix was suggested in 1951 by Linus Pauling and Robert Corey after they studied X-ray diffraction patterns and models of hair and silk proteins. The **α-helix** is a rodlike structure formed by a tightly coiled polypeptide backbone (Figure 5.5). The R side chains branch out from the main chain. The fundamental interactions that hold the polypeptide unit in that arrangement are hydrogen bonds between the N—H group of one amino acid residue and the C═O group of the amino acid four residues ahead in the chain (interactions between amino acids "close together" in the linear sequence). The hydrogen bonds are par-

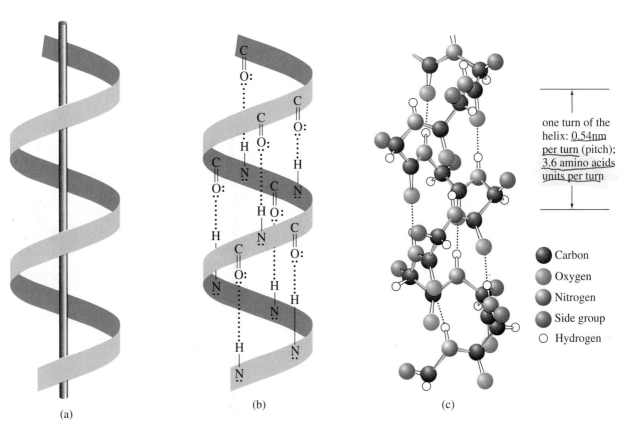

(a)   (b)   (c)

one turn of the helix: 0.54nm per turn (pitch); 3.6 amino acids units per turn

● Carbon
● Oxygen
● Nitrogen
● Side group
○ Hydrogen

Figure 5.5

The α-helix structure shown in three levels of detail. (a) Only the polypeptide backbone in ribbon form around an imaginary axis is shown. (b) All of the atoms involved in hydrogen bonding are shown. Note the coiling of the polypeptide chain and hydrogen bonding (· · ·) between N—H and C═O groups of the backbone. (c) All atoms including the amino acid side chains are shown.

allel to the main axis and, therefore, allow the polypeptide to stretch up to twice its normal length. The hydrogen bonding is weakened by stretching but still holds the chain in the α-helix coil. There are 3.6 amino acid residues for each turn in the coil. *All* of the NH and $C = O$ groups in the main chain of the α-helix region are hydrogen bonded, leading to a large number of favorable interactions and a highly stabilized structural unit. No hydrogen bonding between R side chains is present. An α-helix coil can wind in the right-handed or left-handed direction; however, L-amino acids favor the right-handed form. The length of a single α-helix in a chain can vary from one coil (3.6 residues) to as long as 1000 Å (about 650 residues), as observed in α-keratin, a fibrous protein in hair. However, the average length of an α-helix unit in globular proteins is in the range of only 15 to 30 Å (10–20 residues). The total amount of α-helix also varies from protein to protein. Some proteins like skin collagen have no α-helix character. Chymotrypsin, the digestive enzyme, has about 10% α-helix, whereas myoglobin, an oxygen-carrying protein, has approximately 75% of its amino acid residues in α-helix structures.

The α-helix arrangement is a highly stabilized structure, so why is it not predominant in all proteins? The answer is evident from our earlier discussions. Protein secondary structure is dependent on primary structure. The presence of some amino acids favors the formation of an α-helix, whereas other amino acids destabilize the structure. Some general conclusions about amino acid preferences can be made. The main constraining factor in α-helix structure is the extent of interactions between adjacent R side chains. Recall that these are jutting out from the main backbone and may be near enough to become involved in unfavorable close encounters. The presence of several adjacent, larger side chains such as asparagine, tyrosine, serine, threonine, isoleucine, and cysteine will destabilize the α-helix and make the formation less favorable. [The presence of large numbers of glycine residues (R = H) in a protein actually disfavors the α-helix not because of large size, but because the β sheet, an alternate form of secondary structure, is preferred.] In a similar fashion, several adjacent like-charged R side chains repel each other and destabilize hydrogen bonding in the main chain. For example, three or more adjacent lysine residues or glutamate residues will prevent α-helix formation in that region. An amino acid that severely disrupts α-helix structure is proline. Because of its unique structure incorporating the amide nitrogen in a ring, proline is unable to participate in hydrogen bonding and, in addition, forces a turn in the polypeptide backbone.

β-Sheets

A second important form of protein secondary structure is the **β-sheet** (Figure 5.6). This structural element differs in several ways from the α-helix. In the β configuration, the polypeptide chain is nearly fully extended, hydrogen bonds can be intrachain or interchain, and a β-sheet is constructed by combining two or more regions of the polypeptide. Polypeptide chains in β configuration can combine to form a β-sheet in two possible orientations. If polypeptide chains run in the same direction (both are N $\longrightarrow$ C or C $\longrightarrow$ N), the β-sheet is **parallel.** If the chains alternate in direction, it is described as **antiparallel** (Figures 5.6 and 5.7). The antiparallel arrangement is more common in proteins.

The main limitation in formation of a β-sheet is size and charge of the R side chains, just as in the α-helix. However, the constraints are much more severe in the β-sheet. Proteins that are primarily β-sheet (silk fibroin or proteins of spider webs) have a very high content of alanine and glycine, and repeating patterns in the primary structure.

Bends and Loops

Bends (also called reverse turns) are important elements of secondary structure for two reasons:

1. They reverse direction of the main polypeptide chain.
2. They connect regions of more regular secondary structure (α-helix and β-sheets).

A type of bend common to many proteins is that found connecting antiparallel polypeptides in β-sheet arrangements. This is called a β-bend or sometimes a hairpin turn (Figure 5.8). Four amino acids are usually in the structure with one internal hydrogen bond between the

Spider web protein is in the β-sheet configuration.

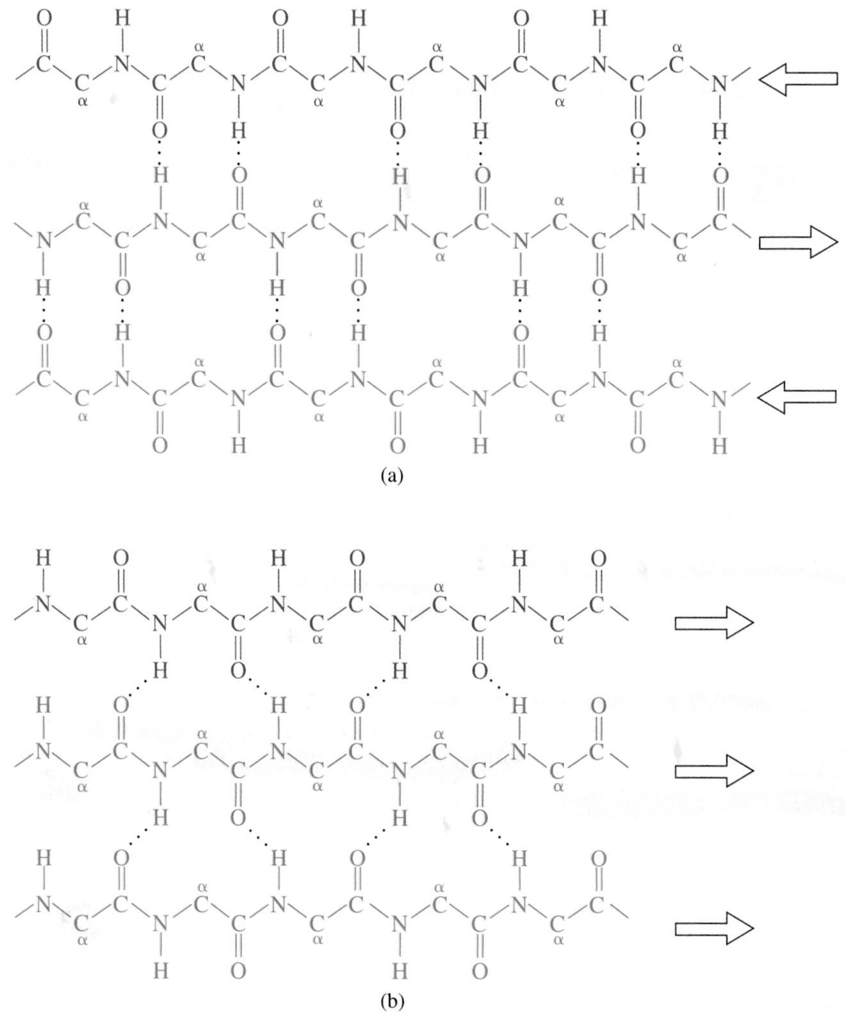

Figure 5.6

The β-sheet secondary structure for proteins. (a) Polypeptide chains are aligned in an antiparallel direction as designated by the arrows. The head of each arrow is pointing toward the C-terminus. Note the hydrogen bonding between N—H and C＝O groups of the polypeptide backbones. (b) Parallel β-sheets are observed in some proteins.

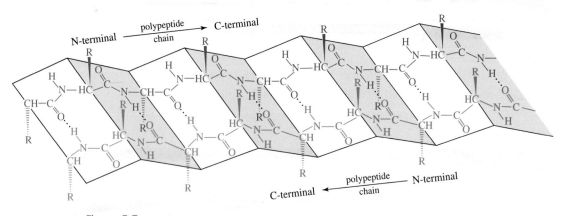

(a)

(b)

first and fourth amino acid. The amino acid residues glycine and proline are prominent in bends; glycine because it is flexible and proline because its cyclic structure forms a natural turn. Bends are considered to be *nonregular* secondary structure elements because they are not periodic. An extended bend is often called a **loop.** A loop is a continuous segment of polypeptide chain that contains between 6 and 16 residues and is usually about 10 Å long.

Figure 5.7

Diagram showing the three-dimensional arrangement of an antiparallel β-sheet. Hydrogen bonding between peptide units is shown with dotted lines. *Source:* Based on Brown and Rogers, 1987.

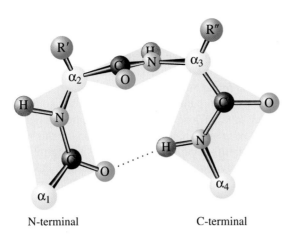

N-terminal   C-terminal

Figure 5.8
Example of a β-bend to form a loop or cause a change of direction in a polypeptide chain. Bends are stabilized by hydrogen bonding as shown with a dotted line. *Source:* Based on Garrett and Grisham, 1999.

Like a bend, a loop contains no periodic secondary structure. Although there are size distinctions between bends and loops, the terms are often interchanged because the two elements play similar roles in protein structure.

Structural Motifs

The individual elements of secondary structure (α-helix, β-sheet, and bends) are often combined into stable, geometrical arrangements. Called **supersecondary structure** or **motifs,** these specific clusters are found many times in the same protein as well as in many different proteins. The clusters are held together by favorable noncovalent interactions between side chains. Probably the simplest such combination found is the α-helix–loop–α-helix or αα motif (Figure 5.9a). This motif is also associated with specific biological functions, such as the ability of the protein to bind to DNA or to sequester a calcium ion. The analogous

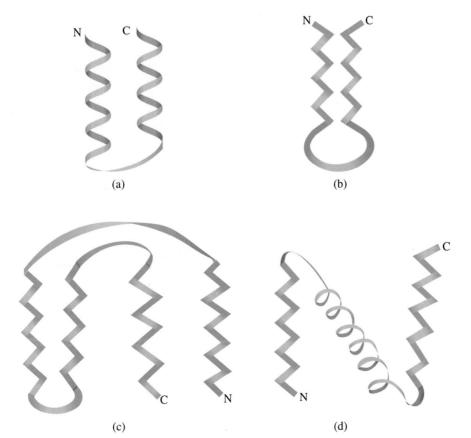

(a)   (b)

(c)   (d)

Figure 5.9
Some common structural motifs of folded proteins: (a) the αα motif; (b) the ββ motif, antiparallel; (c) the Greek key (ββββ) motif; (d) the βαβ motif, parallel.

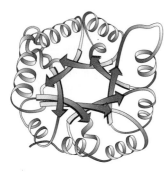

Figure 5.10

Several βαβ motifs combine to form a superbarrel in the glycolysis enzyme triose phosphate isomerase. *Source:* Based on Hein et al, 1993.

β-sheet motif is β–loop–β or ββ, which consists of two antiparallel strands in β configuration connected by a bend or loop (Figure 5.9b). Although it appears quite frequently in proteins, a specific biological function has not yet been discovered for the ββ motif. A ββ motif of parallel strands is rare in protein structure. The Greek key motif (Figure 5.9c), consisting of four adjacent β strands, is found in many proteins including *Staphylococcus* nuclease, an enzyme that degrades DNA. Two parallel β strands are often connected by an α-helix to form a βαβ motif (Figure 5.9d). Triose phosphate isomerase, an enzyme important in carbohydrate metabolism, is made up of repeated combinations of this motif. The combination of a large number of βαβ motifs is called a β barrel or superbarrel (Figure 5.10). One can easily become overwhelmed by looking at a model of a protein in which the atoms of each amino acid residue are shown. Protein structures are now often displayed as schematic diagrams that, instead of showing atomic detail, show elements of secondary structure and outline the general directions of the polypeptide chain. Three symbols are used to define the presence of secondary structure: cylinders or coils are used to represent α-helix regions, arrows for β configurations, and ribbons to represent bends, loops, or random regions. Although this presentation, called a **ribbon diagram,** does not give detailed structure, it does provide an overall view for study (see Figure 5.10).

The structural elements discussed to this point are present in **globular proteins,** water-soluble proteins that play dynamic biological roles in cells. **Fibrous proteins,** those that usually perform a structural role, have their own characteristic secondary structure. One of the most prominent fibrous proteins is collagen, the major protein of skin, bone, and tendons. The primary structure of collagen consists of repeating units, Pro-Gly-*x* or Hyp-Gly-*x,* where Hyp is 4-hydroxyproline and *x* is any amino acid. Because it is rich in proline, collagen is unable to fold into the more common α-helix or β-sheet. The unique three-dimensional structure of collagen consists of three extended helical chains wrapped into a **triple helix** (also called super helix) and held together by hydrogen bonding and covalent

Figure 5.11

The triple helix structure of the important fibrous protein collagen. Each single strand of collagen consists of repeating units Pro-Gly-*x* or Hyp-Gly-*x*. Three extended polypeptide chains are entwined and held together by hydrogen bonding and covalent crosslinks. The triple helix structure is reminiscent of a rope. (a) Artist's conception of collagen. (b) Space-filling model of collagen. *Source:* Part (a) © Irving Geis. Geis Archives Trust. Rights owned by Howard Hughes Medical Institute. Not to be used without permission.

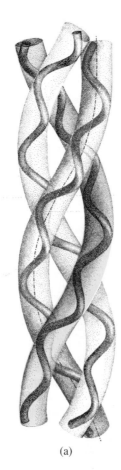

(a)

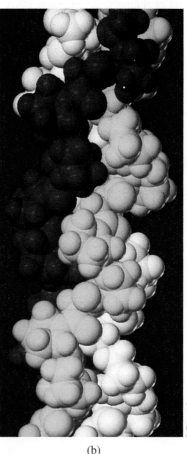

(b)

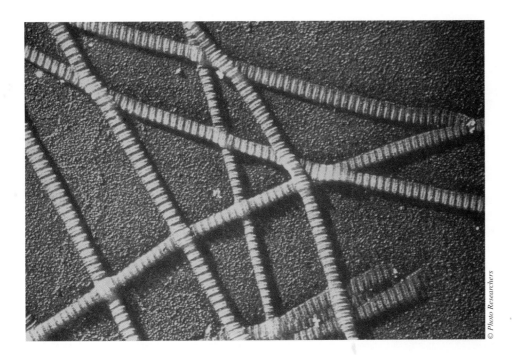

Figure 5.12
Electron micrograph of collagen fibrils.

© Photo Researchers

crosslinks between amino acid side chains (Figure 5.11). The triple helix structure is similar to that of a rope. The resulting structure has very high tensile strength with little ability to stretch, exactly the properties desired for bones and tendons (Figure 5.12). Several metabolic diseases are caused by disorders in the structure and function of collagen. Infants with the genetic disease osteogenesis imperfecta have extremely fragile bones. Collagen in these babies has a cysteine or serine incorporated in a single position usually reserved for glycine. Scurvy, a disease with symptoms of skin rashes and bleeding gums, is caused by a deficiency of ascorbic acid (vitamin C). Ascorbic acid is a cofactor for the enzyme that catalyzes the conversion of proline to hydroxyproline.

5.3 Protein Tertiary Structure

Protein Folding

As described in Section 5.2, the secondary structure of a globular protein defines the arrangement of localized regions of the polypeptide chain into organized units called motifs. These organized units are the result of stabilized interactions between amino acid residues in the polypeptide chain. Interactions stabilizing the units are primarily hydrogen bonds between atoms in the main chain. The **tertiary structure** of a protein describes the positions of all atoms in a protein including the side chains. Tertiary structure is a result of combining several motifs of secondary structure into a compact arrangement, thus causing the close spatial encounter of amino acids that may be far apart in the polypeptide chain. The major forces stabilizing tertiary structure are hydrophobic interactions among nonpolar side chains in the compact core of the protein. Imagine a protein, in aqueous solution, that contains a substantial number of amino acid residues with hydrophobic side chains (group I amino acids: valine, leucine, phenylalanine, etc.). When the protein chain is extended (unfolded), the nonpolar side chains are unable to interact directly with water, thereby forcing the water molecules to form a cage-like boundary (see Section 3.2). This highly ordered arrangement of water molecules is thermodynamically unfavorable for two reasons: (1) The energy level is high (unstable) because no favorable bonding interactions can take place between water molecules and the nonpolar residues, and (2) the entropy (randomness and disorder) of the water molecules is decreased. Now, if the protein chain folds to bury the nonpolar residues inside the structure, water molecules are able to move to the exterior where they can participate in favorable interactions like hydrogen bonding with polar amino acid side chains. The

Interactive Animation: Protein folding

thermodynamic stability arising from the many weak noncovalent interactions more than compensates for the energy required to bring all the atoms into their proper positions (Figure 5.13). Water is excluded from the hydrophobic cores, thus avoiding destabilization caused by encounters between water and nonpolar side chains. In addition to the noncovalent interactions, covalent disulfide bonds (—S—S—) are instrumental in crosslinking cysteine residues that may be far apart in the sequence of the polypeptide chain. Experimental studies on the importance of disulfide bonds in tertiary structure show that they do not directly influence the folding of a protein into its native conformation. Rather, disulfide bonds "lock" the protein in its final form after most of the other stabilizing interactions are in place (hydrophobic, H bonds, etc.).

Several characteristics of tertiary structure are general for all proteins. With freedom of rotation around all bonds (except the amide carbon nitrogen bond), a protein has extensive flexibility and therefore many possible conformations. It has been estimated that a protein of 100 amino acids can assume 5×10^{47} different conformations, yet only one is chosen as the native conformation because it is the most stable (Figure 5.13). However, the final form should not be thought of as static or rigid. There is actually much flexibility in a protein chain, in fact, small conformational changes are a common way to regulate the biological reactivity of some proteins. The hemoglobin molecule, carrying O_2 in your blood, undergoes minor changes in conformation when O_2 is bound or released by the protein.

The **central dogma of protein folding** is now well known: *The primary structure determines the tertiary structure.* This has always been interpreted to mean that a protein with a unique primary structure will fold spontaneously into its distinct tertiary structure. Experimental observation of the protein folding process has provided evidence that it is probably stepwise and begins with the formation of local secondary structure (α-helix, β configuration) to provide a nucleus or seed. The initial structures are likely αα, ββ, αβα, and other simple motifs we have described. The remainder of the protein chain then continues to fold around the initiation nuclei. The folding process is often described as cooperative, implying that each folding step, as it occurs, facilitates the formation of another favorable interaction. In vivo protein folding pathways were for many years thought to be spontaneous self-assembly processes, involving only the folding protein. Although many proteins synthesized in prokaryotic and eukaryotic cells are capable of self-assembly into their native secondary, tertiary, and quarternary structures, some need help. Recently, proteins called

Figure 5.13

Energy diagram of the protein folding process. The completely unfolded protein is thought to be the least stable form. For most proteins, the native conformation is the most thermodynamically stable and the only form that is biologically active.

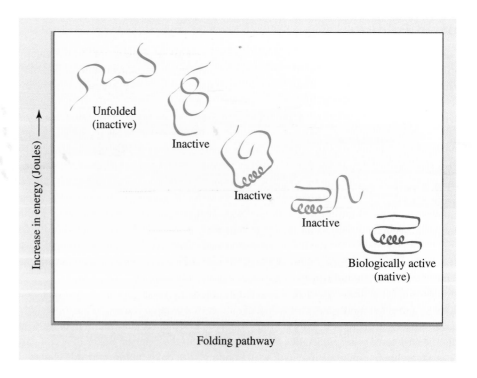

molecular chaperones have been discovered that assist in the folding of other proteins. Molecular chaperones participate in folding by binding to unfolded or partially folded polypeptide chains and preventing the formation of interactions that do not lead to the biologically active structure. Chaperones stabilize interactions that may otherwise be unstable. Chaperones are especially needed for covering exposed nonpolar regions on unfolded protein chains to avoid improper or unfruitful hydrophobic interactions. Improper interactions, which may result from a transient exposure of hydrophobic residues, lead to aggregation and precipitation of the proteins during their synthesis. In general, molecular chaperones function by providing an interior shelter or "safe haven" where other proteins may reside for a brief period as they fold into their native conformations. Chaperones have very little substrate specificity; that is, they accept many different types of unfolded proteins into their protective, interior environments. Molecular chaperones, originally called **heat shock proteins,** were first discovered in *Drosophila* (fruit fly) after exposure to elevated temperatures. In stressed cells, chaperones play a role in preventing the aggregation of heat-denatured polypeptides. Molecular chaperone proteins are found in all prokaryotic and eukaryotic organisms even in cells growing under optimal, nonstressful conditions, where they probably assist in the folding of newly synthesized proteins.

Protein Unfolding

You may be asking yourself the important question, "Can the folding process be reversed; that is, can a protein in its native form be unwound?" Indeed, under various laboratory conditions, it is possible to unravel protein structure (Figure 5.14). The complete loss of organized structure in a protein is called **denaturation** and occurs, for example, during the cooking of an egg. The protein, albumin, present in the egg white, is denatured by heat and changes from a clear, somewhat colorless solution to a white coagulum. Heat can serve as a

Figure 5.14

Denaturation and renaturation of a protein.

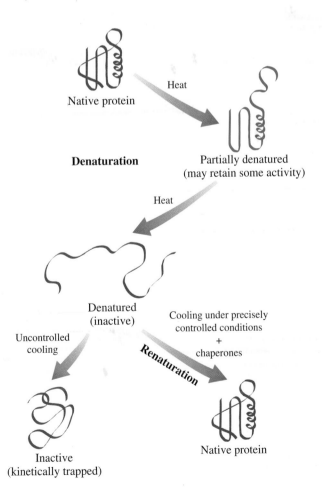

Native protein

Heat

Denaturation

Partially denatured
(may retain some activity)

Heat

Denatured
(inactive)

Uncontrolled
cooling

Cooling under precisely
controlled conditions
+
chaperones

Renaturation

Inactive
(kinetically trapped)

Native protein

denaturant for most proteins; however, the process is often irreversible; that is, the denatured protein cannot be returned to its native biological form. The stabilizing forces holding a protein in its native conformation are relatively weak so they can be disrupted under somewhat mild conditions. Denaturation of a protein involves disruption of secondary and tertiary structure, but the primary structure remains unchanged so the amide bonds remain intact. Common agents used in the laboratory to carry out reversible denaturation include organic solvents (ethyl alcohol or acetone), urea, detergents, and acid or base. The use of the detergent, sodium dodecyl sulfate (SDS) to study protein structure by electrophoresis is described in Window on Biochemistry 4.2. The hydrophobic SDS binds to polypeptide chains and transforms them to random coils. Denaturants disrupt only noncovalent interactions, not the covalent linkages of the primary structure. If the denaturing agent is removed, it is usually possible for the unwound protein to return to its native structure. The renaturation, transforming the random coil of a protein to its active biological form, provides the best experimental proof that the tertiary structure is determined by the primary structure. The information necessary to guide the protein folding process is present in the amino acid sequence.

One point concerning denaturation deserves emphasis. Denaturation of a protein results in the loss of secondary and tertiary structure *and* the loss of biological function. This observation can be used to advantage in the laboratory, if one needs to test the necessity of a protein for a biological process. Substituting a denatured sample of protein in the biological system being studied results in loss of biological activity.

5.4   Protein Quaternary Structure

Monomeric and Oligomeric Proteins

Many proteins are **monomeric;** that is, they are comprised of only one continuous polypeptide chain. A significant number of larger proteins (molecular weight greater than 50,000) are **oligomeric** or comprised of two or more polypeptide chains called **subunits.** Quaternary structures that are the result of associations between identical or nearly identical subunits are called **homotypic.** If the subunits are different, the term **heterotypic** is used to define the quaternary structure. Each individual subunit, of course, possesses a primary, secondary, and tertiary structure. The structure of the complex oligomeric protein can be defined on four levels: **primary, secondary, tertiary,** and **quaternary.** *The fourth architectural level defines the arrangement and position of each subunit in the intact protein molecule.* The quaternary structure details the molecular contacts or interactions among all the subunits. It is not necessary to define new forces stabilizing oligomeric structure since subunits are held together by the familiar, weak, noncovalent interactions: hydrogen bonds, ionic bonds, hydrophobic interactions, and van der Waals interactions. At the contact regions of the subunits, the environment is similar to the hydrophobic core of proteins as we discussed in Sections 5.1 and 5.3. The assembly of individual subunits into a well-defined quaternary structure is not well understood, but presumably the process is spontaneous. The assembly instructions are provided in the amino acid sequence of each polypeptide chain that has been transcribed and translated from DNA via mRNA.

The subunits of a quaternary protein may be identical or different. Glucose oxidase, a bacterial enzyme that metabolizes glucose, is a dimer of identical subunits (MW = 80,000 each), thus its quaternary structure is homotypic. Hemoglobin, an example of a heterotypic protein, is comprised of four subunits (a tetramer), two α type and two β type. Iron storage protein, ferritin, has 24 subunits of two similar but not identical types, H and L, the ratio of which varies with the species and organ of origin. For example, horse spleen ferritin consists of about 85% L and 15% H chains, whereas human heart ferritin consists primarily of H chains.

Multisubunit proteins display very complex biological functions. Many are involved in regulatory processes. As we see in Section 5.5, oxygen binding to hemoglobin is carefully controlled by atomic contacts between α and β subunits. The rate of some enzyme-catalyzed reactions is often regulated by conformational changes in the subunits of an oligomeric

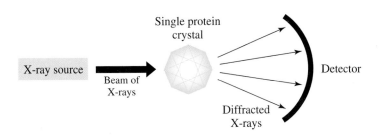

Figure 5.15
The necessary components for structural analysis of a protein crystal by X-ray diffraction.

protein. Very subtle changes in structure at one location of an oligomeric protein can be transferred to a distant site on the protein, causing a change in structure and biological reactivity. The best known protein regulated by long distance or **allosteric interactions** is hemoglobin, which is described in Section 5.5.

The complete three-dimensional structure (secondary, tertiary, quaternary) of a protein can best be determined by X-ray diffraction procedures (Figure 5.15). (Although nuclear magnetic resonance spectroscopy is being developed for protein structural analysis, only proteins with fewer than 250 amino acids are easily analyzed by this method.) An X-ray crystallographic analysis requires three components: (1) a protein crystal, (2) an X-ray source, and (3) a detector, usually a photographic film. A beam of X rays is focused onto the crystal. Many of the X rays passing through the sample are diffracted (scattered) when they encounter electrons associated with atoms. The diffracted beams impinge upon the film detector. The extent of scatter depends on the size and position of each atom in the crystal. By extensive computer analysis of the angle of scatter and of the pattern on the developed film, it is possible to construct an electron density map of a protein molecule showing the arrangement of atoms. Information regarding the number of different types of any subunits present and the molecular weight of each type is measured by the less complex laboratory procedure, SDS-PAGE (see Window on Biochemistry 4.2).

5.5 Protein Structure and Biological Function

The theme that has connected our study of proteins is the relationship between structure and biological function. Each protein is designed for a specific purpose in a cell and organism. Evolutionary changes occurring in the protein primary structure by amino acid substitutions or deletions have resulted in proteins that can direct exquisite and complicated processes.

Crystals of the protein anti-lymphoma intact antibody under polarized light.

Figure 5.16

The structure of hemoglobin, highlighting its secondary, tertiary, and quaternary elements. Hemoglobin has four subunits, two α and two β (labeled α_1, α_2, β_1, β_2). Each subunit contains a heme group with an iron atom. Each heme group can bind an oxygen molecule. The capital letters (A, B, C, . . .) indicate α-helical regions. *Source:* © Irving Geis. Geis Archives Trust. Rights owned by Howard Hughes Medical Institute. Not to be used without permission.

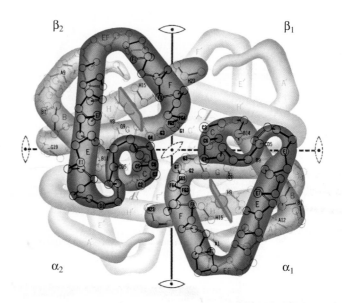

Whether proteins have reached the point of perfection is a matter only for speculation. In this section, the structure and function for three diverse proteins is described. After studying this section, you will be even more in awe of the power of proteins and you will better understand their structure–function relationships.

Hemoglobin

Hemoglobin, the major oxygen-carrying molecule of vertebrates, was the first oligomeric protein to be studied by X-ray diffraction methods. In 1959, Max Perutz and his colleagues in the Cavendish Laboratory at Cambridge University obtained structural data for horse hemoglobin, which contains four polypeptide chains and four heme groups arranged in a compact sphere of about 55 Å diameter (Figure 5.16). The protein chains are comprised of two α-type subunits and two β-type subunits giving a structure of $\alpha_2\beta_2$. (The α and β do not refer to secondary structure but are labels for different kinds of subunits.) The two identical α chains (141 residues each) and two identical β chains (146 residues each) have much sequence homology with over 60 amino acid residues that share identity and position. If the primary structures of α chains, β chains, and the single polypeptide of myoglobin (153 residues) are compared, there are over 20 invariant amino acid residues, suggesting an evolutionary relationship among the chains. As we would predict, the secondary

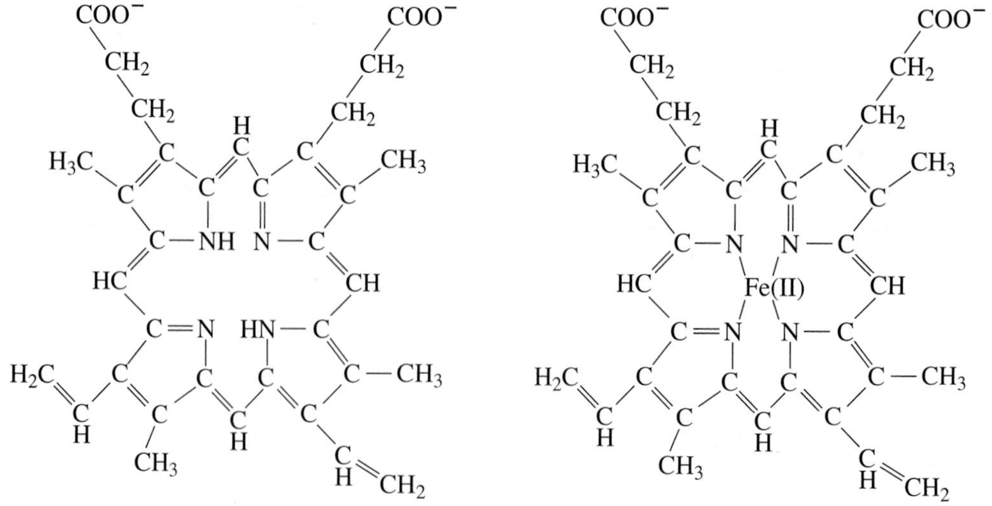

Figure 5.17

(a) Structure of protoporphyrin IX. (b) Structure of the heme prosthetic group, protoporphyrin plus iron.

(a) Protoporphyrin IX

(b) Heme (Fe-protoporphyrin IX)

and tertiary structures of the α and β chains (and the myoglobin chain) are very similar. There are eight major regions of α-helical conformation, comprising over 75% of the polypeptide chain. Each region of α-helix is interrupted by a proline residue and/or a loop. In the quaternary structure of hemoglobin, the four polypeptide chains are in a tetrahedral arrangement. There are important contact points between α and β chains. Hydrophobic interactions and ionic linkages both play a role in stabilizing the quaternary structure. As we note later in this section, the contact regions undergo very small changes during the binding and release of O_2. This brings about conformational changes in the hemoglobin molecule that effect the ease of oxygen binding.

A heme prosthetic group, associated with each chain, is buried in a crevice with a predominance of hydrophobic amino acids. Each heme is held in position by a coordinate covalent bond between the iron atom and a nitrogen atom in the side chain of a histidine residue (see Figures 5.17 and 5.18). The four heme groups serve as the site for oxygen binding; hence, each hemoglobin molecule is able to bind, reversibly, four molecules of O_2. The heme iron must be maintained in the ferrous state (Fe^{2+}) as ferric iron (Fe^{3+}) is unable to bind O_2. The iron atom buried in the protein crevice is shielded by a second histidine residue and thus protected against oxidation.

The biological function of hemoglobin is to transport oxygen over large distances via the blood of aerobic organisms. In contrast, myoglobin is present in muscle and acts as a reserve supply of oxygen within a local region. Their different functions are the result of differences in the process and affinity of oxygen binding. Both myoglobin and hemoglobin bind O_2 reversibly at heme sites, but hemoglobin, with its intricate quaternary structure, can much more precisely regulate oxygen binding and transport.

Let us consider the reactions of oxygen binding to hemoglobin:

$$Hb + O_2 \rightleftharpoons Hb(O_2)$$

$$Hb(O_2) + O_2 \rightleftharpoons Hb(O_2)_2$$

$$Hb(O_2)_2 + O_2 \rightleftharpoons Hb(O_2)_3$$

$$Hb(O_2)_3 + O_2 \rightleftharpoons Hb(O_2)_4$$

The binding of oxygen to myoglobin is similar to the first reaction. The oxygen binding curves of myoglobin and hemoglobin are shown in Figure 5.19. These graphs are obtained

Structure Tutorial:
Myoglobin and hemoglobin

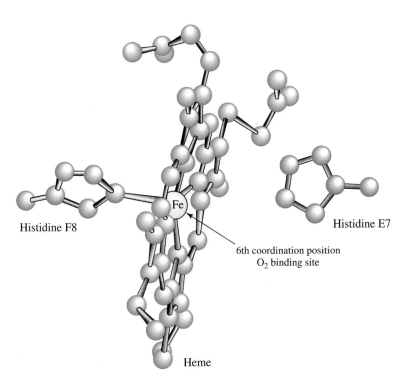

Histidine F8

Fe

Histidine E7

6th coordination position
O_2 binding site

Heme

Figure 5.18

The heme binding site for oxygen in hemoglobin. Note the Fe(II) in the protoporphyrin structure is linked to a histidine in α-helix F by a coordinate covalent bond. The sixth coordination position of iron is the site for O_2 binding. When O_2 is not present, that position is protected from water oxidation with a second histidine residue from α-helix region E. *Source:* Based on Stryer, 1995.

by reacting the protein with increasing amounts of oxygen measured by the partial pressure (pO_2, with units of kilopascals, kPa).[1] The saturation is measured as a percentage of the protein molecule (or heme sites) with a bound oxygen. The curve for myoglobin is hyperbolic and shifted to the left, indicating that it has a greater affinity for oxygen than does hemoglobin. It requires only about 0.2 kPa of oxygen partial pressure to saturate 50% of the myoglobin molecules. Hemoglobin requires about 3 kPa for 50% saturation. Not only is the curve for Hb shifted to the right (lower affinity for O_2), but it is of different shape: S-shaped

---

[1]Kilopascals is a unit of pressure. One atmosphere is equal to 101.33 kPa.

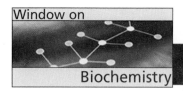

Window on Biochemistry

5.1 The Biochemistry of Fashionable Hair—Do Blondes, Brunettes, or Redheads Have More Fun?

Think about the last time you were comfortably seated in a hair salon. Do you recall the pungent smells of ammonia and other chemicals in the air? There is some serious chemistry (and biochemistry) going on whether clients are having a "perm" or "color."

When you wash your hair, you have probably noticed that hair fibers stretch when wet. The highly coiled structures of α-keratin can unwind even to the point of breaking the hydrogen bonds. The molecular structure of stretched hair resembles the extended structure of a polypeptide in a β-sheet. The disulfide bonds between protein coils also stretch; however, they remain intact and help restore the original helical arrangement when tension is released. For obtaining a permanent wave, wet hair is set in curlers and treated with an ammonia solution that causes the hair cuticle to swell and unpeel exposing the α-keratin. A reducing agent is then added to break disulfide bonds followed by exposure to moist heat to stretch the curled hair. Heat, moisture, and stretching cause extension of the α-helical coils by enhancing cleavage of hydrogen and disulfide bonds. After rinsing out the reducing agent, an oxidizing agent is added to form disulfide bonds between new cysteine partners. Drying the hair brings back the coiling and hydrogen bonds, but the hair is now, it is hoped, in a more fashionable coiffure because of the new disulfide bonds.

Changing your hair color involves even more complex biochemistry. There are three main types of hair coloring agents: (1) temporary and semipermanent dyes, (2) gradual dyes, and (3) permanent dyes that lighten and color. Dyes for temporary color change are usually available in the form of tinted sprays, gels, and rinses. The dyes coat the surface of the hair, but do not penetrate deeply into the hair protein or pigment. They usually wash out after a few shampoos.

Hair dyes like Grecian Formula contain gradual coloring agents. The active ingredient in these products is usually lead acetate, $Pb(CH_3COO^-)_2$. Upon application, these chemicals penetrate the cuticle and Pb^{2+} ions react with protein sulfur to produce dark colored lead sulfide, PbS.

Permanent dyes react chemically with the natural pigments to change hair color. The natural color of hair is due to melanin proteins stored in the cortex beneath the cuticle (Figure 5.21). Permanent hair-coloring products consist of two solutions: (1) a mixture containing 1% ammonia and chemical dyes and agents to link dyes to melanin, and (2) a 6% solution of hydrogen peroxide. The ammonia acts to open the cuticle to expose the melanin pigments. The hydrogen peroxide bleaches the natural melanin color. Chemicals used as coloring agents include o- and p-diaminobenzenes, aminohydroxybenzenes, and nitrodiaminobenzenes depending on the color, tone, and shade desired. The chemical 1,3-dihydroxybenzene produces a yellow shade. For an intense orange-red, the chemical 2-nitro-p-diaminobenzene would be used.

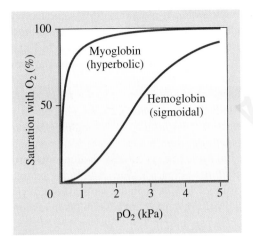

Figure 5.19
The oxygen binding curves of myoglobin and hemoglobin obtained by measuring the percent of heme sites filled with O_2 at varying O_2 concentrations. Myoglobin has greater affinity for O_2 than hemoglobin at all partial pressures (concentrations) of oxygen. The sigmoidal curve for hemoglobin indicates a cooperative binding of O_2.

or sigmoidal. This experimental observation is explained by **cooperative binding** of oxygen molecules. The affinity of deoxyhemoglobin (hemoglobin with no oxygen) for binding O_2 is relatively low. However, once the first O_2 is bound, the second, third, and fourth O_2 molecules bind with increasing affinity. When the first O_2 molecule binds to a heme site, a message is communicated via conformational changes (movements) in the protein chains to the other heme sites. (The heme groups in hemoglobin are not in direct contact; in fact, they are separated by several angstroms in distance.) The subtle conformational changes are triggered by small changes at the contact points between the subunits. The changes include the disruption of ionic linkages at contact points. We call such conformational changes **allosteric interactions,** a term meaning "through space." Although the message of O_2 binding to a heme site appears to be transmitted through space, in reality, it is communicated via movements in the polypeptide chains. The release of O_2 by hemoglobin also displays cooperativity. The first oxygen is removed with some effort, but the others then are released more readily.

The cooperative binding of O_2 to hemoglobin is influenced by H^+ and CO_2, an observation made many years ago by Christian Bohr, therefore, called the **Bohr effect** (Figure 5.20). H^+ and CO_2 bind to the hemoglobin molecule, causing changes in its quaternary structure that stabilize the deoxygenated form and destabilize the oxygenated form. This binding process reduces the affinity of oxygen for the hemoglobin and enhances the release of O_2. The actions of H^+ and CO_2 are caused by allosteric effects. The H^+ and CO_2 bind to hemoglobin (not at the heme sites), causing subtle conformational changes throughout the protein and influencing oxygen affinity at the heme sites.

Hemoglobin functions efficiently to transport oxygen from the lungs to peripheral tissue (muscle cells, brain cells, heart muscle cells, etc.). The partial pressure of oxygen in the

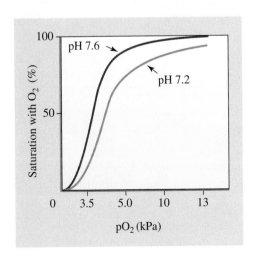

Figure 5.20
The Bohr effect. The affinity of hemoglobin for oxygen decreases with a decrease in pH. This causes enhanced release of oxygen from oxyhemoglobin in muscle.

lungs is about 15 kPa. Hemoglobin is more than 95% saturated with oxygen at these concentrations (see Figure 5.20). The oxygen partial pressure of resting muscle is approximately 5 kPa. Under these conditions, hemoglobin is about 75% saturated; hence, oxygen is released into the muscles for metabolic use. The oxygen partial pressure of working muscle is between 1 and 2 kPa, conditions under which hemoglobin is only 10% saturated. Hemoglobin is able to release needed oxygen very effectively to active tissue. The release of oxygen is even greater than indicated by Figure 5.20. Two important waste products of active muscle, H^+ and CO_2, decrease the affinity of hemoglobin-bound oxygen and cause the release of even more oxygen (the Bohr effect, Figure 5.20).

Now, think about the transport of oxygen from the lungs to peripheral tissue if myoglobin were the carrier. According to Figure 5.19, myoglobin is approximately 100% saturated with oxygen partial pressure of 15 kPa as in the lungs. If the oxygen-saturated myoglobin were then moved to an area of 5-kPa oxygen pressure (as in resting muscle), because of its

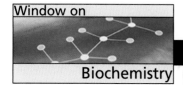

Window on

Biochemistry

5.2 Bacteriorhodopsin—Let the Sun Shine In

A protein very different from oxygen-carrying hemoglobin and fibrous α-keratin is bacteriorhodopsin, a protein found in the membrane of the salt-loving bacterium, *Halobacterium salinarum*. This bacterium thrives in sunny, salt lakes or brine ponds in which NaCl concentrations exceed 3 *M* and temperatures approach 70° C. It is unable to live in seawater, which contains only 0.6 *M* NaCl. An aerobic bacterium, *H. salinarum* uses oxygen to carry out the normal respiratory metabolism of nutrients. That process involves the oxidation of organic fuel molecules with the production of energy in the form of adenosine triphosphate (ATP). However, brine ponds are usually low in oxygen concentration, so the bacteria must find other ways to make ATP for cellular energy. It uses energy from sunlight to synthesize ATP. When these bacteria are grown under low oxygen concentrations, the cell membrane develops purple patches containing the protein bacteriorhodopsin. This protein contains 247 amino acids, and the organic molecule, retinal, is covalently bound to the side chain of a lysine residue. (Retinal is also used as a light-absorbing molecule for animal vision.) The single polypeptide chain of bacteriorhodopsin consists of a bundle of seven α-helical rods of about 25 amino acid residues each. Each rod is perpendicular to the plane of the membrane and spans the internal region of the membrane. The amino acid content of each

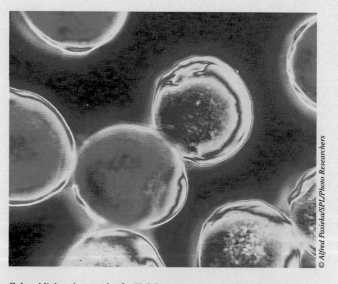

Colored light micrograph of a *Halobacterium,* an organism found in highly saline environments like salt lakes.

high affinity, only about 2% of the oxygen could be released. Therefore, hemoglobin with its intricate regulation of oxygen binding has become adapted to carry oxygen over long distances and release it in an area of lower oxygen concentration, whereas the role of myoglobin is that of storage in a localized region. Hemoglobin and myoglobin have very similar primary, secondary, and tertiary structures. Indeed, it is difficult to distinguish among the α chain, β chain, and the single subunit of myoglobin. The quaternary structure of hemoglobin, which allows it to act allosterically, is able to respond to environmental changes within cells.

Interesting relationships between structure and function can also be studied by noting genetic changes in hemoglobin primary structure. Over 400 varieties of hemoglobin have been discovered. It has been estimated that about 5 in every 1000 individuals have a mutant form of hemoglobin. Table 5.1 lists some mutations that have been discovered and the amino acid substitution. Most of us have the common type, HbA (hemoglobin adult). However,

α-helix unit is largely hydrophobic and is at home in the highly nonpolar environment within the membrane (see Chapters 3 and 9). Hydrophobic interactions between the amino acids and the nonpolar lipids in the membrane stabilize the arrangement. (We will learn in Chapter 9 that a membrane is comprised of nonpolar lipid and protein molecules.) The α-helical units are connected by loops containing polar and charged amino acids. The N-terminus, C-terminus, and other charged residues interact with aqueous solvent on the inner and outer surfaces of the membrane.

The bacteriorhodopsin in the membrane functions as a channel to allow the flow of protons. When cells of *H. salinarum* are illuminated, the bound retinal absorbs some of the light causing the conjugated molecule to isomerize. This triggers a conformational change in the protein. This small movement opens a channel through the membrane and allows the release of protons to the outside. Hence, the membrane protein can be considered a light-driven pump converting light energy into proton movement (chemical energy). Pumping protons through the membrane generates a highly energetic proton concentration gradient across the membrane (high H^+ outside, low H^+ inside). Energy released from the breakdown of the gradient is used by the cell to synthesize ATP by a mechanism discussed in Chapter 17.

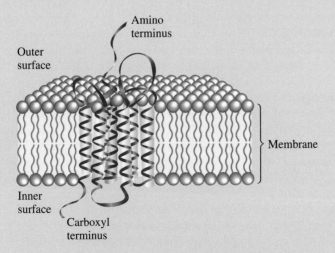

The folding of the protein bacteriorhodopsin through the *H. salinarum* membrane. *Source:* Based on Lehninger et al., 1999.

Table 5.1
Amino acid substitutions in
mutant hemoglobins

Mutant Hemoglobin[a]	Position Number[b]	Normal Residue	Substitutions
α Chain			
G$_{Honolulu}$	30	Glu	Gln
G$_{Philadelphia}$	68	Asn	Lys
I	16	Lys	Glu
M$_{Boston}$	58	His	Tyr
Norfolk	57	Gly	Asp
O$_{Indonesia}$	116	Glu	Lys
β Chain			
C	6	Glu	Lys
D$_{Punjab}$	121	Glu	Gln
G$_{San Jose}$	7	Glu	Gly
E	26	Glu	Lys
S	6	Glu	Val
Zurich	63	His	Arg

[a]The hemoglobins are often named for the cities where they were first discovered.
[b]The numbering for an amino acid position begins at the N-terminus.

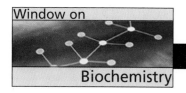

Window on
Biochemistry

5.3   Diagnosis of Lyme Disease by Immunoblotting

The first symptom noted by many victims of the disease is a spreading, ringlike, skin rash. This is followed in a few days by flu-like symptoms—fever and chills, extreme fatigue, nausea, sore throat, muscle and joint pain, and headache. Without medical attention, the disease would progress to a more dangerous stage of memory loss, nerve tremors, and cardiac manifestations. Most experienced clinicians when encountering patients with the above complaints would recognize the classic symptoms of **Lyme disease.**

Lyme disease is caused by a bacterium called *Borrelia burgdorferi* which is contracted by humans and other animals by the bite of an infected deer tick of the *Ixodes* type. *B. burgdorferi,* a relative of the bacterium that causes syphilis, is a large spirochete (0.25 micron × 50 microns, about as long as a human hair is thick) (see figure). The bacterium is unusual in that it is surrounded by a slime layer, a gel-like coat of glycoproteins which helps hide it from the human immune system.

A physician treating a patient with the above symptoms may suspect Lyme disease, but before prescribing the usual treatment of a potent and extended regimen of antibiotic therapy, the doctor must order clinical tests for confirmation. The only way to be absolutely certain of the diagnosis is to isolate the bacterium from blood or tissues. For most early cases of the disease, this is not possible because of extremely low concentrations of the bacterium. Indirect, less sensitive serological tests must be used to indicate the infection. Most indirect tests attempt to detect the presence of serum antibodies produced by the immune system to fight the bacteria. In patients with Lyme disease (and with most diseases caused by infectious agents), the first antibodies to appear in the blood are IgM type (in about 1–3 weeks after infection). However, only about 50% of Lyme disease patients have measurable levels of IgM. Victims of the disease then begin to develop the primary antibodies of the IgG type which last for several weeks and peak in concentration in 3–4 months.

The first antibody test used to diagnose Lyme disease was based on the ELISA (enzyme-linked immunoadsorbent assay). In this test, antigens from *B. burgdorferi* are fixed to a matrix and incubated with the patient's serum. If antibodies to the infectious organism are present in the serum, they will bind to antigens form-

some individuals have a genetic variation of HbA. Most variations are single amino acid changes and display few structural, functional, or clinical effects. We have already discussed one exception to the nonlethality of mutant hemoglobins. HbS, present in individuals with sickle cell anemia, displays very low solubility in the deoxygenated form and precipitates in the red blood cells, causing deformation (sickle-shaped cells) and lysis (causing anemia).

Relatively small changes in the primary structure of hemoglobin can result in significant changes in biological function. Crocodiles can remain underwater without breathing for periods of over one hour. They often use this special ability, which is the result of unusual hemoglobin, to drown their prey. When crocodiles hold their breath, carbon dioxide (or bicarbonate) accumulates in the blood and binds to hemoglobin, causing an increase in the release of oxygen to muscle and brain tissue. (Recall that CO_2 decreases the affinity of hemoglobin-bound O_2.) By changing the identity of just 12 amino acids, researchers have recently modified human hemoglobin so that it displays this enhanced allosteric effect and functions like crocodile hemoglobin.

α-Keratin

The fibrous protein α-keratin is abundant in nature, where it plays a role in constructing protective coatings for organisms. α-Keratin is the major protein constituent of hair, feathers, wool, the outer layer of skin, claws, scales, horns, turtle shells, quills, and hooves. It provides

ing an antigen:antibody complex. This complex may be detected by adding reagents that recognize the complex and trigger the formation of a color. The ELISA test is relatively insensitive because of the presence of many competing antibodies in the patient's blood.

A more reliable antibody test now in use requires a two-step process: (1) fractionation of the patient's serum antibodies by polyacrylamide gel electrophoresis (see Window on Biochemistry 4.2) and (2) transfer of the fractionated proteins from the gel to a nitrocellulose membrane (called an **immunoblot** or **western blot**). The membrane containing blotted proteins is then treated with immunological reagents to detect specific antibodies against the bacterium.

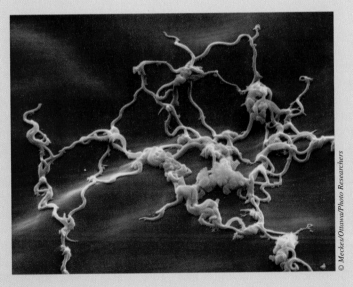

The *Borrelia burgdorferi* bacterium.

© *Meckes/Ottawa/Photo Researchers*

Figure 5.21

Cross-section of a hair showing its beginning with an α-helix of a single α-keratin molecule. *Source:* From Moran et al., 1994.

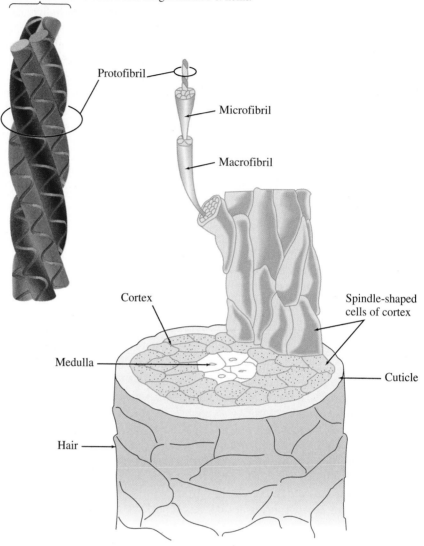

Four α-keratins twisted into a left-handed superhelix.
Each α-keratin is coiled into a right-handed α-helix.

Protofibril

Microfibril

Macrofibril

Cortex

Spindle-shaped cells of cortex

Medulla

Cuticle

Hair

Feathers of the Indigo Bunting contain the protein α-keratin.

© D. Cavagnaro/Visuals Unlimited

the characteristics of strength, water insolubility, and durability to its owners. The amino acid composition of wool, which is representative of all forms of α-keratin, is 12% Glu, 11% Cys, 10% Ser, 8% Gly, 7% Arg, 7% Leu, 6% Asp, 5% Val, 5% Ala, and 29% others. The primary structure of α-keratin has a seven-residue unit that is repeated many times. That repeat unit is represented by a-b-c-d-e-f-g, where a and d are residues with nonpolar side chains that can interact hydrophobically to assist the folding into an alpha helix. The presence of cysteine is important in the packaging of α-keratin into hair and other forms. Two types of α-keratin, "hard" and "soft," are present in nature. Hard α-keratin in turtle shells, hooves, and so on, is more rigid than soft keratin in skin, hair, or wool because of an increase in cysteine and the number of crosslinks formed by disulfide bonds (see Window on Biochemistry 5.1).

A hair is composed of many molecules of α-keratin. The polypeptide chain of α-keratin coils into a right-handed α-helix that is stabilized by hydrogen bonds and hydrophobic interactions between amino acids (Figure 5.21). Extra tensile strength results by wrapping three right-handed helices into a left-handed superhelix called a protofibril (see Figure 5.21). This is similar to the twisting of strings into a rope. Stabilizing factors for the protofibril include intermolecular hydrogen bonds and intermolecular disulfide bridges formed by the oxidation of juxtaposed cysteine residues. Eleven of the protofibrils combine into clusters called microfibrils, hundreds of which then combine to form a protein matrix called a macrofibril. A hair fiber is formed by stacking hair cells made from stacking macrofibrils. Hair fibers can be extended when moist because the highly coiled structure can unwind even to the point of breaking hydrogen bonds. The structure of stretched hair resembles the extended structure of β configuration. The disulfide bonds between protein coils, however, remain intact to help restore the original helical arrangement when tension is released.

Cutting Edge: Prions

Other Proteins

The structures and functions of other proteins including kinesin (in muscle), prions (disease-causing proteins), and antibodies are described in detail on CD-ROM/Website.

Structure Tutorials: Antibodies; Kinesin

Summary

- Each type of protein molecule in its natural environment folds into a unique three-dimensional structure called the **native conformation** defined by secondary, tertiary, and perhaps quaternary levels of organization. The most important determinant in protein structure is the influence of the amino acid sequence (primary structure).

- Weak, noncovalent interactions including hydrogen-bonding, ionic bonds, and hydrophobic interactions stabilize protein structures.

- The most prominent elements of secondary structure are α-helices, β-sheets, triple helices, bends, and loops.

- The tertiary structure of a protein, which describes the position of all atoms in the protein, is the result of combining several motifs of secondary structure into a compact arrangement.

- Protein folding is a stepwise, cooperative process that starts with the formation of local secondary structure.

- Hemoglobin, the oxygen-carrying molecule of vertebrates, is a tetrameric protein that carries up to four dioxygen molecules. The binding of dioxygen is regulated by allosteric interactions that are influenced by H^+, CO_2, and other modulators.

Study Exercises

Understanding Terms

5.1 Using 25 words or less, define each of the following terms.

    a. Fibrous protein
    b. Globular protein
    c. α-Helix
    d. Secondary structure
    e. β-Sheet
    f. Disulfide bond
    g. Supersecondary structure
    h. Structural motif
    i. Superbarrel
    j. Triple helix
    k. Scurvy
    l. Molecular chaperones
    m. Quaternary structure
    n. Oligomeric proteins
    o. Allosteric interactions
    p. Bacteriorhodopsin
    q. Hydrophobic interactions
    r. Central dogma of protein folding
    s. Proteome
    t. Homotypic subunits
    u. Heterotypic subunits
    v. Heat shock proteins
    w. Cooperative binding

Reviewing Concepts

5.2 What is the approximate number of amino acids in an α-helix that is 20 Å long?

5.3 Using a line to represent the polypeptide chain, draw the following combinations of secondary structure.

    a. αα
    b. ββ, antiparallel
    c. Superbarrel
    d. αααα
    e. βαβ, antiparallel

5.4 Draw the hexapeptide Gly-Phe-Asp-Leu-Ala-Glu, and show (using lines) the hydrogen bonding stabilizing an α-helix.

5.5 Describe the forces that stabilize protein quaternary structure.

5.6 Draw the axes for an oxygen binding curve (Figure 5.19). On the graph, draw the curve for hemoglobin at normal physiological pH (about 7.4). Now on the same graph, draw the curve for hemoglobin at pH 7.6; at pH 7.2.

5.7 Select any five amino acids, draw their structures, and circle the atom(s) in each that can serve as hydrogen bond donors and/or acceptors.

5.8 Describe the types of secondary structure present in the myoglobin molecule.

HINT: Study Figure 5.1b.

5.9 Proteins may have up to four levels of structure. For each type, describe the kind of bonding interactions that maintain that type of structure. Select from covalent or noncovalent. If you choose noncovalent, also select specific types (hydrogen bonding, hydrophobic, ionic).

   a. Primary
   b. Secondary
   c. Tertiary
   d. Quaternary

5.10 Compare the characteristics of globular and fibrous proteins. The following items should be discussed and contrasted: water solubility, amino acid composition, types of secondary and tertiary structure, biological function.

5.11 What is the difference between a parallel β-sheet and an antiparallel β-sheet?

5.12 How are the heat shock proteins formed in cells and what are their proposed biological functions?

5.13 What are the differences between a structural motif and a domain?

Solving Problems

5.14 Draw the sequence -Gly-Pro-Leu-Ala- and explain how proline forms a bend in the structure. Use a dotted line to indicate possible intramolecular hydrogen bonding.

5.15 Study the structure of each of the following amino acid residues as given in Chapter 4 and predict the ones on the inside and determine the ones that would be on the outside of a typical globular protein in aqueous solution at pH 7.4.

   Ser   Phe   Asn
   Glu   His   Leu
   Thr   Val   Cys

HINT: Is the side chain polar or nonpolar?

5.16 You have isolated and purified an unknown protein from an insect. Its amino acid composition was determined after acid-catalyzed hydrolysis:

Amino Acid	%
Gly	45
Ala	30
Ser	12
Tyr	5
Val	2
Others	6

Answer the following questions about the unknown protein:

   a. What is the expected secondary structure?
   b. Is the protein fibrous or globular?

   c. Do you expect the protein to be water soluble?
   d. Can you make some predictions about its biological function?

5.17 Study each of the short polypeptide segments below and predict regions of secondary and tertiary structure. Choose from one of the following:

   α-Helix   Random coil
   β-Sheet   Disulfide bond
   ββ motif   αα motif

   a. Gly-Ala-Ala-Gly-Ser-Gly-Ala-Pro-Ala-Gly-Ala-Ala-Ser-Tyr-Gly
   b. Leu-Ala-Lys-Lys-Lys-Lys-Phe-Gly
   c. Val-His-Ala-Thr-Cys-Met-Tyr-Ser
   d. Ala-Phe-Cys-Ser-Gly-Pro-Thr-Ala-Cys-Ala-Phe
   e. Phe-Val-Met-Ala-Thr-Ser-Gly-Pro-Gly-Ala-Phe-Thr-Leu-Phe-Lys
   f. Gly-Glu-Asp-Asp-Glu-Asp-Phe

5.18 A peptide chain of polylysine coils into an α-helix in solution at pH 13; however, it becomes a random coil at pH 7. Explain.

5.19 Some strains of bacteria are able to thrive in hot springs and geysers at 100°C and higher. These thermophilic organisms are of special interest because they contain proteins and nucleic acids that are more thermally stable than those biomolecules from organisms that live under more normal conditions. What differences might you expect in the amino acid composition of proteins from thermophilic organism?

HINT: Ionic interactions and hydrogen bonding become weaker with increasing temperatures; hydrophobic interactions become stronger with increasing temperatures.

5.20 The diagram below shows the results of a sodium dodecyl sulfate polyacrylamide gel electrophoresis (SDS-PAGE) experiment. Identify the protein in each lane using the following list:

   Hemoglobin
   Myoglobin
   Glucose oxidase
   Bacteriorhodopsin

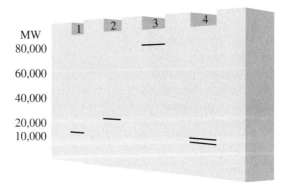

HINT: See Window on Biochemistry 4.2.

5.21 Urea causes denaturation of the 2° and 3° structure of proteins but has no effect on 1° structure. Study the structure of urea below and explain the action of urea as a denaturing agent.

$$H_2N-\overset{\overset{\displaystyle O}{\displaystyle \|}}{C}-NH_2$$

5.22 Predict which of the polypeptides listed below will fold into an α-helix under the conditions given.
  a. Polyalanine, pH = 7.0
  b. Polyaspartic acid, pH = 7.0
  c. Polyaspartic acid, pH = 3.0
  d. Polyproline, pH = 7.0
  e. Polylysine, pH = 7.0

5.23 Below is shown the structure of a hypothetical protein. Define bonding forces that hold the protein in its native conformation. Identify possible amino acid residues involved in interactions.

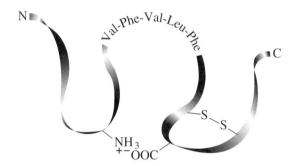

5.24 Study the O$_2$ binding curve in Figure 5.19 and answer the following questions.
  a. At what O$_2$ pressure is myoglobin 50% saturated with O$_2$?
  b. At what O$_2$ pressure is hemoglobin 50% saturated with O$_2$?
  c. At what O$_2$ pressure does hemoglobin have a higher affinity for O$_2$ than does myoglobin?

5.25 Explain why hemoglobin is a better long-distance oxygen transporter than myoglobin.

5.26 A synthetic peptide chain of polyglutamate is found to be a random coil in solution at pH 7. When the pH of the solution is changed to 3, the peptide folds into an α-helix. Explain.

5.27 Each of the following reagents or experimental conditions is capable of disrupting covalent and/or noncovalent bonding in a protein. List the type of bonding broken by each.
  a. Heat at 60°C
  b. HCl + H$_2$O; heat at 100°C
  c. Ethyl alcohol
  d. Urea
  e. NaOH + H$_2$O; heat at 100°C

  f. Sodium dodecyl sulfate
  g. Acetone
  h. Mercaptoethanol
  i. Peptidase + H$_2$O at 37°C

5.28 When a protein is unfolded (denatured), it becomes less water soluble and often precipitates from solution. Explain.

5.29 Identify each of the following proteins as globular or fibrous.
  a. Myoglobin
  b. Insulin
  c. Hemoglobin
  d. α-Keratin
  e. Serum albumin
  f. Collagen

5.30 Bacteriorhodopsin, a protein found in the membrane of a salt-loving bacterium, has seven distinct regions of α-helix. These regions span the cell membrane in the organism. Can you make any conclusions about the amino acid composition in these α-helix regions of bacteriorhodopsin? (See Window on Biochemistry 5.2.)

Writing Biochemistry

5.31 Your biochemistry instructor makes the following challenge in class: "Tell me the primary structure of a protein and I will draw its secondary and tertiary structure." Write a 100-word response to your instructor. Comment especially on the credibility of the challenge.

5.32 The present summary for this chapter has only about 200 words. After reading the chapter, write a more detailed summary that has between 300 and 400 words.

Applying Biochemistry

5.33 The compound 2,3-bisphosphoglycerate (2,3-BPG, see p. 143 for structure) promotes the release of oxygen from oxygenated hemoglobin. The allosteric modulator binds in the central cavity of the protein. Look at the structure of 2,3-BPG and predict what kinds of bonds hold the small molecule into the binding site on hemoglobin. Also, predict what amino acids of hemoglobin hold 2,3-BPG in place.

5.34 There is now substantial evidence that some diseases like bovine spongiform encephalitis (BSE) and chronic wasting disease in deer and elk are transmitted by a protein and nothing more. Why is this a difficult concept for scientists to accept?

5.35 Draw a sketch of an antibody molecule and show the following characteristics and regions: disulfide bonds, variable region, constant (or consistent) region, and antigen binding site.

Further Reading

Atkins, P., 1987. Polypeptides, α-keratin, β-keratin. In *Molecules,* pp. 90–94. New York: W. H. Freeman.

Birge, R., 1995. Protein-based computers. *Sci. Am.* 272(2):90–95.

Branden, C., and Tooze, J., 1998. *Introduction to Protein Structure,* 2nd ed. New York: Garland Publishing.

Brunori, M., 1999. Hemoglobin is an honorary enzyme. *Trends Biochem. Sci.* 24:158–161.

Collins, F. et al., 2001. The human genome. *Nature* 409:745–748.

Craig, E., 1993. Chaperones: Helpers along the pathways to protein folding. *Science* 260:1902–1903.

de Chadarevian, S., 1999. Protein sequencing and the making of molecular genetics. *Trends Biochem. Sci.* 24(5):203–206.

Dill, K., and Chan, H., 1997. From Levinthal pathways to funnels. *Nature Struct. Biol.* 4:10–19.

Dobson, C., 1999. Protein misfolding, evolution and disease. *Trends Biochem. Sci.* 24(9):329–332.

Doolittle, R., and Bork, P., 1993. Evolutionary mobile modules in proteins. *Sci. Am.* 269(4):50–56.

Frydman, J., and Hartl, F., 1996. Principles of chaperone-assisted protein folding. *Science* 272:1497–1502.

Goodsell, D., and Olson, A., 1993. Soluble proteins: Size, shape, and function. *Trends Biochem. Sci.* 18:65–68.

Henry, C., 2000. Protein structure by mass spec. *Chem Eng. News* Nov. 27: 22–26.

Komiyama, H., and Nagai, K., 1995. Transplanting a unique allosteric effect from crocodile into human hemoglobin. *Nature* 373:244.

Lesk, A., 2000. *Introduction to Protein Architecture.* Oxford: Oxford University Press.

Mbewe, B., 2000. Quaternary structure as defined by whom? *Biochem. Mol. Biol. Educ.* 28:240–241.

O'Halloran, T. and Culatta, V., 2000. Metallochaperones, an intracellular shuttle service for metal ions. *J. Biol. Chem.* 275:25057–25060.

Prince, R., and Gunson, D., 1998. Prions are copper-binding proteins. *Trends Biochem. Sci.* 23:197–198.

Prusiner, S., 1995. The prion diseases, *Sci. Am.* 272(1):48–57.

Prusiner, S., 1997. Prion diseases and the BSE crisis. *Science* 278:245–251.

Smith, C., 2000. How do proteins fold? *Biochem. Educ.* 28:76–79.

Venter, J., 2001. The sequence of the human genome. *Science* 291:1304.

Vollrath, F., 1992. Spider webs and silks. *Sci. Am.* 261(3):70–76.

Welch, W., 1993. How cells respond to stress. *Sci. Am.* 268(5):56–64.

Chapter 6

Enzymes I: Reactions, Kinetics, and Inhibition

O f all biomolecules found in cells and organisms, the enzymes, because of their highly specialized characteristics and crucial roles, are among the most well known. The enzymes serve as catalysts for biochemical reactions in living organisms. In this role, they direct and regulate the thousands of reactions providing for energy transformation, synthesis, and metabolic degradation. As catalysts, the enzymes excel all other chemicals in power and specificity. Most of the reactions in the cell would not occur without the presence of catalysts. Prior to 1982 it was common to hear the statement, "All enzymes (biological catalysts) are proteins," because all biological catalysts previously studied were composed of amino acids; however, some forms of RNA that catalyze cellular reactions have been discovered.

In this chapter we focus on protein enzymes and introduce their classification, kinetic characteristics, and mode of action. In addition, enzymes now being developed for use in medicine, food processing, agriculture, and the chemical industry are introduced.

6.1   Enzymes As Biological Catalysts

Introduction to Enzymes

Early research in biochemistry was essentially a study of enzyme and protein chemistry. The first thorough and controlled studies of natural catalytic processes were those in the late 1700s and early 1800s involving the digestion of meat protein by gastric fluids. In other early studies, polysaccharides like starch were found to be changed to glucose by plant cell extracts and animal saliva. In the mid-1800s, Louis Pasteur, whose research was funded by wineries, coined the term "ferments" to define the agents in yeast cells that acted to convert sugars into ethyl alcohol (fermentation). Pasteur believed that only intact, living yeast cells were able to carry out fermentation of sugars; however, in 1897, Eduard and Hans Büchner used extracts of yeast cells to ferment sugar. These experiments began a more sophisticated era in biochemistry because it was demonstrated that the natural catalytic agents, now called **enzymes,** could function independent of a living cell. In 1926, James Sumner at Cornell University isolated and crystallized the enzyme urease from jack bean plant and found the crystals to be composed primarily of protein material. Once it was found that enzymes could be isolated from biological cells, purified, and their reactions studied in a test tube, thousands of reports on enzymes and their properties appeared in the literature. Today, we understand many of the general principles of enzyme function. In this chapter we focus on characteristics that are common for most enzymes.

Cutting Edge: Visionaries of science

PAKISTAN ﭘﺎﮐﺴﺘﺎﻥ

Rs.5

100TH DEATH ANNIVERSARY OF
L O U I S   P A S T E U R
1 8 9 5 - 1 9 9 5

This Pakistani stamp honors the chemist and microbiologist, Louis Pasteur.

Crystals of chitinase under polarized light. This enzyme catalyzes the hydrolysis of chitin, a polysaccharide of invertebrate exoskeletons.

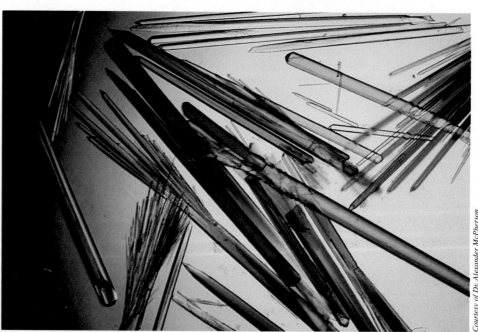

Courtesy of Dr. Alexander McPherson

(a)   (b)   (c)

Figure 6.1
A solution of 3% H_2O_2 in water at 37°C: (a) no catalyst added, (b) with added Fe^{3+} salt, and (c) with added enzyme catalase. The bubbles present in each tube are oxygen gas produced from the breakdown of H_2O_2.

Consider the following reaction that takes place in most plant and animal cells:

$$2 \, H_2O_2 \rightleftharpoons 2 \, H_2O + O_2$$

H_2O_2 (hydrogen peroxide) is a waste product of metabolism, and if left in the cell, it would initiate the formation of free radicals that are detrimental to nucleic acids and other biomolecules. The reaction is thermodynamically favorable but occurs slowly without the assistance of any catalysts. You may have observed bubbles of O_2 gas evolving from a 3% solution of H_2O_2 in water, a common disinfectant present in many medicine chests. Figure 6.1a shows a 3% solution of H_2O_2 at 37°C. The energy diagram for the reaction is shown in Figure 6.2. Before the products H_2O and O_2 can form, H_2O_2 molecules must be lifted to a higher energy

Interactive Animation: Catalysis

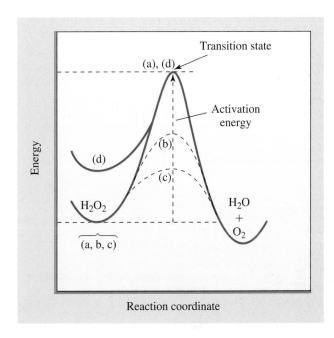

Figure 6.2
The energy diagram for the decomposition of H_2O_2 to H_2O and O_2. H_2O_2 molecules must achieve an energy level equivalent to the activation energy to be converted to product. Curve (a), the activation energy for the reaction in the absence of a catalyst. Curve (b), the activation energy for the breakdown of H_2O_2 is lowered in the presence of an iron catalyst. Curve (c), the energy diagram for the catalase-catalyzed breakdown of H_2O_2. This reaction has the lowest activation energy requirement. Curve (d), the energy diagram for the noncatalyzed breakdown of H_2O_2 at an elevated temperature. The H_2O_2 molecules begin the reaction at a higher energy level.

level. The **activation energy** is the amount of energy that must be acquired by reactant molecules before they can be converted to a product. This level of energy is represented by the top of the "hill" between reactants and products in Figure 6.2. The activation energy can be seen as a "barrier" for the reactant molecules to overcome before conversion to products. At the top of the barrier, the reactant is in its **transition state,** an energetic (unstable), short-lived species with an equal probability of reverting to starting material (H_2O_2) or being transformed to products ($O_2 + H_2O$). At 37°C, only a small fraction of H_2O_2 molecules can gain sufficient energy to react, mainly by collisional processes. If we want to increase the rate of the reaction, one of two things can be done: (Option 1) we can increase the average energy level of the H_2O_2 molecules by increasing the temperature, or (Option 2) we can add a catalyst. Option 1 enhances the rate by increasing the number of molecular collisions and, therefore, the fraction of H_2O_2 molecules that has enough energy to pass over the energy hill (through the transition state). The energy requirements for the reaction remain the same for both temperatures (Figure 6.2a, d). Increasing the temperature is not a realistic option for enhancing the rates of reactions in most organisms. Option 2 is the addition of a catalyst. Ferric (Fe^{3+}) and other metal ions added to solutions of H_2O_2 enhance the rate of breakdown by about 30,000 times compared to no catalyst (compare Figures 6.1a and 6.1b). Recall from your studies in introductory chemistry that a catalyst functions by lowering the activation energy for a reaction (Figure 6.2b, c). The Fe^{3+} guides H_2O_2 molecules through a different decomposition pathway that involves a lower energy barrier; hence the reaction rate is increased. The equilibrium position for the reaction above is not changed in the presence of the catalyst; however, equilibrium is reached much sooner since more H_2O_2 molecules are degraded per time unit. Now, let's consider the addition of an enzyme, a biological catalyst. The enzyme catalase, when added to a solution of H_2O_2, enhances the reaction to about 100,000,000 times faster than with no catalyst (Figure 6.1c). Catalase, a large protein (MW = 250,000 D) with four identical subunits, each associated with a heme unit (iron protoporphyrin), is found in nearly all types of plant and animal cells. Catalase, like all enzymes, has the properties of a true catalyst:

1. It increases the rate of a reaction by lowering the activation energy barrier.
2. It is not "used up" or permanently changed during the catalytic process.
3. It does not change the position of equilibrium only the rate at which equilibrium is attained.
4. It usually acts by forming a transient complex with the reactant, thus stabilizing the transition state.

Some enzymes require, in addition to the protein component, another chemical entity, called a **cofactor,** in order to function properly. A cofactor may be an organic or organometallic molecule (called a **coenzyme**) or a metal ion such as Zn^{2+}, Mg^{2+}, or Cu^{2+} (see Table 7.1 for a list of cofactors and biological function). The term **prosthetic group** is used to define a cofactor or coenzyme that is covalently bonded to the protein. The protein catalase requires the heme cofactor for catalytic activity. The complete molecular package, protein and cofactor, is called the **holoenzyme** form. The term **apoenzyme** is used to define just the protein component, the holoenzyme minus the cofactor.

Table 6.1

Classification and function of enzymes

Classification Number	Enzyme Class	Type of Reaction Catalyzed
1	Oxidoreductases	Transfer of electrons usually in the form of hydride ions or hydrogen atoms
2	Transferases	Transfer of functional groups from one molecule to another
3	Hydrolases	Cleavage of bonds by hydrolysis
4	Lyases	Formation of double bonds by removal of groups or addition of groups to double bonds
5	Isomerases	Transfer of groups within a molecule to yield isomeric forms
6	Ligases	Formation of C—C, C—S, C—O, and C—N bonds by condensation coupled to ATP cleavage

Naming of Enzymes

Thousands of enzymes have now been isolated and studied; confusion would reign without some system for nomenclature and classification. Common names for enzymes are usually formed by adding the suffix *-ase* to the name of the reactant. The enzyme tyrosinase catalyzes the oxidation of tyrosine; cellulase catalyzes the hydrolysis of cellulose to produce glucose. Common names of this type define the substrate but do not describe the chemistry of the reaction. Some early names, such as catalase, trypsin, and pepsin, are even less descriptive and give no clue to their function or substrates. To avoid such confusion, enzymes now have official names that reflect the reactions they catalyze. All known enzymes can be classified into one of six categories (Table 6.1). Each enzyme has an official international name ending in *-ase* and a classification number. The number consists of four digits, each referring to a class and subclass of reaction. Table 6.2 shows an example from each class of enzyme. Detailed chemistry for each class of enzyme is given in Section 14.2.

Oxidoreductase

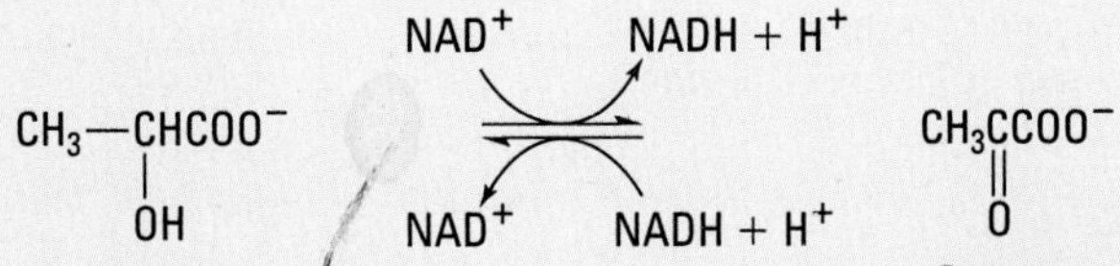

CH₃—CHCOO⁻ with OH below (Lactate) → CH₃CCOO⁻ with O below (Pyruvate)

NAD⁺ → NADH + H⁺ ; NAD⁺ ← NADH + H⁺

Common name: Lactate dehydrogenase
Official name: L-Lactate: NAD⁺ oxidoreductase
Official number: 1.1.2.3

Transferase

$$(d\text{-NMP})_n + d\text{-NTP} \rightleftharpoons (d\text{-NMP})_{n+1} + PP_i$$

$(d\text{-NMP})_n$ = DNA with n nucleotides
$(d\text{-NMP})_{n+1}$ = DNA with $n + 1$ nucleotides
PP_i = Pyrophosphate
Common name: DNA polymerase
Official name: Deoxynucleoside triphosphate: DNA deoxynucleotidyltransferase (DNA directed)
Official Number: 2.7.7.7

Hydrolase

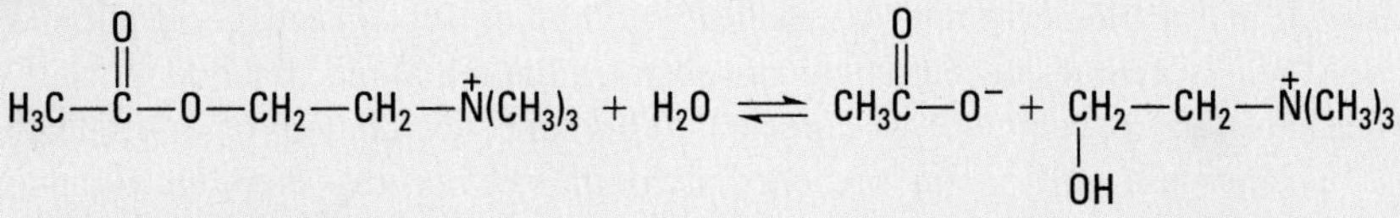

$$H_3C-\overset{O}{\overset{\|}{C}}-O-CH_2-CH_2-\overset{+}{N}(CH_3)_3 + H_2O \rightleftharpoons CH_3\overset{O}{\overset{\|}{C}}-O^- + CH_2-CH_2-\overset{+}{N}(CH_3)_3$$ with OH below

Acetylcholine   Acetate   Choline

Common name: Acetylcholinesterase
Official name: Acetylcholine acetylhydrolase
Official number: 3.1.1.7

Lyase

$$CO_2 + H_2O \rightleftharpoons H_2CO_3$$
Carbonic acid

Common name: Carbonic anhydrase
Official name: Carbonate hydrolyase
Official number: 4.2.1.1

Table 6.2
An example of each class of enzyme

(continued)

Table 6.2—*continued*
An example of each class
of enzyme

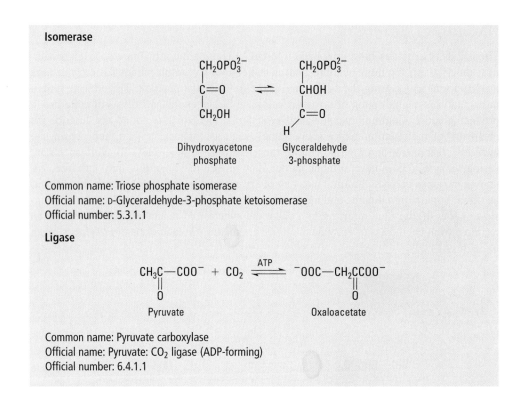

Isomerase

Dihydroxyacetone
phosphate

Glyceraldehyde
3-phosphate

Common name: Triose phosphate isomerase
Official name: D-Glyceraldehyde-3-phosphate ketoisomerase
Official number: 5.3.1.1

Ligase

Pyruvate

Oxaloacetate

Common name: Pyruvate carboxylase
Official name: Pyruvate: CO_2 ligase (ADP-forming)
Official number: 6.4.1.1

6.2 The Kinetic Properties of Enzymes

Michaelis–Menten Equation

Concept Review:
Elementary kinetics

Although enzymes have most of the characteristics of typical organic and inorganic catalysts, they have unique kinetic properties that set them apart from other catalysts. One early observation was the unusual effect of substrate concentration on the rate of the enzyme-catalyzed reaction. The rates of enzyme-catalyzed reactions are studied by mixing substrate and enzyme in an appropriate buffer solution (to maintain a constant pH) and at a constant temperature. The initial rate (v_0, **initial velocity**) is determined during the first few minutes of the reaction by measuring either the decrease of reactant concentration or the increase of product concentration. To study the effect of substrate concentration, an experiment is set up according to Figure 6.3. A set of tubes is prepared, each containing a buffer with increasing amounts of substrate. A constant amount of enzyme is added to each tube, the reaction rate is measured, and a plot or graph of reaction rate versus substrate concentration is constructed. At relatively low concentrations of substrate, the initial reaction rate increases with increasing substrate concentration as expected. At higher concentrations of substrate, the rate increase becomes less until a point is reached where the rate becomes constant no matter how much substrate is present. The resulting curve is hyperbolic in shape. We define the constant rate as the **maximum velocity** or V_{max}. This catalytic behavior, which is observed for most enzymes, can be described by a substrate saturation effect to be discussed in the next section. The first researchers to explain the shape of the rate curve were two biochemists, Leonor Michaelis and Maud Menten. In 1913, they proposed a general theory for enzyme action and derived a mathematical equation to express the hyperbolic shape of the curve and to calculate rate constants. Michaelis and Menten proposed that enzyme molecules, E, and substrate molecules, S, combine in a fast and reversible step to form an **ES complex:**

$$E + S \underset{k_2}{\overset{k_1}{\rightleftharpoons}} ES \underset{k_4}{\overset{k_3}{\rightleftharpoons}} E + P$$

There are two possible fates of the ES complex: (1) it can revert to free enzyme and substrate, or (2) it can proceed by a reversible reaction to form free enzyme and product, P. The terms k_1, k_2, k_3, and k_4 define rate constants for the individual steps.

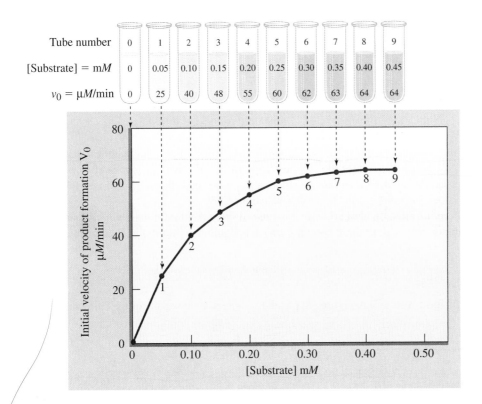

Figure 6.3
Experimental procedure to study the kinetics of an enzyme-catalyzed reaction. An identical amount of enzyme is added to a set of tubes containing increasing amounts of a substrate. The reaction rate or initial velocity is measured for each reaction mixture by determining the rate of product formation. The velocity is plotted against substrate concentration. Most enzymes yield a hyperbolic curve as shown.

The basic equation derived by Michaelis and Menten to explain enzyme-catalyzed reactions is

$$v_0 = \frac{V_{max}\,[\text{S}]}{K_M + [\text{S}]}$$

where

v_0 = initial velocity caused by substrate concentration, [S]
V_{max} = maximum velocity
K_M = Michaelis constant

(You may recognize this equation as the mathematical expression for a hyperbola.) Michaelis and Menten made several assumptions to ease their derivation. They chose to neglect the reaction that reverts product P and free enzyme to the ES complex (defined by k_4 in the reaction sequence). This reaction becomes significant only after relatively high concentrations of P have been produced. When biochemists use the Michaelis–Menten equation in the laboratory, they measure only *initial rates,* when the reaction represented by k_4 is very, very slow (usually during the first few minutes). Another assumption necessary in the derivation of the equation is that the ES complex is a **steady-state intermediate.** That is to say, after mixing E and S, a certain level of ES is formed very rapidly and its concentration remains relatively constant because it is produced at the same rate as it breaks down.

Two important constants in the **Michaelis–Menten equation,** K_M and V_{max}, require further description. The **Michaelis constant, K_M,** is expressed mathematically as a combination of rate constants:

$$K_M = \frac{k_2 + k_3}{k_1}$$

Because it is difficult to gain a real understanding of K_M from this relationship, we will define the Michaelis constant in different terms. If an analysis is completed for K_M in the Michaelis–Menten equation, it is found to have the same units as the substrate concentration, [S]. This implies some relationship between K_M and [S]. What happens to the Michaelis–Menten equation when the value for K_M is equal to the value of [S]?

where

$$K_M = [S]$$

$$v_0 = \frac{V_{max}\,[S]}{[S] + [S]}$$

$$v_0 = \frac{V_{max}\,[S]}{2[S]}$$

$$v_0 = \frac{V_{max}}{2}$$

In words, K_M *is equivalent to the substrate concentration that produces an initial velocity of* $^1/_2$ V_{max}. The K_M values of enzymes range from 10^{-1} M to as low as 10^{-8} M. For those enzymes having more than one substrate, a K_M is defined for each substrate. Table 6.3 lists K_M values for some common enzyme substrate pairs.

More information about K_M can be gleaned from the special case when k_2 is very much greater than k_3. Under these conditions k_3 is insignificant and K_M is defined as

where

$k_2 \gg k_3$

$$K_M = \frac{k_2}{k_1}$$

In this form, you will recognize K_M as the dissociation constant for the ES complex:

$$ES \underset{k_1}{\overset{k_2}{\rightleftharpoons}} E + S$$

Therefore, K_M under special conditions is a measure of the affinity between E and S. When K_M is relatively large (10^{-1} to 10^{-3} M), this implies that k_2, the rate constant for ES dissociation, is relatively large and the ES complex is held together rather weakly. In contrast, a small K_M (less than 10^{-3} M) represents a high affinity between E and S (a strong complex is formed).

We have previously defined maximum velocity, V_{max}, as the maximal rate for an enzyme-catalyzed reaction for a given value of [E]. This term, which can be experimentally measured, is an important constant to help characterize an enzyme and optimize reaction conditions. More importantly, knowing the value for V_{max} allows one to calculate another useful constant, the **turnover number,** k_3, for an enzyme:

$$\text{turnover number} = k_3 = \frac{V_{max}}{[E_T]}$$

where

$[E_T]$ = total enzyme concentration.

Table 6.3

K_M values for some enzyme substrate systems

Enzyme	Substrate	K_M (mM)
Catalase	H_2O_2	0.001
Hexokinase from brain	ATP	0.4
	D-Glucose	0.05
	D-Fructose	1.5
Carbonic anhydrase	HCO_3^-	9
Chymotrypsin	Glycyltyrosinylglycine	108
	N-Benzoyltyrosinamide	2.5
β-Galactosidase	Lactose	4.0
Penicillinase	Benzylpenicillin	0.050
Pyruvate carboxylase	ATP	0.060
	Pyruvate	0.40
	HCO_3^-	1.0
Ribulose 1,5-bisphosphate carboxylase (rubisco)	Ribulose 1,5-bisphosphate	0.028
	CO_2	0.009
Ribulose 1,5-bisphosphate oxygenase (rubisco)	Ribulose 1,5-bisphosphate	0.028
	O_2	0.535

Enzyme	Substrate	k_3 (per second)
Catalase	H_2O_2	40,000,000
Carbonic anhydrase	HCO_3^-	400,000
Acetylcholinesterase	Acetylcholine	25,000
Penicillinase	Benzylpenicillin	2,000
Lactate dehydrogenase	Lactate	1,000
Chymotrypsin	Glycyltyrosinylglycine	100
DNA polymerase	DNA	15
Ribulose 1,5-bisphosphate carboxylase	Ribulose 1,5-bisphosphate + CO_2	3.3
Ribulose 1,5-bisphosphate oxygenase	Ribulose 1,5-bisphosphate + O_2	2.4

Table 6.4
Turnover numbers, k_3, for some enzymes

The turnover number for catalase can be calculated from experimental data as follows:

A 10^{-9} M solution of catalase catalyzes the breakdown of 0.4 M H_2O_2 per second.

$$k_3 = \frac{0.4 \text{ moles/liter of } H_2O_2 \text{ degraded per second}}{10^{-9} \text{ moles/liter catalase}}$$

$k_3 = 40,000,000$ moles of H_2O_2 degraded per mole of catalase per second

In general, the turnover number is the number of moles (or molecules) of substrate transformed to product per mole (or molecule) of enzyme in a defined time period, usually a second. Turnover numbers for several enzymes are listed in Table 6.4.

The constants, K_M, V_{max}, and k_3 tell us much about an enzyme-catalyzed reaction so it is important to obtain them from experiments. Most methods involve graphical analysis of experimental data. The Michaelis–Menten curve can be used to estimate V_{max} and K_M as shown in Figure 6.4. However, an accurate value of V_{max} is difficult to measure because it requires very high levels of substrate that cannot be easily achieved experimentally. V_{max} is usually estimated from the Michaelis–Menten graph and the value for K_M is determined as the concentration of substrate that yields an initial rate of $\frac{1}{2}$ V_{max}.

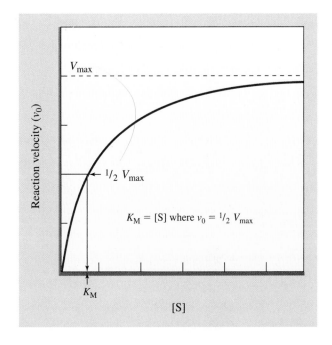

Figure 6.4
Using the Michaelis–Menten curve to estimate V_{max} and K_M. V_{max} is estimated from the graph at the point where the reaction velocity no longer increases with an increase in substrate concentration. K_M is measured on the [S] axis. K_M defines the substrate concentration that yields a velocity of $\frac{1}{2}$ V_{max}.

Figure 6.5

Using the Lineweaver–Burk equation and curve to determine V_{max} and K_M. The graph is obtained by plotting $1/v_0$ versus $1/[S]$ and the points connected in a straight line. The slope of the line is K_M/V_{max}. V_{max} is measured at the intersection of the line with the $1/v_0$ axis. The intersection of the line on the $1/v_0$ axis is $1/V_{max}$. K_M is measured on the $1/[S]$ axis. The intersection point is $-1/K_M$.

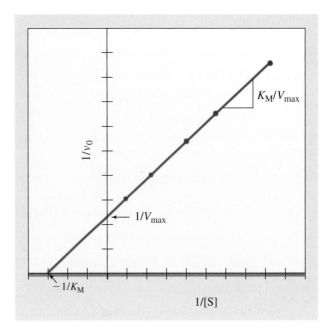

Lineweaver–Burk Equation

In 1934, Hans Lineweaver and Dean Burk reported a method to change the Michaelis–Menten equation into a form that is more amenable to graphical analysis. The **Lineweaver–Burk equation** allows one to plot experimental enzyme rate data in a straight line form (Figure 6.5):

$$\frac{1}{v_0} = \frac{K_M}{V_{max}} \cdot \frac{1}{[S]} + \frac{1}{V_{max}}$$

A graph of $1/v_0$ versus $1/[S]$ (called a Lineweaver–Burk or a double reciprocal plot) yields a straight line with intercepts of $1/V_{max}$ and $-1/K_M$ and a slope of K_M/V_{max}. The Lineweaver–Burk plot is valuable in determining constants for enzyme-catalyzed reactions and also in evaluating the inhibition of enzyme reactions (see Section 6.4).

Characteristics of Enzyme Reactions

To this point we have only discussed the influence of [S] on the enzyme reaction rate. Other experimental factors include enzyme concentration, pH of the reaction solution, and temperature effect on enzyme activity. The result of adding more enzyme should be easily predicted.

Figure 6.6

A plot of initial reaction velocity versus the concentration of enzyme. Note that the velocity increases in a linear fashion with an increase in enzyme concentration.

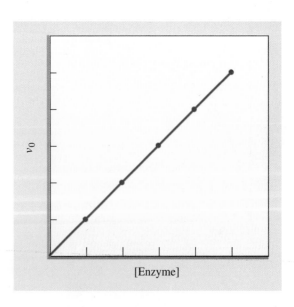

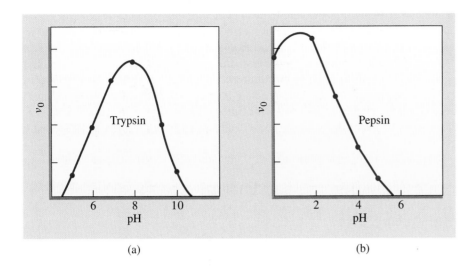

Figure 6.7
The pH–rate profiles for two enzymes obtained by measuring the reaction velocity for a reaction at several pH values: (a) trypsin and (b) pepsin. Note that the enzymes have very different pH optimums.

Since the enzyme is the catalyst, the higher the concentration, the greater should be the initial reaction rate (Figure 6.6). This relationship will hold as long as there is enough substrate present. Enzymes are also sensitive to environmental changes. Most enzymes have an optimal pH at which they function most effectively. This optimum varies from enzyme to enzyme, but most are in the range of pH 6–8. Dependence on pH usually results from the involvement of charged amino acid residues in the enzyme or substrates whose ionic structures will change with variation of pH. Figure 6.7 shows the pH–rate profiles for two enzymes.

Enzymes are also very sensitive to temperature changes as shown in Figure 6.8. At relatively low temperatures, the rate for an enzyme-catalyzed reaction increases proportionally with increasing temperature. At some point, which will vary with the enzyme, the temperature becomes a destructive agent, resulting in a precipitous decline in the rate. (Can you explain the shape of the curve in Figure 6.8? HINT: See Section 5.3.) For most enzymes, the rate decline begins in the temperature range of 50°C to 60°C, although some organisms living in hot springs or ocean thermal vents have enzymes stable to temperatures above 80 to 90°C.

The presence of metal ions and organic cofactors (coenzymes) also influences the action of some enzymes. This topic is covered in detail in Chapter 7.

Although many enzymes display the characteristics of Michaelis–Menten kinetics (hyperbolic rate curves), some do not. A large group of enzymes that are unique in their kinetic behavior are the allosteric enzymes. We have already introduced this term in our discussion of hemoglobin and other multimeric proteins that undergo conformational changes

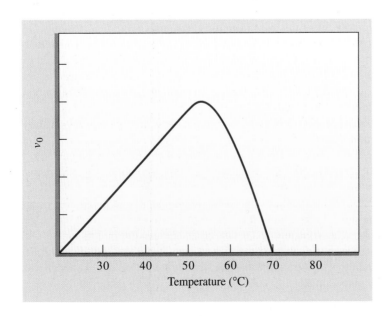

Figure 6.8
The influence of temperature on the rate of an enzyme reaction. The initial rate of the reaction increases with temperature until the enzyme becomes denatured and, therefore, less active.

upon the binding of ligands. In the cell, these enzymes are responsible not only for catalysis but also for regulating the overall rate of metabolic processes. **Allosteric** or **regulatory enzymes** are quite different from the nonregulatory enzymes. Regulatory enzymes are usually larger in molecular weight and are composed of two or more subunits. Their reactivities depend upon some of the same factors as do nonregulatory enzymes: [S], [E], temperature, pH, cofactors, and so on. In addition, they may also display cooperative binding of substrate and/or bind another type of molecule called a modulator or **effector.** Binding of a modulator may enhance or decrease the activity of the regulatory enzyme as discussed in detail in Chapter 7.

6.3   Substrate Binding and Enzyme Action

Enzyme Active Sites

Interactive Animation: Enzyme specificity

As discussed in the previous section, enzymatic catalysis begins with the combination of an enzyme molecule with a substrate molecule to form an ES complex (E + S $\rightleftharpoons$ ES). The substrate molecule, which is usually smaller than the enzyme, binds to a specific region in the enzyme called the **active site.** The idea of the active site was a concept that emerged from the observation of Michaelis–Menten kinetics and studies of protein structure. The active site is a pocket or crevice in the three-dimensional structure of the enzyme where the catalytic event occurs. Each enzyme molecule has only a limited number of active sites, usually no more than one per subunit. In the cell, there is usually a much higher concentration of substrate molecules than enzyme molecules and each substrate must wait its turn to bind to an active site and be transformed to product. This accounts for the "saturation effect" seen in the Michaelis–Menten curve. The limited number of enzyme active sites can only make product at a certain rate (V_{max}) and not even a higher substrate concentration will influence the reaction rate.

The active site has special characteristics that are common for most enzymes:

1. *The active site demonstrates specificity; that is, it is able to discriminate among possible substrate molecules.* The active site has a shape that closely mirrors that of the substrate. There is a "close fit" of a correct substrate into the active site. Enzymes show two types of specificity: absolute or group specificity. Many enzymes will accept only one type of molecule, sometimes being able to discriminate between a D or L isomer of a substrate. Other enzymes can accept any number of closely related substances as long as the reactive functional group is present. Recall from Chapter 4 the action of two enzymes, trypsin and carboxypeptidase. Both catalyze identical reactions: the hydrolysis of peptide bonds. Trypsin shows greater specificity by cleaving only those peptide bonds on the C side of lysine or arginine. Carboxypeptidase, however, shows group reactivity by its catalytic removal of almost any amino acid from the C-terminus.

2. *The active site is a relatively small, three-dimensional region within the enzyme.* A bound substrate interacts directly with perhaps only three to five amino acid residues when it is completely in place at the active site. The amino acid residues need not be contiguous in the linear protein chain because three-dimensional folding may bring amino acids from various locations together into the site.

3. *Substrates are initially held in place at the active site by weak, noncovalent, reversible interactions.* The most important interactions are hydrophobic, ionic, and hydrogen bonding. Functional groups on the enzyme come into close contact with the substrate to allow only certain interactions to take place. The interactions are such as to hold the substrate in a proper orientation to the amino acid residues for the most effective catalytic action. Figure 6.9 shows a substrate molecule bound to the active site of a hypothetical enzyme. The substrate is bound in a way to place strain on the bond to be cleaved by the enzyme action (the **scissile bond**). The bond is partially broken in the transition state. Much of the energy for enzyme reactivity comes from the stabilization energy released upon the binding of substrate. Each noncovalent interaction that forms provides energy to stabilize the transition state.

Scientists collecting microorganisms from a hot spring in Yellowstone National Park.

© American Chemical Society

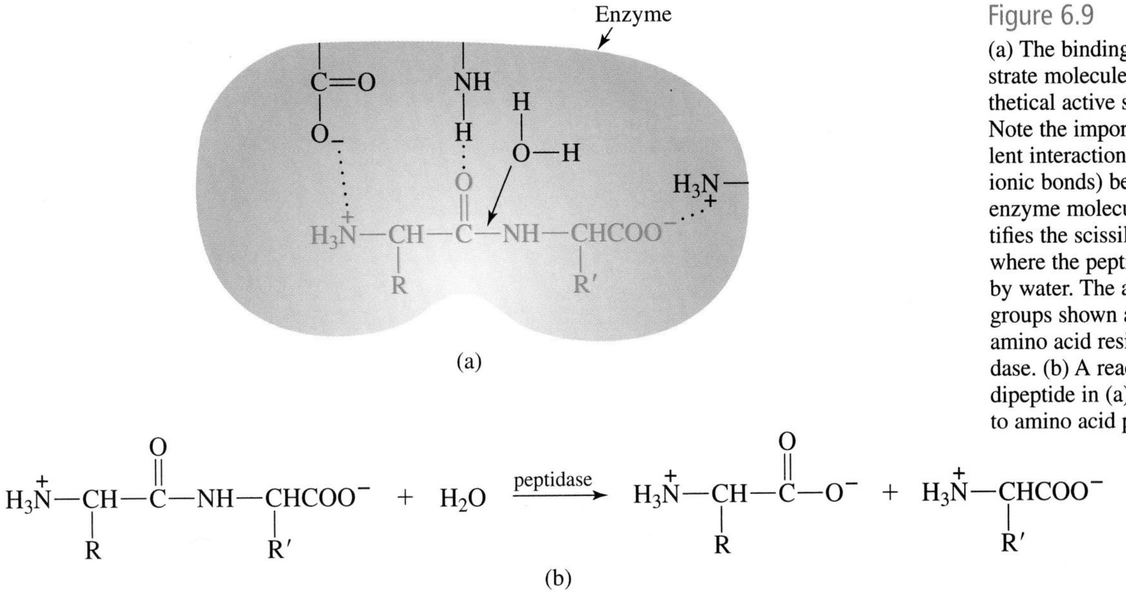

Enzyme

$$C=O \quad NH$$

(a)

Figure 6.9

(a) The binding of a dipeptide substrate molecule (in *red*) to a hypothetical active site in a peptidase. Note the importance of noncovalent interactions (hydrogen bonds, ionic bonds) between substrate and enzyme molecule. The arrow identifies the scissile bond and shows where the peptide bond is attacked by water. The amino and carboxyl groups shown are side chains of amino acid residues of the peptidase. (b) A reaction showing the dipeptide in (a) being hydrolyzed to amino acid products.

$$\overset{+}{H_3N}-\overset{\underset{|}{R}}{CH}-\overset{\overset{O}{\|}}{C}-NH-\overset{\underset{|}{R'}}{CHCOO^-} + H_2O \xrightarrow{\text{peptidase}} \overset{+}{H_3N}-\overset{\underset{|}{R}}{CH}-\overset{\overset{O}{\|}}{C}-O^- + \overset{+}{H_3N}-\overset{\underset{|}{R'}}{CHCOO^-}$$

(b)

The 1890 picture of an ES complex as envisioned by Emil Fischer was that of a **lock and key model** as shown in Figure 6.10. The enzyme active site (lock) is able to accept only a specific type of substrate (key). The lock and key model, however, implies a very rigid, inflexible active site. We now know there is much flexibility in the three-dimensional structure of the enzymes that leads to conformational changes at the active site. A newer idea, the **induced-fit model,** proposed by Daniel Koshland in 1958, assumes continuous changes in active site structure as substrate binds (Figure 6.11). According to Koshland, the active site in the absence of substrate is a rather nondescript region of the enzyme. The process of substrate binding induces specific conformational changes in the protein structure especially in the active site region. The final shape and charge characteristics of the active site are not in place until the substrate is completely bound. X-ray crystallography data of free enzyme molecules and enzyme–substrate complexes now provides evidence for such conformational changes occurring during the formation of the ES complex.

A more sophisticated view of the ES complex continues to evolve. The modern view of the active site shows a region that not only recognizes the substrate but also orients it in such a way as to activate it toward reaction. When the substrate is completely held in the active site, it takes on characteristics of the transition state for the reaction (Figure 6.12). We call

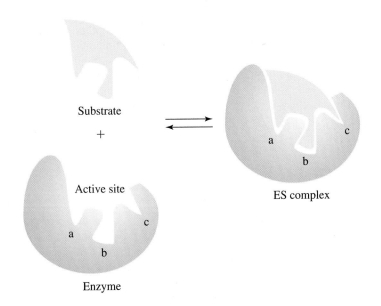

Substrate

+

Active site

Enzyme

ES complex

Figure 6.10

The lock and key model to describe the formation of an ES complex. The substrate has a shape that is complementary or fits into a preformed site on the enzyme. Note that *a, b,* and *c* refer to specific types of interactions that form between substrate and enzyme.

Figure 6.11

The induced-fit model to explain binding of a substrate to an enzyme active site. Initially the enzyme does not have a preformed site for substrate binding. Initial binding of the substrate induces specific conformational changes in the enzyme structure to make it more compatible to the substrate's size, shape, and polarity.

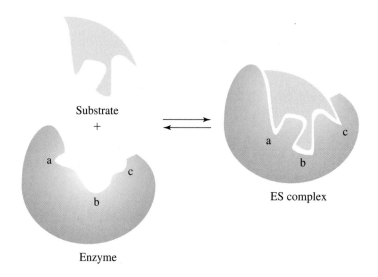

this the **transition-state analog model** for enzyme reactivity. In this picture, we see one of the important functions of the catalyst: to enhance the formation of and stabilize the highly energetic transition state. Binding energy provided by weak, noncovalent interactions in the enzyme–substrate complex accounts for a significant fraction of the large rate accelerations observed for enzyme catalyzed reactions. Once the substrate is bound and its transition state

Figure 6.12

Transition-state analog model of enzyme action. Assume the reaction involves the cleavage of the bond between atoms A and B. When the substrate is bound in the ES complex the bond to be cleaved is already strained and partially broken.

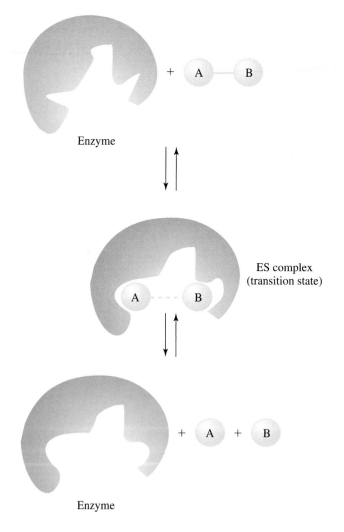

is stabilized, other mechanistic features of the enzyme take over. The most important and best understood mechanistic processes are general acid–base catalysis, metal-ion catalysis, and covalent catalysis.

General Acid–Base Catalysis

Many reactions you studied in organic chemistry including ester and amide hydrolysis involving proton transfers are catalyzed by acids and bases. Figure 6.13 outlines the pathway for general acid catalysis of amide hydrolysis in the absence of enzyme. Enzyme functional groups in the active site region can serve as acids ($—NH_3^+$; $—COOH$) or bases ($—NH_2$; $—COO^-$) to assist in proton transfer reactions and facilitate bond cleavage. The acid and base functional groups of the enzyme are oriented in the active site to interact favorably with the bound substrate. Substrate and proton donors and acceptors are located side by side, therefore facilitating proton transfer. A reaction under these conditions is much more feasible and efficient than relying on random collisions between the substrate and proton donors and acceptors freely diffusing in solution.

Metal-Ion Catalysis

Metal ions associated with the enzyme or substrate molecules often participate in catalysis. Alkali metal ions (Na^+, K^+) and transition metals (Mg^{2+}, Mn^{2+}, Cu^{2+}, Zn^{2+}, Fe^{2+}, Fe^{3+}, Ni^{2+}, and others) assist enzyme reactions in at least three ways:

1. They hold a substrate properly oriented by coordinate covalent bonds; thus, a substrate can be bound to the active site with a very specific geometry (Figure 6.14a).
2. They enhance a reaction by polarizing the scissile bond or by stabilizing a negatively charged intermediate (Figure 6.14b).
3. They participate in biological oxidation–reduction reactions by reversible electron transfer between metal ions and substrate (Figure 6.14c).

Metal ion requirements for an enzyme are usually specific, with little or no enzyme activity observed if the specified metal ion is absent or if another metal ion is substituted for the

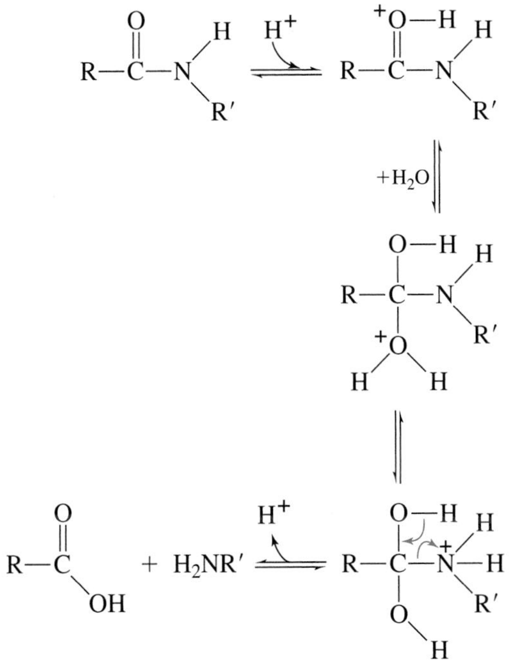

Figure 6.13

The pathway for acid-catalyzed hydrolysis of an amide or peptide bond. In the first step, a proton is added to the carbonyl group. This facilitates attack by water to form a tetrahedral intermediate. In the final step, a base assists by accepting a proton from the intermediate. An enzyme can catalyze such a process by donating a proton in step 1 and accepting a proton in the final step.

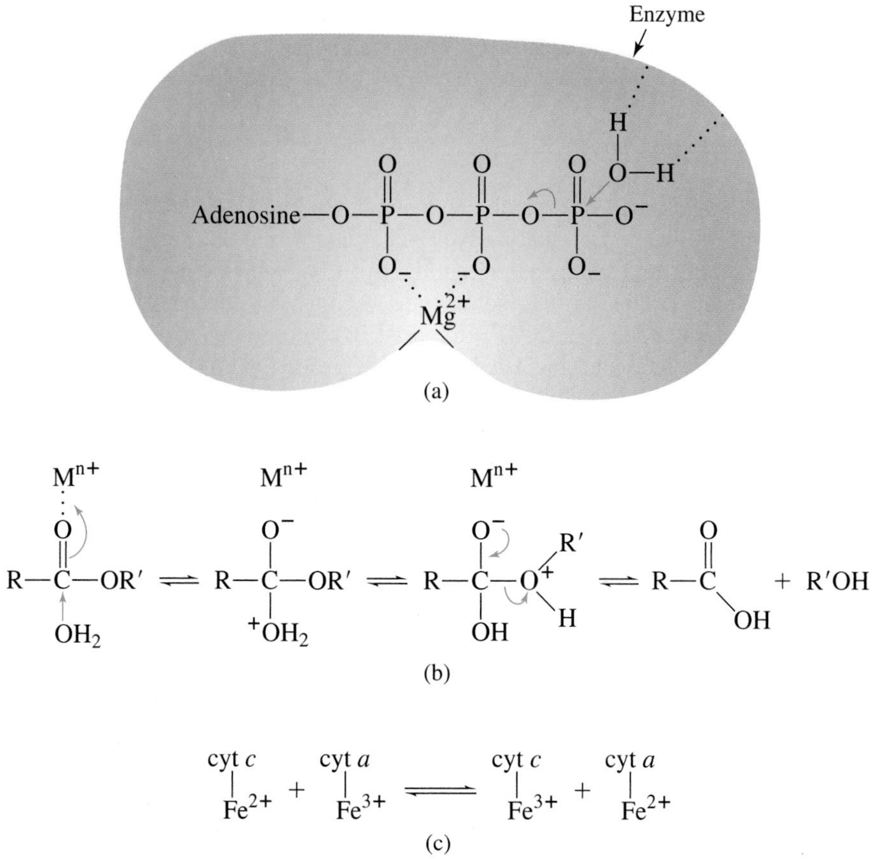

Figure 6.14

Possible roles for metal ions in enzyme-catalyzed reactions. (a) Orienting a substrate into the active site by coordination; the enzyme-bound Mg^{2+} ion assists in holding ATP in the active site. The hypothetical reaction shown is the hydrolysis of ATP to ADP + P_i. (b) Polarizing a bond or stabilizing a negative intermediate. In the reaction of ester hydrolysis, the metal ion (M^{n+}) stabilizes the negative charge on the oxygen anion. (c) Assisting in reversible oxidation–reduction reactions; the iron atoms in the cytochromes promote electron transfer from cytochrome c to cytochrome a. These reactions are important in aerobic metabolism.

(a)

(b)

(c)

natural one. A metal ion is required for about 30% of all known enzymes. Metalloenzymes often combine one or more of the strategies listed above to carry out substrate binding and catalytic action.

Covalent Catalysis

The process of covalent catalysis occurs when a nucleophilic (electron-rich) functional group on an enzyme reacts to form a covalent bond with the substrate. This process leads to a transient intermediate that is especially reactive. Covalent catalysis is observed with the serine proteases, a group of enzymes, including chymotrypsin, that catalyzes cleavage of amide bonds in peptide substrates. Such catalytic processes occur in two steps:

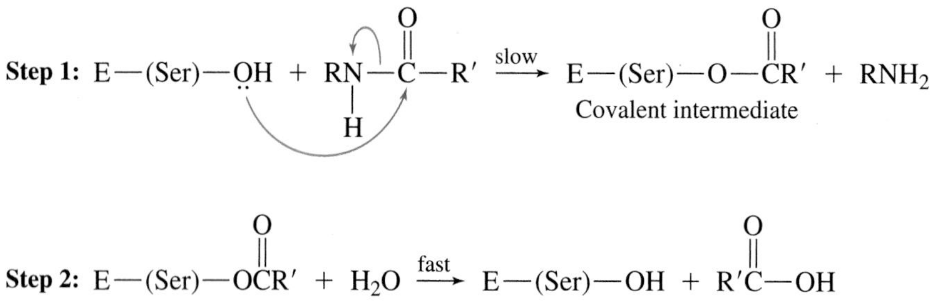

The covalent intermediate involving the enzyme breaks down in step 2. Note that this sequence of reactions does not disobey the principle that catalysts are structurally unchanged because the form of the enzyme is the same at the beginning and the end.

Enzymes use a combination of the different strategies of acid–base catalysis, metal-ion catalysis, and covalent catalysis as described here. As individual enzymes are discussed throughout the book, the mechanisms of catalysis will be described using these processes.

6.4 Enzyme Inhibition

To this point we have only considered the interaction of enzymes with their substrates. In a tissue and cell, an enzyme encounters other chemical agents, including reaction products, substrate analogs, toxins, drugs, metal complexes, and diverse biochemical compounds. Many of these substances will have an inhibitory effect on enzyme action. It may be surprising that biochemists are interested in compounds that lower enzyme reactivity; however, valuable information can be learned from enzyme inhibition studies:

1. Processes for coordinating the thousands of reactions in a cell can be defined because metabolic pathways are regulated by the presence of natural agents that inhibit enzymes.
2. Many drugs and toxins exert their effects through enzyme inhibition.
3. By studying the action of specific inhibitors, enzyme reaction mechanisms, including the role of specific amino acid residues, can be defined.

Reversible and Irreversible Inhibitors

Two broad classes of inhibitors have been identified based on their extent of interaction with enzymes: irreversible and reversible. An **irreversible inhibitor** forms covalent or very strong noncovalent bonds with the enzyme. The site of attack of the inhibitor on the enzyme is at an amino acid functional group that participates in normal substrate binding or catalytic action; therefore, the enzyme is rendered permanently inactive:

$$E—H + R—X \longrightarrow E—R + HX$$

where

E—H = active enzyme
R—X = irreversible inhibitor
E—R = chemically modified enzyme that is catalytically inactive

The action of nerve gases on the enzyme acetylcholinesterase (ACE) provides a good example of irreversible inhibition (Figure 6.15). The compound diisopropylfluorophosphate (DIFP) reacts with the serine residue at the active site of ACE. Since the serine hydroxyl group is essential for the normal reactivity of ACE, the chemically modified form of the enzyme is unable to fulfill its usual role: catalyze the hydrolysis of the neurotransmitter acetylcholine. Individuals exposed to DIFP degrade acetylcholine at nerve junctions at greatly reduced rates,

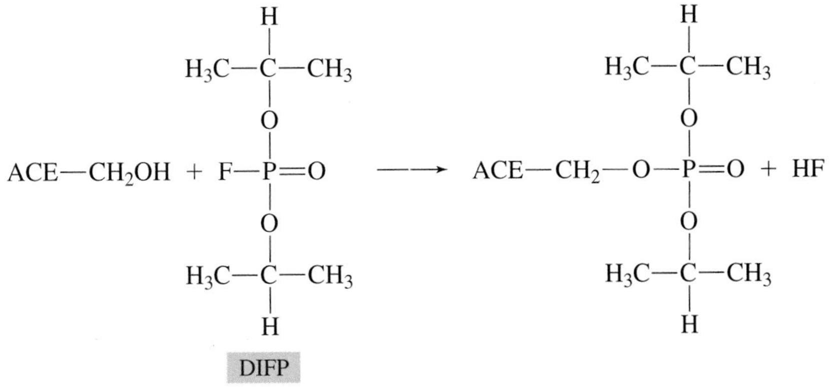

Figure 6.15
An example of irreversible inhibition. Acetylcholinesterase (ACE) is inactivated by diisopropylfluorophosphate (DIFP). A serine residue in ACE, represented as ACE—CH₂OH, reacts to form a covalent bond to DIFP. The hydroxyl group of the serine residue is essential for ACE activity. Therefore, the chemically modified ACE is inactive as an enzyme.

Interactive Animation: Enzyme inhibition

causing continuous nerve impulses to fire. Amino acid residues especially susceptible to chemical modification by irreversible inhibitors are serine, cysteine, and histidine.

Irreversible inhibition can also lead to positive results. The anti-inflammatory, pain-killing drug aspirin (acetylsalicyclic acid) acts by blocking synthesis of pain-producing prostaglandins. Aspirin covalently modifies and thus inactivates prostaglandin synthetase, the enzyme that catalyzes the first step of prostaglandin synthesis (see Chapter 9, Section 9.4).

Reversible inhibitors are those that can readily combine and dissociate with an enzyme and render the enzyme inactive only when bound:

$$E + I \rightleftharpoons EI$$

where

I = inhibitor.

The EI complex is held together by weak, noncovalent interactions similar to the ES complex. Three common types of reversible inhibitors are classified as competitive, uncompetitive, and noncompetitive.

Figure 6.16

Types of reversible inhibition. (a) The ES complex, no inhibitor present. (b) The binding of a competitive inhibitor, I. (c) The binding of a noncompetitive inhibitor. I can bind to E or ES. (d) The binding of an uncompetitive inhibitor to the ES complex.

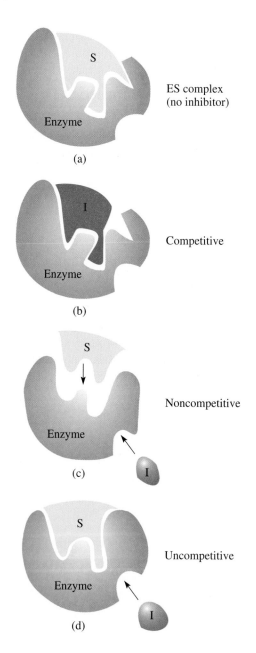

Competitive Inhibition

A **competitive inhibitor** usually resembles the structure of normal substrate and binds to the active site of the enzyme (Figure 6.16b). The binding of substrate and competitive inhibitor to the enzyme active site is a mutually exclusive process: when inhibitor is bound, substrate is unable to bind and vice versa. The kinetic scheme for competitive inhibition is as follows:

$$E + S \rightleftharpoons ES \longrightarrow E + P$$
$$+$$
$$I$$
$$\updownarrow$$
$$EI$$

The quantitative extent of inhibition depends on the concentration ratio of substrate to inhibitor and the binding affinity of each for the enzyme. A very high concentration of substrate lessens the effect of the competitive inhibitor, unless the inhibitor has a much higher binding affinity for the enzyme than does the substrate. The inhibitor molecule cannot be converted to a "product" because it does not have the functional groups normally acted upon by the enzyme. If the inhibitor is capable of binding to the active site, however, it must have some structural features similar to the normal substrate. The inhibition of the citric acid cycle enzyme, succinate dehydrogenase, by malonate, oxalate, and pyrophosphate provides a clear example of the competitive inhibition process (Figure 6.17). These three small molecules have structures and charges similar to the normal substrate succinate, but they cannot be dehydrogenated by the enzyme. These findings have led to the conclusion that the active site of succinate dehydrogenase probably has two appropriately spaced positive charges capable of recognizing and binding the inhibitor molecules and substrate.

The enzyme bisphosphoglycerate mutase provides an example of a competitive inhibition process that regulates a pathway in sugar metabolism (glycolysis). The reaction catalyzed by the enzyme is the isomerization of 1,3-bisphosphoglycerate:

$$
\begin{array}{ccc}
OPO_3^{2-} & & O^- \\
| & & \| \\
\overset{1}{C}=O & & \overset{1}{C}=O \\
| & \rightleftharpoons & | \\
H\overset{2}{C}OH & & H\overset{2}{C}OPO_3^{2-} \\
| & & | \\
H_2\overset{3}{C}OPO_3^{2-} & & H_2\overset{3}{C}OPO_3^{2-}
\end{array}
$$

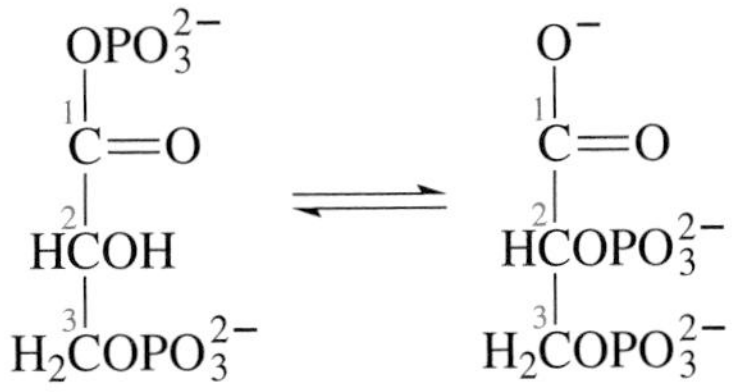

The product 2,3-bisphosphoglycerate is a competitive inhibitor of the enzyme. This process provides the cell with a mechanism for controlling the concentration of 2,3-bisphosphoglycerate.

$$^-OOCCH_2COO^-$$
(a) Malonate

$$^-OOC{-}COO^-$$
(b) Oxalate

$$
\begin{array}{c}
O \quad\quad O \\
\| \quad\quad \| \\
^-O{-}P{-}O{-}P{-}O^- \\
| \quad\quad | \\
\_O \quad\quad O\_
\end{array}
$$
(c) Pyrophosphate

$$
\begin{array}{l}
COO^- \\
| \\
CH_2 \\
| \\
CH_2 \\
| \\
COO\_
\end{array}
\quad + FAD \xrightarrow{\text{succinate dehydrogenase}}
$$
Succinate   (oxidized form)

$$
\begin{array}{c}
H \quad\quad COO^- \\
\diagdown \quad \diagup \\
C \\
\| \\
C \\
\diagup \quad \diagdown \\
^-OOC \quad\quad H
\end{array}
\quad + FADH_2
$$
Fumarate   (reduced form)

(d)

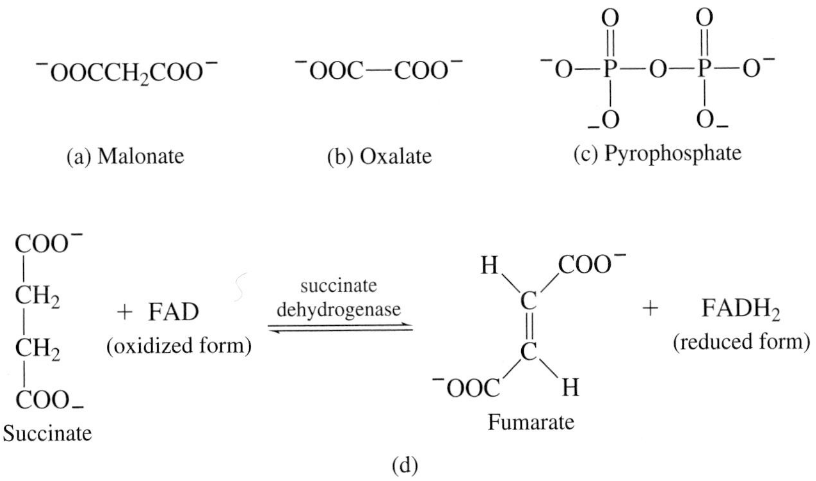

Figure 6.17
The enzyme succinate dehydrogenase is inhibited by substrate analogs: (a) malonate, (b) oxalate, (c) pyrophosphate. The normal substrate, succinate, is shown in (d). The substrate analogs are able to bind to the active site but cannot be dehydrogenated like succinate as shown in the reaction (d). FAD and FADH₂ are the oxidized and reduced form, respectively, of the cofactor, flavin adenine dinucleotide.

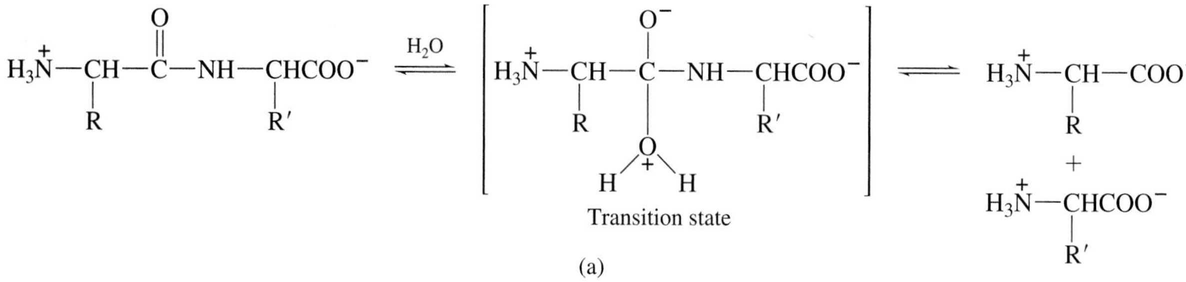

(a)

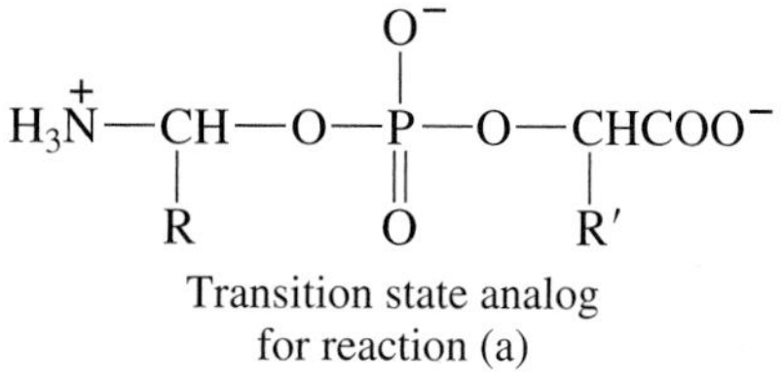

Figure 6.18

Using the principle of transition state analogs to design a competitive inhibitor for a peptidase. (a) The presumed reaction sequence for peptide bond hydrolysis showing the proposed transition state in brackets; this transition state is formed after attack by water on the peptide carbonyl. (b) Shown is a compound that has many characteristics of the presumed transition state, but it does not have an amide bond susceptible to hydrolysis by the enzyme. This compound, a transition state analog, is a possible competitive inhibitor of the hypothetical peptidase.

Transition state analog
for reaction (a)

(b)

Cutting Edge: AIDS therapies

Some of the best competitive inhibitors discovered are **transition state analogs:** those modeled after the structures of substrates in presumed transition states. Recall from our discussion in the previous section that the final binding form of substrate in the ES complex closely resembles the transition state. In fact, the enzyme active site is more complementary to the transition state of the reaction than to the reactants or products. It follows that a molecule designed to resemble the predicted transition state should bind very selectively to the active site and be a potent competitive inhibitor. This approach is now widely used in the design of potential drugs by pharmaceutical companies (see Window on Biochemistry 6.1). Let's look at a hypothetical example to illustrate the use of transition state analogs. Although we do not know with certainty the exact structure of any ES complex, let's assume that the transition state for peptide hydrolysis is similar to the tetrahedral structure shown in brackets in Figure 6.18a. The compound in Figure 6.18b has many characteristics similar to the presumed transition state. It would seem to be a reasonable compound to test as an inhibitor for an enzyme that catalyzes peptide hydrolysis. Note that the proposed competitive inhibitor does not have an amide linkage that would be susceptible to hydrolysis by the enzyme.

Noncompetitive Inhibition

In reversible, **noncompetitive inhibition,** both inhibitor and substrate can bind simultaneously to the enzyme molecule (see Figure 6.16c). Clearly the two molecules must bind to different sites on the enzyme. The presence of inhibitor does not affect substrate binding but does interfere with the catalytic functioning of the enzyme. The actual mechanism of action of the inhibitor varies with each kind of molecule. The kinetic scheme for noncompetitive inhibition is as follows:

A common type of noncompetitive inhibition is the reaction of an enzyme functional group, for example, a sulfhydryl group of a cysteine residue, with a metal ion such as Ag^+, Hg^{2+}, or Pb^{2+} represented by M^{n+}:

$$E—SH + M^{n+} \rightleftharpoons E—S—M^{n+} + H^+$$

The sulfhydryl group in this example is essential for catalytic action of the enzyme but does not participate in substrate binding.

Uncompetitive Inhibition

An **uncompetitive inhibitor** is similar to a noncompetitive inhibitor—it allows substrate to bind to the active site (see Figure 6.16d). It differs, however, because an uncompetitive inhibitor binds only to the ES complex:

$$E + S \rightleftharpoons ES \longrightarrow E + P$$
$$+$$
$$I$$
$$\updownarrow$$
$$ESI$$

Because the inhibitor combines only with ES and not free enzyme, it will influence the activity of the enzyme only when substrate concentrations and, in turn, ES concentrations are high.

The three types of reversible inhibition are distinguishable by completing the proper sets of experiments in the laboratory. Inhibition experiments are performed by using varying levels of substrate (such as experiments designed to determine K_M and V_{max}) and adding a constant amount of enzyme and inhibitor. By analyzing the rate data on a Lineweaver–Burk graph, the inhibitor can be identified as competitive, noncompetitive, or uncompetitive. The family of lines obtained for each type of inhibition is displayed in Figure 6.19. In competitive inhibition, V_{max} is not changed by adding the inhibitor so the lines intersect on the $1/v_0$ axis (Figure 6.19a). In noncompetitive inhibition, the family of lines has a common intercept on the $1/[S]$ axis (unchanged K_M for the lines, Figure 6.19b). For uncompetitive inhibition, both V_{max} and K_M are changed leading to parallel lines (Figure 6.19c).

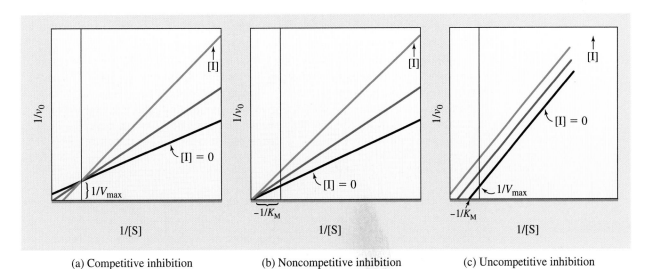

(a) Competitive inhibition   (b) Noncompetitive inhibition   (c) Uncompetitive inhibition

Figure 6.19
Lineweaver–Burk plots used to determine the type of reversible enzyme inhibition. The blue and green lines represent two different inhibitor concentrations. (a) Competitive inhibition: The lines intersect on the $1/v_0$ axis where V_{max} can be calculated. (b) Noncompetitive inhibition: The lines intersect on the $1/[S]$ axis where K_M can be calculated. (c) Uncompetitive inhibition: The lines are parallel. The V_{max} and K_M can be calculated for each line as shown.

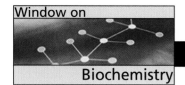

6.1   Protease Inhibitors—A Tale of Two Diseases

Alzheimer's disease (AD) and acquired immunodeficiency syndrome (AIDS); could any two afflictions be more different? AD, a neurological disease characterized by the presence of amyloid protein plaques deposited in the brain, is most often an affliction of the elderly. Although there is some evidence for a genetic link, the cause of AD is currently unknown, but probably not viral. AIDS, a disease that weakens and eventually destroys the immune system, is caused by the human immunodeficiency virus (HIV) and affects mainly the young. Although the differences between the two diseases are stark, an identical approach toward treatment is being used. In both cases, drugs called **protease inhibitors** are being designed, synthesized, and clinically tested.

AD researchers have recently identified the enzyme that makes the final cuts on a protein called amyloid precursor protein (APP). This protease, called **γ-secretase**, generates the Alzheimer peptides, amyloid β-peptides (Aβ), which aggregate to form the telltale plaques in the brains of AD patients. Aβ peptide molecules are produced in two sizes, either 40- or 42-amino-acid residues:

$$APP \xrightarrow[\text{H}_2\text{O}]{\gamma\text{-secretase}} A\beta\text{-40} + A\beta\text{-42}$$

Both sizes are present under normal conditions, but in AD, the Aβ-42 that forms the fibrillar deposits is favored. The protease γ-secretase is an attractive drug target because it generates Aβ peptides. Inhibitors of the enzyme could slow production of the Aβ peptides and potentially be used in treating AD. One of the first compounds tested was L-685,458 (see figure), designed by Merck Research Labs using the transition state analog approach. This approach works on the following premise: Because the enzyme's active site binds substrates by stabilizing the transition state of the reaction, compounds mimicking the transition state, but without reactive functional groups, should fit into the active site and thus act as inhibitors. Indeed, the Merck compound was found to be a potent inhibitor of γ-secretase. Interestingly, the compound was originally designed as a possible treatment for AIDS, but it was found to be ineffective (see below).

The potential for protease inhibition in the treatment of AIDS was first recognized when an important viral enzyme, **HIV protease**, was identified and characterized. The enzyme was found to have a major involvement in viral growth and development. Most proteins necessary for HIV replication are initially synthesized as part of a larger protein called the gag-pol protein. Before the formation of new infectious virions, HIV protease cleaves the gag-pol into smaller polypeptide fragments that form part of the core for new viral particles. Compounds that inhibit the protease should block the assembly of new viral particles. HIV protease, a homotypic dimer, catalyzes the hydrolytic cleavage of several peptide bonds, but prefer-

L-685,458

entially between the amino acids phenylalanine—proline and tyrosine—proline (see natural substrate). Two important HIV protease inhibitors, which are components of the new "triple-combination treatments" now being prescribed for AIDS patients, are also shown. The drugs include **Saquinavir,** developed by Hoffman-LaRoche, and Merck's **Crixivan,** both of which were designed by transition-state analog approaches. Note the similarities (and differences) between the natural substrate (—phe—pro—) and the protease inhibitors. The Merck drug mentioned above, L-685,458, did not inhibit the HIV protease.

It is, of course, beneficial in research that potential treatments for two diseases can use similar approaches as described here. However, it is important to emphasize that even enzymes that catalyze the same kind of reaction do not respond to the same inhibitors. Most likely there will never be found a compound that inhibits both γ-secretase and HIV protease because the enzymes have dissimilar active sites, but we can learn much from overlapping studies. The use of protease inhibitors is a relatively new approach in disease treatment. There is future potential that they will become important in the treatment of other ailments including heart attacks, strokes, influenza, the common cold, rheumatoid arthritis, asthma, and cancer.

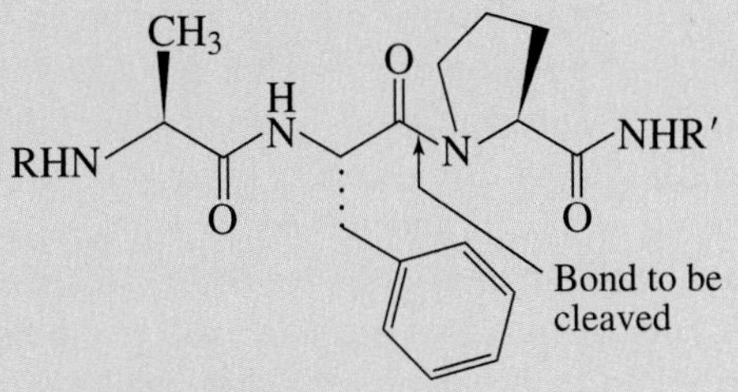

Bond to be cleaved

—Phe—Pro—
HIV protease substrate

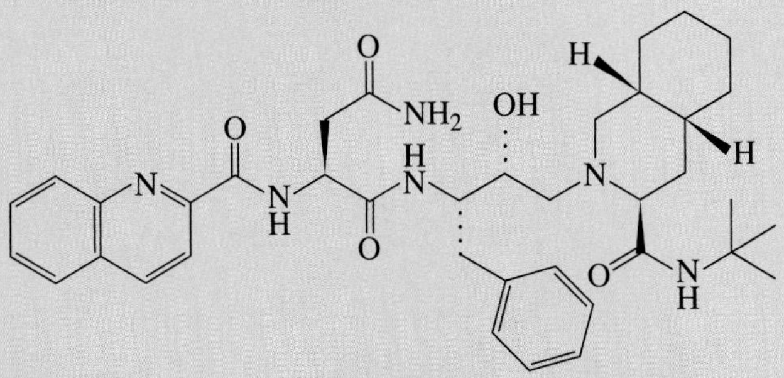

Saquinavir

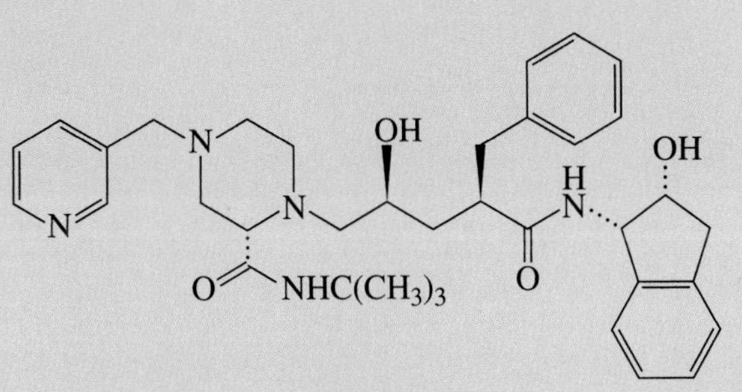

Crixivan

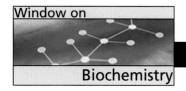

6.2 Practical Catalysts—A Grab Bag of Enzymes

The action of enzymes is not limited to the catalysis of metabolic reactions within biological cells. The specific and robust characteristics of enzymes make them ideal agents for use in medicine, food processing, industrial processes, chemical waste cleanup, and even in the manufacturing of clothing.

The biotechnology industry has developed several enzymes for therapeutic applications. The enzyme **deoxyribonuclease (DNase)** has recently been approved for treatment of cystic fibrosis (CF). A major clinical problem in CF patients is the production in the lungs and airways of large volumes of highly viscous mucus. This often results in respiratory failure, which is the most common killer in CF patients. The mucus contains (a) bacterial organisms, which cause chronic lung infections, and (b) numerous white blood cells to fight the infection. The blood cells and bacteria die, releasing their DNA that greatly increases the viscosity of the mucus. Human DNase is being produced by recombinant methods by the Genentech Company under the brand name **Pulmozyme,** which is administered by inhalation of an aerosol mist. The enzyme catalyzes the hydrolytic cleavage of large DNA molecules to produce smaller nucleotide fragments, thus reducing the viscosity of DNA solutions. The enzyme may also be effective in thinning infected mucus in patients of pneumonia, bronchitis, and emphysema.

Because foods contain natural substrates for enzymes, it is logical to consider the use of enzymes in the processing of agricultural products. The starch-digesting enzyme, **amylase,** is currently used in preparation of low-calorie beers and the clarification of wines and fruit juices. The turbidity present in natural beverages is due primarily to the presence of starch and cellulose molecules too large to be completely water soluble. Added amylase catalyzes the hydrolysis of starch to produce small polysaccharides and glucose that are very water soluble. The enzyme **cellulase** is also added to juices to break down water-insoluble cellulose to glucose.

A cystic fibrosis patient inhaling Pulmozyme®, a form of recombinant deoxyribonuclease.

© Mark Tuschman

Enzymes are also used in the preparation of agricultural feed. The newest and most promising additive is **phytase,** an enzyme that catalyzes the hydrolytic release of phosphorus bound in phytic acid (see structure), the primary storage form of phosphorus in plant seeds and grains. Less inorganic phosphorus, an essential nutrient, needs to be added to the feed because the animals can absorb the phosphorus released from phytic acid. Use of the enzyme in feed reduces the amount of phosphorus in poultry and hog waste by over 30%. Toxin producing microorganisms such as *Pfiesteria* thrive in water contaminated by phosphorus-containing runoff from agricultural waste.

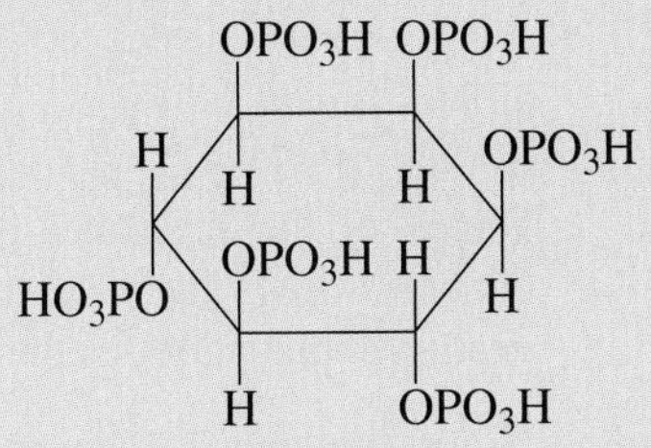

$$OPO_3H \quad OPO_3H$$
H H H OPO$_3$H
OPO$_3$H H
HO$_3$PO H
H OPO$_3$H

Phytic Acid

The bacterial proteinase **subtilisin** is a common additive in household detergents. It acts to remove protein-based stains such as grass, blood, mucus, feces, and foods like egg and gravy. These protein stains are water-insoluble and adhere to the surface of textiles. Subtilisin catalyzes the hydrolysis of protein in the stains to produce smaller peptides that are readily dissolved in the wash water.

A new approach to the treatment of hazardous waste is a process called **bioremediation**—the use of biological organisms and enzymes to clean up oil spills, chemical dumps, and contaminated soils. Recent microbiological searches have identified microorganisms that are able to degrade toxic chemicals such as polychlorinated biphenyls (PCBs), petroleum-based oils, and other wastes. One specific organism that is receiving much attention is the white rot fungus, *Phanerochaete chrysoporium.* This species, which belongs to a family of wood-rotting fungi common in forests, secretes enzymes that degrade lignin in wood to carbon dioxide and water. However, the enzymes are not very selective. When the organisms feed on wood products, their enzymes, especially **oxygenases,** also degrade chemical wastes. These enzymes utilize oxygen as a cosubstrate to degrade xenobiotics (unnatural synthetic chemicals).

Enzymes are being used to make newly purchased jeans have the desired "you've worn them for the last ten years" look. The denim for "stone-washed jeans" was originally prepared by treating the material with the natural mineral pumice. The abrasive action of the pumice wore off the top layer of the denim, releasing indigo dye and producing a used look. Because of environmental concerns about the mining and manufacturing of pumice, jean makers turned to biotech. Denim, a cotton fabric, is now briefly treated with the enzyme **cellulase,** which catalyzes the hydrolysis of surface cellulose in the cotton. This loosens the top layer of the cloth, releasing indigo dye in the process and making the jeans look worn.

Summary

■ **Enzymes** are biomolecules that catalyze and regulate the thousands of chemical reactions occurring in organisms. Most enzymes are proteins, but some forms of RNA display catalytic activity. Enzymes function as catalysts by lowering the activation energy barrier for a reaction.

■ Some enzymes require, in addition to the protein component, another chemical entity called the **cofactor,** which may be an organic molecule or a metal ion.

■ The kinetics for an enzyme-catalyzed reaction are explained by the Michaelis–Menten equation, which defines the Michaelis constant, K_M, and maximum velocity, V_{max}.

■ Enzymatic catalysis begins with the substrate molecule binding to a specific region on the enzyme called the **active site.** Substrates are held by noncovalent interactions. Enzymes display saturation kinetics because there is a limited number of active sites.

■ Enzyme action may be inhibited by the presence of other chemical agents that bind to the enzyme molecule. Reversible inhibition is classified as competitive, noncompetitive, or uncompetitive.

Study Exercises

Understanding Terms

6.1 In your own 25 words or less, define the following terms.

a. Enzyme
b. Activation energy
c. Prosthetic group
d. Apoenzyme
e. K_M
f. Turnover number
g. Active site
h. Induced-fit model
i. Irreversible inhibitor
j. Competitive inhibitor
k. Lineweaver–Burk graph
l. Substrate saturation
m. Lock and key model
n. Holoenzyme
o. Oxidoreductase
p. Initial velocity
q. Michaelis–Menten graph
r. V_{max}
s. Acid–base catalysis
t. Noncompetitive inhibitor
u. Uncompetitive inhibitor
v. Reversible inhibition

Reviewing Concepts

6.2 In an experiment completed in biochemistry laboratory, you found that a 10 μM solution of the enzyme acetylcholinesterase catalyzed the breakdown of 0.5 M acetylcholine in 1 min of reaction time. Calculate the turnover number for acetylcholinesterase in seconds^{-1}.

6.3 While completing the experiment with acetylcholinesterase in Problem 6.2, you also measured the K_M values for two substrates, acetylcholine and butyrylcholine.

$$\begin{array}{c} O \\ \parallel \end{array}$$
$$CH_3C-O-CH_2CH_2\overset{+}{N}(CH_3)_3$$
Acetylcholine
$(K_M = 1 \times 10^{-6}\ M)$

$$\begin{array}{c} O \\ \parallel \end{array}$$
$$CH_3CH_2CH_2C-O-CH_2CH_2\overset{+}{N}(CH_3)_3$$
Butyrylcholine
$(K_M = 2 \times 10^{-4}\ M)$

Which of the substrates has a greater affinity for the enzyme? Study the structure of each substrate and suggest an explanation for the different K_M values.
HINT: $k_2 \gg k_3$

6.4 In Figure 6.9 the dotted lines represent noncovalent interactions between substrate and amino acids at the active site of an enzyme.

a. Identify each type of noncovalent interaction. Choose from hydrogen bonding, ionic bonding, and hydrophobic interactions.
b. Name possible amino acids in the active site of the enzyme.

6.5 Describe the differences among the actions of the three types of reversible inhibitors. Begin by identifying where each type of inhibitor binds to an enzyme. Your answer should include a comparison of the chemical structure of each type of inhibitor to the natural substrate and a discussion of the effect of each inhibitor on the V_{max} and K_M for the reaction.

6.6 Using words or structures, write the reaction catalyzed by each of the following enzymes.

a. Catalase
b. Cellulase
c. Trypsin
d. Acetylcholinesterase
e. Amylase

6.7 Which of the following factors will influence the reaction rate for a typical enzyme? Will the change increase, decrease, or have no effect on the reaction rate?

a. Increase substrate concentration
b. Increase the temperature from 25°C to 37°C
c. Add a competitive inhibitor
d. Increase the pressure from 1 atm to 1.5 atm
e. Change the pH from 7 to 1
f. Increase the temperature from 37°C to 150°C
g. Increase the enzyme concentration

6.8 Which of the following are not general characteristics of enzymes?

    a. All enzymes are proteins.

    b. Enzymes increase reaction rates by lowering the activation energy barrier.

    c. ES complexes are held together by noncovalent interactions (H bonds, hydrophobic interactions, ionic bonds).

    d. Enzymes usually are substrate specific.

    e. The catalytic activity of some enzymes can be regulated.

    f. All enzymes show allosteric properties.

    g. ES complexes are often held together by disulfide bonds, $-S-S-$.

6.9 Describe what kind of enzymes are added to clothes washing powders in order to help dissolve biological stains such as milk and blood.

6.10 Enzymes often use electron-rich, nucleophilic functional groups in their actions. For example, in the process of covalent catalysis, nucleophilic amino acid side chains initiate attack on the substrate. What amino acid side chains are considered nucleophilic?

6.11 Study each of the biochemical reactions below and determine the common name for an enzyme that might catalyze the reaction. Select and match enzyme names from the list below.

Reactions

    a. $CH_3CH_2OCCH_3 + H_2O \rightleftharpoons$
                $\|$
                O

                $CH_3CH_2OH + CH_3COOH$

    b. Ala-Ser + $H_2O \rightleftharpoons$ Ala + Ser
    c. Cellulose + $H_2O \rightleftharpoons$ glucose
    d. Phenylalanine + $O_2 \rightleftharpoons$ tyrosine
    e. Starch + $H_2O \rightleftharpoons$ glucose
    f. $CH_3CH_2OH + NAD^+ \rightleftharpoons CH_3CHO + NADH + H^+$
    g. Phenylalanine $\rightarrow$ 2-phenethylamine + CO_2

Enzyme Names

    1. Amylase
    2. Esterase
    3. Alcohol dehydrogenase
    4. Peptidase
    5. Nuclease
    6. Cellulose hydrolase or cellulase
    7. Phenylalanine hydroxylase
    8. Phenylalanine decarboxylase

6.12 Derive the Lineweaver–Burk equation beginning with the Michaelis–Menten equation.

HINT: Take the reciprocal of the Michaelis–Menten equation and solve for $1/v_0$.

6.13 From studies of the enzyme linkase (Problem 6.20), it has been shown that a positively charged amino acid residue is present at the active site. Linkase functions best at a pH of 7. What amino acid side chains could provide the necessary positive charge?

Solving Problems

6.14 Below are data compiled from kinetic analysis of an enzyme isolated from blueberry plant leaves. The enzyme catalyzes the hydrolysis of esters such as methyl benzoate. Use the Michaelis–Menten and Lineweaver–Burk graphs to determine K_M and V_{max}. Be sure to include the proper units for each constant. Compare your results for each graphical method. The data were obtained by measuring the rate of disappearance of substrate.

[Methyl Benzoate] (μmol/L)	v_0 (nmol/min)
3.7	10
13.0	30
39.0	63
79.0	87
230	116
400	122

6.15 Acetophenone was tested as an inhibitor of the esterase in Problem 6.14. Use the rate data in Problem 6.14 and the following inhibition data to evaluate the effect of acetophenone. Is it a competitive, noncompetitive, or uncompetitive inhibitor? Use a Lineweaver–Burk graph to analyze the data. Rate data for two concentrations of the inhibitor are given.

[Acetophenone] $= 3.7 \times 10^{-4} M$		[Acetophenone] $= 1.58 \times 10^{-3} M$
v_0 (nmol/min)	[S] (μmol/L)	v_0 (nmol/min)
5	3.7	—
20	13.0	—
45	39.0	20
70	79.0	40
102	230	80
122	400	110

6.16 Papain, a protease from the papaya plant, contains an essential cysteine residue at its active site. The compound iodoacetamide has been shown to act as an irreversible inhibitor for the enzyme. Complete the chemical reaction below to describe the action of iodoacetamide on the enzyme.

                            O
                            $\|$
    $E-CH_2-SH + ICH_2CNH_2 \rightarrow$

6.17 Ethyl alcohol is sometimes used in hospital emergency rooms to treat patients who have ingested radiator or gas-line antifreeze. A major chemical component of these products is methanol. The methanol itself is not especially harmful; however, it is transformed by enzymes called dehydrogenases to the very toxic substances, formaldehyde and formic acid. Can you explain the biochemical principles behind this medical treatment?

6.18 Study Figure 6.7a and b and estimate the pH values at which trypsin and pepsin are the most active. Can you predict which of the enzymes is present in the stomach?

6.19 When chemists in the laboratory want to make reactions occur faster they often use heat. To what extent could living organisms use this method to speed metabolic reactions?

6.20 An enzyme called linkase catalyzes the following reaction:

$$A + B \rightleftharpoons A-B$$

Suggest laboratory measurements that could be used to measure the kinetics of the reaction.

6.21 Study the Michaelis–Menten graph below and identify the curve that corresponds to each reaction condition listed below.
  a. No inhibitor
  b. Competitive inhibitor
  c. Noncompetitive inhibitor

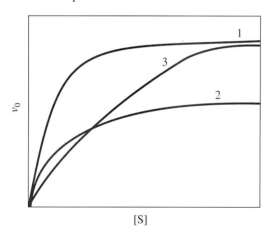

6.22 Which statement about a transition state analog is false?
  a. It usually is a strong inhibitor of the enzyme.
  b. It fits better into the active site than the substrate.
  c. It is a stable molecule and has a structure similar to the presumed transition state.
  d. It would probably function as a noncompetitive inhibitor of the enzyme.

6.23 Assume that an enzyme has the following kinetic constants:

$V_{max} = 50$ µmol substrate transformed per minute per milligram of enzyme
$K_M = 0.001\ M$

Calculate the substrate concentration that will yield a reaction rate of
  a. $\frac{1}{2}\ V_{max}$
  b. $\frac{1}{3}\ V_{max}$

6.24 Describe the biochemical principles behind the use of the peptidases pepsin and papain as meat tenderizers.

6.25 The Environmental Protection Agency (EPA) warns us to avoid the ingestion of materials containing lead. Lead may be present in drinking water and in chips of old paint. What specific and detrimental biochemical actions is lead capable of?

6.26 The structure drawn below is that of the artificial sweetener, aspartame. Predict the reactions involved in its metabolism and give common names for enzymes that may catalyze the reactions.

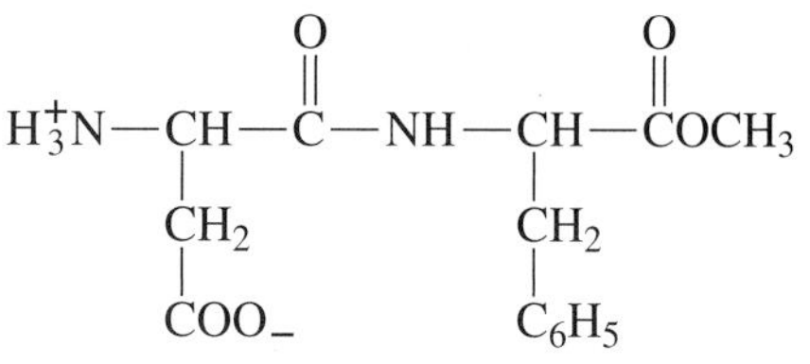

$$\overset{O}{\overset{\|}{H_3^+N-CH-C-NH-CH-\overset{O}{\overset{\|}{C}}OCH_3}}$$
$$\underset{CH_2}{\quad\quad}\quad\quad\underset{CH_2}{\quad}$$
$$\underset{COO^-}{\quad\quad}\quad\quad\underset{C_6H_5}{\quad}$$

6.27 Study the reaction below that is catalyzed by an esterase.

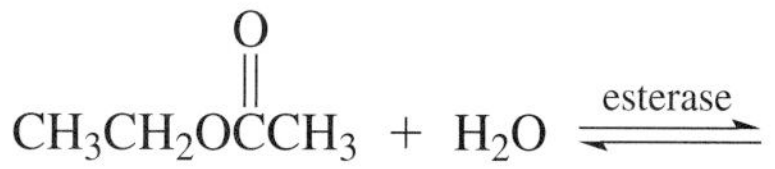

$$CH_3CH_2O\overset{O}{\overset{\|}{C}}CH_3 + H_2O \overset{esterase}{\rightleftharpoons}$$

  a. Complete the reaction by drawing structural formulas for the products.
  b. Draw the structures of two compounds that may serve as competitive inhibitors of the enzyme. What principles did you use to determine these structures?
  c. Draw the structure for a compound that may serve as an irreversible inhibitor of the enzyme. Assume there is an essential serine residue at the active site of the enzyme.

6.28 Complete the following reactions by drawing structures of all organic products. Indicate no reaction by N.R.

  a.
$$C_6H_5\overset{O}{\overset{\|}{C}}OCH_3 + H_2O \overset{esterase}{\rightleftharpoons}$$

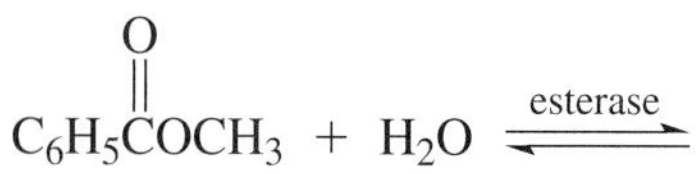

  b.
$$C_6H_5\overset{O}{\overset{\|}{C}}CH_3 + H_2O \overset{esterase}{\rightleftharpoons}$$

  c.
$$CH_3CH_2\overset{O}{\overset{\|}{C}}-N\overset{CH_3}{\underset{CH_3}{\big<}} + H_2O \overset{peptidase}{\longrightarrow}$$

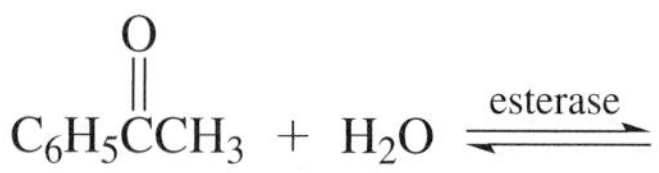

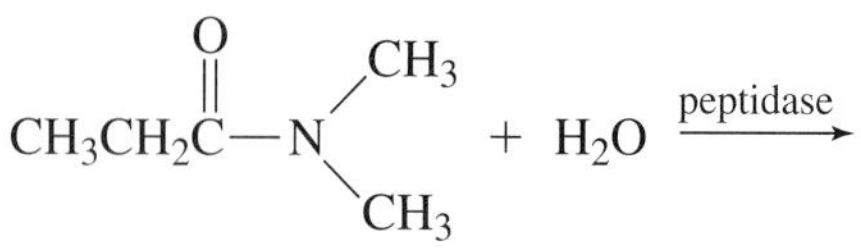

d. $H_3^+N—CHCOO^-$ $\xrightleftharpoons{\text{decarboxylase}}$
     |
     CH_3

e. Phe-Gly + H_2O $\xrightleftharpoons{\text{peptidase}}$

6.29 Kinetic constants for inhibition studies on enzymes are described below. For each part a–c, determine if the numerical values for the constants are identical or different. The first one is worked as an example.

    a. K_M values in the absence and presence of a competitive inhibitor.

    Answer: The K_M values are different.

    b. K_M values in the absence and presence of a noncompetitive inhibitor.

    c. K_M values in the absence and presence of an uncompetitive inhibitor.

6.30 Replace K_M in Problem 6.29 with V_{max} and answer questions a, b, and c again.

6.31 Consider the following sequence of reactions outlining how cholesterol is made in animals.

acetyl CoA $\longrightarrow$ HMGCoA $\longrightarrow$ mevalonate $\longrightarrow$
                                     squalene $\longrightarrow$ cholesterol

The drug lovastatin is often prescribed to individuals diagnosed with hypercholesterolemia and atherosclerosis. Lovastatin is a competitive inhibitor of the enzyme HMGCoA reductase, which catalyzes the reaction HMGCoA $\longrightarrow$ mevalonate. Describe the biochemical effect of this drug and explain why it may help patients with these diseases.

6.32 In a laboratory experiment you completed a study of enzyme kinetics. The following data were collected:

[S] (μmolar)	v_0 (μmolar/min)
50	10
100	19
150	31
200	38
300	55
400	62
800	68
1000	70

Estimate the K_M for this substrate:enzyme combination without graphing the data.

HINT: What is the definition of K_M?

Writing Biochemistry

6.33 Search for an article on the topic of enzymes in a recent science journal. An article from one of the following journals is acceptable: *Scientific American, Science, Nature, Biochemistry and Molecular Biology Education,* or *Journal of Chemical Education.* Write a 200-word summary of the article emphasizing concepts in this chapter.

6.34 Enzymes are becoming so common in our lives that you can often find articles about them in the everyday press. Search for an article on enzymes in a local or national newspaper or a newsmagazine such as *Time* or *Newsweek.* Write a brief (about 100-word) summary of the article. Emphasize biochemical concepts you learned in this chapter.

 ## Applying Biochemistry

6.35 In the discussion on enzyme catalysis in the book and in the accompanying CD/ROM unit, there were several uses of fundamental terms you first learned in your Introductory and Organic Chemistry classes. Briefly explain, in your own words, the meaning of each of the terms below and apply them to enzyme catalysis: Gibbs free energy, reaction coordinate, activation energy, transition state, progress of the reaction, catalyst.

6.36 The concept of "induced fit" is very important in describing enzyme–substrate interactions. Define the term *induced fit,* and explain why this model for substrate binding is better than the older idea of "lock and key."

Further Reading

Arnold, F. et al., 2001. How enzymes adapt: lessons from directed evolution. *Trends Biochem. Sci.* 26:100–106.

Bauhs, D., 1999. The enzyme's party. *Biochem. Educ.* 27(3):135–136.

Boyer, R., 2000. Kinetic properties of enzymes. *Modern Experimental Biochemistry,* 3rd ed., pp. 279–301. San Francisco: Benjamin/Cummings.

Bugg, C., Carson, W., and Montgomery, J. 1993. Drugs by design. *Sci Am.* 269(6):92–98.

Collins, F., 1992. Cystic fibrosis: Molecular biology and therapeutic implications. *Science* 256:774–779.

DeCastro, I., and Alonso, F., 1997. Energy diagrams for enzyme catalyzed reactions. *Biochem. Educ.* 25(2):87–89.

Flam, F., 1994. The chemistry of life at the margins. *Science* 265:471–472.

Fraaije, M., and Mattevi, A., 2000. Flavoenzymes: Diverse catalysts with recurrent features. *Trends Biochem. Sci.* 25(3):126–132.

Garcia-Carreno, F., and Del Toro, M., 1997. Classification of proteases without tears. *Biochem. Educ.* 25(3):161–167.

Johnson, L., and Petsko, G., 1999. David Phillips and the origin of structural enzymology. *Trends Biochem. Sci.* 24(7):287–289.

Price, D., 2001. β-Secretase. *Trends Biochem. Sci.* 26:14.

Ranaldi, F., Vanni, P., and Giachetti, E., 1999. What students must know about the determination of enzyme kinetic parameters. *Biochem. Educ.* 27(2):87–91.

Sutcliffe, M., and Scrutton, N., 2000. Enzyme catalysis: Over the barrier or through the barrier. *Trends Biochem. Sci.* 25:405–408.

Van Wyk, J., Mogale, M., and Moroka, K., 1999. Bioconversion of waste paper materials to sugars. *Biochem. Educ.* 27(4):227–228.

Whiteley, C., 1997. Enzyme kinetics. *Biochem. Educ.* 25(3):144–146.

Whiteley, C., 1999. Enzyme kinetics—Inhibition. *Biochem. Educ.* 27(1):15–18.

Whiteley, C., 2000. Enzyme kinetics—Inhibition. *Biochem. Educ.* 28(3):144–147.

Chapter 7

Enzymes II: Coenzymes, Regulation, Abzymes, and Ribozymes

n the previous chapter, some of the basic principles of enzyme action were introduced including active sites, Michaelis–Menten kinetics, mechanism, and inhibition. Other than the influence of substrates, inhibitors, and some physical changes (pH, temperature), it might be assumed that enzymes act rather independently of their chemical and physical surroundings. In reality, as we discover in this chapter, the catalytic activity of enzymes is precisely regulated by cellular conditions to assure that the thousands of metabolic reactions are properly synchronized. Several modes of regulation are observed for enzymes, including allosteric interactions, covalent modification, isoenzymes, and proteolytic activation. In addition, the presence of coenzymes and essential metal ions controls the reactivity of some enzymes. The intricate roles played by some of these biomolecules and regulatory processes are also discussed in this chapter.

Some of the most recent advances in biological catalysis involves the design of custom-made protein enzymes and catalytic antibodies and the discovery of nonprotein catalysts, the ribozymes (catalytic RNA). The continued development of these nontraditional biological catalysts will add to our repertoire of available reactivities for practical applications in pharmaceuticals, the chemical industry, biotechnology, and agriculture.

7.1 Enzyme:Coenzyme Partners

Vitamins and Coenzymes

Enzymes as protein molecules possess complex primary, secondary, tertiary, and perhaps quaternary structure as discussed in Chapters 4 and 5. In addition, a variety of reactive amino acid side chains is present in enzyme active sites to provide many kinds of catalytic reactivities. For some enzymes however, the protein component is not sufficient for complete reactivity. Certain proteins require the aid of another molecular species to complete their role as an enzyme. The assisting species, which may be organic, organometallic, or a metal ion, is usually a molecule smaller than the protein. As we learned in Chapter 6, the term **coenzyme** is used to define an organic or organometallic molecule that assists an enzyme. During its participation in a reaction process, a coenzyme may be either very tightly or weakly bound to the enzyme. The term **prosthetic group** is used for those coenzymes that are covalently linked or noncovalently bound very tightly to their enzyme partners and, therefore, are always present. Other coenzymes are only temporarily associated with enzymes during the reaction process. They are held into the active site by weak, noncovalent interactions just like substrate molecules. The names of some of the most important coenzymes and prosthetic groups are shown in Table 7.1. Each type of coenzyme, in conjunction with the proper enzyme, has a particular kind of function that it can perform. For example, the coenzyme nicotinamide adenine dinucleotide (NAD^+) linked with a dehydrogenase enzyme, can catalyze special kinds of oxidation–reduction reactions. In fact, almost all enzymes involved in oxidation–reduction reactions require an organic or organometallic coenzyme or a metal ion to assist in electron transfer. Table 7.1 describes the specific function for each coenzyme. Further examples describing the actions of the coenzymes are referenced in the table.

Many years ago during studies on the origin of coenzymes in the cell, it was discovered that most of them were structurally related to the **vitamins,** a group of relatively small, organic molecules essential in small amounts in the diet for proper growth and development. Humans and many other animals require the vitamins in their diets because they have lost their ability to synthesize them in sufficient quantities. The coenzymes of Table 7.1 can be synthesized by higher animals only when the vitamins are present in the diet. Part of each coenzyme structure is contributed by a vitamin. For example, the vitamin niacin is a building block for the synthesis of NAD^+ (Table 7.1).

The vitamins can be divided into two broad classes, water-soluble vitamins and fat-soluble vitamins. The structures, related coenzymes, and deficiency diseases for the vitamins are outlined in Table 7.1. A deficiency of a dietary vitamin usually results in a specific disease that has been identified for each. The clinical links between the deficient coenzyme and symptoms of diseases are poorly understood. For most vitamins, only very

Dorothy Hodgkin is honored in this British stamp for her X-ray diffraction work on vitamin B_{12}.

(Text continues on p. 161)

Table 7.1
Characteristics of vitamins and coenzymes

Name/Structure of Vitamin	Related Coenzyme	Reaction type (page numbers)[a]	Deficiency Disease
Water-Soluble Vitamins			
Niacin (nicotinamide) 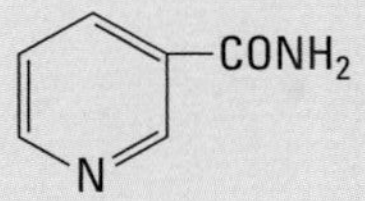	NAD^+, $NADP^+$	Oxidation–reduction (pp. 429–431, 443, 451–456)	Pellagra
Riboflavin (vitamin B_2) 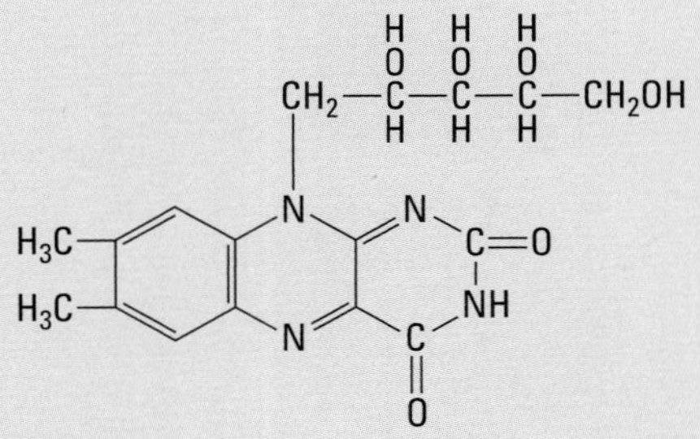	FAD, FMN	Oxidation–reduction (pp. 429–431, 451–456)	Growth retardation
Thiamine (vitamin B_1) 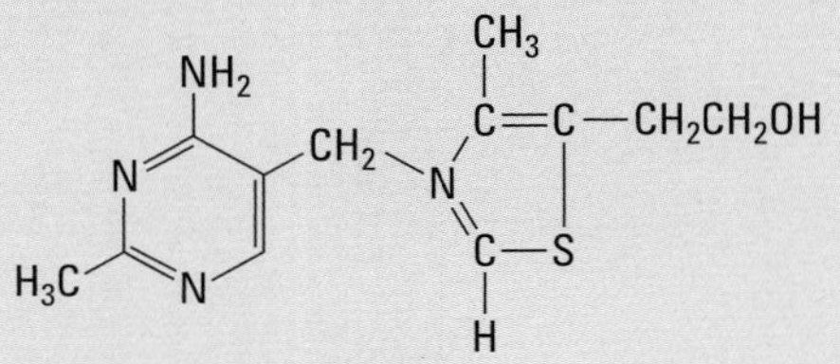	Thiamine pyrophosphate	Decarboxylation (pp. 405, 427–428, 431)	Beriberi
Pantothenic acid 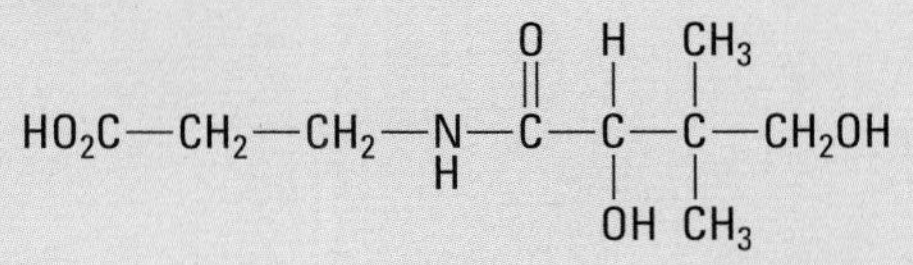	Coenzyme A	Acyl group activation and transfer (pp. 384–385, 427–431, 493–507)	Dermatitis (chickens)

(continued)

157

Table 7.1—*continued*
Characteristics of vitamins and coenzymes

Name/Structure of Vitamin	Related Coenzyme	Reaction type (page numbers)[a]	Deficiency Disease
Water-Soluble Vitamins—*continued*			
Biotin 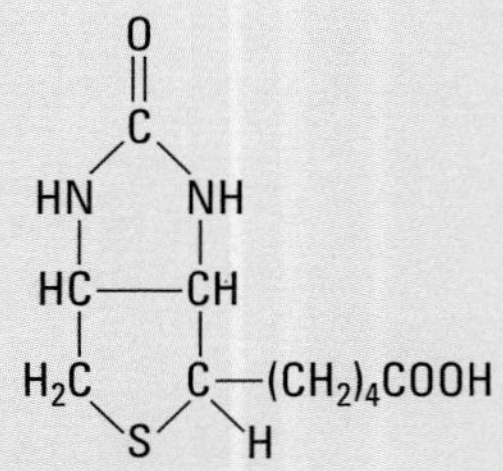	Biotinylated enzymes	CO_2 activation and transfer (pp. 407–408)	Dermatitis (humans)
Pyridoxal (vitamin B_6) 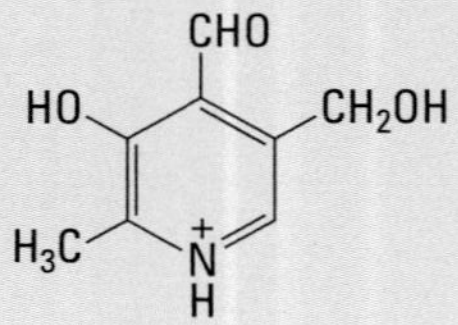	Pyridoxal phosphate	Amino group transfer (pp. 534–535)	Dermatitis (rats): neurological symptoms
Folic acid 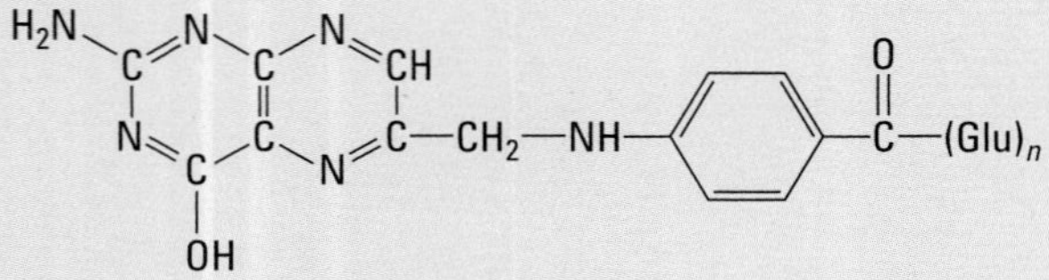	Tetrahydrofolate	Transfer of one carbon unit (pp. 529–532)	Anemias
Lipoic acid (may not be a vitamin) 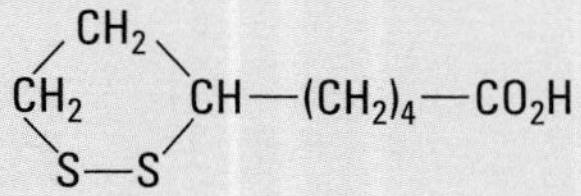	Attached to ε-lysine NH_2 in protein	Acyl group activation and transfer (pp. 427–432)	Growth deficiencies
Choline 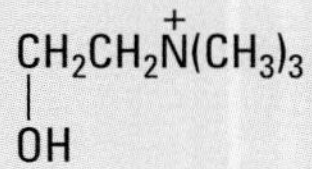	?	(pp. 162–163, 216–218)	Impaired brain development

Pernicious anemia

Methyl group transfer
(pp. 500–501)

Scurvy

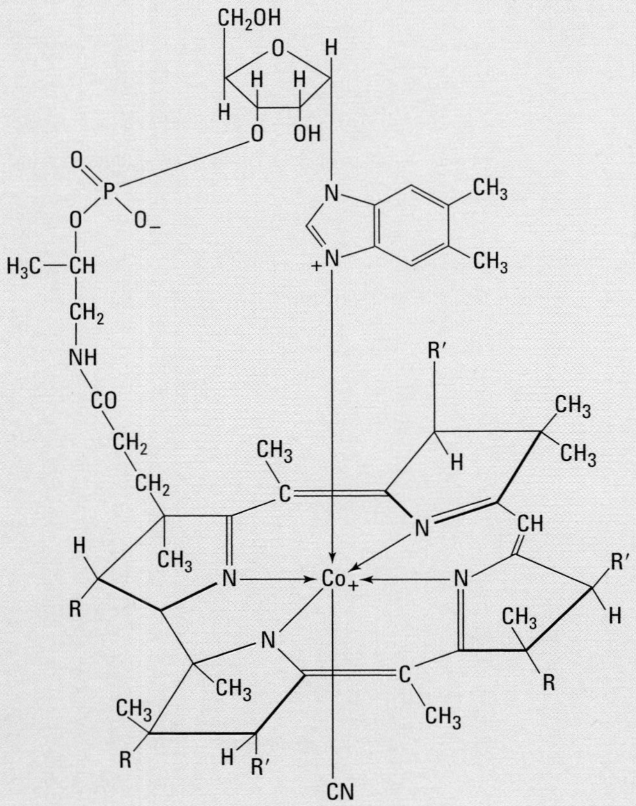

5′-Deoxyadenosyl cobalamin

Hydroxylation
(p. 107, 432)

Cobalamin (vitamin B$_{12}$)

L-Absorbic acid

L-Absorbic acid (vitamin C)

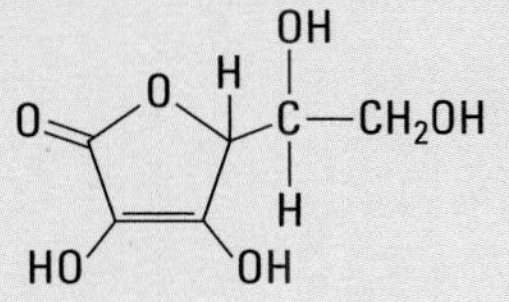

(continued)

159

Table 7.1—*continued*
Characteristics of vitamins and coenzymes

Name/Structure of Vitamin	Related Coenzyme	Reaction type (page numbers[a])	Deficiency Disease
Fat-soluble Vitamins			
trans-Retinol (vitamin A) 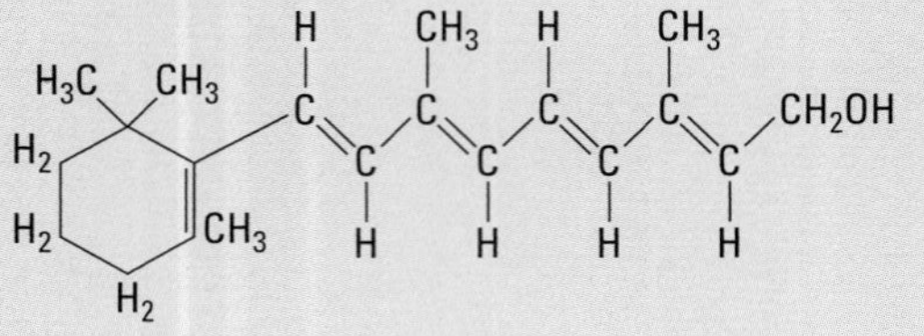	Associated with visual pigment	(pp. 116–117, 224, 473–474)	Night blindness, other effects
Cholecalciferol (vitamin D) 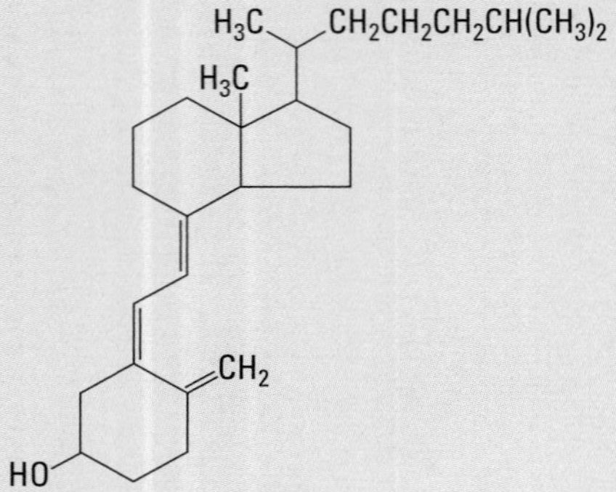	None	(pp. 224, 512–514)	Rickets
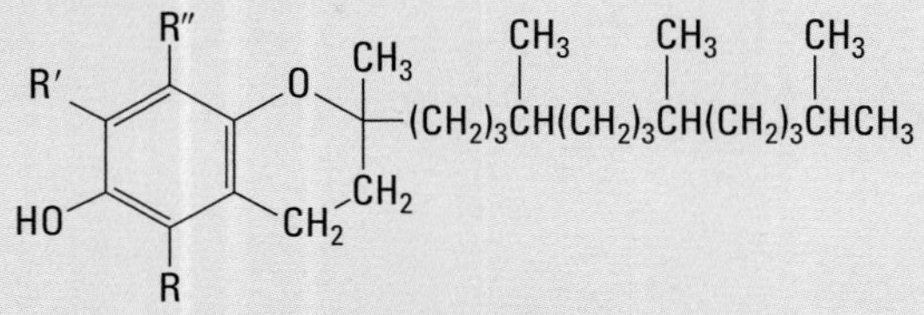 Tocopherol (vitamin E)	None	(p. 224, 432)	Reproductive and other problems in rats; uncertain in humans
(several variants, with R, R', R'' = H or CH₃)			
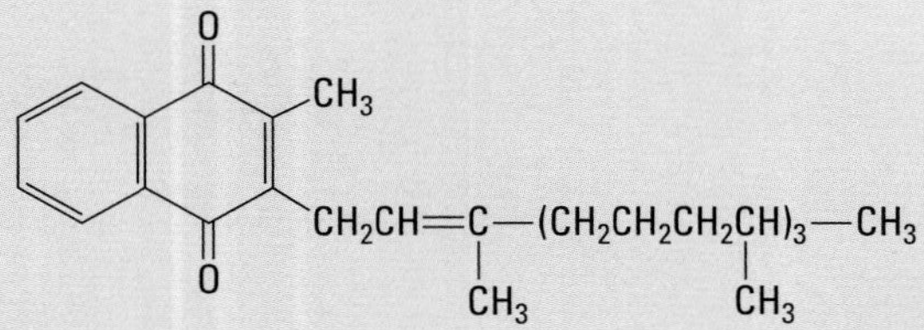 Phylloquinone (vitamin K₁)	None	(pp. 224, 474–476)	Problems in blood clotting

[a]Page numbers listed here refer to page numbers in this book.

160

Enzyme	Metal Ion
Catalase, peroxidase, aconitase, and cytochrome oxidase	Fe^{2+} and Fe^{3+}
Alcohol dehydrogenase, carboxypeptidase A, carboxypeptidase B, and DNA polymerase	Zn^{2+}
Cytochrome oxidase, lysyl oxidase, and superoxide dismutase	Cu^{2+}
Hexokinase and glucose-6-phosphatase	Mg^{2+}
Arginase	Mn^{2+}
Pyruvate kinase	K^+
Urease	Ni^{2+}
Nitrate reductase	Mo^{4+} and Mo^{6+}
Carbonic anhydrase	Zn^{2+}, Cd^{2+}

Table 7.2
Enzymes requiring metal ions as cofactors

small amounts (usually micrograms or milligrams per day) appear to be required in the diets of humans; however, megadoses (several grams) of some, such as vitamin C and niacin, have been recommended for disease prevention. The ingestion of large doses of water-soluble vitamins is usually not harmful; however, since vitamin levels above needed amounts are readily excreted, it does produce "expensive" urine. In contrast, megadoses of the fat-soluble vitamins such as A or D may cause serious effects because of accumulation in fat tissue and membranes. The biological functions of the coenzymes, especially those derived from water-soluble vitamins, are known. The molecular actions of the fat-soluble vitamins are less well understood. It is not entirely clear whether they serve as precursors of coenzymes. Some of the fat-soluble vitamins are known to be converted to an "active" form once inside the cell. For example, vitamin A (the alcohol retinol) is converted to the aldehyde, retinal, a chromophoric coenzyme found in visual proteins.

Metals as Nutrients

The diverse roles played by metal ions as coenzymes have been discussed in Chapter 6, Section 6.3. The ions of magnesium, manganese, iron, cobalt, copper, nickel, molybdenum, and zinc are required for enzyme catalytic action (Table 7.2). As with the organic cofactors, metal ions are required in only very small amounts, so they are considered essential trace elements or **micronutrients.** The biological roles of several metal ions are discussed throughout the book.

7.2 Allosteric Enzymes

The thousands of reactions in cellular metabolism are grouped into sequences each of which has a specific degradative or synthetic function to perform. Figure 7.1 outlines a hypothetical metabolic sequence with biomolecule A being converted in several enzyme-catalyzed steps to a final product, P. For example, the ten enzyme-catalyzed reactions of **glycolysis** transform glucose to pyruvate. The behavior of most enzymes in metabolism can be explained by Michaelis–Menten kinetics. That is, rate curves (v_0 versus [S]) are hyperbolic and the constants K_M, V_{max}, and k_3 can be experimentally measured for each. In a metabolic sequence as shown in Figure 7.1, there is at least one step catalyzed by a **regulatory enzyme** that controls the rate for the entire sequence. The regulatory enzyme(s) may be influenced by the concentration of (1) final product(s) in the metabolic sequence, (2) the beginning substrate in the sequence, (3) an intermediate formed in the pathway, (4) some external factor such as a hormone, or (5) perhaps by all of the above. In many metabolic reaction sequences, the first enzyme in the sequence is the major control enzyme (E_1 in Figure 7.1) and is influenced by the concentration of starting material, A, or the final product, P, or perhaps both. Regulatory enzymes come in many types and are classified according to their mode of action. In this section we concentrate on allosteric enzymes, those regulatory enzymes that are influenced by the reversible, noncovalent binding of a signal molecule. We shall encounter other types of regulatory enzymes in our study of metabolism.

$$A \xrightarrow{E_1} B \xrightarrow{E_2} C \xrightarrow{E_3} D \xrightarrow{E_4} F \xrightarrow{E_5} P$$

Figure 7.1

A hypothetical sequence of reactions comprising a metabolic pathway. Biomolecule A is converted via several intermediates (B, C, D, and F) to final product, P. Enzymes are represented by E_1, E_2, and so on.

Positive and Negative Effectors

Biomolecules that influence the action of an allosteric enzyme are called **effectors** or modulators. These regulatory molecules can serve as stimulants to the enzyme (positive effectors) or inhibitors of the enzyme (negative effectors). For example, a high concentration of A and a low concentration of P in Figure 7.1 are signals that the metabolic sequence should proceed with the production of P. In this case, A may serve as a positive effector *and* substrate. When the product P reaches a certain desired level, it can serve as an inhibitor (negative effector). This mode of regulation is a very logical and practical way to control the synthesis rate of P. With precise regulation, there will be the proper amount of P produced without complete depletion of A.

When the behavior of allosteric enzymes was first studied in detail, they were found in most cases not to display hyperbolic rate plots; that is, they did not obey typical Michaelis–Menten kinetics and inhibition could not be described by traditional Lineweaver–Burk graphs. Rate curves of v_0 versus [S] for allosteric enzymes are sigmoidal (Figure 7.2). The presence of a positive or negative effector also results in a sigmoidal curve, but with an

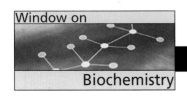

Window on
Biochemistry

7.1   The Newest Vital Amine—Choline

The term "vital amines" (now called vitamins) was first used in 1912 by Casimir Funk, a Polish biochemist, to describe a group of organic compounds that are essential in small amounts in the diet for proper growth and development. Vitamins are required because we have lost the ability to synthesize them in sufficient quantities.

The most recent addition (1998) to the list of essential human nutrients is **choline.** The new vitamin fits the original definition: It is an amine and it is essential for normal growth. When mammalian cells are grown in culture, choline must be added or the cells die. The intracellular sources of choline are primarily from salvage processes: the metabolic turnover of larger biomolecules that contain the choline moiety. The internal sources of choline are insufficient to maintain healthy cells so choline must be present in the diet. Choline is widely distributed in food with an abundant presence in milk, liver, eggs, and peanuts. The actual amount of choline in the diet is difficult to determine because many different dietary compounds contribute to its presence.

At a biochemical level, choline is essential as a precursor for several important molecules (see accompanying scheme). The biosyntheses of the amino acid methionine and the neurotransmitter acetylcholine begin with choline. Choline is also a building block for biomolecules we will encounter in later chapters. Important examples include (a) the phospholipids like sphingomyelin that serve as structural components in cell membranes (Chapter 9) and (b) cell-signaling messengers like diacylglycerol (Chapter 2).

Physiological changes that are the result of choline deficiency are just now being explored. The brain appears to be an organ especially affected by choline deficiency. Studies with pregnant rats provide typical clinical results. If choline is absent in the maternal diet during fetal development, but provided to offspring after birth, the young rats still display impairments in mental attention and memory. The link between choline and brain development is not surprising because brain and nerve cells normally have an abundance of structural phospholipids for membrane construction.

The Recommended Dietary Allowance (RDA) of choline has not yet been determined; however, the Adequate Intake (AI) levels have been set by the Institute of Medicine at the National Academy of Sciences. The AI for women of ages 19–70 years is 425 mg/day. For pregnant women, the AI is increased to 450 mg/day, and for lactating women, 550 mg/day. This increase reflects the special need

enhanced or inhibited rate, respectively. Because Michaelis–Menten kinetics are not obeyed, a K_M cannot be defined as usual. The substrate concentration that yields a v_0 of $\frac{1}{2} V_{max}$ is represented instead by the term $[S]_{0.5}$.

Effector molecules act by reversible, noncovalent binding to a site on the enzyme. All allosteric enzymes so far studied are much larger and more complex than nonallosteric enzymes, and they all have two or more subunits; that is, they are oligomeric. In addition to active sites for reaction (**catalytic sites**), allosteric enzymes have **regulatory sites** for binding specific effectors. *The general principle behind the concept of allosterism is that the binding or catalytic event occurring at one site influences the binding or catalytic event at another site.* Messages are sent from one site to another via conformational changes in the protein structure. The term *allosteric* can be translated to "other shapes." The binding of effector molecules changes the conformation of the protein in a way that tells the other subunits that it is bound. Allosteric enzymes transmit messages, via conformational changes, between binding sites that are spatially distinct.

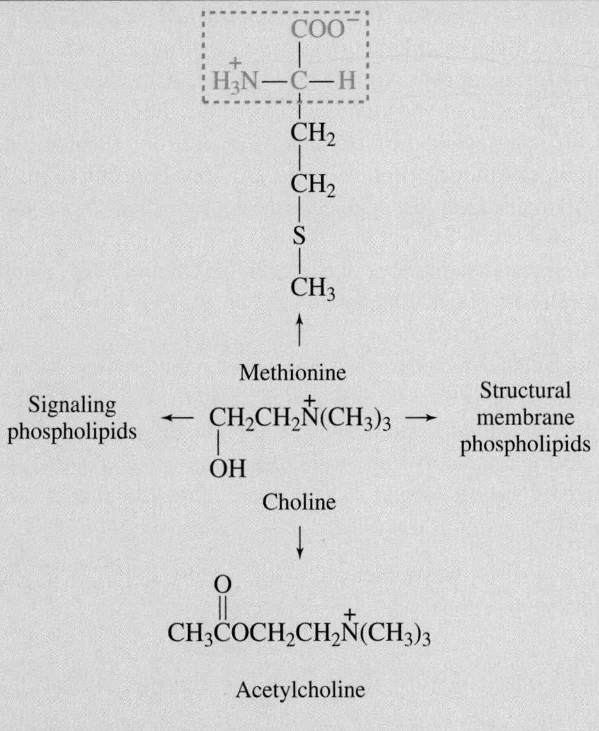

$$
\begin{array}{c}
\fbox{$
\begin{array}{c}
COO^- \\
| \\
\overset{+}{H_3N}-C-H
\end{array}
$} \\
| \\
CH_2 \\
| \\
CH_2 \\
| \\
S \\
| \\
CH_3
\end{array}
$$

Methionine

Signaling phospholipids $\longleftarrow$ $\underset{\underset{OH}{|}}{CH_2CH_2\overset{+}{N}(CH_3)_3}$ $\longrightarrow$ Structural membrane phospholipids

Choline

$$
\begin{array}{c}
O \\
\parallel \\
CH_3COCH_2CH_2\overset{+}{N}(CH_3)_3
\end{array}
$$

Acetylcholine

for choline during fetal development. Human milk, which has a choline content of 160–210 mg/liter, is a primary source of choline for newborns. Men 19 years and older have an AI of 550 mg/day. Most nutritionists agree that, until more complete studies are done, there is no need for supplementary choline for individuals on a balanced diet. The intake of choline at levels above the recommended AI cause hypotension, liver damage, sweating, diarrhea, and a fishy body odor.

Classifying choline as a vitamin will greatly increase nutritional studies on the biological role(s) of this small but interesting molecule. Further studies are needed to identify the molecular mechanisms of choline action, especially its critical role in early brain development.

Figure 7.2

Rate curves for allosteric enzymes. Curve ⓪ is enzyme + substrate, Curve ⊖ is enzyme + substrate + negative modulator and Curve ⊕ is enzyme + substrate + positive modulator. (a) This graph defines the group of allosteric enzymes, where $[S]_{0.5}$ is modulated with no change in V_{max}. The measurement of $[S]_{0.5}$ is shown on the [S] axis for each curve. (b) The group of allosteric enzymes where V_{max} is modulated with constant $[S]_{0.5}$.

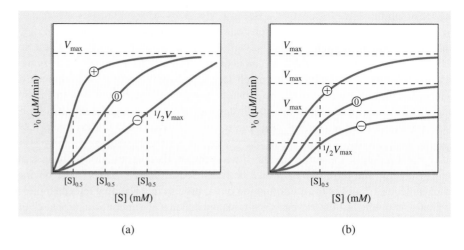

(a)   (b)

Examples of how effectors function will clarify the picture. In Figure 7.3a is shown a hypothetical tetrameric allosteric enzyme comprised of four identical subunits. Each subunit has a catalytic site where a substrate of complementary structure binds and is transformed to product. In a typical allosteric enzyme of this type, when one substrate molecule is bound to an **active site,** a message is transmitted via conformational changes to an active site or another subunit which makes it easier for a substrate molecule to bind and react at that site. This mode of allosteric interaction, where substrate and effector are the same type of molecule, is said to be cooperative and homotropic. Although the example given here shows a reaction enhancement (positive cooperativity), negative cooperativity, although rare, is observed for some allosteric enzymes. Cooperativity in binding and reactivity is identified by sigmoidal kinetics (Figure 7.3b). At low concentrations of substrate, the allosteric enzyme remains in a less active form. As more and more substrate molecules bind (near the midpoint of the curve, $[S]_c$), there is a sharp increase in the rate. The enzyme under these condition works somewhat like an "on–off" switch. Very small changes in substrate concentration lead to large changes in v_0. The enzyme is very sensitive to substrate concentration changes.

Figure 7.4 shows another hypothetical allosteric enzyme, a dimer with nonidentical subunits. Subunit α contains the active site; hence, it is called the catalytic subunit. Subunit β contains the site for effector binding and is called the regulatory subunit. Binding of a specific effector molecule to the regulatory site on the β subunit sends a signal via conformational changes to the catalytic site on subunit α. The message may be one of rate enhancement or

Figure 7.3

(a) A hypothetical, tetrameric, allosteric enzyme showing the cooperativity of substrate binding and reactivity. The binding of S to an active site on one subunit increases the probability that an S molecule can bind to another active site on another subunit. Because S molecules are the same, this allosteric enzyme is defined as cooperative and homotropic. (b) Kinetic curve for allosteric enzyme in (a). At substrate concentrations below $[S]_c$ on the graph, the enzyme displays low activity. At substrate concentrations above $[S]_c$, the enzyme is very active. The sigmoidal curve indicates the enzyme is allosteric and displays cooperativity in substrate binding.

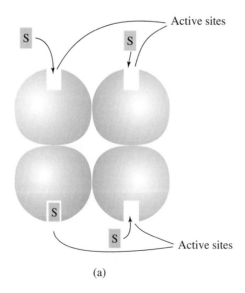 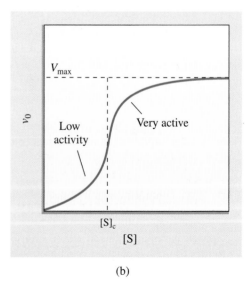

(a)   (b)

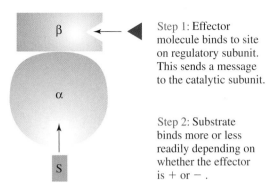

Step 1: Effector molecule binds to site on regulatory subunit. This sends a message to the catalytic subunit.

Step 2: Substrate binds more or less readily depending on whether the effector is + or −.

Figure 7.4
A hypothetical, dimeric, allosteric enzyme consisting of a catalytic subunit (α) and a regulatory subunit (β). Effector molecule binds to a regulatory site on the β subunit; substrate binds to an active site on the α subunit. This is an example of heterotropic allosterism. Regulation begins with binding of an effector to subunit β (Step 1). When the effector is bound, it may be harder or easier for S to bind than in the absence of effector (Step 2).

inhibition depending on the type of effector. Each type of effector for an allosteric enzyme has a specific binding site and sends its own distinct message when bound. Because the substrate and effector are different kinds of molecules, this type of allosterism is heterotropic.

Models to Describe Allosteric Regulation

The discussion above to explain the action of homotropic and heterotropic effectors has been descriptive and does not allow a quantitative analysis of enzyme regulation. Two theoretical models have been devised to explain the molecular and kinetic properties of allosteric enzymes and present the opportunity for a more precise analysis. Most of the experimental data obtained from allosteric enzymes at this time can be rationalized by one or both of these models. The **MWC concerted model** was proposed in 1965 by three French biochemists, Jacques Monod, Jeffries Wyman, and Jean-Pierre Changeux. We will apply this model to a dimer with identical subunits, each with a catalytic site. According to the MWC approach (Figure 7.5), the enzyme molecule can exist in two interconvertible states or conformations, a tense state (T) and a relaxed state (R). Both subunits of a molecule must be in the same conformation. The enzyme conformations RR and TT are possible; but the hybrid RT does not exist in this model. Therefore, transition between the R and T states happens at the same time; the reversible process of interchange is concerted. Subunits in the T state have little affinity for the substrate and thus show only slight catalytic activity. Enzyme molecules in the R state bind substrate molecules with high affinity and are catalytically active. In the absence of the substrate, the equilibrium between the two states is far toward the

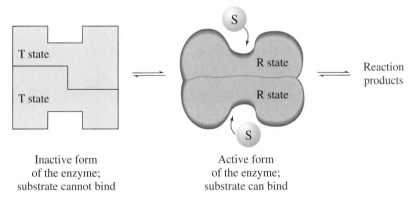

Inactive form of the enzyme; substrate cannot bind

Active form of the enzyme; substrate can bind

Reaction products

Figure 7.5
The MWC concerted model to explain the action of allosteric enzymes. The two forms of the enzyme are tense (T) and relaxed (R). These two forms are in equilibrium. Substrate molecules can not bind to the T state. Substrate molecules can bind to the R form as shown. Once bound, S can be converted to a product. To explain the action of positive and negative effectors, it is suggested that positive effectors bind with more affinity to the R state, thus shifting the equilibrium toward a greater population of enzyme molecules in the R state. Negative effectors bind more readily to the T state, thereby shifting the T ⇌ R equilibrium toward a greater population of enzyme molecules in the T state.

Figure 7.6

The sequential model to explain the action of allosteric enzymes. Substrate cannot bind readily to a subunit in the T state. When a substrate molecule binds, only that subunit shifts to the R state. This conformational change increases the probability that the remaining subunits can shift to the more active R state. Substrate molecules can readily bind to R state subunits. Hybrid dimers, RT, are possible in this model.

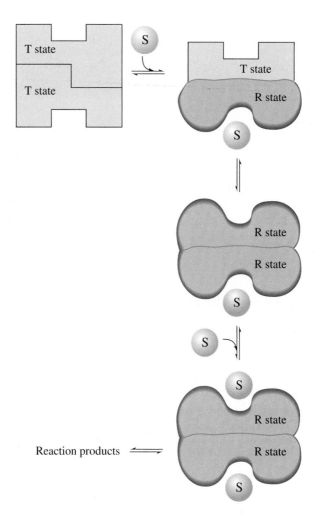

T side. As the substrate concentration increases, more and more substrate molecules bind to R state molecules, shifting the equilibrium toward the R state. The term **cooperative binding** is again useful to explain the process of substrate binding and reactivity.

The action of positive and negative effectors can also be explained by the MWC model. Positive effectors bind preferentially to the R state enzyme molecules, shifting more T molecules into the catalytically active form. In other words, the effector enhances the formation of and stabilizes the R state. Negative effectors prefer to bind to enzyme molecules in the T state, shifting enzyme molecules to this less active conformation.

An alternate explanation of allosteric action, the **sequential model,** was proposed by Daniel Koshland, Jr. at Berkeley in 1966. According to the sequential model, subunits of a single enzyme molecule are allowed to undergo conformational changes individually, so a dimer in the RT state is possible (Figure 7.6). When a substrate molecule binds to the active site, only that subunit shifts to the R state. This increases the probability of the remaining T subunits shifting to the R state. The conformational changes occur in a sequential manner: Each substrate binding step enhances the next step of substrate binding. The sequential model has several features different from the MWC concerted model's: (1) The sequential model does not assume initial equilibrium between R and T conformations, but the change to the R form is *induced* by substrate binding; and (2) the RT form is allowed in the sequential model; however, in the concerted model, an "all or none" principle is in effect for subunit conformation. The action of most allosteric enzymes can be explained by one or the other of these models. Neither one, alone, can explain all the known properties and experimental observations of allosteric enzymes. Because of the diverse characteristics observed for allosteric enzymes, it is unlikely that a single model can be developed to rationalize all observations.

7.3 Cellular Regulation of Enzymes

Covalent Modification of Regulatory Enzymes

In the previous section, we studied allosteric enzymes, those that are made more or less active by the *reversible, noncovalent* binding of small molecules called effectors. In another important kind of regulation, some catalytic activities of enzymes are altered by *reversible, covalent* changes to specific amino acid side chains in the enzyme. Chemical alterations that are commonly used to change the activity of an enzyme are:

1. Phosphorylation of hydroxyl groups in serine, threonine or tyrosine.
2. Attachment of an adenosyl monophosphate to a similar hydroxyl group.
3. Reduction of cysteine disulfide bonds.

Other enzymes serve as catalysts for the chemical changes on the amino acid residues. The attachment of the chemical group or other chemical modifications to specific amino acid residues in some cases transforms an active enzyme to an inactive form or in other cases activates an inactive enzyme. This process of regulation may seem a bit unusual because the enzyme that is covalently altered is actually a substrate for another enzyme that catalyzes the modifying reaction. Although many enzyme substrates are small molecules, we will encounter several examples of large molecules such as proteins and nucleic acids serving as substrates. As we would expect, enzymes that catalyze the chemical changes on other enzymes are themselves under very strict regulatory control.

Control of the regulatory enzyme, glycogen phosphorylase, provides an excellent example of covalent modification (Figure 7.7). As we shall study in Chapter 15, this enzyme catalyzes an important reaction to convert stored carbohydrate (glycogen) to glucose, a form that is readily degraded for energy. A specific serine residue in each of two identical dimers of the enzyme is phosphorylated in a reaction catalyzed by phosphorylase kinase to provide maximal activity to the glycogen phosphorylase:

$$\underset{\substack{| \quad\quad\quad |\\ \text{OH}\quad\text{OH}\\ \text{Less active form}}}{\text{Phosphorylase}} + 2\,\text{ATP} \rightleftharpoons \underset{\substack{| \quad\quad\quad\quad |\\ \text{OPO}_3^{2-}\quad\text{OPO}_3^{2-}\\ \text{More active form}}}{\text{Phosphorylase}} + 2\,\text{ADP}$$

A phosphoryl group ($-\text{PO}_3^{2-}$) is transferred from ATP to each of the serine hydroxyl groups in glycogen phosphorylase. Another enzyme, phosphorylase phosphatase, catalyzes the removal of the phosphate groups, P_i, by hydrolysis:

$$\underset{\substack{| \quad\quad\quad\quad |\\ \text{OPO}_3^{2-}\quad\text{OPO}_3^{2-}\\ \text{More active form}}}{\text{Phosphorylase}} + 2\,\text{H}_2\text{O} \rightleftharpoons \underset{\substack{| \quad\quad\quad |\\ \text{OH}\quad\text{OH}\\ \text{Less active form}}}{\text{Phosphorylase}} + 2\,\text{P}_i$$

Although the primary mode of glycogen phosphorylase regulation is covalent modification, other types of regulation, including allosterism and hormonal control, are also in effect. Several small molecules act as allosteric modulators for the enzyme. When we study regulatory processes in detail, we will note that enzymes at metabolic crossroads, such as glycogen phosphorylase, are controlled at several levels in the cell. Regulation of glycogen synthase is just the opposite of glycogen phosphorylase.

A similar covalent modification reaction occurs with the enzyme glutamine synthetase. This bacterial enzyme plays a major role in nitrogen metabolism (see Chapter 19, Section 19.2). The AMP (adenosyl monophosphate) group from ATP is transferred to the hydroxyl group of a specific tyrosine residue in glutamine synthetase, releasing pyrophosphate, PP_i, as a product. The adenylylated form of the enzyme is inactive.

$$\underset{\substack{|\\ \text{OH}\\ \text{Active form}}}{\text{Enzyme}} + \text{ATP} \rightleftharpoons \underset{\substack{|\\ \text{O}-\text{AMP}\\ \text{Inactive form}}}{\text{Enzyme}} + \text{PP}_i$$

Figure 7.7

Regulation of glycogen metabolism by protein phosphorylation. Two enzymes, kinase and phosphatase, control the activity of two other enzymes, glycogen synthase and glycogen phosphorylase, that regulate glycogen metabolism. Glycogen synthase produces glycogen from activated glucose. Glycogen phosphorylase degrades glycogen to glucose 1-phosphate.

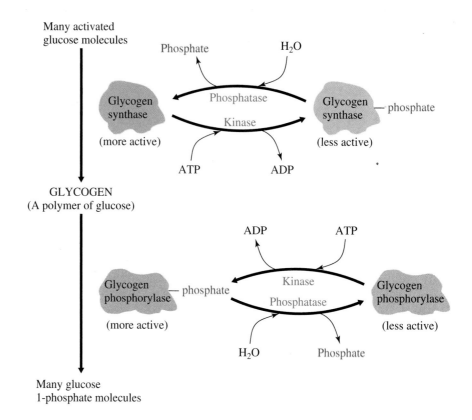

An interesting form of covalent modification that occurs extensively in plants is the reversible reduction of cysteine disulfide bonds by a reducing agent, AH_2:

$$\text{Enzyme} \ + \ AH_2 \ \rightleftharpoons \ \text{Enzyme} \ + \ A$$
$$\begin{array}{cc} | & | \\ S\!-\!S \end{array} \qquad\qquad \begin{array}{cc} | & | \\ SH & SH \end{array}$$

Active form   Inactive form

Many of the light-driven reactions of carbohydrate synthesis and reactions of carbohydrate degradation in plants are catalyzed by enzymes that are regulated by reversible disulfide bond reduction. The plant enzyme of glycolysis, glyceraldehyde-3-phosphate dehydrogenase, is inactivated by disulfide reduction.

Activation by Proteolytic Cleavage

Some enzymes are synthesized initially in an inactive form and require a covalent modification step before they acquire full enzymatic activity. The inactive precursor, called a **zymogen,** is cleaved at one or a few specific peptide bonds to produce the active form of the enzyme. Proteolytic cleavage, at first, may appear to be a type of regulation by covalent modification as discussed in the previous section. However, proteolytic cleavage is an irreversible process and occurs only once in the lifetime of an enzyme molecule. In covalent modification, the enzyme is reversibly altered by a set of regulatory enzymes, a process that may be repeated many times with the same enzyme molecule.

Several important biological processes are regulated by proteolytic cleavage. Many of the digestive peptidases (protein-degrading enzymes) of the stomach and pancreas, including chymotrypsin, pepsin, and carboxypeptidase, are regulated by this mechanism (Table 7.3). The process of blood clotting relies on a series of activation steps, each initiated by proteolytic cleavage. Some protein hormones, such as insulin, are produced in zymogen form (proinsulin) and activated by proteolytic removal of a short peptide.

In this section we consider the mechanistic details of chymotrypsin activation. This enzyme assists in the breakdown of dietary protein in the small intestine. Chymotrypsin cat-

Active Enzyme	Zymogen Form	Site of Zymogen Synthesis
π- or α-Chymotrypsin	Chymotrypsinogen	Pancreas
Pepsin	Pepsinogen	Stomach
Trypsin	Trypsinogen	Pancreas
Carboxypeptidase	Procarboxypeptidase	Pancreas
Elastase	Proelastase	Pancreas

Table 7.3
Digestive enzymes (peptidases) that exist as zymogens

alyzes the hydrolysis of peptide bonds on the carboxyl side of large, hydrophobic amino acid residues, such as phenylalanine, tyrosine, and leucine (see Chapter 4, Section 4.6). Its zymogen, chymotrypsinogen, is synthesized in the pancreas and secreted into the small intestines. Chymotrypsinogen is a single polypeptide chain with 245 amino acid residues and cross-linked by five intrachain disulfide bonds. The zymogen is activated upon trypsin-catalyzed hydrolysis of the peptide bond between arginine 15 and isoleucine 16 (Figure 7.8). The product, called π-chymotrypsin, is a fully active enzyme; however, it does not have a long lifetime because it is susceptible to attack by other π-chymotrypsin molecules. In this cleavage process, two dipeptide fragments (Ser-Arg; Thr-Asn) are removed, resulting in active α-chymotrypsin. The final form of the α enzyme consists of three polypeptides held together by two interchain disulfide bonds. The specific cleavages outlined in Figure 7.8 result in conformational changes in the tertiary structure. The conformational changes uncover amino acid residues at the active sites of α-chymotrypsin and π-chymotrypsin that are buried in the zymogen form.

Regulation by Isoenzymes

Some metabolic processes are regulated by enzymes that exist in different molecular forms. Such multiple forms, called **isoenzymes** or isozymes, have similar but not identical amino acid sequences. All forms demonstrate enzymatic activity and catalyze the same biochemical reaction, but they may differ in kinetics (different K_M and V_{max}), regulatory properties (different effectors), the form of coenzyme they prefer, and even their cellular distribution.

The best-known enzyme, and one of the first enzymes discovered to exist in isoenzymic forms, is lactate dehydrogenase (LDH), which catalyzes a key reaction in muscle metabolism, the reversible conversion of pyruvate to lactate (see Chapters 15 and 16). LDH is a tetramer composed of two possible types of subunits, M and H. The two polypeptide chains, which are made from two separate genes, are similar in amino acid sequence but can be separated by electrophoresis. The predominant form of LDH in skeletal muscle is M_4. In contrast, heart muscle LDH has the subunit formula, H_4. Other tissue such as liver has a mixture of five possible forms including hybrids (M_4, M_3H, M_2H_2, MH_3, H_4). Biochemical

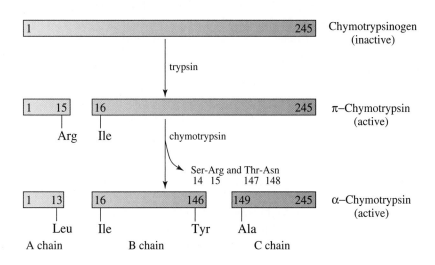

Figure 7.8

Activation of the zymogen, chymotrypsinogen, by proteolytic cleavage. Initial cleavage of the polypeptide chain occurs at Arg to produce active π-chymotrypsin. The final and most stable form, α-chymotrypsin, is formed by removal of two dipeptide fragments from π-chymotrypsin.

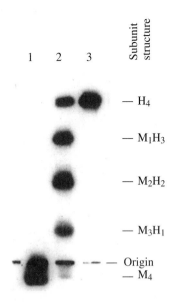

Figure 7.9
Electrophoresis of the isozymic forms of lactate dehydrogenase (LDH). Biological tissue and fluids may be analyzed for isozymic types of LDH present. Position 1 is M_4; position 2 is a sample of all isoenzymes composed from a mixture of M and H; position 3 is H_4. *Source:* Courtesy of C. L. Markert.

reasons for the existence of isoenzymes of LDH and other enzymes are not completely understood; however, it has been noted that the various forms of LDH and other enzymes show strikingly different kinetic and regulatory properties. For example, the M_4 isozyme of LDH has a higher affinity for the substrate pyruvate than the other forms, whereas the H_4 isozyme has a higher affinity for lactate. It is to be expected that the various forms are designed to serve several different regulatory functions. This type of regulation allows different metabolic patterns to function in different organs.

The diagnosis and treatment of some disease conditions have been aided by electrophoretic analysis of LDH isoenzyme forms in blood serum. Upon cell damage caused by myocardial infarction (heart attack) or infectious hepatitis (liver disease) or skeletal muscle diseases, enzymes including LDH leak into the blood. The affected tissue is indicated by the pattern of LDH isoenzymes in the blood. Figure 7.9 shows the electrophoretic pattern for the isoenzymic forms of LDH.

7.4   Site-Directed Mutagenesis and Abzymes

Until 1981, it was assumed that catalysts for biochemical reactions were limited to naturally occurring proteins. Since then, major discoveries have opened new frontiers in biological catalysis:

1. Using a technique called **site-directed mutagenesis,** scientists are able to modify the amino acid sequence of known enzymes and other proteins to change their reactivity, specificity, and even conformation. In fact, proteins of virtually any amino acid composition and sequence now can be created.
2. By using transition-state analogs as antigens, protein antibodies can be prepared that act as biological catalysts; hence they are called **catalytic antibodies** or **abzymes.**

The design of new protein catalysts has led to the development of new classes of catalysts for the biotechnology industry, medicine and basic research. We discuss further details here and in Chapter 13.

Site-Directed Mutagenesis

When studying the structure and activity of an enzyme, it would be of great value to be able to substitute one amino acid residue for another or modify the structure of an existing amino acid. For example, if one surmises that the hydroxyl side chain of a specific serine residue is necessary for catalysis by an enzyme, that serine can be replaced by an alanine residue. The modified protein can be analyzed for catalytic activity. Scientists have recently developed several approaches to modify protein sequences. Early methods included (1) treatment of the protein with chemical reagents that modify amino acid side chains and (2) mutagenization of an organism with ionizing radiation, ultraviolet light, or chemical mutagens. Neither method is especially effective. In the first, all amino acids of a given type are modified, not just the desired one or two. In the second method, it is difficult to focus on the desired gene, so the method is rather random. Techniques that use new recombinant DNA procedures are now available to alter genes at any desired location in the DNA (Chapter 13). Using enzymes that catalyze the cleavage of DNA, synthesize new DNA, and rejoin the DNA strands, the message in the gene can be modified to change amino acid sequences expressed in the protein product. Changes in the DNA that may be made include addition, deletion, rearrangement, or substitution at specific and preselected nucleotide base sites. The modified gene, prepared by chemical or biological synthesis, is introduced into a host cell, where it is cloned and expressed to yield the altered protein in quantities sufficient for further study. Genes can be synthesized that yield proteins of any amino acid sequence. This makes it possible to engineer proteins with new structural and catalytic properties. With only one or two amino acid substitutions, one can find out which are essential for enzymatic reactivity or critical for folding into a native conformation. In addition, entirely new proteins can be designed that may have desired properties not present in native proteins. The protein products may be of value in basic research and in medical and industrial applications.

Abzymes

The protein antibodies in the blood (immunoglobulins) function as protective molecules by tightly binding and thereby neutralizing foreign substances that may do harm to cells and organisms. Antibodies are produced in the blood of higher animals in response to foreign molecules called **antigens.** Many years ago, Linus Pauling proposed that the fundamental difference between enzymes and antibodies is that enzymes selectively bind substrate molecules in the *transition state of a reaction,* whereas antibodies specifically bind antigen-like molecules *in their ground state.* Is it possible, then, to produce an antibody with enzymelike catalytic activity by using a transition-state analog as antigen? If so, this would offer the opportunity to generate antibodies that not only bind molecules but also serve as very selective catalysts for desired chemical reactions.

The first experiments to demonstrate the utility of catalytic antibodies were reported in 1986 by Peter Schultz and Richard Lerner, both at the Scripps Institute, San Diego. These early abzymes were designed to catalyze the hydrolysis of esters and carbonates. The following example outlines the selection of a transition-state analog as antigen to generate antibodies having esterase activity. One reaction targeted for study was the hydrolysis of the methyl ester of p-nitrobenzoate (Figure 7.10). The presumed transition state is shown for the reaction. Hydrolysis is initiated by the attack of the nucleophile H_2O on the carbonyl carbon of the ester to form a tetrahedral center in the transition state. The goal in this study is to produce an antibody that specifically binds this presumed transition state. Since carbon compounds with this structure are very unstable, they cannot be isolated and used as antigens to produce the desired antibodies. However, the analogous tetrahedral phosphonate compound shown in Figure 7.10 is easily synthesized and can be injected into animals for antibody formation. Antibodies produced in this way were found greatly to accelerate the hydrolysis of the target ester, methyl p-nitrobenzoate. The phosphonate antigen, when added to the specific antibody–ester reaction mixture, functioned as a potent competitive inhibitor. This antibody-catalyzed reaction and others are found to obey Michaelis–Menten kinetics. Rate accelerations on the order of 1000- to 1 million-fold over the uncatalyzed rate are observed for catalytic antibodies and K_M and V_{max} values are in the same range as for traditional enzymes.

The possible uses of antibodies as tailor-made catalysts are unlimited and their potentials are just beginning to be tapped. Antibodies can be formed against a vast array of natural and synthetic antigens, making possible the catalysis of reactions that are not feasible with naturally occurring enzymes. Antibodies could be designed to make chiral pharmaceuticals and other expensive chemicals by catalyzing new reactions that are not possible or are very inefficient with current chemical reagents. Future goals for research include

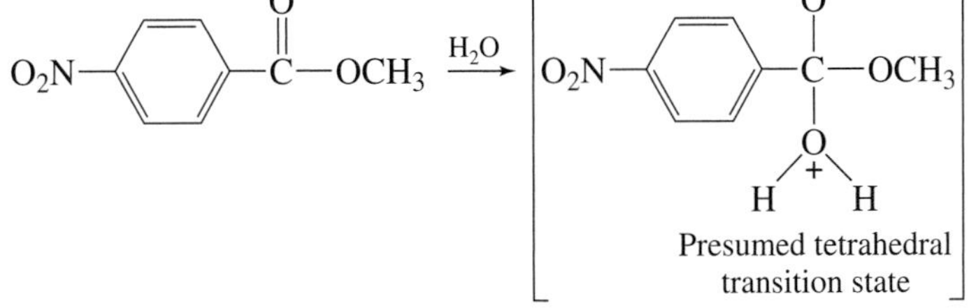

Cutting Edge: Catalytic antibodies

Structure Tutorial: Antibodies

(a)

Transition state analog used
as antigen to prepare
catalytic antibodies

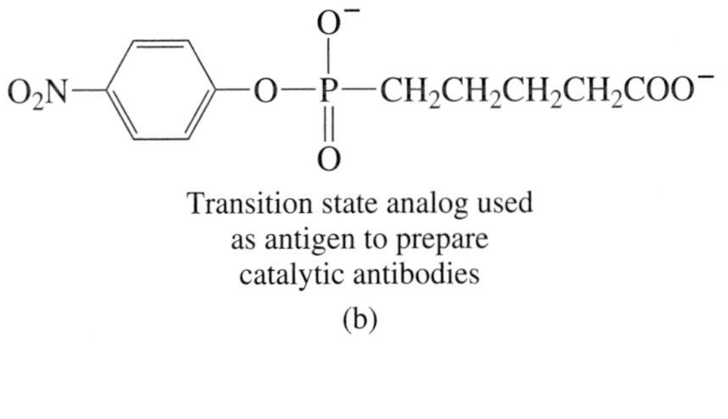

(b)

Figure 7.10
The design of an abzyme.
(a) The transition state for methyl p-nitrobenzoate hydrolysis. (b) A transition-state analog is used as antigen to generate antibodies to catalyze the hydrolysis reaction of methyl p-nitrobenzoate.

(1) antibodies that cleave proteins and nucleic acids in a selective manner to assist in sequence analysis and (2) antibodies that cleave the protein and carbohydrate coats of viruses and cancer cells. The importance of catalytic antibodies in basic research should also be mentioned. Experimental data from catalytic antibody studies have added strong support to the concept of transition-state stabilization in enzyme action and have provided fundamental insight into basic enzyme function.

7.5   Ribozymes

Cutting Edge: Ribozymes

Before 1981, biochemistry students read in textbooks and heard in lectures that "all enzymes are proteins, but not all proteins are enzymes." The study of biological catalysts was active in the mid to late 1800s with investigations on sugar metabolism. During this time, catalysts were called "ferments" and their chemical nature was unknown. In the 1920s, the first enzyme to be purified and crystallized, urease (see Chapter 6, Section 6.1), was found by chemical analysis to consist primarily of protein. From that time until the early 1980s, all enzymes purified and analyzed were found to be proteins. By then, it had become biochemical dogma that all enzymes were proteins. The discovery of RNA molecules that act like enzymes by two independent research groups in 1981 and 1982 began a revolution in biochemistry. Many biochemists were skeptical when research results on **catalytic RNA** were first announced. However, as the laboratory evidence supporting RNA enzymes mounted and several different kinds in different organisms were discovered, even the most skeptical were convinced. The true significance of the discovery that some forms of RNA can serve as a biological catalyst was acknowledged by the awarding of the 1989 Nobel Prize in Chemistry to the codiscoverers of these unique catalysts, Sidney Altman (Yale University) and Thomas Cech (University of Colorado–Boulder). The term **ribozyme** is now used for RNA enzymes. In this section, two examples of RNA molecules as catalysts are presented: (1) the RNA subunit of the bacterial enzyme, ribonuclease P, and (2) self-splicing RNA introns.

Ribonuclease P

The first demonstration of catalytic RNA came from studies on the enzyme **ribonuclease P,** which is found in all organisms. The substrate for the enzyme is a series of inactive, precursor tRNA molecules as shown in Figure 7.11. To prepare functional, mature tRNA, a seg-

Figure 7.11

Substrate for ribonuclease P (RNase P). The precursor tRNA molecule is hydrolyzed at the phosphodiester bond indicated by the arrow to produce active tRNA and a ribonucleotide fragment. Ribonucleotide bases: A = adenine, C = cytosine, U = uracil, G = guanine, ppp = triphosphate 5' end.

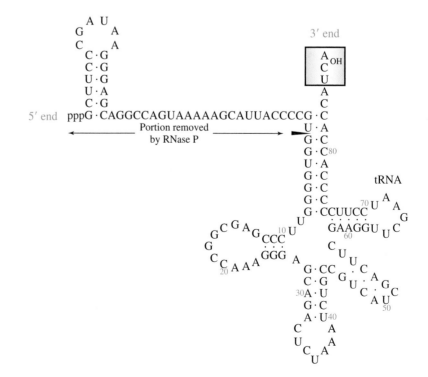

ment of the ribonucleotide is removed by hydrolytic cleavage of the phosphodiester bond, leaving the familiar clover-leaf structure of the tRNA used to select and activate amino acids for protein synthesis. Ribonuclease P is able to recognize and cleave at least 60 different tRNA precursors. Purification and electrophoretic analysis of the enzyme has revealed two components: (1) a relatively small protein subunit of 14,000 molecular weight and (2) an RNA component of 377 nucleotides linked into a single chain. When the enzyme composition was first discovered, it was assumed that the active site resided on the protein subunit and the RNA served as a "cofactor" to help bind the tRNA substrates. However, Altman and co-workers discovered upon separation and study of the components that the RNA subunit in the presence of magnesium ions and in the complete absence of protein was capable of catalyzing the hydrolysis reaction of precursor tRNA. The RNA component functions as a true enzyme: It obeys Michaelis–Menten kinetics, it is needed in only small amounts, it must remain in its native tertiary structure for activity, and it is structurally unchanged in the reaction process.

Self-Splicing RNA Introns

In another laboratory at about the same time, other remarkable capabilities of RNA were being uncovered. T. Cech and his co-workers were investigating biochemical splicing reactions that processed RNA for translation. They were especially interested in how introns or intervening sequences (noncoding sequences) were removed from RNA and how exons were joined together. (For details on RNA processing see Chapter 11). In at least one organism studied (the ciliated protozoan, *Tetrahymena thermophila*), Cech and his co-workers found that splicing of an intron from pre-rRNA was autocatalytic; the RNA cleaved itself without the assistance of protein catalysts. The splicing process cut out an intron, consisting of a sequence of 414 ribonucleotide bases, which was later processed to a product called L-19 IVS RNA. This final product was the original intron (intervening sequence, IVS) lacking 19 nucleotides. Splicing takes place by transesterification steps as shown in Figure 7.12. Although the self-processing of RNA was indeed remarkable, the RNA could not be considered a true

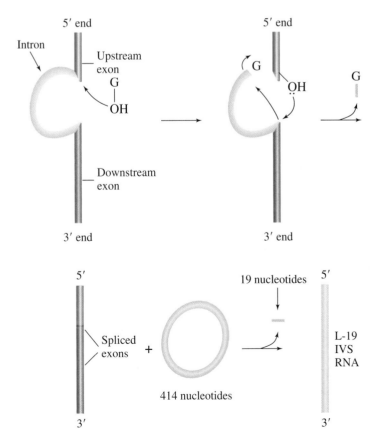

Figure 7.12

Self-splicing of an rRNA precursor from *Tetrahymena*. An intron (*pink*) of 414 ribonucleotide bases is cut from the rRNA. Guanosine (G—OH) acts as a cofactor in splicing the exons together (*red* and *green*). After removal of a short segment of 19 nucleotides from the 414 ribonucleotide intron, the L-19 IVS RNA is produced.

enzyme during the splicing processes because it did not display multiple turnovers; that is, it did not act on more than one substrate molecule like protein enzymes do.

However, upon further study of L-19 IVS RNA, more surprising discoveries were made. This form of RNA can mediate reactions on other RNA molecules, DNA, and other biomolecules. The list of activities for catalytic RNA is growing:

1. Cleavage and rejoining of oligonucleotide substrates
2. Cleavage of DNA phosphodiester bonds
3. Cleavage of RNA at sequence-specific sites
4. Hydrolysis of esters
5. Formation of peptide bonds between amino acids

The kinetic parameters and optimal reaction conditions displayed by ribozymes are similar to those for protein catalysts. Just like protein enzymes, it is essential that the RNA molecules remain in their native secondary and tertiary structures for activity.

The story on ribozymes is only beginning. There is still much to be learned about ribozyme secondary and tertiary structures. It is expected that each type of catalytic RNA molecule will, like a protein, take on its own unique, native conformation that depends on nucleotide sequence. The active site region of a catalytic RNA molecule will most likely be a combination of ionic residues (phosphate groups); polar regions (ribose); and nonpolar, hydrophobic regions (nucleotide bases). We cannot expect as much versatility in reaction type and substrate selectivity with RNA as we have noted with proteins because there are only four different nucleotide building blocks rather than the 20 amino acids used for proteins. Therefore, the prediction has been made that we will not discover nearly as many types of ribozymes as for protein enzymes.

Significance of Ribozymes

Other forms of catalytic RNA are continuously being discovered and studied. Small ribozymes have been found as components of plant RNA viruses. The catalytically active region of this RNA consists of only 19 to 30 nucleotides and because of their characteristic shape and cleavage action have been called "hammerhead" ribozymes.

The discovery of catalytic RNA has very interesting evolutionary implications. With its dual function as an informational molecule and a catalyst, RNA may have played a major role in the early stages of evolution and may have been one of the first large biomolecules. Peptidyl transferase, the ribosomal enzyme responsible for joining amino acids in protein synthesis (p. 315), was one of the first catalysts in the origin of life. It is known to be a ribozyme. During the evolutionary processes, RNA's catalytic function may have been replaced mainly by proteins because they are more effective and diverse catalysts. The informational role of RNA may have been taken over by DNA because it is chemically more stable.

Ribozymes have many potential biomedical applications in addition to their use as tools in specific cleavage of RNA. Many pathogenic viruses contain RNA, which could become a target for cleavage by carefully constructed ribozymes. There is the potential of engineering ribozymes to cleave RNA and DNA molecules only at a predetermined sequence of nucleotides. Infectious mammalian viruses, retroviruses, or cellular mRNAs involved in maintenance of the transformed state of the cell might be inactivated by such treatment. A ribozyme has been designed to catalyze the in vitro cleavage of the RNA of the human immunodeficiency virus (HIV), the virus that causes AIDS. The ribozyme has been approved for use in human clinical trials.

Summary

■ The catalytic activity of enzymes is strongly influenced by cellular activities including allosteric interactions, covalent modification, proteolytic cleavage, isoenzyme formation, and the presence of coenzymes.

■ Allosteric enzymes are influenced by the reversible, noncovalent binding of signal molecules called positive and negative effectors. Two theoretical models, the **Monod, Wyman, and Changeux** model and the **sequential** model as defined by Koshland, attempt to explain the action of allosteric enzymes.

■ The activity of some enzymes is regulated by covalent changes to amino acids such as phosphorylation of hydroxyl groups and reduction of disulfide bonds.

■ Catalytic antibodies (abzymes) may be designed to serve as catalysts for chemical reactions.

■ Some forms of RNA are capable of enzymatic activity. One example of a ribozyme is ribonuclease P, an enzyme that acts on tRNA substrates.

Study Exercises

Understanding Terms

7.1 In 25 words or less, define the following terms.
- a. Vitamin
- b. Niacin
- c. Coenzyme
- d. Prosthetic group
- e. Micronutrient
- f. Negative effector
- g. Regulatory site
- h. Cooperativity
- i. Isoenzymes
- j. Covalent modification
- k. Zymogen
- l. Site-directed mutagenesis
- m. Catalytic antibodies
- n. Ribozyme
- o. $[S]_{0.5}$
- p. Ascorbic acid
- q. Choline
- r. Transition-state analogs
- s. Kinase

Reviewing Concepts

7.2 Although hemoglobin is not an enzyme, its binding of oxygen and effectors can be described using allosteric terms. Which of the following terms can be used to characterize processes a and b below?

Negative cooperativity   Heterotropic
Positive cooperativity   Negative effector
Homotropic   Positive effector

a. Hemoglobin + 4 O_2 ⇌
b. Hemoglobin + 4 O_2 + 2,3-bisphosphoglycerate ⇌

7.3 Predict the coenzyme required for each of the following enzyme-catalyzed reactions.

a. CH_3CH_2OH ⇌ $CH_3-C{\small\begin{smallmatrix}O\\H\end{smallmatrix}}$

Enzyme: alcohol dehydrogenase

b. $CH_3\overset{O}{\overset{||}{C}}COO^- + CO_2 + ATP$ ⇌
Pyruvate

$^-OOCCH_2\overset{O}{\overset{||}{C}}COO^- + ADP + P_i$
Oxaloacetate

Enzyme: pyruvate carboxylase

c. $RCOOH$ ⇌ $CoA-\overset{O}{\overset{||}{C}}R$

Enzyme: acyl CoA synthetase

d. $CH_3\overset{O}{\overset{||}{C}}COO^-$ ⇌ $CH_3\overset{OH}{\underset{H}{\overset{|}{C}}}COO^-$
Pyruvate   Lactate

Enzyme: lactate dehydrogenase

HINT: Use the information in Table 7.1.

7.4 Draw a graph for each of the following kinetic situations.
a. A plot of v_0 versus [E].
b. A plot of v_0 versus [S] for an enzyme that obeys the Michaelis–Menten equation.
c. A plot of v_0 versus [S] for an allosteric enzyme.

7.5 Describe the differences between catalytic sites and regulatory sites on an allosteric enzyme.

7.6 Compare and contrast the characteristics of four types of enzyme regulation described in this chapter: allosterism, covalent modification, proteolytic cleavage, isoenzymes.

7.7 Which of the statements below are true for isoenzymes?
a. They catalyze the same chemical reactions.
b. They have the same quaternary structure.
c. They have the same distribution in organs and tissues.
d. They have the same enzyme classification name and number.

7.8 Describe the general type of biochemical reaction that requires each of the following coenzymes.
a. Thiamine pyrophosphate
b. Pyridoxal phosphate
c. Nicotinamide adenine dinucleotide

7.9 Which of the following statements are not correct for the MWC concerted model for allosteric enzymes?
a. An allosteric enzyme exists in two reversible states, R and T.
b. A dimeric allosteric enzyme may exist as RR.

c. A dimeric allosteric enzyme may exist as TR.

d. Substrates bind with higher affinity to the RR state.

e. Substrates bind with higher affinity to the RT state.

f. Positive effectors bind preferentially to the R state.

7.10 Which of the following statements are not correct for the sequential model for allosteric enzymes?

a. An allosteric enzyme exists in two reversible states, R and T.

b. A dimeric allosteric enzyme may not exist in an RT state.

c. The change of a subunit from T to R is induced by substrate binding.

7.11 Write a reaction catalyzed by each enzyme.

a. Phosphorylase kinase

b. Phosphorylase phosphatase

7.12 Differentiate between the terms *coenzyme* and *prosthetic group*.

7.13 Complete the following reactions that represent modes of regulation by covalent modification of amino acid residues. Assume that the amino acids shown are internal residues in a regulatory enzyme.

a.

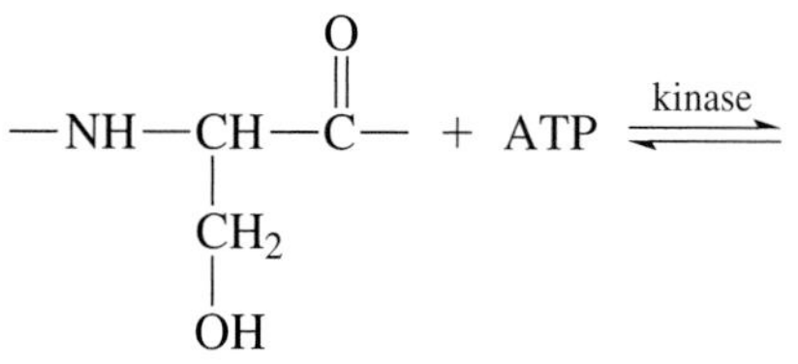

$$-NH-CH-C- + ATP \xrightleftharpoons{\text{kinase}}$$
with CH_2 and OH side chain, C=O

b.

reductase reaction with disulfide bridge:
$$-NH-CH-C-$$
$$CH_2$$
$$S$$
$$S$$ $+ AH_2 \xrightleftharpoons{\text{reductase}}$
$$CH_2$$
$$-NH-CH-C-$$

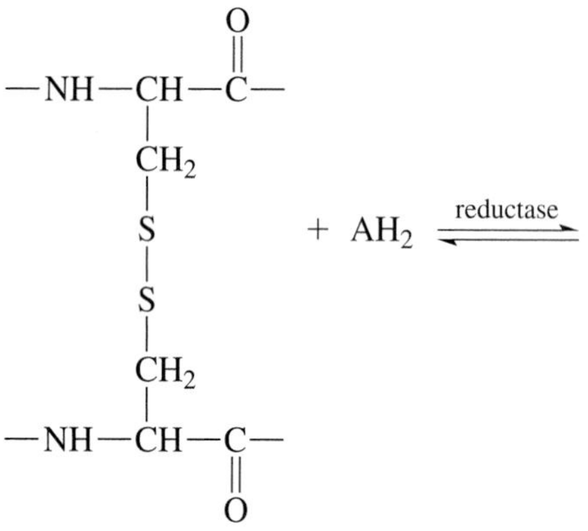

c.

$$-NH-CH-C- + ATP \xrightleftharpoons{\text{kinase}}$$
with CH_2 attached to phenol ring with OH

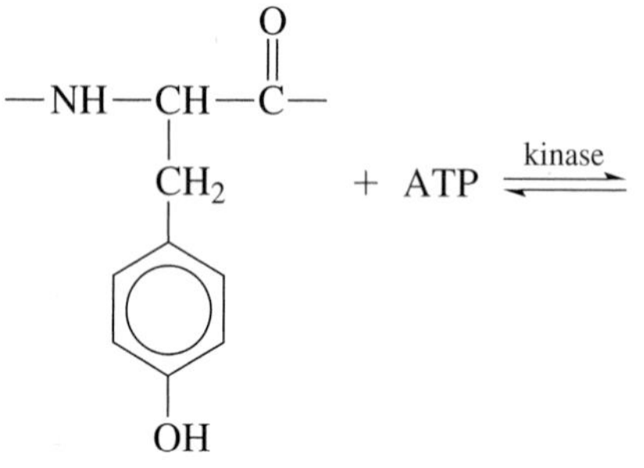

d.

$$-NH-CH-C-$$
$$CH_2$$
$$O$$
$$^-O-P=O + H_2O \xrightleftharpoons{\text{phosphatase}}$$
$$O_-$$

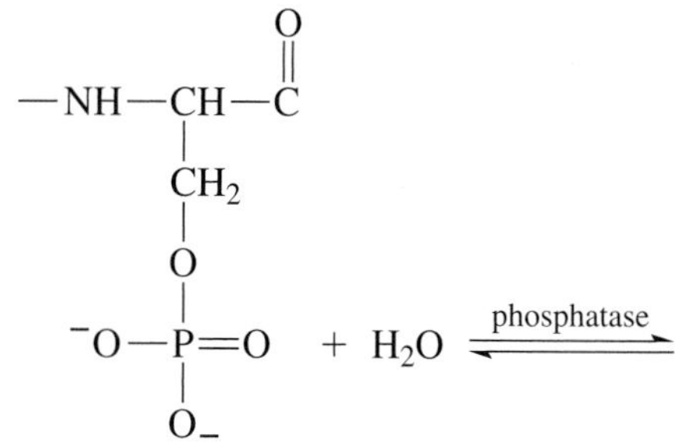

7.14 Match each of the vitamins listed in the first column of the table with its related coenzyme in the second column.

Vitamin	Related Coenzyme
Riboflavin (vitamin B_2)	Pyridoxal phosphate
Vitamin B_6	NAD^+
Niacin	FAD
Folic acid	Coenzyme A
Vitamin B_1	Tetrahydrofolate
	Thiamine pyrophosphate
	Biotin

7.15 Study each reaction in Problem 7.27 and suggest the common name for an enzyme for each part.

7.16 In the list below, draw a circle around those metal ions that are usually considered beneficial and even essential for proper growth and development.

$$Fe^{2+} \quad Mg^{2+} \quad Mo^{6+} \quad Co^{3+}$$
$$Na^+ \quad Hg^{2+} \quad Ni^{2+}$$
$$Pb^{2+} \quad Cu^{2+} \quad Rn$$

HINT: Use Figure 1.3.

Solving Problems

7.17 Why is it logical for the first step in a metabolic sequence of reactions to be the major regulatory step?

7.18 Explain why regulation by proteolytic cleavage is not considered a form of covalent modification.

7.19 Why must the proteolytic enzymes produced in the pancreas and stomach be synthesized in an inactive form?

7.20 Design a molecule (antigen) that could be used to generate antibodies that will catalyze the hydrolysis of the ester, below.

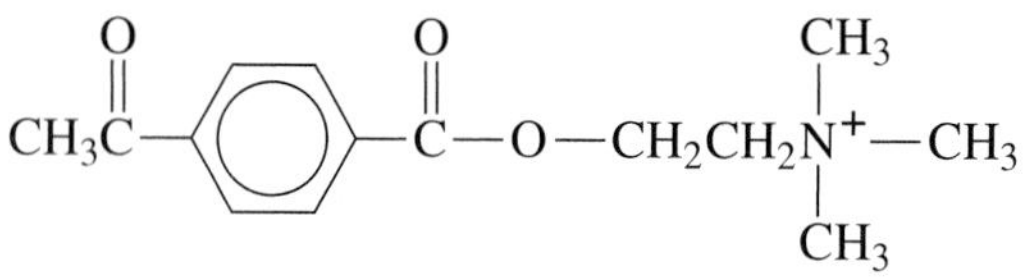

$$CH_3C-\bigcirc-C-O-CH_2CH_2N^+-CH_3$$
with two C=O groups and CH_3 groups on the nitrogen

7.21 An individual suffering chest pains went to the local hospital emergency room for treatment. Analysis of the patient's blood showed elevated levels of the enzyme lactate dehydrogenase (LDH) and a predominance of the H_4 isozyme. Give one possible explanation for the increased levels of LDH. Explain the biochemistry behind your medical diagnosis.

7.22 The addition of 6 M urea to a ribonuclease P reaction mixture reduced the rate of reaction by 50%. Explain the action of the urea.

HINT: Urea can form strong hydrogen bonds.

7.23 The catalytic action of an enzyme occurs with the substrate bound at the active site. How can the binding of another type of chemical at a point distant from the active site influence the catalytic activity at the active site?

7.24 Is it necessary for the amino acid residues at an enzyme active site to be close together in the primary sequence? Explain your answer.

7.25 What is the role of the RNA component in the enzyme system ribonuclease P?

7.26 The pathway for cholesterol synthesis in animals is shown below. Only some of many intermediates and enzymes are given. Use principles of enzyme regulation to predict which reaction is the rate-determining step for cholesterol synthesis. Would you expect the enzyme at that point to be regulatory?

$$HMGCoA \xrightarrow{1} mevalonate \xrightarrow{2} isoprene\ intermediates \xrightarrow{3}$$
$$squalene \xrightarrow{4} lanosterol \xrightarrow{5} cholesterol$$

7.27 Complete the following metabolic reactions by drawing the structure of all organic products derived from substrates. Assume that each reaction is catalyzed by the appropriate enzyme.

a. $CH_3CH_2OH + NAD^+ \rightleftharpoons$

b.
$$\underset{H}{\overset{HOOC}{}}C=C\underset{COOH}{\overset{H}{}} + FADH_2 \rightleftharpoons$$

c. $CH_3\underset{O}{\overset{\|}{C}}COOH + CO_2 + ATP \rightleftharpoons$

7.28 The reaction below is important in converting proinsulin (inactive) to insulin in its active form. Complete the reaction by showing major products.

$$\underset{proinsulin}{-Ala-Arg-} + H_2O \overset{peptidase}{\rightleftharpoons}$$

7.29 Use a term to describe the kind of chemistry occurring in each reaction in Problem 7.27. Select your answer from the following.

Deamination   Oxidation–reduction
Hydrolysis   Carboxylation

7.30 For each of the three parts below, name a biomolecule that has that metal ion as part of its structure.
a. Mg^{2+}
b. Fe^{3+}
c. Co^{3+}

Writing Biochemistry

7.31 Alpha-lipoic acid, a well-known coenzyme (see Table 7.1), has recently received attention as a unique antioxidant. Some researchers and nutritionists recommend that it be taken as a supplement even with a balanced diet. Search your library or the Internet for an article on lipoic acid. Use information in the reference and this book to write a "Window on Biochemistry" that would be appropriate for inclusion in Chapter 7. It should contain 200 to 300 words and include both basic biochemistry and relevant aspects that affect your life.

7.32 The editor of your campus newspaper has asked you to write a brief article on some "hot" topic in biochemistry. Because the newspaper is for general distribution, the article must be written so non-science majors can read and understand it. You have selected to write on homocysteine. Review the CD/ROM unit on this topic and write a 200- to 300-word article for the newspaper. You may want to look for more updated references in the library.

Applying Biochemistry

7.33 Antibodies and enzymes are both classified as proteins; however, they are quite different in their biological functions and mode of action. Scientists are now discovering some common characteristics in the two types of molecules. Compare and contrast the functions and actions of these molecules. Describe how antibodies are now being used as biological catalysts. Do you think it might be feasible to use enzymes as antibodies?

7.34 Peptidyl transferase, the ribosomal enzyme that catalyzes protein synthesis (see Section 12.2), is now known to be a ribozyme. This discovery has an impact on how we think about the origin of life and the early molecules that were present. Is it reasonable to think that RNA might have been present before protein? Discuss.

Further Reading

Arnold, F. et al., 2001. How enzymes adapt: lessons from directed evolution. *Trends Biochem. Sci.* 26:100–106.

Athota, R., and Radhakrishnan, T., 1991. Ribozyme—A novel enzyme. *Biochem. Educ.* 19:72–73.

Blusztajn, J., 1998. Choline, a vital amine. *Science* 281:794–795.

Bugg, C., Carson, W., and Montgomery, J., 1993. Drugs by design. *Sci. Am.* 269(6):92–98.

Danishefsky, S., 1993. Catalytic antibodies. *Science* 259:469–470.

Fraaije, M., and Mattevi, A., 2000. Flavoenzymes: Diverse catalysts with recurrent features. *Trends Biochem. Sci.* 25(3):126–132.

Gerlt, J., 1994. Protein engineering to study enzyme catalytic mechanisms. *Curr. Opin. Struct. Biol.* 4:593–600.

Gesteland, R., Cech, T., and Atkins, J., 1999. *The RNA World.* Cold Spring Harbor, NY: Cold Spring Harbor Press.

Halliwell, B., 1999. Vitamin C: poison, prophylactic or panacea? *Trends Biochem. Sci.* 24:255–259.

Hashim, O., and Adnan, N., 1994. Coenzyme, cofactor and prosthetic groups—Ambiguous biochemical jargon. *Biochem. Educ.* 22:93–94.

Hendrickson, H., 1999. Binding of proflavin to chymotrypsin. *Biochem. Educ.* 27(2):118–121.

Lewis, R., 2000. Homing in on homocysteine. *The Scientist* 14(2):1, 14–15.

Martin, A., and Schultz, P., 1999. Opportunities at the interface of chemistry and biology. *Trends Biochem. Sci.* 24(12):M24–M28.

Pizzi, R., 2000. The discovery of vitamins. *Modern Drug Discovery* 3(7):83–87.

Schultz, P., and Lerner, R., 1995. From molecular diversity to catalysis: Lessons from the immune system. *Science* 269:1835–1842.

Chapter 8

Carbohydrates: Structure and Biological Function

This stamp issued by Redonda honors Emil Fischer for his work on carbohydrates.

Glucose, fructose, sucrose, starch, and cellulose: These chemical names are common words in everyday use. They are among the most abundant members of the class of biomolecules called **carbohydrates,** compounds that have aldehyde or ketone functional groups and multiple hydroxyl groups. In addition to the elements carbon, hydrogen, and oxygen, some carbohydrates also contain nitrogen, sulfur, and phosphorus. Like the other major classes of biomolecules (proteins, nucleic acids, lipids), the carbohydrates are found in all forms of life and serve many different functions:

1. The carbohydrates are probably best known for their roles in energy metabolism. Some compounds in this class (glucose and fructose) serve as fuel for immediate use by organisms, whereas other compounds (starch and glycogen) are chemical stores for future energy needs in plants and animals.
2. Like proteins, some carbohydrates perform structural functions, providing scaffolding for bacterial and plant cell walls, connective tissue such as cartilages in animals, and exoskeleton shells in arthropods.
3. The monosaccharides ribose and deoxyribose, as components of the nucleic acids, serve a chemical structural role (RNA and DNA) and as polar sites for catalytic processes (RNA).
4. Carbohydrates are covalently combined with proteins and complex lipids on cell surfaces to act as markers for molecular recognition by other biomolecules.

In this chapter we first introduce the unique structures and reactions of biologically important carbohydrates and then turn to their participation in biological processes.

8.1   Monosaccharides

The simplest carbohydrates are the **monosaccharides,** compounds with a single aldehyde or ketone unit and multiple hydroxyl groups, which have the empirical formula $(CH_2O)_n$. For most naturally occurring monosaccharides, the range of n is 3 to 7. Although compounds with one or two carbons ($n = 1$ or 2) are not carbohydrates, they are of passing interest. For $n = 1$, the compound is formaldehyde, a toxic gas that has few properties in common with the carbohydrates. When $n = 2$, the compound becomes glycoaldehyde (Figure 8.1). The biological importance of this compound is still unknown; however, its derivative, glycolic acid, is an intermediate in plant and microbial metabolism. Glycolic acid is also an important α-hydroxyacid in skin creams. The smallest compounds considered to be important natural carbohydrates are the trioses ($n = 3$), glyceraldehyde (an aldehyde-based carbohydrate or **aldose**), and dihydroxyacetone (a ketone-based carbohydrate or **ketose**) (Figure 8.2). Most monosaccharides have one hydroxyl group on each carbon atom except for one carbon that has a carbonyl oxygen (aldehyde or ketone). Glyceraldehyde, with four unique groups around carbon number 2, possesses a chiral center; hence, it can exist in two stereoisomers (**enantiomers**) as shown in Figure 8.3. The prefixes D and L are used to distinguish the two structures. (How many chiral centers are present in dihydroxyacetone?) Although the letters D and L were originally used to define the rotated direction of plane-polarized light caused by solutions of each enantiomer (dextrorotatory and levorotatory), the letters now refer to the absolute configura-

Figure 8.1

Glycoaldehyde and glycolic acid. Although these compounds have molecular formulas and some properties similar to carbohydrates, they are not usually considered as members of this family. These two compounds are present especially in plants and microbes.

Glycoaldehyde

Glycolic acid
(an α-hydroxyacid)

tion of each. Naturally occurring glyceraldehyde is in the D configuration. To draw carbohydrate structures, we will use **Fischer projections** (Figure 8.3a). In this system, a tetrahedral carbon atom is represented by two crossed lines. The carbon atom is assumed to be at the intersection of the lines. A horizontal bond to an asymmetric carbon designates bonds in the front plane of the page and vertical bonds behind the plane of the page. For $n = 4$, the tetroses, there are two aldose structures, erythrose and threose, and one ketose, erythrulose (Figure 8.4). At this point we shall concentrate on the aldoses and take up the ketoses later.

Inspection of the aldotetroses in Figure 8.4 reveals two chiral centers, one at carbon number 2 and one at carbon number 3. A molecule with two chiral centers will have four possible stereoisomers (2^n isomers where $n = 2$, the number of chiral centers). This presents a further complication in structural analysis of the tetroses. By convention, we will reserve the terms D and L to designate the stereochemical orientation *at the chiral carbon farthest from the carbonyl carbon*. This will be carbon 3 (C3) for the tetroses. Thus, D-erythrose and D-threose will have the same absolute configuration at C3 as does D-glyceraldehyde at C2. D-Threose and D-erythrose have the opposite stereochemistries at C2; therefore, they are **diastereoisomers** of each other.

The monosaccharides with $n = 5$, 6, and 7 are called **pentoses, hexoses,** and **heptoses,** respectively. Like the smaller monosaccharides, there are also aldoses and ketoses among the larger carbohydrates. As the number of carbon atoms increases, the number of possible isomers increases. Figure 8.5 outlines the stereochemical relationships among the D-aldoses from $n = 3$ to $n = 6$. The same convention to designate D and L isomers is used for both smaller and larger carbohydrates. D-Ribose, with five carbon atoms (an aldopentose), is a component of RNA. The common sugar D-glucose, an aldohexose, has the same stereochemical orientation at C5 (farthest chiral center from the carbonyl group) as does D-glyceraldehyde at C2. Several important monosaccharides are in the D-aldose series. D-Glucose, D-mannose, and D-galactose are the most abundant of the aldohexoses. D-Mannose and D-galactose differ stereochemically from D-glucose at only one chiral center (C2 for mannose; C4 for galactose). Mannose and galactose are **epimers** of glucose.

The stereochemical relationships among the naturally occurring D series of ketoses are shown in Figure 8.6a. These compounds all have a carbonyl oxygen on carbon atom 2 and there is one hydroxyl group on each of the other carbon atoms. The most plentiful ketose in nature is D-fructose. The most abundant ketose with $n = 7$ is D-sedoheptulose (Figure 8.6b). Note from Figure 8.6a that three of the D-ketoses are named by adding -*ul* to the name of the corresponding D-aldose. For example, the five-carbon aldose is D-ribose, whereas the five-carbon ketose is D-ribulose.

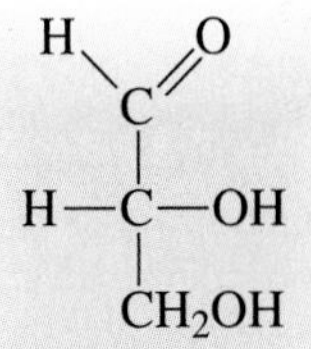

Glyceraldehyde
(aldose)

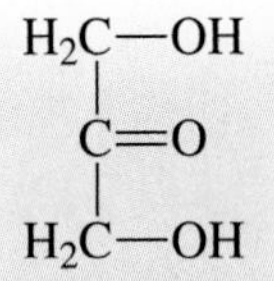

Dihydroxyacetone
(ketose)

Figure 8.2

Monosaccharides containing three carbons: glyceraldehyde, an aldose, and dihydroxyacetone, a ketose. Both are trioses.

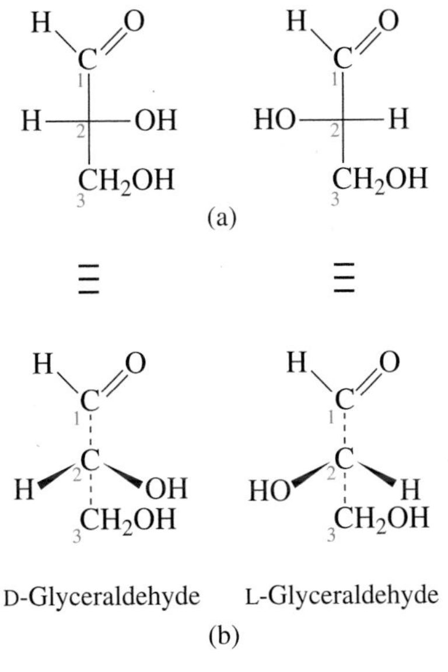

(a)

D-Glyceraldehyde   L-Glyceraldehyde

(b)

Figure 8.3

(a) Fischer projection formulas to show the stereochemistry of carbohydrates. Glyceraldehyde exists in two enantiomeric forms, D and L, because carbon number 2 (C2) is a chiral center. Horizontal bonds are in front of the plane of the page; vertical bonds are behind. The naturally occurring form is D-glyceraldehyde. (b) The Fischer projection structures are compared to another form of representation used for stereochemistry wherein dashed lines indicate bonds behind the plane of the chiral carbon and ⏐ indicates bonds in front.

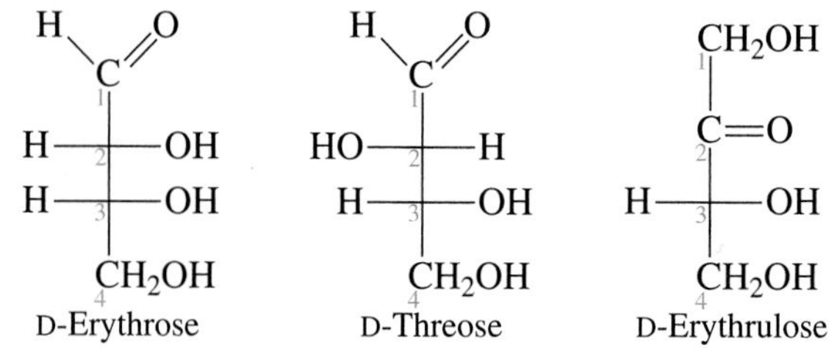

Figure 8.4

Monosaccharides containing four carbons. The stereochemical orientation (D or L) is determined at carbon number 3 (C3), the chiral center farthest from the carbonyl carbon. All three D-tetroses have the same absolute configuration at carbon 3.

D-Erythrose   D-Threose   D-Erythrulose

Names and classifications for the common, naturally occurring monosaccharides are given in Table 8.1.

8.2   Carbohydrates in Cyclic Structures

In Figures 8.1 to 8.6 we have written the structures of the carbohydrates as straight chains or Fischer projections. These structural representations are appropriate for some carbohydrates; however, the monosaccharides with five or more carbon atoms spend most of their

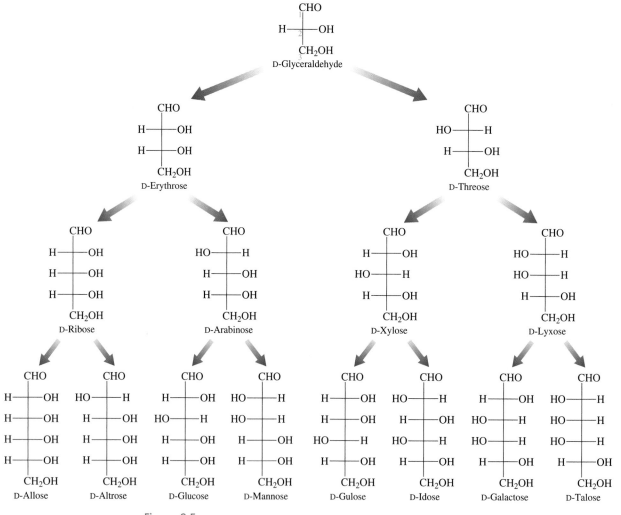

Figure 8.5

The family of D-aldoses containing from three to six carbons. All aldoses shown here have the same absolute configuration at their chiral center farthest from the carbonyl group as does D-glyceraldehyde at carbon 2.

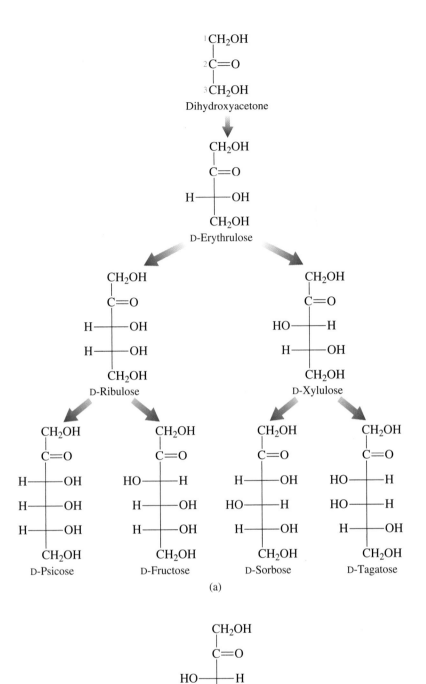

(a)

Figure 8.6

(a) The family of D-ketoses containing from three to six carbons. The D designation defines their absolute stereochemistry at their chiral center farthest from the keto group. All have the same absolute configuration as D-glyceraldehyde at carbon 2. (b) The most important ketoheptose is D-sedoheptulose.

Table 8.1

Names and classification for the common monosaccharides

Monosaccharide	Class
Glyceraldehyde	Aldotriose
Dihydroxyacetone	Ketotriose
Erythrose	Aldotetrose
Erythrulose	Ketotetrose
Ribose	Aldopentose
Ribulose	Ketopentose
Glucose	Aldohexose
Mannose	Aldohexose
Galactose	Aldohexose
Fructose	Ketohexose
Sedoheptulose	Ketoheptose

time in solution as cyclic structures. The ring is formed by reaction of the aldehyde or ketone on one end of the molecule with a hydroxyl group on the other end. The reaction between an aldehyde and a hydroxyl group (alcohol) to form an acetal is shown below:

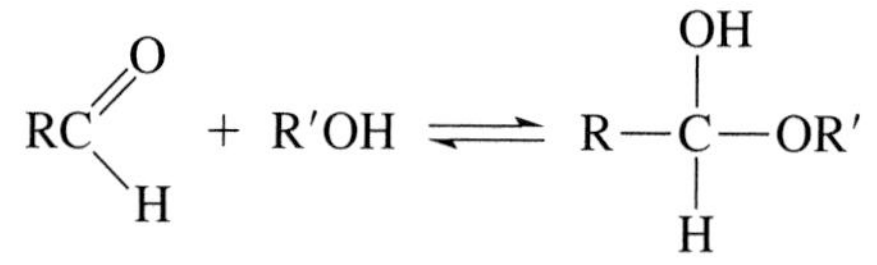

$$RC{\overset{O}{\underset{H}{\diagup}}} \; + \; R'OH \; \rightleftharpoons \; R-\overset{\overset{OH}{|}}{\underset{\underset{H}{|}}{C}}-OR'$$

Aldehyde   Alcohol   Acetal

In Figure 8.7a this chemical reaction is applied to open-chain ribose, which results in a cyclic acetal structure. The five-membered ring is called a furanose because it is similar to the heterocyclic ring, furan (Figure 8.7b). The same reaction applied to D-glucose results in a six-membered ring called a pyranose because it is similar to pyran (Figure 8.7c and Figure 8.8).

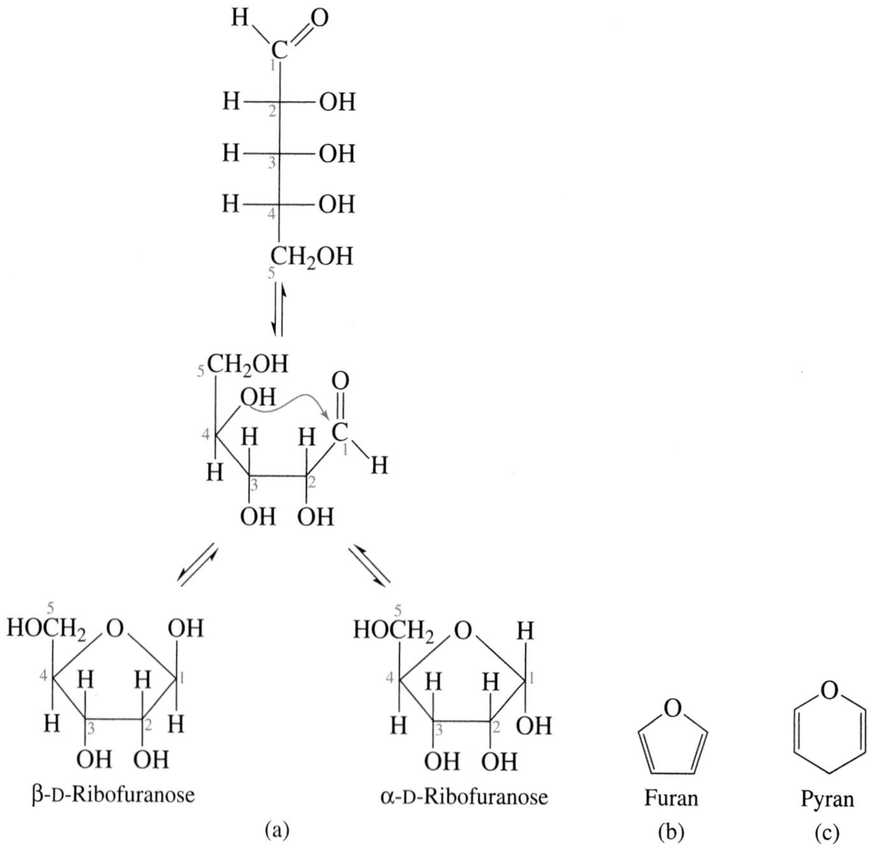

Figure 8.7

Cyclization of the open-chain form of D-ribose. (a) The carbonyl group of D-ribose reacts with the hydroxyl group on C4 to form an acetal. The product consists of a mixture of two acetal compounds, β-D-ribofuranose and α-D-ribofuranose. The two products drawn in the cyclic Haworth projection differ in stereochemistry at C1. (b) The acetal products from ribose have structures similar to furan, and are called furanoses. (c) The acetal products from glucose have structures similar to pyran and are called pyranoses.

β-D-Ribofuranose   α-D-Ribofuranose   Furan   Pyran

(a)   (b)   (c)

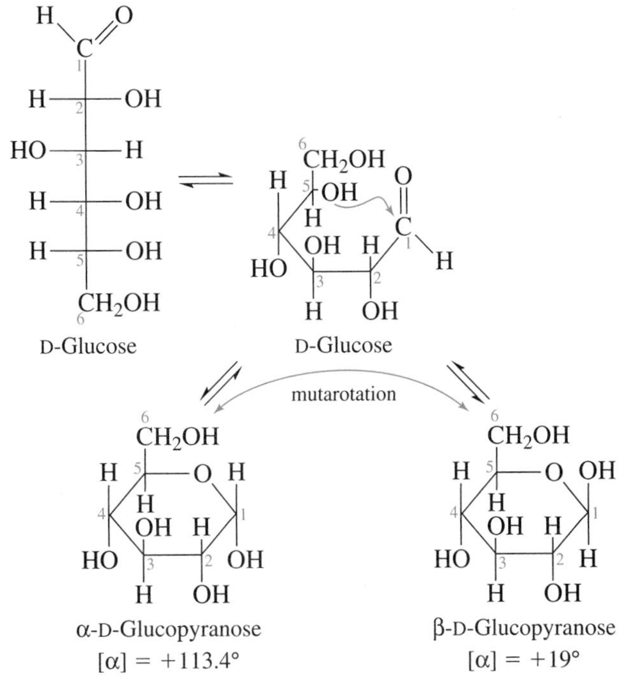

Figure 8.8

Cyclization of D-glucose to form two cyclic acetals. The two acetals drawn in Haworth projection form are α-D-glucopyranose and β-D-glucopyranose. Their six-membered rings are similar to pyran. This figure also shows the interconversion of α and β anomers by mutarotation. The cyclic acetals can undergo rearrangement at C1 by opening to the free aldehyde form and reclosing.

The ketoses with carbon numbers greater than four can react in a similar fashion:

For the ketohexose D-fructose, the carbonyl at C2 reacts with the hydroxyl group on C5 to form a cyclic **ketal** (Figure 8.9).

When the ring closes for each monosaccharide, the former carbonyl carbon (aldehyde or ketone) becomes a chiral center; therefore, two structures can be drawn to represent the stereoisomeric products. In one structure, the hydroxyl group, at carbon 1 (in aldoses) and carbon 2 (in ketoses), is below the plane (α configuration) and in the other structure, the hydroxyl group is above the plane of the ring (β configuration). The α and β forms of sugars are called **anomers** because they differ only by the stereochemistry at the anomeric carbon (C1 in ribose and glucose, C2 in fructose). Note that the configuration at other carbon atoms remains the same for both the α and the β anomer. The full names of the three cyclic sugars above are α- or β-D-ribofuranose, α- or β-D-fructofuranose, and α- or β-D-glucopyranose. The cyclic structures for ribose, glucose, and fructose are drawn in the **Haworth projection form** (see Figures 8.7 to 8.9). This representation shows all hydroxyl groups and hydrogen atoms but does not show carbon atoms in the ring. The plane of the ring is assumed to be parallel to the plane of the page with functional groups indicated above and below the plane of the ring.

When dissolved in aqueous solution, the cyclic forms of the monosaccharides are in equilibrium with the open chains. Thus, the α form of glucose can be readily converted to the β as shown in Figure 8.8. An aqueous solution of pure β-D-glucopyranose has an initial specific rotation of +19° when measured in a polarimeter. If the solution of the β anomer is allowed to sit for several hours, the specific rotation slowly changes, resulting in a final value of +52.2°. A solution of α-D-glucopyranose has an initial specific rotation of +113.4° but, after several hours, the specific rotation changes to +52.2°, the same as with the β anomer. In each case, the same equilibrium mixture of α-D- and β-D-glucose has been produced. This process of interconversion of stereoisomers, called **mutarotation,** has been

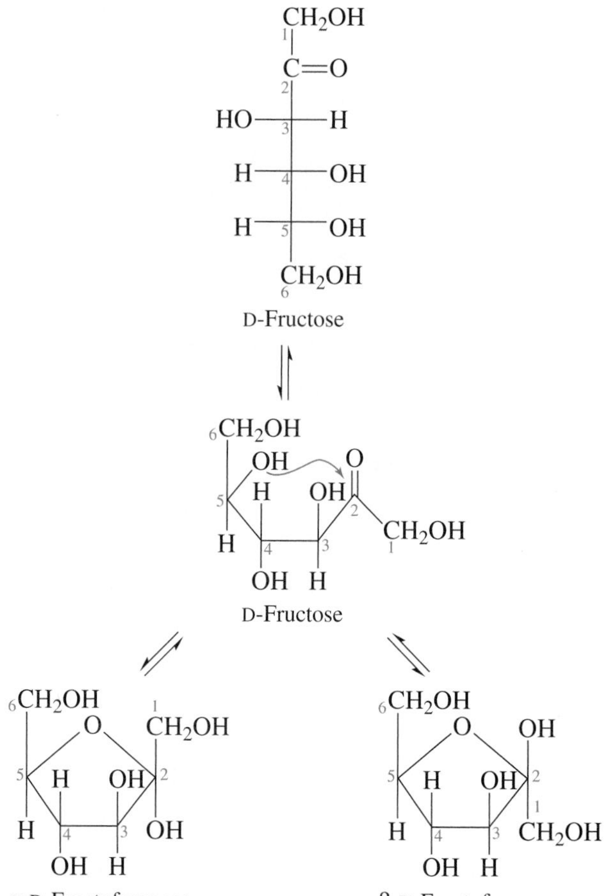

Figure 8.9

Cyclization of D-fructose forming two cyclic ketals, β-D-fructofuranose and α-D-fructofuranose, drawn in Haworth projection form. The two products differ in stereo-chemistry only at C2.

observed for many carbohydrate solutions when the sugar can easily form a five- or six-membered ring (see Figure 8.8). In a D-glucose solution, an equilibrium mixture consists of three forms of the sugar, approximately 33% α anomer, 66% β anomer (the most stable of the three forms), and 1% of the open polyhydroxyl aldehyde chain.

8.3   Reactions of Monosaccharides

With several kinds of functional groups present in an aqueous carbohydrate solution (hydroxyl groups, acetal or ketal groups, and aldehyde or ketone groups), diverse chemical reactions and unique products can be expected. Here we focus on those reactions that are either of biological significance or used for identification and analysis of carbohydrates.

Oxidation–Reduction

The complete metabolic breakdown of glucose to CO_2 and H_2O is a process involving several oxidation steps, the details of which are covered in Chapter 15. Oxidation reactions of carbohydrates can be illustrated with either the open-chain or the cyclic form, which are in equilibrium. The open chain is more straightforward because the functional group most susceptible to oxidation is the aldehyde group, which produces a carboxylic acid group:

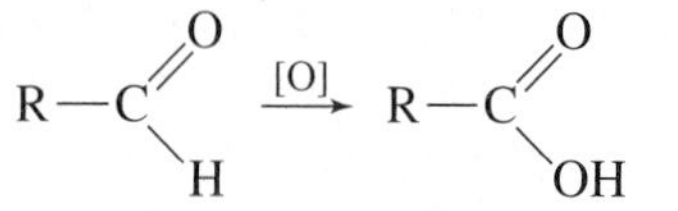

Many oxidizing agents, including Tollens' reagent (silver ammonia complex, $Ag(NH_3)_2^+$) or cupric ion (Cu^{2+}), are used to identify the presence of reducing sugars. **Reducing sugars**

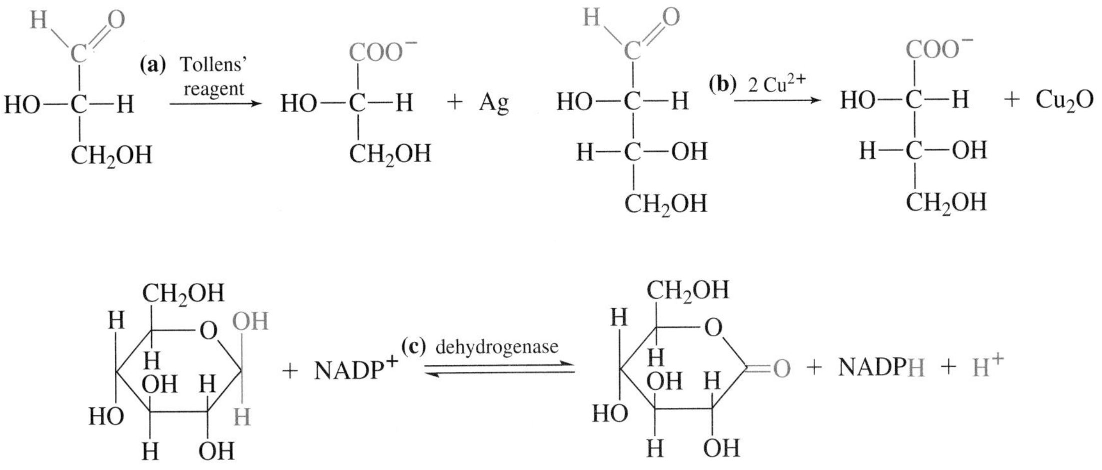

Figure 8.10
Oxidation reactions of carbohydrates. (a) The oxidation of glyceraldehyde by Tollens' reagent (basic silver ammonia). A positive result is indicated by the formation of a silver mirror (Ag). (b) The oxidation of threose by the cupric ion. A positive test is indicated by the formation of a red precipitate of Cu_2O. (c) The oxidation of glucose catalyzed by the enzyme dehydrogenase, assisted by the coenzyme $NADP^+$. The glucose acts as a reducing agent and the $NADP^+$ as an oxidizing agent. The oxidized product of glucose is a lactone, a cyclic ester.

are those that contain a free aldehyde group and are capable of reducing metal ions in solution. The open-chain form of the aldehyde acts as the reducing agent. Figure 8.10 shows three important oxidation reactions of carbohydrates including one catalyzed by an enzyme with the coenzyme, $NADP^+$. Oxidation of the cyclic structure (acetal) results in a carboxylic acid; however, that functional group cyclizes with one of the sugar hydroxyl groups to form a **lactone,** a cyclic ester (Figure 8.10c).

Reduction of the aldehyde or ketone group of a carbohydrate is also of biological importance. The reduction of the aldehyde group in glucose yields sorbitol, a sweetening agent (Figure 8.11). Enzymes that catalyze oxidation–reduction reactions in the cell use the coenzymes NADH or NADPH to carry out carbohydrate reduction reactions. Another form of reduced carbohydrates are the **deoxy sugars,** carbohydrates lacking hydroxyl groups. The most important one is 2-deoxyribose, a component in DNA. 2-Deoxyribose is produced by reduction of the hydroxyl group on C2 of ribose, a process that removes the hydroxyl group and replaces it with hydrogen. Deoxy sugars are synthesized in the cell by reduction using the cofactor NADPH with enzymes called oxidoreductases or more commonly, **dehydrogenases.**

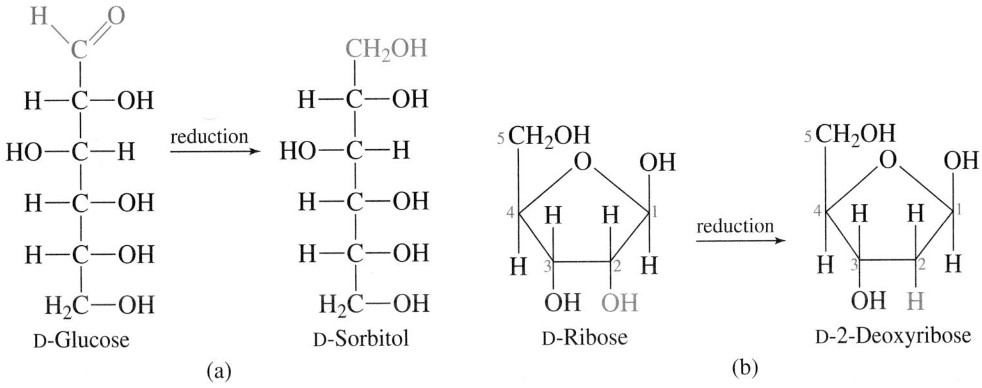

Figure 8.11
Reduction reactions of carbohydrates. (a) Reduction of glucose to sorbitol. The aldehyde group is reduced to an alcohol. (b) Conversion of D-ribose a component in RNA to D-2-deoxyribose, a component in DNA.

Esterification

Esters are compounds formed by reaction of hydroxyl groups (alcohols) with acids:

$$ROH \ + \ HOOCR' \longrightarrow R-O-\overset{\overset{\displaystyle O}{\|}}{C}-R' \ + \ H_2O$$

Alcohol   Acid   Ester

The hydroxyl groups of the carbohydrates also react with acids to produce esters. The most important biological esters of carbohydrates are the phosphate esters made using phosphoric acid:

$$ROH \ + \ HO-\overset{\overset{\displaystyle O}{\|}}{\underset{\underset{\displaystyle OH}{|}}{P}}-OH \longrightarrow HO-\overset{\overset{\displaystyle O}{\|}}{\underset{\underset{\displaystyle OH}{|}}{P}}-O-R \ + \ H_2O$$

Alcohol   Phosphoric acid   Phosphate ester

Important phosphate esters in biology include D-glucose 6-phosphate, D-glyceraldehyde 3-phosphate, D-ribose 5-phosphate, and D-deoxyribose 5-phosphate (Figure 8.12). The first two compounds are formed during the metabolic process of **glycolysis,** the sequence of reactions that generates cellular energy from glucose. The phosphate esters of D-ribose and D-deoxyribose are important precursors of the nucleic acids RNA and DNA, respectively. In the cell, phosphate esters are produced not using the very acidic phosphoric acid but, most often, by transfer of a **phosphoryl group** $-\overset{\overset{\displaystyle O}{\|}}{\underset{\underset{\displaystyle OH}{|}}{P}}-OH$ from adenosine triphosphate (ATP) to a carbohydrate hydroxyl group, a reaction catalyzed by enzymes called **kinases:**

$$\text{D-glucose} \ + \ \text{ATP} \xrightleftharpoons{\text{kinase}} \text{D-glucose 6-phosphate} \ + \ \text{ADP}$$

We will study the importance of this metabolic reaction in Chapter 15.

Amino Derivatives

The replacement of a hydroxyl group on a carbohydrate by an amino group results in an unusual class of compounds, the **amino sugars.** Two amino sugars that occur widely in nature are D-2-aminoglucose (glucosamine) and D-2-aminogalactose (galactosamine) shown in Figure 8.13. Glucosamine and its acetylated derivatives, *N*-acetylglucosamine and *N*-acetylmuramic acid, are found as structural components of bacterial cell walls. As shown in Figure 8.13, the acetyl groups are linked to glucosamine via amide bonds. In addition, *N*-acetylmuramic acid has a three-carbon carboxylic acid (lactic acid) in ester linkage to the 3-hydroxyl group. Galactosamine is a component of **chitin,** a polymer found in the exoskeleton of insects and crustaceans, and is also a major structural unit of chondroitin

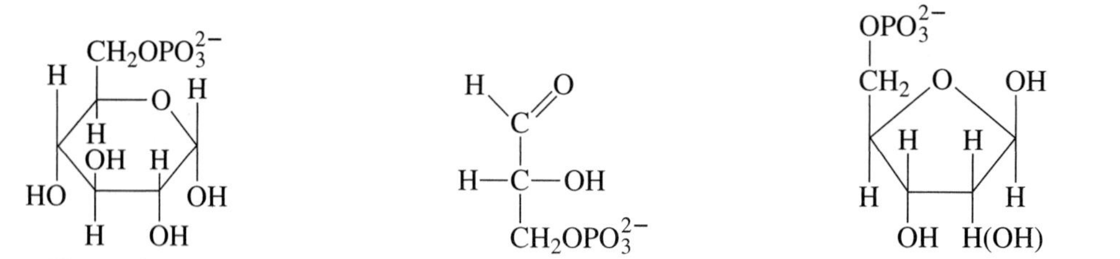

(a) D-Glucose 6-phosphate   (b) D-Glyceraldehyde 3-phosphate   (c) D-Deoxyribose 5-phosphate
                                                                                          (D-Ribose 5-phosphate)

Figure 8.12

Important phosphate esters of carbohydrates: (a) D-glucose 6-phosphate, (b) D-glyceraldehyde 3-phosphate, (c) D-deoxyribose 5-phosphate (or D-ribose 5-phosphate).

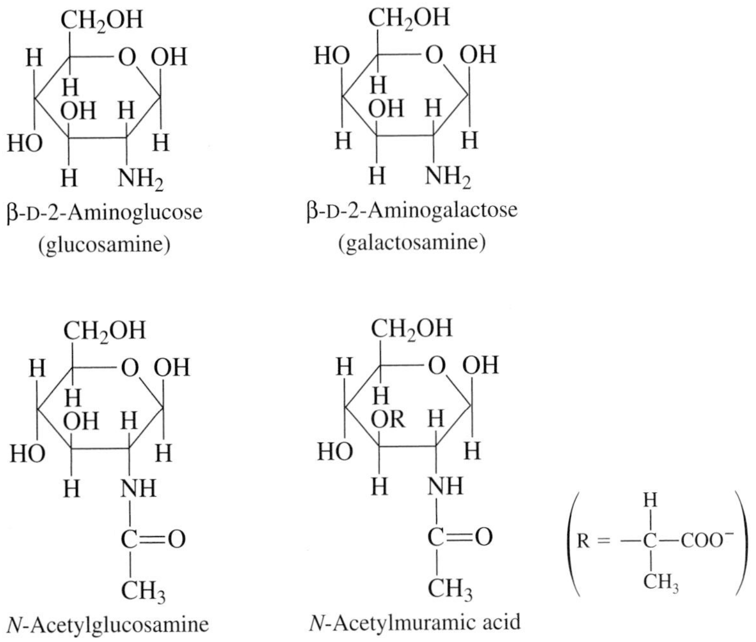

CH₂OH
H ——O OH
H
OH H
HO H
H NH₂

β-D-2-Aminoglucose
(glucosamine)

CH₂OH
HO ——O OH
H
OH H
H H
H NH₂

β-D-2-Aminogalactose
(galactosamine)

CH₂OH
H ——O OH
H
OH H
HO H
H NH
|
C=O
|
CH₃

N-Acetylglucosamine

CH₂OH
H ——O OH
H
OR H
HO H
H NH
|
C=O
|
CH₃

N-Acetylmuramic acid

$$\left(R = -\underset{CH_3}{\overset{H}{\underset{|}{\overset{|}{C}}}}-COO^- \right)$$

Figure 8.13

Important amino derivatives of carbohydrates. These compounds result from the substitution of a hydroxyl group with an amino group. Note that the most common position of substitution is at C2 of the carbohydrate. Two important amino sugars are glucosamine and galactosamine. In some amino sugars, the amino group is acetylated, converted to its amide with the acetyl group. Two common acetylated amino sugars are *N*-acetylglucosamine and *N*-acetylmuramic acid.

sulfate, a polymeric component of cartilage. The acidic amino sugar *N*-acetylneuraminic acid (sialic acid), derived from the carbohydrate rhamnose, is a component of glycoproteins (see Section 8.5) and glycolipids (see Chapter 9, Section 9.3).

Glycoside Formation

When carbohydrates are reacted with hydroxyl groups under anhydrous, mildly acidic conditions, a new linkage, an *O*-glycosidic bond, is formed (Figure 8.14). The alcohol (ROH) reacts with the anomeric hydroxyl group (on C1 or C2) to form a glycosidic bond. In this example, the product, a **glycoside,** would be called methyl-β-glucopyranoside if the alcohol (ROH) is methanol, CH₃OH. The hydroxyl group of another carbohydrate molecule can substitute for the alcohol in the above reaction, linking together two monosaccharides (Figure 8.15). In the glycoside product, one glucose is linked via its anomeric hydroxyl group (on C1) to the hydroxyl group on C4 of the second glucose unit. Because the anomeric configuration is α and the hydroxyl groups linked together are on the 1 and 4 carbon atoms of

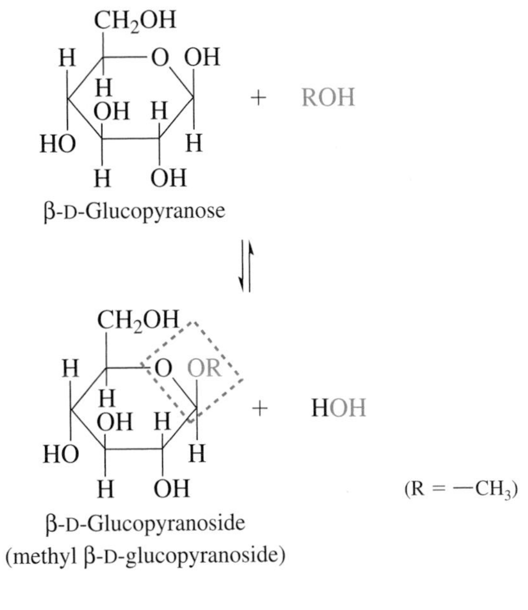

CH₂OH
H ——O OH
H
OH H
HO H
H OH

β-D-Glucopyranose   +   ROH

CH₂OH
H ——O OR
H
OH H
HO H
H OH

β-D-Glucopyranoside   +   HOH
(methyl β-D-glucopyranoside)

(R = —CH₃)

Figure 8.14

A monosaccharide, β-D-glucopyranose, reacts with an alcohol to form a glycoside. The atoms of the glycosidic bond are outlined by the dashed lines. If the alcohol is methanol, the product glycoside is methyl-β-glucopyranoside.

the glucose residues, the new linkage is called an $\alpha(1 \rightarrow 4)$ glycosidic bond. This specific compound is called **maltose.** If the anomeric configuration of the glucose unit connected to the glycosidic bond at C1 is β, the new bond is classified as $\beta(1 \rightarrow 4)$. Two glucose units linked in this manner form the compound **cellobiose** (Figure 8.16), a structural unit of cellulose. The glycosidic linkages in maltose and cellobiose do not equilibrate like the acetal centers of carbohydrates; hence, the alpha and beta configurations are locked into position. *O*-Glycosidic bonds are the basis for linking together carbohydrates (monosaccharides) into the larger carbohydrates, **disaccharides** and **polysaccharides.** Several varieties of *O*-glycosidic bonds are possible in the formation of more complex carbohydrates. In addition to α and $\beta(1 \rightarrow 4)$ bonds, the linkages may be α or $\beta(1 \rightarrow 6)$, α or $\beta(1 \rightarrow 2)$ or even $\alpha\alpha$ or $\beta\beta(1 \rightarrow 1)$. Figure 8.16 shows disaccharides resulting from some of these glycosidic bond variations. The disaccharides shown in the figure are the most abundant in nature. Maltose, a product from the hydrolytic breakdown of starch, is similar to cellobiose, a hydrolysis product of cellulose. Inspection of the structures of maltose and cellobiose provides a preview of the chemical linkages that hold together the polysaccharides starch and cellulose. The disaccharide maltose can be easily digested by humans because of the presence of enzymes that catalyze the hydrolysis of $\alpha(1 \rightarrow 4)$ glycosidic bonds. In contrast, cellobiose is indigestible by humans because we lack enzymes capable of catalyzing the hydrolysis of $\beta(1 \rightarrow 4)$ glycosidic bonds between glucose residues. The disaccharide **lactose** consists of the monosaccharides galactose and glucose in a $\beta(1 \rightarrow 4)$ linkage. (Galactose is a C4 epimer of

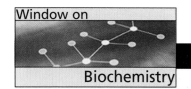

Window on

Biochemistry

8.1 Can You Tolerate Lactose?

Most infants and children have abundant levels of the intestinal enzyme **lactase (β-galactosidase)** so they are able to digest dietary lactose (milk sugar) and use the monosaccharide products, glucose and galactose, which are readily absorbed from the intestines and metabolized. However, in approximately 75% of the world's adults (90% of Asian and African descent; 75% of Mexican and Jewish ancestry; 10–20% of northern/central Europeans), lactase levels decline with age and they are increasingly unable to digest lactose. Lactose in these individuals remains in the small intestines where it is degraded by bacteria, producing the gases, CO_2 and H_2, as well as organic acids that cause gastrointestinal distress including bloating, nausea, cramping, and diarrhea. The medical term for this condition is **lactose intolerance.** Physicians usually recommend that lactose intolerant individuals restrict their intake of the disaccharide that is abundant in milk, yogurt, cheeses, sour cream, ice cream, and other dairy products. For those who enjoy dairy foods, there are now products available that contain lowered levels of lactose. Lactose-reduced products such as milk, ice cream, cottage cheese, and yogurt are prepared by using the enzyme β-galactosidase, which catalyzes the hydrolysis of the $\beta (1 \rightarrow 4)$-glycosidic linkage in lactose. The most popular enzyme is obtained from cultures of the yeast, *Saccharomyces lactis.* This particular enzyme is especially useful because it is active even at low temperatures (4° C); hence, it can be added to milk that is cooled during overnight storage or shipping. Lactose digestion by the enzyme also aids in the production of cheese, yogurt, and ice cream. Lactose is not very soluble and, especially at cold temperatures, tends to crystallize out of solution if concentrations are not reduced. The use of yeast β-galactosidase does have one major disadvantage compared to other enzymes. Complete hydrolysis of all lactose present is not possible because the enzyme is inhibited by the galactose produced in the reaction.

If you are lactose intolerant and want to avoid the extra expense of purchasing already-prepared, lactose-reduced foods, there are other things you can do. For example, you could purchase yogurt that has "active cultures" of live bacteria. The bacteria help digest the lactose by producing β-galactosidase in your digestive system. Also, tablets of β-galactosidase (*Aspergillus niger* or yeast) are available for home use. These may be used by mixing with dairy foods or eating the tablets before ingesting dairy products.

The lactose content in dairy products may be reduced by adding β-galactosidase.

© Leonard Lessin/Peter Arnold, Inc

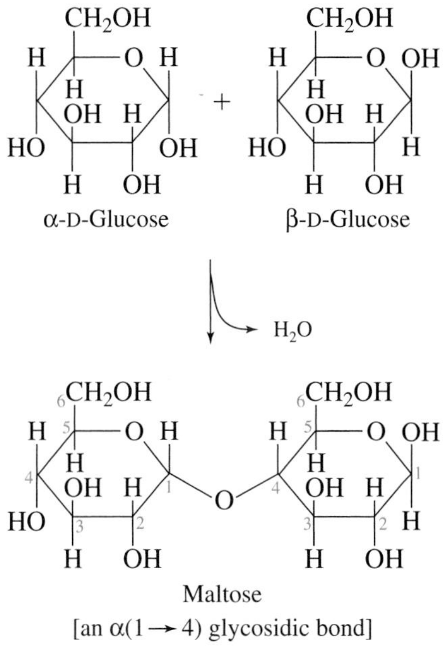

Figure 8.15
Combining two monosaccharide molecules, D-glucose, with a glycosidic linkage to produce the disaccharide maltose. The glycosidic bond is defined as $\alpha(1 \rightarrow 4)$, because the bond is between C1 of a glucose in α form and C4 of the other glucose.

glucose.) Note that galactose provides the anomeric hydroxyl group at C1 and glucose the hydroxyl group from C4 to make the glycosidic bond. The official name for lactose, galactose-$\beta(1 \rightarrow 4)$-glucose, indicates how the monosaccharides are linked together. The disaccharide lactose, also called milk sugar, is found only in dairy products (see Window on Biochemistry 8.1). When lactose is consumed in the human diet, its hydrolysis is catalyzed by the enzyme lactase (β-galactosidase) to produce the free monosaccharide units, glucose and galactose:

$$\text{lactose} + H_2O \xrightleftharpoons{\text{lactase}} \text{galactose} + \text{glucose}$$

Sucrose, common table sugar, is a disaccharide of glucose and fructose linked via an α, $\beta(1 \rightarrow 2)$ glycosidic bond (see Figure 8.16). Both α and β terms are necessary to define this glycosidic bond because the anomeric carbon of glucose (C1) in α configuration is

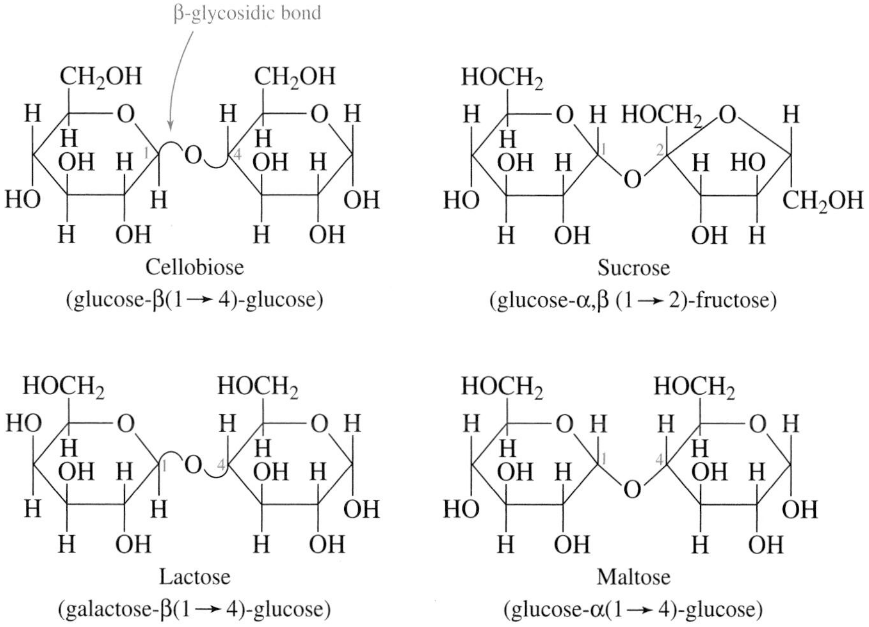

Figure 8.16
Important disaccharides formed by linking monosaccharides with *O*-glycosidic bonds. Cellobiose is a disaccharide composed of two glucose residues linked by a $\beta(1 \rightarrow 4)$ glycosidic bond. Sucrose, common table sugar, is a disaccharide of glu-cose and fruc-tose linked by an α, $\beta(1 \rightarrow 2)$ glycosidic bond. Lactose, milk sugar, is a disaccharide of galactose and glucose linked by a $\beta(1 \rightarrow 4)$ glycosidic bond. Maltose is a disaccharide consisting of two glucose residues linked by an $\alpha(1 \rightarrow 4)$ glycosidic bond.

Table 8.2
Common disaccharides and their structural properties

Name	Monosaccharide Components	Type of Glycosidic Linkage
Maltose	Glucose, glucose	$\alpha(1 \rightarrow 4)$
Cellobiose	Glucose, glucose	$\beta(1 \rightarrow 4)$
Lactose	Galactose, glucose	$\beta(1 \rightarrow 4)$
Sucrose	Glucose, fructose	$\alpha\beta(1 \rightarrow 2)$

linked to the anomeric carbon of fructose (C2) in β configuration. Because both anomeric carbons are tied into a glycosidic bond, neither is able to form an open chain containing a free aldehyde or ketone; therefore, sucrose is a nonreducing sugar. When consumed in the diet of humans, sucrose is hydrolyzed to glucose and fructose. Large amounts of sucrose are ingested in the average American diet. Because of potential health problems and increased caloric intake that can result from excessive sucrose, other sweetening agents are being used. Aspartame, a dipeptide of aspartate and phenylalanine (see Chapter 4, Section 4.2), is widely used as a low-calorie sweetener. Fructose and mixtures of fructose and glucose (invert sugar) are extensively used in soft drinks. Fructose is sweeter than sucrose, so less is required for the same level of sweetness. Names and structural properties for the important disaccharides—maltose, cellobiose, lactose, and sucrose—are given in Table 8.2.

The N—H group of amines can substitute for hydroxyl groups and react at the anomeric carbon center of carbohydrates (Figure 8.17a). The new linkage is called an *N*-glycosidic bond. This type of bond is of paramount importance in the construction of nucleotides such as ATP and in the nucleic acids RNA and DNA. The amine functional group in these cases is a nitrogen atom in the heterocyclic purine and pyrimidine bases. In Figure 8.17b is shown the *N*-glycosidic bond resulting from the carbohydrate ribose combining with the purine base, adenine. (Is the *N*-glycosidic bond shown in Figure 8.17b α or β?)

8.4 Polysaccharides

Serving as monomeric units, the monosaccharides and their derivatives are linked together to form a wide variety of polysaccharides that play diverse biological roles. The stability and variety of *O*-glycosidic bonds make it possible for monosaccharides to combine into structurally distinct and biologically useful polymers. To define the structure of a polysaccharide, several structural features must be recognized:

1. The identity of the monomeric units.
2. The sequence of monosaccharide residues (if more than one kind is present).

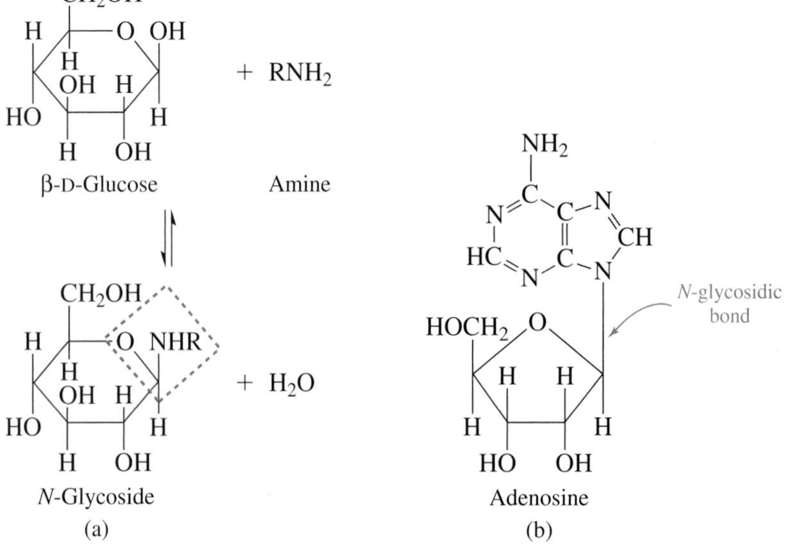

Figure 8.17
(a) The formation of an *N*-glycosidic bond between β-D-glucose and an amine. The atoms in the dashed rectangle comprise the *N*-glycosidic bond. (b) The structure of adenosine, part of ATP. Note the presence of an *N*-glycosidic bond between a nitrogen atom in adenine and the anomeric carbon of the monosaccharide ribose.

3. The types of glycosidic bonds linking the units.
4. The approximate length of the chain (the approximate number of monosaccharide units).
5. The degree of branching.

Homopolysaccharides are composed of a single type of monosaccharide unit, whereas heteropolysaccharides contain two or more types of monosaccharides. The term **oligosaccharide** is used to denote polysaccharides with a small number of monosaccharides (usually fewer than ten). Glucose and its derivatives are the most common monomeric units; however, other monosaccharides, including fructose, galactose, and their derivatives, are found in natural polysaccharides. The polysaccharides differ from proteins in that they occur in variable sizes. They are mixtures of polymers with varying lengths; hence, they have differing molecular weights. Proteins, composed of a specific sequence and composition of amino acids, have definite molecular weights. The roles performed by the polysaccharides in energy storage and biological structure do not require molecules of a reproducible and well-defined size. In some polysaccharides, the individual strands are cross-linked by short peptides. These compounds, the peptidoglycans, are components of bacterial cell walls. In this section, we introduce the polysaccharides in terms of their biological functions, beginning with the storage polysaccharides and then continuing with those serving structural roles.

Storage Polysaccharides

Plants and animals store the energy molecule, glucose, in **starch** and **glycogen,** respectively. These polysaccharides are stored in the cell in cytoplasmic packages called granules (Figure 8.18). Starch is present in the chloroplasts of plant cells, where it is produced by photosynthetic energy (see Chapter 17). Starch is especially abundant in potatoes, corn, and wheat. Glycogen granules are present primarily in liver and muscle cells of animals. Because of the numerous hydroxyl groups on the two polysaccharides, much water is associated by hydrogen bonding. In fact, each gram of glycogen stored in liver or muscle tissue is hydrated with 2 g of water.

Starch is a mixture of two types of polymeric glucose, amylose and amylopectin. Amylose is a linear, unbranched chain of D-glucose units linked by $\alpha(1 \longrightarrow 4)$ O-glycosidic bonds (Figure 8.19a). Because of the nature of the linkages, there are two distinct ends to the polymer: a nonreducing end (the end glucose has a free hydroxyl group on C4) and a reducing end (the end glucose has an anomeric carbon center that is in equilibrium with an open aldehyde chain). The molecular weight of the amylose chain may range from a few thousand to 500,000. Amylopectin has two important structural features: (1) a main backbone composed of glucose units linked by $\alpha(1 \longrightarrow 4)$ glycosidic bonds (exactly like amylose) and (2) branches connected to the backbone via $\alpha(1 \longrightarrow 6)$ glycosidic bonds (Figure 8.19b). The branch points occur on the average of every 25th glucose residue. The average molecular weight of amylopectin is about 1 million. The branching in amylopectin leads to a polymeric molecule that has one reducing end and many nonreducing ends. When starch is ingested in the human diet, its degradation begins in the mouth. The salivary enzyme, α-amylase,

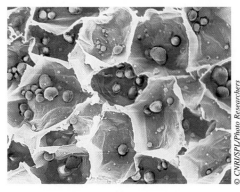

(a)

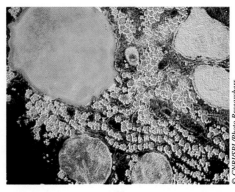

(b)

Figure 8.18

Cellular segregation of storage polysaccharides in cytoplasmic granules shown in electron micrographs: (a) starch granules *(green)* in plant cell chloroplasts and (b) glycogen granules *(pink)* in liver cells.

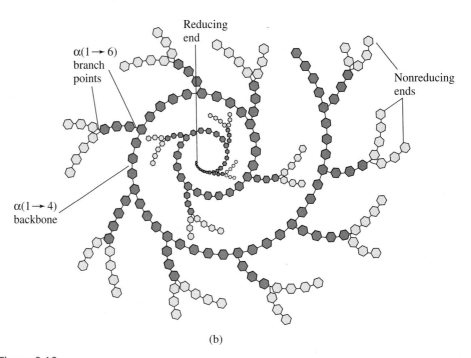

Figure 8.19

(a) The structure of amylose, an unbranched polymer of glucose. The glucose units are held together by $\alpha(1 \rightarrow 4)$ glycosidic bonds. Each amylose molecule has two distinct ends, a nonreducing end (glucose with a free hydroxyl group on C4) and a reducing end (glucose with a free hydroxyl group on C1). (b) The structure of amylopectin and glycogen, branched polymers of glucose. Each hexagon represents a glucose residue. The main chain of glucose units are held together by $\alpha(1 \rightarrow 4)$ glycosidic bonds. Branches from the main chain originate by $\alpha(1 \rightarrow 6)$ glycosidic bonds. Amylopectin has a branch at about every twenty-fifth glucose whereas glycogen is branched at every tenth glucose. Note the presence of one reducing end but several nonreducing ends.

catalyzes the hydrolysis of *O*-glycosidic bonds, yielding the disaccharide maltose and oligosaccharide products. (More detail on starch metabolism is found in Chapter 15).

Animals store glucose for energy metabolism in the highly branched polymer, **glycogen** (Figure 8.19b). This polymer is identical to amylopectin except that it has more numerous $\alpha(1 \rightarrow 6)$ branches (about one every tenth glucose residue in the main chain) and a much higher average molecular weight (several million). Figure 8.20, showing a main chain of $\alpha(1 \rightarrow 4)$ glycosidic linkages and $\alpha(1 \rightarrow 6)$ branches, represents both amylopectin and glycogen. Like amylopectin, glycogen consists of a single reducing end and many nonreducing ends. This is of significance in the mobilization (release) of glucose units for energy metabolism. As we shall discover in Chapter 15, the enzyme glycogen phosphorylase cat-

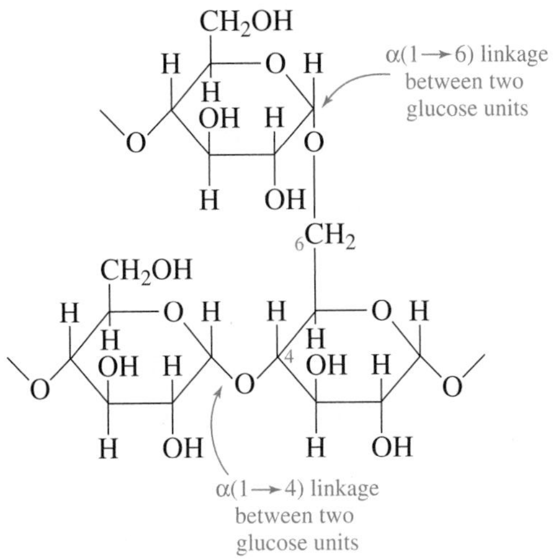

CH₂OH

$\alpha(1 \longrightarrow 6)$ linkage between two glucose units

CH₂OH

$_6$CH₂

$\alpha(1 \longrightarrow 4)$ linkage between two glucose units

Figure 8.20
The chemical structure of glycogen, a storage polysaccharide in animals and amylopectin, a branched, plant polysaccharide. The main chain is composed of glucose linked in $\alpha(1 \longrightarrow 4)$ glycosidic bonds. Branches are formed by an $\alpha(1 \longrightarrow 6)$ glycosidic linkage. Amylopectin has an $\alpha(1 \longrightarrow 6)$ branch from each 25th glucose unit in the main chain. Glycogen is more highly branched, with $\alpha(1 \longrightarrow 6)$ glycosidic bonds at every tenth glucose residue.

alyzes the removal of glucose residues from the numerous nonreducing ends, providing abundant supplies of glucose when necessary. Although glycogen is present both in liver and muscle cells, it is much more abundant in hepatic cells, where it may account for as much as 10% of the wet weight. Only about 1% of the weight of muscle cells is glycogen.

The extended chains and the angle of the $\alpha(1 \longrightarrow 4)$ glycosidic bond of amylose, amylopectin, and glycogen make it possible for them to fold into tightly coiled helical structures as shown in Figure 8.21. Approximately six glucose residues make up one turn of the helix. (Compare this helix to the α-helix structure of proteins.)

Other minor forms of storage polysaccharides are present in nature. **Dextran,** found in yeasts and bacteria, consists of glucose residues connected into a main chain via $\alpha(1 \longrightarrow 6)$ glycosidic linkage with occasional branches formed by $\alpha(1 \longrightarrow 2)$, $\alpha(1 \longrightarrow 3)$, and $\alpha(1 \longrightarrow 4)$ glycosidic bonds. Bacteria growing on teeth produce extracellular dextran that accumulates and becomes an important component of dental plaque. **Inulin,** a homopolymer of D-fructose connected by $\beta(2 \longrightarrow 1)$ glycosidic linkages, is found in artichokes and other vegetables.

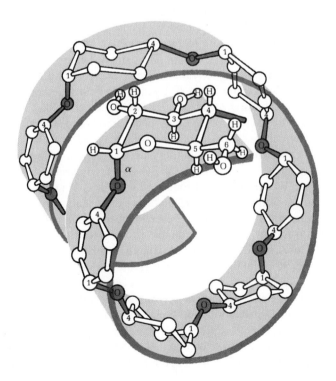

Figure 8.21
The main backbone of amylose, amylopectin, and glycogen made up of $\alpha(1 \longrightarrow 4)$ glycosidic linkages folds into a helical arrangement. One turn of the helix is composed of six glucose units. *Source:* © Irving Geis. Geis Archives Trust. Rights owned by Howard Hughes Medical Institute. Not to be used without permission.

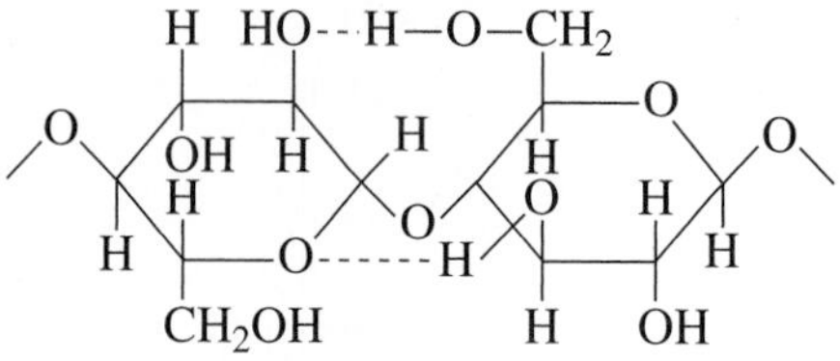

Figure 8.22

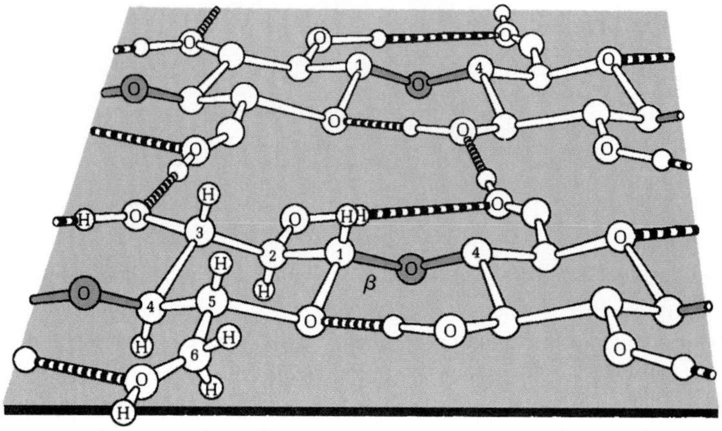

The chemical and three-dimensional structures of cellulose: (a) The glucose units are held together in β(1 → 4) glycosidic bonds. Note the rotation around the β(1 → 4) bond, which brings together glucose functional groups for hydrogen bonding. (b) Two parallel cellulose chains are shown with hydrogen bond crosslinks. These strong and rigid networks provide scaffolding for plant cell walls. *Source:* Part b, © Irving Geis. Geis Archives Trust. Rights owned by Howard Hughes Medical Institute. Not to be used without permission.

(a)

(b)

Structural Polysaccharides

Some polysaccharides such as **cellulose, chitin,** and **mucopolysaccharides** are synthesized inside cells but extruded to the outside to provide a protective wall or lubricative coating to cells. Cellulose is the major structural component of wood and plant fibers. This glucose homopolymer is extremely abundant in nature, making up over 50% of the organic matter in the biosphere. It is an unbranched polymer connecting D-glucose units together by β(1 → 4) glycosidic linkages (Figure 8.22). The average molecule of cellulose contains between 10,000 and 15,000 glucose residues. The β configuration of the glycosidic bonds allows cellulose to form very long and straight chains that differ from the helical coils of starch and glycogen. The extended chains of cellulose can associate into bundles of parallel chains called fibrils. These strong and rigid networks, which provide the scaffolding for plant cell walls, are held together by intra- and intermolecular hydrogen bonding. This strong, stabilized, sheetlike arrangement also makes cellulose a useful product for clothing, paper, cardboard, and building materials. Although a portion of our diet consists of vegetables and fruits containing cellulose, we are unable to extract cellular energy from this glucose polymer. Animals do not produce enzymes that can catalyze the hydrolysis of β(1 → 4) glycosidic bonds between the glucose residues in cellulose. Wood-rot fungi and some bacteria obtain nutrient glucose by synthesizing and secreting enzymes (cellulases) that catalyze the hydrolysis of the β(1 → 4) glycosidic bonds in cellulose. These organisms are largely responsible for the decaying of dead wood in forests. Ruminant animals (cattle, sheep, goats, camels, giraffes) are able to use cellulose as a nutrient source because their second stomachs contain bacteria that secrete cellulases. In a similar fashion, termites can digest wood cellulose because their intestinal tracts contain *Trichonympha,* a symbiotic microorganism that produces cellulase. Cellulose and its derivatives do, however, perform an essential role in the diets of animals as bulk fiber or "roughage" to assist the digestion and absorption of nutrients.

Another important polysaccharide component of plant cell walls is **pectin,** a polymer of D-galacturonic acid (Figure 8.23), a C4 epimer of the glucose derivative, D-glucuronic

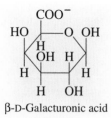

β-D-Galacturonic acid

Figure 8.23

The structure of D-galacturonic acid, the monomeric unit of the plant cell wall polysaccharide, pectin.

Chitin is the major structural polysaccharide in the exoskeleton of the ghost crab (*Ocypode quadrata*).

acid. In galacturonic acid, the —CH_2OH group on C6 of galactose is oxidized to a carboxylic acid (—COOH). Pectin extracted from plants is used as a gelling agent in the preparation of jams and jellies.

The protective exoskeletons of arthropods (insects, crabs, lobsters) are composed primarily of the unbranched homopolysaccharide chitin. This polymer is also found in smaller amounts in the cell walls of yeasts, fungi, and algae. The monomeric building block of chitin is the glucose derivative, *N*-acetylglucosamine, linked in β(1 → 4) glycosidic bonds (Figure 8.24). Like cellulose, chitin exists in extended chains that are associated into fibers by intra- and intermolecular hydrogen bonding. Because humans have no enzymes that can catalyze the hydrolysis of the β(1 → 4) linkages in chitin, it is indigestible.

Some structural polysaccharides are found in connective tissue (cartilages and tendons) or extracellular matrix (ground substance) of higher animals. The extracellular matrix is a gellike material that acts as a glue to hold cells together. A group of polymers, called **mucopolysaccharides,** provide a thin, viscous, jellylike coating to cells. The most abundant in this class of polysaccharides is hyaluronic acid, which has alternating monomeric units of *N*-acetylglucosamine and D-glucuronic acid. In this acidic polymer, C6 of each glucose derivative is oxidized from the level of —CH_2OH (alcohol) to the level of —COOH (carboxylic acid). The connective linkages between the two types of monomeric units are alternating β(1 → 4) and β(1 → 3) glycosidic bonds (Figure 8.25). Hyaluronic acid serves as a lubricant in the synovial fluid of joints and is found in the extracellular matrix of connective tissue. Another component of the extracellular matrix is chondroitin sulfate, a heteropolymer mucopolysaccharide. Like hyaluronic acid, chondroitin sulfate is composed of two alternating monomer units, *N*-acetylgalactosamine sulfate and D-glucuronic acid. They are joined by alternating β(1 → 3) and β(1 → 4) glycosidic bonds (Figure 8.26). The monomer unit, derived from galactose, consists of *N*-acetylgalactosamine (similar to *N*-acetylglucosamine) with a sulfuryl group (—SO_3^{2-}) in ester linkage to the hydroxyl group on C4.

Figure 8.24

The structure of chitin, a polymer of *N*-acetylglucosamine. The monosaccharides are held together by β(1 → 4) glycosidic bonds.

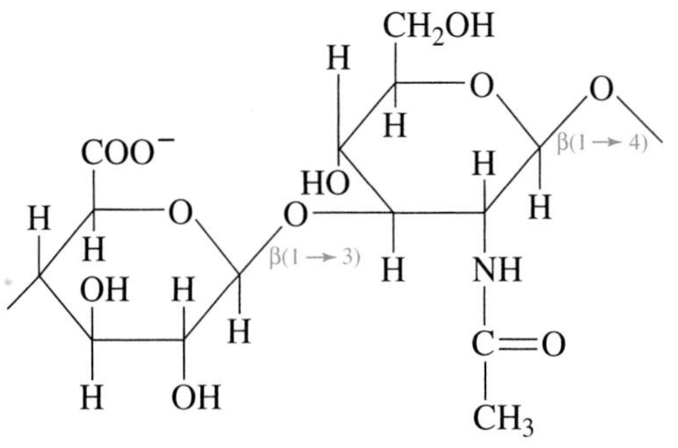

Figure 8.25
The structural unit of hyaluronic acid, a polymer consisting of alternating N-acetylglucosamine and D-glucuronic acid units. The glycosidic bonds alternate between $\beta(1 \rightarrow 3)$ and $\beta(1 \rightarrow 4)$.

Structural Peptidoglycans

The rigid cell walls of bacteria, which provide physical protection, are composed primarily of an unbranched, heteropolymer of alternating N-acetylglucosamine and N-acetylmuramic acid. The monomer units are linked into an extended backbone by $\beta(1 \rightarrow 4)$ glycosidic bonds. The rigidity and strength of bacterial cell walls is the consequence of a network of peptide crosslinks between strands of the linear polysaccharides. The amino acid composition and sequence of the peptide links vary from bacterium to bacterium. In the gram positive bacterium *Staphylococcus aureus,* two sets of peptides, a tetrapeptide and a pentapeptide of five glycine residues, form the crosslinks (Figure 8.27).

Table 8.3 summarizes the composition and biological roles of the polysaccharides introduced in this section.

8.5   Glycoproteins

Proteins that carry covalently bonded carbohydrate units are called **glycoproteins.** These proteins are involved in many biological functions, including immunological protection, cell–cell recognition events, blood clotting, and host–pathogen interactions. In addition, in Chapter 9 we shall discover that many proteins embedded in plasma membrane structures are glycoproteins.

Glycoprotein Structure

The carbohydrate portion of a glycoprotein usually constitutes 1% to 30% of the total weight, although some glycoproteins contain as much as 50% to 60% carbohydrate. The most common monosaccharides found in glycoproteins are glucose, mannose, galactose, fucose, N-acetylgalactosamine, N-acetylglucosamine, and sialic acid (N-acetylneuraminic acid). Structures for these sugars are shown in Figure 8.28. They are attached to proteins as branched oligosaccharide units, usually containing fewer than 15 carbohydrate residues.

There are two important types of linkages involved in the attachment of sugars to proteins: (1) O-glycosidic bonds using the hydroxyl groups of serine and threonine residues in the proteins to react with carbohydrates and (2) N-glycosidic bonds using the side chain

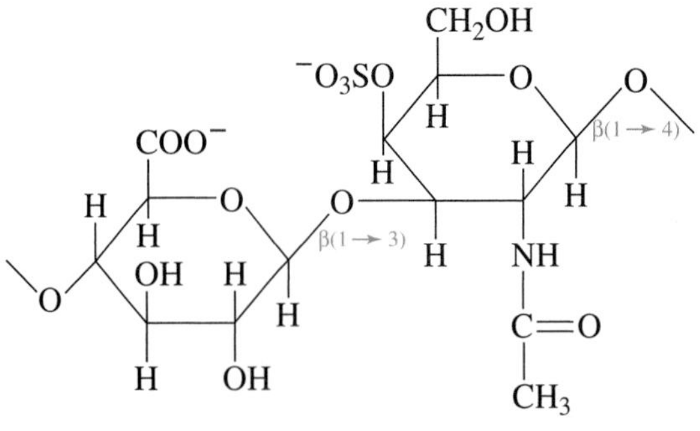

Figure 8.26
The structure of chondroitin sulfate, a component of the extracellular matrix. It is an alternating polymer of N-acetylgalactosamine 4-sulfate and glucuronic acid units held together by alternating $\beta(1 \rightarrow 4)$ and $\beta(1 \rightarrow 3)$ glycosidic bonds.

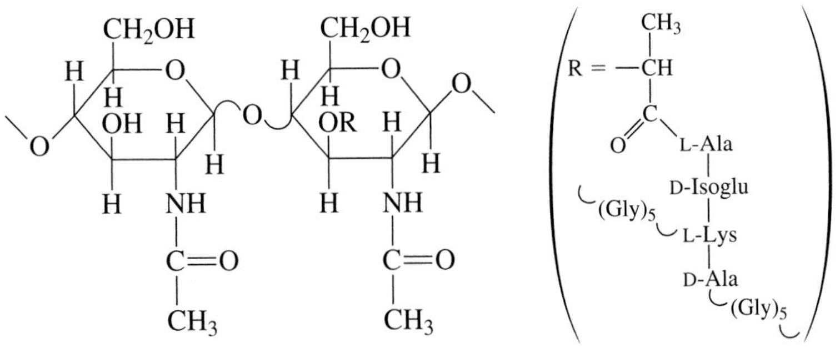

Figure 8.27
Peptidoglycans that form the cell walls of bacteria. The polysaccharide is composed of alternating *N*-acetylglucosamine and *N*-acetylmuramic acid residues linked by $\beta(1 \rightarrow 4)$ glycosidic bonds. Crosslinks between polysaccharide chains are formed by a tetrapeptide and a pentapeptide of glycine residues.

amide nitrogen of the amino acid residue asparagine to link carbohydrates. Examples of common *O*- and *N*-linked carbohydrate units are shown in Figure 8.29. Additional carbohydrates are usually linked to the first monosaccharide on the glycoprotein. Figure 8.30 shows two examples of oligosaccharide groups common in glycoproteins.

Glycoprotein Function

⏺ **Cutting Edge: Molecular recognition**

A wide variety of oligosaccharide units attached to proteins is found throughout the plant and animal kingdoms. As experimental methods for the detection and analysis of oligosaccharides have improved, more and more proteins involved in a diverse array of biological processes have been identified as glycoproteins. Although the function for many glycoproteins is known, the specific role played by the oligosaccharide unit of a glycoprotein is often unknown. It has recently been discovered that oligosaccharides attached to proteins play a major role in cell–cell and cell–molecule recognition. Glycoproteins with unique oligosaccharide compositions and sequences are said to be "information rich." They are frequently found on cell surfaces, where they serve as markers to identify the specific type of cell. The determinants of the common ABO human blood group system are carried by glycoproteins. Each type of cell is marked on its surface with a glycoprotein containing a specific oligosaccharide unit that has

Table 8.3
Structures and properties of common polysaccharides

Name	Type[a]	Components and Linkage	Biological Function
Starch			
Amylose	Homo	Glucose, $\alpha(1 \rightarrow 4)$	Nutrient storage (plants)
Amylopectin	Homo	Glucose, $\alpha(1 \rightarrow 4)$ with $\alpha(1 \rightarrow 6)$ branches	Nutrient storage (plants)
Glycogen	Homo	Glucose, $\alpha(1 \rightarrow 4)$ with $\alpha(1 \rightarrow 6)$ branches	Nutrient storage (animals)
Dextran	Homo	Glucose, $\alpha(1 \rightarrow 6)$ with $\alpha(1 \rightarrow 2)$, $\alpha(1 \rightarrow 3)$, and $\alpha(1 \rightarrow 4)$ branches	Nutrient storage (yeast and bacteria)
Inulin	Homo	Fructose, $\beta(2 \rightarrow 1)$	Nutrient storage (plants)
Cellulose	Homo	Glucose, $\beta(1 \rightarrow 4)$	Structural function in plants
Pectin	Homo	Galacturonic acid	Structural rigidity in plants
Chitin	Homo	*N*-Acetylglucosamine, $\beta(1 \rightarrow 4)$	Structural function in exoskeletons
Hyaluronic acid	Hetero	*N*-Acetylglucosamine; glucuronic acid, $\beta(1 \rightarrow 4)$; $\beta(1 \rightarrow 3)$	Lubricant in synovial fluid, extracellular matrix
Chondroitin sulfate	Hetero	*N*-Acetylgalactosamine sulfate; glucuronic acid; $\beta(1 \rightarrow 3)$ and $\beta(1 \rightarrow 4)$	Lubricant in synovial fluid; extracellular matrix
Peptidoglycans	Hetero, with peptide crosslinks	*N*-Acetylglucosamine; *N*-acetylmuramic acid $\beta(1 \rightarrow 4)$	Structural function in bacterial cell walls

[a]Homopolymeric or heteropolymeric.

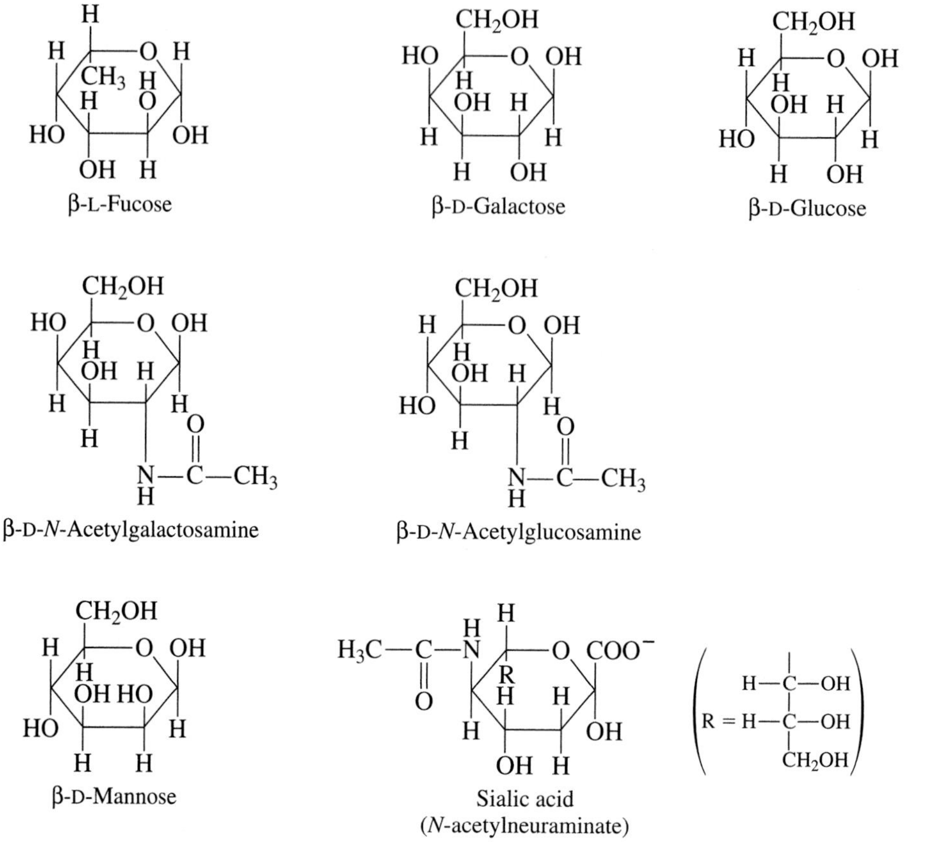

Figure 8.28

Monosaccharides most commonly found in glycoproteins. They are linked to amino acid side chains via *N*- or *O*-glycosidic bonds.

β-L-Fucose

β-D-Galactose

β-D-Glucose

β-D-*N*-Acetylgalactosamine

β-D-*N*-Acetylglucosamine

β-D-Mannose

Sialic acid
(*N*-acetylneuraminate)

a unique nonreducing end. The marker for the A-type blood group determinant is a terminal *N*-acetylgalactosamine on the oligosaccharide branch. α-D-Galactose is the terminal residue of the B-group determinant. Minor differences in the composition and sequence of the sugar units can make the difference between an emergency room patient receiving a transfusion with compatible or incompatible blood type.

When cells transform from a normal state to a malignant state, changes take place in the glycoproteins on the cell surface. These changes are usually caused in the cancerous cell by a deficiency of enzymes, called glycosyltransferases, that attach carbohydrates to membrane proteins by forming *O*- and *N*-glycosidic bonds. Because of a phenomenon called contact

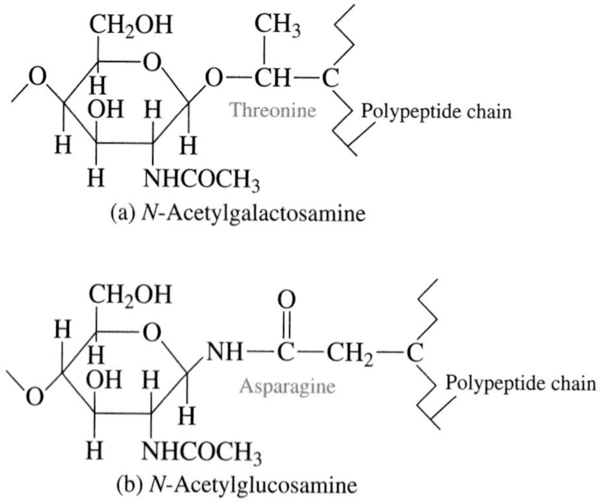

Figure 8.29

Two types of linkages to attach carbohydrates to proteins: (a) *O*-Glycosidic bond between the hydroxyl group on the anomeric carbon (C1) of the carbohydrate and a hydroxyl group of the side chain of a serine or threonine residue of the protein. (b) *N*-Glycosidic bond between the hydroxyl group on the anomeric carbon (C1) of the carbohydrate and the amide nitrogen of an asparagine residue in the protein.

(a) *N*-Acetylgalactosamine

(b) *N*-Acetylglucosamine

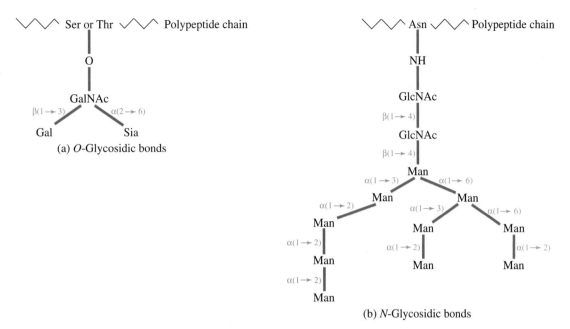

Figure 8.30

Examples of oligosaccharide units in glycoproteins. The amino acid residue in the protein to which the first monosaccharide is attached is shown and the type of glycosidic bond is listed for each carbohydrate added. GalNAc = N-acetylgalactosamine; Gal = galactose; Sia = sialic acid (N-acetylneuraminic acid); GlcNAc = N-acetylglucosamine; Man = mannose. (a) O-Glycosidic bond between protein and sugar. (b) N-Glycosidic bond between protein and sugar. This represents a high-mannose type of oligosaccharide unit.

inhibition, normal cells usually stop growing when their surfaces touch. Cancerous cells with altered cell surfaces do not recognize the message to stop growing.

Oligosaccharides are also used to mark proteins for age. Glycoproteins in the serum are constantly "turned over" with newly synthesized ones taking the place of the aged ones (Figure 8.31). "New" serum glycoproteins usually have terminal sialic acid residues on the carbohydrate portion. For example, the blood protein ceruloplasmin, which catalyzes iron oxidation ($Fe^{2+} \longrightarrow Fe^{3+}$) and transports serum copper, has ten such sialic acid terminal residues. Removing just two of the sialic acid units decreases the serum half-life of ceruloplasmin from 54 h to about 5 min. This appears to be a common pathway for removal of aged serum proteins. In general, removing carbohydrate units from glycoproteins almost always

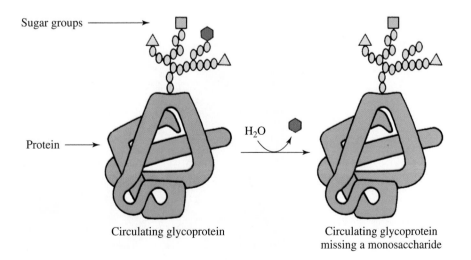

Figure 8.31

Turnover of serum proteins is initiated by removal of carbohydrate residues by hydrolysis.

increases their susceptibility to proteolytic degradation. Liver cells have on their surfaces receptor protein sites that recognize aged glycoproteins lacking sialic acid. Such glycoproteins are taken up by liver cells and degraded by proteolytic enzymes to free amino acids.

Many viruses, including herpes simplex, hepatitis B, influenza, and HIV, contain proteins on their surfaces that have potential sites for linkage of oligosaccharides. Carbohydrates may be linked to these proteins using host cell enzymes to produce glycoproteins that are indistinguishable from those on host cells. This allows the virus to thrive because it avoids

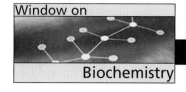

Window on

Biochemistry

8.2 Lectins—Proteins that Recognize Carbohydrates

The identification and characterization of glycoproteins have been greatly facilitated by a group of proteins called **lectins.** Lectins can bind reversibly with high affinity to specific carbohydrates (single or groups) in glycoproteins (see figure). For example, the lectin concanavalin A, from jack bean, has a high specific affinity for mannose but not glucose, its C-2 epimer. The mannose recognized by the lectin can be a nonreducing terminal residue or an internal residue in the oligosaccharide chain linked to the protein. Another lectin, wheat germ agglutinin, is able to recognize and bind to the trisaccharide of N-acetylglucosamine (GlcNAc-GlcNAc-GlcNAc). Protein–carbohydrate interactions involve primarily hydrogen bonds between carbohydrate hydroxyl groups and polar functional groups on amino acid residues.

Several hundred lectins, whose sources are plants, animals, and bacteria, have been isolated and characterized. This large number of lectins with a wide range of unique carbohydrate specificities makes it possible to evaluate the carbohydrate composition of glycoproteins, cell surfaces, and tissues. For example, lectins are used clinically as probes to diagnose changes occurring on the surfaces of cells that are becoming malignant. Mammalian lectins, called **selectins,** are cell adhesion proteins present on the outer surfaces of white blood cells (leukocytes). Here they are involved in regulating the inflammatory response system by binding to glycoprotein ligands on endothelial cells.

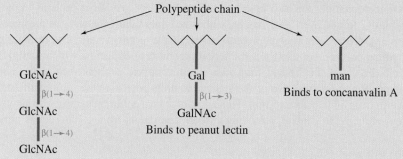

Binds to wheat germ agglutinin

Binds to peanut lectin

Binds to concanavalin A

Crystals of concanavalin B, a lectin that binds specific carbohydrate residues.

Courtesy of Dr. Alexander McPherson

immune surveillance by the host cell and also can attach and fuse to host cell receptors. For example, envelope protein, gp 120 of the HIV virus, has 20 to 25 asparagine residues that are potential sites for attachment, by *N*-glycosidic linkage, of oligosaccharide units. Drugs that influence addition of carbohydrates at these sites may be developed in the future to control the growth of the HIV virus.

Summary

■ Carbohydrates are naturally occurring compounds that have aldehyde or ketone functional groups and multiple hydroxyl groups. They serve in many different functions including energy metabolism, structural roles, and in cell communication.

■ Carbohydrates in aqueous solution form cyclic acetals or ketals by reaction of the aldehyde or ketone group with a hydroxyl group to form a five- or six-membered ring.

■ Monosaccharides can link together via *O*-glycosidic bonds to form disaccharides and polysaccharides.

■ The most abundant carbohydrates in nature are the polysaccharides starch, glycogen, and cellulose.

■ In glycoproteins, carbohydrates are linked to the proteins by two types of covalent bonds: (a) *O*-glycosidic bonds using the hydroxyl groups of serine and threonine and (b) *N*-glycosidic bonds using the side chain amide groups of asparagine. Glycoproteins are involved in many biological functions including immunological protection, cell–cell recognition, and host–pathogen interactions.

Study Exercises

Understanding Terms

8.1 Define the following terms in 25 words or less. Use chemical structures, if appropriate.

a. Monosaccharide	**k.** Chitin
b. Aldose	**l.** Glycoside
c. Glyceraldehyde	**m.** Lactose intolerance
d. Chiral center	**n.** Homopolysaccharide
e. Diastereoisomers	**o.** Glycogen granule
f. Ketohexose	**p.** Mucopolysaccharide
g. Furanose	**q.** Lectin
h. Cyclic hemiacetal	**r.** *O*-glycosidic bond
i. Anomeric center	**s.** Glycoprotein
j. Reducing sugar	**t.** Sialic acid

Reviewing Concepts

8.2 Draw all the possible structures for the triose family of monosaccharides.

8.3 How many chiral centers are present in each of the following monosaccharides?

a. Dihydroxyacetone	**f.** Sedoheptulose
b. Ribose	**g.** 2-Deoxyribose
c. Erythrulose	**h.** 6-Deoxyglucose
d. Glucosamine	**i.** *N*-acetylglucosamine
e. Fructose	**j.** Sialic acid

8.4 Use the Fischer projection method to draw each of the following monosaccharides.

a. D-Glyceraldehyde
b. L-Ribose
c. D-Mannose

8.5 Use the Fischer projection method to draw the D and L enantiomers of glucose.

8.6 The Fehling's reaction is a laboratory analysis used to test the presence of a reducing sugar. To complete the analysis, the unknown sugar is treated with a cupric salt, Cu^{2+}. A positive test for a reducing sugar is the formation of a red precipitate of Cu_2O. Which of the following carbohydrates would give a positive test?

a. Glucose	**d.** Lactose
b. Ribose 5-phosphate	**e.** Sucrose
c. Trehalose	**f.** Maltose

8.7 Study the structures of the compounds written below and list all the organic functional groups present in each molecule.

a. Glyceraldehyde
b. Glucose (Fischer projection)
c. Glucose (Haworth projection)
d. *N*-acetylglucosamine (Figure 8.13)
e. D-Galacturonic acid (Figure 8.23)

8.8 Write the name and draw the structure of a disaccharide or polysaccharide that has at least one of the following types of *O*-glycosidic bonds. Use any monosaccharides to construct the oligosaccharide.

a. β(1 ⟶ 4)
b. α(1 ⟶ 4)
c. α(1 ⟶ 6)

8.9 Which of the following amino acid residues in a protein could be a potential *O*-glycosylation site for attachment of an oligosaccharide unit?

a. Glycine   e. Serine
b. 4-Hydroxyproline   f. 5-Hydroxylysine
c. Asparagine   g. Threonine
d. Valine   h. Cysteine

8.10 Name a functional group in proteins that may serve as a glycosylation site.

8.11 What is the number for the anomeric carbon in each of the following monosaccharides? Assume the standard numbering system is used.

a. Glucose   d. Fructose
b. Ribose   e. Sedoheptulose
c. Galactose

8.12 Define the functional group present at each of the carbon atoms of β-D-fructofuranose.

a. C1
b. C2
c. C3

8.13 Name a specific biomolecule that is a member of each of the following classes.

a. Monosaccharide   d. Homopolysaccharide
b. Disaccharide   e. Heteropolysaccharide
c. Polysaccharide

8.14 Compare amylose starch and glycogen in terms of the following characteristics.

a. Synthesized in what kind of organism?
b. Biological role?
c. Type of saccharide? mono, di, poly, etc.?
d. Chemical structure?
e. Linkage between monosaccharide units?
f. Type of branching?

8.15 Name five monosaccharides that are common in glycoproteins.

8.16 Briefly describe two important functions of glycoproteins.

Solving Problems

8.17 What is the relationship between each pair of compounds listed below?

a. D-Glyceraldehyde:dihydroxyacetone
b. D-Glucose:D-fructose
c. D-Glucose:D-mannose
d. D-Threose:D-erythrose
e. D-2-Glucosamine:D-2-galactosamine
f. α-D-Glucose:β-D-glucose
g. D-Glucose:L-glucose

HINT: Choose from anomers, epimers, enantiomers, aldose–ketose pairs.

8.18 Why do glyceraldehyde and erythrose not exist in cyclic acetal structures, whereas ribose does?

8.19 Use the Haworth projection formula to draw each of the following monosaccharides.

a. α-D-Mannose
b. α-D-Glucose 6-phosphate
c. α-D-Deoxyribose

8.20 The disaccharide trehalose is a major constituent of hemolymph, the circulating fluid of insects, and is abundant in mushrooms and other fungi. It is composed only of two glucose units linked by an αα(1 → 1) glycosidic bond. Draw its structure. Is it a reducing carbohydrate?

8.21 Complete the following reactions of carbohydrates by drawing the structures of all organic products.

a.
$$\text{glyceraldehyde} \ + \ 2\ CH_3\overset{\displaystyle O}{\overset{\displaystyle \|}{C}}-O-\overset{\displaystyle O}{\overset{\displaystyle \|}{C}}CH_3 \ \longrightarrow$$

b. Dihydroxyacetone + $LiAlH_4 \longrightarrow$
c. Erythrose + $Cu^{2+} \longrightarrow$
d. α-D-Glucose + $CH_3OH \longrightarrow$

8.22 Which compound is more soluble in water, 1-hexanol or D-glucose? Explain.

8.23 Why is reaction (c) in Figure 8.10 classified as an oxidation–reduction reaction?

8.24 Write a reaction catalyzed by each of the following enzymes.

a. Cellulase   c. Lactase
b. α-Amylase   d. Ceruloplasmin

8.25 Why are all monosaccharides and disaccharides soluble in water?

8.26 Write structures to show the chemistry of each of the following reactions.

a. D-Glucose + ATP $\rightleftharpoons$ glucose 1-phosphate + ADP
b. Lactose + $H_2O \rightleftharpoons$ galactose + glucose
c. Glyceraldehyde 3-phosphate $\rightleftharpoons$ dihydroxyacetone phosphate
d. Glucose + NADH + $H^+ \rightleftharpoons$ sorbitol + NAD^+

8.27 Use one of the terms below to describe the chemistry occurring in parts a–d of Problem 8.26.

Oxidation–reduction
Hydrolysis
Phosphorylation
Isomerization

8.28 List a common name for an enzyme that would catalyze each of the reactions in Problem 8.26.

8.29 Suggest a treatment for lactose intolerance.

8.30 Draw the structure of D-galacturonic acid under the following conditions.
  a. at pH 1.0
  b. at pH 7.0
  c. at pH 12.0

8.31 Peanut lectin binds specifically to the disaccharide, galactose *N*-acetylgalactosamine. The two monosaccharides are linked by a β(1 ⟶ 3) glycosidic bond. Draw the structure of this disaccharide.
HINT: The nonreducing end is galactose

8.32 To model how lectins might bind to carbohydrates, use molecular structures to show the possible bonding interactions between glucose and the amino acid serine.

8.33 Draw molecular structures that show an *O*-glycosyl bond between serine and the C-1 of glucose.

8.34 Draw molecular structures that show an *N*-glycosyl bond between the C-1 of glucose and the amino acid asparagine.

Writing Biochemistry

8.35 Search the literature for a recent article on the general topic of lectins and selectins. Use one of the following journals: *Science, Nature, Scientific American, Journal of Chemical Education, Biochemistry and Molecular Biology Education,* or one similar. Write a 500-word summary of the article emphasizing concepts from this chapter.

8.36 Write a two-minute paper immediately after your professor's lecture on this chapter. Answer the following questions in the paper: (1) What were the two to three most important concepts? (2) What concepts were a review from organic chemistry? (3) Were there concepts you found particularly confusing?

 ## Applying Biochemistry

8.37 Carbohydrates such as galactose and mannose are often involved in molecular recognition processes (lectins and ricin, for examples). The recognition processes are dependent on the formation of favorable, noncovalent interactions between proteins and the carbohydrates. Draw the structure of a monosaccharide and label those functional groups that could form noncovalent interactions with amino acids in proteins. Define types of interactions (choose from hydrogen bonding, ionic, and hydrophobic) and select some amino acids that could participate in the bonding interactions.

Further Reading

Bentley, R., 1998. The transformation of linear to cyclic monosaccharide structures. *Biochem. Educ.* 26(3):205–210.

Boyer, R., 2000. Glycoproteins. In *Modern Experimental Biochemistry,* 3rd ed., pp. 321–331. San Francisco: Benjamin/Cummings.

Cole, C., and Smith, C., 1989. Glycoprotein biochemistry (structure and function). *Biochem. Educ.* 17:179–189.

Cole, C., and Smith, C., 1990. Glycoprotein biochemistry (biosynthesis). *Biochem. Educ.* 18:110–122.

Elgavish, S., and Shaanan, B., 1997. Lectin-carbohydrate interactions. *Trends Biochem. Sci.* 22:462–467.

Gahmberg, C., and Tolavanen, M., 1996. Why mammalian cell surface proteins are glycoproteins. *Trends Biochem. Sci.* 21:308–311.

Jentoft, N., 1990. Why are proteins O-glycosylated? *Trends Biochem. Sci.* 15:291–294.

Ruiz-Chica, A., et al., 1999. Stereochemistry of carbohydrates. *Biochem. Educ.* 27(1):7–8.

Ryan, M., 1998. Living with lactose intolerance. *Today's Chemist At Work,* Dec:49–52.

Sanchez, M., and Cabezas, J., 1998. Lectins in the laboratory. *Biochem. Educ.* 26:309–312.

Sharon, N., and Lis, H., 1993. Carbohydrates in cell recognition. *Sci. Am.* 268(1):82–89.

van Wyk, J., et al., 1999. Bioconversion of waste paper materials to sugars. *Biochem. Educ.* 27(4):227–228.

Zatta, P., and Cummings, R., 1992. Lectins. *Biochem. Educ.* 20:2–9.

Chapter 9

Lipids, Biological Membranes, and Cellular Transport

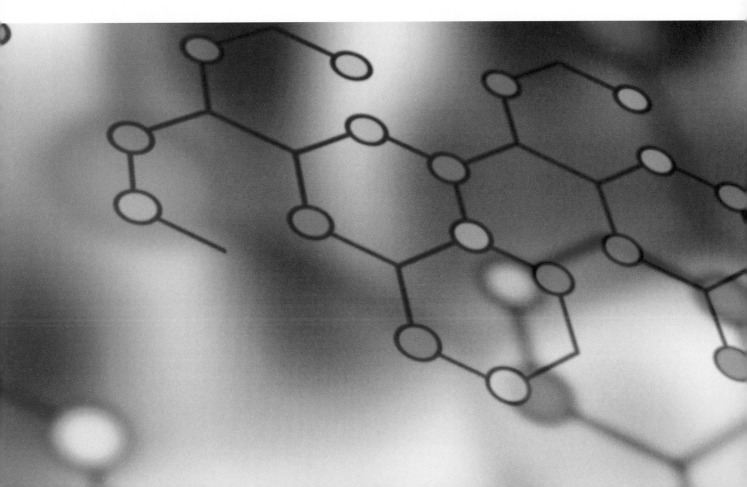

Lipids are a class of biomolecules whose distinctive characteristic is their solubility behavior. Because of their hydrophobic nature, they are more soluble in nonpolar solvents such as diethyl ether, methanol, and hexane than in water. Because they are defined by their physical *behavior* rather than their chemical *structure* like amino acids and carbohydrates), we may expect to find a great variety of chemical structures among the lipid class of molecules. In addition to the expected elements of carbon and hydrogen, which lend nonpolar character, lipids may also contain oxygen, nitrogen, and phosphorus. The chemical functional groups most common to the lipids are carbon–carbon single and double bonds, carboxylate esters, phosphate esters, amines and amides. We have discussed in earlier chapters how chemical diversity among biomolecular structure always leads to a diversity of biological function. This is true also for the lipids. The lipids are probably best known for their role in energy metabolism. In most organisms, the principal molecules for energy storage are the nonpolar lipids called fats.

Polar lipids, some nitrogen and phosphorus containing, are important components of biological membranes. Biological membranes, which are composed of lipids and proteins, form molecular boundaries around all cells and cell organelles. They give the cell its shape and form and protect its contents from the outside environment. Some protein molecules in the membrane act as channels and gates to regulate the transport of biomaterials in and out of the cell. The steroid class of lipids is represented by cholesterol, which is found in biological membranes and serves as a precursor for many hormones. Miscellaneous lipids present in only minor quantities in the cell are involved as light-absorbing pigments (β-carotene, retinal), enzyme cofactors (vitamin K), hormones (estrogens, testosterone), signal molecules (prostaglandins), and electron carriers (ubiquinone).

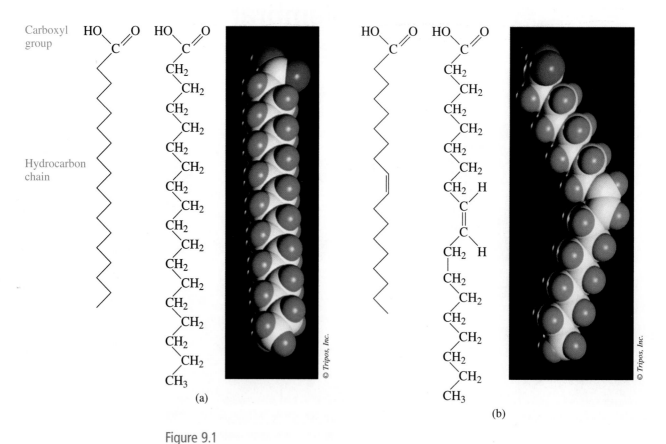

Figure 9.1

The structures of two fatty acids: (a) octadecanoic acid, a saturated acid, and (b) 9-octadecenoic acid, an unsaturated acid. The fatty acids are shown in three forms, an abbreviated structure, where a zigzag line represents the hydrocarbon chain; a structure showing all carbon atoms; and space-filling models showing the actual shape of each molecule.

Table 9.1

Structures and names of common, naturally occurring fatty acids

Number of Carbons[a]	Common Name	Systemic Name	Abbreviated Symbol[b]	Structure[c]
12	Lauric acid	n-Dodecanoic acid	12:0	$CH_3(CH_2)_{10}COOH$
14	Myristic acid	n-Tetradecanoic acid	14:0	$CH_3(CH_2)_{12}COOH$
16	Palmitic acid	n-Hexadecanoic acid	16:0	$CH_3(CH_2)_{14}COOH$
16	Palmitoleic acid	n-Hexadecenoic acid	$16:1^{\Delta 9}$	$CH_3(CH_2)_5CH == CH(CH_2)_7COOH$
18	Stearic acid	n-Octadecanoic acid	18:0	$CH_3(CH_2)_{16}COOH$
18	Oleic acid	n-Octadecenoic acid	$18:1^{\Delta 9}$	$CH_3(CH_2)_7CH == CH(CH_2)_7COOH$
18	Linoleic acid	—	$18:2^{\Delta 9,12}$	$CH_3(CH_2)_4CH == CHCH_2CH == CH(CH_2)_7COOH$
18	Linolenic acid	—	$18:3^{\Delta 9,12,15}$	$CH_3CH_2CH == CHCH_2CH == CHCH_2CH == CH(CH_2)_7COOH$
20	Arachidonic acid	—	$20:4^{\Delta 5,8,11,14}$	$CH_3(CH_2)_4CH == CHCH_2CH == CHCH_2CH == CHCH_2CH == CH(CH_2)_3COOH$
20	EPA	Eicosanpentaenoic acid	$20:5^{\Delta 5, 8, 11, 14, 17}$	
22	DHA	Docosahexaenoic acid	$22:6^{\Delta 4, 7, 10, 13, 16, 19}$	

[a]Note that all have an even number of carbons.
[b]Indicates the number of carbon atoms and the position of the carbon–carbon double bonds.
[c]All double bonds are *cis*.

In this chapter we first introduce the structures and functions of the various lipid families. Second, we turn to a discussion of the chemical architecture and transport activities of biological membranes, with emphasis on the roles of lipids and proteins.

9.1 Fatty Acids

Fatty Acid Structure

Fatty acids are biomolecules containing a (polar) carboxyl functional group (—COOH) connected to an unbranched aliphatic chain (Figure 9.1). These structural features give them a split personality: One end is polar and sometimes ionic (the carboxyl group), whereas the opposite end (the hydrocarbon chain) has nonpolar properties. (In Chapter 3, Section 3.3, we called such molecules **amphiphilic.**) Fatty acids are rarely found in a free form in cells and tissues but are most often bound in fats (triacylglycerols). The number of carbon atoms in a fatty acid can range from 4 (as found in butter) to as many as 36 (found in the brain). Most fatty acids found in nature however, contain between 12 and 24 carbon atoms, with those containing 16 and 18 carbons the most prevalent. Table 9.1 provides a list of names and structures of the most common naturally occurring fatty acids.

Several general rules are followed for the cellular construction of fatty acids (Table 9.2). Because fatty acids are synthesized by combining the C_2 units of acetic acid (CH_3COOH), almost all contain an even number of carbon atoms. The hydrocarbon chain, which is almost

1. Most fatty acids have an even number of carbon atoms.
2. The hydrocarbon chain is almost always unbranched.
3. Most carbon–carbon bonds are single; however, acids may contain one, two, or more carbon–carbon double bonds.
4. Double bonds are most often cis.
5. For monounsaturated fatty acids, the double bond is usually between carbons 9 and 10.
6. If more than one carbon–carbon double bond is present they are not conjugated but are separated by a methylene unit.

Table 9.2

General rules for the structures of naturally occurring fatty acids

always unbranched, can consist of all carbon–carbon single bonds (saturated) or one or more carbon–carbon double bonds (unsaturated) (see Figure 9.1). For monounsaturated acids, the double bond is usually between carbons 9 and 10 (C1 is the carboxyl carbon). For diunsaturated acids, the second double bond is most often between carbons 12 and 13. Note in Table 9.2 that multiple double bonds are not conjugated ($-CH=CH-CH=CH-$) but are separated by one methylene group ($-CH=CH-CH_2-CH=CH-$). Fatty acids with two or more double bonds are called **polyunsaturated.** Double bonds in naturally occurring fatty acids are almost always of the cis configuration. Because of the many varieties of fatty acids present in nature, a special, abbreviated nomenclature system has been developed. This system indicates the total number of carbon atoms, the number of carbon–carbon double bonds, and the position of each double bond. Fatty acids also have common names, which are more often used than their official names. Lauric acid (dodecanoic acid), a fatty acid with 12 carbon atoms and no carbon–carbon double bonds, is designated in the abbreviated form as 12:0. Linoleic acid, with 18 carbon atoms and two carbon–carbon double bonds (at C9–C10 and C12–C13), is $18:2^{\Delta 9,12}$. [This is read 18:2 delta (Δ) 9,12.] This nomenclature is used to define each fatty acid in Table 9.1.

Physical and Chemical Properties of Fatty Acids

The physical properties of fatty acids can be predicted from a knowledge of their structures. They are all soluble in organic solvents, such as alcohols, hexane, and diethyl ether. The smaller chain acid butanoic (4:0) is infinitely soluble in H_2O; however, solubility decreases with increasing chain length. Lauric acid (12:0) is soluble in water to the extent of 0.06 g per gram of water. Fatty acids greater than 12:0 are increasingly insoluble in water. The saturated acids with ten or more carbons are waxy solids at room temperature. All saturated acids with fewer than ten carbons and all unsaturated acids are oily liquids at room temperature. The double bonds put "kinks" in the hydrocarbon chains of the unsaturated fatty acids (see Figure 9.1b). These molecules cannot line up in a fully extended form to pack together in a crystalline array as is the case for saturated fatty acids. Although larger chain fatty acids are insoluble in water, they appear to dissolve in dilute aqueous NaOH or KOH solutions. This is because an acid–base reaction takes place, forming the Na^+ or K^+ salt of the acids:

$$CH_3(CH_2)_{10}COOH + NaOH \longrightarrow CH_3(CH_2)_{10}COO^-Na^+ + H_2O$$

Lauric acid   Sodium laurate
(a soap)

The Na^+ or K^+ salts of fatty acids are called **soaps.** The soaps appear to dissolve in water, but they do not form true solutions. The amphiphilic molecules aggregate into molecular arrangements called **micelles** (see Chapter 3, Section 3.2). Soaps are able to remove grease and other stains from cloth and skin by coating the oily residues with their hydrophobic tails and at the same time extending their ionic salt heads into the water. Free fatty acids under physiological conditions are present as carboxylate salts (ions) since their pK_a values are between 4 and 5. Therefore, the names of fatty acids under physiological conditions should end with -*ate,* indicating salts: laurate, myristate, oleate, and so on.

The chemical reactivity of the hydrocarbon chains of fatty acids depends on the extent of unsaturation. Saturated chains of fatty acids are relatively unreactive. Unsaturated chains display the usual reactivity of carbon–carbon double bonds. For example, halogen or hydrogen atoms may be added:

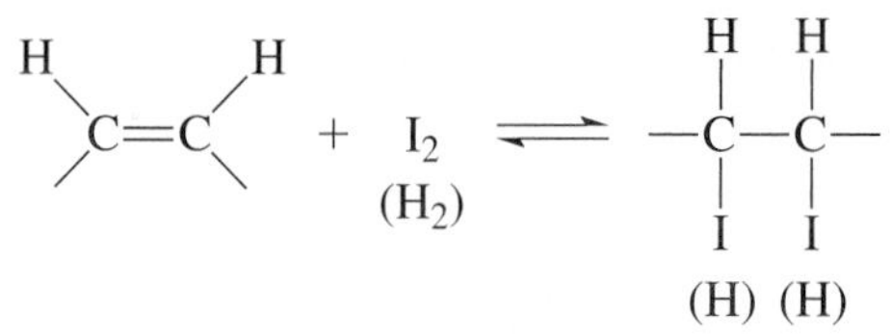

The iodine addition reaction shown above has been used to measure experimentally the number of double bonds in a fatty acid sample and the hydrogenation process is used to produce solid fats from oils. The double bonds of unsaturated fatty acids are attacked also by oxygen. The process, called **autooxidation,** results in complex, yellow products such as smaller, odoriferous aldehydes and acids.

9.2 Triacylglycerols

Triacylglycerol Structures

The primary biological role of the fatty acids is to serve as metabolic fuel for cells. The acids are ingested and stored for future energy use in **triacylglycerols.** When energy is needed, they are released in a free form by enzyme-catalyzed hydrolysis reactions, bound to serum albumin in the blood, and circulated throughout an organism. In heart and skeletal muscle, fatty acids, which have hydrocarbon structures similar to fossil fuels, are oxidized to CO_2 and H_2O with release of large amounts of energy. The stored fatty acids are especially efficient for energy production because the hydrocarbon chains are in a highly reduced state. As we shall see in Chapters 15 and 18, the complete oxidative metabolism of fatty acids produces twice as much energy per gram as for carbohydrates. Almost all fatty acids present in nature are found as constituents of the nonpolar lipids called **triacylglycerols.** The basic foundation molecule of the triacylglycerols is the trihydroxyl compound **glycerol.** Each hydroxyl group can be linked to a fatty acid by esterification (Figure 9.2). The esterification process can occur in a stepwise manner to produce intermediate monoacylglycerols and diacylglycerols (either 1, 2 or 1, 3). Simple triacylglycerols, which are rare in nature, have three identical fatty acids, whereas the more common mixed triacylglycerols have two or three different fatty acids. The polar hydroxyl groups of glycerol and the polar (and sometimes ionic) carboxyl group of each fatty acid are tied up in neutral ester linkages, so triacylglycerols are nonpolar, hydrophobic molecules. As

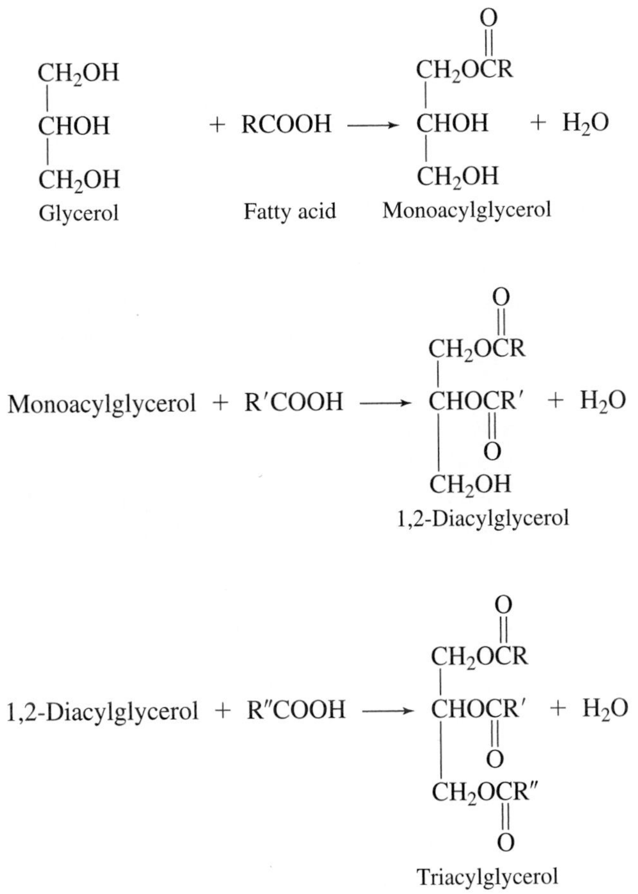

Figure 9.2
The stepwise addition of fatty acids to glycerols to the final stage of a triacylglycerol. Ester bonds hold the acyl groups to the glycerol backbone. R, R′, R″ are aliphatic saturated or unsaturated hydrocarbon chains of variable lengths.

we can predict by inspection of the structures, triacylglycerols are insoluble in water but soluble in nonpolar solvents. [Recall that oil and vinegar salad dressings separate into two phases, a lower layer of vinegar (4% acetic acid in H_2O) and an upper layer of vegetable oil (mixtures of triacylglycerols).] Triacylglycerols are extracted from plant and animal tissue using solvent systems such as chloroform–methanol and hexane–isopropanol mixtures. Triacylglycerols isolated from animal tissues are called **fats** and are solids at room temperature because they contain predominately saturated fatty acids. Triacylglycerol mixtures from plant seeds are termed **oils** and contain mainly unsaturated acids. Table 9.3 compares the fatty acid content of several plant and animal triacylglycerol mixtures. Nutmeg oil is the source of trimyristin (contains three molecules of myristic acid, 14:0), one of the few simple triacylglycerols found in nature.

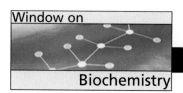

Window on

Biochemistry

9.1   Fish Oils—Omega-3 Fatty Acids for a Healthy Heart

Private health organizations and many world governments recommend that individuals should reduce their dietary fat intake and, when possible, use polyunsaturated oils rather than saturated fats. Extensive research by physicians and scientists has proven that individuals who favor unsaturated fats in the diet show decreased incidence of atherosclerosis, heart disease, cancer, and other significant health problems. A study of Table 9.3 leads one to the general conclusion that plant oils contain more unsaturated fatty acids than animal fats. (Coconut oil is an exception to this generality.) We cannot completely restrict fats from the diet. Three **essential fatty acids**—linoleic, linolenic, and arachidonic—are required in the human diet for proper growth and development (see Sections 9.4 and 18.4).

Now, nutritionists are suggesting an even greater health benefit with a special class of highly unsaturated fatty acids, the omega-3 (ω-3) fatty acids. These acids are present in seafood, especially cold-water fish like salmon, tuna, trout, and mackerel. The two most abundant omega-3 acids in fish oils are eicosapentaenoic acid (EPA) $20:5^{\Delta 5, 8, 11, 14, 17}$ and docosahexaenoic acid (DHA) $22:6^{\Delta 4, 7, 10, 13, 16, 19}$. The label "omega-3" is derived from the position of a double bond, three carbons from the last (omega, ω) carbon of the chain. The amounts of EPA and DHA combined make up about 30% of the total fat in the oil of the Canadian tuna, the Atlantic mackerel, the Chinook salmon, and other fish. Fish are unable to synthesize these acids, but obtain them from their primary food source, the planktons. Therefore, it is possible to increase omega-3 fatty acids in commercial, farm-raised fish by supplementing their food with the fatty acids.

Many years ago, it was noted that cultures in Greenland, Japan, and Alaska that live from the sea have a relatively low incidence of heart disease even though they have high-fat diets and high levels of serum cholesterol. Upon further study it was found that their diets were low in saturated fats and high in the omega-3 acids, EPA and DHA. Nutritionists have since uncovered many health benefits obtained from dietary and supplemental use of the omega-3 acids. Some include: (1) prevention of cardiovascular disease, lessening of angina (chest pain), and lowering of blood pressure. The benefits are thought to be due to inhibition of platelet formation which has the effect of thinning the blood and thus reducing the risk of blood clotting. We now know that the highly unsaturated omega-3 fatty acids inhibit the synthesis of prostaglandins that enhance blood clotting (see Section 9.4); (2) prevention of some cancers including colon, breast, prostate, and others; and (3) treatment of joint stiffness and swelling caused by rheumatoid arthritis. Because omega-3 acids enhance blood-thinning (as does aspirin), supplemental fish oil and aspirin should be avoided 1–2 weeks before surgery.

Oil from these canola plants is a rich source of polyunsaturated fatty acids.

Table 9.3
Fatty acid content of common oils and fats. The fatty acids are present in triacylglycerol form

	Fatty Acids[a]				
	Saturated				Unsaturated
Source	C_4–C_{12}	C_{14}	C_{16}	C_{18}	C_{16} + C_{18}
Canola oil	—	—	5	1	94
Olive oil	2	2	13	3	80
Butter	10	11	29	10	40
Beef fat	2	2	29	21	46
Coconut oil	60	18	11	2	8
Corn oil	—	2	10	3	85
Palm oil	—	2	40	6	52
Nutmeg oil	7	90	3	—	—

[a]The numbers represent percentage of each fatty acid.

Triacylglycerol Reactivity

The chemical behavior of triacylglycerols depends mainly on the reactivity of the ester linkages. In an important commercial process called **saponification,** each ester is hydrolyzed in a reaction catalyzed by NaOH to produce glycerol and soaps:

$$
\begin{array}{c}
\text{CH}_2\text{OCR} \\
| \quad \| \\
\text{O} \\
\text{CHOCR}' \\
| \quad \| \\
\text{O} \\
\text{CH}_2\text{OCR}'' \\
| \quad \| \\
\text{O}
\end{array}
+ \text{H}_2\text{O} \xrightarrow[\Delta]{\text{NaOH}}
\begin{array}{c}
\text{CH}_2\text{OH} \\
| \\
\text{CHOH} \\
| \\
\text{CH}_2\text{OH}
\end{array}
+
\begin{array}{c}
\text{RCOO}^-\text{Na}^+ \\
\text{R}'\text{COO}^-\text{Na}^+ \\
\text{R}''\text{COO}^-\text{Na}^+
\end{array}
$$

Triacylglycerols   Glycerol   Soaps

In ancient times soaps were prepared by boiling animal fats with wood ash, which contains lye (NaOH). Soaps have been replaced largely by synthetic detergents, such as sodium dodecyl sulfate (SDS; see Window on Biochemistry 4.2), which do not form precipitates or a bathtub ring in hard water the way sodium salts of fatty acids do.

Hydrolysis of triacylglycerols also occurs under physiological conditions; however, the catalysts are enzymes called **lipases.** These enzymes are present in the intestines and fat cells (adipocytes), where they release fatty acids for energy metabolism:

$$
\begin{array}{c}
\text{CH}_2\text{OCR} \\
| \quad \| \\
\text{O} \\
\text{CHOCR}' \\
| \quad \| \\
\text{O} \\
\text{CH}_2\text{OCR}'' \\
| \quad \| \\
\text{O}
\end{array}
+ 3\,\text{H}_2\text{O} \underset{}{\overset{\text{lipases}}{\rightleftharpoons}}
\begin{array}{c}
\text{CH}_2\text{OH} \\
| \\
\text{CHOH} \\
| \\
\text{CH}_2\text{OH}
\end{array}
+
\begin{array}{c}
\text{RCOOH} \\
\text{R}'\text{COOH} \\
\text{R}''\text{COOH}
\end{array}
$$

In Chapter 18 we discuss this important cellular reaction in detail. Reactions of triacylglycerols do not involve only the ester linkages. The double bonds of unsaturated fatty acids in triacylglycerols may add halogen or hydrogen or be attacked by oxygen, just as with free acids. The preparation of oleomargarine, a butter substitute that is high in desired polyunsaturated fatty acids, involves partial hydrogenation of liquid vegetable oil (usually corn), thereby changing some double bonds to single bonds and transforming the oil to a firm, but spreadable solid. The commercial hydrogenation process often leads to the rearrangement of cis carbon–carbon double bonds to trans double bonds (Figure 9.3). Trans fatty acids may increase blood cholesterol levels. Highly pure vegetable oils are clear,

The primary component of beeswax is a nonpolar ester of palmitic acid.

Figure 9.3
The commercial hydrogenation of
vegetable oils often leads to trans
acids.

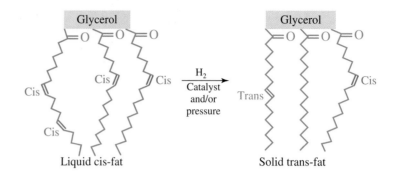

Liquid cis-fat → Solid trans-fat

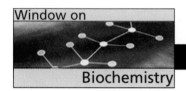

Window on
Biochemistry

9.2 Anandamide and Chocolate Cravings

Arachidonic acid ($20:4^{\Delta 5, 8, 11, 14}$) has recently received attention for its involvement in neurochemistry. **Anandamide,** a product of the metabolism of the polyunsaturated, C-20 acid, has been found in the brains of humans, pigs, and other animals. It has also been isolated from cocoa powder and chocolate products. Anandamide, produced by the combination of the C-20 acid with ethanolamine (see structure), is a neurotransmitter involved in the regulation of movement. Researchers have found that stimulating nerve cells in the dorsal striatum, a region of the brain that participates in the control of movement, caused the cells to release anandamide. The newly discovered neurotransmitter is a natural ligand for the cannabinoid receptor, CB-1. The CB-1 receptor is present in high concentrations in the cerebellum and basal ganglia where motor effects are mediated and in the hippocampus, which involves memory. Other functions of the CB-1 receptor may involve mood, cognition, coordination, sleep, and appetite. The CB-1 receptor is part of the G-protein family of receptors (see Section 2.5). The signal transduction pathway (Section 2.5) involves calcium and cyclic AMP as messengers. CB-1 is also the protein receptor that binds and mediates the CNS effects of the cannabinoids, especially Δ9-tetrahydrocannabinol (THC), the active ingredient in marijuana. The discovery of anandamide in cocoa powder may explain the craving for chocolate that many people experience. Research on anandamide and THC and their interactions with the CB-1 receptors may eventually lead to new and better treatments for movement disorders such as Parkinson's disease and Tourette's syndrome as well as for drug addiction.

© P. Motta/Dept. of Anatomy/Univ. La Sapienza Rome/SPL/Photo Researchers

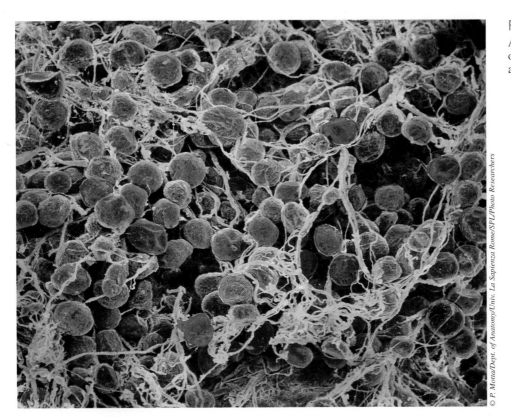

Figure 9.4
A scanning electron micrograph of adipocytes *(red)*. Adipocytes are specialized fat storage cells.

colorless, and almost odorless. The yellow color and strong odor of old or rancid fats and oils result from autooxidation of unsaturated fatty acid chains in the triacylglycerols. Addition of special preservatives, such as butylated hydroxytoluene (BHT), can slow oxidation by trapping free radical intermediates, which are produced by reactions with O_2.

Biological Properties of Triacylglycerols

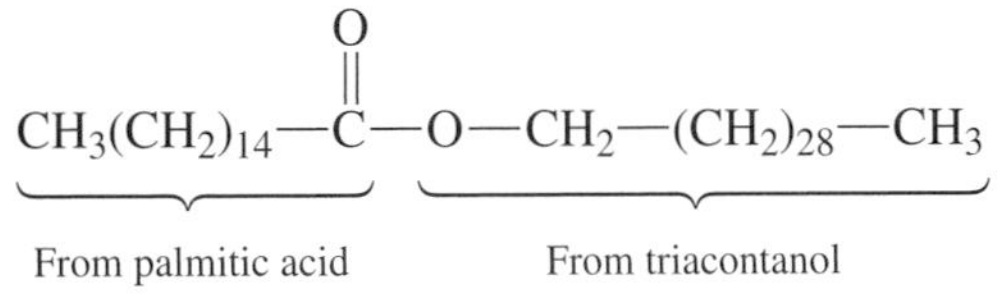

Cutting Edge: Fat blockers

The triacylglycerols have two primary biological roles, energy metabolism and insulation. They serve as the molecular form for storage of the important fuel molecules, fatty acids. (In comparison, the fuel molecule glucose is stored in the form of starch and glycogen in plants and animals, respectively). Triacylglycerols are present in oily droplets in the cytoplasm of plant and animal cells. **Adipocytes** are animal cells specialized for fat storage (Figure 9.4). Almost the entire volume of each cell is filled by a fat droplet. These cells serve as storage depots of metabolic fuel. Enzymes called **lipases** are present in adipocytes to catalyze the release of fatty acids. Triacylglycerols are also present below the skin layer of animals, especially those living in polar regions, to insulate against temperature extremes.

The fatty acids, in addition to being found in triacylglycerols are also present in other derivative forms. Chemically related to the triacylglycerols are nonpolar lipids called **waxes,** which serve many biological roles such as protective coatings for plant leaves, lubrication of skin, and water repellency for the feathers of waterfowl. Representative of the waxes is beeswax, an ester composed of palmitic acid (16:0) and the alcohol triacontanol, a compound containing an unbranched saturated chain of 30 carbons (Figure 9.5).

O
CH3(CH2)14—C—O—CH2—(CH2)28—CH3
From palmitic acid
From triacontanol

Figure 9.5
Beeswax, an important derivative of fatty acids is a nonpolar lipid composed of palmitic acid in ester linkage to the alcohol triacontanol.

9.3 Polar Lipids

The nonpolar class of lipids represented primarily by the triacylglycerols serve as storage molecules for metabolic fuel. The polar lipids are a class of lipids with chemical structures similar to the triacylglycerols, but quite different in biological function. These biomolecules are combined with protein molecules for the construction of biological membranes, which provide a protective shield around cells and cellular organelles. The membrane forms a permeable barrier that selects some molecules for passage, but not others. The polar lipids are found almost exclusively in membranes and not stored as in the case for triacylglycerols. In this section we discuss the chemical structures for the polar lipids and the properties that make them well suited for membrane structure. As a preview for this section, Figure 9.6 compares the structural features of the nonpolar lipids (triacylglycerols) with polar lipids found in membranes. Two new classes of lipids are introduced: **glycerophospholipids** and **sphingolipids.** (The steroid cholesterol is also present in membranes; however, because of its very different chemical structure and additional biological properties, it is explored in a separate section.)

Glycerophospholipids

The foundation molecule for the glycerophospholipids is 1,2-diacylglycerol 3-phosphate, which has the generic name **phosphatidic acid** (Figure 9.7a). The fatty acids linked by ester bonds to the glycerol molecule are the same variety as found in triacylglycerols. Fatty acids with 16 and 18 carbons are most prevalent. Both saturated and unsaturated hydrocarbon chains are found; however, an analysis of many glycerophospholipids has uncovered a preference for saturated fatty acids at position 1 and for unsaturated fatty acids at position 2. (Note that the term "phosphatidic acid" actually refers to a group of many molecules that differ by the identity of the two fatty acids.) The third hydroxyl group of glycerol is esterified by phosphoric acid. Because phosphoric acid is triprotic, it is able to react with up to three alcohol moieties to form mono-, di-, and triesters. In the glycerophospholipids, a second alcohol is esterified with the phosphate group. That alcohol is usually one of the following: the amino alcohols ethanolamine or choline, the amino acid serine, or the polyhydroxyl compound inositol (Figure 9.7b–e). The four resulting polar lipids are named by combining "phosphatidyl" with the alcohol name, for example, phosphatidylethanolamine. With the exception of inositol, all of the alcohols have an amino or other functional group that becomes ionic at physiological pH. Except for phosphatidylethanolamine and phosphatidylcholine, all of the glycerophospholipids are electrically charged at cellular pH. We should by now see a major structural difference when comparing triacylglycerols and glycerophospholipids. Triacylglycerols are nonpolar and hydrophobic. The glycerophospholipids, on the other hand, have highly polar and sometimes electrically charged regions on the molecule in addition to nonpolar regions. Therefore, we can distinguish two structural features in the glycerophospholipids: a polar head and nonpolar tails. These characteristics are essential for membrane structure.

Figure 9.6

Structural features of storage and membrane lipids: (a) The storage lipids are composed of nonpolar triacylglycerols. (b) The membrane lipids are composed of the glycerophospholipids and sphingolipids, which have polar and nonpolar regions. Specific members of the polar lipid class that have various X substituents are shown in later figures.

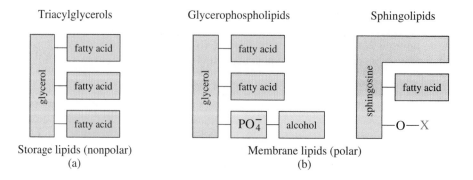

Sphingolipids

A second group of polar lipids found in membranes are the **sphingolipids.** This major class of lipids is represented by three subclasses: ceramides, sphingomyelins, and glycosphingolipids. Here the foundation molecule is the 18-carbon amino alcohol **sphingosine** (Figure 9.8a), rather than the simple glycerol molecule. Sphingosine has two functional groups (amino and hydroxyl) that can be chemically modified to make a variety of sphingolipids. The amino group at C2 of sphingosine may be linked via an amide bond with a fatty acid. The resulting molecule is called a ceramide and has a polar head (hydroxyl group on C1 of sphingosine) and two nonpolar tails (Figure 9.8b). Another class of sphingolipids uses the amino and hydroxyl groups to link additional units. The sphingomyelins, the only phosphorus-containing sphingolipids, contain a fatty acid on the amino group (like the ceramides) and a phosphocholine unit esterified with the hydroxyl group (see Figures 9.6 and 9.8c). This results in a molecule with a polar (in fact, ionic) head and two nonpolar tails similar to the glycerophospholipids. Although the sphingomyelins are found in plasma membranes, they are probably best known and named for their presence in the myelin sheath, where they insulate nerve axons. A third subclass of the sphingolipids is represented by the glycosphingolipids. These carbohydrate-containing lipids use the ceramides as foundation molecules. The cerebrosides consist of ceramide with a monosaccharide unit in glycosidic linkage at the hydroxyl group at C1 of sphingosine (Figure 9.8d). The carbohydrates commonly found are glucose, galactose, and N-acetylgalactosamine (see Chapter 8, Section 8.3). Cerebrosides, as the name implies, are abundant in the membranes of the brain and nervous system. The most complex sugar-containing lipids are the gangliosides, which contain a polar head composed of several carbohydrate units linked by glycosidic bonds

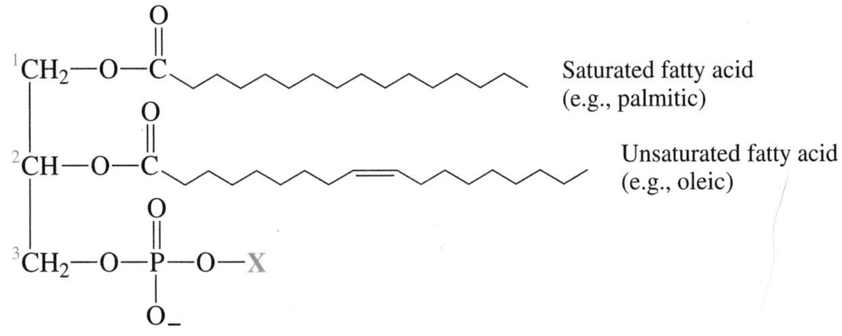

X	Structure of X	Name of Glycerophospholipids
(a) hydrogen	—H	phosphatidic acid
(b) ethanolamine	$-CH_2-CH_2-\overset{+}{N}H_3$	phosphatidylethanolamine
(c) choline	$-CH_2-CH_2-\overset{+}{N}(CH_3)_3$	phosphatidylcholine
(d) serine	$-CH_2-\underset{\underset{COO_-}{\mid}}{CH}-\overset{+}{N}H_3$	phosphatidylserine
(e) inositol	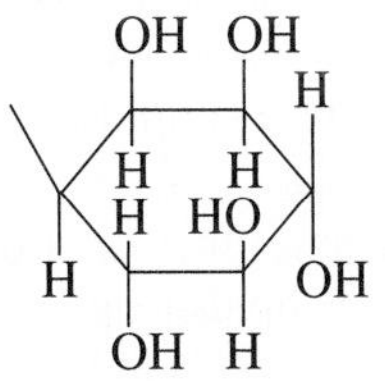 (inositol ring structure)	phosphatidylinositol

Figure 9.7

Phosphatidic acid is the foundation molecule for the glycerophospholipids. It contains two fatty acids in ester linkage to glycerol (at carbons 1 and 2) and a phosphate in ester linkage to glycerol (C3). The X moiety in (a) phosphatidic acid is H. A variety of saturated or unsaturated fatty acids may be present. The four major glycerophospholipids are: (b) phosphatidylethanolamine, (c) phosphatidylcholine, (d) phosphatidylserine, and (e) phosphatidylinositol.

Figure 9.8

The foundation molecule for the sphingolipids is sphingosine, which is derivatized at the amino group on C2 with a fatty acid and at the hydroxyl group on C1 with an X substituent. (a) The arrows point to the amino and hydroxyl groups of sphingosine that are derivatized to make sphingolipids. The four major sphingolipids are: (b) ceramide, (c) sphingomyelin, (d) cerebroside, and (e) ganglioside; NAN = *N*-acetylneuraminic acid.

Sphingolipid (general structure)

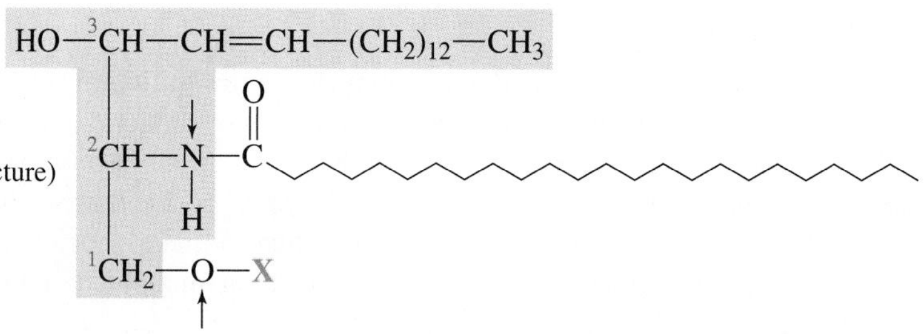

(a)

X	Structure of X	Name of Sphingolipid
(b) hydrogen	—H	ceramide
(c) phosphocholine		sphingomyelin
(d) glucose	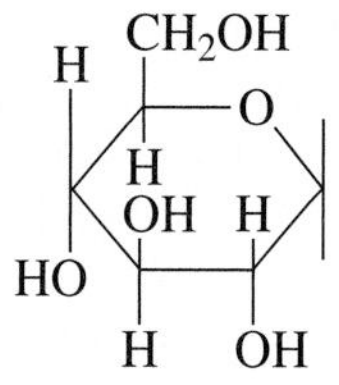	glucosylcerebroside
(e) complex oligosaccharide	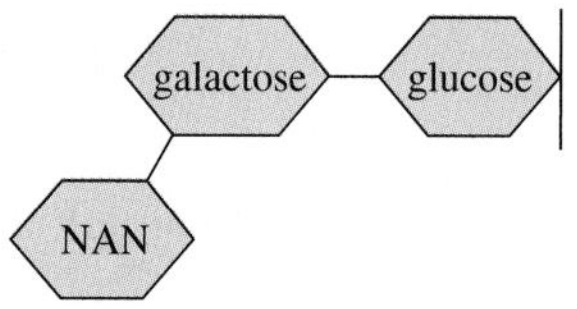	ganglioside

(Figure 9.8e). Like the cerebrosides, these are abundant in brain and nerve membranes. In addition to their role in biological membrane structure, the glycosphingolipids are also involved in other specialized cellular functions, including recognition events at cell surfaces (see Chapter 8, Section 8.5, glycoproteins), tissue specificity of cell association, and the transmission of nerve impulses. These important functions explain some of the major medical and clinical consequences of improper metabolism of the glycosphingolipids. In Tay-Sachs disease, for example, a specific ganglioside accumulates in the brain and spleen because the enzyme that is responsible for its degradation is lacking. This genetic disease results in slow development, paralysis, blindness, and finally, death by the age of 3 to 4 years.

We have observed a variety of chemical structures for the polar lipids we call glycerophospholipids and sphingolipids. In spite of chemical differences, these molecules serve a similar biological role as structural units in membranes. Their similar features, a polar region and a nonpolar region, endow them with their common biological characteristic. We have observed that salts of fatty acids with a polar (ionic) head and a single nonpolar tail assemble spontaneously into spherical structures called **micelles.** The polar lipids represented by the glycerophospholipids and sphingolipids each has a polar (and sometimes ionic) head and *two* hydrophobic tails. Because of the extra space taken by the nonpolar tails, these lipids are unable to assemble into micelles. Instead, they form **bilayers,** composed of two monolayers or sheets of polar lipids (Figure 9.9). The nonpolar side of each sheet combines by hydrophobic interactions to exclude water in the central region of the bilayer. The bilayer provides a structural framework for membrane assembly.

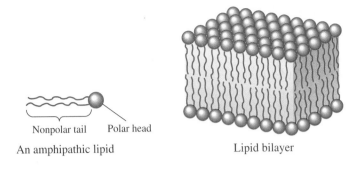

Nonpolar tail   Polar head

An amphipathic lipid   Lipid bilayer

Figure 9.9
Assembly of polar lipids with ionic heads and nonpolar tails into bilayer sheets. The nonpolar tails cluster in the interior of the bilayer. Hydrophobic interactions provide stabilizing energy to hold the bilayer together.

9.4 Steroids and Other Lipids

The two large classes of lipids previously covered, the triacylglycerols and the polar lipids, perform functions in energy storage and biological membrane construction, respectively. A lipid extraction of cells and tissues contains many more lipid classes in addition to these two. In this section a potpourri of lipids and their functions are explored.

Steroids

One of the most well known and best studied of the lipid groups is the **steroid** class. The steroids all have the characteristic fused-ring system of three six-membered rings labeled A, B, C and one five-membered ring called the D ring (Figure 9.10a). Ketones, alcohols, double bonds, and hydrocarbon chains decorate the ring system in various types of steroids. **Cholesterol,** the best known steroid, has a hydroxyl group on the A ring, a double bond in ring B, and hydrocarbon chains attached at several locations (Figure 9.10b). Although the chemical structure is quite different from lipids previously studied, the cholesterol molecule is amphiphilic with a polar head (the —OH group) and an extensive nonpolar region (the fused

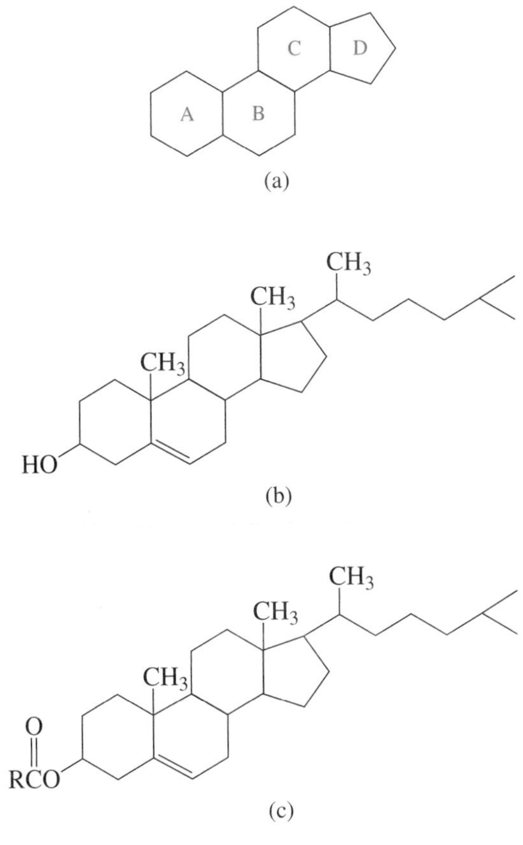

Interactive Animation:
Cholesterol

(a)

Figure 9.10
(a) The molecular structure common to all steroids showing the four fused rings, A, B, C, and D. (b) Cholesterol has a polar head (hydroxyl group) and a nonpolar tail (hydrocarbon skeleton). (c) A cholesteryl ester formed between the cholesterol hydroxyl group and a fatty acid with a long aliphatic side chain, R.

(b)

(c)

British stamp issued to acknowledge the synthesis and analysis of the steroids. Note the outline of the steroid carbon skeleton.

rings and hydrocarbon tails). Indeed, cholesterol and some of its derivatives accompany the glycerophospholipids and sphingolipids in biological membranes. The most chemically reactive portion of the cholesterol structure is the hydroxyl group. Under physiological conditions, it is common for a fatty acid to be esterified at this position (Figure 9.10c). Cholesterol and its ester derivatives are abundant in plasma proteins called **lipoproteins** whose function it is to transport the cholesterol to peripheral tissue for use in construction of membranes and as a biosynthetic precursor for steroid hormones and other biologically active products. We discuss the roles of cholesterol and lipoproteins in the development of **atherosclerosis** in Chapter 18. The cholesterol molecule, which is found almost exclusively in animal tissue, is derived from units of five carbons called **isoprene** (2-methyl-1,3-butadiene, Figure 9.11a). Multiples of these C_5 building blocks are combined to make compounds with 10, 15, and eventually 30 carbons to synthesize the characteristic fused-ring system of the steroids. Cholesterol is the starting point for biosynthesis of the steroid hormones and the bile acids. Several steroid hormones, including estradiol (female sex hormone), testosterone (male sex hormone) and cortisol (glucose metabolism regulator), are pictured in Figure 9.11b–d. Enzyme-catalyzed oxidation reactions on the cholesterol ring system lead to the bile acids (Figure 9.11e,f). Two important bile acids in humans are cholic acid and its glycine deriva-

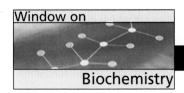

Window on

Biochemistry

9.3 Phytosterols—Fighting Disease with Plants

Nutrition experts constantly remind us that one of the best ways to achieve good cardiovascular health is to keep serum cholesterol levels low. Reducing the intake of this animal sterol is the method most often suggested. For many reasons, especially genetic and lifestyle, this is not always effective or practical, so individuals often turn to prescription, cholesterol-lowering drugs like Mevinolin or Lipitor (Chapter 18). However, more and more people are seeking natural methods to fight diseases. One current strategy is the use of "functional foods"—foods that contain ingredients that may provide a health benefit beyond the traditional nutrients present.

Functional foods targeted toward good cardiovascular health include rice bran, wheat germ, soybeans, corn oils, and some vegetables. One class of natural ingredients in these foods that provide health benefits are the **phytosterols:** β-sitosterol, stigmasterol, and campesterol (see structures on facing page). These compounds have all been shown to lower serum cholesterol by blocking dietary cholesterol and cholesterol secreted in the bile from being absorbed by the intestine. The plant sterols may prevent cholesterol from forming micelles that are more readily absorbed. The phytosterols themselves are very poorly absorbed. Note that the phytosterols have structures very similar to cholesterol. They differ by having methyl and ethyl groups and unsaturation on the side chain. Relatively high doses (500 mg to 10 g) of the plant sterols are required to show a beneficial effect. A daily serving of the specialized margarine (Benecol or Take Control, for example) is three tablespoons (42 g, enough for three slices of bread), which contains about 4.5 g of phytosterols. An individual on this schedule can expect to achieve an average 10–15% reduction in serum cholesterol after a few weeks or months of use. No significant side affects or interactions have yet been reported.

β-Sitosterol is also known to benefit sufferers of benign prostatic hyperplasia (BPH, enlarged prostate). Daily phytosterol doses of 60–130 mg relieve the discomfort of frequent and painful urination. Very recent research has also provided evidence that β-sitosterol may inhibit the growth of human prostate cancer cells. The phytosterol acts by replacing some of the cholesterol in the cell membranes which apparently inhibits cell growth.

Phytosterols have been used for years in Europe both for cholesterol management and for BPH. We can expect to see an increase in the use of these natural remedies throughout the world.

tive, glycocholic acid. Note that these compounds have acidic functional groups with protons that dissociate at physiological pH to produce ionic structures called **bile salts.** The bile salts are stored in the gallbladder and secreted into the intestines to help solubilize, digest, and absorb dietary fats. (Can you explain how the bile salts function?) At one time, steroids were thought to be present only in animals. Sterols with structures similar to cholesterol have recently been isolated from plants. The plant steroids are derived from isoprene, just like cholesterol (see Section 18.4). We do not yet know the roles played by the steroids in plants, but they have been shown to be of benefit to human health (see Window on Biochemistry 9.3).

Terpenes

The lipid class of **terpenes** includes all molecules biosynthesized from the isoprenes. (According to this definition, cholesterol and its derivatives are also in this class.) Important terpenes in plants and animals include limonene, β-carotene, gibberellic acid, **squalene,** and lycopene (Figure 9.12). Many of these compounds provide the colors and odors associated with plants. Squalene is an intermediate in the biosynthesis of plant and animal steroids (see Section 18.4).

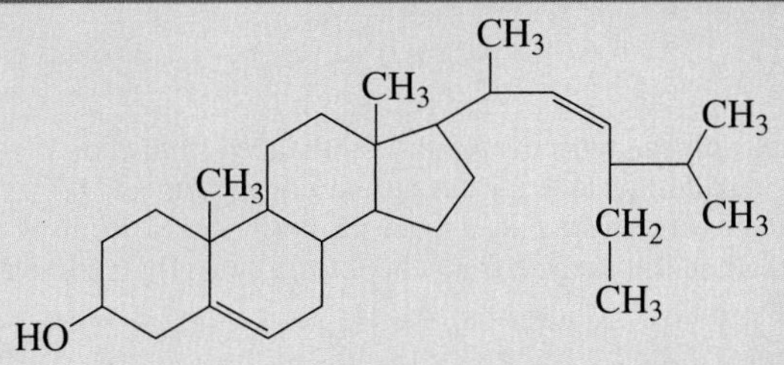

Stigmasterol

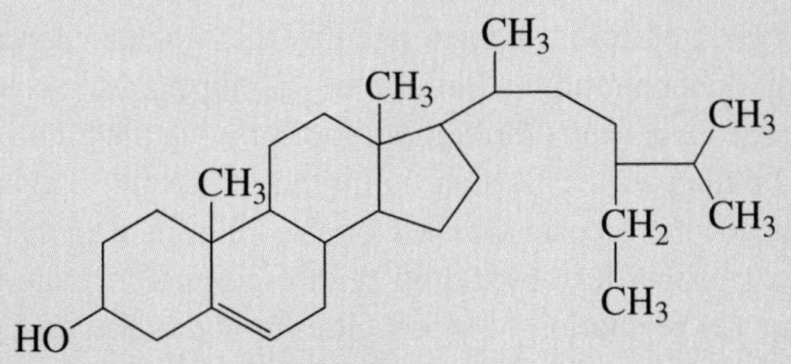

β-Sitosterol

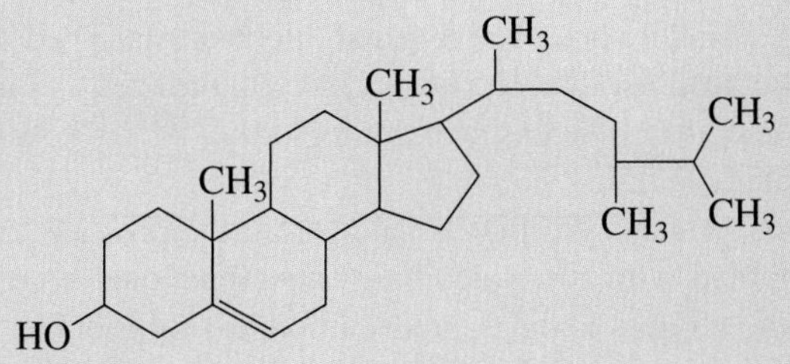

Campesterol

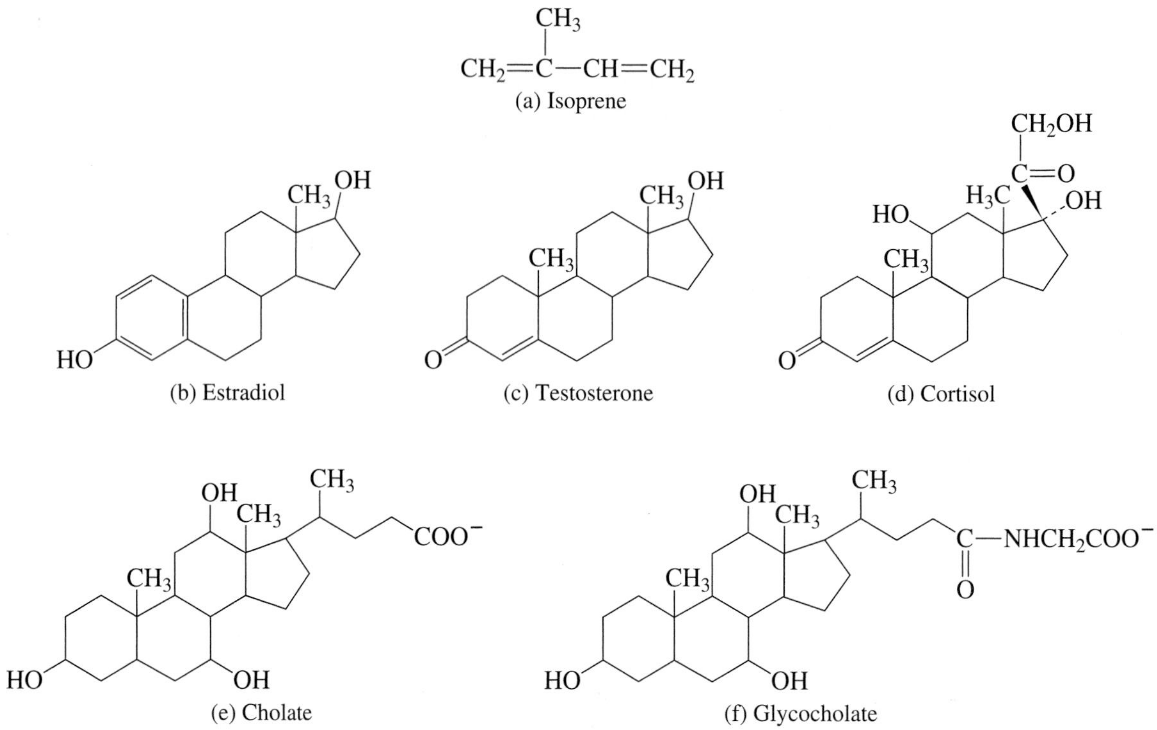

Figure 9.11

Structures of the isoprene unit and bioactive products produced from cholesterol: (a) isoprene, or 2-methyl-1,3-butadiene, a building block for the steroid ring system and for other terpenes; (b) estradiol, a female sex hormone; (c) testosterone, a male sex hormone; (d) cortisol, regulator of glucose metabolism; (e) cholate, a bile salt derived from cholic acid; and (f) glycocholate, a bile salt derived from glycocholic acid.

Eicosanoids

The **eicosanoids** are a class of lipids characterized by their localized, hormonelike activities and very low cellular concentrations. Better known hormones, such as adrenaline and insulin, are transported to target cells throughout the body via the blood. They influence cellular metabolism by binding to the plasma membrane, thereby relaying a message to the cell's interior. In contrast, the eicosanoids act on the cells in which they are synthesized. Some of the biological properties associated with eicosanoids include: (1) influence on reproductive functions; (2) regulation of blood clotting and blood pressure; (3) the generation of inflammation, fever, and pain associated with injury and diseases; and (4) regulation of temperature and the sleep–wake cycle in humans and other animals. The initial discovery and many subsequent studies of eicosanoid action were carried out by Ulf von Euler, Sune Bergstrom, and Bengt Samuelsson, three Swedish biochemists. There are three subclasses of eicosanoids: **prostaglandins,** thromboxanes, and leukotrienes. All are derived from the 20-carbon, polyunsaturated fatty acid arachidonate ($20:4^{\Delta5,8,11,14}$), but each has a unique chemical structure (Figure 9.13).

The **prostaglandins,** which were first isolated from the prostate gland, contain a five-membered ring substituted with two side chains and functional groups, including a carboxylic acid, hydroxyl groups, ketones, and carbon–carbon double bonds. The prostaglandins are now known to be distributed in virtually all mammalian tissue and organs. These compounds exhibit numerous and diverse biological effects on many physiological and pathological activities. The compounds PGE_2 and PGD_2 have recently been discovered in the brain of mammals, including humans. They are especially concentrated in the preoptic area, the sleep center of the brain. Experiments in rats, monkeys, and humans now show that

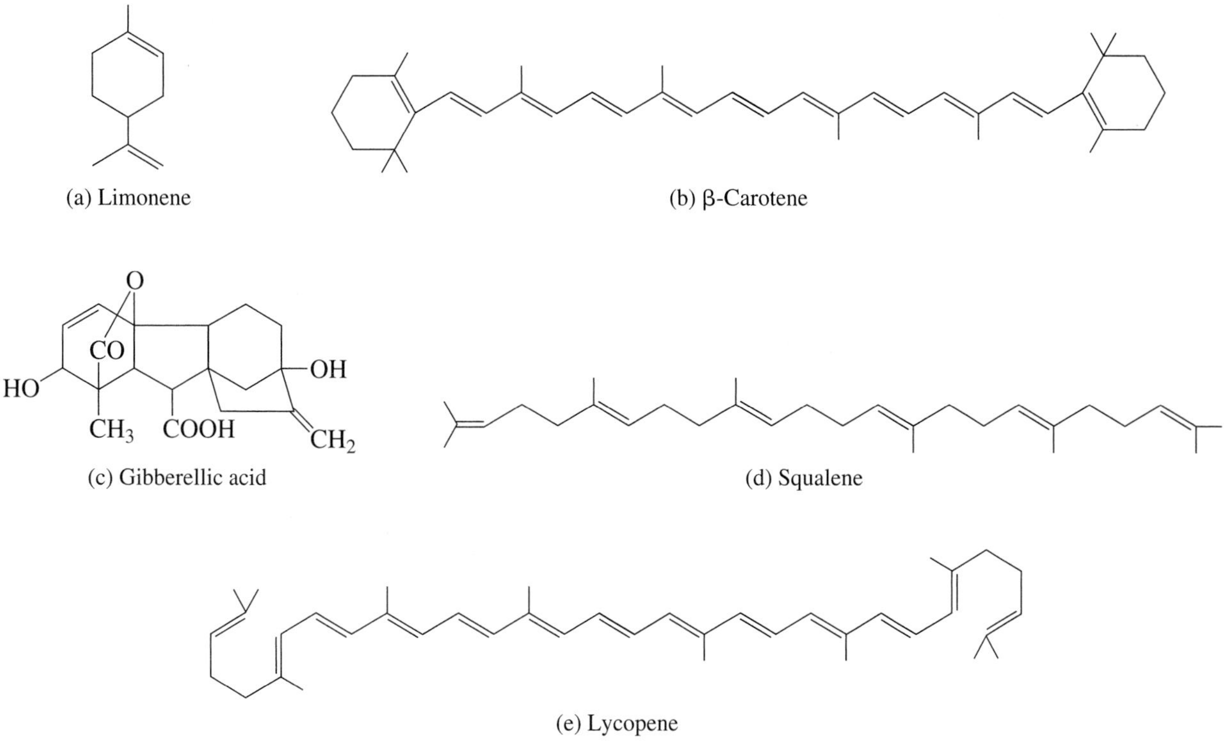

Figure 9.12

Important terpenes found in plants and animals: (a) limonene, found in citrus fruits, where it provides the distinct odor; (b) β-carotene, the source of the orange color in carrots (the precursor of vitamin A); (c) gibberellic acid, a plant growth hormone; (d) squalene, an intermediate in cholesterol synthesis; and (e) lycopene, the red pigment of tomato skin.

PGD_2 promotes physiological sleep and PGE_2 induces wakefulness. The structures for PGE_2 and PGD_2 are shown in Figure 9.13. The prostaglandins that cause fever, inflammation, and pain are diminished in cellular concentration by aspirin and ibuprofen. These over-the-counter pain killers inhibit an enzyme, prostaglandin synthase, which transforms arachidonate to the prostaglandin structures.

The thromboxanes, also derived from arachidonate, are characterized by a six-membered ring containing oxygen. The thromboxanes were first isolated from blood platelets (thrombocytes), where they are thought to facilitate the formation of blood clots. (See Window on Biochemistry 9.1.)

A third group of eicosanoids synthesized from arachidonate are the leukotrienes. They are named after white blood cells (leukocytes), from which they were first extracted. The structural characteristics that separate them from the other eicosanoids are the linear chain (no rings) and three conjugated double bonds. The leukotrienes cause the contraction of smooth muscle, especially in the lungs. Asthmatic attacks and allergic reactions may be caused by overproduction of leukotrienes.

Lipid-Soluble Vitamins

In Chapter 7, Section 7.1, we identified two classes of vitamins, those that are water soluble and those that are fat soluble. Here we only briefly mention the biological properties of the fat-soluble vitamins and wait until later chapters to study their roles in nutrition and metabolism. This lipid-soluble group of vitamins can be classified as terpenes (derived from isoprenes), but because of their importance in human health they are usually considered a separate category of lipids. The most important compounds to consider

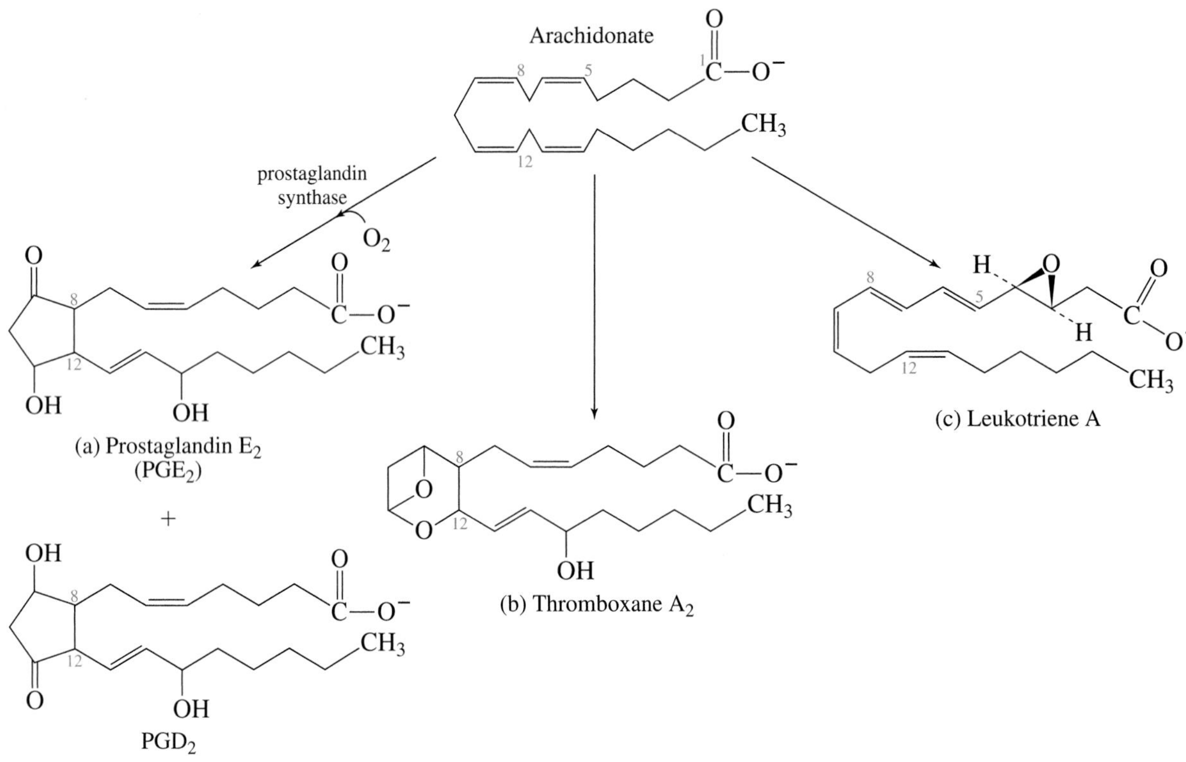

Figure 9.13
Structures of important eicosanoids and route of synthesis from arachidonate: (a) prostaglandins, (b) thromboxane A_2, and (c) leukotriene A.

here are vitamins A, D, E, and K. Table 9.4 provides the names, chemical characteristics, and the most significant biological functions for the fat-soluble vitamins.

Pheromones

Some organisms release chemical substances into the environment that alter the behavior of other organisms of the same species. Most often the behavior is linked to sexual attraction, but it may also involve trail marking and alarms. These hormonelike substances are called **pheromones.** Although all organisms, including humans, may release pheromones, the best studied are those from insects. The compound muscalure, which is a long-chain hydrocarbon (Figure 9.14a), is secreted by the common female housefly to attract a male partner. The sex attractant used by a honeybee queen in her nuptial flight is 9-ketodecenoic acid (Figure 9.14b).

Table 9.4
Common fat-soluble vitamins and their biological functions

Vitamin	Common Name	Chemical Characteristics	Biological Function
A	Retinol	A terpene with 20 carbons	Absorption of light in vision
D	Several forms; one is D_3-cholecalciferol	Formed from cholesterol by ultraviolet radiation	Regulation of calcium and phosphorus metabolism
E	α-Tocopherol	Aromatic ring with long hydrocarbon chain	Antioxidant; prevents oxidation damage to cellular membranes
K	Vitamin K	Bicyclic ring system with long hydrocarbon chain	Regulates blood clotting; bone formation

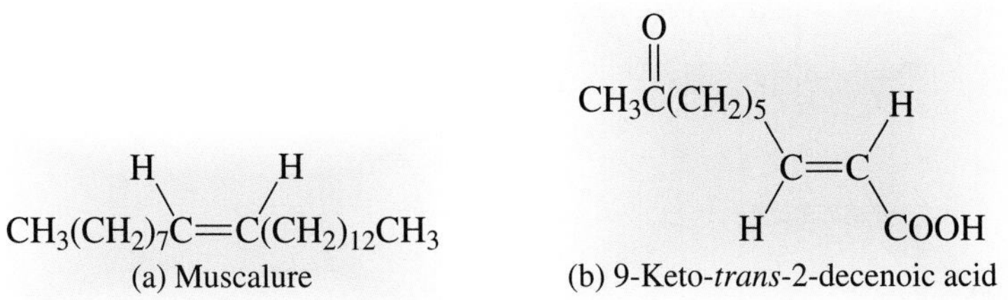

$$CH_3(CH_2)_7C=C(CH_2)_{12}CH_3$$

(a) Muscalure

$$CH_3\overset{\overset{\displaystyle O}{\|}}{C}(CH_2)_5 \quad H$$
$$C=C$$
$$H \quad\quad COOH$$

(b) 9-Keto-*trans*-2-decenoic acid

Figure 9.14
The molecular structures for two insect pheromones: (a) muscalure (housefly) and (b) 9-keto-*trans*-2-decenoic acid (honeybees).

Electron Carriers

The final group of lipids to be considered here are those that serve as electron carriers for the oxidation–reduction reactions of energy metabolism. Ubiquinone (coenzyme Q), a primary component of the mitochondrial electron transport chain in animals, is named for its ubiquitous presence in cells. Plastoquinone, a compound in plant chloroplasts, serves as an electron carrier for the production of adenosine triphosphate (ATP) generated by light absorption (photosynthesis). Menaquinone is the primary electron carrier of bacteria. All these compounds function as redox cofactors, cycling between oxidized and reduced states as they shuttle electrons from one biomolecule to another (see Chapter 17).

9.5 Biological Membranes

Biological Roles of Membranes

All biological cells are surrounded by a boundary layer called a **plasma membrane** that consists primarily of proteins and polar lipids. (Eukaryotic cell organelles, such as the nucleus, mitochondria, and chloroplasts, also have their own membrane systems that are similar in structure and function to plasma membranes.) At first glance these barriers may be thought of only as protective shields to isolate the cell's and organelle's sensitive interiors from their exterior environments. Although this role of membranes is important, they perform other functions critical to the cell's survival. As a physical barrier, the membrane provides structural integrity to a cell, giving it shape and form and literally "packaging" the cellular components. The membrane structure is responsible for the organization and compartmentation of biochemical activities within tissues and cells. However, a cell cannot live in a completely isolated state. The membrane also must act as a selective filter to allow the entry of nutrients necessary for the cell's growth and development and the exit of metabolic waste products. The selectivity of membrane transport brought about by protein channels, pumps, and gates regulates the flow of biomolecules and the ionic composition of the cellular milieu. Biochemically, this results in unidirectional or vectorial membrane processes such as the transport of some nutrients and ions from the outside to the inside of the cell, but not the reverse.

The cell also must communicate with its surroundings. Embedded in the exterior side of plasma membranes are protein receptors that bind such hormones as adrenaline and insulin, thereby relaying a chemical signal that regulates metabolic processes in the interior of the cell (see Chapter 2; Section 2.5, and Chapter 8, Section 8.5). Thus, chemical stimuli outside the cell can influence the cell's inner biochemistry.

Finally, some specialized membranes contain protein assemblages that act as **energy transduction** systems. Animal mitochondrial membranes contain enzymes and other proteins that convert energy released by the oxidation of fats and carbohydrates to the chemical form of ATP. In photosynthetic organisms, energy from light is trapped by pigments and transformed to chemical energy (to drive glucose synthesis) by proteins in chloroplast membranes. The details of these important and interesting membrane processes must be postponed for later chapters. First, we shall explore the components, structures, and dynamic properties of biomembranes that make such processes possible.

Table 9.5
The lipid and protein compositions
of several membranes

Percentage by Weight[a]		
Membrane Source	Lipid	Protein
Myelin	80	18
Mouse liver	52	45
Human erythrocyte (plasma)	43	49
Corn leaf	45	47
Mitochondria (outer)	48	52
Mitochondria (inner)	24	76
Escherichia coli	25	75

[a]If the total is under 100%, the balance is made up by carbohydrate.

Membrane Components and Structure

Biomembranes are supramolecular assemblies of lipids, proteins, and carbohydrates. The composition ratio varies depending on the source of the membrane. Each type of cell and organelle has unique functions that depend on the properties of their membranes. Diversity in membrane composition can lead to unique membrane functions which are necessary for cell individuality. Table 9.5 compares the compositions of several diverse membranes. It is evident that membranes exist that have a wide range of protein and lipid compositions. The ratio of protein to lipid can range from 80:20 to 20:80. It is surprising that membranes have so many characteristics in common even though their compositions are so varied. Free carbohydrates are not present in membranes. The carbohydrates of membranes are covalently bound to lipid and protein molecules. The carbohydrate content is usually 5% or under and is represented in the glycolipids (see Section 9.3) and glycoproteins. The carbohydrate tails perform important functions as cell recognition sites (see Chapter 8, Section 8.5).

The basic structural framework of membranes is provided by a double layer of lipids (lipid bilayer, Figure 9.15a). Membrane lipids possess common structural features favoring

Figure 9.15

Lipid bilayer folding into a closed spherical vesicle: (a) The nonpolar components of membranes can minimize their exposure to water by forming a bilayer. The hydrophobic tails of the lipids cluster in the nonpolar region and the hydrophilic heads of the lipids face water on both sides of the bilayer. (b) The bilayer closes to form a vesicle, thereby segregating fluid inside the vesicle from fluid surrounding it.

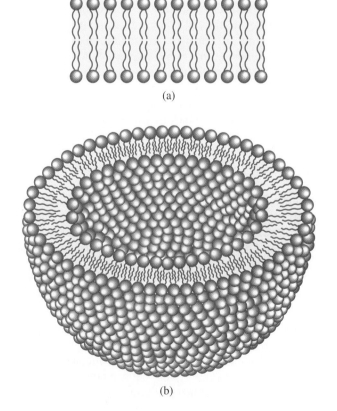

(a)

(b)

their assembly in aqueous environments into a stable bilayer arrangement that is held together by noncovalent, hydrophobic interactions. Indeed, the first picture proposed for biomembranes in 1925 consisted of two lipid monolayers combined into a bilayer. This model accounted for many of the properties known for membranes at that time. The bilayer of polar lipids is formed in a spontaneous process; therefore, this must represent a low free energy arrangement. The bilayer in water, upon further study, was found to fold into a closed spherical vesicle, thus covering the "edges" of the bilayer and reducing their exposure to water (Figure 9.15b). Such vesicles were used for years to study membrane structure and function. They display two important features of membranes: unique transport properties and fluidity. Transport studies of closed spherical vesicles composed of polar lipids show that small, nonpolar molecules, such as CO_2 and hydrocarbons, diffuse through the bilayer, but most polar biomolecules, such as amino acids, sugars, proteins, and nucleic acids are prohibited from transport by simple diffusion. Thus, one function of the membrane bilayer is to act as a selective, permeable barrier for transport of molecules. Another feature of the bilayer vesicles that is important for membrane structure is fluidity. The bilayer is liquid with a consistency of vegetable oil. Using radioactively labeled lipids, it has been shown that the polar lipids of vesicles and biomembranes are mobile and capable of rapid lateral or sideways movement, but polar lipids in one monolayer rarely exchange with lipids in the other monolayer.

More detailed study of lipid bilayers in membranes has shown a nonrandom distribution of lipids in the two monolayers. In the erythrocyte plasma membrane, the outer layer has a predominance of phosphatidylcholine and sphingomyelin. The opposite monolayer (cytoplasmic side) has more phosphatidylethanolamine, phosphatidylserine, phosphatidylinositol, and glycolipids. The two distinct sides display elements of asymmetry in membrane structure (Figure 9.16).

Cholesterol, which has a very different chemical structure than other membrane lipids, has great influence on animal membranes. Its presence can vary from as low as 3% of the lipids in mitochondrial membrane to as high as 38% in typical plasma membrane. It is never found in plants. The cholesterol molecule is stiff and more rigid than other polar lipids, making membranes less fluid. Because they are devoid of cholesterol, plant membranes tend to be more fluid than animal membranes.

Membrane Proteins

Although spherical, closed vesicles formed from polar lipids serve well as models for membrane structures, they explain poorly the more dynamic actions and distinctive characteristics of membranes. If biomembranes were composed only of lipids, then only very small molecules and nonpolar molecules could be transported and all cell membranes would have very similar biological properties. In addition, cells of different types of tissues and organisms could not display such diverse and unique functions. The dynamic activities of the cell membrane are carried out by membrane proteins. The proteins are of several different

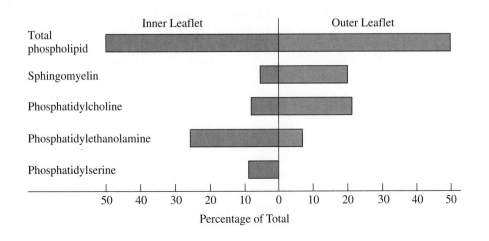

Figure 9.16

Distribution of lipids in the two monolayers (leaflets) of the human erythrocyte. *Source:* Redrawn from Voet & Voet, 1996.

Figure 9.17

Examples of two types of membrane proteins: (a–c) integral proteins and (d and e) peripheral proteins.

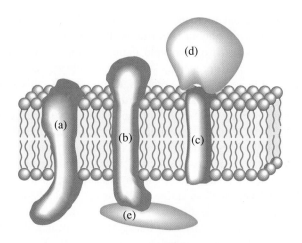

Figure 9.17

Examples of two types of membrane proteins: (a–c) integral proteins and (d and e) peripheral proteins.

functional types including enzymes, receptor proteins and transport proteins. All membrane proteins are classified into one of two general groups according to their ease of extraction from membranes (Figure 9.17). The removal of proteins from membranes depends on the polar or nonpolar characteristics of the protein. Some proteins are removed simply by treating the membrane with a dilute aqueous salt solution (1 M NaCl). Because of the ease of their removal, these proteins are presumably on the surface and are relatively polar and very water soluble. They are called **peripheral proteins.** The other class of membrane proteins requires harsher treatment for removal. Extraction of membranes with aqueous detergents such as SDS (see Chapter 4, Section 4.6) releases **integral proteins** that are buried inside the membrane. They can be released only by disrupting the lipid bilayer and interfering with the hydrophobic interactions between protein and lipid. Peripheral proteins are usually bound to the membrane by noncovalent, ionic, and hydrogen bonding interactions between their amino acid residues with complementary amino acid residues in integral proteins that are exposed on the membrane surface. The majority of peripheral proteins probably function as receptor sites and enzymes. Membranes also show asymmetry in terms of protein location since different peripheral proteins are located on the inside and the outside. This allows cells to have distinctive biological activities on each side, outside and cytoplasmic. Integral proteins contain large portions (domains) of hydrophobic amino acid residues that allow them to become buried in the nonpolar region of the lipid bilayer. Since they are usually transmembrane (span the membrane), the integral proteins function primarily as transport proteins, facilitating the movement of solute molecules from one side of the membrane to the other. A substantial portion of the protein is within the membrane, where it often acts as a channel or gate.

The Fluid-Mosaic Model for Membranes

The picture that best represents what we know about the chemical structure and biological function of membranes is the **fluid-mosaic model** (Figure 9.18). S. Jonathan Singer and Garth Nicholson in 1972 suggested this model, which consists of a lipid bilayer embedded with proteins, with some on the surface (peripheral) and others passing through the entire bilayer (integral). This model presumes intimate contact between lipid and protein. The proteins float somewhat freely in and on the bilayer, creating a mosaic pattern. The Singer–Nicholson model is supported by X-ray diffraction data, which reveal two areas of high electron density, represented by the two sides of the bilayer, and a central area of low electron density, represented by the hydrophobic tails of the lipids. The dynamic transport properties can also be explained by the model. Nonpolar molecules presumably diffuse through the membrane because of the hydrophobic nature of the central region. Polar, hydrophilic molecules must be assisted in transport by specific proteins located in the membrane. The fluid-mosaic model is also consistent with the experimental observation of asymmetric organization of membrane compounds. Some peripheral proteins are located on the exterior surface of the membrane, whereas others are on the interior surface. Biochem-

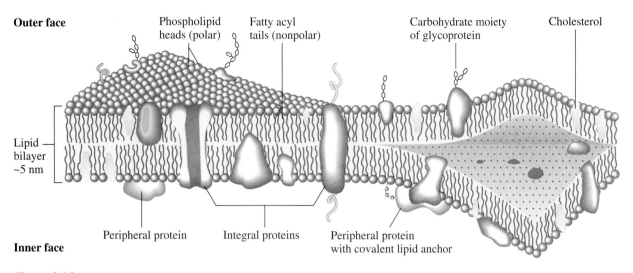

Outer face

Phospholipid heads (polar)

Fatty acyl tails (nonpolar)

Carbohydrate moiety of glycoprotein

Cholesterol

Lipid bilayer ~5 nm

Peripheral protein

Integral proteins

Peripheral protein with covalent lipid anchor

Inner face

Figure 9.18

The fluid-mosaic model for biological membranes. This model is composed of a lipid bilayer that is embedded with integral and peripheral proteins. The lipids provide the structural framework and the proteins function as enzymes or as transport proteins. Peripheral proteins are usually rather hydrophilic, whereas integral proteins contain hydrophobic regions to anchor them in the lipid bilayer. Some integral proteins form channels for transport of solutes. Carbohydrates in membranes are usually covalently linked to lipids and to proteins. *Source:* Based on Lehninger et al., 1999.

ically this results in unidirectional or vectorial membrane processes such as the transport of nutrients from outside to the inside of the cell, but not the reverse. The fluid-mosaic model portrays a rather static picture, but the true picture is more complex and dynamic. The membrane is fluid and constantly changing by lateral movements of proteins and lipids.

9.6 Membrane Transport

For a cell to survive, it must be constantly supplied with nutrients and metabolic waste products must be removed. All biomolecules that enter or exit the cell encounter the plasma membrane barrier. (A similar situation exists for the entry and exit of biomolecules through the membranes of cell organelles.) Certain small, nonpolar molecules can readily diffuse through the membrane. Other molecules, specifically those that are large and polar, require the assistance of protein channels, carriers, gates, and pumps to achieve transport. Membrane proteins, especially the integral proteins, regulate the flow of molecular traffic. All forms of membrane transport can be classified into one of two major groups, **passive transport** or **active transport.** The key distinction between these groups is their dependence on energy. In **passive transport,** a biomolecule (solute) moves from a region of higher concentration through a permeable divider to a region of lower concentration of that solute. Because this is thermodynamically favorable (entropy is increasing), no expenditure of energy is required (Figure 9.19). In this section, we explore the characteristics of two kinds of passive transport: **simple diffusion** and **facilitated diffusion** (a protein-mediated process). In the second form of solute transport, **active transport,** expenditure of energy by the cell is required because solutes are moved from a region of lower concentration through a permeable barrier to a region of higher concentration. This results in a net accumulation of solute on one side of the membrane, a process that is energy dependent (entropy is decreasing). The transport of polar biomolecules and ions like amino acids, carbohydrates, Na^+, K^+, and Ca^{2+} usually requires an energy source. The energy sources for active transport are varied, but the most common method is to couple transport to energy-producing systems such as a chemical reaction (cleavage of ATP) or light absorption. The forms of active transport to be discussed include ion pumping (ATPase-dependent processes) and cotransport.

Interactive Animation: Cellular transport

Figure 9.19

Passive transport of solute molecules through a permeable membrane, C_L and C_R refer to the solute concentrations on the left side and right side, respectively, of the membrane. The net rate of movement of a solute through a permeable membrane depends on the relative concentrations of solute on the two sides (C_L and C_R). When $C_L >> C_R$, net diffusion of molecules is from left to right. When $C_L = C_R$ solute molecules continue to diffuse in both directions but the net flow is zero.

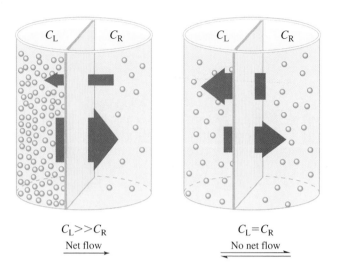

Passive Transport: Simple and Facilitated Diffusion

The process of **simple diffusion** is depicted in Figure 9.19. Imagine two compartments, divided by a permeable barrier (a membrane), with unequal concentrations of a solute, C_L and C_R. C_L refers to the solute concentration on the left side and C_R to solute concentration on the right side. The net direction of solute molecule flow through the barrier is from the C_L to the C_R compartment. Although some solute molecules will diffuse from C_R to C_L, more will flow in the opposite direction. This trend continues until the concentration of solute molecules is equal on both sides. There will at this point still be diffusion of molecules, but no *net* flow; that is, an equal number of solute molecules will be moving in each direction (equilibrium will have been reached). The process of simple diffusion requires no expenditure of energy because the molecules are moving "downhill" on an energy scale. Biological examples of molecules that move through membranes by simple diffusion include fatty acids and the important gases CO_2, N_2, O_2, and CH_4. Although the transport mechanism of these small molecules through biological membranes is not completely understood, we assume they pass through the lipid bilayers because they are small and/or nonpolar. Water molecules, at one time, were thought to be transported by simple diffusion. However, recent studies have implicated the involvement of membrane proteins called **aquaporins.** Several such water transport systems have been discovered in human cells, especially the kidney.

Some solute molecules such as sugars and amino acids are too large and polar for simple diffusion through a lipid bilayer. The transport of these molecules is assisted by specific membrane proteins. We call this **facilitated diffusion.** There is no requirement for an energy input so it is still classified as passive transport. Membrane proteins, sometimes called carrier or transport proteins, assist by forming transmembrane channels or perhaps by binding and "carrying" solute molecules. Some proteins create transmembrane channels by folding into α-helical conformations. We would expect these to be integral proteins spanning the membrane. Proteins that bind molecules and thus enhance their transport are sometimes called **permeases** because their actions are similar to enzymes. Just because the transport of solutes is facilitated, we must not assume that molecules can be moved, without energy input, from a region of lower concentration to a region of higher concentration. The same thermodynamic rules are in effect as for simple diffusion. The net flow of solute molecules is from an area of higher concentration to an area of lower concentration. In Section 9.7 we study an important example of facilitated diffusion: the transport of glucose through the erythrocyte membrane. This may also be classified as uniport, where proteins carry only one type of molecule (Figure 9.20).

Facilitated diffusion sometimes involves the transport of two solute molecules (cotransport). Two types of facilitated diffusion by cotransport have been studied: symport, where the two solute molecules transported move in the same direction, and antiport, where the two mol-

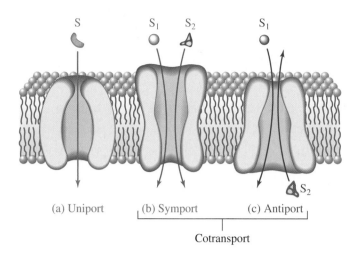

(a) Uniport (b) Symport (c) Antiport

Cotransport

Figure 9.20
Three types of membrane transport are based on the number of solute molecule types and the direction of net flow: (a) Uniport: only one type of solute molecule, S, is transported; (b) symport: two types of solute molecules, S_1 and S_2, are transported in the same direction; (c) antiport: two types of solute molecules, S_1 and S_2, are transported in opposite directions. Cotransport refers to the transport of two solutes. Classification of a transport process into one of these groups does not indicate whether the process is passive or active. This can be determined only by the need for energy. *Source:* Based on Lehninger et al., 1999.

ecules move in opposite directions. Amino acids and sodium ions (Na^+) are transported through intestine and kidney membranes by symport processes. An example of antiport is the exchange of chloride (Cl^-) and bicarbonate (HCO_3^-) by proteins in erythrocyte membrane.

Active Transport: Ion Pumping

Since energy is expended by the process, active transport can achieve the net accumulation of solute on one side of a permeable membrane. Even though $C_L >> C_R$, solute molecules are transported from an area of lower concentration to one of higher concentration (Figure 9.21). This is so alike pumping water uphill that membrane systems are called transport pumps. Just as an energy source (electricity, gas, etc.) is required for pumping water uphill, energy must be supplied for active transport of solute molecules. Most active transport processes are driven by chemical reactions or by light absorption. Active transport processes are crucial for the survival of organisms. The entry of nutrients into *Escherichia coli* cells is brought about by a membrane permease that acts by cotransport. For example, sugars, such as the disaccharide lactose are transported by a symport process requiring the simultaneous entry of protons. In higher plant cells, K^+ and H^+ are actively transported by an antiport process. The active transport of Na^+ and K^+ by the Na^+–K^+ ATPase pump, which is present in the plasma membrane of virtually every animal cell, is described in the next section.

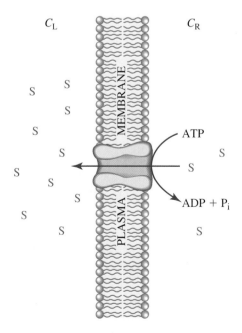

Figure 9.21
Active transport of solutes. Transported solute molecules (S) move from C_R, a region of lower concentration to C_L, a region of higher concentration. Because this net flow is against the concentration gradient, energy input is required. The source of energy is often from the cleavage of a phosphoanhydride bond in ATP. In the reaction shown, ATP is cleaved to ADP and P_i, with release of energy that is coupled to the transport of S molecules.

9.7 Examples of Membrane Transport

In this section the structure and function of several important membrane proteins are explored.

Glycophorin A of the Erythrocyte Membrane

One of the most studied and best understood membrane systems is that of the red blood cell (erythrocyte). Glycophorins, a family of glycoproteins, are extracted from erythrocyte membrane by the use of detergents. One specific protein in this group is glycophorin A, which has 131 amino acid residues and 16 oligosaccharide groups containing a total of

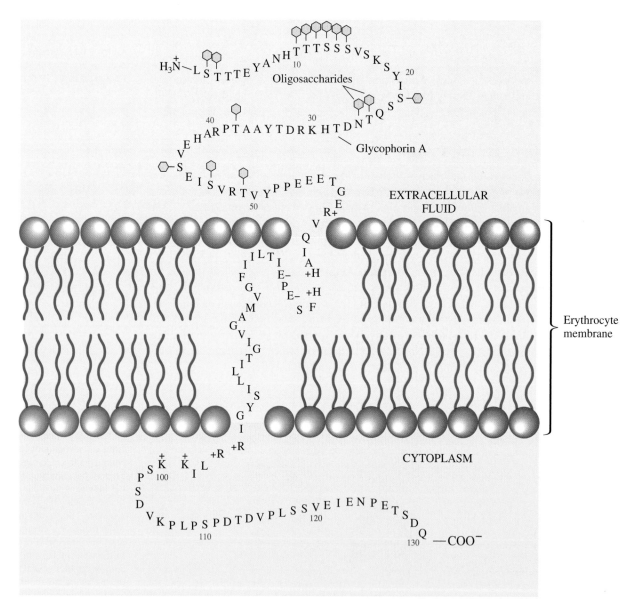

Figure 9.22

The proposed arrangement of glycophorin A in the erythrocyte membrane. A nonpolar region of about 30 amino acids in the protein spans the membrane. Relatively polar regions of the protein are inside and outside the membrane. The hexagon rings show the covalent linkage of carbohydrates. The symbols for the amino acids are: Q, glutamine; I, isoleucine; A, alanine; H, histidine; F, phenylalanine; S, serine; E, glutamate; P, proline; T, threonine; L, leucine; G, glycine; V, valine; M, methionine; Y, tyrosine; R, arginine; N, asparagine; D, aspartate; C, cysteine; K, lysine; and W, tryptophan.

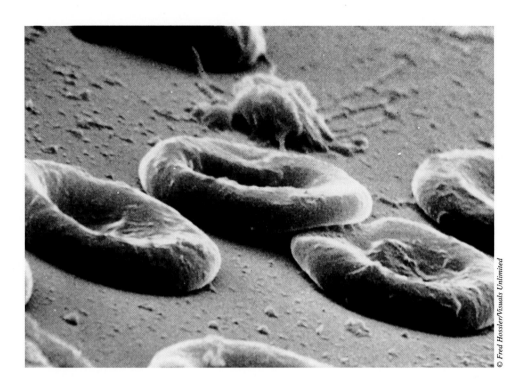

Scanning electron micrograph of human erythrocytes (×7500). One type of protein in the membrane is glycophorin.

© Fred Hossler/Visuals Unlimited

almost 100 monosaccharides. (About 60% of the mass of the protein is carbohydrate, with abundant sialic acid present.) Glycophorin A is an integral protein with part of the molecule on the outside of the plasma membrane and part on the cytoplasmic side (Figure 9.22). This arrangement requires that the protein span the membrane; that is, it must pass through the lipid bilayer, which suggests a hydrophobic domain in addition to the hydrophilic ends extending on either side of the membrane. Indeed, amino acid sequencing of glycophorin A uncovered a region of about 30 hydrophobic amino acid residues that anchors the protein in the membrane. Furthermore, the geometric conditions and amino acid sequence are right for this hydrophobic region to fold into an α-helix structure that spans the membrane. Figure 9.22 shows the proposed position of glycophorin A in the erythrocyte membrane. Note that almost one-half of the protein (the N-terminus) is outside the membrane in the extracellular fluid. With so much known about its primary and secondary structure and its membrane location, it is surprising that the biological function has not yet been found. However, much has been learned from study of the glycophorins. We now know of several transmembrane systems that use α-helical protein spans as channels for transport of solutes. One of the best understood is the chloride–bicarbonate antiport process previously mentioned. In Window on Biochemistry 5.2, we introduced another integral protein, bacteriorhodopsin, from the purple membrane of *Halobacterium;* this protein traverses the lipid bilayer with seven hydrophobic, α-helical regions and pumps protons across the membrane.

Glucose Permease of Erythrocyte Membrane

The primary fuel molecule for red blood cell metabolism is glucose. Blood glucose is provided to erythrocytes by a facilitated diffusion process carried out by a specific glucose transporting protein called **glucose permease** (Figure 9.23). The permease is a single polypeptide chain containing almost 500 amino acid residues, with about 12 hydrophobic domains spanning the membrane. The transport mechanism has many characteristics similar to enzyme action. A graph of initial velocity of glucose entry versus external glucose concentration resembles the Michaelis–Menten curve for enzymatic catalysis (Figure 9.24). Apparently at higher concentrations of glucose, all permease molecules become saturated. The maximum rate of entry (V_{max}) occurs at this point and can be graphically determined.

Figure 9.23

The glucose permease of erythrocyte membrane. This transport protein is a single polypeptide chain that spans the membrane. The permease accepts glucose from the extracellular fluid and assists its transport through the membrane. Thus, this is a passive, facilitated process. Cytoplasmic glucose concentrations can never exceed the external glucose concentrations (about 5 mM).

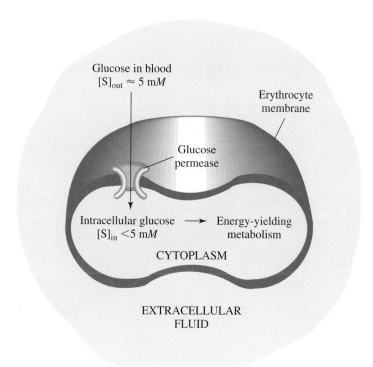

Glucose in blood
[S]$_{out}$ ≈ 5 mM

Erythrocyte membrane

Glucose permease

Intracellular glucose ⟶ Energy-yielding
[S]$_{in}$ <5 mM metabolism

CYTOPLASM

EXTRACELLULAR
FLUID

The external glucose concentration that yields an entry rate of $\frac{1}{2} V_{max}$ is represented by $K_{transport}$ (K_t instead of the familiar K_M for enzymes). Another observed similarity between permeases and enzymes is substrate specificity. The permease displays the highest transport rate with the monosaccharide glucose. The glucose epimers, galactose and mannose, are transported by the permease at a rate approximately 20% that for glucose. The transport of glucose can be competitively inhibited by glucose analogs. The transport process for the glucose permease involves the following steps:

1. Binding of the glucose to a specific site on the extracellular fluid side of the transport protein.
2. Passage of glucose through the membrane, perhaps through a channel formed by α-helical regions in the permease.
3. Release of the glucose on the cytoplasmic side.

Figure 9.24

The transport of glucose through the erythrocyte membrane by permease. From the graph of initial rate of glucose entry versus external glucose concentration, the values of V_{max} and K_t can be determined. The kinetics of the transport process are similar to the kinetics of enzyme action.

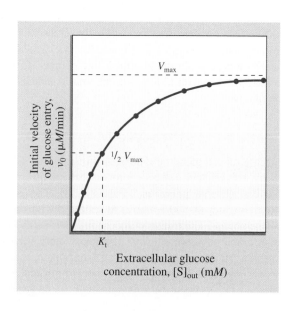

V_{max}

$\frac{1}{2} V_{max}$

K_t

Extracellular glucose
concentration, [S]$_{out}$ (mM)

Initial velocity of glucose entry, v_0 (μM/min)

The amino acid sequence of human erythrocyte glucose permease has been determined. It consists of 492 amino acids that can be arranged into 25 domains: 12 hydrophobic and 13 hydrophilic. The polar and nonpolar segments alternate with each other. The picture that develops for the permease in the membrane is shown in Figure 9.25. The permease may weave back and forth at least 12 times across the membrane. It has been suggested that the protein folds into a helical arrangement to create a pore. The glucose permease, which has characteristics of solute specificity, saturation kinetics and inhibition, is representative of facilitated transport systems in other membranes.

This Yanomami Indian hunts with poison darts in the Brazilian jungle.

Na$^+$–K$^+$ ATPase Pump

Inside eukaryotic cells, the concentration of sodium and potassium ions is maintained at different levels than in extracellular fluid:

$$[Na^+]_{inside} < [Na^+]_{outside}$$
$$12 \text{ m}M \qquad 145 \text{ m}M$$

$$[K^+]_{inside} > [K^+]_{outside}$$
$$140 \text{ m}M \qquad 4 \text{ m}M$$

The balanced flow of ions (antiport) requires an active transport process that pumps Na$^+$ outside and K$^+$ inside. Each is pumped from a region of lower concentration to one of higher concentration. The plasma membrane of all eukaryotic cells contains an integral membrane protein pump called the **sodium–potassium (Na$^+$–K$^+$) ATPase pump.** The enzyme word ending, *ase,* is used because the transport of Na$^+$ and K$^+$ has characteristics

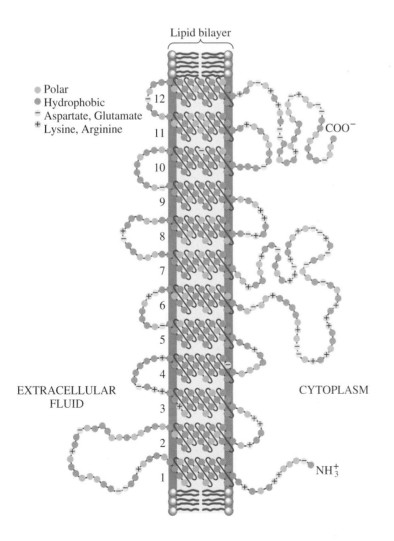

Lipid bilayer

- Polar
- Hydrophobic
- Aspartate, Glutamate
- Lysine, Arginine

COO$^-$

EXTRACELLULAR
FLUID

CYTOPLASM

NH$_3^+$

Figure 9.25

The proposed arrangement of glucose permease in the erythrocyte membrane. The 492 amino acids are folded into segments, 12 hydrophobic regions and 13 hydrophilic regions. Each hydrophobic region spanning the membrane is coiled into an α-helix to provide a channel for transport of glucose. *Source:* From Leinhard et al., 1992.

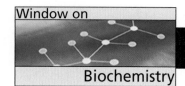

9.4 Patch Clamps and Membrane Transport—Low Tech with Surprising Results

An important method of communication between cells is the transmission of chemical signals through **ion-channels,** pores in transmembrane proteins that can be opened and closed like gates. Ion-channels are present in all types of cell membranes, including the plasma membrane of neurons and muscle cells. The best-studied example of gated channels is the acetylcholine receptor of synapses. The transmission of nerve impulse to a neuron causes the release of acetylcholine into the synapse. Acetylcholine diffuses through the synaptic cleft and binds to a specific receptor protein on the adjoining neuron. The binding triggers a conformational change in the receptor, which results in the opening of an ion channel in an adjacent protein. Sodium ions move from the synapse into the channel and adjoining neuron, thereby depolarizing that cell and continuing the nerve impulse.

Much of our current knowledge of nerve transmission and ion-channels comes from the use of a relatively simple procedure that allows the isolation and study of individual ion-channels. Erwin Neher and Bert Sakmann, of Germany, the developers of the **patch clamp technique** to study ion-channels, were awarded the Nobel Prize for Physiology or Medicine in 1991. Ion-channels can be isolated by pressing a thin, specially shaped glass pipette directly against an enzymatically cleaned surface of membrane as shown to the left. A tight seal is produced by gentle suction. Chemical stimuli can be applied to the membrane and resulting chemical and physical changes are monitored through the pipette.

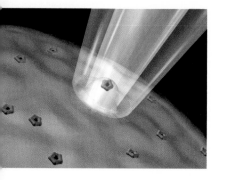

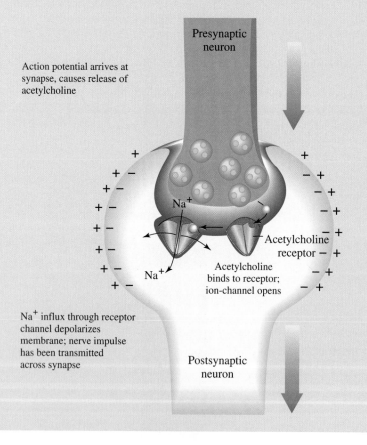

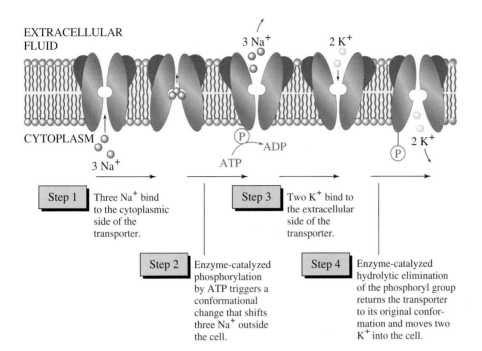

EXTRACELLULAR FLUID

CYTOPLASM

3 Na$^+$

ATP

ADP

2 K$^+$

2 K$^+$

| Step 1 | Three Na$^+$ bind to the cytoplasmic side of the transporter. |

| Step 2 | Enzyme-catalyzed phosphorylation by ATP triggers a conformational change that shifts three Na$^+$ outside the cell. |

| Step 3 | Two K$^+$ bind to the extracellular side of the transporter. |

| Step 4 | Enzyme-catalyzed hydrolytic elimination of the phosphoryl group returns the transporter to its original conformation and moves two K$^+$ into the cell. |

Figure 9.26
Transport of sodium and potassium ions by the Na$^+$–K$^+$ ATPase pump of plasma membrane. Three Na$^+$ are transported out of the cell for every two K$^+$ that move inside. Since this is an active transport process, energy is required. The energy source is the cleavage of ATP to ADP + P$_i$. *Source:* Based on Devlin, 1999.

of enzyme action and is coupled directly to the chemical cleavage of a phosphoanhydride bond in ATP. ("Coupled" refers to a linkage between the transport of Na$^+$ and K$^+$ and cleavage of ATP.) The ATPase is composed of two subunits, both of which span the plasma membrane and are involved in ion transport and ATP cleavage. The detailed mechanism of ATPase action is not completely known, but the overall process can be outlined as shown in Figure 9.26. Note that three Na$^+$ are transported out of the cell for every two K$^+$ moving inward. This results in a net electrical potential difference across the membrane (negative inside, positive outside). In neuronal cells this potential difference is crucial for nerve impulse generation. It has been estimated that between 25% and 30% of the ATP generated by humans is used to support the Na$^+$–K$^+$ ATPase pump, implying a significant importance attached to the active transport of these ions. Ouabain (wä′ bān), a naturally occurring steroid found in the seeds of some jungle plants and used as a toxin on the tips of hunting arrows, is a strong inhibitor of the Na$^+$–K$^+$ ATPase.

Summary

■ The lipids are biomolecules that are extracted from cells using nonpolar solvents such as hexane, methanol, and diethyl ether.

■ The most abundant constituents of lipids are the fatty acids. Most fatty acids contain between 12 and 24 carbon atoms and are present as esters linked to glycerol. The primary biological role of fatty acids is to serve as metabolic fuel.

■ Polar lipids, including the glycerophospholipids and sphingolipids, are combined with protein molecules for the construction of biological membranes.

■ Steroids, a special class of lipids, all have a characteristic fused-ring system. The best-known steroid, cholesterol, is abundant in membranes but is also an important biosynthetic precursor for steroid hormones, bile salts, and other biomolecules.

■ All biological cells are surrounded by a boundary layer called a plasma membrane. It consists of a double layer of protein and polar lipids. The lipids provide the structural framework and the proteins the dynamic activities of membranes.

■ The **fluid-mosaic model** provides the best picture for the membrane.

■ Membrane transport processes are classified as **passive** or **active** depending on the need for energy input.

Study Exercises

Understanding Terms

9.1 Give a brief definition for each of the following terms.

a. Polyunsaturated fatty acids
b. Micelle
c. Lipid bilayer
d. Inositol
e. Gangliosides
f. Tay-Sachs disease
g. Atherosclerosis
h. Isoprenes
i. Eicosanoids
j. Plastoquinone
k. Integral proteins
l. Facilitated diffusion
m. Antiport
n. Na^+–K^+ ATPase
o. Omega-3 fatty acids
p. Soaps
q. Triacylglycerols
r. Saturated fats
s. Prostaglandins
t. Phytosteroids

Reviewing Concepts

9.2 Which of the following molecules are in the lipid family of compounds?

a. 1-Decanol
b. Alanine
c. Fructose
d. Palmitic acid
e. Trimyristin
f. Glycerol
g. Adenine
h. β-Carotene
i. Aspartame
j. Insulin
k. Ubiquinone
l. Ethanol

9.3 Work out the shorthand nomenclature for each of the following fatty acids. The first problem is worked as an example.

a. $CH_3(CH_2)_5CH = CH(CH_2)_7COOH$
(shorthand nomenclature = $16:1^{\Delta 9}$)
b. $CH_3(CH_2)_5CH = CHCH_2CH = CH(CH_2)_4COOH$
c. $CH_3(CH_2)_4CH = CHCH_2CH = CHCH_2CH = CHCH_2CH = CH(CH_2)_3COOH$

9.4 Draw the chemical structures for the following fatty acids.

a. $10:1^{\Delta 4}$
b. $18:2^{\Delta 9,12}$
c. $18:3^{\Delta 9,12,15}$

9.5 Some scientists and physicians now recommend the ingestion of fish oil to reduce the risk of heart disease. The two major fatty acid components in a fish oil capsule are listed below. Draw the structures of these compounds.

a. Eicosanpentaenoic acid
$20:5^{\Delta 5,8,11,14,17}$

b. Docosahexaenoic acid
$22:6^{\Delta 5,8,11,14,17,20}$

9.6 Complete the following reactions.

a. $CH_3(CH_2)_5CH = CH(CH_2)_7COOH + Br_2 \longrightarrow$

b.

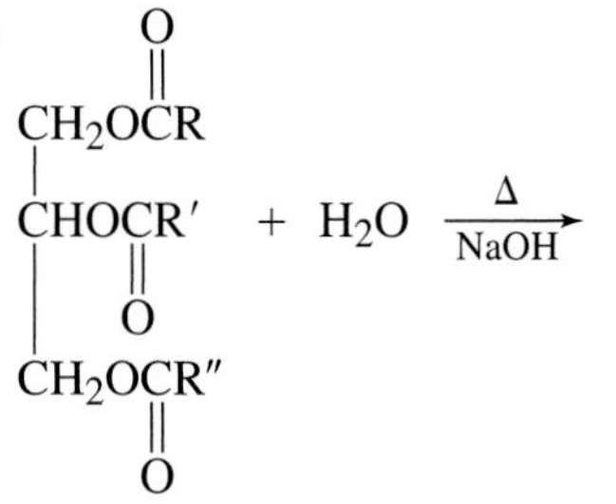

c.
$$CH_3(CH_2)_{14}-\overset{O}{\overset{||}{C}}OCH_2(CH_2)_{28}CH_3 + H_2O \xrightarrow[NaOH]{\Delta}$$

d.
cholesterol

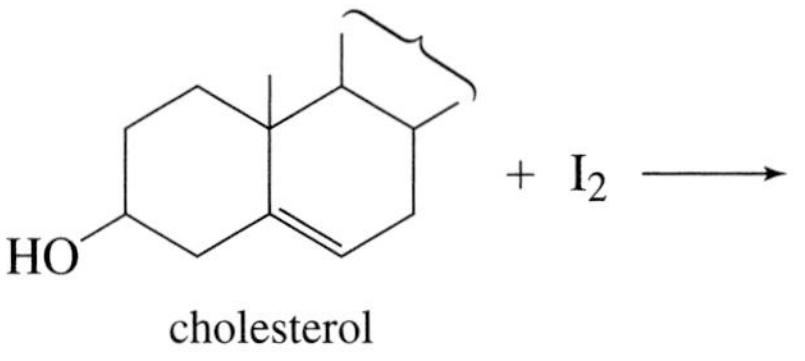

e.

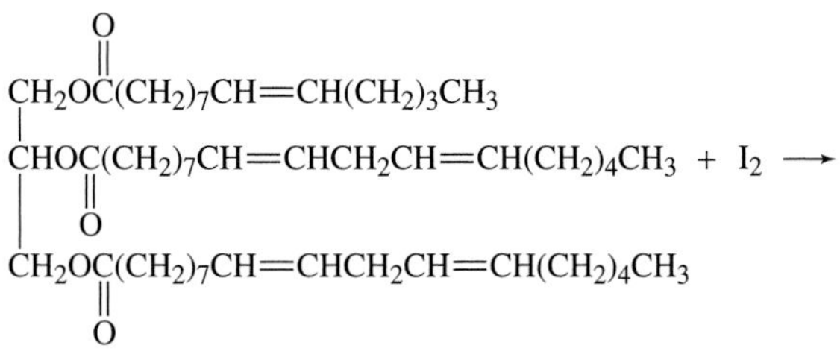

9.7 Explain why soaps in aqueous solutions assemble into micellar structures.

9.8 How do bile salts function to help digest fats?

9.9 Write a reaction or describe the action of each enzyme below.

a. Lipase
b. Prostaglandin synthase
c. Permease

9.10 In enzyme kinetics K_M is an important constant for substrate binding, whereas in transport $K_{transport}$ is used to define the process. How are K_M and $K_{transport}$ similar and how are they different?

9.11 Draw the structure of butanoic acid as it would exist under the following conditions.

a. pH 1.0
b. pH 7.0
c. pH 12.0

9.12 The major ingredient in many ointments for treatment of athlete's foot is the zinc salt of undecylenic acid ($11:1^{\Delta 10}$). Draw the structure of this potent fungicide. Would you expect this acid to be naturally occurring in plants or animals?

9.13 How do the prostaglandins and leukotrienes differ in chemical structure?

9.14 The lipid bilayer was a very early model for biological membranes. Explain how the lipid bilayer differs from a membrane as envisioned by the fluid-mosaic model.

9.15 Describe the patch clamp technique and explain how it may be used to study ion-selective channels.

Solving Problems

9.16 Why do some cooking oils, such as canola oil and olive oil, become rancid sooner than solid shortenings?

9.17 Arrange the following four compounds in the order of increasing solubility in water. When acyl groups are present, assume they are equivalent.

Glycerol
1,3-Diacylglycerol
2-Monoacylglycerol
1,2,3-Triacylglycerol

9.18 The chemical process of hydrogenation is used to convert oils to solids. The reaction below shows how hydrogenation takes place. Complete the reaction by writing the structure of the major organic product.
$$CH_3CH=CHCH_2CH=CHCH_2COOH + 2\,H_2 \xrightarrow{\text{Pt catalyst}}$$

9.19 The reactions listed below are important in the chemistry of lipids. Complete the reactions by drawing the structure for each product.

a. $CH_3(CH_2)_{10}COOH + NaOH \longrightarrow$

b.
$$CH_3(CH_2)_{10}\overset{\displaystyle O}{\overset{\displaystyle \|}{C}}OCH_2CH_3 + H_2O \xrightarrow[\Delta]{\text{NaOH}}$$

c.
$$\underset{H}{\overset{CH_3}{>}}C=C\underset{\overset{|}{C}-OH}{\overset{H}{<}} + H_2O \underset{\rightleftarrows}{\xrightarrow{\text{hydratase}}}$$
with $\overset{\|}{O}$ on the C—OH group

d.
$$\underset{H}{\overset{CH_3}{>}}C=C\underset{\overset{|}{C}-OH}{\overset{H}{<}} + FADH_2 \underset{\rightleftarrows}{\xrightarrow{\text{reductase}}}$$
with $\overset{\|}{O}$ on the C—OH group

9.20 Predict the mode of transport for each of the following molecules through the erythrocyte plasma membrane.

a. Phenylalanine   **e.** Cl^-
b. Lactose   **f.** K^+
c. H_2O **g.** Glucose
d. CO_2 **h.** Galactose

HINT: Choose from simple diffusion, passive–facilitated diffusion, or active transport.

9.21 Membrane lipids are said to have a polar head and non-polar tail(s). Assign these two regions in each of the following molecules.

a. Cholesterol
b. Phosphatidylcholine
c. Sphingomyelin
d. Cerebrosides

9.22 In a laboratory project your goal is to determine $K_{transport}$ for glucose transport through the erythrocyte membrane. The following data are collected:

[Glucose]$_{outside}$ (mM)	Rate of glucose entry (μm/min)
0	0
0.5	7
1.0	14
2.0	26
3.0	35
4.0	37
5.0	38

Calculate the experimental value for $K_{transport}$. What are the proper units for $K_{transport}$?

HINT: Prepare a graph of [glucose]$_{outside}$ versus rate of glucose entry or a plot of 1/[glucose] versus 1/rate of entry (see Chapter 6, Section 6.2).

9.23 Explain the following observations of plasma membrane transport.

a. Some H_2O may be transported by simple diffusion. Why is Na^+ not?
b. CO_2 can pass through a membrane by simple diffusion. HCO_3^- (bicarbonate) cannot.
c. Nitric oxide (NO) is transported by simple diffusion. Nitrite ion (NO_2^-) is transported by active transport.
d. Glucose is transported by facilitated diffusion. Glucose 6-phosphate cannot be transported.

9.24 Aspirin is an inhibitor of the enzyme prostaglandin synthase. What is the effect of aspirin on the cellular concentrations of the prostaglandins? How does aspirin reduce the inflammatory pain and cause blood "thinning"?

9.25 What are the common structural features of phosphatidylserine, phosphatidylethanolamine, phosphatidylcholine, and phosphatidylinositol?

9.26 Assume you have triacylglycerol samples containing the sets of fatty acids listed below. Is each sample a liquid (oil) or solid (fat) at room temperature?

a. palmitic, stearic, palmitic
b. oleic, linoleic, arachidonic
c. oleic, oleic, oleic
d. myristic, myristic, myristic

9.27 What solvents would be most effective in extracting terpenes from plant tissue?

HINT: Choose from water, diethyl ether, chloroform, hexane, methanol, and salt water.

9.28 Why do most naturally occurring fatty acids contain an even number of carbon atoms?

9.29 Why do fat-soluble vitamins tend to accumulate in cells rather than being flushed out in urine?

9.30 Compare and contrast the characteristics of membrane transport with those of enzyme action.

Writing Biochemistry

9.31 Write a brief summary for this chapter in paragraph form. The summary should be between 100 and 200 words. You may use the existing summary as a guide, but yours should be different.

9.32 The biochemistry of the phytosteroids is expanding rapidly. Find a recent article in your library or on the Internet on the topic of β-sitosterol or stigmasterol (see Window on Biochemistry 9.2). Write a 200-word summary of the article.

Applying Biochemistry

9.33 Membrane transport proteins are often given the suffix "-ase" used also for enzymes. For example, we discussed glucose permease in this chapter. Is this practice of naming transport proteins as enzymes "biochemically correct"? Do you find it confusing to think about transport proteins as enzymes or does it help in your understanding of membrane transport?

Further Reading

Agosta, W., 1994. Using chemicals to communicate. *J. Chem. Educ.* 71:242–246.

Bayley, H., 1997. Building doors into cells. *Sci. Am.* 277(3):62–67.

Borgnia, M. et al. 1999. Cellular and molecular biology of the aquaporin water channels. *Ann. Rev. Biochem.* 68:425–458.

Boyer, R., 2000. Purification and characterization of triacylglycerols. In *Modern Experimental Biochemistry*, 3rd ed., pp. 303–319. San Francisco: Benjamin/Cummings.

Bretshcer, M., 1985. The molecules of the cell membrane. *Sci. Am.* 253(4):100–109.

Changeux, J., 1993. Chemical signaling in the brain. *Sci. Am.* 269(5):58–62.

Costa, M., 1999. The natural nature of biomembrane lipids. *Biochem. Educ.* 27:207–208.

Dominguez, E., and Heredia, A., 1998. Waxes: a forgotten topic in lipid teaching. *Biochem. Educ.* 26:315–316.

Engelman, D., 1996. Crossing the hydrophobic barrier: Insertion of membrane proteins. *Science* 274: 1850–1851.

Guterman, L., 1999. The joy of soy. *Modern Drug Discovery* Nov./Dec.: 69–73.

Hajjar, D., and Nicolson, A., 1995. Atherosclerosis. *Am. Scientist* 83: 460–467.

Hanahan, D., 1997. *A Guide to Phospholipid Chemistry.* New York: Oxford University Press.

Jacobson, K., Sheets, E., and Simson, R., 1995. Revisiting the fluid mosaic model of membranes. *Science* 268:1441–1442.

Lawn, R., 1992. Lipoprotein(a) in heart disease. *Sci. Am.* 266(6):54–60.

Leinhard, G., Slot, J., James, D., and Mueckler, M., 1992. How cells absorb glucose. *Sci. Am.* 266(1):86–91.

Neher, E., and Sakmann, B., 1992. The patch clamp technique. *Sci. Am.* 266(3):44–51.

Ritter, S., 1999. Anabolic steroids. *Chem. Eng. News* Sept. 6:42–52.

Weissman, G. 1991. Aspirin. *Sci. Am.* 264(1):84–90.

Zeuthen, T., 2001. How water molecules pass through aquaporins. *Trends Biochem Sci.* 26:77–79.

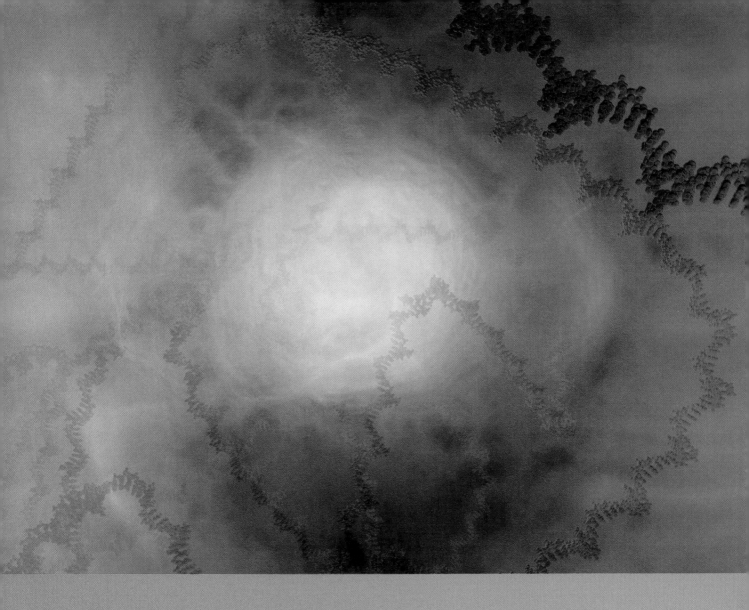

Part III

**Storage and Transfer
of Biological Information**

DNA played a major role in the development and growth of this rat snake.

n Part III we shall focus on the **nucleic acids,** biomolecules important for their roles in the storage, transfer, and expression of genetic information. Two fundamental types of nucleic acids participate as genetic molecules: (1) deoxyribonucleic acid (DNA) and (2) ribonucleic acid (RNA). DNA is found primarily in the chromosomal form in the cell's nucleus, where it serves as the repository of genetic information. Additional DNA functions in mitochrondia and chloroplasts. Directions for the synthesis of a cell's constituents reside in the cell's DNA. RNA is present in three major types: ribosomal, messenger, and transfer. Each of these three forms plays a role in the expression of the genetic information in DNA. Some forms of RNA (ribozymes) serve as splicing catalysts in the processing of RNA (see Chapter 7, Section 7.5). You may wish to review Chapter 2, in which the flow of biological information was described by the sequence, DNA → RNA → proteins → cell structure and function.

In Chapter 10, the chemical and biological properties of RNA and DNA are explored. As we study the molecular characteristics of the nucleic acids we note some similarities to the proteins. Proteins are linear polymers built from 20 different amino acids. Nucleic acids are linear polymers constructed from four different monomers called **nucleotides.** We learned in Chapters 4 and 5 that the sequence of amino acids in a protein contains directions for how the protein should fold into 2°, 3°, and 4° structure and how it functions biologically. In a similar fashion, the sequence of nucleotides in nucleic acids carries important information about three-dimensional structure and cell function. We first study the molecular characteristics of RNA and DNA from the perspectives of primary, secondary, and tertiary structures. After a discussion of the chemical and physical properties of the nucleic acids, we conclude the chapter with an introduction to functional combinations of proteins and nucleic acids.

10.1   RNA and DNA Chemical Structures

Components of Nucleotides

In Chapters 1 and 2 nucleic acids were described as sequence variable, linear polymers of monomeric units called **nucleotides.** A nucleotide is composed of three chemical parts:

1. An aromatic cyclic compound containing carbon and nitrogen atoms; because of their chemical properties, these compounds are called **nitrogenous bases.**
2. A five-carbon carbohydrate (an aldopentose).
3. One, two, or three phosphate groups.

The general structure for a nucleotide is shown in Figure 10.1. All of the nitrogenous bases in DNA and RNA are derivatives of the two heterocyclic compounds, **purine** and **pyrimidine.** The purines in DNA are adenine (A) and guanine (G); the pyrimidines in DNA are thymine (T) and cytosine (C). Similarly, the predominant purines in RNA are adenine and guanine; however, the pyrimidines in RNA are cytosine and uracil (U). Structures of the heterocyclic bases and official numbering systems are shown in Figure 10.2. Nucleic acids, especially RNA, contain small quantities of methylated bases, including 5-methylcytosine, 5-hydroxymethylcytosine, 2-methyladenine, and 1-methylguanine. The bases are usually modified after they have been incorporated in the nucleic acids.

Figure 10.1

General structure of a nucleotide showing the three fundamental units: a purine or pyrimidine nitrogenous base, an aldopentose (ribose or deoxyribose), and phosphate.

(a) *Major Bases*　　　　　　　　　　　　　　(b) *Minor Bases*

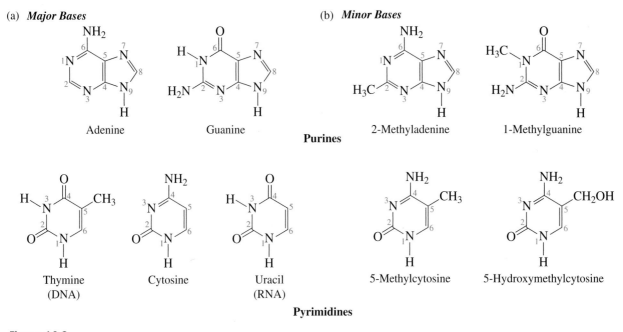

Adenine　　　　　　　Guanine　　　　　　　2-Methyladenine　　　　1-Methylguanine

Purines

Thymine　　　　　Cytosine　　　　　Uracil　　　　　5-Methylcytosine　　5-Hydroxymethylcytosine
(DNA)　　　　　　　　　　　　　　　(RNA)

Pyrimidines

Figure 10.2

The major and some minor heterocyclic bases in RNA and DNA. All are derived from purine or pyrimidine.

Two types of aldopentoses are found in the nucleic acids. **Ribose** occurs in RNA; **2-deoxyribose** in DNA (Figure 10.3). 2-Deoxyribose, a derivative of ribose, lacks an oxygen atom at C2. (The chemical and biological characteristics of the pentoses are described in Chapter 8). The linking of a nitrogenous base and an aldopentose through the *N*-glycosidic linkage results in a **nucleoside.** The covalent linkage forms, by elimination of water, between N9 of purines or N1 of pyrimidines and C1, the anomeric carbon of ribose or 2-deoxyribose (Figure 10.4). The β configuration exists for all *N*-glycosidic bonds in naturally occurring nucleosides. Notice that two numbering systems are necessary to identify carbon and nitrogen atoms in the rings of the nucleoside. Primed numbers (1′, 2′, etc.) are used for the pentose ring. The nomenclature for nucleosides is given in Table 10.1. Many synthetic nucleosides have physiological activity and are used for the treatment of cancers and other diseases. The nucleosides AZT (3′-azidodeoxythymidine) and DDI (2′,3′-dideoxyinosine) shown in Figure 10.5 are important drugs for treating patients with acquired immune deficiency syndrome (AIDS).

A **nucleotide** forms when a phosphoryl group ($-PO_3^{2-}$) is linked to a carbohydrate hydroxyl group on the nucleoside. Three hydroxyl groups are available on ribose for this esterification and two on deoxyribose; however, the most common site for ester linkage is the hydroxyl group on carbon 5′ of the pentoses (Figure 10.6). The resulting product is called a nucleoside 5′-monophosphate or more commonly, 5′-mononucleotide, indicating

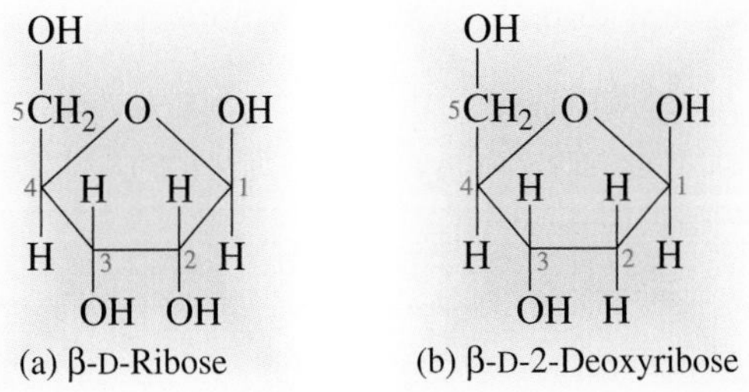

Interactive Animations: Nucleotides game

Cutting Edge: AIDS therapies.

Figure 10.3

The aldopentoses in RNA and DNA: (a) β-D-Ribose, (b) β-D-2-deoxyribose.

(a) β-D-Ribose　　　　　(b) β-D-2-Deoxyribose

Figure 10.4

A nucleoside consists of a purine or pyrimidine base linked to a carbohydrate (ribose or deoxyribose) by an N-glycosidic bond. Two numbering systems (primed and unprimed) are necessary to distinguish the two rings.

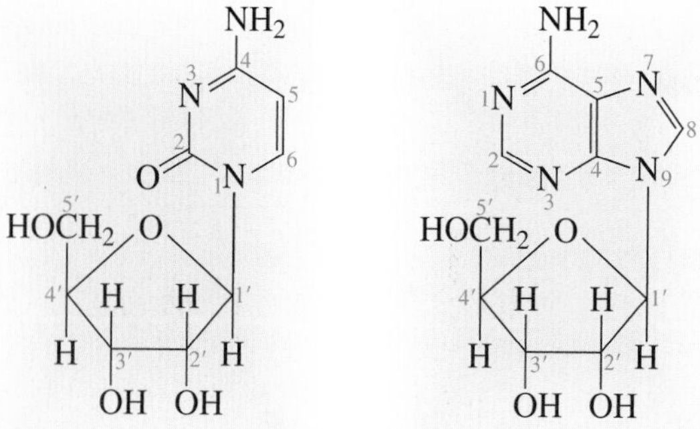

Pyrimidine nucleoside   Purine nucleoside

the presence of *one* phosphate. The most important and most abundant mononucleotide in the cell is adenosine 5′-monophosphate (5′-AMP). AMP and other 5′-mononucleotides can add another phosphoryl group to produce nucleoside diphosphates and nucleoside triphosphates. Note that when the source of the phosphoryl group is a phosphoric acid dianion (HPO_4^{2-}) a water molecule is eliminated to form an ester bond:

$$\text{nucleoside}-\text{OH(5′)} + \text{HO}-\overset{\overset{\displaystyle O}{\|}}{\underset{\underset{\displaystyle O_-}{|}}{P}}-O^- \longrightarrow \text{nucleoside}-\text{O}-\overset{\overset{\displaystyle O}{\|}}{\underset{\underset{\displaystyle O_-}{|}}{P}}-O^- + H_2O$$

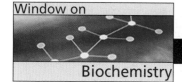

Window on
Biochemistry

10.1 Purines and Pyrimidines with Physiological Activity

Section 10.1 described the chemical components of the nucleotides and their importance in nucleic acid biochemistry. The section emphasized nucleosides and nucleotides that occur naturally in DNA and RNA and also some synthetic ones that are used as drugs (AZT, DDI). There are also many purine and pyrimidine derivatives, lacking ribose and phosphate groups, that have physiological activity.

The popular stimulant **caffeine** is a trimethylated derivative of the fundamental purine ring. It occurs naturally in the plants *Coffea arabica, Camellia thea, Cola acuminata,* and *Cola nitida.* The beverages made from the seeds of these plants provide the major sources of dietary caffeine. A typical 6-oz. serving of coffee, tea, or cola contains about 100 mg, 70 mg, and 25 mg, respectively, of caffeine. Caffeine stimulates the cerebral cortex by inhibiting the enzyme phosphodiesterase, which is involved in cell-signaling processes (see Window on Biochemistry 2.1).

The cancer treatment drug **6-mercaptopurine** (6-MP or Purinthol) is an analog of the nucleosides and nucleotides so it functions by interfering with their normal biochemical reactions. The primary action of 6-MP is to block the synthesis of nucleic acids. It is especially effective for treatment of leukemia. The drug inhibits the uncontrolled DNA replication that is associated with proliferation of white blood cells. 6-MP cannot distinguish between normal and cancerous cells, but because the more rapidly growing cancerous cells take up the drug faster, they are more greatly affected.

Acyclovir (Zovirax) is an antiviral agent. It is currently used in the treatment of herpes viral infections. The drug acts by inhibiting the DNA polymerase of the herpes simplex virus. Acyclovir achieves its active form in the cell after phosphorylation of the side chain hydroxyl group catalyzed by a kinase.

Two biologically active pyrimidines are **5-bromouracil** (5-BU) and **5-fluorouracil** (5-FU). (5-BU is discussed in Section 11.3). 5-FU (Adrucil) is an example of the delib-

Table 10.1
Nomenclature for nucleosides and nucleotides in DNA and RNA

Base	Nucleoside	Nucleotide (Abbreviation)	Nucleic Acid
Purine			
Adenine	Adenosine,	Adenosine 5′-monophosphate (5′-AMP)	RNA
	deoxyadenosine	Deoxyadenosine 5′-monophosphate (5′-dAMP)	DNA
Guanine	Guanosine,	Guanosine 5′-monophosphate (5′-GMP)	RNA
	deoxyguanosine	Deoxyguanosine 5′-monophosphate (5′-dGMP)	DNA
Pyrimidine			
Cytosine	Cytidine,	Cytidine 5′-monophosphate (5′-CMP)	RNA
	deoxycytidine	Deoxycytidine 5′-monophosphate (5′-dCMP)	DNA
Thymine	Deoxythymidine	Deoxythymidine 5′-monophosphate (5′-dTMP)	DNA
Uracil	Uridine	Uridine 5′-monophosphate (5′-UMP)	RNA

Phosphoryl groups can also be transferred from other phosphorylated compounds as we will see in Chap. 14. Table 10.1 lists the names of nucleotides derived from the major nitrogen bases. The first phosphoryl group added (α in Figure 10.6) forms a phosphate ester linkage. Continued addition of the β and γ phosphoryl groups are by phosphoanhydride linkages as shown in Figure 10.6. Differences between phosphate ester bonds and phosphoanhydride bonds are discussed in Chapter 14. With AMP, the resulting sequence of compounds is adenosine 5′-diphosphate (ADP) and adenosine 5′-triphosphate (ATP). ATP is the principal carrier of chemical energy in the cell. The interconversion of ADP and ATP is linked to energy transfer in metabolism. P_i refers to inorganic phosphate (HPO_4^{2-}):

$$ATP + H_2O \rightleftharpoons ADP + P_i + energy$$

erate use of carbon–fluorine bonds in drugs. Although C—F bonds are quite rare in natural compounds, they can be of value in drug design because (1) the fluorine atom is very small and can be used to replace hydrogen atoms with little change in steric factors, (2) the fluorine atom is very electronegative and provides some polarity to the C—F bond, and (3) the C—F bond is quite stable and unreactive so it remains intact under cellular conditions. 5-FU, an analog of thymine, is a potent inhibitor of DNA synthesis and has been widely used in cancer treatment. It inhibits the enzyme thymidylate synthase (see Section 19.5).

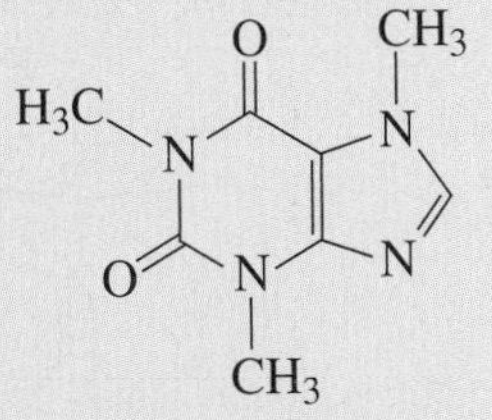

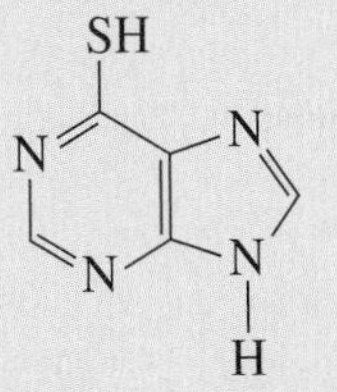

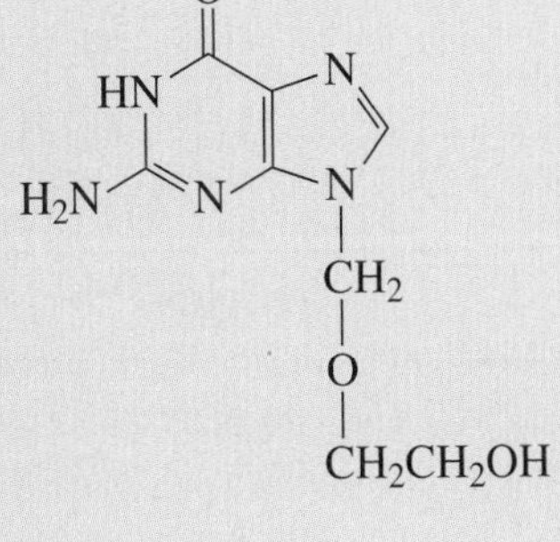

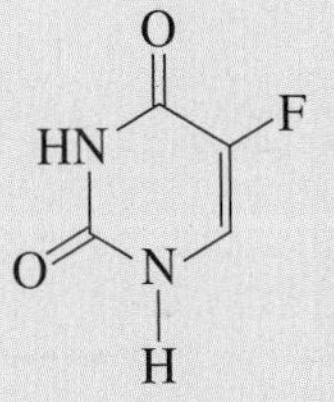

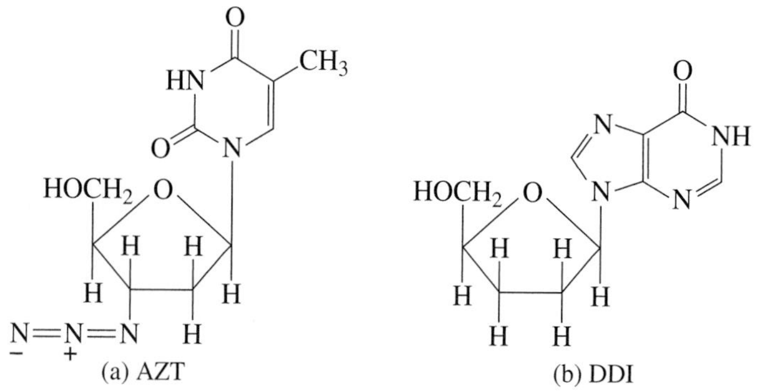

Figure 10.5

Structure of (a) 3′-azidodeoxy-thymidine (AZT) and (b) 2′,3′-dideoxyinosine (DDI). These compounds are used for the treatment of AIDS.

(a) AZT

(b) DDI

Other nucleoside triphosphates, guanosine triphosphate (GTP), uridine triphosphate (UTP), deoxyguanosine triphosphate (dGTP), and deoxyuridine triphosphate (dUTP), participate to a lesser extent in energy metabolism but are important in biosynthetic processes including the synthesis of nucleic acids. Nucleoside phosphates are also important as signal molecules involved in cellular communication. GTP, cyclic AMP, and cyclic GMP are transient intermediates that send messages via signal transduction through cell membranes (see Chapter 2, Section 2.6). We have also encountered the nucleotides as constituents of the important cofactors **coenzyme A** (CoA), flavin adenine dinucleotide (FAD), and nicotinamide adenine dinucleotides (NAD and NADP) (see Chapter 7, Section 7.1).

The nucleotides are relatively acidic compounds because of the presence of phosphoric acid moieties. At physiological pH values, all protons are dissociated from the phosphates, leading to compounds with several negative charges. Note that ATP has a net charge of -4 (see Figure 10.6). Cellular nucleotides are associated with metal ions such as Mg^{2+} and Mn^{2+} to neutralize the charges.

Nucleic Acids

To form nucleic acids, nucleotides are linked together through their phosphate groups. Specifically, the 5′-phosphate group of one nucleotide is linked to the 3′-hydroxyl group of the next nucleotide. The bridge between each monomeric unit is a **3′,5′-phosphodiester bond** (Figure 10.7). Several important characteristics of the covalent, primary structure of nucleic acids must be introduced. The covalent backbone of DNA (and RNA) consists of alternating deoxyriboses (or riboses) and phosphate groups. This is a *common, invariant region* found in all nucleic acids. This part of DNA, which remains the same throughout

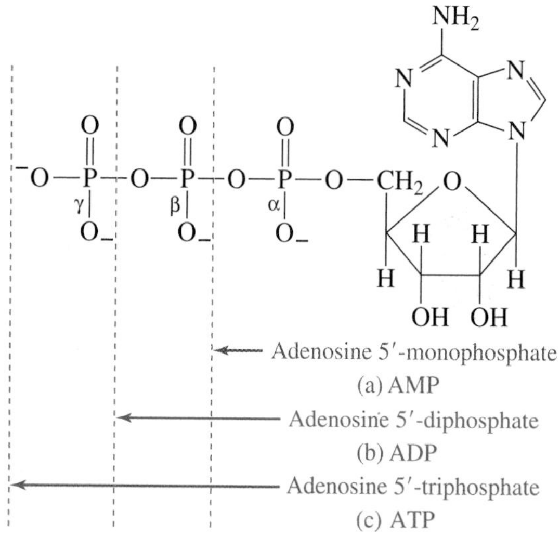

Figure 10.6

Structures for three types of nucleotides, which are composed of a nitrogenous base, pentose, and one, two, or three phosphates: (a) AMP, (b) ADP, (c) ATP.

Adenosine 5′-monophosphate
(a) AMP

Adenosine 5′-diphosphate
(b) ADP

Adenosine 5′-triphosphate
(c) ATP

Figure 10.7
Phosphodiester bonds linking mononucleotides into nucleic acids. The phosphodiester bonds are between the 3′ carbon of one nucleotide and the 5′ carbon of the second nucleotide. This gives direction to the nucleic acids. One end has a free 5′ OH; the other end has a free 3′ OH. The 3′,5′-phosphodiester bonds are high-lighted with green.

the molecule, plays an important structural role. Another important characteristic of the covalent backbone is its directionality. One end of the chain has a 3′-hydroxyl group; at the other end is a 5′-hydroxyl group. (We have also encountered directionality in proteins that have an N-terminus amino acid residue and a C-terminus amino acid residue.) The heterocyclic bases linked to the covalent backbone via *N*-glycosidic bonds protrude from the backbonelike side chains. This structural feature portrays the *variable region* of DNA and RNA. The sequence of the four kinds of bases (A, G, C, T in DNA; A, G, C, U in RNA) carries the genetic message in a specified manner. The sequence of nitrogenous bases serves as a template to specify a particular order that can be varied from molecule to molecule. In Chapter 11 we discuss methods for the sequential analysis of bases in DNA and RNA.

It is very cumbersome to draw the nucleic acids as shown in Figure 10.7. A simplified version can be used by drawing vertical lines to represent sugars and letters to indicate bases. Thus, nucleosides are represented as

Structure Tutorial: DNA structure

The top of the line indicates C1′ of the sugar where the base is linked by an *N*-glycosidic bond. The bottom of the line is labeled C5′. The phosphate group is inserted as P and the 3′,5′-phosphodiester bonds are represented by connecting lines:

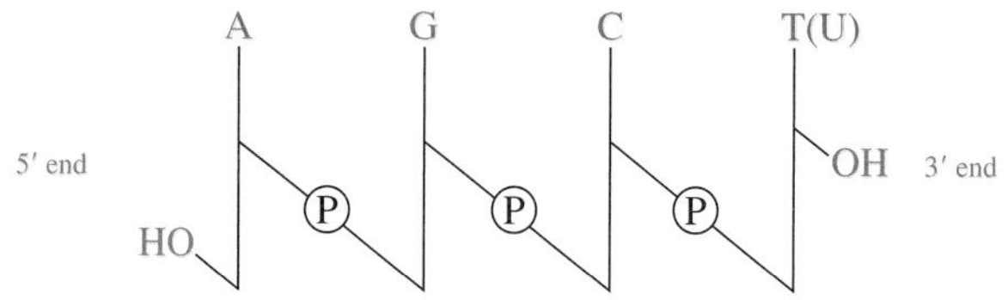

This structure represents an **oligonucleotide,** a short sequence of a nucleic acid. The standard format for drawing an oligonucleotide begins with the 5′ end on the left. The very abbreviated nomenclature, AGCT(U), can also be used to represent this sequence of DNA (or one of RNA).

DNA and RNA

In native DNA, many nucleotides are linked together, producing an extremely long molecule with a definite molecular size and sequence depending on the species of origin. DNA from simple prokaryotic cells such as *Escherichia coli* (which has a single chromosome) is a huge molecule with a mass of 2.6 billion daltons and 4.7 million base pairs. Its length is about 1.4 mm. Eukaryotic DNA as found in humans has approximately 3 billion base pairs and a total, unfolded length of at least 1 to 2 million μm (1–2 meters). DNA from eukaryotic cells is combined in the nucleus with basic proteins called **histones** (Section 10.5). Because methods required to extract cellular DNA are rather harsh, it is difficult to isolate intact DNA. The physical properties of DNA from different species are compared in Table 10.2.

RNA comprises about 5% to 10% of the total weight of a cell (DNA is about 1%). RNA molecules are not as homogeneous as DNA but exist in three major forms. Each type of RNA has a characteristic base composition and molecular size. These are compared in Table 10.3. **Ribosomal RNA** (rRNA), the most abundant form, is associated with the protein-synthesizing organelles, **ribosomes.** Three important kinds of prokaryotic rRNA, labeled 5S, 16S, and 23S, can be separated by centrifugation. (S refers to the svedberg, the basic unit of measure of the sedimentation coefficient of a molecule. The sedimentation coefficient defines the rate at which particles or molecules sediment in a centrifugal field. S is dependent on a molecule's weight, shape, and density. In general, the higher the S value, the larger the molecule.) **Messenger RNA** (mRNA) carries the transient message for protein synthesis from nuclear DNA to the ribosomes. Each molecule of mRNA carries the instructions for a single gene, which usually codes for one type of polypeptide product. Thus, mRNA molecules of many different base sequences are present in the cell. **Transfer RNAs** (tRNA), the smallest nucleic acids (between 74 and 93 nucleotides), form esters with specific amino acids for use in protein synthesis. There are several types of tRNA molecules used to select and activate amino acids.

Table 10.2

Comparison of DNA from different species

Organism	Number of Base Pairs	Length (μm)	Conformation
Viruses			
SV40	5100	1.7	Circular
Adenovirus	36,000	12	Linear
λ phage	48,600	17	Circular
Bacteria			
E. coli	4,700,000	1400	Circular
Eukaryotes			
Yeast	13,500,000	4600	Linear
Fruit fly	165,000,000	56,000	Linear
Human	3,000,000,000	$1–2 \times 10^6$	Linear

Type	Relative Amount (%)	S^a Value	MW (Daltons)	Average Number of Nucleotides
Ribosomal RNA (rRNA)	80	23S	1.2×10^6	3700
		16S	0.55×10^6	1700
		5S	36,000	120
Transfer RNA (tRNA)	15	4S	25,000	74–93
Messenger RNA (mRNA)	5	4S	Heterogeneous mixture	

^{a} The S value refers to the sedimentation coefficient and is related to the size of the molecule.

Table 10.3
RNA molecules in *E. coli*

Before we move on to a discussion of the secondary and tertiary structures of DNA and RNA, we describe important chemical characteristics of the covalent structure (1° structure) that affect nucleic acid structure and function. Nucleic acids have acidic and basic regions. The phosphate groups of the backbone derived from phosphoric acid have negative charges because of proton dissociation at physiological pH. The ionic character of the many negative charges and the polar sugar components establish a hydrophilic region (the covalent backbone) that interacts favorably with water. In contrast, the variable region of purine and pyrimidine bases is hydrophobic. In addition, the weakly basic nitrogen atoms and amino groups are not protonated and hence carry no charge at physiological pH. The polarity differences between the backbone region and the bases are important in the folding of DNA and RNA into three-dimensional structures. In the aqueous environment of the cell, the polar backbone becomes the surface of the molecule and the nonpolar bases move to the inside, where hydrophobic interactions between bases are favored. Hidden inside the molecular structure, the nonpolar bases have little contact with water but instead interact with each other.

Other important interactions influence nucleic acid folding. Several functional groups on the nitrogenous bases form hydrogen bonds. Ring N—H groups and amino groups ($—NH_2$) are potential hydrogen bond donors. Carbonyl groups ($\overset{\backslash}{\underset{/}{C}}=O$) and ring nitrogen atoms ($=N—$) are potential hydrogen bond acceptors. Specific hydrogen bonding occurs between the nitrogenous bases of one strand and the bases of another strand. Two strands of deoxyribonucleic acid will therefore associate into the well-known double helix for DNA as proposed by Watson and Crick. This hydrogen bonding between base pairs is important in the folding of single-stranded RNA molecules.

10.2 DNA Structural Elements

DNA was first extracted from biological material in 1868, but its participation in the transfer of genetic information was not fully recognized and proved until the mid-1940s (Table 10.4). Since that time DNA has been the subject of thousands of physical, chemical, and biological investigations. The landmark discovery that had special significance in our understanding of DNA was the elucidation of its regular, three-dimensional structure (2° structure) by Watson and Crick in 1953. It consists of two polynucleotide strands folded into a double helix. The **double helix** as envisioned by Watson and Crick is now recognized as the predominant structural form of DNA in vivo.

Nucleic acid structure can best be understood using the same concepts introduced for protein structure. The *primary* structure of DNA and RNA, consisting of the invariant, covalent backbone and sequence-variable nitrogen bases, is described in Section 10.1. The *secondary* structure of the nucleic acids describes the regular, periodic arrangement of the polynucleotide strands. In this section we consider the 2° and 3° structural features of DNA.

The DNA Double Helix

After a thorough study of the X-ray diffraction data and other experimental results of Franklin, Wilkins, Chargaff, Avery, MacLeod, McCarty, and others as shown in Table 10.4, Watson and Crick postulated a unique, double-stranded helical structure for the DNA

Cutting Edge: Visionaries

Interactive Animation: Central dogma of biochemistry

Structure Tutorial: DNA structure

Table 10.4

Historical landmarks in early DNA biochemistry

Date	Researcher(s)	Discovery
1868	Friedrich Miescher	Isolated and studied a phosphorus-containing substance from cell nuclei and salmon sperm cells. Miescher called the substance "nuclein." It was suspected to be associated with cellular inheritance.
1928	Frederick Griffith	Defined a "transforming principle" in the pneumococcus bacterium.
1944	Oswald Avery Colin MacLeod Maclyn McCarty	Published experimental evidence that DNA is a component in chromosomes and the principal agent involved in the transfer of genetic information.
1950	Erwin Chargaff	Studied the composition of DNA from different species and found the ratios of adenine to thymine and of guanine to cytosine to be 1.0.
Early 1950s	Rosalind Franklin Maurice Wilkins	Studied X-ray diffraction of DNA crystals and found periodic patterns.
1953	James Watson Francis Crick	Formulated a three-dimensional structure (double helix) for DNA that accounted for X-ray diffraction and A-T, G-C equivalence data.

molecule. In brief, the double helix consists of two complementary polynucleotide chains twisted to form a circular staircase-type structure. The detailed features of the DNA double helix, as shown in Figure 10.8a, are as follows.

1. Two right-handed, helical, polynucleotide chains are coiled around a common axis to form a double helix. Two characteristic topographic features, a major groove and a minor groove, are created in the highly organized arrangement.
2. The two strands run in opposite directions; they are antiparallel. That is, their 3′, 5′-phosphodiester bridges run in opposite directions.
3. In an aqueous environment, the polar, charged, covalent backbone of alternating deoxyribose and phosphate groups is on the outside of the helix, where interaction with H_2O is likely; the hydrophobic purine and pyrimidine bases avoid water by turning toward the inside of the structure.
4. The double helix is stabilized by two types of forces:
   a. Hydrogen bonds between pairs of complementary bases on opposite strands (A-T; G-C). There are two strong hydrogen bonds between A and T and three between G and C (Figures 10.8b and 10.9). Each base in one strand can form specific bonds to the complementary base directly across from it in the other strand.
   b. Van der Waals and hydrophobic interactions between "stacked bases." The planes of the nucleotide bases are nearly perpendicular to the common axis and take on a stacking arrangement like stairs in a circular staircase. The bases are brought close enough together to allow attractive forces between the electrons in their π orbitals (van der Waals interactions).

Perhaps the most important feature of the double helix that allows it to function in storage and transfer of genetic information is the *complementary base pairing*. In each case the pair consists of a purine (A or G) with a pyrimidine (T or C). This combination leads to the maximum possible number of hydrogen bonds and also allows the base pairs to give maximum stability. In addition, the internal space between the polynucleotide strands is just enough to allow for a pyrimidine–purine pair. The distance between the two strands does not accommodate two purines, and two pyrimidines are too far apart for strong hydrogen bond interaction.

The Watson and Crick model for the double helix, called B-DNA, is one of several conformational varieties that have been experimentally observed. B-DNA is the structure of the nucleic acid form that is crystallized from water and retains water molecules within the crystal structure. Biochemists believe that the B-DNA form is the predominant one under physiological conditions. If DNA in the B form is treated with reagents or conditions that dehydrate the crystals, the structure rearranges to a similar form called A-DNA. This form has 11 bases per helix turn and therefore is more compact (shorter) with a greater diameter

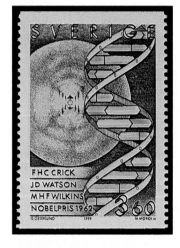

Swedish stamp honoring Watson, Crick, and Wilkins for their DNA structural analysis by X-ray diffraction.

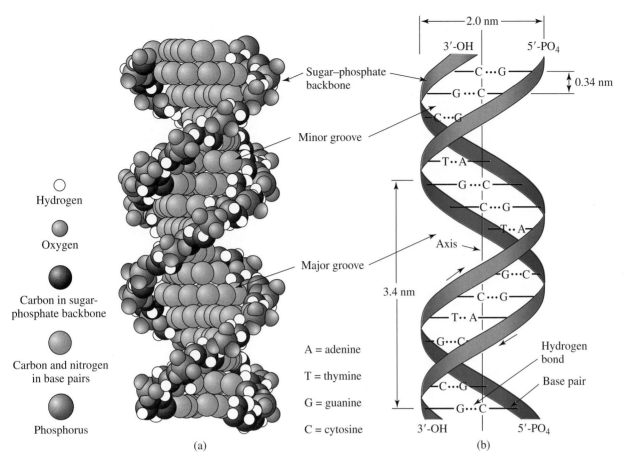

Figure 10.8

(a) The Watson–Crick double helix. (b) Pairing of A-T and G-C bases in the DNA double helix. Extensive hydrogen bonding occurs between the base pairs.

(26 Å). (B-DNA has 10.5 bases per turn and a diameter of 20 Å.) The structural features common to B- and A-DNA are a right-handed helix, antiparallel strands, and base complementarity (A-T; G-C). Figure 10.10 contrasts the forms.

A new structural form, Z-DNA, has recently been observed in short strands of synthetic DNA. It differs considerably from A- and B-DNA. The Z form, which is a left-handed helix, has 12 bases per turn and a diameter of 18 Å. The Z refers to a zigzag arrangement of the covalent backbone. The Z form has also been found in short stretches of native DNA. It may play a role in the regulation of gene expression.

Physical and Biological Properties of the Double Helix

Upon inspection of the DNA double helix, a possible mechanism for duplication (replication) and information transfer (expression) is readily apparent. Separation of the two strands by disruption of the hydrogen bonds and van der Waals interactions could lead to two single polynucleotide strands that may be used as templates for new DNA synthesis (Figure 10.11). Indeed, such a mechanism did not escape the attention of Watson and Crick. In a 1953 manuscript describing the genetic implications of the double helix, the following statement was made by Watson and Crick:

> Now our model for deoxyribonucleic acid is, in effect, a *pair* of templates, each of which is complementary to the other. We imagine that prior to duplication the hydrogen bonds are broken and the two chains unwind and separate. Each chain then acts as a template for the formation onto itself of a new companion chain, so that eventually we shall have *two* pairs of chains, where we only had one before. Moreover, the sequence of the pairs of bases will have been duplicated exactly. (Watson, J. and Crick, F., 1953. *Nature* 171:964–967)

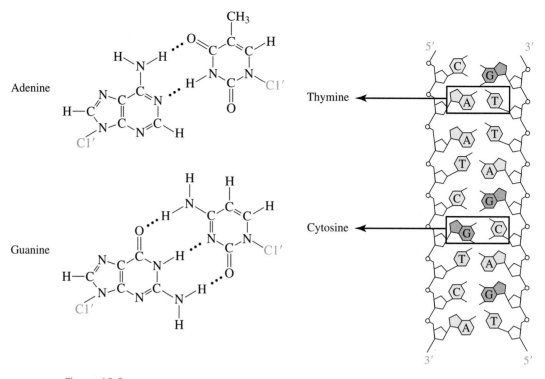

Figure 10.9
Hydrogen bonding between Watson–Crick base pairs.

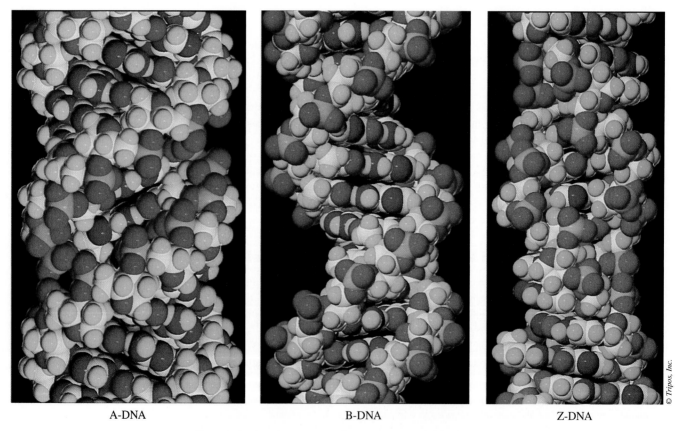

A-DNA   B-DNA   Z-DNA

Figure 10.10
Comparison of the secondary structures of A-, B-, and Z-DNA.

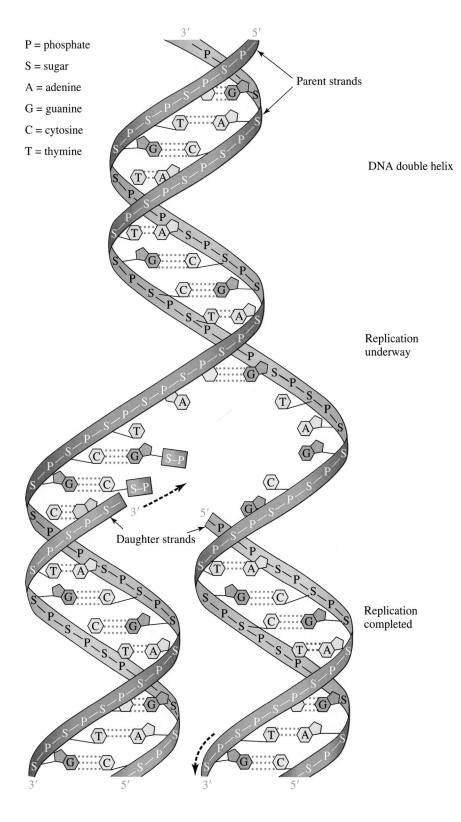

P = phosphate
S = sugar
A = adenine
G = guanine
C = cytosine
T = thymine

Parent strands

DNA double helix

Replication
underway

Daughter strands

Replication
completed

Figure 10.11
Postulated scheme for the replication of the DNA double helix.
Watson and Crick suggested that
each strand of the double helix
(parent strands) was used as a
template to make complementary
daughter strands. This general
idea of DNA replication was later
found to be correct; however,
specific details of the process
were not.

Is it reasonable to consider such a process for DNA duplication and expression? Does the double helix of DNA unwind (are the H bonds easily broken) without causing damage to the primary (covalent) structure? When DNA molecules in the double helix are subjected to relatively mild conditions such as heat, acids, bases, or organic solvents, the two strands come apart (Figure 10.12). The double helix is **denatured.** When increased temperature is the dena-turing agent, the process is called melting because it occurs in a small temperature range

Figure 10.12
DNA during denaturation and renaturation.

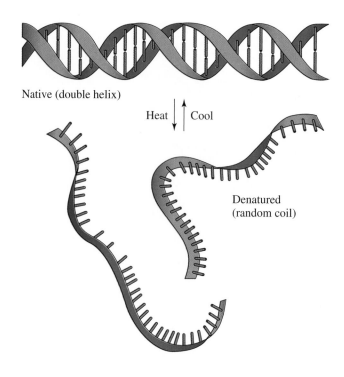

Native (double helix)

Heat ↓ ↑ Cool

Denatured
(random coil)

(85–90°C for many DNA molecules). Accompanying the melting is a significant increase in the absorption of ultraviolet (UV) light by the DNA molecule. This provides an experimental technique for quantitatively characterizing denaturation, and it reveals important information about the strength of hydrogen bonding and stacking interactions between base pairs. The melting curves for some DNA molecules are shown in Figure 10.13. The origin of the UV absorption increase, called a **hyperchromic effect,** is well understood. The absorption changes are those that result from the transition of an ordered double helix DNA structure to a denatured state or random, unpaired DNA strands. Native DNA in solution exists in the double helix held together primarily by hydrogen bonding between complementary base pairs on each strand. Hydrophobic and stacking interactions between base pairs also strengthen the double helix. Agents that disrupt these forces (hydrogen bonding, hydrophobic and stacking interactions) cause dissociation or unwinding of the double helix. In a random coil arrangement, base–base stacking interactions are at a minimum; this alters the arrangement of valence (π) electrons in the aromatic rings, causing an increase in absorption of UV light.

It is possible to observe reassociation of the two free strands into the original double helix if solutions of heat-denatured DNA are slowly cooled. This process is called **annealing** or

Figure 10.13
Melting curves for DNA molecules from four microorganisms. These are obtained by heating DNA solutions and measuring the absorbance changes at 260 nm. When DNA is denatured, more light is absorbed. The actual melting temperature (T_m) is measured at the midpoint for each curve. T_m for *Pneumococcus* is about 83°C.

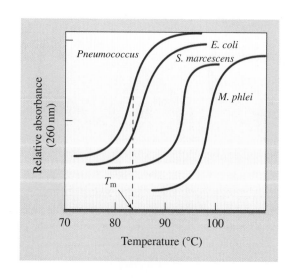

renaturation (see Figure 10.12). The melting experiment showing reversible dissociation supports Watson and Crick's assertion about the overall process of DNA replication. However, although their general idea was correct, the specific details were not.

Tertiary Structure of DNA

The 2° structure for DNA has been described as a double helix. Is it possible for this already highly organized structure to twist into more complex arrangements, as we have observed for the 3° structure of proteins? Genetic studies and electron microscopy on native, intact DNA have revealed two distinct forms, linear and circular. Many diagrams of DNA illustrate the linear nature of the double helix and imply that DNA molecules are extended rods with two ends. The chromosomal DNA of many microorganisms, however, is a closed circle, a result of covalent joining of the two ends of a linear double helix. Circular DNA has been discovered in simian virus 40 (SV40), bacteriophages (for example, φX174), bacteria (*E. coli,* for example), and several species of animals. On the other hand, DNA in T2 phage is linear, while λ phage DNA exists in linear and circular form, depending upon the time of life cycle. Closed, circular, duplex DNA has a unique structural feature. It is found twisted into a new conformation, **supercoiled DNA.** Figure 10.14 illustrates both relaxed (part a) and supercoiled (part b) DNA. Form (b) is not simply a coiling of the covalent circular DNA but is the result of extra twisting in the linear duplex form just before covalent joining of the two double strands. The supercoiled form is less stable than the relaxed form; however, there is sufficient evidence to conclude that supercoiled DNA exists in vivo. The most convincing evidence is the isolation and characterization of proteins that catalyze the interconversion of relaxed and supercoiled DNA. These proteins are called **topoisomerases** because they catalyze changes in the topology of DNA.

The cause of supercoiling is not completely understood, but it may be the presence of more than the standard number of bases per unit length of a DNA helix, forcing tension in the molecule. Supercoiling may have biological significance in that the DNA molecule becomes more compact; hence, it is more easily stored in the cell. Supercoiling may also play a regulatory role in DNA replication.

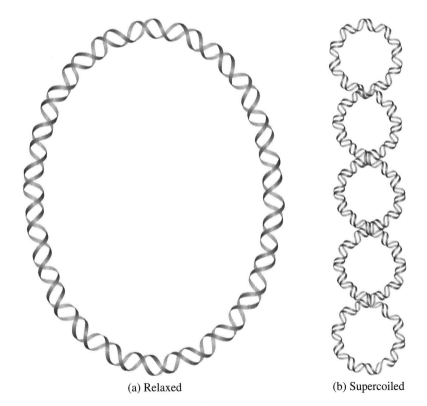

(a) Relaxed

(b) Supercoiled

Circular, duplex DNA

Figure 10.14
Tertiary forms of DNA: (a) relaxed and (b) supercoiled. The supercoiled form is less stable than the relaxed form. Enzymes, called topoisomerases, which catalyze interconversion of relaxed and supercoiled DNA, are present in the cell.

10.3   RNA Structural Elements

We turn now to structural details of the other nucleic acid, RNA. Several classes of RNA are organized by biological functions; hence, we expect variety also in chemical structure. Recall that RNA and DNA are both long, unbranched polymers composed of nucleotide monomers.

There are two fundamental differences in the primary, covalent structures of RNA and DNA: (1) RNA contains the carbohydrate ribose rather than 2-deoxyribose and (2) one of the major bases in RNA is uracil (U) instead of thymine (T) in DNA. The presence of an extra hydroxyl group in RNA (on the 2′ carbon next to the 3′-phosphodiester bridge to the next nucleotide) makes RNA more susceptible to hydrolysis than DNA. (This may be the most important reason that the more stable DNA is the ultimate repository of genetic information.) These differences in the primary structures of RNA and DNA result in important differences in 2° and 3° structural features.

Figure 10.15

General features of RNA secondary and tertiary structure. Important structural elements of single-stranded RNA include hairpin turns, right-handed double helixes, and loops. The double helixes are formed by complementary base pairing between A-U and G-C. Note that the RNA strands in double-helix regions are antiparallel. *Source:* Redrawn from Wolfe, 1993.

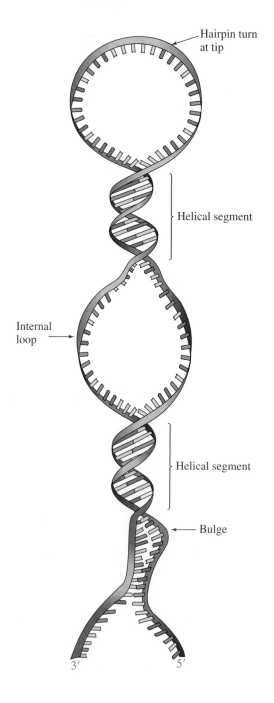

Hairpin turn at tip

Helical segment

Internal loop

Helical segment

Bulge

3′   5′

Most (but not all) cellular DNA is double stranded and folded into the Watson–Crick helix as previously described. All classes of RNA, no matter what their size or biological function, are synthesized as single-stranded molecules. Their primary structures consist of a phosphodiester backbone of alternating ribose and phosphate and side-arm nitrogenous bases that are complementary to the DNA template (AGCT in DNA leads to UCGA in RNA). Although RNA molecules do not take on extended, periodic 2° arrangements like DNA does, RNA is not totally devoid of regular structure. Rather than folding into a uniform, periodic pattern, as in DNA, single strands of RNA arrange themselves into conformations containing several different structural elements. Elements of periodic structure are found scattered throughout RNA molecules. The most important structural elements are shown in Figure 10.15 and described as follows.

1. *Hairpin turns* are loops in the single chain that bring together complementary stretches for base pairing. These sequences in RNA are often long enough for the formation of a double helix region similar in structure to A-DNA.
2. *Right-handed double helixes* in RNA are the result of intrastrand folding. A hairpin turn can be the trigger that initiates double helix regions, but other structural features also cooperate. Double-helical regions in RNA are antiparallel and are stabilized by the same interactions as in DNA. Complementary hydrogen bonding and stacking interactions hold the double helix together. The same rules for base pairing in RNA are in effect for DNA with the substitution of uracil for thymine (A-U; G-C) (Figure 10.16). Because of steric interference of the extra hydroxyl groups, RNA single strands cannot fold into the analogous B-DNA arrangement. The extent of base-pairing interactions and double helix formation in RNA varies with the class of RNA.
3. *Internal loops* and *bulges*, which are relatively common in RNA molecules, are structural features that disrupt the formation of continuous double helix regions (see Figure 10.15).

There are features that are common in each class of RNA, including tRNA and rRNA. Each mRNA molecule, which carries the message for a gene, has a definite size, but because of its instability and transient existence we know little about the three-dimensional structure.

tRNA Structure

The smallest types of RNA are the highly structured tRNA molecules. These molecules are carriers of specific amino acids used for protein synthesis. Each of the 20 amino acids has at least one corresponding tRNA. All tRNAs contain between 74 and 93 nucleotides in a single chain. They often consist of several unusual purine or pyrimidine bases that are chemical modifications of the basic four. tRNAs have been crystallized and their structures determined by X-ray analysis. All tRNA molecules have a common 2° and 3° structure, as shown in Figures 10.17 and 10.18. The basic pattern is that of a cloverleaf. Structural features include hairpin turns, regions of double helix, and loops (see Figure 10.15). Figure 10.18 displays the three-dimensional structure of a specific tRNA for the amino acid phenylalanine.

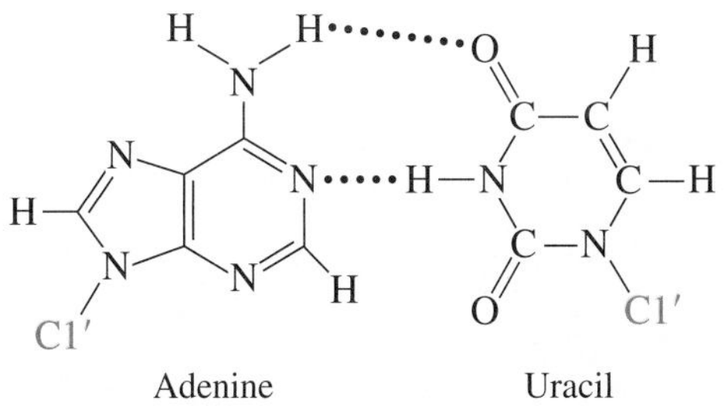

Structure Tutorial: tRNA

Figure 10.16
Base pairing of A-U in RNA.

Figure 10.17

The cloverleaf structure for tRNA molecules. Structural elements include three loops formed by hairpin turns and double-helix regions stabilized by hydrogen bonding between complementary bases. Modified nucleotides are abbreviated: methylguanosine (m^1G); dimethylguanosine (m^2G); inosine (I); methylinosine (m^1I); dihydrouridine (D); and pseudouridine (ψ).

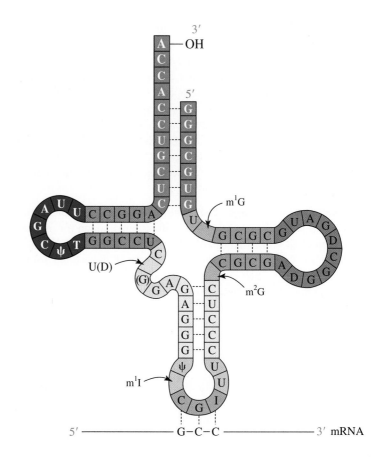

rRNA Structure

The 2° and 3° structures of rRNA molecules display many of the same elements as tRNAs; however, they are much larger. In most rRNA molecules, there are many extensive regions of double-helix formation. Of the three types of prokaryotic rRNA (5S, 16S, and 23S), we have the greatest understanding of the 16S structure. The proposed 2° structure of 16S rRNA from *E. coli* is shown in Figure 10.19. Current experimental work in X-ray diffraction crystallography is providing evidence that 16S rRNA from a wide variety of species is similar in 2° and 3° structure to that of *E. coli*. The relationships between RNA structures and function are considered in more detail in Chapters 11 and 12.

Figure 10.18

The three-dimensional structure of yeast tRNA for phenylalanine. Views from two angles are shown.

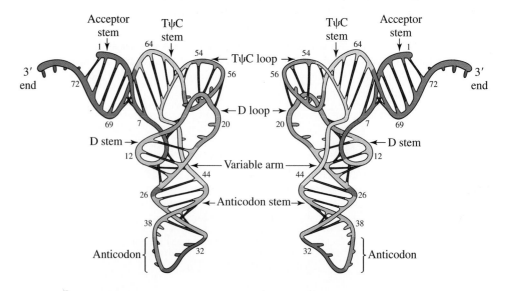

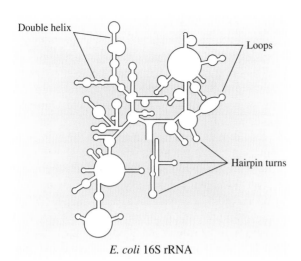

Figure 10.19

Proposed secondary structure for *E. coli* 16S rRNA. Note several types of structure, including double helix regions, loops, and hairpin turns. *Source:* From Sadava, 1993.

10.4   Cleavage of DNA and RNA by Nucleases

Nucleases

The characterization of nucleic acid structure and function often requires breaking the usually long, complex molecules into smaller fragments more amenable to analysis. This is especially important for (1) primary structural analysis of DNA and RNA and (2) for cleavage at specific sites for recombinant DNA manipulation (see Chapter 13). Historically, the most important procedures for nucleic acid fragmentation have involved hydrolysis reactions catalyzed by acids, bases, and enzymes. RNA withstands treatment with dilute acid, but DNA in 1 *M* HCl solution undergoes degradation by removal of purine bases. On the other hand, DNA is relatively resistant to alkaline hydrolysis, whereas RNA is hydrolyzed under basic conditions at random phosphodiester bonds along the polynucleotide chains to produce polynucleotide fragments of varying lengths. For careful study of nucleic acid structure, ideally one desires a mild cleavage procedure that results in selective, predictable, and reproducible cuts in the chain.

All cells contain enzymes called **nucleases** that catalyze the hydrolysis of phosphodiester bonds. Their normal biochemical role is to catalyze degradation of damaged or aged nucleic acids by processes that are necessary for general housekeeping in the cell. Many nucleases have been isolated from cells and are used to achieve specific fragmentation of nucleic acids under laboratory conditions. We must introduce terminology to understand the action of nucleases. Some enzymes work on both RNA and DNA substrates, whereas others have substrate specificity (**deoxyribonucleases,** DNases; **ribonucleases,** RNases). **Exonucleases** catalyze the hydrolytic removal of terminal nucleotides (Figure 10.20). The exonucleases

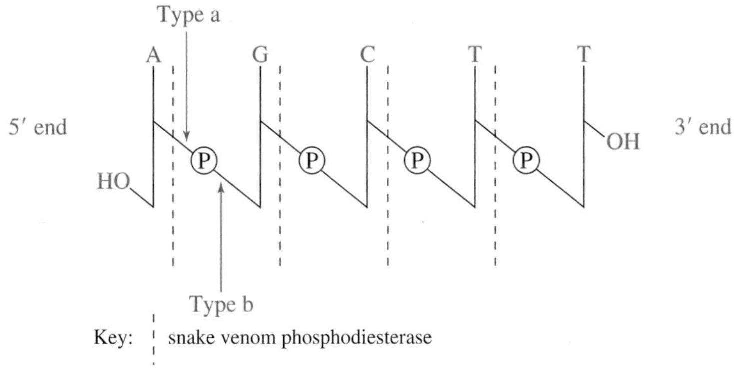

Figure 10.20

Specificity of snake venom phosphodiesterase, an exonuclease that catalyzes the hydrolysis of type-a bonds in the phosphodiester linkages of RNA and DNA. The dashed lines indicate the bonds broken by hydrolysis.

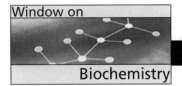

Window on
Biochemistry

10.2   Ribonuclease—A New Look at an Old Enzyme

Pancreatic ribonuclease A (Rnase A) has often been cited as the most-studied enzyme of the 20th century. This enzyme, which is produced in the pancreas, is secreted into the cell's endoplasmic reticulum and also transported to the gut, where it plays a major role in food digestion. **RNase A,** a type endo, b, nuclease, catalyzes the hydrolysis of RNA. The human form of the nuclease has no measurable cytotoxic activity. A ribonuclease with human sequence homology has been found in the Northern leopard frog *(Rana pipiens)* that displays strong in vitro antitumor activity with selective cytotoxicity toward tumor cell lines. The frog enzyme in a drug form called **onconase** (see ribbon structure below) is currently being tested for the treatment of malignant mesothelioma, an asbestos-related cancer, and for metastatic renal carcinoma. These forms of cancer have no current reliable treatment, and average patient survival after diagnosis is 1–2 years. Results from initial clinical trials using onconase are not yet available, but side effects appear to be minor and involve only fatigue and flu-like symptoms. Hair loss has not been observed. The mild side effects are probably because of the similarity between onconase and the human enzyme. Another advantage of the new drug is its unusual stability. It is especially resistant to proteolysis, which means slow breakdown under in vivo conditions. Earlier clinical trials using onconase on patients with pancreatic cancer showed no advantage over present drugs.

There is experimental evidence that the cytotoxic action of onconase involves more than just the nuclease activity; in fact some forms of the enzyme that have very low catalytic activity show high cytotoxic activity. It has recently been observed that the human enzyme, which alone shows no toxic action, becomes highly cytotoxic when linked to the iron protein transferrin. The nuclease activity of the Rnase A was not changed with addition of transferrin. The effect of transferrin is currently being studied. Even though it has been thoroughly investigated for years, there is much yet to be learned about Rnase A.

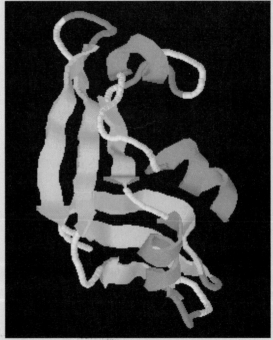

© Alphacell Corp

Ribbon structure of onconase.

are organized into two broad categories, 5′-exonucleases and 3′-exonucleases depending on their site of action. Hydrolytic cleavage of *internal* phosphodiester bonds is carried out by **endonucleases** (Figure 10.20). Close inspection of a phosphodiester bridge reveals two different types of ester bonds: (1) those connecting the 3′-hydroxyl group of a nucleotide with the phosphorus (called type a) and (2) those connecting the 5′-hydroxyl group of a nucleotide with the phosphorus (called type b). All phosphodiester bridges between nucleotides have one type a ester bond and one type b bond.

The most-studied and best-understood nucleases are rattlesnake venom phosphodiesterase, spleen phosphodiesterase, pancreatic ribonuclease A, and spleen deoxyribonuclease II. The properties of these enzymes are compared in Table 10.5. Although these enzymes are useful for cutting DNA or RNA into more manageable sizes, they do not have a high degree of site specificity determined by base content or sequence. That is, each phosphodiester bond has almost equal probability of hydrolytic attack.

DNA Restriction Enzymes

The most specific enzymes available for DNA cleavage, the **restriction endonucleases** or restriction enzymes, were discovered in the early 1970s. Bacterial cells produce these enzymes that act to degrade or "restrict" DNA molecules foreign to the cell. Restriction enzymes recognize specific base sequences in double-stranded DNA and catalyze hydrolytic cleavage of the two strands in or near that specific region. Host DNA is protected from hydrolysis because some bases near the cleavage sites are methylated.

Several hundred restriction enzymes have been isolated and characterized. Nomenclature for the enzymes consists of a three-letter abbreviation representing the source, a letter representing the strain, and a roman numeral designating the order of discovery. *Eco*RI is the first restriction enzyme to be isolated from *E. coli* (strain R). The cleavage specificity of the restriction enzyme *Eco*RI is shown in the reaction:

$$
\begin{array}{l}
5'...G-A-A-T-T-C...3' \\
3'...C-T-T-A-A-G...5'
\end{array}
\xrightarrow[\text{2 H}_2\text{O}]{\textit{Eco}\text{RI}}
\begin{array}{l}
5'...G \overset{\text{OH}}{} \\
3'...C-T-T-A-A
\end{array}
+
\begin{array}{l}
A-A-T-T-C...3' \\
\underset{\text{HO}}{G...5'}
\end{array}
$$

The site of action of *Eco*RI is a specific hexanucleotide sequence. Two phosphodiester bonds are hydrolyzed (see red arrows), resulting in fragmentation of both strands of the DNA double helix. Note the twofold rotational symmetry feature at the recognition site and

The venom from this red diamond-back rattler contains RNA and DNA nucleases.

© Tom McHugh/Photo Researchers

💿 **Interactive Animation: Restriction digestion and electrophoresis of DNA**

Table 10.5
Properties of nucleases

Enzyme	Substrate	Type	Specificity
Standard Nucleases			
Rattlesnake venom phosphodiesterase	DNA, RNA	exo (a)	Begins at 3′ end; no base specificity
Spleen phosphodiesterase	DNA, RNA	exo (b)	Begins at 5′ end; no base specificity
Pancreatic ribonuclease A	RNA	endo (b)	Preference of pyrimidine on 3′ side
Spleen deoxyribonuclease II	DNA	endo (b)	Internal ester bonds; no base specificity
Restriction Endonucleases			
*Eco*RI	DNA	endo (a)	5′ GAATTC[a]
*Bal*I	DNA	endo (a)	5′ TGGCCA
*Taq*I	DNA	endo (a)	5′ TCGA
*Hinf*I	DNA	endo (a)	5′ GANTC

[a]The recognition sequence for each restriction endonuclease is shown with a red arrow denoting the actual site of cleavage.

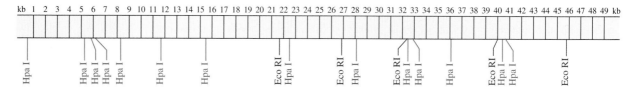

Figure 10.21
Restriction enzyme cleavage sites in λ phage DNA. The DNA is marked by kilobase pairs (kb) written above. The action of two restriction enzymes, *Eco*RI and *Hpa*I is shown.

the formation of ends with complementary bases. The weak pairing between the four bases (cohesive ends) is not sufficient to hold the two product fragments together. Table 10.5 lists several restriction enzymes and their recognition sequence for cleavage.

Restriction enzymes are used extensively in nucleic acid chemistry. They are especially useful to catalyze the cleavage of large DNA molecules into smaller fragments that are more amenable to analysis. For example, λ phage DNA, a linear, double-stranded molecule of 48,502 base pairs (molecular weight 31×10^6 daltons), is hydrolyzed into six fragments when catalyzed by *Eco*RI or into 15 fragments when *Hpa*I *(Haemophilus parainfluenzae)* is the enzyme (Figure 10.21). The base sequence recognized by a restriction enzyme is likely to occur only a very few times in any particular DNA molecule; therefore, the smaller the DNA molecule, the fewer the number of specific sites. The λ phage DNA is cleaved into two or more fragments, depending on the restriction enzyme used, whereas larger bacterial or animal DNA will most likely have many recognition sites and be cleaved into hundreds of fragments. Smaller DNA molecules, therefore, have a much greater chance of producing a unique set of fragments with a particular restriction enzyme. It is unlikely that this set of fragments will be the same for any two different DNA molecules, so the fragmentation pattern is unique and can be considered a "fingerprint" of the DNA substrate. The fragments are readily separated and sized by agarose gel electrophoresis. Restriction enzyme cleavage is a widely used technique for the analysis of DNA (DNA fingerprinting) found in body fluids remaining at crime scenes (see Chapter 13, Section 13.3).

No analogous enzymes have yet been found that catalyze the cleavage of RNA into well-defined fragments by cutting at predictable sites. However, ribozymes (catalytic RNA) are now being designed that display the action of RNA restriction endonucleases (see Chapter 7, Section 7.5). The availability of these specific catalysts aids greatly in elucidating RNA structure and function and may provide new pharmaceuticals that could control viral and bacterial infections.

10.5   Nucleic Acid–Protein Complexes

We have introduced the term "supramolecular assemblies" to describe organized clusters of macromolecules that display biochemical function (see Chapter 1, Section 1.4). Prominent examples of these higher levels of organization are **nucleoproteins,** complexes of nucleic acids and proteins. Combined assemblies of nucleic acids and proteins have complex structures and biological activities not observed for their individual components. Several examples that have important biological activities are introduced in this section.

Viruses

Viruses are stable, infective particles composed of a nucleic acid, either DNA or RNA, and protein subunits. In the basic viral package, the protein molecules form a protective shell around the nucleic acid core. In large, more complex viruses, the protein shell is coated with an envelope of glycoproteins (see Chapter 8, Section 8.5) and membrane lipids (see Chapter 9, Section 9.3). Viruses cannot exist independently and are usually not considered as forms of life. Rather, they are deemed parasitic because they survive by infecting a host cell and pirating the metabolic machinery, which is then used to synthesize the molecular components for new viral particles. Many different virus types have been isolated and studied.

They vary greatly in size, shape, and chemical composition (see Figure 1.7). Viruses average about 100 nm in length. Each viral form is usually specific for a type of host cell. Viruses selective for bacterial, plant, or animal cells have been discovered and characterized.

Viruses that are specific for bacteria are called bacteriophages or just **phages** for short. The majority of phages are DNA viruses; that is, their genome is in the form of DNA. A well-studied phage is φX174, which infects *E. coli* cells. The DNA genome of φX174 is a relatively small, single-stranded, circular molecule with just 5386 nucleotide bases.

One of the best-known plant viruses is tobacco mosaic virus (TMV), which infects the leaves of the plant. Its single strand of genomic RNA contains 6390 nucleotides and carries the message for over 2100 identical protein subunits. Some RNA viruses, such as TMV, replicate in their host cells by the action of RNA-directed RNA polymerases. RNA templates carry the instructions for the synthesis of complementary RNA by RNA polymerases.

Viruses that infect animal cells have also been isolated and characterized. Some are highly pathogenic in humans, including those causing influenza, the common cold, some kinds of cancer, poliomyelitis, herpes, and AIDS. These are often called retroviruses because they require a special enzyme to decode their RNA-based genetic information. These viruses direct their host cells to synthesize the enzyme reverse transcriptase, which catalyzes the synthesis of DNA complementary to the RNA genome. The Rous sarcoma virus, which causes tumors, is a well-known variety. Of greatest current concern is the **human immunodeficiency virus, HIV,** shown in Figure 10.22, which causes acquired immune deficiency syndrome (AIDS). The core of HIV, a retrovirus, contains two copies of a single-strand RNA genome. The two RNA molecules have a total of only 9700 nucleotides. The RNA is replicated by reverse transcriptase to double-strand DNA that becomes incorporated into the host cell genome. The core is surrounded by a lipid bilayer membrane containing two specific glycoproteins, gp41 and gp121. The human host cell, which is taken over and destroyed by the virus, is the helper T cell; loss of this cell causes victims to have greatly weakened immune systems. Even the most minor infections result in death. The development of an AIDS vaccine is extremely challenging because the virus is constantly changing the amino acid sequence of the membrane proteins, gp41 and gp121. Other methods for combating the virus include the design of drugs to inhibit the enzymes HIV-1 protease and RNase H, which play key roles in the maturation of the virus. The drugs DDI and AZT (see Figure 10.5) slow the growth of the virus by inhibiting HIV reverse transcriptase.

Cutting Edge: AIDS therapies and Window on Biochemistry 6.1.

Chromosomes

The genomic DNA in the nucleus of eukaryotic cells is packaged in functional units called **chromosomes.** The number of chromosomes varies from species to species; fruit flies have a total of 8 chromosomes per cell, whereas humans have 46 per cell. The chromosomes of eukaryotic cells are packages composed of DNA and protein. The packaging must be highly ordered and compact in order to fit the huge DNA molecules into the cell's nucleus. The total length of all DNA in a human cell is estimated at 1–2 m. The nucleus of the cell, which must find room for all the DNA, is only 5 μm in diameter.

The nucleoprotein complex in the eukaryotic cell nucleus is called **chromatin.** The primary biochemical components of chromatin are the DNA molecules and two classes of proteins, **histones** and nonhistone chromosomal proteins. The histones are a family of small proteins that contain relatively large numbers of the basic amino acid residues arginine and lysine. Five types of histone proteins have been identified. They vary in molecular size (10,000–20,000 daltons) and in Lys/Arg ratios. The nonhistone proteins are relatively few in number.

Wrapping the genomic material into a package small enough to fit into the cell nucleus requires a well-organized and compact arrangement of the DNA molecule (Figure 10.23). The packaging process begins with the formation of DNA–histone complexes. Histone cores consisting of eight protein molecules are tightly coiled with DNA. The spooled complexes, called **nucleosomes,** are held together by ionic bonds between positively charged

Figure 10.22

The human immunodeficiency virus: (a) schematic diagram and (b) electron micrograph. In part (a), the glycoproteins in the membrane, gp41 and gp121, are shown in two shades of green. The viral core contains two kinds of protein subunits shown in tan and orange; an RNA genome in red; and reverse transcriptase in light blue.

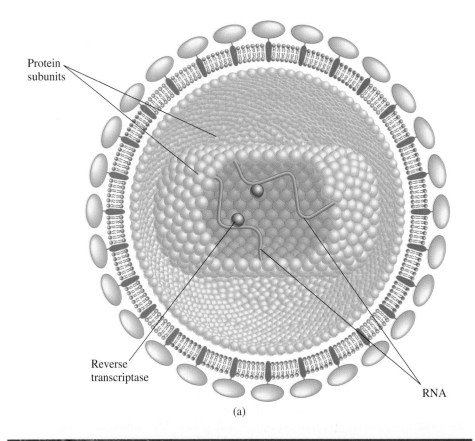

Protein subunits

Reverse transcriptase

RNA

(a)

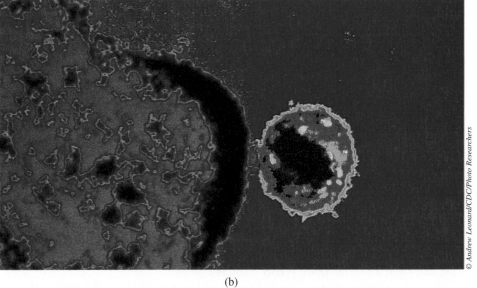

(b)

© *Andrew Leonard/CDC/Photo Researchers*

arginine and lysine protein residues with the negatively charged phosphate groups in the DNA backbone. Each human chromosome requires approximately 1 million nucleosomes to spool the DNA. At this point, the structure takes on the appearance of "beads on a string." The histone packets (beads) are aligned on the DNA molecule (string). Nucleosomes then continue to wind tightly into a structure reminiscent of a filament or fiber (called chromatin fiber). Wrapping continues with the formation of loop clusters until a chromatid is produced. Chromatids then combine, resulting in the chromosome unit. This entire process in human cells represents a compacting factor of 10^5. The biochemical functioning of the chromosome is introduced in Chapter 11.

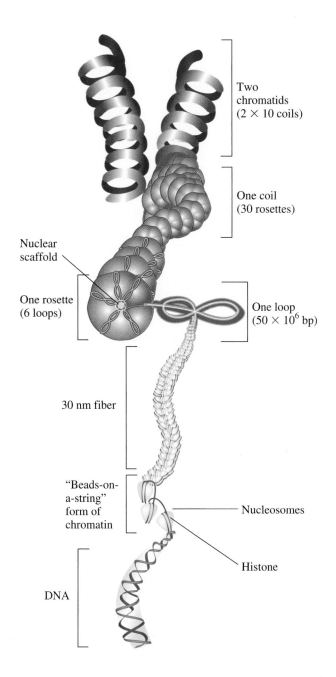

Figure 10.23
The packaging of eukaryotic DNA into chromosomes.

Two chromatids
(2×10 coils)

One coil
(30 rosettes)

Nuclear scaffold

One rosette
(6 loops)

One loop
(50×10^6 bp)

30 nm fiber

"Beads-on-a-string" form of chromatin

Nucleosomes

Histone

DNA

snRNPs

Biochemists and molecular biologists have recently discovered a new class of functional ribonucleoprotein complexes that play a role in RNA processing. These active clusters composed of small nuclear ribonucleic acids (snRNAs) and proteins are called **small nuclear ribonucleoprotein particles** (snRNPs, pronounced "snurps"). Analogous particles found in the cytoplasm are called scRNPs ("scurps"). The RNA molecules are relatively small with about 100 to 200 nucleotide bases, an intermediate size between tRNA and small, 5S rRNA. The stable ribonucleoprotein complexes have been found only in eukaryotic cells. The snRNPs play very important roles in the preparation (processing) of mRNA. They catalyze specific splicing reactions that transform gene transcripts (heterogeneous nuclear RNA) into mature mRNA before it is transported from the nucleus to ribosomes in the cytoplasm. Specifically, the RNA transcripts must be spliced; that is, regions of intervening sequences **(introns)** are removed and coding sequences **(exons)** are joined into active mRNA (see Chapter 2, Section 2.4; Chapter 7, Section 7.5; and Chapter 11, Section 11.5).

Ribosomes

Ribosomes are supramolecular assemblies of RNA and protein that function as the intracellular sites for protein synthesis (translation of RNA). Ribosomes are composed of about 65% RNA and 35% protein. They have two major subunits that dissociate during the translation process. The detailed structure and function of ribosomes are discussed in Chapter 12, Sections 12.1 and 12.2.

Ribonucleoprotein Enzymes

Important enzymes that are composed of protein and nucleic acid parts include ribonuclease P (see Chapter 7, Section 7.5), telomerase (see Chapter 11, Section 11.2), and peptidyl transferase (Chapter 12, Section 12.2).

Summary

■ The two types of nucleic acids, DNA and RNA, are important for their roles in the storage and expression of genetic information. Both DNA and RNA are polymers of nucleotides that consist of a nitrogen base, a pentose, and a phosphate group.

■ Watson and Crick in 1953 discovered that the DNA molecule is a double helix consisting of two complementary polynucleotide chains twisted together to form a circular staircase-like structure. The two chains are held together by hydrogen bonds between pairs of complementary bases, A-T, and G-C.

■ RNA exists in three forms: ribosomal (r), messenger (m), and transfer (t).

■ Large complex molecules of DNA are sequence-analyzed by using enzymes to break them into smaller fragments. The most important enzymes are the restriction endonucleases, which recognize specific nucleotide sequences.

■ Viruses are nonliving but infective particles composed of a nucleic acid (DNA or RNA) and protein subunits.

Study Exercises

Understanding Terms

10.1 Define the following terms in 25 words or less.

a. Purine bases	**j.** Restriction
b. Pyrimidine bases	endonuclease
c. Nucleoside	**k.** Bacteriophages
d. Oligonucleotide	**l.** Histones
e. Complementary	**m.** snRNPs
base pairs	**n.** Chromatin
f. B form of DNA	**o.** Caffeine
g. Denaturation	**p.** 6-Mercaptopurine
of DNA	**q.** Chemical mutagen
h. Supercoiled DNA	**r.** Topoisomerase
i. Hairpin turns	**s.** Nuclease

Reviewing Concepts

10.2 Draw molecular structures for the following compounds. Draw the structures as they would exist at physiological pH.

a. 2′-Deoxythymidine 5′-monophosphate
b. Guanosine
c. Uridine 5′-diphosphate
d. 3′,5′-Cyclic AMP

10.3 Name the following nucleosides and nucleotides.

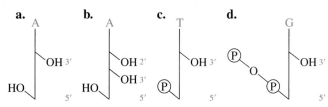

10.4 Assume that the short polynucleotide strands below are part of a DNA or RNA molecule. Draw a complementary strand for each.

a. 5′ AGCTTACGTCC
b. 5′ UAGGUACUUCG

10.5 Draw the structure of an mRNA strand that would be transcribed by the following DNA strand.

5′ ATGTACCGATA

10.6 Use arrows to show the bond cleaved by each of the following nucleases.

a. Rattlesnake venom phosphodiesterase
b. Spleen deoxyribonuclease II

10.7 Write the structure of the charged form of AMP at pH 7. Show how a magnesium ion (Mg^{2+}) could bind to AMP.

10.8 Compare the three structural forms of DNA (A, B, and Z) by preparing a table containing the following characteristics.

Form	Direction of Helix	Number of Base Pairs per Helix Turn	Diameter
A			
B			
Z			

10.9 List the differences between DNA and RNA. Answer this in terms of primary, secondary, and tertiary structure.

10.10 Draw the structures for the following purines and pyrimidines. We will encounter some of these in later chapters.
  a. 5-Bromouracil
  b. 5-Methyluracil (what is an alternative name for this?)
  c. 2-Aminopurine
  d. 2,6-Diaminopurine

10.11 How many phosphodiester bonds are present in a polynucleotide of 10 nucleotide units?

10.12 Which of the following base sequences are probably not recognition sites for cleavage by restriction endonucleases? Why not?
  a. 5′ GAATTC 3′ **c.** 5′ CATATG 3′
  b. 5′ CATTAG 3′ **d.** 5′ CAATTG 3′

10.13 Draw the Fischer projection formula for each carbohydrate listed below.
  a. D-Ribose
  b. D-2-Deoxyribose
  c. D-Glucose

10.14 Draw the Haworth projection for each carbohydrate in Problem 10.13.

10.15 Name three coenzymes that have nucleotide constituents in their overall structures.

10.16 The terms listed below are associated with various aspects of mammalian genetic material. Arrange the terms in order of increasing complexity (and increasing size).

nitrogenous base, nucleotide, double-strand DNA, genome, nucleoside, chromosome, gene

HINT: Begin with nitrogenous base < etc.

10.17 AZT, an important drug for the treatment of AIDS has the complete name, 3′-azido-2′,3′-dideoxythymidine. Draw its structure.

HINT: The azido group is N_3.

10.18 Another drug that is being tested for use in AIDS treatment is 2′,3′-dideoxycytidine (DDC). Draw its structure.

Solving Problems

10.19 Draw the structure for cytidine 4′-monophosphate. Is it likely that such a 4′-monophosphate nucleotide would be found naturally occurring? Why or why not?

10.20 How many fragments of DNA are produced by the action of the restriction enzyme *Taq*I on the following strand of DNA?

5′ AACTCGATTCTCGAACCG

HINT: You will need to start this problem by writing the sequence for the complementary strand.

10.21 Draw the covalent backbone for a short strand of DNA and show how arginine and lysine residues in histones could interact by ionic bonds.

10.22 The nitrogenous bases found in nucleic acids can exist in tautomeric forms (keto and enol). These chemically distinct forms are produced by rearrangements of protons. The base uracil exists in several tautomeric forms, some drawn below. Identify functional groups that act as hydrogen bond donors and acceptors in each tautomeric form.

What will be the effect of tautomerization in base pairing processes?

10.23 Assume that the structure of RNA shown below is part of the genome of a virus. Draw the structure of double-stranded DNA that would be produced by the enzyme reverse transcriptase using the RNA as template.

5′ AGGGCUAACUCUAAG

10.24 The following enzymes have been discussed in this chapter. Write a reaction catalyzed by each or describe its action in words.
  a. Topoisomerase
  b. Ribonuclease
  c. *Hinf*I

10.25 DNA with a high content of G-C base pairs has a higher melting temperature (T_m) than DNA with a high content of A-T base pairs. Explain why.

10.26 Which of the following reagents and conditions will denature double-stranded DNA?

    a. Heat

    b. Urea

    c. Extreme pH changes

    d. Ethanol

10.27 The DNA from bacteriophage ϕX174 cannot be degraded by rattlesnake venom phosphodiesterase. Why not?

10.28 Study the structures you drew in Problem 10.2 and follow the directions below.

    a. Identify an *N*-glycosidic bond in each structure a, b, c, and d.

    b. Identify a phosphoester bond in a, c, and d.

    c. Identify a cyclic ester bond in a, b, c, or d.

10.29 We often describe DNA and RNA as being "information-rich molecules." Explain what this description means.

10.30 Give a brief explanation for how AZT and DDC function in inhibiting the growth of the HIV.

Writing Biochemistry

10.31 Read the article on ancient DNA by S. Paabo listed in Further Reading (*Sci. Am.* 269(5):86–92). Write a "Window on Biochemistry" on the topic. Include interesting and practical aspects from the article, but limit your writing to 200 words.

10.32 Your instructor will probably present the material from this chapter in one or two class meetings. Immediately after all the concepts have been presented, write a 100-word response to the following questions: (1)What concepts were a review for you? (2) What concepts were new to you? (3)What concepts were confusing?

Applying Biochemistry

10.33 Describe the possible normal function(s) of the restriction enzymes found in bacterial cells.

10.34 Study the CD/ROM unit on "The Central Dogma of Biochemistry." Describe in your own words what is meant by the phrase, "the central dogma of biochemistry." Explain how it is possible that a single type of molecule (DNA) can possibly be involved in directing cellular processes such as the synthesis of carbohydrates.

Further Reading

Bartel, D., and Unrau, P., 1999. Constructing an RNA world. *Trends Biochem. Sci.* 24:M9–M12.

Boyer, R., 2000. DNA structure and function. In *Modern Experimental Biochemistry,* 3rd ed., pp. 399–414. San Francisco: Benjamin/Cummings.

Cavanee, W., and White, R., 1995. The genetic basis of cancer. *Sci. Am.* 272(3):72–79.

Cohen, J., and Hogan, M., 1994. The new genetic medicines. *Sci. Am.* 271(6):74–82.

Collins, F. et al. 2001. The human genome. *Nature* 409:745–748.

Lebowitz, J., 1990. Through the looking glass: Discovery of supercoiled DNA. *Trends Biochem. Sci.* 15:202–207.

Neidle, S., 1994. *DNA Structure and Recognition.* Oxford: IRL Press.

Notomista, E., et al., 2000. Onconase: An unusually stable protein. *Biochemistry* 39:8711–8718.

Nowak, M., and McMichael, A., 1995. How HIV defeats the immune system. *Sci. Am.* 273(2):58–65.

Paabo, S., 1993. Ancient DNA. *Sci. Am.* 269(5):86–92.

Piper, A., 1998. Light on a dark lady: Rosiland Franklin. *Trends Biochem. Sci.* 23:151–154.

Rennie, J., 1993. DNA's new twists. *Sci. Am.* 266(3):88–96.

Venter, J., 2001. The sequence of the human genome. *Science* 291:1304.

Volker, E., 1993. An attack on the AIDS virus: inhibition of the HIV-1 protease. *J. Chem. Educ.* 70:3–9.

Watson, J., 1968. *The double helix.* New York: Atheneum Publishers.

Weintrab, H., 1990. Antisense RNA and DNA. *Sci. Am.* 262(1):40–46.

Chapter 11

DNA Replication and Transcription: Biosynthesis of DNA and RNA

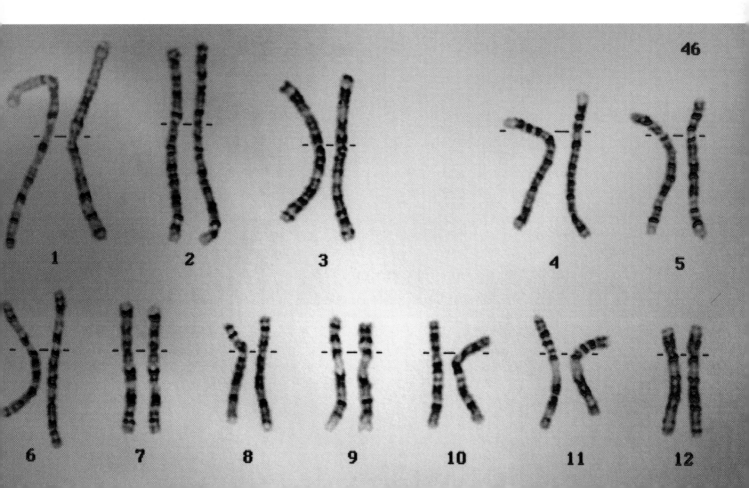

Spanish stamp honoring Severa Ochoa, who studied the enzymes of transcription.

By the early 1950s, soon after Watson and Crick's proposal for DNA's secondary structure, our molecular understanding about the transfer of hereditary traits from one generation to the next could be summarized in two wide-ranging statements:

1. The biomolecule that is ultimately responsible for the storage and transfer of genetic information is DNA.
2. The structure of the DNA molecule is described as a double helix of two polydeoxyribonucleotide strands formed by pairing of complementary bases.

Both of these statements had a revolutionary impact on biochemistry and genetics. As is often the case in science, important discoveries such as those above have led to new sets of questions that must be answered. The elucidation of the DNA structure by Watson and Crick, especially, spurred interest in the design of experiments that better explained the molecular events occurring in genetic information transfer. Important questions still to be answered were:

1. What is the molecular mechanism for transfer of DNA-based information from one generation of cells to the next? DNA must be copied and the duplicate transferred to daughter cells so the cell line can continue. To maintain the species in a viable form, the duplication process must be reproducible and accurate.
2. How is the sequence of bases in DNA decoded to make proteins that maintain the cell's structure and function? Biological information in the language of DNA must be translated to the language of proteins.

We have answered these questions briefly in Chapters 2 and 10, but here we shall add depth and breadth to our biochemical understanding of nucleic acid function. In this and later chapters, we shall change our focus from structural matters to a description of nucleic acid function. Transfer of genetic information requires faithful duplication of the information, a process we call **replication.** The double-helix concept, which greatly enhanced our understanding of DNA structure, suggested a process for replication. Unwinding of the double helix results in single polynucleotide strands that act as **templates** or "surfaces" for copying. When DNA is replicated, the new copy of DNA for the daughter cell must be identical to the parent DNA. The complex replication process, however, is not always 100% error-free; mistakes, although very rare, are made. In addition, chemicals unnatural to the cell and extreme environmental conditions may lead to chemical and sequence changes in DNA. Specialized enzyme systems recognize mistakes in DNA base sequence and catalyze repair of those mistakes. If errors are allowed to be transcribed into RNA and translated, protein products change in amino acid sequence and have altered properties.

Next, we shall turn to the process of DNA **transcription** or biosynthesis of RNA. The RNA initially formed must be biochemically modified before it becomes biologically active. Post-transcriptional modification of RNA is carried out by enzymes and ribozymes.

We conclude the chapter by describing experimental techniques for studying DNA and RNA base sequence. These help us answer the question, How do we know that the new molecule of DNA or RNA has a base sequence complementary to the template?

11.1   Replication of DNA

At first sight, DNA's seemingly rigid, double-helical conformation gives the impression that it is a static, unchanging structure. On the contrary, the existence of at least three types of secondary conformations (A, B, Z) and extensive tertiary structure (supercoiling) tells us that DNA has a dynamic structure capable of assuming various arrangements. In replication and transcription, the DNA double helix can be observed being wrapped into supercoils, unwound to single strands, and bent into forks for decoding. These processes must be accurate, precise, gentle, and carefully monitored because any changes in the DNA primary structure result in the transfer of incorrect information to future generations. A concept weaving through this and later chapters is the required faithful transmission of genetic information through all the transfer steps.

Although there are exceptions, most naturally occurring DNA molecules are double-stranded helixes that exist in circular or linear forms. We have a relatively complete understanding of DNA replication and transcription as it occurs in prokaryotic cells, especially *Escherichia coli.* Our understanding of replication in eukaryotic cells is incomplete. Therefore, our study of DNA metabolism focuses on the double-stranded, circular DNA of *E. coli.* The fundamentals of replication and transcription for eukaroyotic cells appear to be similar except for the more complex, linear DNA structure and a greater number of required enzymes and other protein factors.

Characteristics of the Replication Process

The model of the DNA double helix by Watson and Crick aided the interpretation of many experiments done on replication. Early laboratory studies of DNA replication uncovered fundamental characteristics that appeared to be universal in all organisms studied. We begin our study of replication with a preview of these characteristics:

DNA Replication Is Semiconservative

If Watson and Crick's hypothesis for replication is true, then each original strand of DNA is used as a template for synthesis of a new strand. The "new" DNA according to their proposed **semiconservative replication** model would be a hybrid: Each new duplex would be composed of an original strand and a newly synthesized strand (Figure 11.1). In an alternate model, called conservative replication, each single strand of DNA is replicated and the newly synthesized strands combine to form one new DNA duplex while the original strands reassociate into another duplex.

Evidence supporting the semiconservative model was obtained in 1957–1958 by Matthew Meselson and Franklin Stahl while they were working at Harvard University. They

🔘 **Interactive Animation: DNA replication**

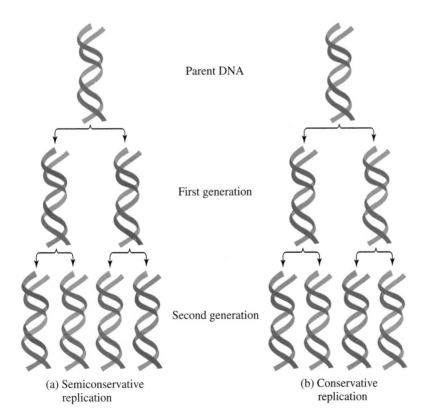

(a) Semiconservative replication

(b) Conservative replication

Figure 11.1

Two possible models for DNA replication: (a) In semiconservative replication each first generation DNA will have one parental strand (green) and one new strand (red). (b) In conservative replication the two parental strands remain together (green) and the first generation daughter DNA has two new strands (red).

Parent DNA

First generation

Second generation

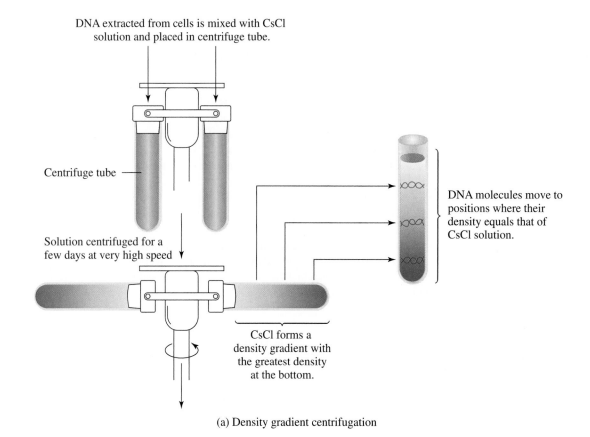

DNA extracted from cells is mixed with CsCl solution and placed in centrifuge tube.

Centrifuge tube

Solution centrifuged for a few days at very high speed

DNA molecules move to positions where their density equals that of CsCl solution.

CsCl forms a density gradient with the greatest density at the bottom.

(a) Density gradient centrifugation

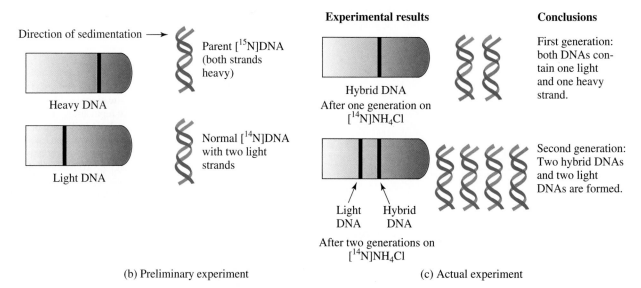

Direction of sedimentation →

Heavy DNA

Light DNA

Parent [^{15}N]DNA (both strands heavy)

Normal [^{14}N]DNA with two light strands

Experimental results

Hybrid DNA
After one generation on [^{14}N]NH$_4$Cl

Light   Hybrid
DNA   DNA

After two generations on [^{14}N]NH$_4$Cl

Conclusions

First generation: both DNAs contain one light and one heavy strand.

Second generation: Two hybrid DNAs and two light DNAs are formed.

(b) Preliminary experiment

(c) Actual experiment

Figure 11.2

(a) The use of density gradient centrifugation to study DNA replication. The Meselson–Stahl experiment was designed to distinguish between conservative and semiconservative replication of DNA. (b) The original parent DNA molecule grown in ^{15}N (red) sediments near the bottom of the density gradient (shown in blue), whereas normal [^{14}N] DNA (green) sediments nearer the top of the gradient. (c) The first generation of daughter molecules obtained from the "heavy" DNA consisted of a single band of "hybrid" DNA, one strand of each contained ^{15}N (shown in red) and one strand contained ^{14}N (shown in green). The second generation of daughter DNA molecules consisted of two types of DNA, DNA molecules that had a density equal to "light" DNA (shown in green) and DNA molecules that had a density equal to "hybrid" DNA (shown in green and red).

grew many generations of *E. coli* cells in a culture medium containing NH$_4$Cl labeled with ^{15}N, the heavy isotope of nitrogen, instead of the normal lighter isotope, ^{14}N. Thus, nitrogen-containing biomolecules synthesized in the cells, including DNA, contained ^{15}N rather than ^{14}N. The ^{15}N-containing DNA is heavier than ^{14}N-DNA. Meselson and Stahl then transferred some of the *E. coli* cells with ^{15}N-DNA to a fresh medium containing ^{14}NH$_4$Cl. After allowing time for cell growth of one and two generations (doubling of cell population each time), Meselson and Stahl isolated the DNA from the cells and analyzed it by density gradient centrifugation (Chapter 1, Section 1.5). This experimental technique allowed the separation, identification, and characterization of three kinds of DNA, (1) "heavy" DNA, containing two strands of ^{15}N-DNA from the original labeling experiment; (2) "light" DNA, containing two strands of ^{14}N-DNA; and (3) "hybrid" DNA, containing one strand with ^{15}N and one with ^{14}N (Figure 11.2). Experimental results supported the semiconservative model for replication of DNA in *E. coli*. Meselson and Stahl–like experiments, repeated with other prokaryotic cells and with eukaryotic cells (including plants), have demonstrated that the semiconservative mechanism for replication is universal.

Replication Begins at a Discrete Point on DNA

The process of DNA replication was first visually observed in the late 1950s by John Cairns. He grew *E. coli* cells in a medium containing radioactively labeled nucleoside, ^{3}H-thymidine. The ^{3}H-labeled base was incorporated into newly synthesized DNA. The isolated, radioactive DNA was then analyzed by autoradiography. The DNA was spread on a photographic emulsion plate in order to prepare a photographic image (Figure 11.3). Cairns observed forklike structures or loops that he interpreted to be DNA molecules undergoing replication. He named these structures **replication forks.** In a circular DNA molecule, replication was observed to start at a discrete point, the origin, and proceed in *both* directions, thereby copying *each* polynucleotide strand. Further experiments based on Cairns' observations showed that two replication forks were present in each DNA molecule. The overall process of replication as we now understand it is outlined in Figure 11.4. Formation of two replication forks results in the development of a replication bubble, which increases in size as replication proceeds. The replicating forks advance in opposite directions, terminating by meeting on the other side of the circular

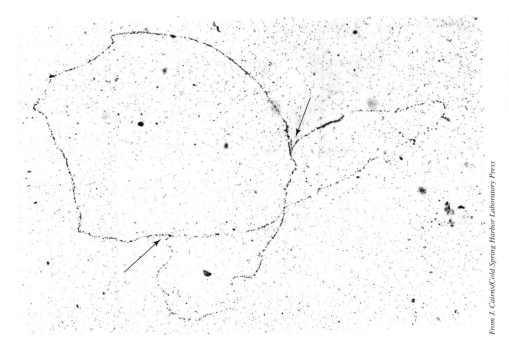

Figure 11.3

A photographic image (obtained by autoradiography) of replicating *E. coli* chromosome. The DNA contained ^{3}H-labeled thymidine. Note the presence of two replicating forks (*arrows*).

From J. Cairns/Cold Spring Harbor Laboratory Press

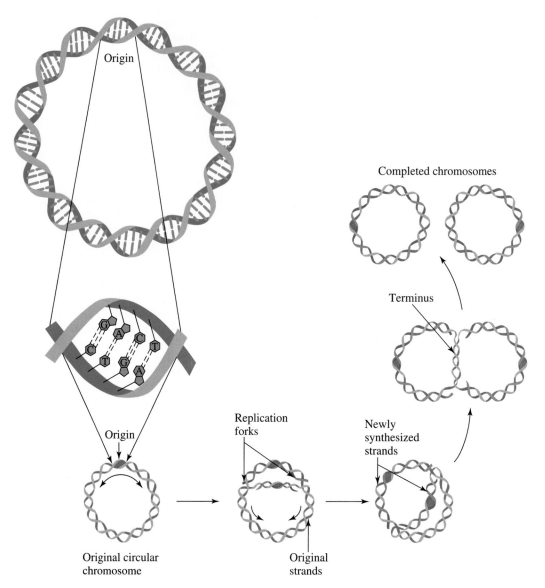

Figure 11.4

The theta (θ) model for replication of a circular DNA molecule. Note that the DNA structures resemble the Greek letter θ. The two replicating forks advance in opposite directions. *Source:* Redrawn from Ingraham and Ingraham, 1995.

DNA. This mechanism is often called the theta model for replication because the intermediates resemble the Greek letter theta (θ).

Eukaryotic DNA molecules, which are linear, replicate in a similar fashion (Figure 11.5). The major difference is the presence of several initiation sites; these result in several replication bubbles eventually forming two separate double-stranded DNA molecules.

The experiments by Meselson and Stahl, Cairns, and others provided general and descriptive principles of DNA replication but uncovered little information at the molecular level. Questions remaining unanswered were: What is the direction of DNA synthesis ($5' \longrightarrow 3'$ or $3' \longrightarrow 5'$)? How are nucleotides selected for incorporation? and How are the phosphodiester bonds between mononucleotides formed? Answers to these questions awaited completion of biochemical searches for enzymes and other molecular factors necessary for replication.

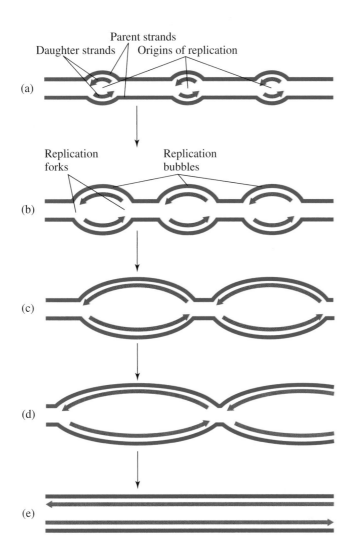

(a)

(b)

(c)

(d)

(e)

Figure 11.5
Proposed pathway for replication of eukaryotic DNA. There are several origins of replication (a) and a pair of replication forks begins at each origin (b). As the forks advance in opposite directions, the bubbles coalesce to form two double-stranded DNA molecules (c, d, e).

11.2   Action of DNA Polymerases

DNA Polymerase I

The first enzyme discovered that could catalyze the synthesis of DNA, now called **DNA polymerase I,** was isolated from *E. coli,* purified, and characterized by Arthur Kornberg and his colleagues in 1957. The general reaction catalyzed by the enzyme is:

$$\text{dNTP} + (\text{dNMP})_n \rightleftharpoons (\text{dNMP})_{n+1} + \text{PP}_i$$

dNTP = deoxyribonucleoside triphosphates, dATP, dGTP, dCTP, dTTP
(dNMP) = preformed DNA with n or $n + 1$ mononucleotides
PP_i = pyrophosphate

Specific requirements for the reaction process include magnesium ion (Mg^{2+}) for nucleotide complexation and the presence of "preformed DNA," which serves two purposes (Figure 11.6). First, the DNA acts as a **template** that carries the message to be copied. The message is read according to Watson–Crick base pairing rules (adenine in the template adds thymine to the newly synthesized strand, etc.). Second, the preformed DNA must have a **primer** segment with a free 3′-hydroxyl group. This site provides access for covalent attachment of the entering nucleotide. DNA polymerase I is unable to start a new polynucleotide chain without such a primer. In chemical terms, the reaction proceeds with nucleophilic attack by the free 3′-hydroxyl group of the primer on the entering nucleotide,

Figure 11.6

The action of DNA polymerase I. Preformed DNA performs two roles, one as a template (red), which carries the message to be copied, and one as a primer (purple) for attachment of added nucleotides. The incoming deoxythymidine triphosphate (dTTP, shown in blue) is held in position by complementary hydrogen bonds to adenine in the template strand. The new phosphoester bond is formed adding a base at the 3′ end of the growing strand. Extra energy is provided by hydrolysis of pyrophosphate.

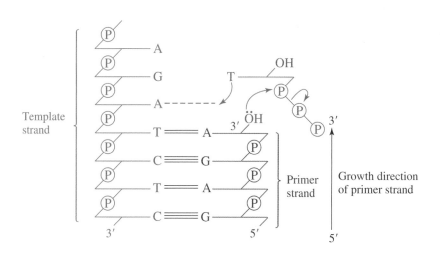

which is selected and held in place by specific base pairing. Thus, elongation occurs at the 3′ end with net synthesis of DNA proceeding in the 5′ → 3′ direction. Since the chains are antiparallel, the message in the DNA template is read in the 3′ → 5′ direction. Energy for the reaction is supplied by release of pyrophosphate, PP_i a reactive molecule that is readily hydrolyzed, with release of energy, to form two molecules of orthophosphate, P_i ($PP_i + H_2O \rightarrow 2 P_i$).

DNA Polymerases II and III

Further studies on *E. coli* mutant cells have uncovered two additional DNA polymerases, called **DNA polymerase II** and **DNA polymerase III.** These enzymes have many of the same reaction requirements described for DNA polymerase I. Properties of the three polymerases are compared in Table 11.1. We now believe that the primary replicating enzyme in all *E. coli* cells is DNA polymerase III. Note in the table that DNA polymerase III has a much higher polymerization rate, but that it also has a very complex structure with ten polypeptide subunits and a molecular weight of 900,000. What, then, are the functions of DNA polymerase I and II? They probably serve editing functions, such as proofreading and repair to ensure fidelity in DNA synthesis. Specifically, these enzymes act as nucleases to catalyze hydrolysis of phosphodiester bonds. DNA polymerase I is a 3′ → 5′ exonuclease (Chapter 10, Section 10.4). Its action as a hydrolytic enzyme is shown in Figure 11.7. In addition, DNA polymerase I can catalyze the hydrolysis of DNA beginning at the 5′ end; that is, DNA polymerase I is also a 5′ → 3′ exonuclease. Thus, the DNA polymerase I molecule, which is rather small, has three known catalytic activities associated with its single polypeptide chain: (1) DNA polymerase, (2) 3′ → 5′ exonuclease, and (3) 5′ → 3′ exonuclease. The 5′ → 3′ exonuclease activity of DNA polymerase I serves an important function in the replication process.

Table 11.1

Comparison of *E. coli* DNA polymerases

	DNA Polymerase		
Characteristic	**I**	**II**	**III**
Molecular weight (daltons)	103,000	88,000	900,000
Polypeptide subunits[a]	1	4	10
Polymerization rate (nucleotides added per second)	16–20	7	250–1000
3′ → 5′ Exonuclease activity	Yes	Yes	Yes
5′ → 3′ Exonuclease activity	Yes	No	No

[a]The number of polypeptide subunits defines the quaternary structure.

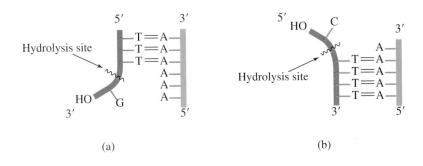

Figure 11.7
Exonuclease activity of DNA polymerase I: (a) $3' \rightarrow 5'$ exonuclease activity; (b) $5' \rightarrow 3'$ exonuclease activity. DNA polymerase I acts as a proofreading and repair enzyme by catalyzing hydrolytic removal of mismatched bases.

Okazaki Fragments

All known DNA polymerases catalyze chain elongation in the $5' \rightarrow 3'$ direction. A closeup of a replication fork shows two strands of parental DNA serving as templates for synthesis of new DNA (Figure 11.8). The two template strands are oriented in an antiparallel fashion; therefore, DNA polymerases can only catalyze continuous nucleotide addition by reading the $3' \rightarrow 5'$ parental DNA strand. The complementary strand is synthesized in the $5' \rightarrow 3'$ direction. How is the $5' \rightarrow 3'$ parental strand copied if synthesis can only proceed in the $5' \rightarrow 3'$ direction? This question was answered when it was discovered that the $5' \rightarrow 3'$ parental strand was replicated in smaller, discontinuous fragments. The two growing strands at the replication fork elongate in different directions and can be defined as the leading strand (molecules are added continuously in the $5' \rightarrow 3'$ direction) and the lagging strand (proceeds in short pieces in the reverse direction of the leading strand). However, note that the actual direction of DNA polymerase action is always $5' \rightarrow 3'$. Experimental support for the concept of discontinuous replication was provided by Reiji Okazaki and his colleagues, who found short DNA fragments (1000–2000 nucleotides) during replication. The discontinuous **Okazaki fragments** are covalently joined in later steps of replication to form the completed daughter strand.

Several eukaryotic DNA polymerases have been discovered and characterized. DNA polymerase α, found only in the cell nucleus, has the same reaction features as the prokaryotic enzymes, requirement for a DNA template and primer, and synthesis in the $5' \rightarrow 3'$ direction. Additional DNA polymerases β, δ, and ε are found in the nucleus and γ is in mitochondria.

DNA Replication

Now with an understanding of some of the biochemical reactions and mechanics involved, the complex process of prokaryotic DNA replication can be outlined in detail. The sequence of steps beginning with unwinding the parental double helix and ending with covalent joining of the discontinuous Okazaki fragments is outlined in Figure 11.9. A list of major proteins necessary for DNA replication in *E. coli* is given in Table 11.2. The steps of *E. coli* replication are as follows:

1. Helicase (also called unwinding protein and rep protein) recognizes and binds to the origin for replication and catalyzes separation of the two DNA strands by breaking

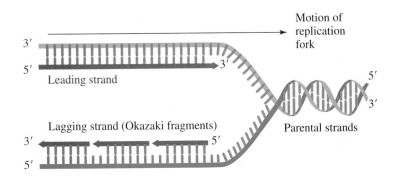

Figure 11.8
Closeup of a replication fork showing initiation of the continuous leading strand and the discontinuous, lagging strand (Okazaki fragments).

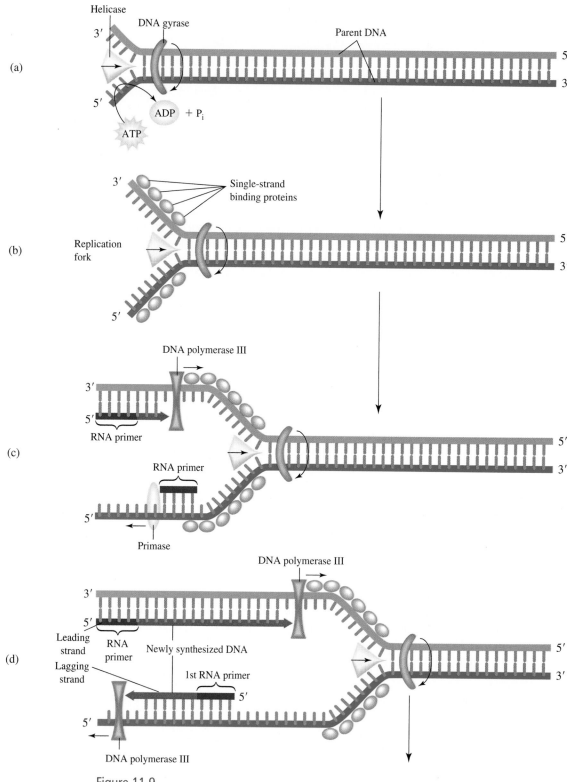

Figure 11.9
Complete scheme showing sequential steps of the replication process.

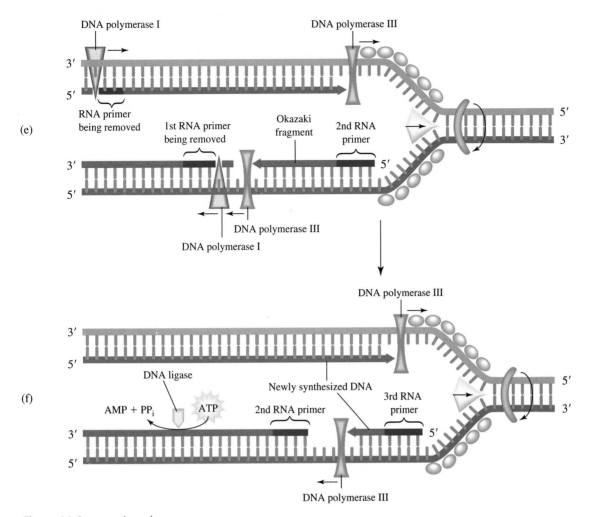

Figure 11.9—*continued*

hydrogen bonding between base pairs (Figure 11.9a). This endothermic reaction is coupled to the cleavage of ATP (ATP + H_2O → ADP + P_i + energy). **DNA gyrase,** a topoisomerase, assists unwinding of DNA strands by inducing supercoiling. These combined processes result in the formation of a replication fork.

2. The exposed single strands of DNA must be stabilized and protected from hydrolytic cleavage of phosphodiester bonds. The **single-stranded DNA binding proteins (SSB proteins)** perform this protective role. The now separated polynucleotide strands are used as templates for the synthesis of complementary strands (Figure 11.9b).

3. The next step, initiation of DNA synthesis, brings in a new twist for DNA polymerase III. The action of all DNA polymerases, including DNA polymerase III, requires a primer with a free 3′-hydroxyl end. Since this is not present at the replication fork, synthesis of

Protein	Function
Helicase	Begins unwinding of DNA double helix
DNA gyrase	Assists unwinding
SSB proteins	Stabilize single strands of DNA
Primase	Synthesis of RNA primer
DNA polymerase III	Elongation of chain by DNA synthesis
DNA polymerase I	Removal of RNA primer and fill in gap with DNA
DNA ligase	Closes last phosphoester gap

Table 11.2
Proteins necessary for DNA replication in *E. coli*

DNA cannot begin immediately. This problem is resolved by the synthesis of a short stretch of RNA (four to ten nucleotides) complementary to the DNA template. RNA synthesis is catalyzed by an enzyme called **primase** (Figure 11.9c). The action of primase is required only once for the initiation of the leading strand of DNA. On the other hand, each Okazaki fragment must be initiated by the action of primase. After the few ribonucleotides are added, DNA synthesis catalyzed by DNA polymerase III can proceed from the 3'-hydroxyl group. In later steps, the RNA primer is removed from the DNA by the 5' → 3' nuclease action of DNA polymerase I and by RNase H.

4. Now, with the availability of the free 3'-hydroxyl group, DNA synthesis begins, catalyzed by DNA polymerase III. Both the leading strand and the lagging strand are extended in this way. The leading strand proceeds in the direction of the advancing replication fork. Synthesis of the lagging strand continues in the opposite direction until it meets the fragment previously synthesized (Figure 11.9d).

5. At this time, the RNA primers are removed by the 5' → 3' nuclease action of DNA polymerase I. Small gaps remain that are filled in by the synthesizing action of DNA polymerase I (Figure 11.9e).

6. DNA polymerase activity can bring in all of the deoxyribonucleotide bases needed to fill in the gaps; however, the final phosphoester bond to close completely the last gap must be formed by an additional enzyme, **DNA ligase.** This enzyme catalyzes ATP-dependent phosphate ester formation between a free hydroxyl group at the 3' end of one fragment with a phosphate group at the 5' end of the other (Figures 11.9f and 11.10). Termination of the replicating process occurs when the two replication forks meet on the circular *E. coli* chromosome (see Figure 11.4).

Eukaryotic Chromosomes and Telomeres

Cutting Edge: Telomeres, eternal life, and cancer

The picture of DNA replication in the prokaryotic cell as we have outlined here is well understood. However, there is much we do not understand about the more complicated process in eukaryotic cells. One of the most important differences between prokaryotic and eukaryotic DNA is the presence of specialized ends called **telomeres** in eukaryotic DNA. Sequential base analysis of these ends reveals hundreds of repeats of a hexanucleotide sequence. In human DNA, the 3' end is made up of the repeating sequence AGGGTT. The enzyme **telomerase** catalyzes the synthesis of these DNA ends. Telomerase is an unusual enzyme that contains an RNA molecule (ribozyme) that serves as a template to guide the addition of the right nucleotides.

Varying activity levels of telomerase may serve to regulate cell division, thus having an influence on the aging of cells. Telomeres usually shorten during the normal cell cycle. If telomeres become too short (telomerase is inactive), chromosomes (DNA molecules) become unstable and cell division is inhibited. The progressive shortening of telomeres may function as a biological clock that limits the number of times a cell can replicate. New research has shown that telomerase becomes activated in human cancer cells. Telomerase

Figure 11.10

The DNA ligase–catalyzed reaction to close the final phosphodiester bond in newly synthesized DNA. ATP is required as a source of energy.

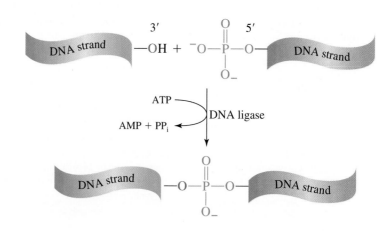

activity was found in cancer cells from women with late-stage ovarian cancer but absent from nonmalignant body cells from the same cancer patients. Large numbers of tumor cell lines have now been studied and over 85% display telomerase activity. A similar survey of normal tissues shows the absence of telomerase activity except in sperm cells. These studies suggest that inhibitors of telomerase may be effective anticancer agents.

11.3   DNA Damage and Repair

DNA is the instruction manual for the cell. It contains all the information necessary to make proteins that perform cellular functions. Therefore, it is imperative that cells maintain the integrity of their DNA. Errors committed during the replication of DNA must be repaired so they are not passed on to future generations. Furthermore, DNA when damaged by extreme environmental conditions, must be restored to its original form. Changes in the base sequence of DNA are called **mutations.** Cell DNA repair systems prevent mutations from being replicated and transferred to daughter cells, which would lead to permanent changes in future cells. Most mutations are harmful and can even be lethal to cells. Silent mutations are changes in DNA base sequence that do not affect the function of protein products.

Mutations are classified into one of two categories depending on the cause of the damage. The change may occur by spontaneous processes or by induced processes.

Spontaneous Mutations

Spontaneous mutations are those changes that occur during normal genetic and metabolic functions in the cell. The two major types of spontaneous mutations are (1) mistakes in the incorporation of deoxyribonucleotides (mismatching of base pairs) during DNA replication and (2) base modifications caused by hydrolytic reactions. Final mistakes in *E. coli* replication occur about once for every 10^{10} base pairs incorporated, which is a very rare event. The *actual error rate* of base incorporation during replication may be as high as one in every 10^4 to 10^5 bases inserted; however, repair systems correct most mismatched bases. Most mutations lead to undesirable changes in genes. Occasionally, spontaneous changes are favorable in that they confer a selective advantage to an organism. Perhaps these changes are the causes of evolutionary processes. Replication errors are of three types: (1) substitution of one base pair for another (point mutation), (2) insertion of one or more extra base pairs, and (3) deletion of one or more base pairs. Substitution is the most common type of spontaneous mutagenesis. It is often caused by tautomerism in the nitrogen bases, which leads to inappropriate hydrogen bond pairing (see Study Problem 10.22 in Chapter 10). Tautomerism can lead to nonstandard base pairing. It has been estimated that the standard bases may spend about 0.01% of their time in a less stable tautomeric form.

Escherichia coli cells have relatively complex systems to detect and repair mismatched bases. The general repair mechanism proceeds in four steps: (1) endonuclease-catalyzed cleavage of a phosphoester bond holding the mismatched nucleotide in position, (2) removal of the mismatched base by an exonuclease, (3) incorporation of the correct base by DNA polymerase I or III, and (4) closure of the final gap by DNA ligase (Figure 11.11).

Another common type of spontaneous damage to DNA is hydrolytic. Nucleotides containing purine bases can undergo spontaneous hydrolysis at the *N*-glycosidic bond to remove the purine ring (Figure 11.12a). It has been estimated that a normal human cell may lose up to 10,000 purine bases per day by hydrolysis. In addition, heterocyclic bases can be deaminated by hydrolytic reactions. The most common deamination reaction is the conversion of cytosine to uracil (Figure 11.12b). A rate of 100 base deaminations per day may be common in the human cell. Damage by deamination of bases is detected and repaired by special enzymes called DNA glycosidases. These enzymes catalyze removal of damaged bases by hydrolysis of the *N*-glycosidic bond to the deoxyribose. An endonuclease removes the sugar by cleavage of both phosphodiester bonds. DNA polymerase I incorporates the proper nucleotide and the final gap is closed by DNA ligase.

Figure 11.11

The basic repair mechanism for mismatched base pairs in *E. coli* DNA: (a) An endonuclease catalyzes the hydrolysis of a phosphoester bond to the mismatched base. (b) An exonuclease catalyzes the hydrolytic removal of the mononucleotide, dGMP. (c) A DNA polymerase catalyzes the addition of the correct nucleotide to the free 3′ OH. (d) The final gap is closed by DNA ligase.

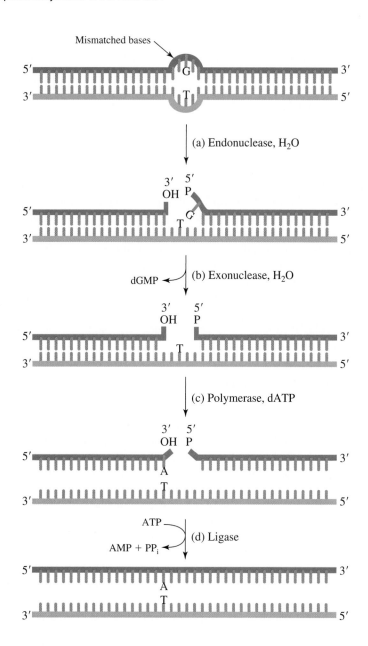

Induced Mutations

Organisms are often exposed to environmental agents that produce mutations. These agents are called **mutagens.** Common mutagens include ionizing radiation, some chemicals, and ultraviolet light.

Ionizing Radiation

Natural sources of ionizing radiation include cosmic rays from outer space and radioactivity that is released from the ground. Humans and other organisms have been exposed to natural levels of radiation for millions of years; in fact, this exposure may have played a role in evolutionary development. But with nuclear weapons testing, nuclear power plants, more airplane travel, and increased medical use of X-rays, our exposure to higher radiation levels is increasing. Extensive research has been done to study the molecular events occurring when DNA is exposed to ionizing radiation. When bombarded by ionizing radiation, molecules readily dissociate to produce ions and highly reactive free radicals. In secondary reactions, the free radical products attack DNA (Figure 11.13). For example, ionizing radiation

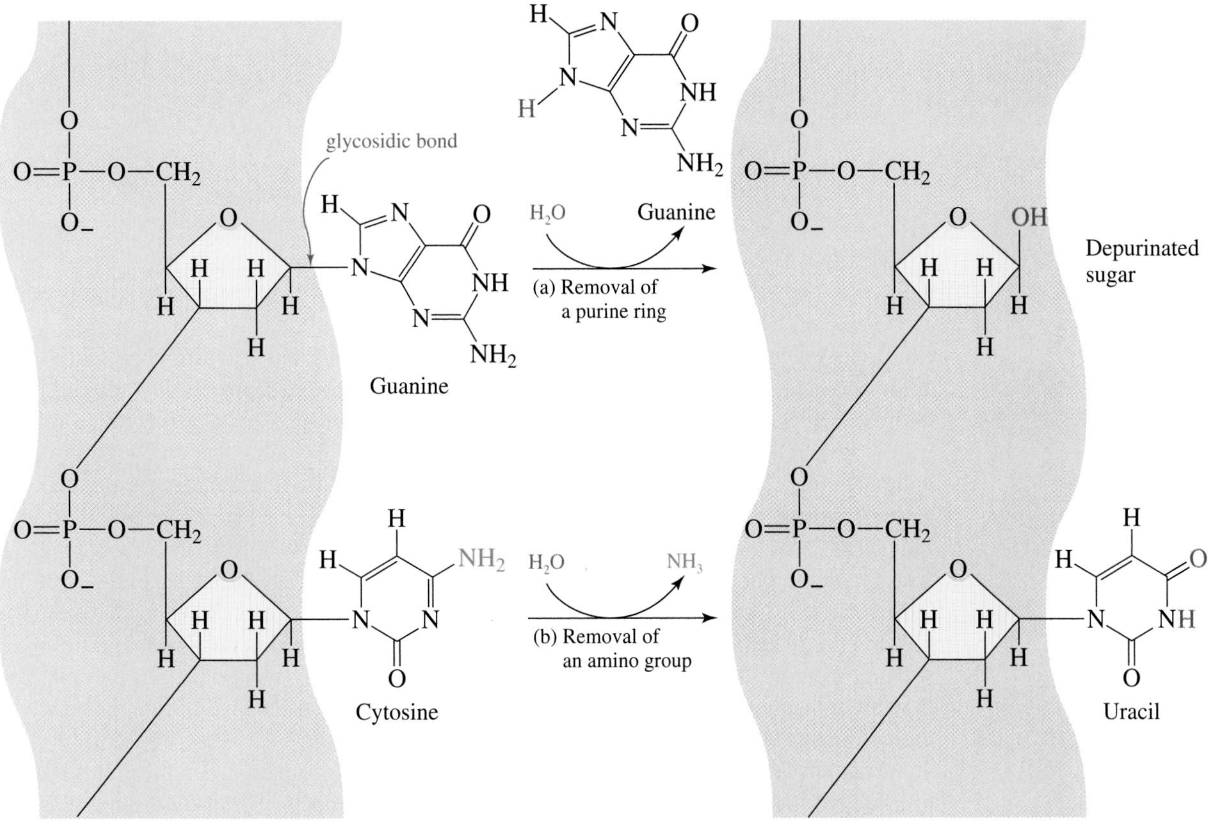

Figure 11.12

Mutagenic damage to DNA caused by spontaneous processes: (a) removal of a purine base by hydrolysis of a *N*-glycosidic bond; (b) removal of an amino group by hydrolysis. In (b), the base cytosine is converted to uracil. *Source:* Based on Becker, 1986.

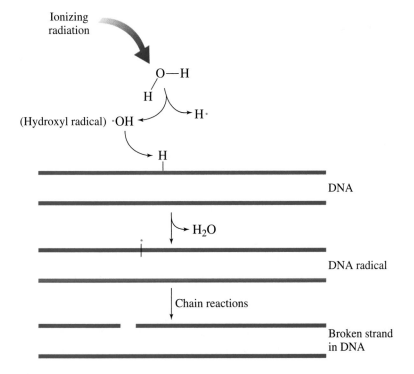

Figure 11.13

Secondary or indirect damage to DNA caused by the hydroxyl radical, ·OH. The ·OH is formed by interaction of ionizing radiation with H_2O. The ·OH removes a hydrogen atom from DNA, forming H_2O and a reactive DNA radical. This results in a broken DNA strand.

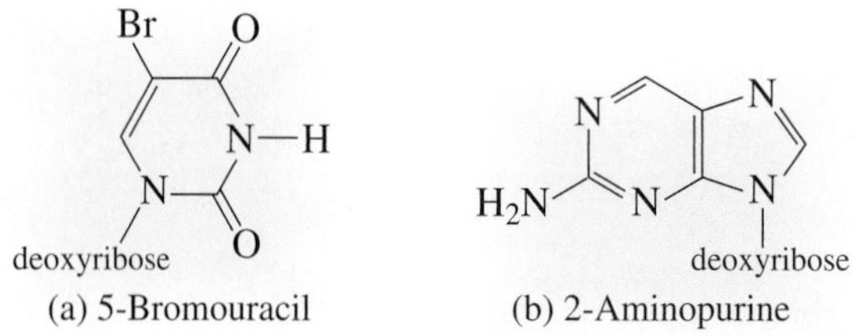

Figure 11.14

Chemical mutagens. Heterocyclic base analogs like (a) 5-bromouracil and (b) 2-aminopurine are incorporated into replicating DNA and induce mutations by altering base-pairing characteristics.

(a) 5-Bromouracil

(b) 2-Aminopurine

can dissociate a water molecule into a hydrogen atom (H·) and a hydroxyl radical (·OH). With water so abundant in biological cells, hydroxyl radicals are readily formed. If a hydroxyl radical reacts with a DNA molecule, it can remove a hydrogen atom to reform water (H· + ·OH $\longrightarrow$ H$_2$O). However, the second product, a free radical on the DNA molecule, begins a chain reaction resulting in DNA strand cleavage. It has been estimated that strand breakage occurs as soon as a millionth of a second after ·OH attack on the DNA. This kind of DNA damage is by "indirect" processes, compared to a "direct" interaction of DNA with radiation. For most radiation sources up to 80% of the DNA damage results from the indirect process. An important source of ionizing radiation that is receiving current attention is radon gas, which in certain geographical regions of the United States is released from the soil and accumulates in homes and offices. Radon emits highly energetic alpha particles and is more damaging than other radiation sources because DNA damage occurs primarily by direct processes.

Most cells can repair damage from low levels of ionizing radiation. Single-stranded damage to DNA is readily repaired. Double-stranded DNA is more difficult for the cell to repair and may lead to cell death or cancer. The supercoiled loops of plant and animal DNA (called nucleosomes; see Chapter 10, Section 10.5) seem to be better protected than the more open DNA in bacteria.

Chemicals

Many chemicals are now known that cause genetic mutations and more will be identified in the future. **Chemical mutagens** are classified according to their mode of action. Several classes and examples of chemical mutagens are shown in Figures 11.14 through 11.16 and are discussed as follows.

1. Heterocyclic base analogs can be incorporated into replicated DNA and induce mutations by altering base-pairing characteristics. 5-Bromouracil (Figure 11.14a), an analog of thymine, becomes incorporated in DNA at sites normally taken by thymine. The keto tautomer of 5-bromouracil pairs with adenine (as thymine would do). The enol tautomer of

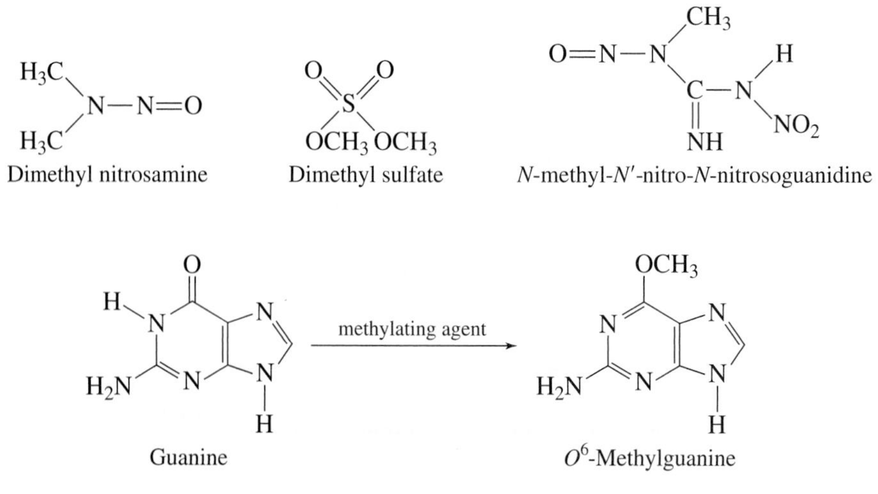

Figure 11.15

Alkylating agents as chemical mutagens. Reactive chemical mutagens change functional groups in the standard DNA bases. Reactive methylating agents shown here can convert guanine (which pairs with cytosine) to O^6-methylguanine (which pairs with thymine).

Dimethyl nitrosamine

Dimethyl sulfate

N-methyl-N′-nitro-N-nitrosoguanidine

Guanine

methylating agent

O^6-Methylguanine

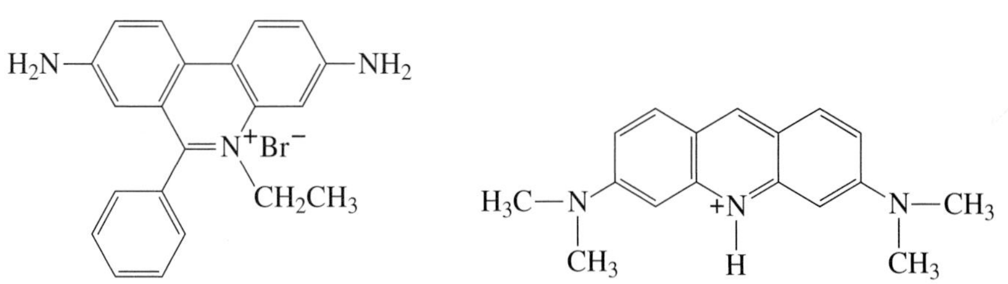

(a) Ethidium bromide   (b) Acridine orange

Figure 11.16

Intercalating agents as chemical mutagens: (a) ethidium bromide and (b) acridine orange insert between stacked base pairs in DNA.

5-bromouracil, which is quite frequently present, pairs with guanine. Thus, 5-bromouracil causes a substitution-type mutation of an A-T pair to a G-C pair. A similar change is made by the incorporation of the base analog 2-aminopurine (Figure 11.14b).

2. Reactive chemical mutagens are reagents that chemically change functional groups on the DNA bases. For example, nitrous acid (HNO_2) deaminates cytosine residues with formation of uracil. The cytosine base residue that normally forms hydrogen bonds with guanine is now changed to uracil and would form a base pair with adenine. Alkylating agents are mutagenic because they add methyl or ethyl groups to base residues, thus altering their base-pairing characteristics (Figure 11.15). Well-known alkylating reagents include nitrosamines, nitrosoguanidines, and alkyl sulfates (dimethyl sulfate). A typical reaction induced by an alkylating reagent is the methylation of guanine. The product O^6-methylguanine forms a stable base pair with thymine.

3. Intercalating agents are flat, hydrophobic, usually aromatic molecules that insert between stacked base pairs in DNA (Figures 11.16 and 11.17). This interaction increases the distance between bases and causes mutations by insertion or deletion of bases. Known intercalating agents are ethidium bromide and acridine orange. Ethidium bromide has the unique characteristic of enhanced fluorescence when intercalated into double-stranded DNA. This interaction yields a beautiful orange color and has led to the extensive use of ethidium bromide as a fluorescent dye to detect small quantities of DNA.

Exposure of human and other animal cells to mutagens often results in the initiation of tumors and other forms of cancer. Indeed, there is a strong positive correlation between exposure to mutagens and cancer. Therefore, it is important to identify those chemicals that may be mutagenic so that their use can be eliminated. Bruce Ames at the University of California at Berkeley has developed a simple, sensitive, and rapid test for chemical mutagens. The **Ames test** uses a special *Salmonella* mutant that is unable to grow in the absence of histidine. A culture plate of the bacterial strain is prepared with no histidine present. If the plate is spotted with a mutagenic reagent, many new mutant cells are produced. A small but measurable proportion of the mutants revert to a mutant that can synthesize histidine. Therefore, they grow on the plate. The extent of new mutant growth is an indication of the potency of the chemical as a mutagen. The test is completed in 2 days, compared to months or years to test mutagenic activity in animals. Of the thousands of compounds tested by this procedure, about 90% of those that give a positive Ames test (are mutagenic) are carcinogenic (cancer-causing) in animal tests.

11.4 Synthesis of RNA

DNA, the blueprint for life present in every cell, contains the information to make the 3000–4000 proteins in an *E. coli* cell or the 30,000–40,000 proteins in a human being. However, the genetic information in DNA language (A, G, C, T) must be deciphered and translated into the language of proteins (amino acids). RNA is the molecular vehicle that carries the genetic information from DNA to protein synthesis. We have already been introduced to the roles of RNA in the cell (see Chapter 2, Sections 2.2 and 2.3) and to the molecular structures of three types (tRNA, mRNA, and rRNA; see Chapter 10, Section 10.3). A catalytic role for RNA was also introduced in Chapter 7, Section 7.5. With the exception of RNA viruses, all organisms synthesize RNA that is complementary to a DNA template

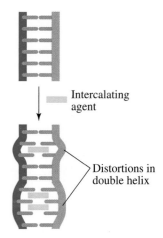

Figure 11.17

Binding of intercalating agents to DNA, which causes structural distortions.

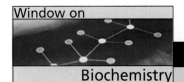

11.1   Mutagenic Effects of UV Light—No Fun In the Sun

Excessive exposure of human skin to the sun greatly increases the risk of skin cancer. Over one million cases of the disease are diagnosed each year in the United States. All three major types of skin cells—basal cells, squamous cells, and melanocytes—may be affected. **Malignant melanoma,** the most lethal skin cancer type, kills over 7000 individuals in the United States each year.

UV light is a potent mutagenic agent. Research on the interaction of UV radiation with DNA and other biomolecules is rapidly expanding. The UV rays make up just a small wavelength range of sunlight (200–400 nm); however, they are energy intense and are able to initiate deleterious biochemical reactions. The exposure of DNA to UV light results in the covalent crosslinking of adjacent pyrimidine bases (T or C) to form cyclobutane and other photoproduct dimers (see figure). Dimer formation forces an abnormal kink in the DNA double helix that causes mismatched pairing of the linked bases during replication. This creates a genetic mutation in the DNA. A majority of the mutations are C to T substitutions. Cancer may develop if repair enzymes fail to spot and correct the mistakes. *Escherichia coli* cells have an enzyme called **DNA photolyase** that, upon absorption of light, catalyzes reversal of the dimer formation. The enzyme, which has two cofactors, reduced flavin adenine dinucleotide ($FADH_2$) and folate, absorbs light in the near UV and visible range to provide energy to cleave covalent bonds between pyrimidine bases. The photolyase has not been detected in human cells, but other dimer repair enzyme systems are present in normal cells. Deficiencies of these enzymes cause the rare human skin disease called **xeroderma pigmentosum.** Individuals with this genetic disease suffer skin dryness, atrophy, keratoses, and eventual development of skin cancers upon exposure to sunlight.

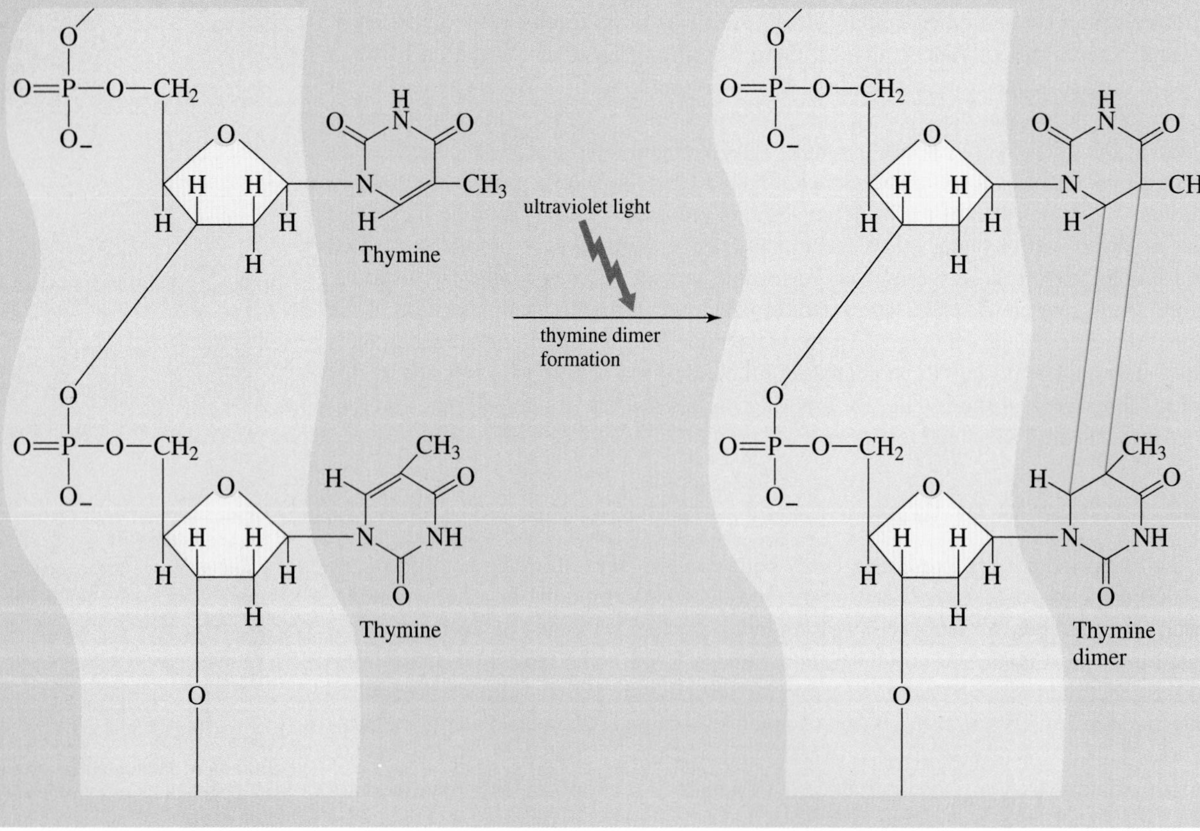

Thymine

ultraviolet light

thymine dimer formation

Thymine

Thymine dimer

(the transcription process). The newly synthesized RNA is then biochemically modified to produce an active molecule able to carry out its specific cellular function.

Before we begin our detailed look at transcription we should define terms in common use to describe genetic information flow (Figure 11.18). The strand of duplex DNA used as a template for RNA synthesis is called the **template (sense) strand** and is read in the $3' \rightarrow 5'$ direction. Therefore, the RNA product molecule, which is called a **transcript,** is synthesized in the $5' \rightarrow 3'$ direction and has a sequence complementary to the template strand. The other, noncopied DNA strand is called the coding strand because it has the same sequence as the RNA transcript except T is substituted for U. By convention, the numbering for a gene to be transcribed begins at the nitrogen base where transcription starts. The start base is numbered $+1$ and continuing bases to the right are $+2$, $+3$, and so on. The base just to the left of $+1$ is -1. There is no 0 in the numbering system. Negative numbers indicate bases upstream from the start and positive numbers are downstream. We shall describe the process of transcription in the *E. coli* cell and compare it with that in eukaryotic organisms. In addition to DNA-directed RNA synthesis, which is the most common mode, we also describe RNA-directed RNA synthesis, which occurs in RNA-based viruses. In prokaryotic cells, transcription takes place in the cytoplasm, whereas the machinery for eukaryotic transcription is in the cell nucleus.

DNA-Directed RNA Synthesis

Prokaryotic organisms have an enzyme, DNA-directed RNA polymerase (**RNA polymerase**), that directs and catalyzes all biochemical steps in the transcription process. RNA synthesis occurs in three stages: initiation, elongation, and termination. To preview the process, RNA polymerase seeks and binds to gene initiation sites on the DNA template. A short segment of double-helix DNA is unwound and the template strand is complemented in an RNA transcript. RNA polymerase brings in the appropriate ribonucleoside triphosphate and links it to an adjoining ribonucleotide by a phosphoester bond. The elongation process continues until an entire gene is transcribed. At this point, RNA polymerase detects termination sites in the DNA template, stops the synthesis process, and releases the transcript. The elongation reaction catalyzed by RNA polymerase proceeds as follows:

$$\text{NTP} + (\text{NMP})_n \xrightarrow{\text{DNA template}} (\text{NMP})_{n+1} + \text{PP}_i$$
$$\text{RNA} \qquad\qquad\qquad \text{Lengthened}$$
$$\text{RNA}$$

where

NTP = ribonucleoside triphosphates, ATP, GTP, CTP, UTP
(NMP) = preformed RNA with n or $n+1$ mononucleotides

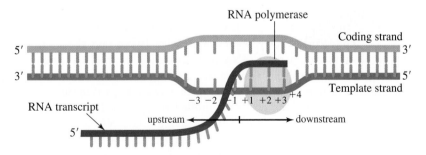

Figure 11.18

Common terms used to define the flow of information from DNA to RNA (transcription). The template strand of DNA is used for RNA synthesis. It is read in the $3' \rightarrow 5'$ direction. The RNA transcript has a base sequence complementary to the template strand but identical to the DNA coding strand. The gene region on the template strand begins at base $+1$ and proceeds downstream (increasing $+$ numbers). RNA polymerase catalyzes the synthesis of RNA transcript.

Requirements for the reaction are (1) the four ribonucleoside triphosphates, (2) a DNA template, and (3) Mg^{2+}. The new phosphodiester bond is formed by nucleophilic attack of the 3′-hydroxyl group of the growing RNA chain on the α-phosphate of the entering ribonucleotide triphosphate. The product PP_i does not accumulate since this reactive anhydride is readily hydrolyzed ($PP_i + H_2O \rightarrow 2 P_i$). We have noted similar chemistry in the process of DNA replication (see Figure 11.6).

The *E. coli* RNA polymerase is a large, complex enzyme system (465,000 daltons) containing four different kinds of subunits. The **holoenzyme,** with subunit structure $\alpha_2\beta\beta'\sigma$, initiates RNA synthesis, but once started the σ subunit dissociates, leaving the core enzyme ($\alpha_2\beta\beta'$) that contains the catalytic and substrate binding sites.

The steps of *E. coli* transcription are described below and shown in Figure 11.19.

1. Transcription of a gene begins with recognition of the promoter region and initiation of RNA synthesis. RNA polymerase holoenzyme binds to most regions of template DNA; however, the affinity of interaction increases in certain areas, called **promoter regions** (Figure 11.19a). These are short segments (20–200 base pairs; average is 40 base pairs) of DNA that are upstream from the initiation site (+1). The entire *E. coli* DNA molecule of 4.7×10^6 base pairs has approximately 2000 such promoter sites. There are two important subregions within a promoter region that are found in all prokaryotic DNA; (1) the −10 region (also called the Pribnow box), which has the common sequence TATAAT, and (2) the −35 region, which has the sequence TTGACA. RNA polymerase holoenzyme binds nonspecifically to DNA template and migrates downstream (3′ → 5′ of template strand) until it recognizes a promoter region. Stronger binding in the promoter sequence, especially in the −10 region, causes slower movement of the RNA polymerase and results in unwinding of the DNA double helix. Approximately 12 to 15 base pairs are separated, leading to the formation of a transcription bubble. Synthesis of the RNA begins with entry of the first ribonucleotide selected by hydrogen bonding to the DNA template and binding to the RNA polymerase active site. This incoming base is complementary to the +1 base of DNA. The next ribonucleotide, complementary to +2 base, enters in a like manner. RNA polymerase catalyzes the formation of a phosphodiester bond between the first two ribonucleotides. Two important characteristics of this process should be noted: (1) No RNA primer is required as in DNA synthesis and (2) a 5′-triphosphate tail remains on the first ribonucleotide. Regulation of the initiation process by catabolite gene activation protein (CAP) and protein repressors will be discussed in Chapter 12, Section 12.4.

2. Elongation of the RNA chain proceeds in the 5′ → 3′ direction. RNA polymerase continues to move along the DNA template, adding ribonucleotides to the growing RNA chain. After approximately ten bases have been added, the σ subunit of RNA polymerase dissociates, leaving the core enzyme. Apparently the σ subunit is required for recognition of promoter regions and initiation of synthesis but is not essential for elongation. Elongation catalyzed by core RNA polymerase continues, producing RNA transcript complementary to the DNA template strand. A short stretch of RNA–DNA transient hybrid is a characteristic of this synthetic period (Figure 11.19b). RNA polymerase catalyzes elongation of the chain with a high degree of accuracy. Insertion of a mismatched base occurs about once for every 10^4 to 10^5 nucleotides incorporated. RNA polymerase, unlike the DNA polymerases, is unable to recognize and correct errors. Because many copies of RNA are made, an error is of less consequence than in DNA replication, where only one copy is made.

3. Termination of RNA synthesis is controlled by specific DNA sequences. RNA polymerase core enzyme continues to move along the DNA template until base sequences signaling termination are encountered (Figure 11.19c). Two termination methods in *E. coli* have been identified and studied. One type requires an accessory termination factor called rho (ρ) protein; the other type is ρ protein independent. The ρ protein–independent type relies on a specific DNA base sequence that ends the RNA transcript with a stable hairpin structure. This stop signal in DNA begins with a GC-rich region, followed by an AT-rich region and a poly A region. The self-complementary residues in the transcript fold into a hairpin loop, ending with a poly U sequence at

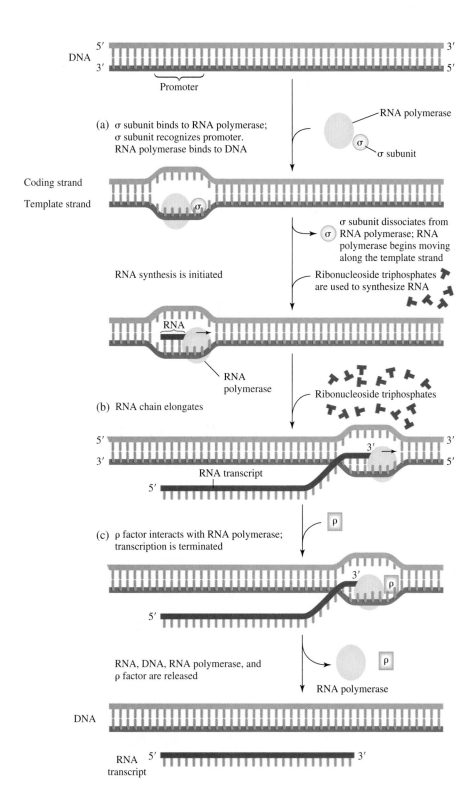

Figure 11.19
Mechanism of action of DNA-directed RNA polymerase in phosphoester bond formation in *E. coli:* (a) Initiation of transcription; (b) elongation of the RNA chain in transcription; (c) termination of transcription.

DNA 5′ 3′ / 3′ 5′

Promoter

(a) σ subunit binds to RNA polymerase; σ subunit recognizes promoter. RNA polymerase binds to DNA

RNA polymerase

σ subunit

Coding strand

Template strand

σ

σ subunit dissociates from RNA polymerase; RNA polymerase begins moving along the template strand

RNA synthesis is initiated

Ribonucleoside triphosphates are used to synthesize RNA

RNA

RNA polymerase

(b) RNA chain elongates

Ribonucleoside triphosphates

5′ 3′ / 3′ 5′

RNA transcript

5′ 3′

(c) ρ factor interacts with RNA polymerase; transcription is terminated

ρ

3′

5′ ρ

RNA, DNA, RNA polymerase, and ρ factor are released

ρ

RNA polymerase

DNA

RNA transcript 5′ 3′

the 3′ end (Figure 11.20). The ρ protein acts by disrupting the RNA–DNA hybrid, thereby dissociating the transcript from the template and terminating synthesis.

Eukaryotic cells differ in the transcription process by having three classes of RNA polymerases (compared to one in prokaryotic cells). Each eukaryotic polymerase (designated I, II, III) has a specific role in RNA synthesis. Polymerase I transcribes large ribosomal RNA genes; II transcribes protein-encoding genes; III transcribes small RNAs, including tRNA and 5S rRNA.

Figure 11.20

Folding of RNA transcript into a hairpin loop to terminate synthesis.

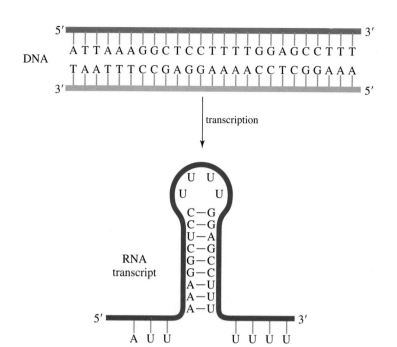

Figure 11.21

Inhibitors of transcription: (a) Actinomycin D inhibits elongation by intercalation into the DNA. (b) α-Amanitin, a toxin in mushrooms, inhibits formation of mRNA.

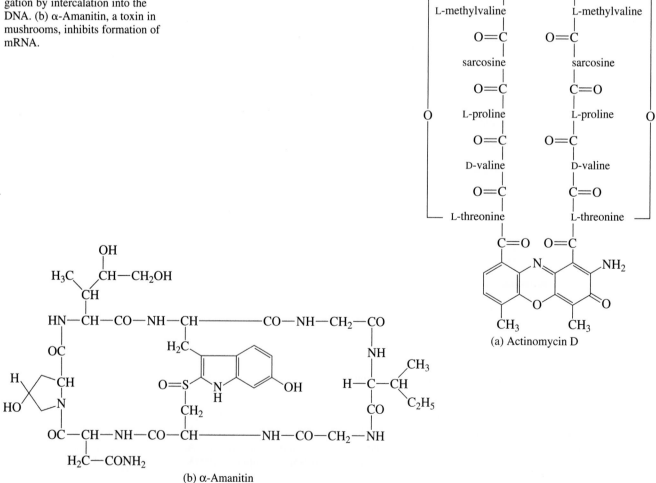

(a) Actinomycin D

(b) α-Amanitin

Figure 11.22

Mechanism of RNA-directed RNA polymerase. The 3′-hydroxyl group (in red) of the RNA primer strand attacks at the α-phosphate of the entering ribonucleoside triphosphate to form a new phosphoester bond. Pyrophosphate (in green) is produced by the cleavage of the phosphoanhydride bond.

The mechanistic details of transcription have been elucidated by the use of specific inhibitors (Figure 11.21). **Actinomycin D,** an antibiotic from a strain of *Streptomyces,* inhibits RNA elongation in prokaryotic and eukaryotic organisms by intercalation into the DNA template thereby blocking transcription. RNA synthesis in animal cells is inhibited by **α-amanitin,** a toxic compound in the *Amanita* species of mushrooms that causes the death of about 100 persons each year. α-Amanitin binds to RNA polymerase II blocking formation of mRNA.

RNA-Directed RNA Synthesis

An alternate mode of RNA synthesis is found in RNA viruses, those that have an RNA genome. Well-studied RNA viruses include the *E. coli* bacteriophages, Qβ, MS2, TMV, and R17. These viruses induce the formation of the enzyme RNA-directed RNA polymerase **(RNA replicase)** in the host cell. The reaction catalyzed by RNA replicase has characteristics similar to the reaction catalyzed by DNA-directed RNA polymerase:

1. Direction of synthesis is 5′ ⟶ 3′.
2. The mechanism is the same with nucleophilic attack on the incoming ribonucleoside triphosphate by the 3′ end hydroxyl and release of pyrophosphate (Figure 11.22).
3. The RNA transcript is complementary to the RNA template.
4. No editing, proofreading, or repair activities are associated with RNA replicase.

The major difference between the DNA-directed and RNA-directed RNA syntheses is the requirement for a single-stranded RNA template by RNA replicase.

11.5 Post-transcriptional Modification of RNA

Newly synthesized RNA molecules, called primary transcripts, are usually biologically inactive. They must be processed into mature, biologically functional molecules. The actual steps required in post-transcriptional modification of the primary transcript depend on the type of cell and kind of RNA.

tRNA and rRNA Processing

Eukaryotic tRNA molecules such as in yeast are biochemically altered by four processes: (1) trimming of the ends by phosphoester bond cleavage, (2) splicing to remove an intron,

This *Amanita muscaria* (var. *Formosa*) mushroom contains the toxin α-amanitin.

© Matt Meadows/Peter Arnold, Inc.

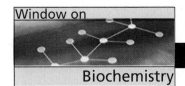

11.2   RNA Drugs—Making Sense Out of Antisense

In the transcription of DNA, RNA polymerase catalyzes the synthesis of mRNA molecules with a sequence complementary to the template (sense) strand (see Figure 11.18). The deoxyribonucleotide sequence of the coding (antisense) strand is identical to that of the mRNA being synthesized except T residues are in DNA where U residues are in RNA. A new RNA molecule transcribed from the coding strand would be complementary to the RNA (transcript) synthesized by RNA polymerase using the template strand. Because the antisense strand is used to make the new RNA, it is called **antisense RNA.** The two RNA molecules (transcript or sense, and antisense) are complementary and should bind together to form double-stranded hybrids. This specific and selective interaction between transcript and antisense RNA has the potential for application to the treatment of cancer and other diseases. The argument to support this prediction goes like this: It is often possible to identify a protein whose presence either initiates or enhances a disease. That protein's action can be prevented by interfering with its synthesis. Perhaps that can be done with a strip of antisense RNA that will bind to the transcript mRNA that carries the message for the protein's production. Antisense RNA drugs developed so far interfere by binding strongly to the targeted transcript mRNA. Most current antisense RNA drugs are relatively small oligonucleotides from 17 to 25 ribonucleotides long. There are two barriers that still need to be overcome before antisense drugs can complete in the mainstream of other pharmaceuticals: (1) Normal RNA molecules are negatively charged and do not readily pass through cell membranes. Antisense molecules that have nonionic backbones are currently being synthesized. (2) RNA molecules are rapidly degraded by nucleases. To slow down metabolic degradation, the new RNA drugs are modified by replacing one of the phosphate oxygens in each backbone by sulfur. One antisense compound currently in clinical trials is GEM 92, produced by Hybridon. It is a 25-ribonucleotide compound that targets the mRNA of the *gag* gene of HIV. The *gag* gene product is essential for viral replication. A space-filling model of Hybridon's GEM 92 binding to a complementary portion of the target mRNA is shown below.

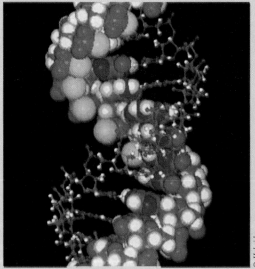

Space-filling model of Hybridon's anti-AIDS antisense drug, GEM, 92 forms a double helix with a complementary portion of the target messenger RNA.

(3) addition of terminal sequences, and (4) heterocyclic base modification, usually methylation. Prokaryotic tRNA undergoes fewer of the above modification steps: Some tRNAs have introns that must be removed; some prokaryotic tRNAs have terminal sequences added and some heterocyclic and ribose rings are modified. In *E. coli,* an endonuclease called ribonuclease P contains a catalytically active RNA molecule (see Chapter 7, Section 5) and catalyzes removal of a section from the 5′ end of pre-tRNA by a hydrolytic cleavage (Figure 11.23).

rRNA molecules in prokaryotes and eukaryotes are processed by trimming to the appropriate sizes and methylation of bases.

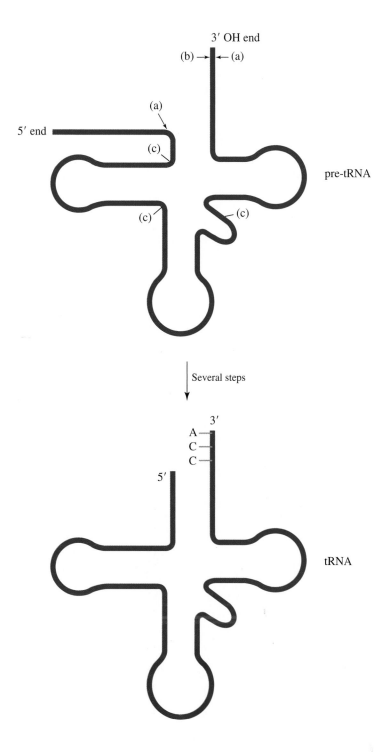

Figure 11.23

Post-transcriptional modification of tRNA in *E. coli*: (a) Ribonuclease P catalyzes the hydrolysis of a phosphoester bond near the 5′ end of pre-tRNA. This releases a small section of RNA. An endonuclease catalyzes the removal of a small section from the 3′ end of the tRNA. (b) CCA is added to the 3′ end. (c) Some bases are chemically modified.

mRNA Processing

Most prokaryotic mRNA requires little or no post-transcriptional alteration before it is translated into protein structures. In fact, it is sometimes translated while it is being transcribed. In contrast, eukaryotic mRNA molecules are chemically modified in the nucleus before transport to the ribosomes for translation. Three biochemical processes occur for most eukaryotic primary mRNA transcripts (sometimes called pre-mRNAs). These enzyme-catalyzed changes include capping at the 5′ end, addition of polyadenylate tails at the 3′ end, and splicing.

Capping

Almost immediately after synthesis, the 5′ end of the mRNA is modified by hydrolytic removal of a phosphate from the triphosphate functional group. Next, guanosine triphosphate (GTP) is used to attach a GMP residue to the 5′ end, resulting in an unusual 5′–5′-triphosphate covalent linkage (Figure 11.24). The end guanine residue is methylated at N7 to complete one kind of 5′ cap. Additional capping reactions may include methylation at ribose hydroxyl groups. Capping is probably necessary for marking mature mRNAs and to protect against cleavage catalyzed by exonucleases.

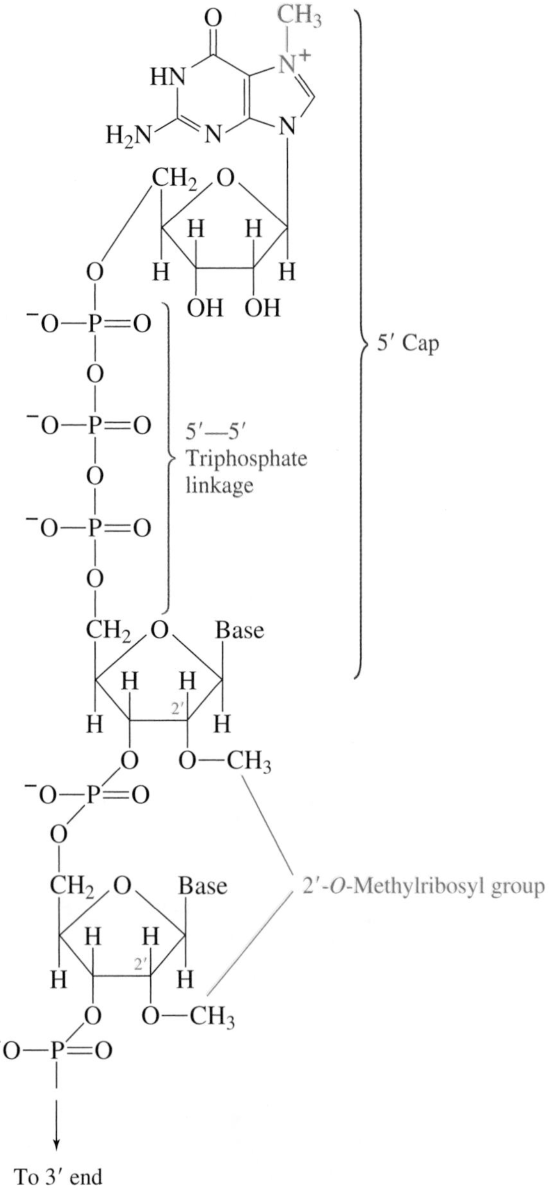

Figure 11.24

Capping reactions important in processing mRNA. Two capping processes occur: attachment of a GMP residue at the 5′ end and methylation at N7 (in *red*) of guanine and at the ribose 2′ hydroxyl groups.

Poly A Addition

The 3′ end of mRNAs is modified by the addition of a polyadenylate (poly A) tail. Most mRNA molecules have a 3′ poly A tail 20 to 250 nucleotide residues long. Before the poly A tail is added, an endonuclease catalyzes removal of a few 3′ base residues from the transcript. Addition of adenine residues is catalyzed by polyadenylate polymerase. The poly A tail is thought to stabilize the mRNA by increasing resistance to cellular nucleases.

Splicing of Coding Sequences

We initially introduced the concept of discontinuous genes in eukaryotes. Coding regions on the gene, called **exons,** are interrupted by noncoding regions, called intervening sequences or **introns.** The final form of mRNA used for translation is the result of extensive and complicated chemical processing events called splicing. Introns are cut from the gene and exon fragments are spliced or joined to produce the mature, functional mRNA. In chemical terms, these processes require the cleavage and reformation of phosphodiester bonds (transesterification). A summary of mRNA processing using the chicken ovalbumin gene is shown in Figure 11.25. The ovalbumin gene DNA consists of approximately 7700 base pairs. The primary transcript, which contains approximately the same number of base pairs, is first capped at the 5′ end. The 3′ poly A tail is added after a short segment of "extra RNA" is excised. Removal of introns by phosphoester bond cleavage and splicing of exons produces the mature mRNA, which has only 1872 base pairs not including the poly A tail. The mature mRNA codes for ovalbumin, which has just over 600 amino acids.

The splicing events in post-transcriptional processing of mRNA involve complicated chemical reactions that have added a whole new dimension to biochemistry in the last few years. Many of the enzymes required for cleavage of phosphoester bonds at exon–intron boundaries are the usual protein type; however, some of the reactions involve autocatalytic or self-splicing processes. That is, some RNA molecules catalyze their own processing (splicing) without the assistance of proteins. These **catalytic RNA** molecules, also called **ribozymes,** were introduced in Chapter 7, Section 7.5. Figure 7.12 outlined the self-catalyzed processing of RNA from *Tetrahymena.* In chemical terms, splicing by RNA involves transesterification reactions. Small nuclear RNA molecules combine with specific proteins to form snRNPs (see Chapter 10, Section 10.5) that catalyze some splicing reactions in the nucleus.

Cutting Edge: Ribozymes

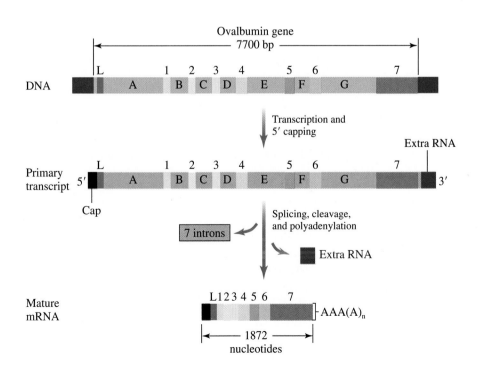

Figure 11.25

Processing of the chicken ovalbumin gene transcript. Modification of the primary transcript begins by capping at the 5′ end. At the 3′ end, a small section of RNA is removed and the 3′ end capped with a poly A tail. Splicing processes remove introns A to G and combine exons 1 to 7 to produce the mature mRNA. L represents an untranslated leader sequence.

11.6   Base Sequences in DNA

When it was determined in the 1950s that genetic information in DNA was coded in the form of nucleotide base sequence, a primary focus of research shifted to the development of experimental procedures for determining the base order in DNA. Early sequence analysis of nucleic acids was done using acid-, base-, and nuclease-catalyzed hydrolysis (see Chapter 10, Section 10.4). These techniques, applied even to short polynucleotides, were complicated, labor intensive, and difficult to reproduce. Since essentially every experiment done with replication and transcription depended on knowing DNA and RNA sequences, the development of rapid and accurate sequence techniques was imperative. Two methods that rely on relatively simple procedures and inexpensive equipment are now available. Both methods are the result of advances in two areas:

1. Development of new chemical cleavage and enzymatic reactions for nucleic acids that generate a predictable series of fragments of varying sizes.
2. Discovery of new electrophoretic techniques that allow separation of nucleic acid fragments that differ in size by only one nucleotide.

The Maxam–Gilbert chemical cleavage method of sequencing was developed at Harvard University. Specific bases in the target DNA are chemically modified, which promotes cleavage at selected phosphodiester linkages. Cleavage products, which have one common end and vary in length, are analyzed by polyacrylamide gel electrophoresis (see Window on Biochemistry 4.2).

The Sanger chain-termination sequencing method was developed by Frederick Sanger at Cambridge University. (This is the same Sanger who developed the protein sequencing method discussed in Chapter 5, Section 5.6.) This more commonly used procedure for DNA sequencing requires the use of:

1. Preformed DNA template (strand to be sequenced) and short primer strand at the 3′ end.
2. DNA polymerase.
3. The four standard deoxyribonucleoside triphosphates (dATP, dGTP, dCTP, dTTP). One is radioactively labeled, usually ^{32}P-dATP.
4. Four 2′,3′-dideoxyribonucleoside triphosphates (ddATP, ddGTP, ddCTP, ddTTP) (the ddNTPs). (Figure 11.26).

Four individual reaction tubes are prepared, all containing reagents 1 to 3 above (Figure 11.27). One ddNTP, dideoxyribonucleoside triphosphate, is added to each reaction; that is, reaction A contains only ddATP, reaction G only ddGTP, and so on. When all reagents are present in a reaction mixture, DNA synthesis of a strand complementary to the template begins at the 3′ end of the primer. The radioactive dATP is added to label the DNA fragments synthesized so they can be detected by autoradiography after separation by electrophoresis. The key to the Sanger method is the presence of the ddNTPs. Because these compounds have base structures identical to the standard dNTPs, they are incorporated into the synthesized strand. However, with no 3′-hydroxyl group, synthesis is terminated when a ddNTP is incorporated into the newly synthesized DNA strand. Since there is a mix of dNTPs and ddNTPs, termination does not occur at every elongation step. The ratio of dNTP to ddNTP in each reaction mixture is carefully adjusted so that a ddNTP is incorporated one time for every 100 nucleotides incorporated (1% of the time a ddNTP is added; 99% of the time a dNTP is added).

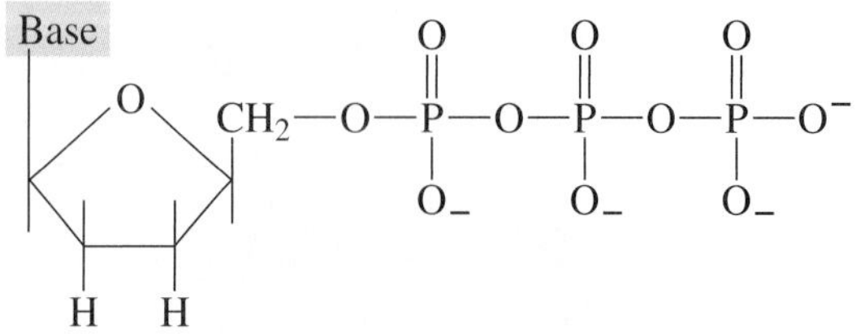

Figure 11.26

Four 2′,3′-dideoxyribonucleoside triphosphates (ddATP, ddGTP, ddCTP, ddTTP) are used to terminate the reactions in the Sanger chain-termination sequencing method.

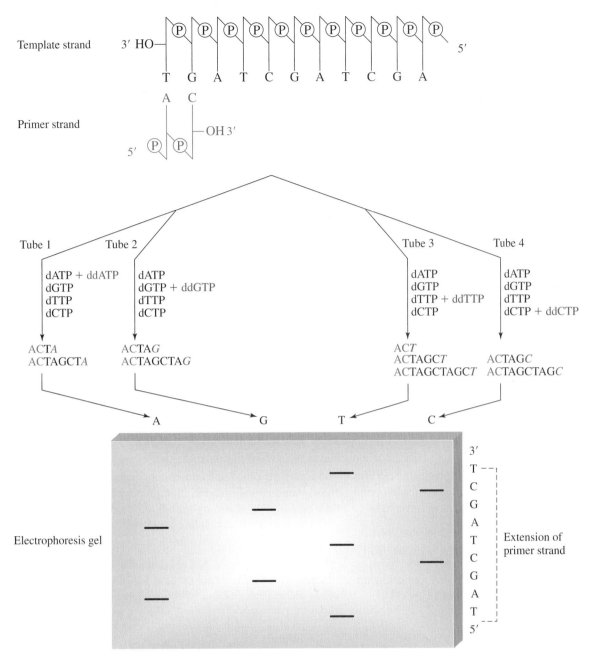

Template strand   3′ HO—

Primer strand

Tube 1

dATP + ddATP
dGTP
dTTP
dCTP

ACT*A*
ACTAGCT*A*

Tube 2

dATP
dGTP + ddGTP
dTTP
dCTP

ACTA*G*
ACTAGCTA*G*

Tube 3

dATP
dGTP
dTTP + ddTTP
dCTP

AC*T*
ACTAGC*T*
ACTAGCTAGC*T*

Tube 4

dATP
dGTP
dTTP
dCTP + ddCTP

ACTAG*C*
ACTAGCTAG*C*

A   G   T   C

Electrophoresis gel

3′
T
C
G
A
T   Extension of
C   primer strand
G
A
T
5′

Resulting X-ray image of primer strand sequence

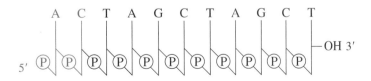

Figure 11.27

Sequence analysis of DNA by the Sanger chain-termination method. Each tube contains the four standard deoxyribonucleoside triphosphates and one of four dideoxyribonucleoside triphosphates (in *blue*). DNA synthesis begins at 3′ of the primer strand (in *red*). Incorporation of dideoxyribonucleotides terminates DNA synthesis and produces DNA fragments analyzed by electrophoresis.

To understand the basis of the Sanger method, let's focus on one reaction mixture. Mixture G contains preformed DNA, DNA polymerase, four dNTPs, and ddGTP. Every time the polymerase encounters a C on the template, a G is inserted into the growing complementary strand. For every 100 guanine residues incorporated, 99 are from dGTP, thus allowing synthesis to continue. However, one reaction out of 100 will result in the incorporation of ddGTP, which terminates elongation. Because there are millions of DNA template molecules in the reaction mixture, G will contain a family of DNA fragments that end with 3′ dideoxyguanine (3′ ddG) and have varying lengths. Consider how many fragments are produced for a DNA template. Assume a newly synthesized oligonucleotide (complementary to template) has the sequence 3′-TAAGCAAGT. In this reaction, two fragments are generated, one with four nucleotides and one with eight nucleotides. (Write the sequence for each fragment.)

The reactions in the four tubes are quenched by the addition of a chemical reagent (denaturing agent) that dissociates the synthesized fragments from the template. Electrophoresis of each reaction mixture in polyacrylamide gels separates the fragments according to size, with the smaller fragments moving more rapidly to the bottom of the gel. Because the fragments contain radioactively labeled ^{32}P, the gel can be placed on an X-ray film to determine the position of fragments. The base order of the complementary, newly synthesized strand of DNA is read in sequence from the bottom to the top of the gel.

Summary

■ Transfer of genetic information from one generation to another requires the exact duplication of the DNA molecule. The duplication process, called **replication,** is semiconservative; that is, the new strand is composed of an original strand and a newly synthesized one.

■ The replication process, which is performed by the enzyme DNA polymerase III, requires the presence of preformed DNA to serve as a template and primer.

■ Errors called mutations are classified as spontaneous (those that occur during normal function) and induced (those caused by environmental agents such as chemicals, radioactivity, or UV light). Most errors committed during DNA replication are recognized and repaired by special enzymes.

■ The process of RNA synthesis is called **transcription.** RNA may be synthesized from DNA templates using the enzyme DNA-directed RNA polymerase or from RNA templates using viral RNA-directed RNA polymerase.

■ Primary transcripts of RNA are usually biologically inactive and must be processed into mature functional molecules by post-transcriptional modification processes. tRNA and rRNA are processed by trimming of ends, addition of base sequences, base modification, and splicing. mRNA is processed by capping, poly A addition, and splicing.

■ Base sequencing of DNA is usually carried out by the Sanger chain-termination method using dideoxyribonucleoside triphosphates as terminating reagents.

Study Exercises
Understanding Terms

11.1 Define the following terms in 25 words or less.
- **a.** Semiconservative replication
- **b.** DNA polymerase I
- **c.** Leading strand
- **d.** Lagging strand
- **e.** Helicase
- **f.** Mutation
- **g.** Alkylating agents
- **h.** Intercalation
- **i.** RNA polymerase holoenzyme
- **j.** ρ(rho) protein
- **k.** Capping
- **l.** Splicing
- **m.** Intron
- **n.** Exon
- **o.** DNA polymerases II and III
- **p.** Template strand
- **q.** Coding strand
- **r.** Antisense RNA
- **s.** Sanger base sequencing

Reviewing Concepts

11.2 Before the Meselson–Stahl experiments were done to determine the mode of DNA replication, two possible mechanistic options were considered: conservative or semiconservative. Assume that the actual mechanism is conservative. Using the same type of illustration as Figure 11.2, draw the centrifugation results obtained from such a process.

11.3 Assume that the following polynucleotide is part of a longer DNA molecule. Write out the product that would be obtained from its replication.

3′ AAGCTTTCCG

11.4 Many enzymes and other protein factors are required for the replication process. Briefly describe the role for each of the following in replication.

a. Helicase
b. SSB proteins
c. DNA ligase
d. Primase
e. DNA polymerase III
f. DNA polymerase I

11.5 What is the RNA product obtained from the transcription of this DNA template strand?

3′ GGCATATCG

11.6 What is the RNA product obtained from transcription of the DNA in Problem 11.5 if it is assumed to be the coding strand?

11.7 Compare DNA polymerase III and RNA polymerase in terms of the following characteristics.

a. Need for primer
b. Direction of new chain elongation
c. Proofreading functions
d. Form and identity of entering nucleotides

11.8 Below is written a short strand of double-helix DNA. The start of RNA transcription is shown with the number +1. Identify each of the following characteristics.

$$\begin{array}{c} +1 \\ 3'\ \text{AGTACGCAAGTT} \\ 5'\ \text{TCATGCGTTCAA} \end{array}$$

a. Template strand
b. Coding strand
c. Identity of base number +3
d. Identity of base number −4
e. What is the identity of the base five nucleotides upstream from +1?
f. What is the base two nucleotides downstream from +1?

11.9 Compare DNA-directed RNA synthesis and RNA-directed RNA synthesis in terms of the following characteristics.

a. Major elongation enzyme
b. Direction for new chain elongation
c. Form of entering nucleotides
d. Mechanism of phosphoester bond formation
e. Proofreading
f. Template
g. Need for primer

11.10 Briefly describe each type of mutation listed below. Your description should clarify any differences and identify examples of causes if known.

a. Lethal mutation
b. Silent mutation
c. Spontaneous mutation
d. Induced mutation

11.11 Briefly describe the role of each of the following proteins in the transcription process.

a. σ subunit
b. ρ protein
c. RNA polymerase

11.12 List the functional differences between DNA and RNA.

11.13 Explain the difference between *continuous* and *discontinuous* DNA replication.

11.14 What is the reaction catalyzed by telomerase?

11.15 Which of the following features are characteristic for the process of DNA transcription?

a. Conservative
b. Semiconservative
c. Phosphodiester bonds are formed
d. Catalyzed by polymerases
e. Final product is RNA
f. Newly formed polymers are complementary to a template
g. Final product is a polypeptide

Solving Problems

11.16 The compounds whose structures are drawn below are known to be mutagenic. Predict how each acts as a mutagen.

a. CH_3CH_2
   $\diagdown$
   $N{-}N{=}O$
   $\diagup$
   CH_3CH_2
   Diethylnitrosamide

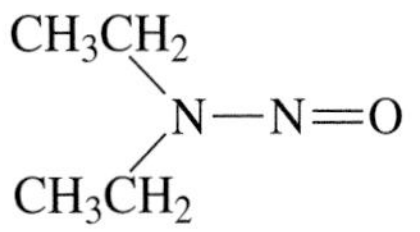

b. 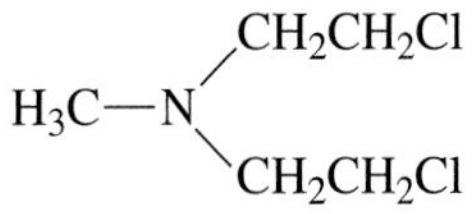
   $\phantom{H_3C{-}N}CH_2CH_2Cl$
   $\phantom{H_3C{-}N}\diagup$
   $H_3C{-}N$
   $\phantom{H_3C{-}N}\diagdown$
   $\phantom{H_3C{-}N}CH_2CH_2Cl$
   Nitrogen mustard

c.

   3,4-Benzpyrene

d.

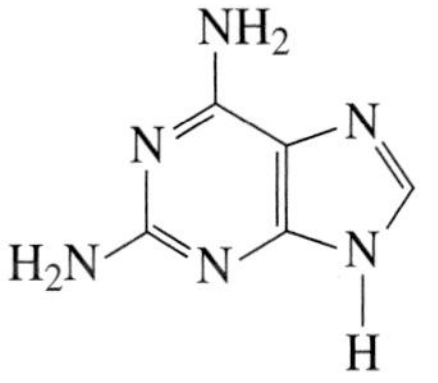

   2,6-Diaminopurine

11.17 Show how the base pairing properties of 2-amino-purine differ in the keto and enol forms.

11.18 Cordycepin, also called 3'-deoxyadenosine, is an inhibitor of prokaryotic transcription. Draw its structure and try to predict how it acts as an inhibitor.

HINT: Compare to the structure for adenosine.

11.19 5-Bromouracil is a base analog mutagen that is incorporated into synthesized DNA in place of thymine. Explain how this substitution could occur.

11.20 What is the relationship in terms of base sequence between the RNA transcript and the DNA coding strand?

11.21 Classify the mutagenic agents below as spontaneous or induced.

   a. Errors in replication
   b. Intercalculating agents
   c. UV light
   d. Removal of nucleotide bases by hydrolysis
   e. Radon gas

11.22 Draw the structure of the AIDS drug, 3'-azido-2',3'-dideoxythymidine (AZT) and explain how it blocks transcription.

HINT: Compare the structure of AZT to thymidine.

11.23 Complete the following reactions.

   a.

   b. $PP_i + H_2O \xrightarrow{\text{pyrophosphatase}}$

   c. H_2O + ionizing radiation $\longrightarrow$

11.24 Describe how the Ames test saves time and money in detecting chemicals that are potential carcinogens.

11.25 Which of the following features are characteristic for the process of DNA replication?

   a. Conservative
   b. Semiconservative
   c. Catalyzed by nucleases
   d. Final product is DNA
   e. Final product is RNA
   f. Newly formed polymers are complementary to a template
   g. Phosphodiester linkages are formed
   h. Catalyzed by polymerases

11.26 Describe the molecular interactions that hold together a single strand of DNA and a single strand of RNA in a hybrid during the transcription process. Are the interactions covalent or noncovalent or both?

11.27 Which of the following features are characteristic of ribozymes?

   a. Are related to ribosomes
   b. Are DNA molecules
   c. Are RNA molecules
   d. Involved in post-transcriptional processing
   e. Catalysts like enzymes
   f. Cleavage of RNA

11.28 Explain the possible link between telomerase activity and cancer.

11.29 Show how a single RNA molecule can have base paired regions.

11.30 A chemical, X, gives a positive result when studied in the Ames test. Is it valid to state categorically that X is a human carcinogen? Explain.

Writing Biochemistry

11.31 Read the section on telomerase on page 282 of this book and study the CD/ROM unit on telomers. Use the information to write a 300-word paper on the possibility that inhibitors of telomerase may be effective anti-cancer agents. Begin the paper with an introduction to telomers and telomerase and continue by describing how an inhibitor could be designed. Use articles in the Further Reading section or search the Internet to find the structure of a possible telomerase inhibitor.

11.32 Write a 200-word paper describing how 5-bromouracil acts as a mutagenic agent. Define and draw the enol and keto forms of the chemical and explain their action on DNA replication.

Applying Biochemistry

11.33 In this chapter, you were introduced to the Sanger chain-termination DNA sequencing method (dideoxy method). In the CD/ROM unit, "Shotgun Sequencing," another procedure for DNA sequencing was described. Compare the two sequencing methods in terms of experimental procedure, final results, and limitations.

11.34 The structure, synthesis, and functional role of tRNA were introduced in this chapter and in the CD/ROM unit on "tRNA." tRNAs are called "adaptor molecules." They read the message of nucleotide sequence in mRNA and convert it into the language of amino acids in proteins. Explain in molecular terms how tRNAs carry out this translation process.

Further Reading

Apirion, D., and Miczak, A., 1993. RNA processing in prokaryotic cells. *BioEssays* 15:113–120.

Collins, F. et al., 2001. The human genome. *Nature* 409:745–748.

Greider, C., and Blackburn, E., 1996. Telomeres, telomerase, and cancer. *Sci. Am.* 274(2):80–85.

Hubscher, U., Nasheuer, H., and Syvaoja, J., 2000. Eukaryotic DNA polymerases, a growing family. *Trends Biochem. Sci.* 25:143–147.

Koshland, D., 1994. The DNA repair enzymes. *Science* 266:1925–1928.

Leffell, D., and Brash, D., 1996. Sunlight and skin cancer. *Sci. Am.* 275(1):52–59.

Pindur, U., Haber, M., and Sattler, K., 1993. Antitumor active drugs as intercalators of DNA. *J. Chem. Educ.* 70:263–272.

Proudfoot, N., 2000. Connecting transcription to messenger RNA processing. *Trends Biochem. Sci.* 25:290–293.

Santesmases, M., 2001. Severo Ochoa (1905–1993): the changing world of biochemistry. *Trends Biochem. Sci.* 26:140–142.

Shore, D., 1997. Telomerase and telomere-binding proteins: controlling the endgame. *Trends Biochem. Sci.* 22:233–235.

Soultanas, P., and Wigley, D., 2001. Unwinding the 'Gordian knot' of helicase action. *Trends Biochem. Sci.* 26:47–54.

Sousa, R., 1996. Structural and mechanistic relationships between nucleic acid polymerases. *Trends Biochem. Sci.* 21:186–190.

Tanaka, K., and Wood, R., 1994. Xeroderma pigmentosum and nucleotide excision repair of DNA. *Trends Biochem. Sci.* 19:83–86.

Venter, J., 2001. The sequence of the human genome. *Science* 291:1304.

Verdine, G., and Bruner, S., 1997. How do DNA repair proteins locate damaged bases in the genome? *Chem. Biol.* 4:329–334.

Wallace, D., 1997. Mitochondrial DNA in aging and disease. *Sci. Am.* 277(2):40–47.

Watson, J., 1968. The Double Helix. New York: Atheneum Publishers.

Zimm, B., 1999. One chromosome: one DNA molecule. *Trends Biochem. Sci.* 24:121–123.

Chapter 12

Translation of RNA: The Genetic Code and Protein Metabolism

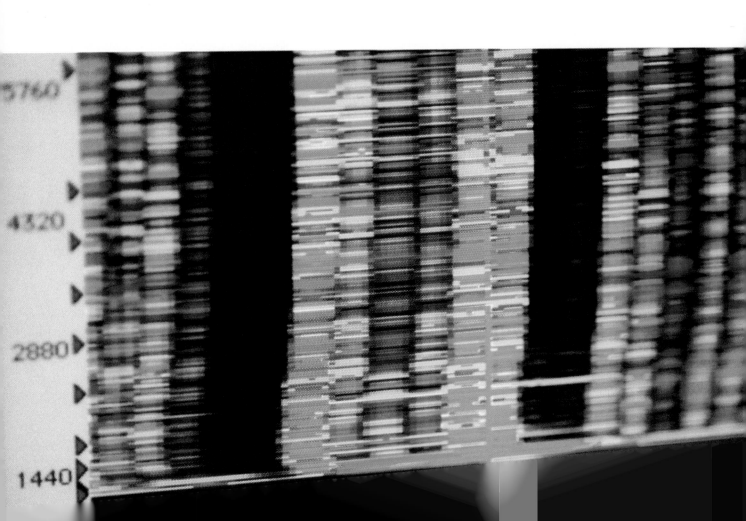

We turn now to the final step in the flow of genetic information, the synthesis of proteins. We have previewed the transfer pathway, DNA → RNA → proteins; we began our detailed journey with DNA, the storehouse of all genetic information. The term "replication" describes the duplication of DNA for future generations of cells. The message in DNA is in the form of a linear sequence of four nucleotide bases (ATGC). Within a cell the 4-letter language of DNA is transcribed into a similar language for mRNA (AUGC). In the transcription process we have noted a one to one correspondence between the nucleotide bases. This correspondence has its origin in the formation of hydrogen bonded complementary base pairs (A-U, A-T, G-C). The final step, transfer of sequential information in mRNA to the amino acid sequence of proteins, is much more complicated. A **translation** process is required to convert the 4-letter language of RNA into the 20-letter language of protein molecules (amino acids). Amino acids and nucleotide bases are quite different chemically, and no direct biochemical correlations or correspondence have been observed. We describe in this chapter how Francis Crick in 1958 suggested that some kind of "adaptor" molecules carried amino acids to the mRNA template for protein synthesis. Those molecules that act as translators are now identified as tRNAs. These small nucleic acids are linked together with the appropriate amino acid.

Other important concepts of protein synthesis introduced in this chapter include **ribosomes,** the RNA molecular machines that make proteins, the workings of the **genetic code,** the chemistry and mechanics of protein synthesis, and **post-translational modification** of protein products.

The chapter concludes with a description of the regulation of protein synthesis and gene expression. It would be wasteful for a cell to synthesize all proteins at all times. Only a small percentage of the cell's proteins are required continuously. The cell uses intricate mechanisms to regulate and coordinate protein concentration.

12.1 The Process of Protein Synthesis

Research on protein synthesis, which began in earnest in the 1950s, was spurred by the excitement of related discoveries, including the DNA double helix and DNA replication and transcription. The early experiments on protein synthesis uncovered fundamental concepts that appeared to be universal in all organisms studied. We begin our study of protein synthesis with a review of these characteristics:

Characteristics of Protein Synthesis

Protein Synthesis Occurs on Ribosomal Particles

Paul Zamecnik and his colleagues in the United States attempted to study the cellular location of protein synthesis by injecting rats with radioactive amino acids. At various time intervals, livers of the rats were removed, homogenized, centrifuged to separate organelles, and analyzed for radioactivity. Within minutes from injection, it was found that the radioactivity accumulated on cellular particles composed of ribonucleic acids and proteins. Zamecnik suggested that these particles were the sites of protein synthesis. Subsequent studies of the ribonucleoprotein particles, which we now call **ribosomes,** showed that they were present in the cytoplasm of all types of cells as well as in the matrix of mitochondria and the stroma of plant chloroplasts. The *Escherichia coli* ribosomes, which are representative of those in prokaryotic cells, have a diameter of 25 nm and a mass of about 2500 kilodaltons. Approximately 15,000 ribosomes are present in a single *E. coli* cell. Each one is composed of two subunits of unequal size that dissociate from each other in solutions of low Mg^{2+} concentration. The sizes of intact ribosomes, their individual subunits, and their component biomolecules are correlated with their **sedimentation coefficients** *(S)*. This is a measure of how rapidly a molecule or particle sediments during gradient centrifugation (see Window on Biochemistry 1.3; Chapter 10, Sections 10.1 and 10.3). The higher the *S* value the larger the particle and the more rapidly it settles. The intact *E. coli* ribosome has a sedimentation coefficient of 70S. The two subunits are defined as 50S and 30S particles. The *E. coli* ribosome consists of approximately 66% RNA and 34% protein. Its structural organization is

Spanish stamp celebrating the 1969 European Biochemistry Congress. The structure of DNA is shown with the genetic code.

Characteristic	Intact Ribosome	Large Subunit	Small Subunit
Sedimentation coefficient	70S	50S	30S
Molecular mass (kilodaltons)	2520	1590	930
RNA content	66%	23S (2904 bases) 5S (120 bases)	16S (1542 bases)
Protein content	34%	34 different proteins	21 different proteins

Table 12.1
Structural organization of the *E. coli* ribosome

described in Table 12.1 and Figure 12.1a. New X-ray crystallographic, high-resolution structures of the subunits of prokaryotic ribosomes were published in 2000. The new pictures confirmed the long-accepted structure as described here and also provided updated information about the catalytic mechanism of protein synthesis (Section 12.2). Eukaryotes have larger, more complex ribosomes than prokaryotes (Figure 12.1b). The intact mammalian ribosome is an 80S particle with 60S and 40S subunits. At least 80 distinct proteins are present in the 80S particle. Functionally, ribosomes are molecular structures that facilitate all the steps of protein synthesis: (1) They move along mRNA templates deciphering the code for conversion from nucleotide to amino acid sequence; (2) they bring to the template the proper adaptor molecule "charged" with the proper amino acid; and (3) they catalyze the formation of peptide bonds between amino acids using energy from ATP or GTP. During protein synthesis the two ribosomal subunits combine together so as to form a channel through which the mRNA moves. If the purified RNA and protein components of ribosomes are mixed under proper conditions, they self-assemble into intact functional ribosomes, capable of directing protein synthesis.

Protein Synthesis Begins at the Amino Terminus

An early view of protein synthesis portrays a polymerization process whereby amino acids are added one at a time to a growing end. But what is the direction of elongation? Does it proceed from the amino terminus to the carboxyl terminus or vice versa (Figure 12.2)? Howard Dintzis and his colleagues in 1961 answered this question by incubating tritium-labeled leucine with rabbit reticulocytes (immature red blood cells) that were synthesizing hemoglobin. The location of radioactive leucine in product hemoglobin allowed Dintzis to conclude that amino acids are added to the growing carboxyl

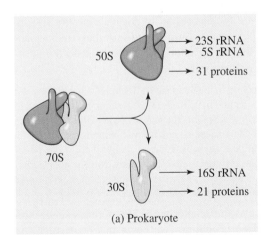

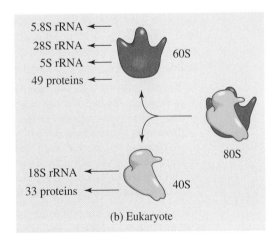

(a) Prokaryote

(b) Eukaryote

Figure 12.1
(a) Model for the *E. coli* ribosome showing the individual and combined subunits. The individual 30S and 50S subunits reversibly associate into the 70S complex, which functions in protein synthesis. The 70S complex moves along the mRNA to be translated. (b) Model for the eukaryotic ribosome made up of an 80S particle.

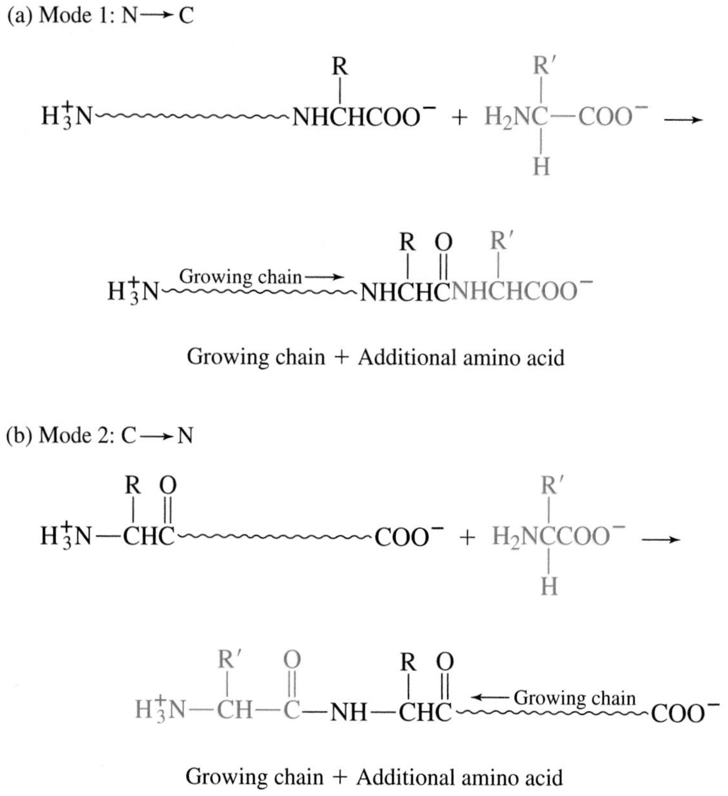
(a) Mode 1: N⟶C
Growing chain + Additional amino acid
(b) Mode 2: C⟶N
Growing chain
Growing chain + Additional amino acid

Figure 12.2
Possible directional modes of protein synthesis: (a) Mode 1, from the amino terminus toward the carboxyl terminus; (b) Mode 2, from the carboxyl terminus toward the amino terminus. Protein synthesis occurs by mode 1, from the amino terminus toward the carboxyl terminus.

terminus of the polypeptide. Thus, protein synthesis proceeds from the amino terminus to the carboxyl terminus (mode 1 in Figure 12.2).

The experiments performed by Dintzis had an additional benefit in that they provided useful information about the time required for synthesis of a polypeptide. A single ribosome in the rabbit reticulocytes completed the α-chain of hemoglobin (146 amino acids) in about 3 min at 37°C, a rate of slightly less than one residue added per second. An *E. coli* ribosome can construct a 100-amino acid polypeptide at 37°C in about 5 s (20 residues per second).

Structure Tutorial: tRNA

Amino Acids Are Activated and Combined with Specific tRNAs

We come now to what was perhaps the most difficult concept to elucidate in protein synthesis: How is the 4-letter language of mRNA sequence translated into the 20-letter language of proteins? Francis Crick suggested that an "adaptor" molecule spanned the information gap between the two languages. In Crick's words:

> One would expect, therefore, that whatever went on to the template (mRNA) in a *specific* way did so by forming hydrogen bonds. It is therefore a natural hypothesis that the amino acid is carried to the template by an "adaptor" molecule, and that the adaptor is the part which actually fits on to the RNA. In its simplest form one would require twenty adaptors, one for each amino acid.
>
> What sort of molecules such adaptors might be is anybody's guess. . . . But there is one possibility which seems inherently more likely than any other—that they might contain nucleotides. This would enable them to join on to the RNA template by the same "pairing" of bases as is found in DNA, or in polynucleotides. (Crick, F., 1966. The genetic code. *Sci. Am.* 215:55–62.)

We now know that the tRNA molecules serve as the adaptor molecules. Crick's suggestion that nucleotides might be involved was indeed insightful. As shown in Figure 12.3, the amino acid is covalently linked by an ester bond to the 2′ or 3′-hydroxyl end of a specific tRNA. At the other end of the tRNA, three adjacent nucleotide bases (the anticodon) bind to the mRNA template via hydrogen bonding between complementary base pairs.

Amino acids are linked to tRNAs by enzymes called **aminoacyl-tRNA synthetases.** It is critical that these enzymes display two kinds of specificity: They must recognize

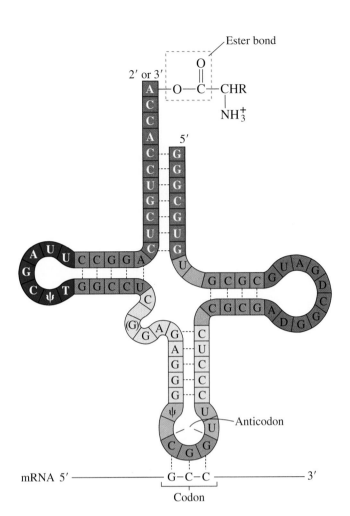

Figure 12.3

tRNA as an adaptor molecule. At one region of the tRNA molecule three adjacent bases (anticodon) form complementary hydrogen bonds with the mRNA codon for the mRNA. The amino acid is linked at the 2′ or 3′ end of the tRNA by an ester bond highlighted by dotted lines.

both the proper tRNA and the proper amino acid for linkage. This is a difficult task for the enzymes because all tRNA molecules have very similar primary, secondary, and tertiary structures except at the anticodon, where each is unique. In addition, the 20 amino acids, which have similar structural characteristics, must be distinguished from each other. With some exceptions cells are unable to correct the wrong pairing of an amino acid and tRNA. Most organisms have 20 aminoacyl-tRNA synthetases, one for each amino acid. As many as 60 different tRNA molecules have been isolated from organisms, so some amino acids have more than one tRNA. (We explain the significance of this later in this section.)

Aminoacyl-tRNA synthetases catalyze a reaction sequence that requires the energy from cleavage of two phosphoanhydride bonds in ATP:

$$\text{amino acid} + \text{ATP} \rightleftharpoons \text{aminoacyl adenylate} + \text{PP}_i$$

$$\text{aminoacyl adenylate} + \text{tRNA} \rightleftharpoons \text{aminoacyl-tRNA} + \text{AMP}$$

$$\text{PP}_i + \text{H}_2\text{O} \rightleftharpoons 2\,\text{P}_i$$

The aminoacyl adenylate formed by the reaction of an amino acid with ATP is an unstable, enzyme-bound intermediate that contains a reactive anhydride linkage (Figure 12.4). The overall equilibrium constant for the first two reactions is approximately 1; therefore, a thermodynamic driving force is necessary to pull the reaction to completion. That extra energy comes from the hydrolysis of the pyrophosphate product (PP_i). Although we have emphasized the importance of this reaction process in bringing together an amino acid with its correct tRNA, the reaction has another important function. By linkage of the amino acid via an ester bond, the amino acid becomes activated for later peptide bond formation in protein synthesis.

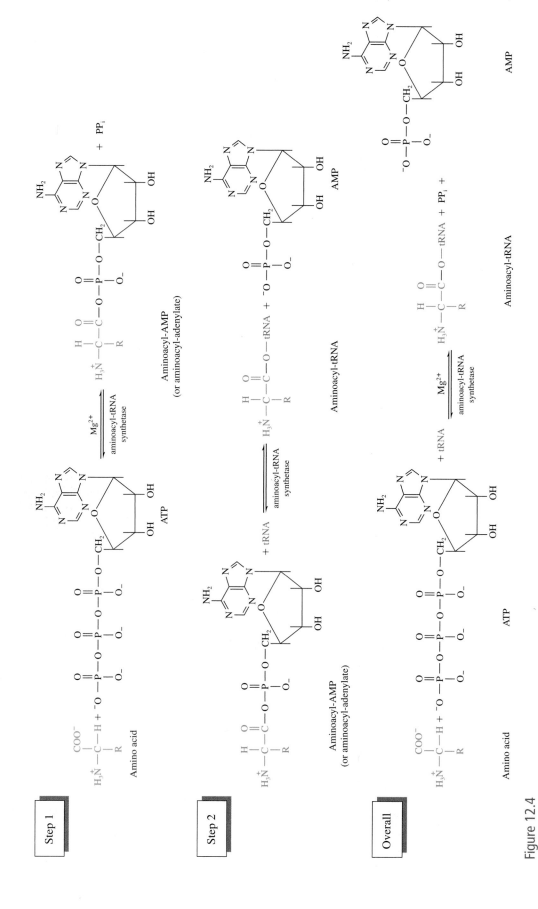

Figure 12.4

Mechanism of aminoacyl-tRNA synthetase action with formation of an aminoacyl adenylate intermediate. Energy from the cleavage of two phosphoanhydride bonds (ATP and PP_i) drives the reaction to completion.

Relationships Between mRNA Base Sequence and Protein Amino Acid Sequence

Several terms have been used to describe the functional characteristics of the genetic code (Chapter 2, Section 2.3):

- *Triplet.* A set of three nucleotide bases on mRNA code for one amino acid.
- *Nonoverlapping.* A set of three adjacent bases is treated as a complete group. The set of bases, called the **codon,** is used once for each translation step.
- *No punctuation.* There are no punctuation marks (intervening bases) between triplets. Therefore, an mRNA is read from start to finish without commas or other interruptions. (Perhaps the argument could be made that termination signals are periods.)
- *Degenerate.* A single amino acid may have more than one triplet code. There is usually a sequential relationship between degenerate codes.
- *Universal.* The same genetic code is used in all organisms except for a few exceptions, including mitochondria and some algae.

The first experiments designed to "break" the genetic code were reported by Marshall Nirenberg and Heinrich Matthaei in the early 1960s. They incubated cell-free extracts containing ribosomes, tRNA, amino acids, and aminoacyl-tRNA synthetases with synthetic mRNA of known sequences. For example, when poly U (U-$(U)_n$-U) was used as template, the polypeptide product was polyphenylalanine, indicating that the base codon U-U-U was translated as the amino acid phenylalanine (Table 12.2). Using similar experiments with other synthetic polyribonucleotides allowed researchers to elucidate completely the genetic code.

Synthetic polyribonucleotides were also used to determine the direction mRNA is read, $5' \rightarrow 3'$ or $3' \rightarrow 5'$. The template $5'$ A-A-A-$(A-A-A)_n$-A-A-C was synthesized and used for protein synthesis. The triplet AAA codes for lysine and AAC for asparagine. The experimental polypeptide product from this template was $H_3\overset{+}{N}$-Lys-$(Lys)_n$-Asn-COO^-, indicating that the direction of translation was $5' \rightarrow 3'$. It is significant that this is the same direction that the mRNA is formed. Hence, the ribosomes may begin translating the mRNA immediately after or perhaps even during its transcription.

As indicated earlier, some amino acids have degenerate codons; that is, they have more than one codon. Phenylalanine uses the two triplets UUU *and* UUC; threonine has four,

Table 12.2
The genetic code

		Second Base of Codon				
		U	C	A	G	
First Base of Codon	U	UUU } UUC } Phe UUA } UUG } Leu	UCU } UCC } UCA } Ser UCG }	UAU } UAC } Tyr UAA } UAG }	UGU } UGC } Cys UGA } UGG Trp	U C A G
	C	CUU } CUC } CUA } Leu CUG }	CCU } CCC } CCA } Pro CCG }	CAU } CAC } His CAA } CAG } Gln	CGU } CGC } CGA } Arg CGG }	U C A G
	A	AUU } AUC } Ile AUA } AUG Met	ACU } ACC } ACA } Thr ACG }	AAU } AAC } Asn AAA } AAG } Lys	AGU } AGC } Ser AGA } AGG } Arg	U C A G
	G	GUU } GUC } GUA } Val GUG }	GCU } GCC } GCA } Ala GCG }	GAU } GAC } Asp GAA } GAG } Glu	GGU } GGC } GGA } Gly GGG }	U C A G

Note: AUG is the start codon and UAA, UAG, and UGA are stop codons as highlighted in table.
Source: Redrawn from Wolfe, 1993.

ACU, ACC, ACA, and ACG. From a study of Table 12.2, it is apparent that when degeneracy exists, the first two bases are usually the same; only the third differs. It has been concluded that the first two bases of each codon are primary determinants for specificity, and one would expect strong interactions between these bases and two complementary bases on the tRNA called the **anticodon.** The third base interaction between the codon and anticodon is probably weaker and is sometimes called the wobble base. The wobble factor provides three benefits: (1) It adds flexibility to the genetic code; (2) the weaker interaction causes faster dissociation of the tRNA from the mRNA, thereby speeding protein synthesis; and (3) it minimizes the effects of mutations.

12.2 The Three Stages of Protein Synthesis

We turn now to details of the protein synthesis process. As with DNA replication and transcription, we focus on the process as it occurs in prokaryotic cells, especially *E. coli,* because it is much better understood than the process in eukaryotic cells. It is likely that over 100 different proteins and several types of RNA molecules are required for protein synthesis and processing in prokaryotic cells. The number of protein factors and RNA molecules required in eukaryotic cells may be several hundred. To add to the complexity, protein synthesis is more than a chemical polymerization process; it also involves the mechanical motions of ribosomal particles along the mRNA template and the specific binding of aminoacyl-tRNA. The steps for protein synthesis in *E. coli* are described as follows and shown in Figure 12.5a–f. For eukaryotic protein synthesis, specific factors may be different, but the overall scheme is the same as for prokaryotes.

Initiation, Elongation, and Termination

Stage 1

Protein synthesis begins with ribosomal recognition of the starting point on the mRNA and entry of tRNA-activated *N*-formylmethionine (fMet). The full cast of characters in the initiation step includes: (a) the dissociated 30S and 50S ribosomal subunits; (b) the mRNA to be translated; (c) protein **initiation factors** (IF); (d) the beginning amino acid in activated form, fMet-tRNA; and (e) GTP. The process starts with formation of a complex containing the 30S ribosomal subunit, the mRNA, the initiation factors, and GTP (Figure 12.5a).

The 30S subunit and mRNA combine to place the two aminoacyl-tRNA binding sites (P for peptidyl; A for aminoacyl) directly over the codons for the first two amino acids. The ribosomal subunit is directed to this placement by the Shine–Dalgarno sequence of bases on the mRNA. This initiation region of approximately five to ten bases is centered at about ten bases upstream from the start codon. This purine-rich region pairs with a complementary pyrimidine-rich region of the 16S RNA in the 30S subunit. This correct alignment facilitates binding of the first activated amino acid, fMet-tRNA as shown in Figure 12.6, which will become the N-terminus. The binding of tRNA-activated amino acid is specified by codon–anticodon base pairing. All proteins in bacterial cells begin with the amino acid formylmethionine at the amino terminus. The start codon on the mRNA is AUG (see Figure 2.7). The complete package of participants at this point is called the 30S initiation complex (see Figure 12.5a). Initiation proceeds with joining of the 50S subunit to the 30S complex. Formation of the new 70S initiation complex is accompanied by the hydrolysis of GTP (GTP + $H_2O \longrightarrow$ GDP + P_i + energy) (Figure 12.5b). The 70S complex is now ready to begin the elongation stage of protein synthesis.

Stage 2

The participants in the elongation stage include the 70S initiation complex, the next aminoacyl-tRNA, and proteins called **elongation factors** (EF). This stage begins with entry of the second amino acid that is guided into the A ribosomal binding site by elongation factors. As

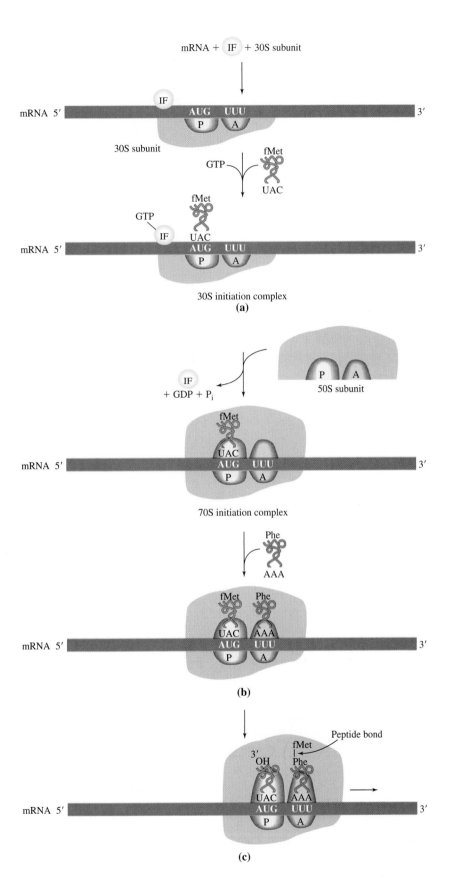

(continued)

Figure 12.5

Steps of protein synthesis: (a) initiation, (b) formation of 70S complex, (c) elongation by peptide bond synthesis.

Figure 12.5—*continued*
(d) translocation, (e) continued
elongation to termination codon,
and (f) termination. IF = initiation factors; RF = release factor.

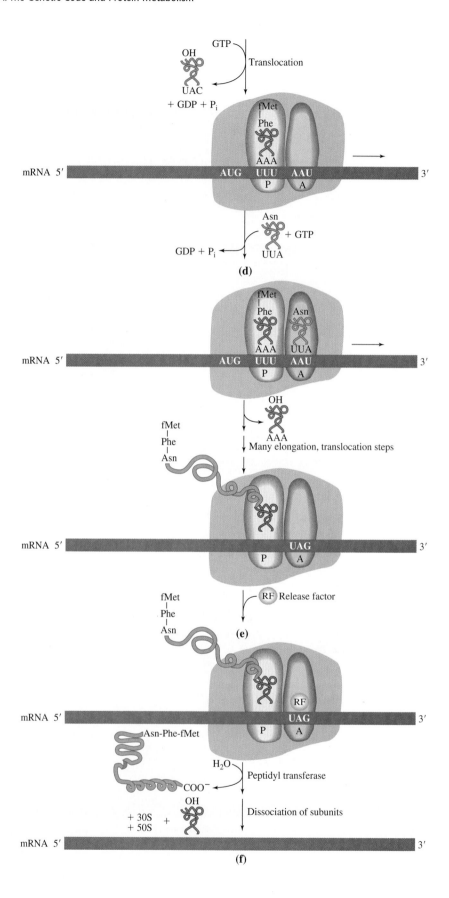

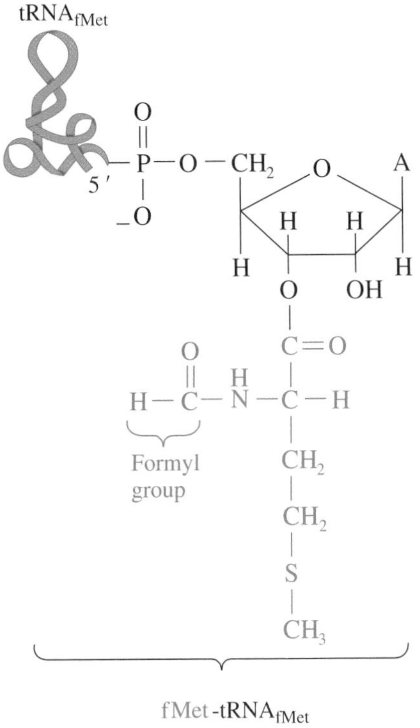

tRNA$_{fMet}$

Formyl group

fMet -tRNA$_{fMet}$

Figure 12.6
The structure of the amino acid fMet activated by its tRNA.

Cutting Edge: Ribozymes

with fMet-tRNA, the new aminoacyl-tRNA is selected by codon–anticodon interactions that specify binding. The complex is set for formation of the first peptide bond that links the carboxyl group of fMet to the amino group of the second amino acid (Figure 12.5c). This reaction produces a dipeptidyl-tRNA that is now positioned in the A ribosomal binding site. An expanded view of peptide bond formation is shown in Figure 12.7. The formation of the peptide bond is catalyzed by **peptidyl transferase,** an RNA enzyme or **ribozyme** associated with the 50S ribosome. Highly resolved structures of the prokaryotic 70S ribosomal subunits have recently been produced by X-ray crystallography (Figure 12.8). These updated pictures show that there are no proteins in the vicinity of the peptidyl transferase active site where catalysis occurs and new covalent bond formation takes place. Only ribosomal RNA molecules are present at the active site. This new evidence confirms that the protein synthesis activity of the ribosome has its basis in RNA. RNA molecules in the subunits function as catalysts by orienting the two substrates, mRNA and aminoacyl-tRNA, close together so that peptide bond formation can occur. Apparently, proteins play predominately a structural role in ribosomes.

Formation of the peptide bond is a thermodynamically favorable reaction because of the reactive ester bond between fMet and its tRNA. The stage is now set for one of the mechanical motions of the ribosomal particles. The 70S ribosomal unit moves the distance of one codon toward the 3′ end of the mRNA (Figure 12.5d). This action, called **translocation,** is accompanied by displacement of deacylated tRNA from the codon region and P site and eventually releases the free tRNA into the cytoplasm. At the same time the dipeptidyl-tRNA is shifted from the A site to the P site, leaving the A site vacant. Each translocation step is coupled to the hydrolysis of GTP. The third aminoacyl-tRNA of the peptide that is bound at the vacant P site is selected by codon–anticodon interactions and facilitated by elongation factors. The binding of each additional aminoacyl-tRNA is coupled to GTP hydrolysis. In fact, two GTPs are necessary for each amino acid, one for delivery and one for each translocation step. Protein synthesis proceeds with formation of the new peptide bond between the second and third amino acids, resulting in a tripeptide still linked to the tRNA of the third amino acid (Figure 12.5e).

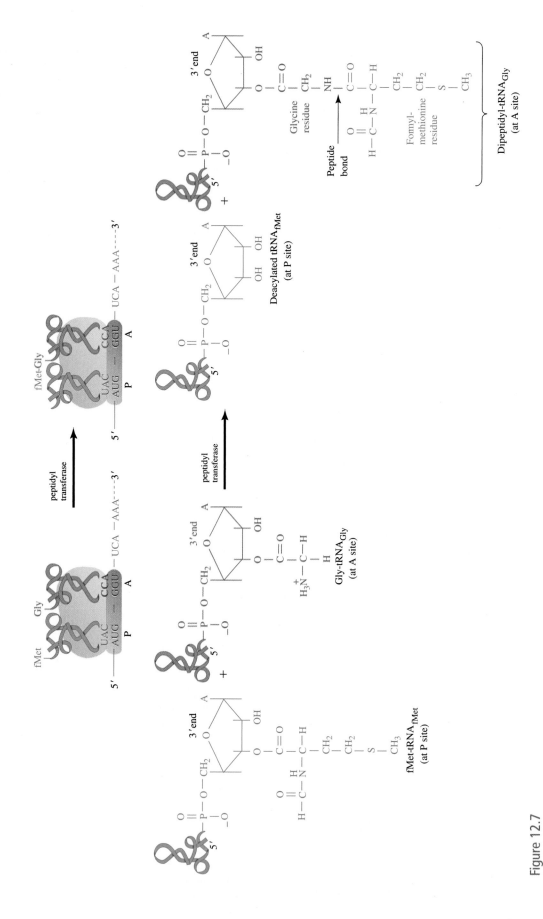

Figure 12.7

Expanded view of peptide bond formation. The amino acid (Gly) to be added to the growing polypeptide chain is in the A site of the ribosome. The growing chain is in the P site. Peptidyl transferase, an RNA enzyme, catalyzes formation of the new peptide bond. Peptide bond formation is initiated by the removal of a proton from the amino group of bound glycine. The nitrogen atom then directs a nucleophilic attack on the ester bond holding fMet.

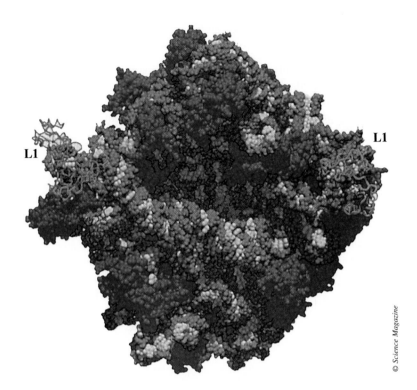

Figure 12.8

This space-filling model represents a large subunit of the ribosome looking down the cleft into the active site, identified by the presence of a red inhibitor molecule. RNA's sugar–phosphate backbone is shown in orange; its bases are white. Proteins, shown in blue, surround the perimeter of the complex but are not present at the active site. The structures of two proteins, labeled L1 and L11, are shown only at lower resolution.

Stage 3

Termination is signaled by special codons on the mRNA. The elongation process continues as described above until the ribosomal A site moves to one of the codons for termination, UAG, UGA, or UAA. No aminoacyl-tRNAs exist with complementary anticodons for binding to these codons. Instead, protein **release factors** (RF) bind to the A site, thus activating peptidyl transferase. In place of peptide bond formation, the transferase catalyzes the hydrolysis of the ester bond linking the carboxyl group of the newly synthesized protein to the tRNA in the P site (Figure 12.5f). The protein product is released from the complex. tRNA and other factors diffuse into the cytoplasm while the 70S ribosome dissociates into its subunits to initiate translation of another mRNA molecule.

Polyribosomes

We have demonstrated the process of protein synthesis using only one ribosome on a mRNA molecule. In reality, clusters of ribosomes called **polyribosomes** can simultaneously translate a mRNA molecule, thereby making many identical protein molecules from a single copy of mRNA (Figure 12.9). Each ribosomal particle along the mRNA is at a different stage of translation. Those ribosomes closer to the 3′ end of the mRNA are nearer to completion of the polypeptide product than those closer to the 5′ end. The actual number of ribosomes on a single mRNA depends on the size of the mRNA. The maximum number is probably about 1 ribosome per 80 nucleotides. The presence of polyribosomes, of course, adds immensely to the efficiency of protein synthesis. mRNA molecules have a relatively short lifetime, especially in prokaryotes, so it is important to use each one as efficiently as possible.

Protein Synthesis and Energy

The total energy requirements for protein synthesis are quite high (Table 12.3). To summarize energy needs:

1. Two anhydride bonds in ATP are cleaved for the activation of each amino acid and the synthesis of a specific aminoacyl-tRNA by aminoacyl-tRNA synthetase.

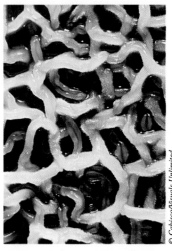

The mold *Streptomyces venezuelae* produces the antibiotic puromycin.

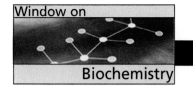

Window on

Biochemistry

2. One GTP is required for entry of each amino acid into the ribosomal A site.

3. One GTP is required during each translocation step.

Therefore, the total requirement is four anhydride bonds cleaved for each amino acid incorporated into a protein. As we shall see in Chapter 14, cleavage of each anhydride bond releases approximately 30 kJ/mole. This makes protein synthesis a very expensive, energy-demanding process. However, the high level of accuracy needed to synthesize functional proteins justifies this high cost.

Inhibition of Protein Synthesis

Because protein synthesis is such an important function for life, it is a target for drug design. Prokaryotic or bacterial protein synthesis is different enough from the eukaryotic process that it is possible to design drugs that inhibit bacterial protein synthesis but have little effect on the process in eukaryotes. We have learned much about drug design by studying the natural antibiotics synthesized by some microorganisms as toxins to prevent the growth of other microorganisms. Probably best known among antibiotics is puromycin, produced by a *Streptomyces* mold. This inhibitor of prokaryotic protein synthesis mimics aminoacyl-tRNA molecules (Figure 12.10). Its structure is similar enough to the amino acid tRNA ester that it binds to the ribosomal A site. The

12.1   Triclosan—An Antibacterial Agent in Your Toothpaste

When medical doctors encounter patients suffering from diseases caused by bacterial infections, the treatment often involves a prescription of a strong antibiotic such as tetracycline, erythromycin, or vancomycin. As you just read in the section on antibiotics, these compounds kill bacteria by inhibiting specific steps in the process of prokaryotic protein synthesis. But what should we do, if anything, about the many germs that are always present on our bodies and in our homes? Manufacturers of personal care products and household cleaning agents are adding antibacterial reagents to limit the spread of these microorganisms. These germ-killing agents are now being incorporated into dishwashing detergents, hand creams, sponges, hand soaps, and cosmetics, and even toothpaste, toys, mattress pads, and food cutting boards. The most widely used biocide is the chlorinated chemical **triclosan** (see structure). The action of triclosan and other biocides like alcohol and detergents has always been thought to be nonspecific—they simply kill all microorganisms by an extensive disruption of the cell membranes. If the biocides have no specific biological targets, then it could be assumed that bacteria could not develop a resistance to the chemical. This is different from the antibiotics that target specific biochemical events and thus allow bacteria to develop resistant strains by mutations.

New studies on triclosan now show that its biocide action does involve a specific biological target. The chemical inhibits enoyl-ACP reductase, an important enzyme in the biosynthesis of fatty acids (see Section 18.4). The research also shows that bacteria could possibly develop resistance to triclosan by mutations in the gene that encodes the reductase. Manufacturers of products containing triclosan and other antibacterial additives maintain that their use is safe and approved by the government. Further research needs to be done concerning the widespread use of the antibacterial agents.

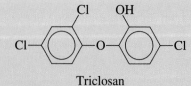

Triclosan

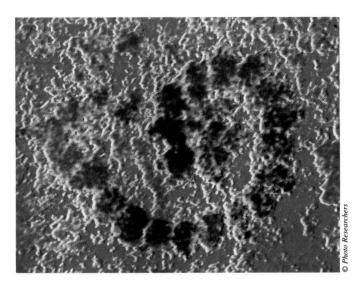

Figure 12.9
The action of several ribosomes on a single mRNA molecule. Electron micrograph of *E. coli* polyribosome. Each ribosome represented by a green circle, synthesizes a copy of the polypeptide by moving along the mRNA (see Figure 2.5).

amino group on the antibiotic participates in peptide bond formation, producing a protein molecule with puromycin at the carboxyl end. Because the peptidyl-puromycin cannot be translocated to the P site, prematurely terminated, inactive protein is hydrolyzed by peptidyl transferase and dissociates to the cytoplasm. The structure and function for several other antibiotic inhibitors of protein synthesis are given in Figure 12.11 and Table 12.4 on page 321. Two interesting nonantibiotic inhibitors of eukaryotic protein synthesis are (1) ricin, a toxic protein present in castor beans, and (2) diphtheria toxin, an enzyme secreted by *Corynebacterium diphtheriae* bacteria carrying the phage corynephage β. Many of the antibiotics shown here have been used medically for years, but the problem of antibiotic resistance is becoming increasingly serious. Bacteria can become resistant to antibiotics either through genetic mutations or by transfer of resistance genes from other bacteria via plasmids. A new antibiotic currently undergoing clinical trials is linezolid. It functions by preventing the initial interaction of the 30S ribosome subunit, mRNA, and *N*-formyl met. Protein synthesis is stopped even before it starts.

Table 12.3
Energy requirements for protein synthesis

Process	Reaction	Number of Phosphoanhydride Bonds Broken
1. Selection and activation of amino acid	Amino acid + tRNA ⟶ aminoacyl-tRNA	
	ATP ⟶ AMP + PP$_i$	1
	PP$_i$ + H$_2$O ⟶ 2 P$_i$	1
2. Entry of aminoacyl-tRNA into ribosomal site A	GTP + H$_2$O ⟶ GDP + P$_i$	1
3. Translocation	GTP + H$_2$O ⟶ GDP + P$_i$	1
		Total: 4

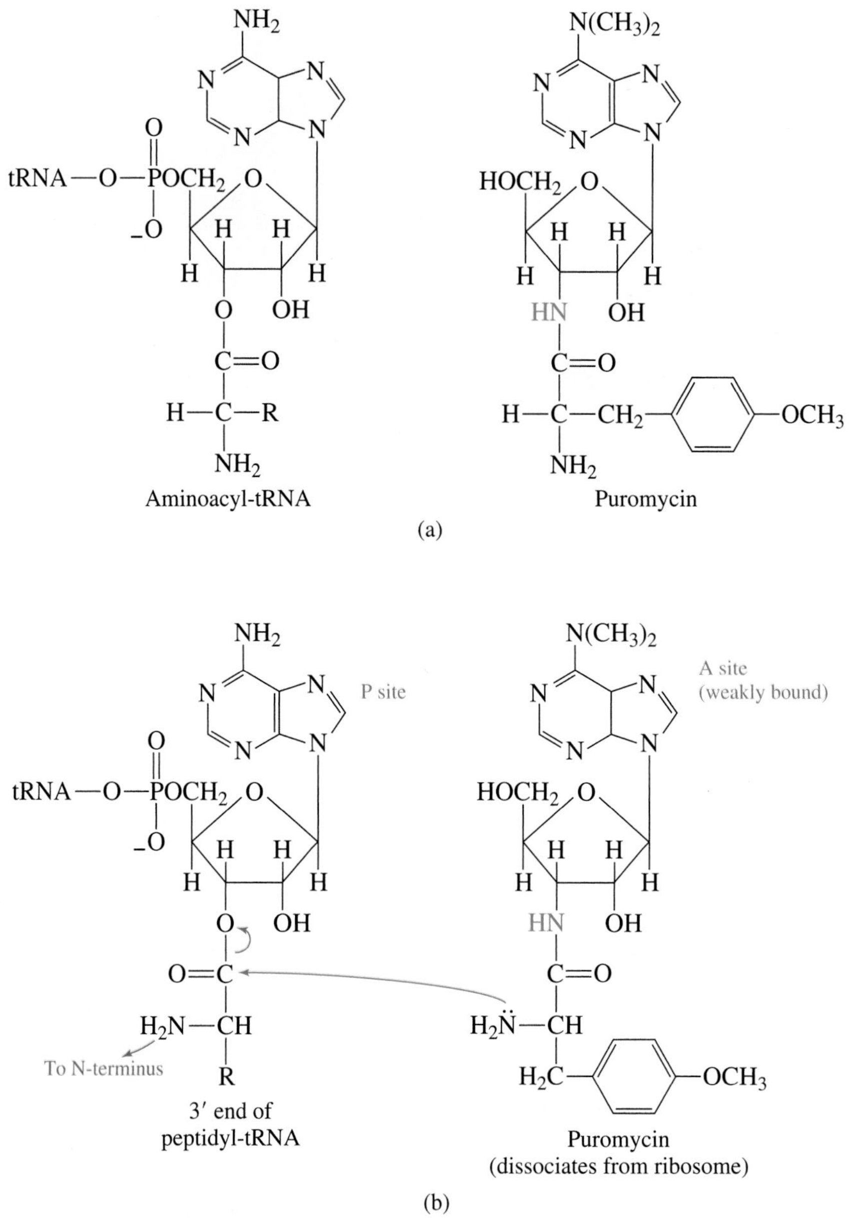

Figure 12.10

The structure of the antibiotic puromycin and mechanism for its action as an inhibitor of prokaryotic protein synthesis. (a) Puromycin has a structure similar to an aminoacyl-tRNA so it is able to bind weakly to the A site of ribosomes. (b) When bound to ribosomes, puromycin can participate in peptide bond formation. But the peptidyl–puromycin binds weakly and dissociates from the ribosome, thus terminating protein synthesis. The growing peptide is linked to puromycin by a very strong amide bond (in *red*) and cannot be broken by normal protein chain elongation.

Aminoacyl-tRNA   Puromycin

(a)

3′ end of
peptidyl-tRNA

Puromycin
(dissociates from ribosome)

(b)

12.3   Post-Translational Processing of Proteins

Most polypeptide translation products are not yet in their biologically active form (native conformation) immediately after synthesis. Several steps of folding and biochemical modification are often required before a protein can perform its essential and specific function. In addition, most proteins (except those made in mitochondria and chloroplasts) are synthesized in the cytoplasm and they must be transported to other cellular regions where they serve their specific biological purpose(s).

Protein Folding

The first action for most proteins after synthesis is to fold into their native conformation. In some cases the protein molecules begin the folding process before synthesis is complete. As discussed in Chapter 5, Section 5.3, the final tertiary structure of a protein is determined by

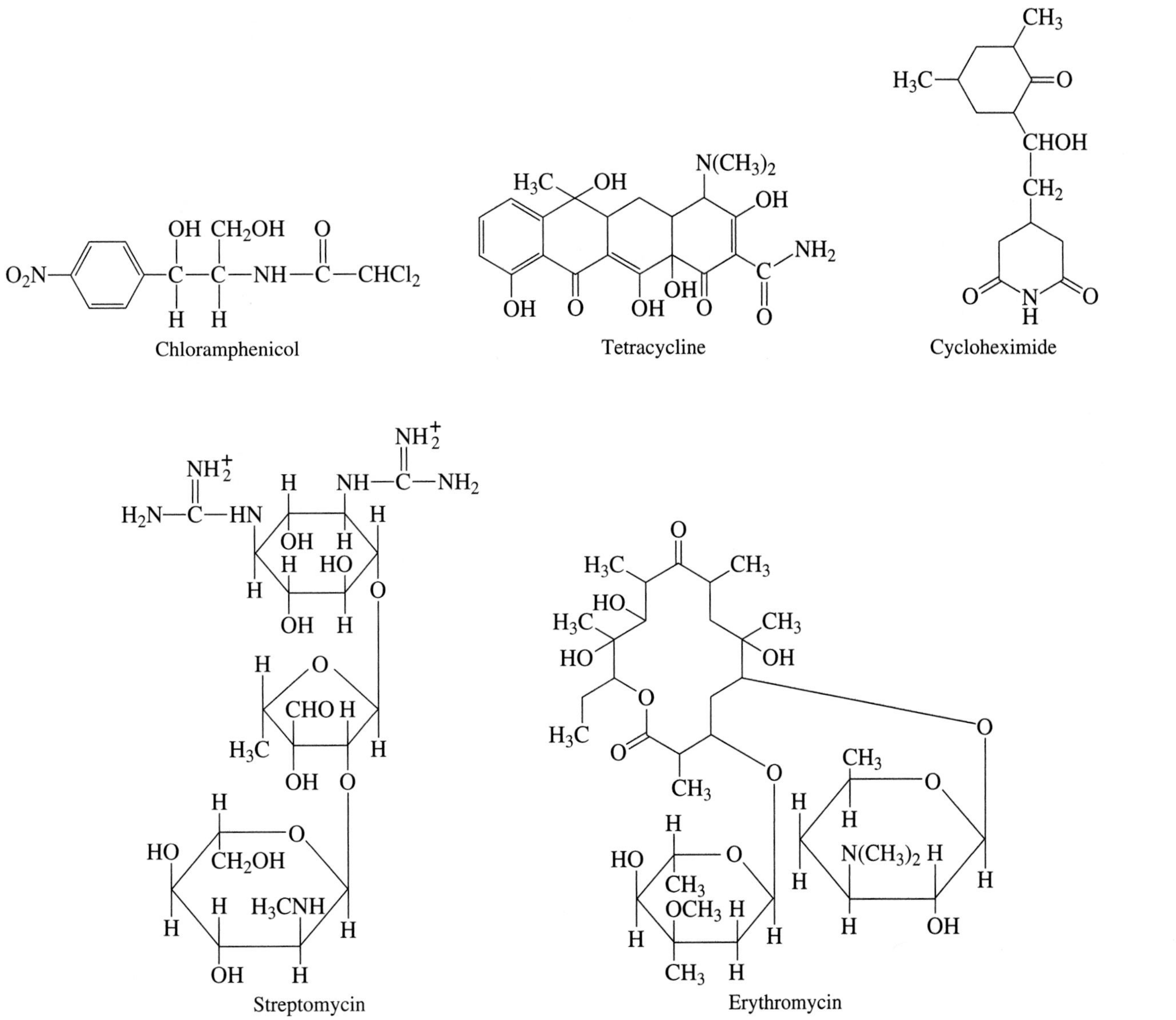

Figure 12.11
Structures of antibiotic inhibitors of prokaryotic protein synthesis.

Table 12.4
Antibiotic inhibitors of protein synthesis

Antibiotic	Mode of Action
Puromycin	Causes early termination by mimicking the action of an aminoacyl-tRNA; acts on prokaryotes and eukaryotes
Streptomycin	Causes misreading of mRNA and inhibits initiation; acts on prokaryotes
Tetracycline	Binds to the A site of ribosomes and blocks entry of aminoacyl-tRNAs; acts on prokaryotes
Erythromycin	Binds to ribosome and inhibits translocation; acts on prokaryotes
Chloramphenicol	Binds to 50S subunit and inhibits peptidyl transferase; acts on prokaryotes
Cycloheximide	Inhibits translocation of eukaryotic peptidyl-tRNA

The protein synthesis inhibitor ricin is present in castor beans, *Ricinus communis.*

the primary structure (amino acid sequence). The folding process probably begins with the formation of local secondary structure (α-helix, β-conformation) to provide a nucleus or seed. The remainder of the protein chain then continues to fold around the initiation nucleus. The process often has the characteristics of cooperativity indicating that each folding step facilitates the formation of other favorable interactions. The goal of protein folding is to form the maximum number of strong interactions (hydrophobic, hydrogen bonding, van der Waals, ionic) so a stable, three-dimensional arrangement can result. Many polypeptides receive assistance in the folding process. Proteins called **chaperones** act as catalysts to guide and facilitate folding. Some chaperones are enzymes that couple ATP hydrolysis to the protein folding process. The chaperones also bind to folding proteins in order to hide exposed hydrophobic amino acid residues so interactions do not occur out of order.

Most favorable interactions holding a protein in its native conformation are noncovalent in nature; but, covalent disulfide bonds (S—S) often crosslink cysteine residues that may be far apart in the sequence of the polypeptide chain. Experimental studies on the importance of disulfide bonds in tertiary structure show that they do not directly influence the folding of most proteins into their native conformations. Instead, disulfide bonds lock the protein into its final form after most other stabilizing, noncovalent interactions are in place. Some proteins are unable to fold completely into their native conformations until further biochemical modifications occur.

Biochemical Modifications

Many proteins must undergo covalent changes of their primary structures before they are biologically functional. Most of these biochemical changes involve chain cleavage processes, amino acid residue alterations, and addition of other factors.

Proteolytic Cleavage

All prokaryotic proteins begin with the amino acid *N*-formylmethionine; however, for about 50% of the proteins synthesized in *E. coli,* the N-terminus amino acid is removed by hydrolysis. In a similar fashion, the methionine residue at the N-terminus of eukaryotic proteins is removed. In addition, approximately one-half of eukaryotic proteins are modified by addition of an acetyl group to their amino end:

$$CH_3\overset{\displaystyle O}{\overset{\displaystyle \|}{C}}\!-\!NH\sim\!\!\sim\!\!\sim$$

In Chapter 7 the covalent modification of enzymes was described as a means to regulate activity. Some enzymes are initially constructed as inactive precursors, called **zymogens,** that are cleaved at one or more specific peptide bonds to produce the active form of the enzyme. Chymotrypsin is a protease that is regulated by such covalent modification (see Figure 7.8).

Amino Acid Modification

The activity of enzymes and other proteins can be altered by biochemical modification on amino acid residues (see Chapter 4 and Chapter 7). The chemical processes of phosphorylation and hydroxylation are the most common. The hydroxyl groups of serine, threonine, and tyrosine residues are modified by transfer of a phosphoryl group, $-PO_3^{2-}$, from ATP. The activity of the enzyme glycogen phosphorylase is regulated by phosphorylation of specific serine residues. Side-chain hydroxylation of proline and lysine residues are common alterations in the structural protein collagen. Chemical modifications occur not on the free amino acid molecules but after they are incorporated into proteins.

Attachment of Carbohydrates

Glycoproteins, those proteins that have covalently bonded carbohydrates, are involved in many biological functions including immunological protection, cell–cell recognition, and blood clotting. Carbohydrate chains up to 15 residues long are covalently bonded to proteins at the hydroxyl groups of serine and threonine or on the side chain amide nitrogen of asparagine (see Section 8.5).

Addition of Prosthetic Groups

Many proteins depend on the presence of a covalently bound cofactor or prosthetic group for biological activity. During our studies in metabolism we will encounter proteins with prosthetic groups such as heme, FAD, biotin, and pantothenic acid. Recall that most cofactors and prosthetic groups are derived from vitamins.

Protein Targeting

Except for a few proteins made in mitochondria and chloroplasts, essentially all proteins are synthesized by ribosomes in the cytoplasm of cells. However, proteins are needed not only in the cytoplasm but also in other cellular regions and organelles. Bacteria typically need proteins in four basic compartments: outer membrane, plasma (inner) membrane, periplasmic space (between membranes), and cytoplasm. Eukaryotic cells have several compartments needing proteins: membranes, mitochondria, chloroplasts, nucleus, lysosomes, and others. Each region, whether it is membrane or the nucleus, has the requirement for a unique set of proteins. How are proteins sorted and transported to their final destinations? These are the subjects of **protein targeting.**

In general, proteins that must be transported from the cytoplasm across plasma or vesicle membranes are synthesized with a short sequence of extra amino acid residues called the signal sequence. Except for those transported to the nucleus, all proteins are labeled with a **signal sequence** of 14 to 26 amino acids at the amino terminus. The signal sequence is usually removed by hydrolysis when the protein reaches its final destination. (For nuclear proteins such as DNA polymerase I, the signal sequence is internal and not cleaved.) The general features of a prokaryotic *(E. coli)* signal sequence are shown in Figure 12.12. The label consists of (1) a short sequence of positively charged amino acids, followed by (2) a hydrophobic region of 7 to 13 amino acids, (3) a nonhelical region containing proline or glycine, and (4) a cleavage site for peptidase action to remove the signal sequence.

Signal sequences for eukaryotic proteins are very similar to prokaryotic proteins; however, more complicated processes are necessary for transport to cellular organelles and other regions or for export to other cells. The best understood targeting system in eukaryotic cells involves the **endoplasmic reticulum** (ER). Proteins bound for membranes, lysosomes, or export are synthesized on ribosomes attached to the ER. Following synthesis, their signal sequences mark them for transport through the ER membrane. Inside the ER, the signal

Figure 12.12
General features of a prokaryotic signal sequence. The sequence consists of a basic region of positively charged amino acids, a hydrophobic region, and a nonhelical region. The cleavage site is for removal of the signal sequence by the action of a peptidase.

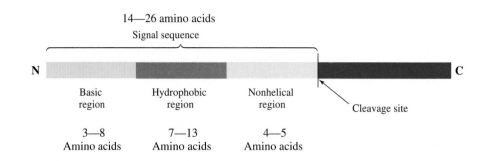

sequence is cut off and chemical modification on the proteins occurs, usually by addition of carbohydrates. Proteins are then moved from the ER to the Golgi apparatus, where they are sorted on the basis of carbohydrate markers, packaged in vesicles, and transported to their final destinations.

The Proteasome and Protein Degradation

Cellular and extracellular proteins are continuously being degraded and replaced by newly synthesized protein molecules. Although this process appears wasteful, it serves many purposes including removal of damaged or misfolded proteins and destruction of regulatory proteins that are not needed at the time. It also gives an organism the chance to adapt to changing conditions. The rate of turnover varies greatly from protein to protein. For example, two rat liver enzymes, RNA polymerase I and cytochrome c, have half-lives of 1.3 and 150 minutes, respectively. Human hemoglobin molecules may exist in erythrocytes for as long as 100 days.

Extracellular proteins taken up by cells are usually destroyed by the actions of lysosomal proteases. Degradation of unwanted intracellular proteins in prokaryotes and eukaryotes is carried out by ATP-dependent proteases associated with large protein complexes called the **proteasomes.** Proteins destined for destruction are marked in various ways. In eukaryotic cells, the **ubiquitin pathway** is important for protein labeling and degradation. Ubiquitin is a small protein of 76 amino acid residues found in all eukaryotic cells. Its amino acid sequence is highly conserved. For example, the sequence is identical in the ubiquitin of humans and fruitflies. By an ATP-dependent process, ubiquitin becomes covalently attached to unneeded or damaged proteins via an unusual peptide bond between the carboxyl terminus of ubiquitin and the ε-amino group of lysine residues in the targeted protein (Figure 12-13). The labeled protein is then degraded by the proteolytic actions of the proteasome (Figure 12.14).

The intact proteasome complex in eukaryotic cells is the **26S proteasome.** The barrel-shaped, 1700-kD complex is about 45 nm long and is composed of a 20S core plus two additional structures known as 19S caps (Figure 12.14). A cavity that has an opening diameter of about 1.3 nm runs down the inside center of the barrel. Entry to the central chamber, where proteolytic destruction occurs, is regulated by the 19S caps. Intracellular proteins that are labeled with ubiquitin are sought by the 19S caps, unfolded, and directed

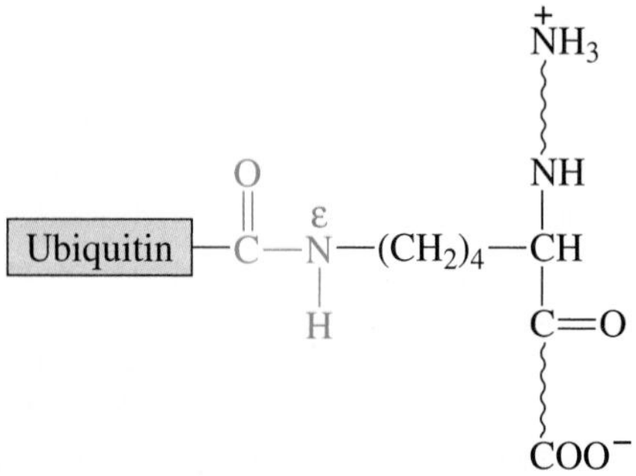

Figure 12.13
Linkage of ubiquitin to a defective protein. Note the unusual amide bond (in *red*) to the ε-amino group of a lysine residue in the defective protein. Several ubiquitin molecules may be attached to a protein in order to mark it for destruction. Ubiquitin is a small polypeptide of 76 amino acid residues.

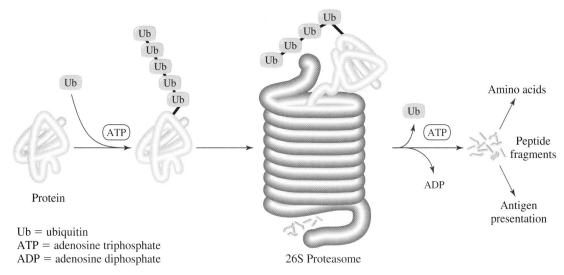

Figure 12.14
Functioning of the 26S proteasome in protein degradation.

into the core. Once inside, the marked proteins are cleaved by proteolytic action. Several types of catalytic activity appear to take place including trypsin-like cleavage at basic amino acid residues, chymotrypsin-like action at aromatic acid residues, and cleavage at acidic amino acid residues. The initial peptide products range in size from 3 to 23 amino acid residues. These are further degraded to free amino acids after release from the proteasome.

12.4 Regulation of Protein Synthesis and Gene Expression

The typical bacterial cell (*E. coli,* for example) has about 4000 genes in its DNA genome that upon transcription and translation could potentially lead to a like number of polypeptide products. The situation in eukaryotic cells is even more overwhelming. With an estimated 30,000 to 40,000 genes in the human genome, there is the potential for a similar number of protein products in the human organism. But all proteins in prokaryotic or eukaryotic cells are not required at all times, so it would be a waste of energy and supplies to synthesize all proteins continuously. In fact, only a fraction of genes is expressed at any given time. Hence, very careful regulation of gene expression is required. Some proteins and enzymes must be present in a cell at all times. Others are needed in only small amounts and at specific times. Some types of cells require specific proteins that are not present in all cells. The environments of some cells, especially prokaryotic, are constantly changing. Genes must be turned on and off during different development stages. The mechanisms by which organisms respond to these situations and balance protein synthesis and degradation are extremely complicated, and we have much less than a complete understanding of regulatory processes. As with many other biological processes, our understanding of the regulation of gene expression is at a higher level for prokaryotes than for eukaryotes.

There are many steps from DNA to protein that can potentially be regulated (Figure 12.15). A cell can control the amount of protein by regulating (a) the rate of transcription, (b) the rate of post-transcriptional processing, (c) the rate of mRNA degradation, (d) the rate of protein synthesis (translation), (e) the rate of post-translational processing, and (f) the rate of protein degradation. Although examples of regulation at all of these stages are known, most gene expression is controlled at the level of transcription initiation. Control at this stage limits the number of mRNA molecules synthesized since the rate of synthesis of any protein is most closely related to the quantity

Structure Tutorial: Protein-DNA interactions

Cutting Edge: Molecular recognition

Figure 12.15

Several steps in protein synthesis at which regulation can influence the concentration of proteins in a cell. Protein synthesis could be regulated at (a) the transcription step, (b) during post-transcriptional processing, (c) by controlling mRNA degradation, (d) by regulating translation, (e) during post-translational processing, or (f) by degrading final protein products. In process (e), proteins are activated by addition of prosthetic groups or carbohydrates and by modification of amino acid residues. Most protein synthesis is regulated at the transcription step (a). *Source:* Based on Lehninger et al., 1999.

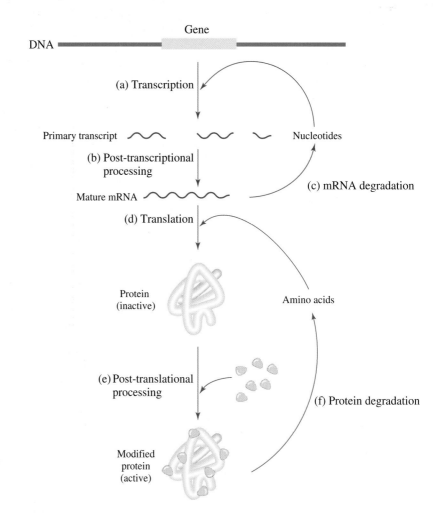

of its corresponding mRNA. The quantity of mRNA is determined as the rate of mRNA synthesis minus the rate of mRNA breakdown. Here we will concentrate on the regulation of transcription initiation.

There are two fundamental types of gene expression:

1. **Expression of constitutive genes** refers to continuous transcription, resulting in a constant level of certain protein products. These housekeeping genes maintain a constant supply of their products for general cell maintenance and central metabolism.
2. **Expression of inducible or repressible genes** refers to those that can be activated (induced) to increase the level of mRNA and protein product or can be deactivated (repressed) to decrease the level of mRNA and protein. These genes are regulated by the action of RNA polymerase and molecular signals such as **regulatory proteins,** hormones, and metabolites.

Figure 12.16

The operon model for regulation of gene transcription in prokaryotes. An operon is a regional unit on the chromosome that carries the genes for related proteins. The operon is comprised of structural genes, a promoter region, a binding site for activators, and a binding site for repressors (operator).

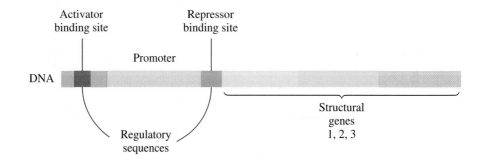

Principles of Regulating Gene Expression

Several fundamental principles of regulating gene expression are presented here.

1. In many prokaryotic cells, the genes for proteins that are related in function are clustered into units on the chromosome called **operons** (Figure 12.16). For example, the enzymes for a single metabolic pathway would be expected in such a cluster. The components within these units are:

   a. The genes to be transcribed and translated (called the structural genes).

   b. The promoter region, which is responsible for RNA polymerase binding to the initiation site (see Chapter 11, Section 11.4).

   c. A binding site for activators.

   d. A binding site for repressors (called an **operator**).

   All structural genes are regulated by nucleotide sequences upstream from the start site ($+1$). The operon model, which was postulated by Jacques Monod and François Jacob in 1960, serves as a paradigm for prokaryotic gene regulation; however, other regulatory processes also are important.

   The key participant in transcription is RNA polymerase. Recall that transcription is initiated when RNA polymerase binds to a promoter region on the DNA. The nucleotide sequences in promoter regions vary considerably, which changes the affinity of RNA polymerase binding. The number of mRNA molecules made is influenced by the affinity of RNA polymerase binding. For constitutive genes, RNA polymerase binding to promoter leads to transcription without further regulation. For inducible or repressible genes, other levels of control are superimposed onto the transcription process. Expression of these genes is controlled by regulatory proteins and other molecular signals. Regulation of eukaryotic gene expression is a more complicated process in several ways:

   i. Complex sets of regulatory elements are present in promoter regions.

   ii. Three classes of RNA polymerases with different modes of regulation are present.

   iii. The DNA is much more complex in size and structure.

2. RNA polymerase activity is mediated by **regulatory proteins.** Two major types of proteins that influence the action of RNA polymerase have been identified. Both types recognize and bind to specific DNA sequences on the chromosome that turns the polymerase "on" or "off":

   a. **Activators** are regulatory proteins that bind next to the promoter regions and assist the binding of RNA polymerase to the adjacent promotor (Figure 12.16). This increases the rate of gene transcription.

   b. **Repressors** are proteins that bind to specific base sequences in the promoter regions (called operator in prokaryotic cells). When bound, these proteins prevent the RNA polymerase from gaining access to the promoter; thus, transcription is blocked.

   The primary mode of action of the regulatory proteins is to recognize and bind to specific base sequences on the chromosome. The proteins have discrete DNA binding domains within their polypeptide chains. The regulatory proteins can "read" DNA sequences and bind to specific regions. Additionally, some regulatory proteins also have protein binding domains that allow them to bind to RNA polymerase, other regulatory proteins, and with themselves to form dimers, trimers, and so on. Binding between protein molecules most likely involves hydrophobic interactions and hydrogen bonding. The action of a repressor or activator protein is influenced by other molecular signals such as small metabolites or hormones. The binding of these small molecules to regulatory proteins may increase or decrease the affinity of the protein–DNA interaction. Before we discuss the details of regulatory protein structure and function we will pause to consider the common types of gene regulation.

3. Most regulatory processes can be classified into one of four mechanistic types (Figure 12.17). Positive regulation refers to the action of an activator on transcription. Two modes are possible. In the first type, transcription proceeds until a specific molecular

Figure 12.17

Several modes of gene regulation at the transcription level. (a) Positive regulation; binding of molecular signal causes dissociation of activator from DNA. (b) Positive regulation; binding of molecular signal causes strong binding of activator to DNA. (c) Negative regulation; binding of molecular signal causes dissociation of operator from DNA. (d) Negative regulation; binding of molecular signal causes stronger operator binding to DNA and inhibition of RNA polymerase action.

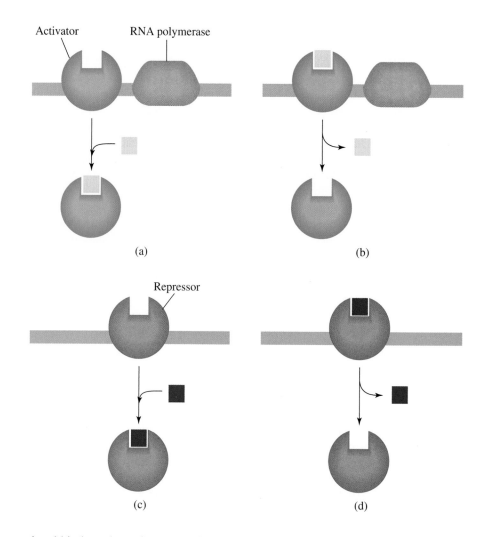

signal binds to the activator, causing it to dissociate from the DNA (Figure 12.17a). In the second positive mode, only the activator–molecular signal complex binds to DNA. If the molecular signal dissociates, the activator falls off the DNA stopping transcription (Figure 12.17b). In negative regulation, binding of repressor causes inhibition of transcription. Some repressors dissociate from the DNA when a specific signal molecule is present (Figure 12.17c), whereas other repressors require the binding of a signal molecule, which is necessary for repressor binding and transcription inhibition (Figure 12.17d). All four types of regulation are found in prokaryotic cells; positive regulation is an especially common feature of eukaryotic gene expression. At the conclusion of this section several examples of some of these common modes of gene regulation are provided.

4. Regulatory proteins have common structural features and binding characteristics. Now that many regulatory proteins have been discovered and studied, it is becoming clear that they have much in common. Regulatory proteins have discrete binding domains that allow them to recognize and bind to specific DNA sequences. The DNA binding domains on the proteins are relatively small, consisting of 20 to 100 amino acid residues. How do the proteins interact with DNA? Molecular recognition is the result of an exact fit between the surfaces of two molecules. Exact fit implies favorable interactions that hold the molecules together. At the outside edge of the DNA double helix is the major groove where the nucleotide bases are sufficiently exposed for possible hydrogen bonding to proteins. Hydrogen bonding between amino acid residues and nucleotide bases is possible without disrupting the base pairing and without unwinding the double helix (Figure 12.18). The amino acid residues

(a) Thymine:adenine

(b) Cytosine:guanine

of the regulatory proteins that participate in hydrogen bonding are the side chains of lysine, arginine, glutamate, asparagine, and glutamine. In each DNA–protein complex many possible contacts lead to specific and relatively tight binding.

Three Classes of Regulatory Proteins

The structures of many regulatory proteins have been elucidated, making it possible to search for similar characteristics. It has been discovered that about 80% of the currently known regulatory proteins can be classified into one of three classes based on the presence of common structural motifs (see Chapter 5, Sections 5.2 and 5.3):

1. The helix–turn–helix motif
2. The zinc finger motif
3. The leucine zipper motif

The **helix–turn–helix motif,** which has a length of about 20 amino acid residues, consists of two short α-helical regions of 7 to 9 amino acids connected by a β turn, often caused by the amino acid glycine. It is the most common DNA binding domain in prokaryotic regulatory proteins. One of the α-helixes, the recognition helix, can bind to DNA by fitting snugly into the major groove (Figure 12.19). Two of the best known regulatory proteins with this structural motif are *lac* repressor and *trp* repressor.

A second structural motif common in regulatory proteins is the **zinc finger motif.** To date, this DNA binding domain has been found only in eukaryotic regulatory proteins. This interesting motif is a region composed of about 30 amino acids. The key features are four amino acid residues whose combined side chains form a binding site for a single Zn^{2+}. Three families of zinc finger proteins have been discovered. They are distinguished by the zinc binding amino acid residues: (a) Cys-Cys-His-His, (b) Cys-Cys-Cys-Cys, and (c) Cys-Cys-His-Cys (Figure 12.20). The sulfur and/or nitrogen atoms in each of the four amino acid side chains coordinate to the zinc ion. Two well-studied zinc finger regulatory proteins are the transcription factor TFIIIA from *Xenopus laevis* (the African clawed toad) and glucocorticoid receptor protein. Proteins with zinc fingers probably bind to DNA in the major groove and wrap around the double-helix axis (Figure 12.21).

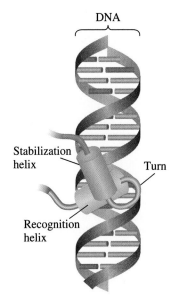

Figure 12.19

The helix–turn–helix structural motif for regulatory proteins. The recognition helix may be able to bind into the major groove of DNA, thus blocking transcription.

One of the first zinc finger regulatory proteins studied was found in the African clawed toad, *Xenopus laevis.*

The third structural motif found in regulatory proteins is the **leucine zipper motif,** which allows the proteins to interact with each other (Figure 12.22). The characteristic feature is an α-helix region of approximately 30 amino acids, with the amino acid leucine occurring about every seventh residue. The leucine residues protrude from one side of the protein. This structural motif allows two molecules of the regulatory protein to combine by forming hydrophobic interactions between the leucine-rich areas. The combined proteins with alternating leucine residues take on a "zipperlike" appearance. Hence, this structural motif serves to provide protein–protein interactions and functions in protein dimerization. Regulatory proteins with leucine zipper motifs often have a DNA binding domain, composed of an α-helix with a high content of lysine and arginine, that is adjacent to the protein binding region. Thus, leucine zipper proteins have the distinction of two possible types of interactions, one with the DNA and one with another regulatory protein molecule.

Examples of Gene Regulation

With some of the fundamental principles of gene regulation described, we can apply them to actual regulation systems. Several examples of regulatory mechanisms are listed in Table 12.5 and described next.

The *lac* Operon

The *lac* operon regulatory system, discovered in 1960 by Monod and Jacob, was the first regulatory mechanism understood in detail. The structural genes in this operon code for three enzymes that are required for lactose metabolism. In the absence of substrate lactose in the growth medium for *E. coli* cells, the *lac* operon is repressed. The specific *lac* repressor molecule binds to the operator, thereby inhibiting transcription of the structural genes by RNA polymerase and slowing the synthesis of enzymes for lactose metabolism (see Figure 12.17c). When lactose is present, it binds to the *lac* repressor, thus causing its dissociation from the operator site. Under these conditions the lactose-metabolizing enzymes are synthesized.

The *lac* operon is also regulated by glucose concentrations in the growth medium. Glucose is the preferred nutrient for *E. coli* growth and its presence represses the metabolism of other carbohydrates, including lactose. Glucose acts through the regulatory protein, catabolite activator protein (CAP). When lactose is present and glucose is absent, CAP binds near the *lac* promotor and enhances transcription of lactose metabolism enzymes. When lactose and glucose both are present, CAP dissociates from the DNA and decreases the production of lactose metabolism enzymes. The action of glucose on CAP is mediated by the second messenger (cAMP) (see Chapter 2, Section 2.5).

Figure 12.20

The zinc finger motif for regulatory proteins, showing possible binding sites for the Zn^{2+}. Sulfur atoms in Cys residues and nitrogen atoms in the His side chains coordinate to the metal ion. *Source:* Redrawn from Wolfe, 1993.

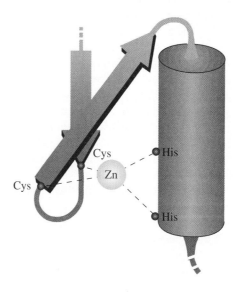

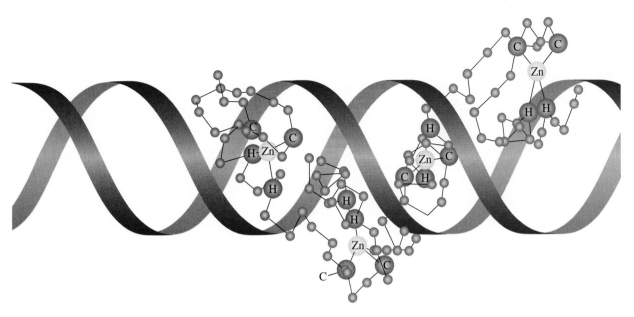

Figure 12.21

A zinc finger regulatory protein interacting with DNA. C and H represent cysteine and histidine, respectively. *Source:* From Annual Reviews, Inc., 1990.

The *trp* Operon

The *trp* operon regulatory system controls the synthesis of tryptophan in *E. coli*. The *trp* operon is composed of a regulatory region (promotor, operator, etc.) and five structural genes that code for enzymes needed to synthesize tryptophan from chorismate (see Chapter 19, Section 19.2; Figure 19.9). If the amino acid tryptophan is present in growth medium,

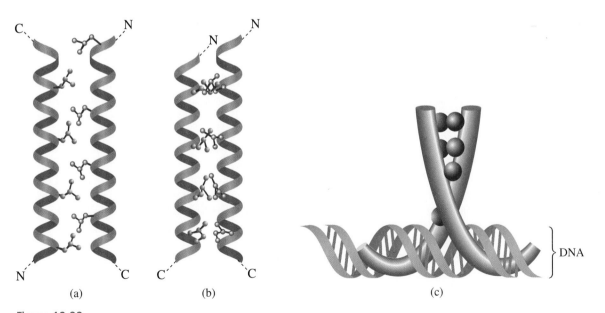

(a) (b) (c)

Figure 12.22

(a) Regulatory proteins in the leucine zipper class have α-helical regions of about 30 amino acid residues. About every seventh residue is the amino acid leucine with side chains shown in red.
(b) Two proteins can combine by hydrophobic interactions between leucine residues to form dimers.
(c) The dimer can open like a zipper to bind to DNA. Leucine side chains are shown as blue spheres. *Source:* Parts (a) and (b) based on *Science,* 1989, 246:911; Part (c) based on Stryer, 1995.

Table 12.5
Examples of gene regulation

Regulatory System	Regulatory Protein and Structural Motif	Molecular Signal	Mechanism of Action
lac operon (*E. coli*)	Tetramer of identical subunits, helix–turn–helix	Lactose or allolactose	Negative; signal causes dissociation of regulatory protein
lac operon (*E. coli*)	Catabolite activator protein, dimer of identical subunits, helix–turn–helix	Glucose and cAMP	Glucose presence represses *lac* genes; glucose absence stimulates *lac* genes
trp operon (*E. coli*)	Dimer of identical subunits, helix–turn–helix	Tryptophan	Negative; signal binds to repressor causing binding to operator; inhibits synthesis of tryptophan
Steroid hormone response elements (eukaryotes)	Glucocorticoid receptor protein, dimer, two zinc fingers	Glucocorticoid	Regulation of carbohydrate metabolism
Metal response elements (eukaryotes)	MTF-1 (fish), six zinc fingers	Heavy metal ions, Zn^{2+}, Hg^{2+}, etcetera	Synthesis of metallothionein, a protein to complex toxic metal ions

the biosynthetic enzymes are not needed and transcription of the structural genes is repressed. Their regulation is mediated by Trp repressor. Tryptophan present in the growth medium binds to protein repressor forming a complex that associates with the operator region. The trp:repressor complex binds to the operator region thus blocking the transcription of the five structural genes by RNA polymerase.

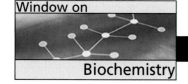

Window on

Biochemistry

12.2 Toxic Metals and Metallothionein Gene Regulation

Organisms must be protected against the ubiquitous presence in the environment of toxic heavy metals such as mercury, lead, cadmium, zinc, and copper. Most organisms require for normal growth very low concentrations of some of the metals (see Section 1.2). However, they are very toxic in the concentrations often found in rivers, streams, and watershed areas. In eukaryotes, ingested metals are detoxified by the protein **metallothionein (MT)**. MT is a low-molecular-weight, cysteine-rich protein that has a high affinity for binding metal ions. It is present in the kidney, liver, and many other cell types. The water-soluble, complexed metal ions are readily removed from cells. MT is always present in cells in relatively low concentrations, but the concentration of MT increases in response to an increase in metal ion concentration. The levels of MT in an organism are often used as a biomarker to evaluate the extent of heavy metal contamination in the local environment.

The MT gene promoter region is structurally complex. It consists of several elements including a TATA box, at least four regions called metal response elements (MRE), and one glucocorticoid response element. The MREs are designed to accept specific transcription factor proteins. These transcription factors must be complexed with metal ions in order to bind to the specific regions in DNA. This binding induces RNA polymerase to transcribe the structural gene for MT, thus producing increased MT.

The complex regulation of MT synthesis is used to an advantage in the treatment of Wilson's disease. This is a rare genetic disorder characterized by excessive storage of copper in the body especially in the liver, kidney, and brain. One experimental drug is zinc acetate. The presence of zinc ions stimulates the synthesis of MT which can then bind and remove the excess copper ions. Copper ions bind to MT with 100,000 times the affinity of zinc so copper can readily displace bound zinc.

Steroid Hormone Response Elements (HRE)

Steroid hormone control of metabolism in humans is mediated through a complex regulatory system. These hydrophobic hormones, such as estrogen, progesterone, and glucocorticoids, present in the bloodstream, are able to diffuse through the plasma membranes of target cells (see Chapter 9, Section 9.4). In the cell nucleus, the hormones form complexes with regulatory proteins and bind to specific promotor regions in the DNA called hormone response elements. Binding of the hormone–regulatory protein complex to DNA may enhance or repress transcription depending on the hormone. The glucocorticoid receptor system is described in Table 12.5.

Summary

■ Protein synthesis (translation) in all organisms has the following fundamental characteristics: (a) The process occurs on ribosomal particles; (b) a protein is synthesized beginning at its amino terminus and proceeding to its carboxyl terminus; (c) amino acids are activated by covalent linking to tRNA molecules; (d) the relationships between mRNA base sequence and protein amino acid sequence is defined by the genetic code.

■ Protein synthesis proceeds in three stages: initiation, elongation, and termination.

■ Protein synthesis is energy consuming, requiring the cleavage of four phosphoanhydride bonds in ATP and GTP.

■ Initial translational products are not biologically active and must be modified by proteolytic cleavage, amino acid functional group change, and attachment of prosthetic groups and carbohydrates.

■ Proteins synthesized on ribosomes are later sorted by the presence of signal sequences and transported to other cell compartments where they are needed (protein targeting).

■ Proteins that have served their useful purposes are labeled with ubiquitin to mark them for destruction by the proteolytic action of the proteasome.

■ Regulation of gene expression is controlled at the level of transcription initiation. Regulatory proteins recognize and bind to specific base sequences on the chromosomes, which turns RNA polymerases "on or off."

■ Regulatory proteins have common structural features and are classified accordingly: helix–turn–helix proteins, zinc finger proteins, and leucine zipper proteins.

Study Exercises

Understanding Terms

12.1 Define the following terms in 25 words or less.

    a. Translation
    b. Codon
    c. Amino acyl adenylate intermediates
    d. 70S initiation complex
    e. Polyribosome
    f. Protein targeting
    g. Ubiquitin
    h. Regulatory proteins
    i. Operon
    j. Zinc fingers
    k. Repressors
    l. Anticodon
    m. Antibiotics
    n. Peptidyl transferase
    o. *Lac* operon
    p. Post-translational processing

Reviewing Concepts

12.2 Write the sequence of amino acids in a polypeptide translated from the following mRNAs.
    a. 5′ UUUCUAGAUAGAGUU
    b. 5′ GGAGGAGUAAGUUGU
HINT: See Figure 2.7

12.3 Write out the sequence of bases in mRNA that encodes the message for the peptide His-Asn-Pro. Is there more than one possible mRNA sequence that answers this question? Explain.

12.4 Compare the following characteristics for the DNA binding domain in helix–turn–helix regulatory proteins and zinc fingers.

    a. Amino acid composition
    b. Number of amino acid residues
    c. Sequence of amino acids
    d. Presence of α-helices, β-sheets, turns
    e. Presence of metal ions

12.5 Differentiate between the A and P binding sites on the ribosome–mRNA complex.

12.6 Assume that a mRNA molecule has the following sequence:

5'AUGCUCACUUCAGGGAGAAGC

a. What polypeptide sequence would be translated from this mRNA?

b. Assume that nucleotide 4 (C) is deleted from the mRNA. What polypeptide sequence would now be translated from this modified mRNA?

12.7 The word "translation" describes which of the following steps in the flow of genetic information?

a. DNA → DNA

b. DNA → RNA

c. RNA → protein

12.8 Briefly describe the function of each of the following in protein synthesis.

a. Ribosomes

b. Codon

c. Anticodon

d. Aminoacyl-tRNA synthetases

e. Peptidyl transferase

12.9 Which of the following statements about protein synthesis are true?

a. Protein synthesis occurs by adding amino acids to the amino terminus of the growing polypeptide chain.

b. The first amino acid to be incorporated into a protein is usually *N*-formylmethionine.

c. Amino acids are selected and activated by a special form of rRNA.

d. Peptidyl transferase is a ribozyme.

12.10 Complete the following questions by identifying characteristics of the aminoacyl-tRNA synthetase–catalyzed reaction.

a. Source of energy (ATP or GTP)? _____

b. Fate of ATP molecule? _____

c. Enzyme-bound intermediate? _____

12.11 Differentiate between constitutive and inducible genes.

12.12 Which of the following features are characteristic of the translation process?

a. Occurs at the surfaces of mitochondria.

b. The first amino acid incorporated is the amino terminus.

c. Amino acids are activated by linkage to cyclic AMP.

d. Aminoacyl-tRNA synthetases catalyze activation of amino acids.

e. The process occurs on ribosomal particles.

12.13 Which of the following biochemical processes are part of posttranslational modification events?

a. Modification of amino acid residues

b. Addition of FAD and other prosthetic groups to proteins

c. Formation of disulfide bonds

d. Proteolytic cleavage

12.14 Many types of molecules and molecular associations are required for the translation process. Circle those listed below that are *not* directly involved in translation.

a. mRNA

b. Aminoacyl-tRNA

c. DNA

d. Peptidyl transferase

e. DNA polymerase III

f. Ribosomes

12.15 Compare and contrast the structures and functional characteristics of the two types of regulatory proteins, activators and repressors.

Working Problems

12.16 Approximately how many phosphoanhydride bonds must be hydrolyzed during translation to synthesize a 300-amino acid protein? Assume that you are starting with a functional mRNA, ribosomal subunits, tRNA, and free amino acids.

12.17 Recall from Chapter 11 that mutations are caused by base changes in DNA. Which of the following amino acid substitutions are caused by a single base mutation?

a. Ser → Pro **d.** Ser → Phe

b. Val → Ser **e.** Lys → Glu

c. Leu → Phe **f.** Gly → Ala

12.18 Show how the amino acid residues asparagine and glutamine in regulatory proteins could hydrogen bond to the thymine:adenine pair in DNA without disrupting complementary base pairing.

12.19 Write the reaction catalyzed by each of the following enzymes and describe physical properties of the enzyme.

a. Aminoacyl-tRNA synthetases

b. Peptidyl transferase

c. RNA polymerase

12.20 Could the 20 amino acids be incorporated into proteins using a genetic code consisting of one nucleotide base = one amino acid? How many different types of amino acids could be coded with this ratio?

12.21 Explain how metal ions are bound in the protein metallothionein. What atoms in cysteine residues complex with metal ions?

12.22 Name the type of bond that links the following molecules.

    a. Amino acid to amino acid in proteins

    b. Nucleotide to nucleotide in DNA

    c. Amino acid to tRNA

    d. Nucleotide to nucleotide in RNA

    e. Codon in mRNA to anticodon in aminoacyl-tRNA

12.23 What is the advantage to the cell of having polyribosomes working on a single mRNA molecule rather than just one ribosome?

12.24 The amino acids hydroxylysine and hydroxyproline are not incorporated into a protein using the genetic code and therefore there is no triplet code listed for them in Figure 2.7. Explain how these amino acids become present in proteins.

12.25 Could the 20 amino acids be incorporated into proteins using a genetic code consisting of two nucleotide bases = one amino acid? Explain.

12.26 Complete the following biochemical reactions of amino acid residue modification that are described in this chapter.

    a.

    b.

12.27 Describe the kind of chemical bonding that holds together a codon with its anticodon.

HINT: Is it covalent, noncovalent, or both?

12.28 Circle those amino acids in the list below that are altered by biochemical reactions when they are present as residues in proteins.

    Ala   Pro
    Tyr   Ser
    Lys   Thr
    Val

12.29 Describe the kind of chemical bond that binds ubiquitin to proteins targeted for destruction.

12.30 Describe the action of the *trp* operon in 25 words or less.

Writing Biochemistry

12.31 Your biochemistry instructor has asked each student in the class to prepare and hand in two essay questions on the material covered in Chapter 12. Some of the questions submitted will be selected by your instructor for use on the next hour exam. The only rule is that the questions should require a 25- to 50-word answer. Prepare two questions for submission and also write answers to the questions for your instructor to use as key.

12.32 Select one of the antibiotics in Figure 12.11 for study. Use an advanced biochemistry textbook or a literature reference to learn all you can about the compound. Then, write a 200-word review of the antibiotic. Your writing should include a discussion of the source of the antibiotic and how it functions to control the growth of microorganisms.

Applying Biochemistry

12.33 Protein synthesis is controlled by the interactions of regulatory proteins with specific regions on DNA. The most important types of bonding between the proteins and DNA include hydrogen bonding, hydrophobic interactions, and ionic interactions.

    a. Describe these three types of bonding and show how amino acid residues in proteins can interact with nucleotides in DNA.

    b. Interactions between regulatory proteins and DNA are said to be **reversible** and **noncovalent.** Explain why both of these characteristics are essential in regulatory protein–DNA interactions.

12.34 Large amounts of energy are required for protein synthesis. The energy from the cleavage of four phosphoanhydride bonds in ATP and GTP are necessary. Outline the steps of protein synthesis to show how the energy from ATP and GTP are used.

Further Reading

Borman, S., 2000. Intricacies of the proteasome. *Chem. Eng. News* March 20:43–48.

Cohen, J., and Hogan, M., 1994. The new genetic medicines. *Sci. Am.* 271(6):76–82.

Collins, F., et al., 2001. The human genome. *Nature* 409:745–748.

Crick, F., 1966. The genetic code. *Sci. Am.* 215(4):55–62.

Grunstein, M., 1992. Histones as regulators of genes. *Sci. Am.* 267(4):68–74.

Ibba, M., et al., 2000. The adapter hypothesis revisited. *Trends Biochem. Sci.* 25:311–316.

Henry, C., 2000. Antibiotic resistance. *Chem. Eng. News,* March 6:41–46.

Kirschner, M., 1999. Intracellular proteolysis. *Trends Biochem. Sci.* 24: M42–M45.

Levy, S., 1998. The challenge of antibiotic resistance. *Sci. Am.* 278(3):46–53.

Macay, J., and Crossley, M., 1998. Zinc fingers are sticking together. *Trends Biochem. Sci.* 23:1–4.

McKnight, S., 1991. Molecular zippers in gene regulation. *Sci. Am.* 264(4):54–64.

Nirenberg, M., 1963. The genetic code II. *Sci. Am.* 209(3):80–94.

Nissen, P. et al., 2000. Structural basis of ribosome activity in peptide bond synthesis. *Science* 289:920–930.

Nomura, M., 1997. Reflections on the days of ribosome reconstitution research. *Trends Biochem. Sci.* 22:275–279.

Rawls, R., 2000. RNA holds key to protein synthesis. *Chem. Eng. New* Aug. 14, pp. 9–10.

Rhodes, D., and Klug, A., 1993. Zinc fingers. *Sci. Am.* 268(2):56–65.

Schimmel, P., and Ribas de Pouplana, L., 2000. Footprints of aminacyl-tRNA synthetases are everywhere. *Trends Biochem. Sci.* 25:207–209.

Thieffry, D., and Sarkar, S., 1998. Forty years under the central dogma. *Trends Biochem. Sci.* 23:312–316.

Tijan, R., 1995. Molecular machines that control genes. *Sci. Am.* 272(2):54–61.

Venter, J., 2001. The sequence of the human genome. *Science* 291:1304.

Weintrab, H., 1990. Antisense RNA and DNA. *Sci. Am.* 263(1):40–46.

Woodson, S., and Leontis, N., 1998. Structure and dynamics of ribosomal RNA. *Curr. Opin. Struct. Biol.* 8:254–300.

Chapter 13

**Recombinant DNA
and Other Topics
in Biotechnology**

The techniques and products of biotechnology are highlighted in this Canadian stamp.

Throughout the development of biochemistry as an academic discipline, there have always been "practical" or "applied" aspects of the science. Recall that the historical origins of biochemistry involved the practical technologies of fermentation (wine and beer making), baking, nutrition, and treatment of human maladies. Discovery of the practical uses of a process was usually followed by an attempt to gain a more fundamental or molecular understanding of the process. Because much of what we have read in this text appears to be derived from "basic" research, modern biochemistry may appear to have lost its "practical" side. However, a major new set of experimental tools is bringing together biochemistry, biology, chemistry, physics, computer science, and engineering: **Biotechnology** seeks to apply our understanding of the intricate workings of the cell to the solution of practical problems. The general definition of biotechnology is "the practical use of biological cells and cellular components." Examples of current biotechnology projects include the following:

1. The making of cheese, wine, and other food commodities.
2. The use of bacterial cells to produce large quantities of scarce proteins needed to treat diseases.
3. Using enzymes to catalyze reaction steps in the industrial production of speciality chemicals or biochemicals.
4. Modifying the genes of a plant so it can grow under adverse conditions.
5. Production of fuel alcohol from plant starch.
6. Using bacteria and plants for cleanup of chemical waste sites.
7. Mining of metals, including silver and gold.
8. Gene replacement in individuals with genetic disorders (gene therapy).
9. Identification of biological specimens at crime scenes.
10. Production of recombinant bovine growth hormone, which when injected into cows increases milk production.

This long list provides evidence for the current advancement of practical biochemistry. We have also observed during the past decade the establishment of hundreds of biotechnology firms that have specific goals to improve medicine, agriculture, and other human endeavors.

Perhaps the most rapid and most controversial advances in biotechnology depend on our ability to change the genetic characteristics of fundamental life forms. This has been made possible by the development of techniques for manipulating the nucleic acids, especially DNA. The generic term to describe this new activity is **recombinant DNA** technology. Today we are not satisfied with the slow, natural, evolutionary processes that change or "improve" organisms. We seek to "engineer" changes in an organism's genome that will then perform tasks for us and provide a perceived benefit. At this point, our knowledge and activities have provided both an extensive understanding of life and the ability to alter it. New laboratory procedures have made DNA one of the easiest molecules with which to work. Using the relatively new techniques of recombinant DNA, the DNA from one species can be cut into well-defined fragments and spliced into the DNA of another species. The new hybrid or recombinant DNA, when introduced into a host cell, results in the production of proteins not normally made by that cell. Protein products of this recombinant technology are now showing practical value in medicine, industrial processes, and agriculture.

We must become familiar with a new set of terms to understand recombinant DNA technology and its application. These terms include **vectors, plasmids, cloning, transformation, genomic libraries, hybridization probes,** and the **polymerase chain reaction**.

13.1   Recombinant DNA Technology

Our knowledge of DNA structure and function has increased at a somewhat gradual pace over the past 100 years, but some discoveries have had a special impact on the progress and direction of DNA research. Although DNA was first discovered in cell nuclei in 1868, it was not confirmed as the carrier of genetic information until 1944. This major discovery was closely followed by the announcement of the double-helix model for DNA structure in 1953 and by the discovery of the biomolecular participants and mechanisms in DNA replication, transcription, translation, and regulation. The deciphering of the genetic code in the 1960s was a major milestone in the development of biochemistry and molecular biology. The more recent development of technology that allows genetic manipulation by insertion of "foreign" DNA fragments into the natural, replicating DNA of an organism may well have a greater impact on the direction of DNA research than the previous discoveries. In the short time since the first construction and replication of plasmid recombinant DNA in the mid-1970s, many scientific, industrial, and medical applications of the new technology have been announced. Some of the predicted achievements in recombinant DNA research have been slow in coming; however, future workers in biochemistry and related fields must be familiar with the principles, techniques, and applications, of biotechnology.

Molecular Cloning

DNA recombination or **molecular cloning** consists of the covalent insertion of a DNA fragment from one type of cell or organism into the replicating DNA of another type of cell. Essentially any segment of DNA can be selected and prepared for copying. Many copies of the hybrid DNA may be produced by the progeny of the recipient cells; hence, the DNA molecule is cloned. If the inserted fragment is a functional gene carrying the code for a specific protein and an upstream promoter region is present in the DNA, many copies of that gene and translated protein are produced in the host cell. This process has become important for the large-scale production of proteins (insulin, somatotropin, tissue plasminogen activator, mutant hemoglobins, bovine growth hormone, and others) that are of value in medicine, agriculture, and basic science but are difficult and expensive to obtain by other methods.

The basic steps involved in constructing a recombinant DNA molecule are listed below and are outlined in Figure 13.1.

1. Select and isolate a DNA molecule to serve as the carrier (called the vehicle or **vector**) for the foreign DNA.
2. Cleave the deoxyribonucleotide strands of the vector (DNA carrier) with a restriction endonuclease.
3. Prepare and insert the foreign DNA fragment into the vector. This produces a **recombinant** or hybrid DNA molecule.
4. Introduce the hybrid DNA into a host organism (often a bacterial cell, but animal and plant cells may also be used), where it can be replicated. This process is called **transformation.**
5. Develop a method for identifying and screening the host cells that have accepted and are replicating the hybrid DNA.

Our current state of knowledge of recombinant DNA is the result of technological advances made in the 1970s by Paul Berg, Stanley Cohen, and Herbert Boyer at Stanford University. The first major breakthrough was the isolation of mutant strains of *Escherichia coli* that did not have restriction endonucleases that degraded foreign DNA. These strains are now used as host organisms for the replication of recombinant DNA. The most important organisms for cloning hybrid DNA are various strains of *E. coli* K12. The use of yeast cells, higher plant cells, and mammalian cell cultures as host cells is still in its infancy, but advances in this area will be rapid in the future. Since *E. coli* cells are presently the most versatile host organisms, techniques for their use are emphasized here.

The second advancement was the development of bacterial extrachromosomal DNA (plasmids) and bacteriophage DNA as cloning vectors. The final, necessary advance was the design of laboratory methods for insertion of the foreign DNA into the vector. The discovery of **restriction endonucleases,** enzymes that catalyze the hydrolytic cleavage of DNA at selective sites, provided a reproducible method for opening (linearizing) circular vectors. Methods for covalent joining of the foreign DNA to the vector ends and final closure of the circular hybrid plasmid were then developed. The enzyme, **DNA ligase,** which catalyzes the ATP-dependent formation of phosphodiester linkages at the insertion sites, is used for final closure (Figures 11.10 and 13.2).

Interactive Animations: Cloning; Restriction digestion and electrophoresis of DNA

Figure 13.1

Outline of a typical experimental procedure for constructing recombinant DNA.

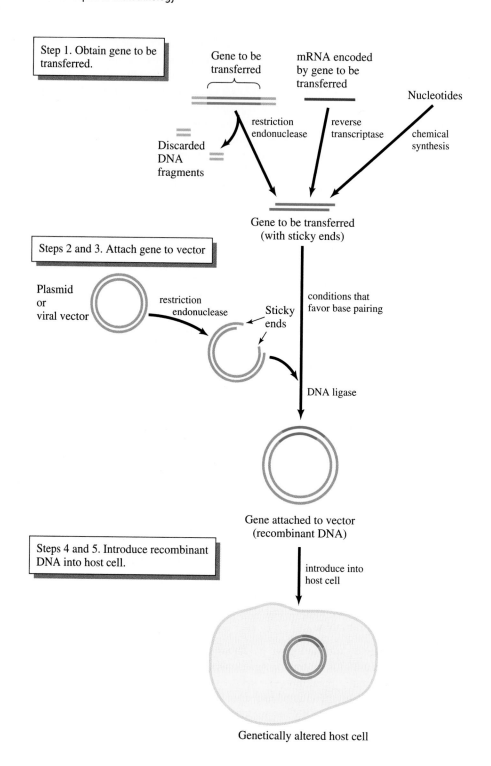

Cloning Vectors

Two types of molecular vectors are in common use for the preparation and replication of hybrid DNA in *E. coli* cells: plasmids and bacteriophage DNA.

Plasmids

Many bacterial cells contain self-replicating, extrachromosomal DNA molecules called **plasmids.** This form of DNA is closed circular, double-stranded, and much smaller than chromosomal DNA; its molecular weight ranges from 2×10^6 to 20×10^6 daltons, which

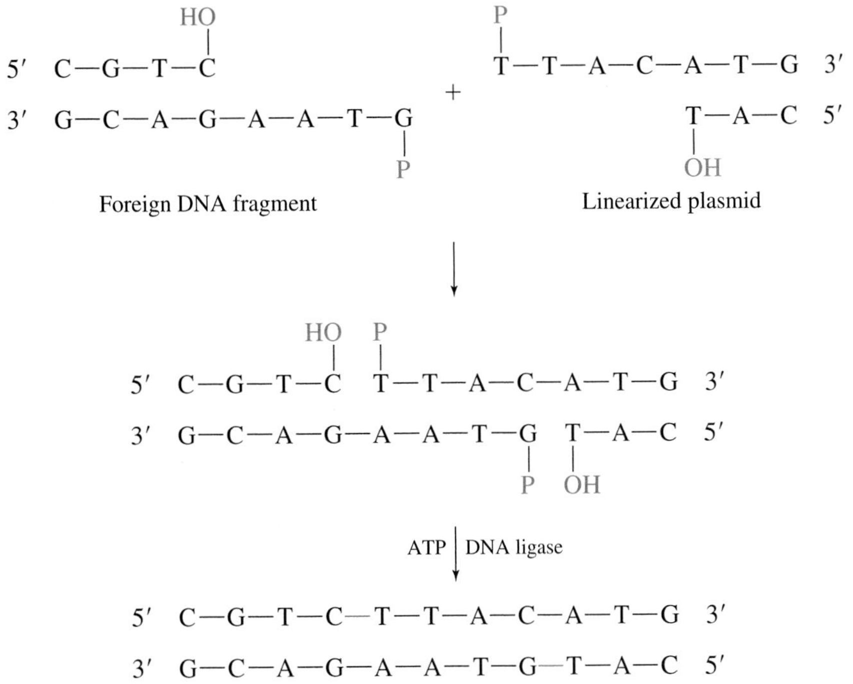

HO
|
5′ C—G—T—C

3′ G—C—A—G—A—A—T—G
                    |
                    P

Foreign DNA fragment

            +

P
|
T—T—A—C—A—T—G 3′

              T—A—C 5′
               |
              OH

Linearized plasmid

Figure 13.2
Insertion of a DNA fragment into
a linearized plasmid using DNA
ligase. DNA ligase catalyzes the
formation of phosphoester bonds.

↓

            HO P
            | |
5′ C—G—T—C T—T—A—C—A—T—G 3′

3′ G—C—A—G—A—A—T—G T—A—C 5′
                    | |
                    P   OH

ATP | DNA ligase

↓

5′ C—G—T—C—T—T—A—C—A—T—G 3′

3′ G—C—A—G—A—A—T—G—T—A—C 5′

corresponds to between 3000 and 30,000 base pairs. Bacterial plasmids normally contain genetic information for the translation of proteins that confer a specialized and sometimes protective characteristic (phenotype) on the organism. Examples of these characteristics are enzyme systems necessary for the production of antibiotics and other toxins and enzymes that degrade antibiotics. Plasmids are replicated in the cell by one of two possible modes. Stringent replicated plasmids are present in only a few copies and relaxed replicated plasmids are present in many copies, sometimes up to 200. In addition, some relaxed plasmids continue to be produced even after the antibiotic chloramphenicol is used to inhibit chromosomal DNA synthesis in the host cell (see Chapter 12, Section 12.2; Table 12.4; and Figure 12.11). Under these conditions, many copies of the plasmid DNA may be produced (up to 2000 or 3000) and may accumulate 30% to 40% of the total cellular DNA. The typical plasmid will accept foreign DNA inserts up to 15,000 base pairs.

The ideal plasmid cloning vector has the following properties:

1. The plasmid should replicate in a relaxed fashion so that many copies are produced.
2. The plasmid should be small; then it is easier to separate from the larger chromosomal DNA, easier to handle without physical damage, and probably contains only a few sites for attack by restriction endonucleases.
3. The plasmid should contain identifiable markers for screening progeny for the presence of the plasmid. Resistance to antibiotics is a convenient type of marker.
4. The plasmid should have only one cleavage site for a restriction endonuclease. This provides only two "ends" to which the foreign DNA can be attached. Ideally, the single restriction site should be within a gene, so that insertion of the foreign DNA will inactivate the gene (called insertional marker inactivation).

Among the most widely used *E. coli* plasmids are derivatives of the replicon plasmid ColE1. This plasmid carries a resistance gene against the antibiotic colicin E. The plasmid is under relaxed control and up to 3000 copies may be produced when the proper *E. coli* strain is grown in the presence of chloramphenicol. One especially useful derivative plasmid of ColE1 is pBR322. It has all the properties previously outlined; in addition, its nucleotide sequence of 4363 base pairs is known, and it contains several different restriction endonuclease cleavage sites where foreign DNA can be inserted. For example, pBR322 has a single restriction site for the restriction enzyme *Eco*RI.

Bacteriophage DNA

The λ phage has been the most widely used cloning vector in this category. λ phage DNA is a double-stranded molecule with approximately 50,000 base pairs. It has several advantages as a cloning vector:

1. Many copies of recombinant phage DNA can be replicated in the host cell.
2. The recombinant phage DNA may be efficiently packaged in the phage particle, which can be used to infect the host bacteria.
3. Recombinant phage DNA is easily screened for identification purposes.
4. λ phage DNA is larger than plasmid DNA; therefore, it is especially useful for insertion of larger fragments of eukaryotic DNA. λ phage accepts DNA fragments up to 23,000 base pairs.

The vectors for use in *E. coli* cells are not especially suitable for insertion of eukaryotic genes and other large DNA fragments. A typical eukaryotic gene may have as many as 100,000 to a few million base pairs. Large fragments of DNA can be cloned in **yeast artificial chromosomes** (YACs). Eukaryotic DNA is fragmented by restriction enzymes, inserted into laboratory-designed YACs, and cloned in yeast cells. YACs containing DNA inserts of 100,000 to 1,000,000 base pairs may be prepared for cloning.

13.2   Preparing Recombinant DNA

Design of Recombinant DNA

🔘 **Cutting Edge: Shotgun sequencing**

The foreign double-stranded DNA that is to be inserted into the vector and replicated may be prepared by one of several methods, including chemical synthesis, restriction endonuclease action on a larger DNA fragment, or reverse transcription of mRNA. For preparation of prokaryotic DNA, the most widely used method is cleavage of DNA into multiple fragments by restriction enzymes. The entire mixture of fragments may be inserted into the plasmid (shotgun method), or the fragments may be separated and purified by chromatography or electrophoretic techniques before insertion into the vector. If only a specific gene is to be inserted, it is essential to purify the DNA fragment before joining it to the vector. Isolation and cloning of a specific gene is described later in this section.

DNA fragments prepared for plasmid insertion by restriction endonuclease action have either cohesive (sticky) ends or blunt ends, as shown in Figure 13.3. The unpaired regions of cohesive ends, which may be up to five nucleotide bases in length, can be joined to a plasmid that also has been cleaved by the same restriction enzyme. This process is shown in Figure 13.2. However, this leads to an overlap region of only a few bases and may not yield a stable linkage. Stability can be given by using DNA ligase to catalyze the formation of phosphodiester linkages at the insertion sites. The reaction catalyzed by DNA ligase is shown in

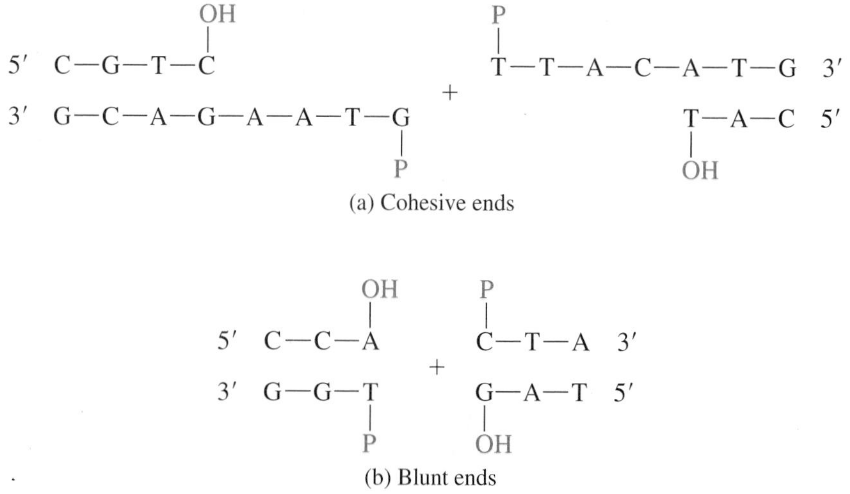

Figure 13.3

Results of the cleavage of DNA by a restriction endonuclease. Two types of ends after DNA cleavage: (a) Cohesive ends may remain weakly associated by hydrogen bonds between complementary bases and (b) blunt ends do not overlap.

(a) Cohesive ends

(b) Blunt ends

Figure 13.2 (see also Chapter 11, Section 11.2 and Figure 11.10). Blunt ends are also fused in the ATP-dependent reaction catalyzed by DNA ligase. These ends can be joined provided that the 5′ ends have phosphate groups and that a free hydroxyl group is present at the 3′ end.

Joining cohesive or blunt ends with ligase is time consuming and not especially versatile. A more widely used procedure for joining DNA fragments is the use of homopolymer tails, as shown in Figure 13.4. Here, a segment of 50 to 150 poly A (or G) residues is added to one fragment (usually the cleaved plasmid vector) and an equal length of poly T (or C) residues is added to the DNA fragment to be inserted. The tails are added by the action of

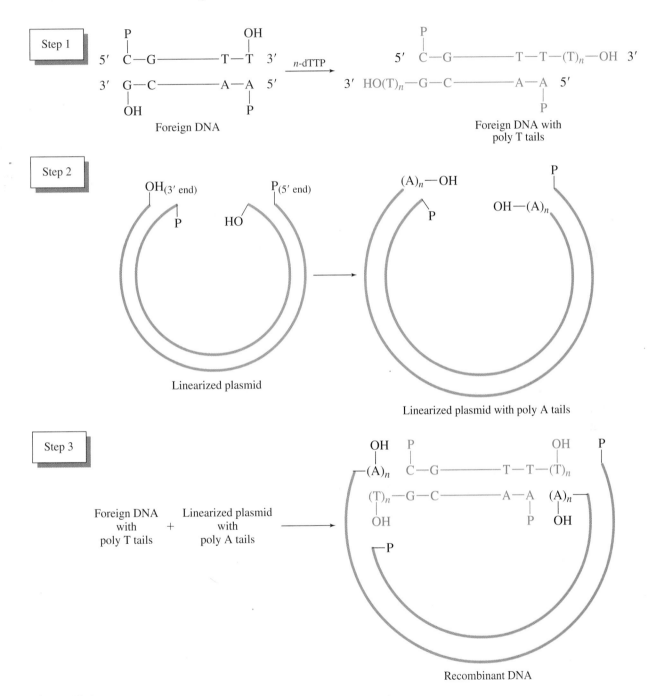

Figure 13.4

Insertion of a DNA fragment into a linearized plasmid using homopolymer tails. In Step 1, poly T tails are added to the 3′ ends of the foreign DNA. Poly A tails are added to the linearized plasmid (Step 2). When the foreign DNA with poly T tails is combined with the linearized plasmid, the A-T poly tails form complementary hydrogen bonding (Step 3).

deoxynucleotidyl transferase, an enzyme that catalyzes the addition of deoxyribonucleotides to the unblocked 3′-hydroxyl ends of single- or double-stranded DNA. When the two products with homopolymer tails are incubated together, the insertion is completed by hydrogen bond formation between the complementary homopolymer tails. The hybrid DNA formed by the insertion then takes the form of circular, double-stranded DNA. Not all of the phosphodiester bonds are intact at the joints, but the long segment of complementary base pairing at the joints is sufficient to hold the insert intact.

Transformation

Now that the recombinant DNA molecule has been prepared, it must be introduced into a host cell where it can be replicated. Current methods for incorporation of the recombinant DNA into the host cell are still rather inefficient, with only one DNA molecule in 10,000 being successfully replicated. This low efficiency, however, usually leads to a sufficient quantity of hybrid DNA and protein products for analytical purposes.

The cloning of plasmid vectors is accomplished by the **transformation** of E. coli cells. It has been demonstrated that the introduction of hybrid plasmid and phage DNA into E. coli cells is promoted by calcium chloride. The general technique for bacterial transformation consists of washing the recipient E. coli cells with calcium chloride solution and incubating the washed cells with a solution of recombinant DNA. Upon introduction into the bacterial cell, the hybrid DNA is replicated and the selective markers are expressed. If the markers express drug resistance, the successfully transformed cells have the ability to grow in the presence of a particular antibiotic.

The cloning of bacteriophage DNA is accomplished by infection of host bacterial cells. For instance, bacterial cells are grown to a midlog phase and infected with virus containing the recombinant DNA. The viral particles enter the host cell, where the recombinant DNA is replicated with expression of its selective markers.

Because the efficiency of successful hybrid DNA transplant and replication is so low, it is essential to be able to select and isolate those cells replicating the hybrid DNA. A method for selection is illustrated here using the plasmid vehicle pBR322 (Figure 13.5). Plasmid pBR322 contains two identifiable markers; both are antibiotic-resistance genes. The location of these resistance genes is shown in Figure 13.5. The section of the plasmid labeled Amp^r confers resistance to ampicillin or penicillin, and that labeled Tet^r gives resistance to tetracycline. Assume that a DNA fragment, obtained in the reaction catalyzed by the restriction enzyme BamHI, is inserted into the BamHI restriction site of the plasmid. Cohesive ends are produced on the fragment and plasmid DNA. The final step in circularization is completed by DNA ligase. Because the insertion process is not highly efficient, the product is a mixture of recircularized pBR322 with no inserted DNA and hybrid pBR322 containing the foreign DNA. When E. coli cells are incubated with the vector mixture in the presence of calcium chloride, three types of host cells are obtained: (1) E. coli cells containing neither pBR322 nor the hybrid pBR322, (2) cells containing pBR322, and (3) transformed cells containing hybrid plasmid. The three types of cells show different drug resistances. The normal E. coli cells do not survive in the presence of ampicillin; the cells containing unmodified pBR322 survive in the presence of ampicillin *and* tetracycline; and the transformed cells survive in the presence of ampicillin but not tetracycline. Because the passenger DNA is inserted within the tetracycline-resistance gene, the hybrid cells are not able to produce proteins that protect them against this antibiotic. The cells are screened for antibiotic resistance using a technique called replica plating (Figure 13.6). Cells are streaked and allowed to grow on agar plates containing ampicillin, where only host cells containing pBR322 and hybrid pBR322 can survive. Some of the viable colonies are transferred to agar plates containing ampicillin or tetracycline. These two plates are incubated for several hours at 37°C; colonies are identified that are growing on the ampicillin plate, but not on the tetracycline plate. Those colonies are most likely made up of cells containing the hybrid pBR322 plasmid. The ampicillin-resistant, tetracycline-sensitive E. coli colonies are removed from the plate and grown in liquid culture to replicate many copies of the hybrid plasmid.

Several methods are now available for the rapid isolation and analysis of small amounts of recombinant DNA. One ultimate goal of plasmid isolation and analysis is to construct a

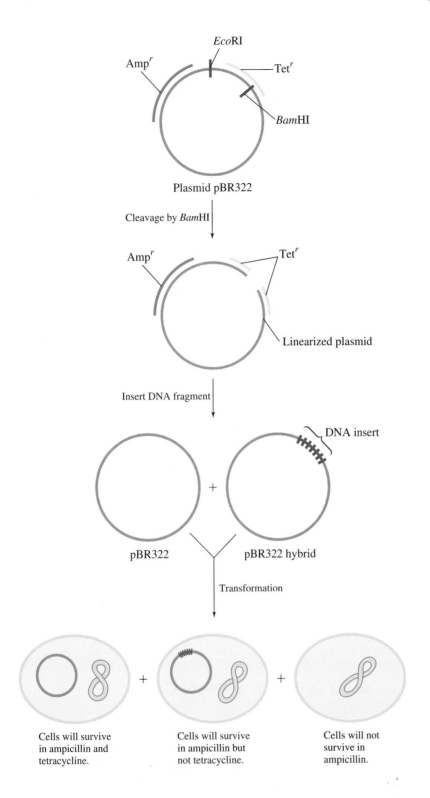

Figure 13.5
Procedure for selecting bacterial cells that contain hybrid DNA.

Plasmid pBR322

Cleavage by *Bam*HI

Linearized plasmid

Insert DNA fragment

DNA insert

pBR322   +   pBR322 hybrid

Transformation

Cells will survive in ampicillin and tetracycline.   +   Cells will survive in ampicillin but not tetracycline.   +   Cells will not survive in ampicillin.

restriction enzyme map for the plasmid. Such a map, shown in Figure 13.7, displays the sites of cleavage by several restriction endonucleases and the number of fragments obtained after digestion with each enzyme. At the center of the circle is shown the position of the two resistance genes, Amp^r and Tet^r. The innermost circle shows the site of binding by restriction enzymes that catalyze cleavage at one location. The action of each of these enzymes on pBR322 will linearize the plasmid for insertion of a DNA fragment. A zero point is defined by the site cleaved by *Eco*RI. Each concentric circle represents the action of an individual restriction enzyme that produces one or more breaks in the plasmid. For example, the

Figure 13.6
Replica plating to screen for antibiotic resistance.

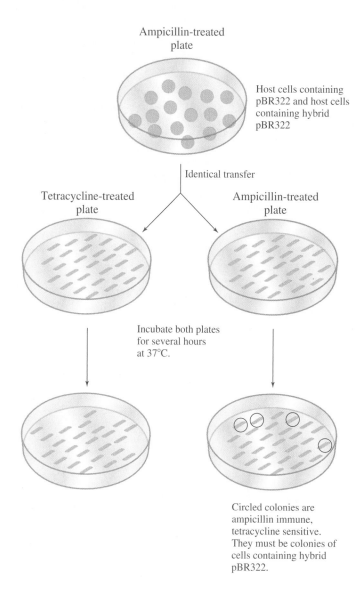

Ampicillin-treated plate

Host cells containing pBR322 and host cells containing hybrid pBR322

Identical transfer

Tetracycline-treated plate

Ampicillin-treated plate

Incubate both plates for several hours at 37°C.

Circled colonies are ampicillin immune, tetracycline sensitive. They must be colonies of cells containing hybrid pBR322.

restriction enzyme *Taq*I acts at seven sites, producing fragments labeled A, B, C, D, E_1, E_2, and F. The map is of value in the selection of plasmids for cloning and for characterization of recombinant DNA molecules by fragment analysis.

Isolation and Cloning of a Single Gene

The major goal of recombinant DNA technology is often to identify, locate, and sequence a specific gene that occurs only once in a chromosome. One begins this selective isolation process by constructing a genomic library. First, genomic DNA is cut into many thousands of large fragments using restriction endonucleases, yielding a random population of overlapping DNA fragments. To ease the preparation of recombinant DNA molecules, the fragment mixture is subjected to gel electrophoresis or density gradient centrifugation to isolate a fragment population of similar molecular size. After attachment of homopolymer tails, the fragments are inserted into linearized vectors (plasmids or phage DNA) that have been prepared to accept the DNA fragments. The DNA recombinant molecules are then cloned in suitable cells, usually *E. coli*. The final result is large population groups of cells, each containing different fragments of DNA in recombinant DNA molecules. It is hoped that all of the original genomic DNA is represented in this mixture, which is called a **genomic library** (Figure 13.8). This library must now be screened to find those cells containing the

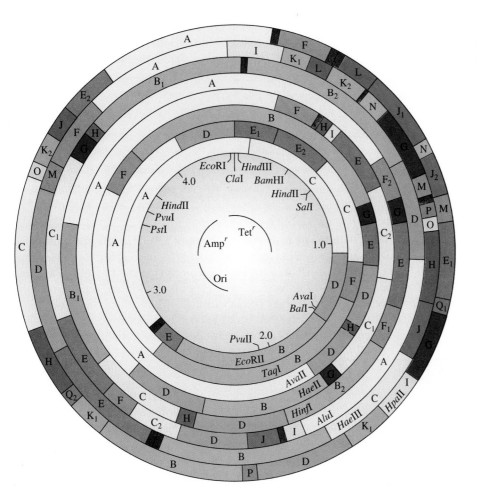

Figure 13.7
Restriction enzyme map for the plasmid pBR322.

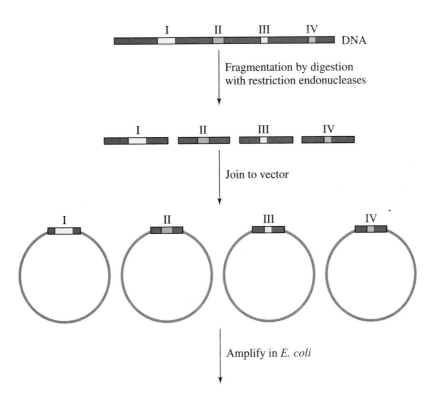

Figure 13.8
Creating a genomic library. I, II, III, and IV represent genes within the fragments.

Figure 13.9

Screening a genomic library for a specific gene (see Window on Biochemistry 13.1).

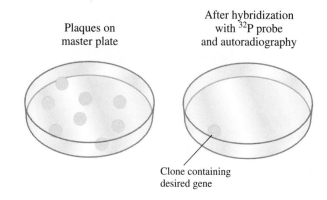

Plaques on
master plate

After hybridization
with ^{32}P probe
and autoradiography

Clone containing
desired gene

recombinant DNA with the gene of interest (Figure 13.9). The bacterial cells containing recombinant DNA are grown on an agar plate. They grow in colonies, each colony containing a different recombinant DNA. Samples of each cell type can be isolated and localized by a technique called **blotting.** Nitrocellulose paper is applied and pressed onto the agar plate to produce a replica or imprint. The paper is treated with dilute NaOH solution to lyse the cells and release recombinant DNA, which remains localized but denatured on the paper. The DNA fragment of interest is then usually identified with the use of a hybridization

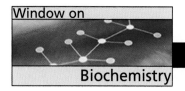

Window on

Biochemistry

13.1   Biochips—Microarray Analysis of DNA and Other Molecules

Early results from the Human Genome Project estimate that the human genome contains between 30,000 and 40,000 protein-coding genes. Using the traditional experimental approach of "one gene at a time" makes further analysis and characterization of the genome long and tedious. In the past few years, a new technology called **microarray analysis** has been introduced that promises to revolutionize the study of gene expression by monitoring the activities of thousands of genes in a single experiment. Other terms used for this technology include DNA chips, biochips, gene arrays, and genome chips.

A microarray is an orderly arrangement of experimental samples. Microarrays for nucleic acids consist of ordered sets of DNA (or DNA fragments) fixed at discrete locations on the solid surface of a glass (silicon) chip. Thousands of samples may be robotically deposited on a single one-inch-square chip. This provides a medium for using base-pairing rules (A-T and G-C for DNA; A-U and G-C for RNA) to match known DNA samples with complementary, but unknown, DNA or RNA. Two specific experimental goals, among many options, are possible: (1) identification of paired sequences in complementary DNA (cDNA) and mRNA and (2) determination of the expression level (abundance) of genes. For example, cDNA spots on the chip may be analyzed for base-pair association (hybridization) by annealing (reacting) the chip with a complex sample of mRNA or DNA. Most microarrays use RNA or DNA with fluorescent tags as a method of detection. Formation of a hybrid generates a fluorescent spot at a definite site on the chip. The color or brightness represents the relative abundance of the hybrid (see figure).

Microarrays are not limited to nucleic acid analysis. **Protein microarrays** have recently been introduced. A protein spotted thousands of times on a chip may be reacted with unknown protein mixtures and used to identify specific protein–protein interactions. Another possible format is to treat the deposited protein with drugs and study protein–drug interactions. This technology could assist pharmaceutical companies in identifying potential drug targets. The National Institutes of Health (NIH) is investigating **tissue microarrays.** These allow life scientists to survey the presence of one gene at a time but in thousands of tissue specimens. It should be possible to study the relationships between individual genes and disease and apply that information to the treatment of an individual. Specifically,

probe. A DNA or RNA molecule (probe) is prepared that is complementary to the sequence of the gene. For detection purposes, the probe is labeled with radioactive phosphorus, ^{32}P. Then the nitrocellulose paper with recombinant DNA is treated with the radioactive probe that binds only to the gene of interest. A diagram of the nitrocellulose paper reveals the location of the desired bacterial colony and recombinant DNA. This method of detection, of course, depends on at least a partial knowledge of the nucleotide sequence of a gene of interest, often the major limiting factor in isolating and cloning a gene. The nucleotide sequences of a gene can often be determined by sequence analysis of its protein product and use of the genetic code relationship between amino acid sequence and nucleotide sequence.

13.3 DNA Amplification by the Polymerase Chain Reaction

Fundamentals of the Polymerase Chain Reaction

Producing multiple copies of a particular DNA fragment need not always require the tedious and time-consuming procedures of molecular cloning, as described above. If at least part of the sequence of a DNA fragment is known, it is possible to make many copies using the **polymerase chain reaction** (PCR). The PCR method was conceived by Kary Mullis (Nobel laureate in chemistry, 1993) during a moonlit drive through the mountains of northern

**Interactive Animations:
Polymerase chain reaction**

one could study why certain drugs work better in some patients than in others and why some drugs may be highly toxic to certain patients.

Obtaining experimental data from microarrays is becoming routine, possibly too easy. The problem is that the data are too abundant. Very complex biostatistical methods, many of them still not developed, are needed to interpret completely all the information available.

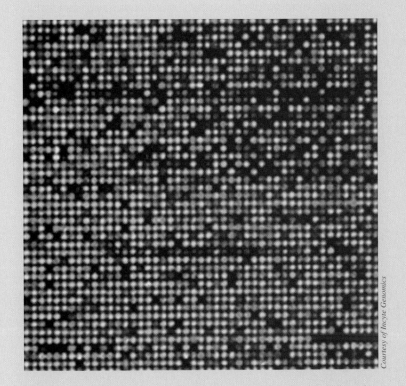

Courtesy of Incyte Genomics

Immobilized cDNA was treated with mRNA. Spot intensities and colors indicate the abundance of particular gene specific mRNAs.

California. The fundamentals of this relatively simple process are outlined using a fragment of double-stranded DNA. The DNA template is divided into five regions designated I to V in Figure 13.10. The complementary regions are I′ to V′. The region targeted for amplification is III (and III′), which can be reproduced by PCR if the nucleotide sequences of the flanking regions, II and IV, are known. Requirements for the PCR include:

1. Two synthetic oligonucleotide primers about 20 base pairs each, which are complementary to the flanking sequences II and IV.
2. A heat-stable DNA polymerase.
3. The four deoxyribonucleoside triphosphates, dATP, dGTP, dCTP, and dTTP.

The PCR is performed in cycles of three steps. Each cycle consists of:

1. Denaturation to achieve template DNA strand separation is done by heating a mixture of all components at 95°C for about 15 s.
2. Abrupt cooling of the mixture to the range of 37°C to 55°C allows the primers to hybridize with the appropriate flanking regions. The primers are oriented on the template so their 3′ ends are directed toward each other. Synthesis of DNA extends across regions III and III′. Note the dual roles played by the complementary oligonucleotides:

Figure 13.10

The polymerase chain reaction has three steps: (a) strand separation by heating at 95°C, (b) hybridization of the primers, and (c) extension of the primers by DNA synthesis. Segments are labeled I, II, III, IV, V on the original DNA strand and I′, II′, III′, IV′, V′ on the complementary strand. Primer II is in yellow and primer IV′ is in blue. Newly synthesized DNA is shown in red.

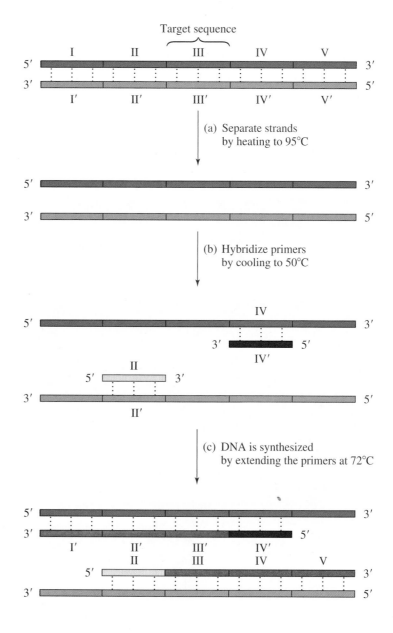

They locate the starting points for duplication of the desired DNA segment and they serve as 3′-hydroxy primers to initiate DNA synthesis. A large excess of primers to DNA is added to the reaction mixture to favor hybridization and prevent reannealing of the DNA template strands.

3. Synthesis of the targeted DNA is catalyzed by *Taq* DNA polymerase. The temperature is raised to 72°C to enhance the rate of the polymerization reaction. The enzyme extends both primers, producing two new strands of DNA, II-III-IV and II′-III′-IV′. The chosen polymerase is from a thermophilic bacterium, *Thermus aquaticus,* an organism originally discovered in a Yellowstone National Park hot spring. Other useful polymerases have been isolated from bacteria found in geothermal vents on the ocean floor. These enzymes are heat stable, so the reaction can be carried out at a high temperature, leading to a high rate of DNA synthesis. The DNA synthesis reaction is usually complete in about 30 s.

The usefulness of the PCR lies in the three steps—denaturation, hybridization, and DNA synthesis—that can be repeated many times simply by changing the temperature of the reaction mixture. Each newly synthesized strand of DNA can serve as a template, so the target DNA concentration increases at an exponential rate. In a process consisting of 20 cycles, the amplification for the DNA fragment is about a millionfold. Thirty cycles, which can be completed in 1 to 3 hours, provide a billionfold amplification. Theoretically, one could begin with a single molecule of target DNA (not enough for sequence analysis or characterization) and produce in 1 hour enough DNA for the Sanger dideoxy chain-termination sequence procedure (see Chapter 11, Section 11.6). Another benefit of the PCR is its simplicity. Instruments called thermocyclers are now commercially available that allow the laboratory technician to mix the reagents and insert the reaction mixture in the cycler, which then automatically repeats the reaction steps by changing temperatures.

The PCR has become a routine tool in basic research carried out in universities, hospitals, and pharmaceutical companies. It is especially useful to amplify small amounts of DNA for sequencing. It has been used in the **Human Genome Project,** which recently announced a rough draft for the sequence of the 3 billion base pairs in the human genome. In addition, PCR has become a powerful technique when applied to diagnostic medicine, forensics, and analysis of ancient DNA (see Window on Biochemistry 13.2).

Applications of the Polymerase Chain Reaction

PCR may be used to detect the presence of infectious bacteria and viruses in an individual even before symptoms begin to appear. HIV infection is usually confirmed using tests that detect serum antibody proteins made against the viral proteins. However, there may be a period of at least 6 months between infection and when the anti-HIV antibodies reach a detectable concentration. PCR can be used to detect proviral DNA. An especially important application is the identification of HIV-infected infants. The antibody-based test cannot be done on infants because they have maternal antibodies for a period up to 15 months after birth. Other medical applications of PCR include early detection of tuberculosis and cancers, especially leukemias, and analysis of small samples of amniotic fluid to detect genetic abnormalities in fetuses.

Forensics

Recovery of fingerprints from a crime scene has been a long-standing and well-accepted method in forensics. A new procedure, **DNA fingerprinting,** is being developed and achieving widespread use. DNA fingerprinting is the biochemical analysis of DNA in biological samples (semen, blood, saliva) remaining at a crime scene. Since every individual possesses a unique hereditary composition, each individual has a characteristic phenotype that is reflected in his or her DNA sequence. DNA fingerprinting attempts to show these genetic variations that are found from person to person. These differences can be detected by analysis of an individual's DNA. Two experimental methods are currently used in DNA fingerprinting.

🔴 **Cutting Edge: DNA fingerprinting**

Restriction Fragment Length Polymorphisms (RFLPs)

In this method, DNA samples are digested with restriction enzymes. The resulting DNA fragments are separated by gel electrophoresis, blotted onto a membrane, hybridized to a DNA probe, and analyzed by autoradiography. Results vary from individual to individual (except identical twins, who have identical DNA) because of their differences in DNA sequence. Since each person's DNA has a unique sequence pattern, the restriction enzymes cut differently and lead to different sized fragments.

PCR-Based Analysis

The second method of DNA fingerprinting involves the use of PCR-amplified DNA. Allele-specific oligonucleotides (ASO) probes are used for hybridization to complementary sequences. PCR-based DNA fingerprinting has many advantages over RFLP methods, including speed, simplicity, no requirement for radioactive probes, and greater sensitivity. (DNA from a single hair or minute samples of blood, saliva, and semen can be analyzed.)

13.4   Applications of Recombinant DNA Technology

The cloning of a gene or DNA fragment is often a preliminary step to achieving a larger goal or yielding a genetic "product." The product of a cloning experiment may range from a specific protein to a genetically altered organism. The hybrid DNA may be constructed so it can be expressed in the form of a unique and beneficial product. Likewise, the "product" of cloning may be a newly engineered organism that has received new, desirable genetic traits

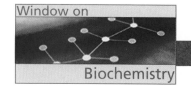
Window on
Biochemistry

13.2   Ancient DNA—When Is Old Too Old?

What do the following organisms have in common: a 5000-year-old Egyptian mummy, the 5000-year-old Iceman found frozen in the Alps, bones of the 9500-year-old Kennewick Man found in the state of Washington, a 40,000-year-old skeleton of the Siberian woolly mammoth, a chloroplast gene from 17-million-year-old magnolia leaves, a 65-million-year-old dinosaur bone, and a 120-million-year-old amber-preserved weevil? Attempts have been made to extract DNA samples from each so that sequence analysis may be completed.

Biochemists have discovered that DNA can be isolated from archeological and paleontological specimens up to several million years old. Several laboratories around the world have attempted to amplify ancient DNA by the polymerase chain reaction so that base sequence analysis can be done. This work has started the new science of "molecular paleontology." It is hoped that evolutionary relationships can be established by comparing DNA sequences from ancient specimens to the sequences from present day samples. However, DNA molecules extracted from fossils and other ancient items are often degraded in size, and the bases become chemically modified by natural oxidation. To add to the problems, there are difficulties with small sample size, poor efficiency of PCR with damaged nucleotide bases, and contamination with contemporary DNA, especially that of workers who handle the specimens. The ability to extract DNA from ancient bone is dependent on many factors including the manner of preservation, humidity, and temperature. The best samples are those that were protected from air and moisture. Many scientists are now concerned about the reliability of data obtained from old DNA. The National Center for Biotechnology Information (NCBI), the storehouse for sequence data in GenBank, has set an outer limit for acceptable DNA at about 100,000 years of age.

Not all of the attempts to extract intact and undamaged DNA for PCR amplification has been successful in the examples listed above. The most noteworthy and recent attempt is with the Kennewick Man found along the Columbia River in Washington State in 1996 (see photo, page 34). The age of the skeleton has been established at about 9500 years, making it one of the

or has had undesirable traits deleted. This section presents some specific applications of recombinant DNA, especially those that impact medicine and agriculture.

Recombinant Protein Products

Recombinant DNA molecules can be designed that when expressed in a host cell yield proteins that are of commercial or medical importance. Bacteria can serve as "factories" to produce a wide range of prokaryotic and eukaryotic proteins if the hybrid genes are appropriately constructed. The first recombinant protein to be marketed commercially was human insulin. The polypeptide chains for insulin are produced from two synthetic DNAs, one coding for the A chain and one for the B chain. The appropriate DNAs are fused into separate vectors and expressed in *E. coli* cells. Mixing polypeptides A and B results in active insulin (see Figure 4.11).

The first major recombinant protein product of agricultural interest was bovine growth hormone, which was approved for sale in 1994. Use of the growth hormone in milk cows yields an average 15% increase in milk production. Currently, only about 25% of the milk produced in the United States is from hormone-injected cows. There is resistance to the use of the hormone from consumer groups in Canada and Europe. Opponents are concerned that use of the hormone will cause increased mastitis and lead to greater use of antibiotics, which will lead to higher levels of antibiotic residues in milk products. The European Union currently has a ban on the use of the hormone and on the import of dairy products from hormone-treated cows.

oldest found in North America. Some anthropologists believe that further study of the skeleton could prove whether or not people who once lived in Europe or southern Asia were early inhabitants of the North American continent. However, because the skeleton was found in wetlands and extensive bone mineralization had occurred, scientists have been unable to extract enough undamaged DNA for PCR amplification.

© Steven Kagan/Gamma Liaison

The DNA from these amber-preserved insects can be isolated, reconstructed by the PCR, and sequence analyzed.

Light micrograph of human insulin crystals produced by recombinant DNA techniques.

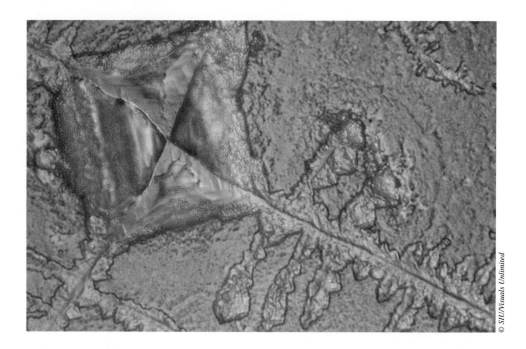

Bacterial cells are not able to translate many human and other eukaryotic genes because they are unable to carry out important exon splicing reactions and other post-translational modifications (see Chapter 12, Section 12.3). Eukaryotic host cells must be used for the correct expression of most eukaryotic genes. Animal cells have been especially popular host cells because their use can lead to important mechanistic and technological information regarding eukaryotic translation and the potential for human gene therapy. Several methods are currently available for introducing recombinant DNA molecules into animal cells. All methods involve the integration of the DNA into a host cell's chromosomes.

1. *Endocytotic uptake of calcium phosphate–precipitated DNA.* This method has an efficiency of uptake of under 1%; however, the experimental method is relatively simple.
2. *Electroporation.* Host cells that are exposed to a brief high-voltage pulse become more permeable to recombinant DNA.
3. *Microinjection.* Recombinant DNA solution is injected directly into individual cells using a fine-tipped glass micropipette. This tedious, time-consuming method results in a cell transformation of about 2%.

In spite of the difficulties outlined above, there are several therapeutic gene products that are now being mass produced or are in various stages of development. Many of the biotechnology companies that have been created during the past several years have as their goal the production of proteins useful for therapeutic or medical diagnostic purposes. Table 13.1 lists some of these products and their potential uses.

It has been assumed in this discussion that the structure of a recombinant protein product consists of its "normal" amino acid sequence. Although this is the case for many protein products, the synthesis of mutant proteins is often highly desirable. A distinct advantage afforded by recombinant DNA technology is the ability to create specific mutations. By altering the sequence of nucleotides in the recombinant DNA, the structure of the expressed protein can be changed. Using methods of **site-directed mutagenesis,** one or several amino acids may be exchanged in the protein product. The design of such novel proteins has greatly enhanced our understanding of the roles of specific amino acids in enzyme catalytic activity, protein folding, and molecular recognition. For example, an enzyme may be suspected to utilize the hydroxyl group of an active site serine residue in its catalytic action. This hypothesis can be tested by ex-

Recombinant Protein	Therapeutic Use
Human insulin	Treatment of diabetes
Somatotropin	Human growth hormone; used to treat dwarfism
Tissue plasminogen activator (TPA)	Treatment of heart attack, stroke victims; dissolves blood clots
Erythropoietin	Stimulates erythrocyte production in anemia
Interferons	Treatment of cancers
Superoxide dismutase	Destroys reactive radical species derived from O_2
Atrial natriuretic factor	Reduces high blood pressure
Mutant hemoglobins	Eliminate risks associated with blood transfusions
Leptin	Treatment of obesity

Table 13.1

Examples of recombinant DNA products of therapeutic use

changing the serine in the normal enzyme with alanine. If the mutant enzyme is catalytically inactive, the original hypothesis is supported.

Genetically Altered Organisms

Recombinant DNA technology makes it possible to alter life forms rapidly and precisely by deleting, adding, or changing specific nucleotide bases in gene structure. Changes in nucleotide sequence can be as small as one base pair to substitute one amino acid for another in a protein or as large as a gene that affords an organism a new genetic trait, for example, the ability to synthesize a new protein. Organisms that have been produced in genetically, altered forms include bacteria, plants, and animals (see Window on Biochemistry 13.3).

Bacteria

Paul Berg and his colleagues at Stanford University were the first to insert foreign DNA into a plasmid and express the hybrid DNA product in bacterial cells. Twenty-five years later, producing a genetically altered bacterium is a routine laboratory experiment. We read daily of genetic changes being made in bacteria to achieve the presence of a desired hereditary trait that may be the technological solution to some problem. Initially there was much concern regarding the safety of releasing genetically engineered bacteria into the environment. Very strict rules for handling modified bacterial strains were put into practice. For example, it was imperative that they be mutated so they could not survive outside the laboratory. Although safety concerns have lessened with each new discovery, environmental impact statements must be completed before modified organisms can be introduced into humans or the environment. Three specific examples of genetically altered bacteria are presented here.

The common bacterium *Pseudomonas syringae* lives on the leaves and stems of many crop plants. Proteins present on the bacterial cell surface enhance the formation of ice crystals, thus making plants more susceptible to frost damage. Some *P. syringae* cells were altered by removal of the gene expressing the surface protein. The "ice-minus bacteria" were sprayed onto a California strawberry field just before a frost. Although physical damage to the treated plants appeared to be less than normal, good statistical data could not be collected because environmental activists destroyed some of the strawberry plants.

Microbial strains are currently being developed to have novel biodegradative capabilities. In particular, genetically modified *Pseudomonas* bacteria are being constructed with genes to express enzymes that catalyze the degradation of complex chlorinated hydrocarbons such as dioxin, crude oil components (mixtures of hydrocarbons), TNT, PCBs, and trichloroethylene. Many of these hybrid organisms are being tested for bioremediation of toxic waste sites and oil spills.

The iron- and sulfur-oxidizing bacterium *Thiobacillus ferrooxidans* is being genetically modified to enhance its ability to desulfurize coal. Sulfur in coal and other fossil fuels is in the form of inorganic sulfur (pyrite, FeS_2) and organic sulfur (benzothiophenes). Removal of these compounds produces a cleaner fuel that produces less sulfur dioxide (SO_2) when burned. *T. ferrooxidans* is also used to degrade arsenopyrite and pyrite in mineral ores. Extraction of gold and silver metals from these ores is much more efficient if sulfides are oxidized first.

Plants

Recombinant DNA molecules are introduced into plant cells with the aid of the common soil bacterium *Agrobacterium tumefaciens.* This parasitic organism infects plants and introduces plasmids (called Ti for tumor inducing) into the plant genome. The site of infection on the plant usually develops a lump of tumor tissue called a crown gall. Ti plasmids are used as vectors to introduce foreign genes into plants.

Genetically engineered plants will have a major impact on agriculture by allowing the food industry to grow higher yielding crop plants under adverse conditions and to produce more

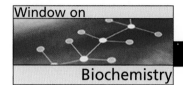

Window on

Biochemistry

13.3   Products from Genetically Modified Organisms on the Grocery Shelf

Take a stroll down the aisles of almost any supermarket in the world and you will pass by dozens of food products containing ingredients derived from **genetically modified organisms (GMO).** Any food item with products from corn, soybeans, and canola (rapeseed) has about a 50% chance of containing genetically engineered ingredients. A short list of specific grocery items would include margarines, corn chips, vegetable oils, corn starch, and bakery items. If we add to the list products from animals that are fed genetically modified grains, then we must include dairy products, eggs, beef, pork, poultry, and even fish. Transgenic versions of the important commodities, wheat and rice, which are currently under development, will greatly increase the number of genetically engineered items on the shopping list.

Major changes in our food products have occurred since the introduction of the first genetically engineered seed, Monsanto's Roundup Ready soybean, in 1996. The soybeans, which were herbicide-tolerant, were quickly followed by insect-resistant corn. This "first-generation" of agricultural biotech products was designed primarily to help the grower by providing a higher yield. The next generation of products currently in development will focus on "output" characteristics, those traits that improve the quality of crops for human and animal food. These promise nutritional enhancement such as increased protein, amino acid, starch, and vitamin content for the consumer.

All of these developments in agricultural biotechnology are occurring in an atmosphere of controversy and concern on the part of consumer groups and the general public. Questions are being raised about food safety, antibiotic resistance, allergic reactions, and other health effects, as well as possible ecological damage. Nowhere is this controversy more intense than in Europe. This is partly due to recent experiences with bovine spongiform encephalopathy (BSE, "mad cow disease"), even though there is no connection between BSE and genetically engineered food. However, it raises concerns about food safety in general. Two recent announcements have added to the controversy: (1) Rats fed transgenic potatoes show changes in immune response and reduced sizes of liver, heart, and brain; and (2) the pollen from genetically modified corn is toxic to monarch butterflies.

The future result of this controversy will probably be a separation into two markets, namely, genetically modified food products and genetically unmodified food products. Hence, there will be a need for careful labeling of the products. European governments have proposed that labels are required for food products in which at least one ingredient contains more than 1% of genetically modified material. The current position of the U.S. Food and Drug Administration (FDA) is that genetically modified foods are not significantly different from conventional foods. In the United States, labeling is required only if there is an inherent health or safety concern or if the product is substantially different in nutritional content and promotes a special health benefit. In order to segregate the foods into their proper category, they must be tested for the presence of ingredients from GMO. Inexpensive test kits are now available that use immunoassays to detect the presence of proteins that are indicative of specific genetic traits.

attractive and nutritious fruits and vegetables with longer shelf-lives. For example, the Flavr Savr® tomato developed by Calgene had its genes altered to inhibit rotting and allow it to ripen longer on the vine. Production was halted because of slow sales. Soybean and cotton plants have been developed that are herbicide protected. Fields of these plants may be treated with weed killing chemicals without damage occurring to the crop plants. Squash, corn, potatoes, and cotton (Figure 13.11) have been developed that are protected from pests such as beetles, blight, and viruses. About 50% of the current U.S. soybean crop is genetically altered to resist the herbicide Round-up®. About 33% of the U.S. corn acreage is modified to be insect resistant. Plants are also being developed that can grow under adverse conditions such as high salinity, drought, and extreme cold. Although these applications may hold great promise, critics of these developments fear they may harm natural ecological systems. Environmentalists are especially concerned that plants altered to increase resistance to herbicides can pass that trait on to other plants including weeds. Furthermore, pollen from insect resistant corn may be lethal to monarch butterflies.

Animals

Genetic engineering in animals is very difficult in terms of technology and also raises serious ethical concerns. In multicellular organisms with different kinds of tissue, genes can be transferred either into germ cells (sperm or eggs) or somatic cells (those not destined to become sperm or egg). Germ cell changes are passed from generation to generation, whereas genetic changes in somatic cells are present only in that particular individual. (Genetic therapy experiments in humans are allowed only with somatic cells.)

Perhaps the best known examples of genetically engineered animals are the creation of the cloned sheep, Dolly, and the production of giant mice (Figure 13.12). To produce oversized mice, the gene for rat growth hormone (somatotropin) was inserted into a plasmid and microinjected into mouse eggs. The mice that were born grew to approximately twice the normal size. Analysis of their genomic DNA showed the presence of the rat somatotropin gene. Animals such as these mice that carry genetically engineered heritable genes are called **transgenic.**

Human Gene Therapy

Gene therapy may be defined as the attempt to correct a genetic defect by inserting the normal gene into the cells of an organism. If it is possible to alter mouse cells genetically as described in the previous paragraph, then it is technologically feasible to do the same with

Figure 13.11
Cotton genetically modified to be protected from insects (left) may produce larger bolls than unprotected cotton.

Courtesy USDA

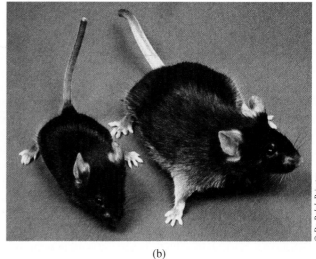

(a)

(b)

Figure 13.12
Genetically engineered animals: (a) Dolly, the cloned sheep, and (b) a giant mouse (*left*) created by inserting a plasmid containing the gene for rat growth hormone into a mouse egg.

Cutting Edge: Hot careers in biochemistry

human cells. In fact, experiments are already underway to test the efficacy of gene therapy in humans. Approximately 4000 diseases are caused by genetic damage in single genes, so there are many potential targets for gene therapy studies. Gene therapy procedures are currently under development for several diseases, including AIDS, brain cancers, obesity, multiple sclerosis, whooping cough, and Kaposi's sarcoma (a cancer found in AIDS patients). Results at this time, however, are less than encouraging. Because of the potential for abuse with these powerful techniques, government commissions have set very strict guidelines and all experiments must be approved by the appropriate federal agencies. For human-related studies approval must come from the Recombinant DNA Advisory Committee of the National Institutes of Health. In general, human gene therapy must be restricted to somatic cells, the genetic defect must be in a single gene, the genetic disease must be life threatening, and the experiments must first be modeled in other animals. The potential for human gene therapy is greatly enhanced by recent advances in locating genes on chromosomes (see Section 13.3).

Table 13.2
Gene therapy projects[a]

Disease	Defective Protein, Gene, or Inserted DNA
Lesch-Nyhan syndrome	Hypoxanthine-guanine phosphoribosyl transferase (Chapter 19, Section 19.5)
Amyotrophic lateral sclerosis (ALS, Lou Gehrig's disease)	Superoxide dismutase
Adrenoleukodystrophy (ALD)	Very long chain fatty acid synthetase transporting protein
Severe combined immunodeficiency (SCID)	Adenosine deaminase
β-Thalassemia	β-Globin, a polypeptide of Hb
Familial hypercholesterolemia	Liver receptor for low density lipoprotein (LDL) (Chapter 18, Section 18.5)
Hemophilia	Blood-clotting factors
Duchenne's muscular dystrophy	Dystrophin
AIDS	The gene to produce a ribozyme that cleaves HIV RNA
Inherited emphysema	α_1-Antitrypsin
Cystic fibrosis	A product that unclogs lung mucus is inhaled in a nasal spray
Cancer	E1A and p53 tumor suppressor genes

[a]Currently in or preparing for clinical trials.

Clinical trials for the treatment of several genetic diseases are currently ongoing or being planned. Delivery of recombinant genes into host cells is usually by viral transduction, calcium phosphate precipitation, or microinjection. In general, scientists remove somatic cells from a patient, insert a normal gene, and reintroduce the host cells into the patient. Host cells usually include bone marrow, skin, and liver. Some current studies in the clinical stage are listed in Table 13.2.

Summary

■ Biotechnology is defined as the use of biological cells and cell components to solve practical problems. Current examples of biotechnology include (a) the use of enzymes to synthesize chemicals for industrial and pharmaceutical application, (b) gene therapy to treat genetic diseases, and (c) the production of cheese, wine, and other food items.

■ Biotechnology is made possible by the design of recombinant DNA procedures: The DNA from one species can be cut into fragments and spliced into the DNA of another species. The hybrid DNA is introduced into a host cell where new proteins are made.

■ An alternative way to make multiple copies of a DNA fragment is with the polymerase chain reaction (PCR).

■ Recombinant DNA technology is now being used for the mass production of specific products, sequence analysis of the human genome, and the creation of genetically altered organisms.

Study Exercises

Understanding Terms

13.1 Define the following terms in 25 words or less.

a. Biotechnology
b. Recombinant DNA
c. Plasmid
d. Cohesive ends
e. Transformation
f. Genomic library
g. Polymerase chain reaction
h. RFLPs
i. Gene therapy
j. Blotting
k. Vector
l. Microarray analysis
m. DNA ligase

Reviewing Concepts

13.2 Which of the following gaps in DNA can be closed by the action of DNA ligase and ATP?

a.

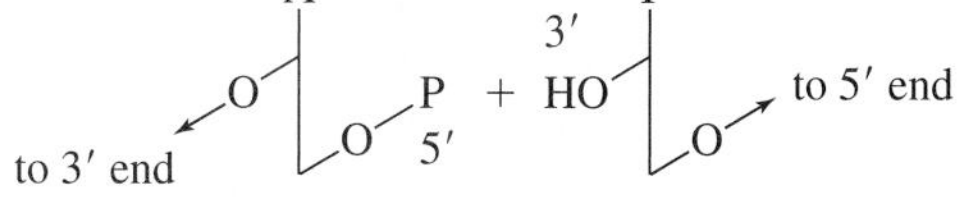

b.

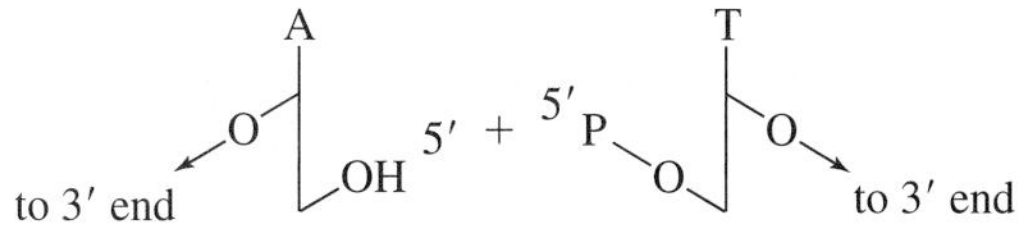

c.

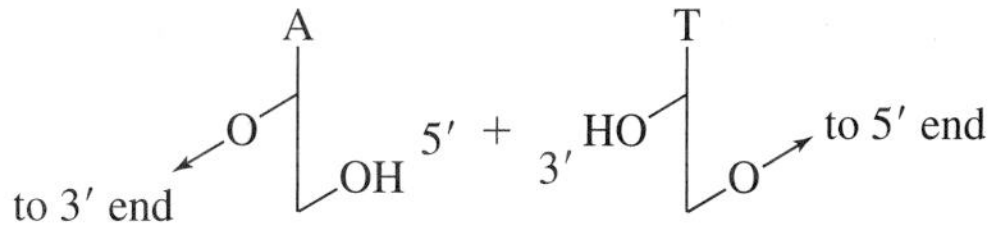

d.

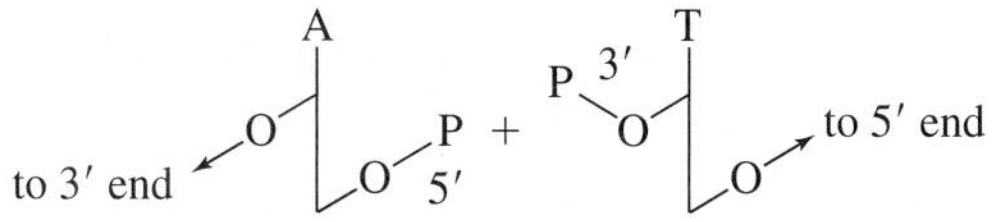

e.

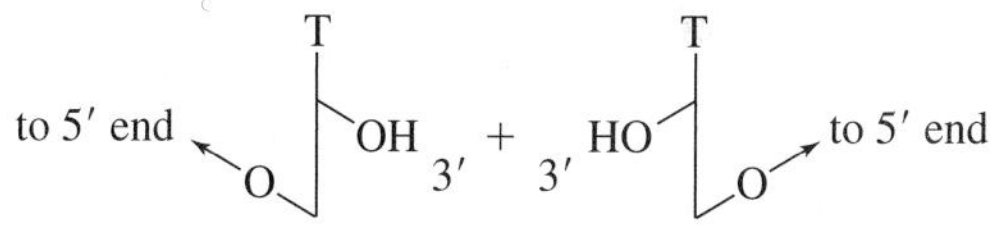

13.3 You have accepted a job as laboratory technician with Biotech Inc. Your first assignment is to construct a recombinant DNA that contains the gene for the hormone vasopressin. You have devised a plan to synthesize the gene by chemical methods. Write out the structure of nucleotide sequences in the gene for vasopressin. The amino acid sequence for vasopressin is given below.

Cys-Tyr-Phe-Gln-Asn-Cys-Pro-Arg-Gly

13.4 Which of the following segments of duplex DNA may be restriction enzyme cleavage sites?

a. 5′ ACGA
   3′ TGCT
b. 5′ CGAT
   3′ GCTA
c. 5′ ACGT
   3′ TGCA

13.5 Differentiate between stringent replicated plasmids and relaxed replicated plasmids.

13.6 Which of the following reagents and biomolecules are necessary to make a recombinant DNA?

a. λ phage DNA
b. Restriction enzyme
c. Chymotrypsin
d. DNA fragment to be cloned
e. Glucose
f. DNA ligase
g. Calcium chloride
h. Starch
i. Ampicillin

13.7 The following enzymes are useful in recombinant DNA preparation and analysis. For each, write a reaction catalyzed by the enzyme and describe its usefulness.
  a. Restriction endonuclease
  b. Deoxynucleotidyl transferase
  c. *Taq* DNA polymerase
  d. DNA ligase

13.8 Discuss the advantages and disadvantages of using plasmids and bacteriophage DNA as cloning vectors.

13.9 Use line structures of DNA to show what happens to DNA during the first step of the PCR process (heating to 95°C for 15 s).

13.10 As your next project in the molecular biology laboratory, you are attempting to change a serine residue in an enzyme to an alanine residue. The amino acid sequence near this serine residue is -Leu-Glu-Ser-Val-Gly-.
  a. What is the nucleotide sequence of DNA that codes for this segment of the enzyme?
  b. How would you change the nucleotide sequence in the gene to insert Ala in place of Ser?

13.11 The term "palindrome" is often used to describe the site of attack on DNA by a restriction enzyme. If you are not familiar with the term, use a dictionary to find its definition. Explain why this term is appropriate here.

13.12 Is there any danger of genetic diseases becoming contagious? Explain.

13.13 Which of the following are characteristics of bacterial plasmids?
  a. Useful cloning vectors
  b. DNA molecules that are linear
  c. Self-replicating, extrachromosomal DNA molecules
  d. Closed circular in structure
  e. They have all of the bacteria's genetic information

13.14 What kind of chemical reactivity best describes the action of DNA ligase?
  a. DNA ligase is a polymerase
  b. DNA ligase catalyzes oxidation–reduction reactions
  c. DNA ligase catalyzes the formation of a phosphoester bond
  d. DNA ligase catalyzes the hydrolysis of peptide bonds

13.15 Compare the similarities and differences of a restriction endonuclease with the nucleases discussed in Chapter 10, Section 10.4.

13.16 What single word best describes the reaction catalyzed by a restriction endonuclease? Choose from the list below.
  a. Redox
  b. Hydrolysis
  c. Phosphoryl group transfer
  d. Isomerization

13.17 What process best describes the action of *Taq* DNA polymerase?
  a. Catalyzes hydrolysis of a phosphoester bond.
  b. Catalyzes an oxidation–reduction reaction.
  c. Catalyzes the formation of a phosphoester bond.

Solving Problems

13.18 A plasmid such as pBR322 is treated with the restriction endonuclease *Eco*RI. Write out the nucleotide sequences at the two ends of the linearized DNA.
HINT: See Table 10.5.

13.19 A linear fragment of eukaryotic DNA has two *Eco*RI cleavage sites as shown below with arrows. Write out the nucleotide sequences at the two ends of the center segment, A, after reaction with *Eco*RI.

13.20 Draw the structure of the product obtained after incubating the linearized plasmid from Problem 13.18 with fragment A from Problem 13.19.

13.21 Write out the nucleotide sequence of a hybridization probe that could be used to locate a gene of the following structure. Assume the gene is inserted into a plasmid.

5′ ATTGGTACAA

13.22 You have prepared the eukaryotic gene shown below and wish to insert it into a recombinant DNA for expression in animal cells. Indicate on the drawing those nucleotide sequences or regions that must be part of the recombinant DNA so it will be properly transcribed, translated, and regulated.

HINT: Review Section 12.4

13.23 Use Figure 13.7 to determine the number of DNA fragments made by the action of each restriction endonuclease on pBR322.
  a. *Eco*RI   **c.** *Taq*I
  b. *Hind*III   **d.** *Hinf*I

13.24 Why is it advantageous to do the last step of the PCR process at 72°C rather than 37°C?

13.25 Would you drink milk from a cow that had been injected with recombinant bovine growth hormone? Why or why not? Should such milk sold in a grocery store have a label describing its origin?

13.26 It is now possible to determine susceptibility to many genetic diseases by analysis of a person's DNA. Will you be first or last in line to find out this information about yourself? Discuss.

13.27 Discuss the pros and cons of gene therapy procedures on a newborn infant diagnosed with a life-threatening genetic disease.

13.28 A sheep has recently been cloned and many world leaders have denounced the work. Even the President of the United States expressed concern and stated that no federal or private money should ever be used to research human cloning. Do you agree or disagree? Explain.

13.29 What are the pros and cons of allowing DNA evidence to be admitted in a court of law?

Writing Biochemistry

13.30 Search for an article in a current newspaper or news magazine that is based on the general topic of biotechnology. Write a 100-word summary of the article, emphasizing the important biochemistry and ethical implications.

13.31 Determine the three most confusing concepts covered in this chapter. List the concepts in the form of questions, for example, what are the differences between blunt ends and cohesive ends? After writing each question, try to prepare a written answer for each.

Applying Biochemistry

13.32 The materials listed below are required to prepare recombinant DNA. Describe the role of each in molecular cloning experiments.

  a. Vector
  b. Restriction endonuclease
  c. Plasmid
  d. DNA ligase
  e. Ampicillin

Further Reading

Anderson, W., 1995. Gene therapy. *Sci. Am.* 273(3):124–128.

Backman, K., 2001. The advent of genetic engineering. *Trends Biochem. Sci.* 26:268–270.

Brown, K., 2001. GM Foods: are they safe? *Sci. Am.* 284(4):50–56.

Burachik, M., Vazquez, M., and Koss, A., 1997. The Biotechnology Enterprise Project. *Biochem. Educ.* 25:31–34.

deLange, T., 1998. Telomeres and senescence. *Science* 279:334–335.

Gerhold, D., Rushmore, T., and Caskey, C., 1999. DNA chips: Promising toys become powerful tools. *Trends Biochem. Sci.* 24:168–173.

Kornberg, A., 1997. Centenary of the birth of modern biochemistry. *Trends Biochem. Sci.* 22:282–283.

Lander, E., 1996. The new genomics: global views of biology. *Science* 274:536–539.

Lyon, J., and Gorner, P., 1995. *Altered Fates: Gene Therapy and the Retooling of Human Life.* New York: Norton Publishers.

Miller, R., 1998. Bacterial gene swapping in nature. *Sci. Am.* 278(1):67–71.

Mullis, K., 1990. The unusual origin of the polymerase chain reaction. *Sci. Am.* 263(4):56–65.

Paabo, S., 1993. Ancient DNA. *Sci. Am.* 269(5):86–92.

Palevitz, B., 1999. Ancient DNA—When is old too old? *The Scientist* 13(13):10.

Service, R., 1998. Microchip arrays put DNA on the spot. *Science* 282:396–399.

Thompson, G., and Esposito, M., 1999. The genetics of complex diseases. *Trends Biochem. Sci.* 24:M17–M20.

Vogel, G., 1998. Possible new cause of Alzheimer's disease found. *Science* 279:174.

Zimm, B., 1999. One chromosome: one DNA molecule. *Trends Biochem. Sci.* 24:121–123.

Part IV
Metabolism and Energy

Chapter 14

Basic Concepts of Cellular Metabolism and Bioenergetics

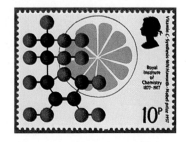

This British stamp celebrates the discovery of vitamin C, a necessary component in our diet.

In Parts I, II, and III of this book, we have focused on the structures and functions of important biomolecules and their assembly into membranes, cellular organelles, and cells. We have paid little attention to the maze of reactions that degrades and/or produces the molecules or to the energy requirements of biological processes. In Part IV our studies target the overall process of **metabolism:** how cells acquire, transform, store, and use energy. Metabolism is often described as the "sum total of all chemical reactions in an organism"; however, this definition is incomplete and a great oversimplification. Metabolism is indeed a study of the thousands of reactions in a cell, but the study also includes exploring their coordination, regulation, and energy requirements.

It is not uncommon for chemists to expose molecules to an environment of high temperatures, several atmospheres of pressure, strong chemical catalysts, and extreme pH values to achieve desired chemical reactions. By chemical standards, the internal conditions in most cells and organisms are relatively mild: one atmosphere pressure, a moderate temperature, and neutral pH. (Some exceptions exist, for example, very acidic pH in the stomach and intestines of animals; bacteria living in hot springs and ocean vents.) We have already explained part of the reason cells can carry out reactions under these mild conditions: the presence of catalysts, called enzymes, that function by bringing a reaction more rapidly to equilibrium. Even with catalysts, many metabolic reactions could not take place without a source of energy and mechanisms for converting energy from one form to another. Organisms and cells are similar to machines in that they require an input of energy to function. The machine changes one form of energy into other forms. In humans and many other nonphotosynthetic organisms, the energy input is in the form of organic molecules: fats and carbohydrates. In photosynthetic organisms, light is the source of energy. Both types of organisms then convert the energy to a different form for use in biosynthesis, transport of nutrients and other energy-requiring processes. Thus we begin to explore the myriad of integrated biochemical reactions whereby cells extract useful energy from foodstuffs and use that energy to drive thermodynamically unfavorable reactions and carry out other critical functions. Specific questions to answer include: How do organisms transform the potential energy in carbohydrates and fats into useful energy for biosynthesis, membrane transport, and muscle contraction? How much energy is associated with the important energy transfer molecule ATP? How do cells regulate the metabolic processes so as not to deplete some resources or produce too much of others? How do photosynthetic organisms harness energy from the sun to make carbohydrates? In addition, we shall discover the answers to practical questions such as, How much energy is

Figure 14.1

Most autotrophs use energy from the sun and CO_2 to synthesize biopolymers. Heterotrophs obtain energy by ingesting and degrading complex carbon compounds. Autotrophs and heterotrophs break down biopolymers by similar pathways.

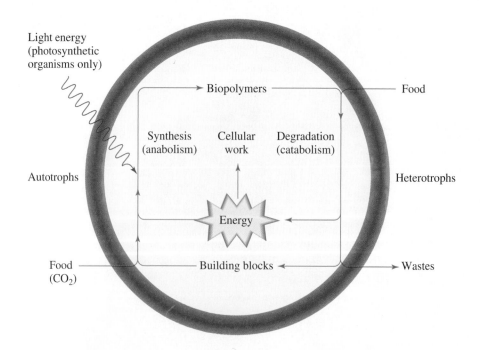

available in glucose? Why do fats provide more energy per gram than carbohydrates? What are the roles of the vitamins in metabolic reactions?

Fortunately for us, most organisms use the same general pathways for extraction and utilization of energy. The reactions of glycolysis (glucose degradation) in humans are the same as those in bacteria and other organisms, including higher plants. However, we will encounter differences in the relative importance of metabolic pathways, regulation, enzyme structure, and reaction mechanisms. Also, we will observe compartmentation of metabolism within an organism; that is, some types of cells are used for energy production and others for energy utilization.

Metabolic differences among organisms can best be studied by dividing all living organisms into two major classes (Figure 14.1). **Autotrophs** ("self-feeding" organisms) can use atmospheric carbon dioxide as the sole source of carbon, from which they build complex carbon-containing biomolecules. Most autotrophs, including photosynthetic bacteria and higher plants, use the sun's energy for biosynthetic purposes and are therefore relatively self-sufficient (see Chapter 17, Sections 17.5 and 17.6). **Heterotrophs** ("feeding on others") obtain energy by ingesting complex carbon-containing compounds such as carbohydrates and fats. The heterotrophs, which include higher animals and most microorganisms, are dependent on the autotrophs for these compounds. Heterotrophic organisms can be further divided into two subclasses depending on their need for molecular oxygen. Aerobes live in air and use molecular oxygen for metabolic reactions. The anaerobes do not require oxygen for survival; in fact, for some, the strict anaerobes, molecular oxygen is toxic.

This introductory chapter focuses on the heterotrophs. We do this because the important metabolic paths have been discovered using microorganisms and animal tissue and in some ways are less complicated than in plants. Figure 14.2 summarizes the aspects of metabolism in a heterotrophic cell.

Trees and other autotrophs use the sun's energy and CO_2 to synthesize carbohydrates.

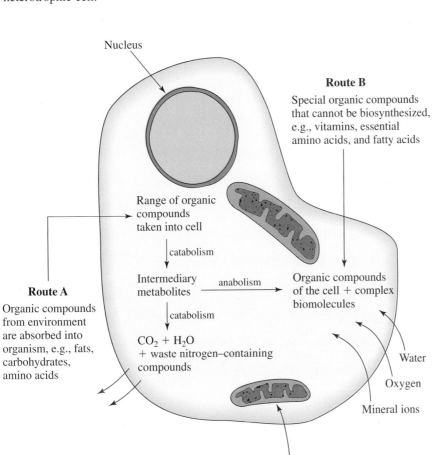

Figure 14.2

Metabolic processes in heterotropic cells. Route A is for entry of all carbon, nitrogen, and sulfur; most organic hydrogen and oxygen; and some phosphorus. Route B is for entry of specific organic compounds that are required by the cell but cannot be synthesized by the cell (vitamins, essential amino acids, essential fatty acids). Catabolism refers to degradation processes with release of energy; anabolism refers to biosynthetic processes that require energy.

14.1 Intermediary Metabolism

In a typical *Escherichia coli* cell, there are at least a thousand biochemical reactions that make up metabolism. A human cell may contain as many as 3000 enzymes, each catalyzing a specific reaction. The process of metabolism in all organisms takes place via sequences of consecutive enzyme-catalyzed reactions. Each step is usually a single, highly specific, chemical change leading to a product that becomes the reactant for the next step. This process is sometimes called **intermediary metabolism** because each reaction leads to a stable product or intermediate. (Metabolic intermediates, called **metabolites,** should not be confused with the unstable variety of intermediates often used in detailed reaction mechanisms.) A sequence of reactions that has a specific purpose, for example, the degradation of glucose to pyruvate (glycolysis), is called a **metabolic pathway.** A pathway may be linear (glycolysis), branched (biosynthesis of amino acids), cyclic (citric acid cycle), or spiral (fatty acid degradation). These possibilities are outlined in Figure 14.3.

Two Paths of Metabolism

Metabolic pathways can be grouped into two paths depending on their biochemical purpose. **Catabolism** is the degradative path, whereby complex organic molecules such as fats, carbohydrates, and proteins are degraded to simpler molecules such as lactate, pyruvate,

Figure 14.3

Possible sequential arrangements for metabolic pathways: (a) linear; (b) branched, converging; (c) branched, diverging; (d) cyclic; (e) spiral, A, B, C, D, F, G, H, and I refer to metabolic intermediates. Primed letters refer to modifications of A, B, C, and D. E_n refers to enzymes that catalyze the reactions.

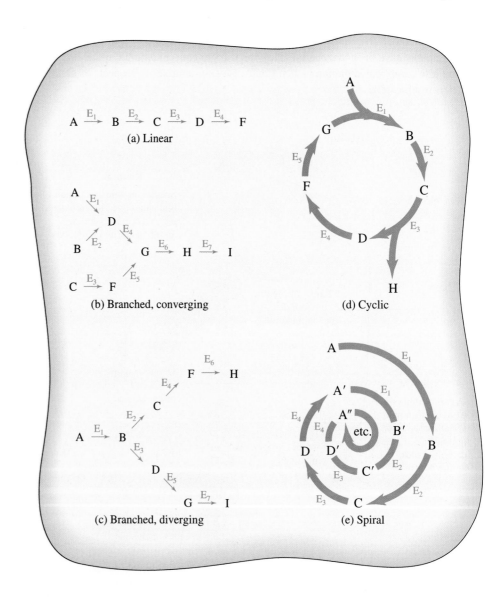

ethanol, CO_2, H_2O, and NH_3 (see Figures 14.1 and 14.2). Catabolism is characterized by oxidation reactions and, because foodstuffs contain potential energy, by release of free energy, which is captured eventually in the form of ATP. The other part of metabolism is **anabolism** or biosynthesis, which is characterized by the construction of large, complex biomolecules from smaller precursor molecules. The synthesis of proteins from individual amino acids, the formation of glucose from two pyruvate molecules, and the synthesis of DNA from nucleotides represent anabolic processes. In contrast to catabolism, anabolism is generally characterized by reduction reactions and energy input. Energy is supplied by ATP, NADH, and NADPH molecules, which contain energy released from catabolism. It is not possible to label every individual reaction as catabolic or anabolic because some individual steps in metabolism neither release nor require energy. In addition, not all reactions are oxidations or reductions. (The first step in the catabolic process of glycolysis involves a phosphoryl group transfer from ATP to glucose. This reaction is not an oxidation and it requires the input of energy; however, the *overall* process of glycolysis is one of oxidation and energy release.) The division into the two parts is not meant to separate metabolism into unrelated processes but to facilitate the learning of each pathway and defining its role in the overall process. The general characteristics of catabolism and anabolism are compared in Table 14.1. In biological cells, anabolic and catabolic processes are regulated to be nonsynchronous; that is, there is a flow or **flux** of metabolites in one direction or the other depending upon the cell's energy state. For example, when a muscle cell is in a resting state, there is surplus energy and therefore no need for catabolism of glucose to CO_2 and H_2O, which would produce even more potential energy. Instead, glucose is stored in its polymeric form, glycogen.

Anabolism and catabolism are linked together by their contrasting but coordinated energy requirements (bioenergetics): *Catabolic processes release the potential energy from food and collect it in the reactive intermediate, ATP; anabolic processes use the free energy stored in ATP to perform work.* The two processes are coupled by the **ATP energy cycle** (Figure 14.4). ATP serves as the *universal carrier of biochemical energy.* In a later section of this chapter we explain the special chemical and thermodynamic properties of ATP and how it effectively performs as an agent of energy transfer. From an energy standpoint, the *recycling of ATP is the central theme of metabolism.* The molecular structure for ATP is shown in Figure 14.5.

Most ATP formation in a cell is not linked directly to metabolic reactions. Instead, it is produced by oxidation of the cofactor, NADH, which is generated during catabolic reactions. NADH, a reactive metabolic intermediate, is therefore involved directly in the acquisition of energy from the degradation of nutrients.

Stages of Metabolism

Catabolism and anabolism form a coordinated network of reactions, but for practical and pedagogical reasons, we divide each into three major stages (Figure 14.6). This is done at the risk of taking pathways and individual steps "out of context" from the whole picture of metabolism. Stage I of catabolism is the breakdown of macromolecules (proteins, fats

Table 14.1
Contrasting characteristics of catabolism and anabolism

Catabolism
Leads to degradation of biomolecules
Overall process of chemical oxidation and formation of reduced cofactors of NADH, NADPH, $FADH_2$
Release of chemical energy (exothermic) and production of ATP from ADP
Convergence of pathways

Anabolism
Synthesis of biomolecules
Overall process of chemical reduction and formation of oxidized cofactors NAD^+, $NADP^+$, FAD
Requirement for energy input (endothermic) and use of ATP
Divergence of pathways

Figure 14.4

The ATP energy cycle, which links together anabolic and catabolic processes of metabolism. Catabolism provides the primary source of energy by producing ATP by the phosphorylation of ADP. (a) Anaerobic metabolism (no use of oxygen) leads to fermentation, with formation of lactate or ethanol or other products. (b) Aerobic metabolism (use of oxygen) leads to more complete degradation of organics ($CO_2 + H_2O$). The cellular activities that can be driven by ATP are also listed.

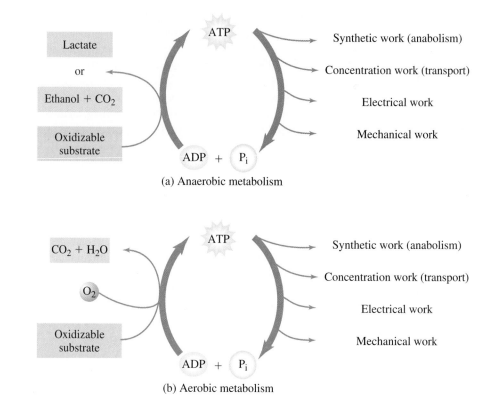

(a) Anaerobic metabolism

(b) Aerobic metabolism

[triacylglycerols], polysaccharides) into their respective building blocks. All proteins are hydrolyzed to their component amino acids; triacylglycerols are hydrolyzed to fatty acids plus glycerol; and polysaccharides are hydrolyzed to monosaccharides, primarily glucose. Essentially all reactions in stage I of catabolism involve the cleavage of chemical bonds by hydrolysis with no release of useful energy in the form of NADH or ATP. Stage I is preparation for the next level of reactions, where the diverse amino acids, fatty acids, and monosaccharides are oxidized to a common metabolite, acetyl CoA (stage II). Some energy is released and captured in the form of NADH and ATP during stage II. In stage III acetyl CoA enters the **citric acid cycle,** where it is oxidized to CO_2, the end product of aerobic carbon metabolism. The oxidation of citric acid cycle intermediates leads to the formation of reduced and reactive cofactors, NADH and $FADH_2$, which eventually give up their electrons. The electrons are transported via the **respiratory assembly** to molecular oxygen, ultimately producing energy and water. The energy released in electron transport is coupled directly to ATP synthesis from ADP and P_i. Stage III of catabolism is the location of most ATP production by a process called **oxidative phosphorylation.** Catabolism is characterized by a convergence of three major routes toward a final common pathway. Many differ-

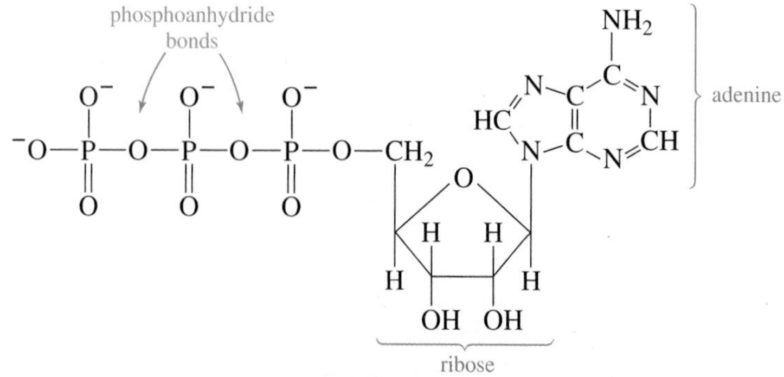

Figure 14.5

The structure of adenosine triphosphate (ATP). ATP is a typical nucleotide consisting of an adenine ring, a ribose, and three phosphate groups.

Adenosine triphosphate (ATP)

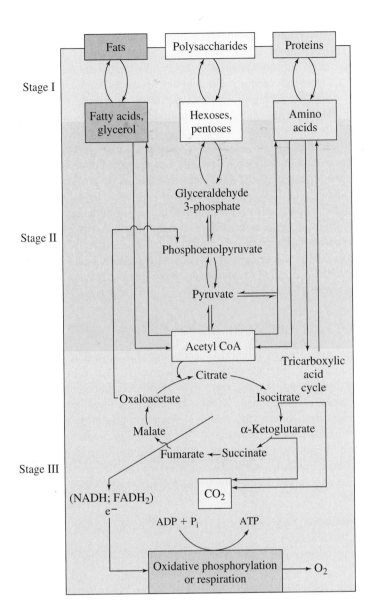

Figure 14.6
The stages of catabolism and anabolism.

ent kinds of proteins, fats, and polysaccharides enter unique pathways that converge at the citric acid cycle. Convergence of pathways is efficient and economical for the cell because it greatly reduces the number of different enzymes necessary for degradation of foods.

Anabolism can also be divided into three stages; however, in contrast to catabolism, the pathways are characterized by divergence. Monosaccharide and polysaccharide syntheses may begin with CO_2, oxaloacetate, pyruvate, or lactate. The amino acids for protein synthesis are formed from acetyl CoA and by the amination of pyruvate and α-keto acids from the citric acid cycle. The triacylglycerols are constructed using fatty acids synthesized from acetyl CoA. Several metabolites such as oxaloacetate, acetyl CoA, and pyruvate are common precursors for proteins, triacylglycerols, and polysaccharides; however, they branch from their starting points, leading to hundreds more of these diverse proteins, triacylglycerols, and polysaccharides.

Many anabolic pathways require energy in the form of ATP and NADPH. By noting the use of special dual arrows in Figure 14.6, it can be seen that anabolism and catabolism are not just the reverse of each other. The two processes are *similar* in terms of intermediates and enzymes, but they are not *identical*. Catabolism and anabolism of glucose illustrate the relationships. Glucose is degraded by glycolysis to pyruvate, a pathway requiring a sequence of 10 enzyme-catalyzed steps. The reversal of glycolysis or synthesis of glucose from pyruvate requires 11 steps, of which 8 are the exact reverse of those in glycolysis. Three steps in glycolysis are thermodynamically irreversible and their reverse requires three different steps in glucose synthesis. It does not seem economical to have even slightly different pathways for anabolism and catabolism, but distinct routes are a thermodynamic and regulatory necessity. Catabolic processes are energetically like a boulder taking the most direct route as it rolls downhill. Carrying the same boulder uphill, as in anabolism, requires some changes in the route; however, portions of the downhill pathway may be retraced. Distinct but coordinated paths for anabolism and catabolism also allow for metabolic regulation. Regulation of metabolic pathways often occurs at thermodynamically irreversible steps. Since irreversible steps in catabolism are replaced by different steps in anabolism, sites of regulation will differ in the two processes.

Anabolic and catabolic pathways also differ in types of cells, intracellular location, and regulation. For example, glucose degradation predominates in active muscle cells, whereas glucose synthesis takes place primarily in liver cells. Fatty acid synthesis occurs in the cytoplasm of adipose cells; the enzymes for fatty acid degradation occur in the mitochondria of resting muscle cells. A detailed discussion on tissue specificity and integration of pathways is presented in Chapter 20.

14.2 The Chemistry of Metabolism

A cell, with its networks of integrated reactions, may appear to be a chemical maze. There are hundreds of enzyme-catalyzed reactions in the most simple single-celled organism, and as many as 3000 reactions in a human cell. This raises questions about the chemical principles of metabolism: What is the chemical logic behind the evolution of a sequence of reactions? Is each biochemical reaction unique, or do different metabolic pathways use similar chemical processes? Is it possible to predict the metabolic fate of a new compound by studying known pathways? Based on thousands of cellular reactions that have now been studied, we conclude that organisms use only a few types of chemical reactions to accomplish metabolism. Six categories of biochemical reactions have been identified:

1. Oxidation–reduction reactions
2. Group-transfer reactions
3. Hydrolysis reactions
4. Nonhydrolytic cleavage reactions
5. Isomerization and rearrangement reactions
6. Bond formation reactions using energy from ATP

These six types of reactions correlate with the six classes of enzymes introduced in Tables 6.1, 6.2, and 14.2. It may surprise you to learn that metabolic pathways have so few different reaction types. However, by now we should recognize that cellular reaction processes are often characterized by simplicity and economy (see Window on Biochemistry 14.1).

Table 14.2
Types of chemical reactions in metabolism correlated to enzyme classes

Type of Reaction	Enzyme Class	Description of Reaction
1. Oxidation–reduction	Oxidoreductases (dehydrogenases)	Transfer of electrons
2. Group transfer	Transferases	Transfer of a functional group from one molecule to another or within a single molecule
3. Hydrolytic cleavage (hydrolysis)	Hydrolases	Cleavage of bonds by water (transfer of functional groups to water)
4. Nonhydrolytic cleavage	Lyases	Splitting a molecule by nonhydrolytic processes
5. Isomerization and rearrangement	Isomerases	Rearrangement of functional groups to form isomers
6. Bond formation using energy from ATP	Ligases	Formation of carbon–carbon and other bond with energy from ATP

Oxidation–Reduction Reactions

Oxidation–reduction or redox reactions are among the most common of all reactions in metabolism. Redox reactions are those in which electrons are transferred from one molecule or atom to another. There must always be two reactant molecules, an electron donor and an electron acceptor. Redox reactions are most readily identified by observing transfer of hydrogen atoms:

Concept Review: Redox reactions

$$AH_2 + B \rightleftharpoons A + BH_2$$

where

AH_2 = electron donor (or reducing agent or reductant)
B = electron acceptor (or oxidizing agent or oxidant)

Enzymes catalyzing these reactions are **oxidoreductases,** or more commonly, **dehydrogenases.**

Carbon atoms in substrates are usually the sites of biochemical redox processes. During the degradation of fatty acids, carbon atoms at the $—CH_2—$ oxidation level (methylene) are transformed, by stepwise reactions, to carbon atoms at the $—COOH$ level (carboxyl). This is clearly an oxidative process because hydrogen atoms are replaced by oxygen atoms; however, it is not always a simple matter to recognize oxidation–reduction at a carbon center. The most straightforward means to assess oxidation level is to count the number of oxygen or hydrogen bonds to a carbon center. The larger the number of oxygen bonds to carbon, the higher the state of oxidation. Table 14.3 shows the relative oxidation levels at a carbon center in important functional groups in biochemistry. A carbon atom can exist in many oxidation states with the extremes being CH_4 (most highly reduced) and CO_2 (most highly oxidized). Note that the carbon atom of methane has no bonds to oxygen atoms, but all four bonds are to hydrogen. Carbon dioxide with all four bonds linked to oxygen atoms is the highest possible oxidation level of a carbon atom. When a carbon center is transformed from a methylene to a methene, electrons and hydrogens are removed from the carbon atom, resulting in an oxidation process. It is important

Functional Group[a,b]	Name	Geometry of Carbon Center	Relative Oxidation Level
C H$_4$	Methane	Tetrahedral	Lowest
C H$_4$	Methyl	Tetrahedral	
C H$_4$	Methylene	Tetrahedral	
RHC$=$CHR′	Methene	Planar	
RC HR′ | OH	Alcohol	Tetrahedral	
RC H || O	Aldehyde	Planar	
RC R′ || O	Ketone	Planar	
RC X || O	Carboxylic acid or derivative	Planar	
O$=$C$=$O	Carbon dioxide	Planar	Highest

[a]The carbon center undergoing oxidation is in red.
[b]R, R′ = alkyl or aryl group; X = —OH (acid), —OR (ester), —NH$_2$ (amide).

Table 14.3
Relative oxidation levels of carbon in functional groups

to recognize that redox reactions also involve changes in the geometry of bonding at the carbon center. For example, some carbon atoms, such as methyl, are tetrahedral (*sp^3* hybridized), whereas carbon in other oxidation levels like methene involve planar bonding (*sp^2* hybridized). Table 14.3 is to assist you in recognizing and defining metabolic reactions that involve redox processes.

When a carbon center or other functional group or atom is oxidized, some other molecule or atom must accept those electrons and therefore be reduced. Many oxidation–reduction reactions in biochemistry are coupled to the coenzyme redox pairs, NAD^+/NADH; $NADP^+$/NADPH; and FAD/$FADH_2$ (see Chapter 7, Section 7.1). The catabolism of ethanol, catalyzed by two dehydrogenases, shows the use of NAD^+ as an electron acceptor (oxidizing agent):

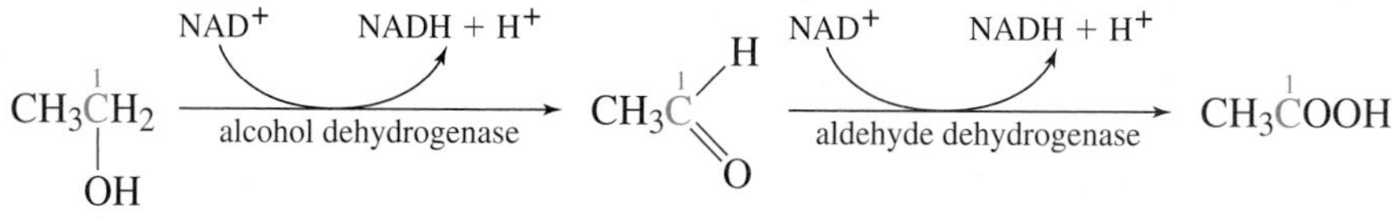

Ethanol · Acetaldehyde · Acetic acid

Ethanol and acetaldehyde act as electron donors (reducing agents) by transferring electrons and hydrogen to the electron acceptor, NAD^+, thereby generating the reduced form of the cofactor, NADH. Two electrons are transferred at each step. The oxidative nature of ethanol catabolism is clearly evident by observing the chemical changes occurring at carbon atom number 1. An oxidation process requires a coupled reduction process. In this example NAD^+ is reduced to NADH.

Metal ions, especially the copper and iron redox couples (Cu^+/Cu^{2+}; Fe^{2+}/Fe^{3+}), often associated with enzymes and other proteins, also function as oxidizing and reducing agents (see Chapter 6, Section 6.3).

Group-Transfer Reactions

Group-transfer reactions involve the transfer of a chemical functional group from one molecule to another (intermolecular) or group transfer within a single molecule (intramolecular) (Table 14.4). One of the most important transferred groups in biochemistry is the **phosphoryl group,** $—PO_3^{2-}$, which usually has its origin in ATP (Figure 14.7). This example shows the transfer of a phosphoryl group from ATP to glucose in a reaction catalyzed by a phosphotransferase better known as hexokinase.[1] This is the first step to activate glucose for

[1]A **kinase** is a subclass of the transferases that catalyzes the transfer of a phosphoryl group from ATP to an acceptor molecule.

Table 14.4
Groups transferred in metabolic reactions

Name	Structure
Phosphoryl	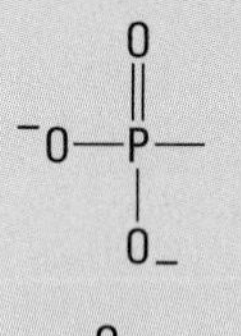
Acyl	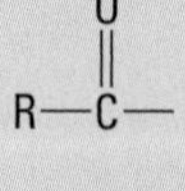
Glycosyl	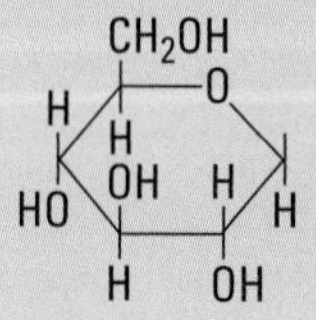

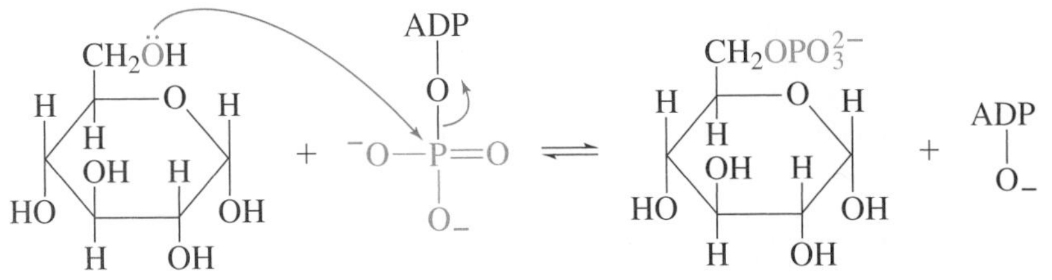

Figure 14.7
Example of a group-transfer reaction, the transfer of a phosphoryl group from ATP to glucose. This is the first step in the pathway of glucose degradation by glycolysis.

entry into oxidative degradation by glycolysis. Other important groups transferred in biochemical reactions include the acyl group and the glycosyl group (carbohydrates). The acyl group, derived from a carboxylic acid derivative, is often made more reactive by transfer to coenzyme A (CoASH):

$$\underset{\substack{\text{Carboxylic acid} \\ \text{derivative}}}{\text{RC}-\text{X}} + \underset{\text{Coenzyme A}}{\text{CoASH}} \rightleftharpoons \underset{\text{Thioester}}{\text{CoAS}-\overset{\text{O}}{\overset{\|}{\text{C}}}-\text{R}} + \text{XH}$$

The reactive —SH group of CoASH forms a thioester linkage with acyl groups. (See Window on Biochemistry 14.2.) Glycoproteins and polysaccharides are synthesized by the transfer of glycosyl units, derived from glucose, mannose, and other monosaccharides (see Chapter 8, Section 8.4).

Hydrolysis Reactions

In hydrolysis reactions, water is used to split a single molecule into two distinct molecules. Such reactions are numerous in metabolism, with cleavage of esters, amides, and glycosidic bonds being the most common.

1. **Ester hydrolysis.** The hydrolytic release of a fatty acid from a triacylglycerol represents the mobilization of metabolic fuel from stored fat (Figure 14.8a).
2. **Amide hydrolysis.** An important example of amide bond hydrolysis is the peptidase-catalyzed reaction. Proteins ingested in foodstuffs are hydrolyzed to free amino acids by enzymes in the stomach and intestines. These reactions occur in stage I of catabolism. The example shown here is the hydrolysis reaction catalyzed by carboxypeptidase (CP) (Figure 14.8b).
3. **Glycoside hydrolysis.** Glycosidic bonds in oligosaccharides are hydrolyzed by glucosidases. For example, the disaccharide lactose (milk sugar) is cleaved by hydrolysis into its component monosaccharides galactose and glucose (Figure 14.8c).

Nonhydrolytic Cleavage Reactions

In this class of reactions, molecules are split without the use of water. The most prevalent reaction in this category is carbon–carbon bond cleavage. Enzymes catalyzing these reactions are called **lyases.** A representative illustration of nonhydrolytic carbon–carbon bond cleavage is the splitting of fructose 1,6-bisphosphate catalyzed by fructose 1,6-bisphosphate glyceraldehyde 3-phosphate lyase, commonly called **aldolase** (Figure 14.9a). Nonhydrolytic cleavage reactions may also include the addition of functional groups to double bonds or the removal (elimination) of functional groups to form double bonds. In the enolase-catalyzed reaction from glycolysis (Figure 14.9b), the elements of a water molecule are removed to form a carbon–carbon double bond. Note that the reverse reaction catalyzed by aldolase is actually the addition of the C_3 unit of dihydroxyacetone phosphate to the carbonyl double bond of glyceraldehyde 3-phosphate; thus, it also fits into the second definition.

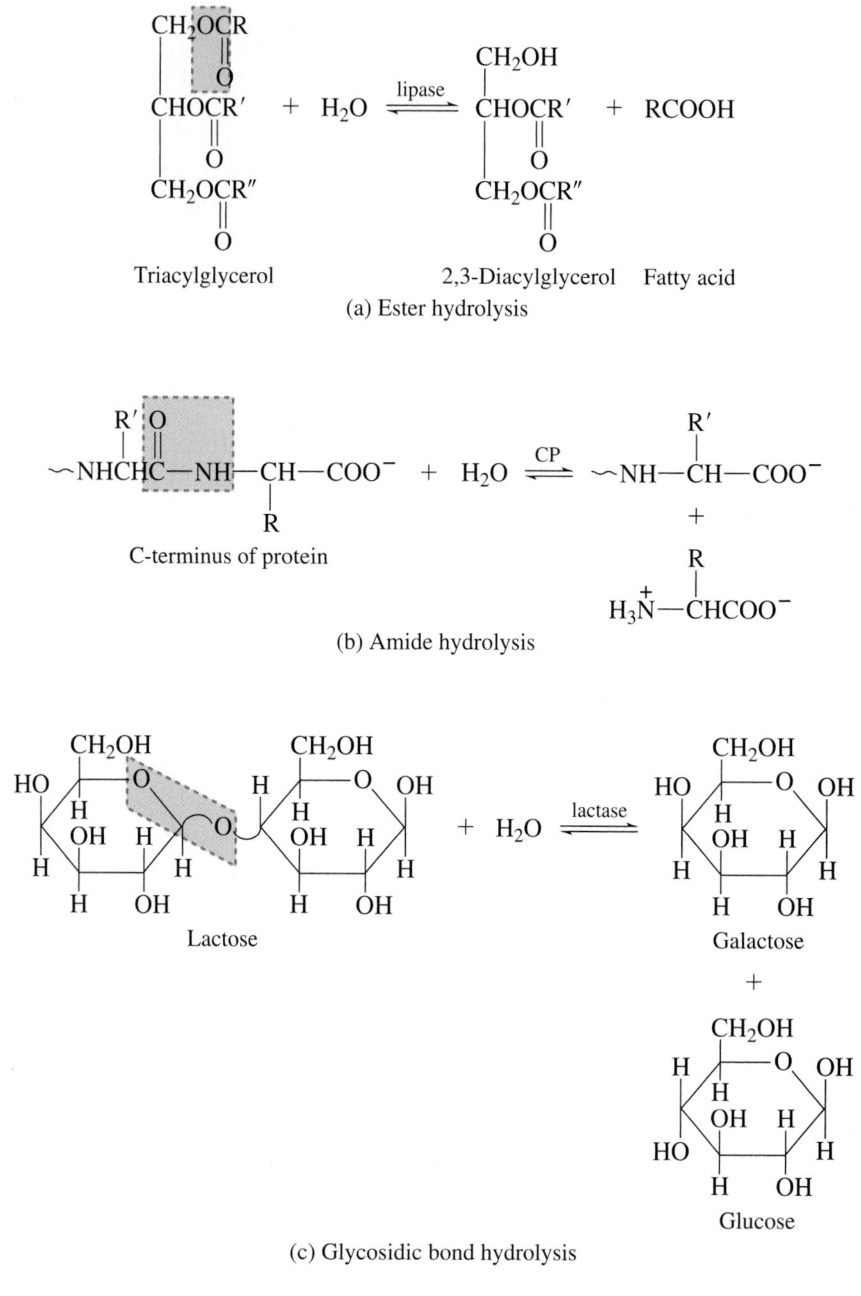

Figure 14.8

Examples of hydrolysis reactions. (a) Hydrolysis of ester bonds in triacylglycerol (fat) molecules. This releases fatty acids for energy metabolism. (b) Hydrolysis of amide bonds is important in the digestion of proteins. CP refers to the enzyme carboxypeptidase. (c) The β(1 → 4) glycosidic bond in lactose (milk sugar) is hydrolyzed to yield two monosaccharides.

Triacylglycerol   2,3-Diacylglycerol   Fatty acid

(a) Ester hydrolysis

C-terminus of protein

(b) Amide hydrolysis

Lactose   Galactose

Glucose

(c) Glycosidic bond hydrolysis

Isomerization and Rearrangement Reactions

Although isomerization and rearrangement reactions are sometimes undesired in organic chemistry, they play important roles in metabolism. Biochemical reactions in this category involve two kinds of chemical transformations:

1. Intramolecular hydrogen atom shifts changing the location of a double bond.
2. Intramolecular rearrangements of functional groups.

Both types transform a substrate molecule into its isomer. The most prominent reaction type of group 1 is the aldose–ketose isomerization (Figure 14.10a). This example is an important reaction in glycolysis, but we shall observe many other uses of this type of chemistry.

Intramolecular rearrangements of functional groups are relatively scarce in metabolism. One example from glucose metabolism is the intramolecular rearrangement of a phosphoryl group catalyzed by phosphoglucomutase (Figure 14.10b). **Mutases** belong to a subclass of isomerases that catalyze intramolecular rearrangements.

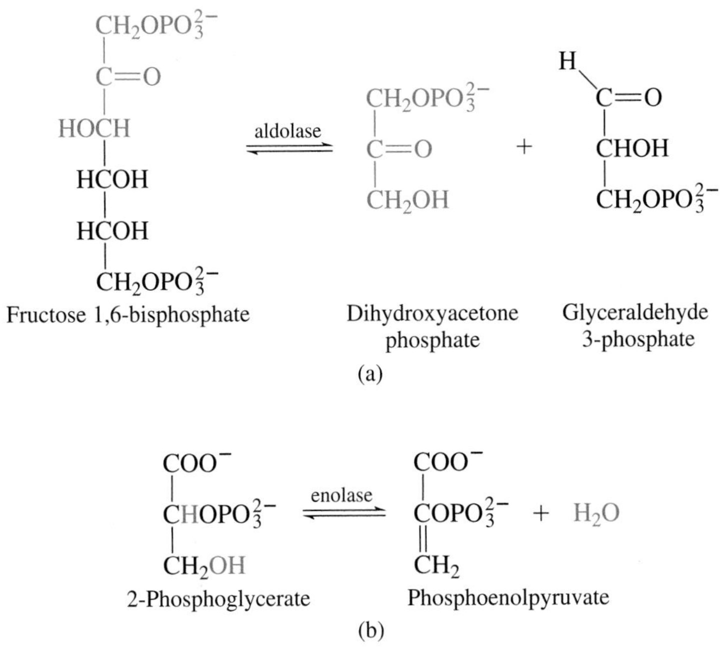

Figure 14.9
Examples of nonhydrolytic cleavage reactions: (a) cleavage of fructose 1,6-bisphosphate to dihydroxyacetone phosphate and glyceraldehyde 3-phosphate and (b) the dehydration of 2-phosphoglycerate. Both reactions are from carbohydrate metabolism by glycolysis.

Bond Formation Reactions Using Energy from ATP

A broad category of biochemical bond formation reactions use energy in ATP to form new bonds between molecules. Enzymes catalyzing the joining of two separate molecules are called **ligases** or **synthetases.** Carbon–carbon bonds in living organisms are usually formed by the reaction of a stabilized carbanion with the carbonyl functional groups of ketones, esters, or CO_2:

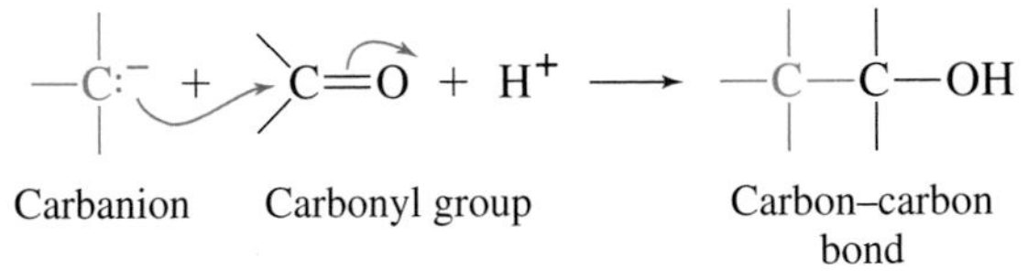

A carbanion is stabilized by the presence of electron-withdrawing substituents such as acyl groups. These groups provide stabilization by inductive withdrawal of electrons or

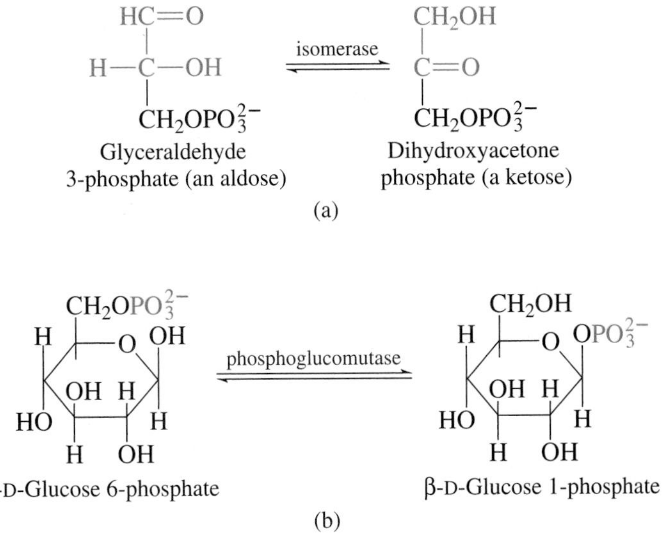

Figure 14.10

Examples of isomerization and rearrangement reactions: (a) rearrangement of an aldose to a ketose and (b) rearrangement of glucose 6-phosphate to glucose 1-phosphate. Both reactions are in the glycolytic pathway of carbohydrate degradation.

resonance stabilization of the negative charge. Two examples of enzyme-catalyzed reactions that proceed through stabilized carbanions are (1) the carboxylation of pyruvate, a reaction important for glucose synthesis (Figure 14.11a), and (2) the synthesis of citroyl CoA in the citric acid cycle (Figure 14.11b). In each example, a resonance-stabilized carbanion is generated on the methyl group of the substrate (pyruvate or acetyl CoA; Figure 14.11c).

Metabolic pathways display only a limited variety of chemical reactions. However, the six categories outlined here can lead to thousands of different reactions in the cell simply by modifying the structures of the reactants (see Window on Biochemistry 14.1). Most of the metabolic steps we have encountered in this section demonstrate a single type of chemical reaction. Organisms can also carry out unusual and varied chemistries by combining two or more of these reaction types in a single metabolic step. The decarboxylation of isocitrate provides an example:

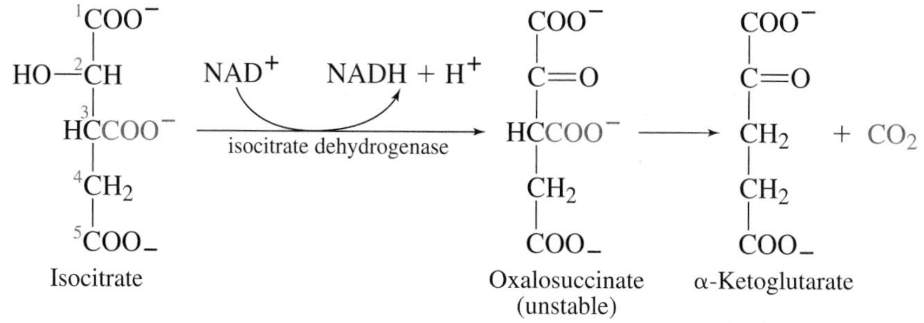

In the first reaction, an alcohol functional group on C2 of isocitrate is oxidized to a keto group. This oxidation reaction is coupled to the reduction of NAD^+ to NADH. The intermediate formed in this reaction, oxalosuccinate, is unstable and spontaneously loses

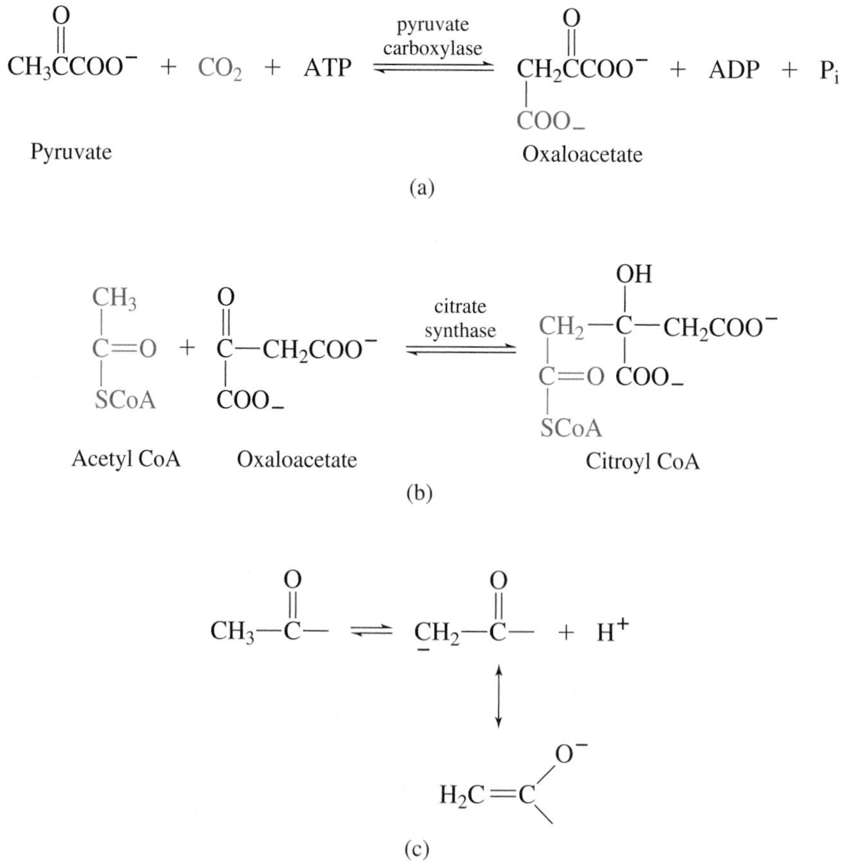

Figure 14.11

Examples of bond-forming reactions that use energy from ATP or reactions that proceed through stabilized carbanions. (a) Carboxylation of pyruvate to synthesize oxaloacetate. This reaction is important for the synthesis of glucose. (b) The combination of acetyl CoA and oxaloacetate to produce citroyl CoA. This reaction replenishes intermediates from the citric acid cycle. (c) Formation of a resonance-stabilized carbanion in an acetyl group.

a carboxyl functional group (decarboxylation). Thus, two types of chemistry occur in this metabolic step: oxidation–reduction and nonhydrolytic cleavage of a carbon–carbon bond. The overall process is described as **oxidative decarboxylation.**

14.3 Concepts of Bioenergetics

Standard Free Energy Change

From a thermodynamic point of view, metabolism is an energy-transforming process whereby catabolism provides energy for anabolism. (In this book, "energy" is defined as the capacity to cause or undergo change.) Cells and organisms are able to harness forms of energy (oxidative degradation of nutrients; absorption of light) and convert them to other forms suitable to support movement, active transport, and biosynthesis. As noted in Section 14.1, the medium of energy exchange is ATP, which is the universal carrier of energy. The process of anabolism (as well as active transport and muscle contraction) consumes ATP, which is then regenerated by catabolism (the ATP cycle; see Figure 14.4). We can combine these introductory statements into a fundamental concept of metabolism: *The overall process of catabolism releases energy (it is spontaneous); the overall process of anabolism requires energy input.* In this section we focus on the bioenergetics of the ATP energy cycle, concentrating on quantitative aspects. More specifically, we need to determine the *amount* of energy associated with ATP. The most useful thermodynamic term for this purpose is ΔG, the **free energy change** for a chemical process. This term, defined by Josiah Gibbs (1839–1903), is a measure of the energy available to do useful work, such as driving an unfavorable reaction to completion. The free energy changes for living systems are very difficult to determine experimentally. The term ΔG is a constant value for a reaction under a defined set of conditions. The form of ΔG that we will use is $\Delta G°$, the **standard free energy change.** Practically

A waterfall has both kinetic and potential energy.

Concept Reviews:
Logarithms; Thermodynamics

speaking, ΔG° is the energy change occurring when a reaction, under standard conditions, proceeds from start to equilibrium. In your introductory chemistry courses, standard conditions for solute reactions were defined as 1 atm of pressure, 25°C, and an initial concentration of reactant(s) and product(s) of 1 M. For biochemical processes, we will add the condition of a pH of 7 and define the modified standard free energy change as $\Delta G^{\circ\prime}$.

The standard free energy change for a reaction ($\Delta G^{\circ\prime}$) is related to its equilibrium constant. Consider the general reaction:

$$A + B \rightleftharpoons C + D$$

The equilibrium constant for this reaction is defined by the ratio of the concentrations of products to the concentrations of reactants:

$$K'_{eq} = \frac{[C][D]}{[A][B]}$$

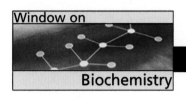

Window on

Biochemistry

14.1   Metabolism—How Do We Know the Pathways?

To understand a metabolic pathway, it is essential to know all the details of the reaction steps: (a) biochemical characteristics including enzymes, coenzymes, substrates, products, intermediates, kinetics, and mechanisms; (b) control of the pathway including the regulation at the reaction level (presence of allosteric enzymes) and the control by cellular conditions (regulation at the gene level), and (c) the physiological function of the pathway and how it coordinates with the overall metabolism of the organism. Several approaches for gathering this information will be outlined here and described in greater detail in later chapters.

Some of the earliest studies of metabolism were done in whole organisms by feeding various foods and collecting and analyzing waste products in urine and feces. These kinds of studies only identified final products and did not provide details of individual metabolic pathways. The introduction of radioisotopes near the middle part of the 20th century allowed biochemists to feed labeled biochemicals to animals, sacrifice the animals, and analyze tissue for the presence of the isotopes. Radioisotopes were used in the 1940s to elucidate the pathway for cholesterol synthesis in rats (Section 18.5). Other metabolic pathways and details were uncovered by the use of tissue slices. For example, the citric acid cycle was elucidated in the late 1930s by incubating thin slices of pigeon liver and heart tissue with the biochemical acids, citrate, malate, oxaloacetate, and succinate. The sequence of the citric acid cycle reactions was determined by adding metabolic inhibitors to the tissues and analyzing the buildup of certain intermediates.

Today, biochemists are more likely to use genetics and even sophisticated instrumentation to study metabolism. When we use genetic engineering, it is possible to prepare specific bacterial mutants that lack an enzyme because of an inactivated or deleted gene. Higher organisms that lack a particular gene can also be produced. These so-called **"gene knockouts"** are particularly useful if a single missing gene results in a metabolic defect that can be characterized.

An experimental approach still used by today's biochemists is to prepare **cell-free extracts.** The Büchners in their studies of glucose metabolism in yeast in the 1890s were the first to use cell-free extracts. Animal and plant cells are homogenized in a buffer to release cell components and **differential centrifugation** is used to separate organelles (see Window on Biochemistry 1.3). For example, these methods were used to locate the enzymes of the citric acid cycle, respiration, and fatty acid catabolism in the mitochondria. Automated chromoatography and electrophoresis procedures may be applied to cell-free extracts in order to purify and characterize enzymes and other molecules essential in metabolism (see Window on Biochemistry 4.2).

where

K'_{eq} = the equilibrium constant under standard conditions:

Pressure = 1 atm
Temperature = 25°C
pH = 7.0
Initial concentrations of A, B, C, D = 1 M

K'_{eq} is determined experimentally by (1) mixing 1 M concentrations of all reactants and products under standard conditions, (2) allowing the reaction to come to equilibrium (this step can be hastened by adding the appropriate enzyme), and (3) measuring the equilibrium concentrations of A, B, C, and D and calculating K'_{eq}. The standard free energy change ($\Delta G^{\circ\prime}$) is then calculated from the expression:

$$\Delta G^{\circ\prime} = -2.303RT \log K'_{eq}$$

A modern method for monitoring the fate of metabolism in whole cells or tissues is **nuclear magnetic resonance (NMR).** This instrumental technique can monitor biochemicals containing ^{1}H, ^{13}C, ^{15}N, and ^{31}P. The nuclei are identified by their characteristic nuclear spins. Energy metabolism in muscle has been studied using ^{31}P NMR (see figure). **Magnetic resonance imaging (MRI),** the medical equivalent of NMR, is a noninvasive procedure to study metabolism in normal and diseased tissue. MRI can be applied to the whole body of humans and other animals and to individual organ study.

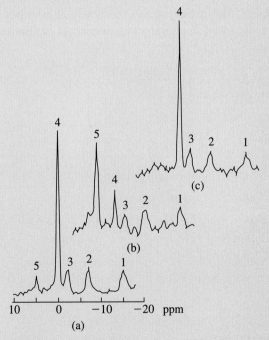

^{31}P NMR spectra of human forearm muscle showing the effect of exercise. (a) Before exercise; (b) after 19 minutes of exercise; (c) 5–6 minutes after spectrum b. Peak assignments: 1, β-phosphorus of ATP; 2, α-phosphorus of ATP; 3, γ-phosphorus of ATP; 4, phosphocreatine; 5, Pi. Phosphocreatine is used as a major source of energy during exercise. It is hydrolyzed to creatine and Pi. Note that the level of ATP remains relatively constant during exercise because it is produced and used at about the same rate. *Source:* G. Radda, *Science* 233:641 (1986), Reprinted with permission from American Association for the Advancement of Science.

where

$\Delta G^{\circ\prime}$ = free energy change under standard conditions
R = the gas constant, 8.315 J/mol
T = the absolute temperature, 273° + 25°C = 298 K
K'_{eq} = the equilibrium constant under standard conditions

We should insert a word of caution about the use of $\Delta G^{\circ\prime}$. Standard conditions do not normally exist in living cells. Therefore, absolute $\Delta G^{\circ\prime}$ values measured in a test tube reaction do not correspond directly to energy considerations for that reaction in a cell. We should not put great significance in the actual value of $\Delta G^{\circ\prime}$. However, $\Delta G^{\circ\prime}$ values are very useful in comparing energy requirements among the many reactions of metabolism.

$\Delta G^{\circ\prime}$ *defines the difference between the energy content of products and the energy content of reactants under standard conditions.* By knowing the sign and value for $\Delta G^{\circ\prime}$, we can predict whether a reaction under standard conditions is favorable (energy releasing) in the direction written and the approximate amount of energy released or required. The possible relationships between the sign of $\Delta G^{\circ\prime}$ and the thermodynamic consequence are given in Table 14.5. The larger the numerical value of $\Delta G^{\circ\prime}$ (whether + or −), the greater the amount of free energy released or required for a reaction. A reaction with $\Delta G^{\circ\prime}$ of −100 kJ/mol is more favorable (releases more energy) than one that has a $\Delta G^{\circ\prime}$ of −10 kJ/mol. A reaction that has a $\Delta G^{\circ\prime}$ of +100 kJ/mol requires a greater energy input than one with a $\Delta G^{\circ\prime}$ of +10 kJ/mol.

Experimental Measurement of $\Delta G^{\circ\prime}$

Now, let's use this information to calculate the standard free energy change for a reaction important in glycolysis, the isomerization of glucose 6-phosphate to fructose 6-phosphate:

glucose 6-phosphate $\rightleftharpoons$ fructose 6-phosphate

To begin the experiment, 1 M concentrations of each reagent are mixed under standard conditions with the enzyme glucose-6-phosphate isomerase. At equilibrium the concentrations measured are

$$K'_{eq} = \frac{[\text{fructose 6-phosphate}]}{[\text{glucose 6-phosphate}]} = \frac{0.67 \, M}{1.33 \, M} = 0.50$$

where

[glucose 6-phosphate] = 1.33 M
[fructose 6-phosphate] = 0.67 M

Using the expression for standard free energy change:

$$\Delta G^{\circ\prime} = -2.303 RT \log K'_{eq}$$
$$\Delta G^{\circ\prime} = (-2.303)(8.315 \text{ J/mol})(298 \text{ K}) \log 0.5$$
$$\Delta G^{\circ\prime} = +1718 \text{ J/mol} = +1.718 \text{ kJ/mol}$$

According to the definitions in Table 14.5, this reaction requires energy input to proceed in the direction written from left to right. When the glucose 6-phosphate and fructose 6-phosphate

Table 14.5
Significance of $\Delta G^{\circ\prime}$ values

Value and Sign of $\Delta G^{\circ\prime}$	Thermodynamic Consequences
$\Delta G^{\circ\prime} = 0$	The reactants and the products are at the same energy level. The reaction under standard conditions is at equilibrium. No release of or requirement for energy.
$\Delta G^{\circ\prime} < 0$ (negative values)	The reaction releases energy as it approaches equilibrium. The reactants are at a higher energy level than products. Useful energy is released and available to do work.
$\Delta G^{\circ\prime} > 0$ (positive values)	The reactants are at a lower energy than products. The reaction requires an input of energy to proceed as written.

were mixed with enzyme under standard conditions, glucose 6-phosphate molecules were converted to fructose 6-phosphate and vice versa. However, the *net flow* of reaction was from fructose 6-phosphate to glucose 6-phosphate; the reaction actually proceeded in the reverse direction.

Now that we have been introduced to some basic concepts of bioenergetics, let's turn to a discussion of the approximate amount of useful energy associated with the ATP energy cycle. The recycling of ATP as it occurs in metabolism is best represented by a phosphoryl transfer reaction:

$$\underset{\text{ATP}}{\text{AMP}-\text{O}-\overset{\overset{\text{O}}{\|}}{\underset{\underset{\text{O}_-}{|}}{\text{P}}}-\text{O}-\overset{\overset{\text{O}}{\|}}{\underset{\underset{\text{O}_-}{|}}{\text{P}}}-\text{O}^- + \text{H}_2\text{O} \rightleftharpoons \underset{\text{ADP}}{\text{AMP}-\text{O}-\overset{\overset{\text{O}}{\|}}{\underset{\underset{\text{O}_-}{|}}{\text{P}}}-\text{O}^-} + \underset{\text{P}_i}{{}^-\text{O}-\overset{\overset{\text{O}}{\|}}{\underset{\underset{\text{O}_-}{|}}{\text{P}}}-\text{OH}} + \text{H}^+$$

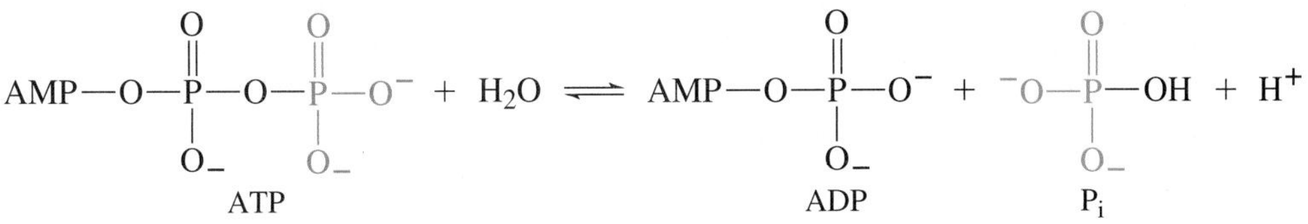

or

$$\text{ATP} + \text{H}_2\text{O} \rightleftharpoons \text{ADP} + \text{P}_i$$

$$K'_{eq} = \frac{[\text{ADP}][\text{P}_i]}{[\text{ATP}][\text{H}_2\text{O}]}$$

In chemical terms, this reaction involves the transfer of a phosphoryl group $(-\text{PO}_3^{2-})$ from ATP to water. The reaction may also be defined as hydrolysis of a **phosphoanhydride bond.** If $\Delta G^{\circ\prime}$ for this reaction is determined by the method discussed above (mix 1 M reactants and products under standard conditions), the equilibrium concentrations of ATP and H_2O are found to be too low to be experimentally measurable, thus making it impossible to calculate K'_{eq}. We have learned one important piece of information even though we could not measure K'_{eq}: the reaction of phosphoryl group transfer from ATP to H_2O is thermodynamically very favorable.

An important thermodynamic concept that will help us measure the standard free energy change for the above reaction is that $\Delta G^{\circ\prime}$ *values for coupled reactions are additive.* To use this concept, we must choose two reactions that have a common intermediate (the reactions are coupled) so that when the reactions are added, the result is the desired phosphoryl transfer reaction for ATP. We will choose the following two reactions: (1) a phosphoryl transfer reaction and (2) hydrolysis of a phosphoester bond. When added, the two reactions represent the transfer of a phosphoryl group from ATP to water.

	$\Delta G^{\circ\prime}$ kJ/mol
glucose + ATP $\rightleftharpoons$ glucose 6-phosphate + ADP	−16.7
glucose 6-phosphate + H_2O $\rightleftharpoons$ glucose + P_i	−13.8
Sum: ATP + H_2O $\rightleftharpoons$ ADP + P_i	−30.5

The standard free energy change of −30.5 kJ/mol for the phosphoryl group transfer from ATP to water indicates a spontaneous reaction. This can be stated in different terms: There is a high potential for ATP in water solution to be converted to ADP and P_i. A relatively large amount of useful chemical energy is available to do work.

ATP and Phosphoanhydride Bonds

Now, we need to explain on a chemical basis the large amount of energy available in ATP. The bond cleaved during the transfer reaction is a **phosphoanhydride bond:**

$$\overset{\overset{\text{phosphoanhydride bond}}{\overbrace{\qquad\qquad}}}{\text{AMP}-\text{O}-\underset{\underset{\text{O}_-}{|}}{\overset{\overset{\text{O}}{\|}}{\text{P}}}-\text{O}-\underset{\underset{\text{O}_-}{|}}{\overset{\overset{\text{O}}{\|}}{\text{P}}}-\text{O}^-}$$

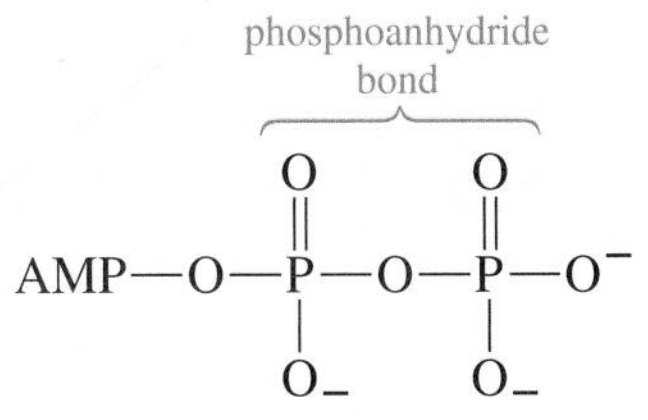

You may recall from organic chemistry that anhydride bonds are relatively reactive; that is, energy is released during their cleavage. For the hydrolysis of ATP, we can state this in other words: *the products (ADP and P_i) are at a lower energy level than the reactants (ATP and H_2O); therefore, energy is released* (Figure 14.12). The products ADP and P_i are made less reactive (more stable) by **resonance stabilization.** Several more resonance hybrid structures may be drawn for the products of the reaction (ADP and P_i) than for the reactants (ATP and H_2O); therefore, the products are at a lower energy level (are more stable) (Figure 14.13). A second reason for the large amount of energy associated with ATP is **charge repulsion.** The large number of negative charges in close proximity on ATP increases the potential for bond cleavage (see Figure 14.5). When phosphoanhydride bonds are broken, the negative charge repulsion is relieved. Note that a phosphoanhydride bond that can undergo phosphoryl group transfer to H_2O also exists in ADP:

$$\text{Adenosine}-\text{O}-\overset{\overset{\text{O}}{\|}}{\underset{\underset{\text{O}_-}{|}}{\text{P}}}-\text{O}-\overset{\overset{\text{O}}{\|}}{\underset{\underset{\text{O}_-}{|}}{\text{P}}}-\text{O}^- + \text{H}_2\text{O} \longrightarrow \text{Adenosine}-\text{O}-\overset{\overset{\text{O}}{\|}}{\underset{\underset{\text{O}_-}{|}}{\text{P}}}-\text{O}^- + \text{P}_i + \text{H}^+$$

$$\qquad\qquad\qquad\text{ADP}\qquad\qquad\qquad\qquad\qquad\qquad\qquad\text{AMP}$$

where

$$\Delta G^{\circ\prime} = -30.5 \text{ kJ/mol}$$

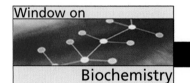

Window on

Biochemistry

14.2   Thioesters—Ancient, But Still Lots of Energy

In Section 14.3 we studied several phosphorylated biomolecules that were defined as "highly reactive" or "energy rich" because they have high phosphoryl group transfer potential. Reactive biomolecules used for energy transfer are not limited to those that are phosphorylated. Another category of important reactive molecules are the **thioesters**. A thioester is formed when a thiol (R—SH) joins with a carboxylic acid (R′COOH):

$$\text{R}-\text{SH} + \text{R}'-\text{COOH} \rightleftharpoons \text{R}-\text{S}-\overset{\overset{\text{O}}{\|}}{\text{C}}-\text{R}' + \text{H}_2\text{O}$$

The most common reactive thioesters in the cell are produced from the thiol **coenzyme A (CoASH,** where SH denotes the sulfhydryl group). This coenzyme, which is derived from the vitamin pantothenic acid (see figure below), functions as a carrier of activated acetyl and other acyl groups:

$$\text{CoA}-\text{SH} + -\overset{\overset{\text{O}}{\|}}{\text{C}}-\text{CH}_3 \rightleftharpoons \text{CoA}-\text{S}-\overset{\overset{\text{O}}{\|}}{\text{C}}-\text{CH}_3$$

$$\quad\text{Coenzyme A}\qquad\quad\text{Acetyl}\qquad\qquad\quad\text{Acetyl CoA}$$

The reactive acetyl group is linked to the SH group of CoASH.

Acetyl CoA is an important metabolite formed in the stage II catabolism of carbohydrates, fats, and amino acids (Figure 14.6). The thioester bond is of high energy, and that energy is released upon transfer of the acetyl group to an acceptor molecule—for example, water:

$$\text{CoA}-\text{S}-\overset{\overset{\text{O}}{\|}}{\text{C}}-\text{CH}_3 + \text{H}_2\text{O} \rightleftharpoons \text{CoA}-\text{SH} + \text{CH}_3\text{COOH} \quad \Delta G^{\circ\prime} = -31.4\,\text{kJ/mol}$$

Other Reactive Biomolecules for Energy Transfer

ATP and ADP are often defined as "reactive" or "energy-rich" molecules. In other words, this means that they have a high tendency to transfer phosphoryl groups. ATP and ADP are not the only reactive molecules in biological cells. In fact, some phosphorylated compounds have even greater potential energy than ATP; that is, they have a greater tendency than ATP to transfer a phosphoryl group. Consider the transfer of a phosphoryl group from the glycolysis intermediate phosphoenolpyruvate (PEP) to H_2O:

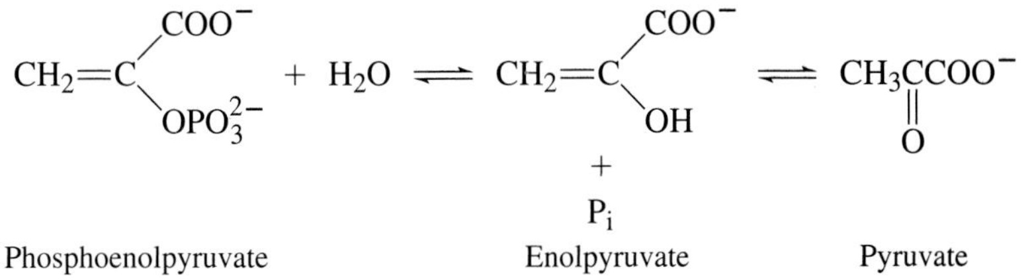

$$CH_2\!=\!C\!\!\begin{array}{c}COO^-\\OPO_3^{2-}\end{array} + H_2O \rightleftharpoons CH_2\!=\!C\!\!\begin{array}{c}COO^-\\OH\end{array} \rightleftharpoons CH_3CCOO^-$$

$$+\ P_i$$

| Phosphoenolpyruvate | Enolpyruvate | Pyruvate |

The $\Delta G^{\circ\prime}$ for this phosphoryl group transfer reaction is -61.9 kJ/mol, just about twice that of phosphoryl group transfer from ATP to H_2O. The phosphoryl group in PEP is linked not via a phosphoanhydride bond, as in ATP, but as a phosphoester bond, which is usually less reactive. The reason for the larger amount of energy in PEP is that the first product, enolpyruvate, is very unstable (high energy level) and releases energy during its isomerization to the more stable pyruvate (sometimes called ketopyruvate).

The cleavage of a thioester bond is more exergonic than that for an oxyester because the thioester is less stabilized by resonance. The energy derived from the exergonic transfer of the acetyl group linked via a thioester may be coupled to a thermodynamically unfavorable reaction. We will encounter the use of CoASH as an acetyl and acyl group carrier in a number of biochemical reactions including pyruvate catabolism, fatty acid catabolism, and the citric acid cycle.

Acetyl CoA and other thioesters are participants in many reactions where ATP is used and recycled. Biochemists who research molecules possibly involved in the origin of life often cite thioesters as important primeval molecules. The thioesters could have actually played the role of ATP in a prebiotic world where phosphate and phosphorylated compounds were scarce. Note that the energy available upon the transfer of the acetyl group from acetyl CoASH ($\Delta G^{\circ\prime} = -31.4$ kJ/mol) is almost identical to that released when a phosphoryl group is transferred from ATP ($\Delta G^{\circ\prime} = -30.5$ kJ/mol).

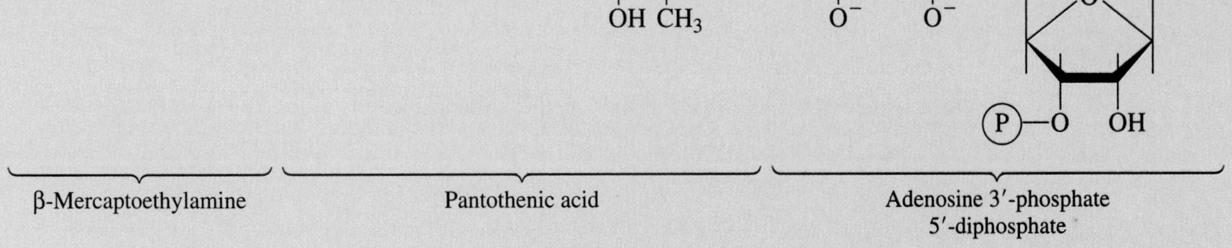

$$HS\!-\!CH_2\!-\!CH_2\!-\!NH\!-\!\overset{O}{\overset{\|}{C}}\!-\!CH_2\!-\!CH_2\!-\!NH\!-\!\overset{O}{\overset{\|}{C}}\!-\!\overset{H}{\underset{OH}{\overset{|}{C}}}\!-\!\overset{CH_3}{\underset{CH_3}{\overset{|}{C}}}\!-\!CH_2\!-\!O\!-\!\overset{O}{\overset{\|}{\underset{O^-}{P}}}\!-\!O\!-\!\overset{O}{\overset{\|}{\underset{O^-}{P}}}\!-\!O\!-\!CH_2 \quad Adenine$$

| β-Mercaptoethylamine | Pantothenic acid | Adenosine 3′-phosphate 5′-diphosphate |

CoASH structure

Figure 14.12

Energy diagram for the reaction $ATP + H_2O \longrightarrow ADP + P_i$. The $\Delta G^{\circ\prime}$ (standard free energy change) is a measure of the amount of energy that is released. That value is -30.5 kJ/mol. ATP is at a higher energy level than $ADP + P_i$; therefore, energy is released when the phosphoryl group is transferred in this case to H_2O.

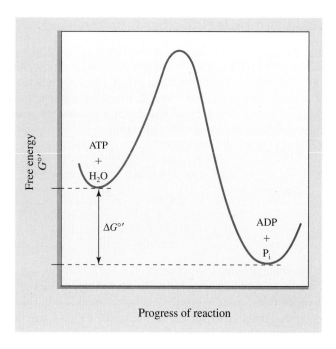

It is useful to rank PEP, ATP, and other phosphorylated compounds in terms of **phosphoryl group transfer potential.** This is a measure of their relative tendency to transfer a phosphoryl group to an acceptor molecule such as H_2O (Table 14.6). The phosphoryl group transfer potential is related directly to the $\Delta G^{\circ\prime}$ for the transfer reaction. The more negative the $\Delta G^{\circ\prime}$, the greater the tendency to transfer the phosphoryl group. ATP is important in energy metabolism not only because of a relatively high phosphoryl group transfer ability. In addition, ATP has *a position in the middle of the phosphoryl group transfer scale* (Figure 14.14). ATP can act as a common intermediate to link two reactions energetically. Consider the following enzyme catalyzed reactions:

$$\text{phosphoenolpyruvate} + \text{ADP} \underset{}{\overset{\text{pyruvate kinase}}{\rightleftharpoons}} \text{pyruvate} + \text{ATP}$$

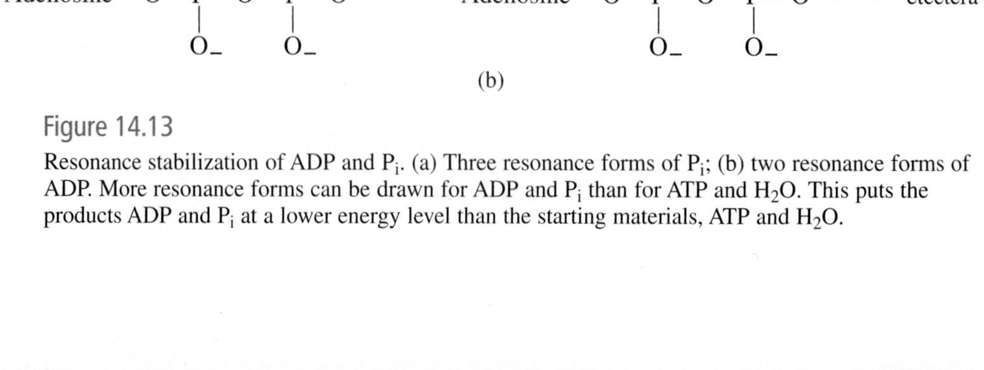

Figure 14.13

Resonance stabilization of ADP and P_i. (a) Three resonance forms of P_i; (b) two resonance forms of ADP. More resonance forms can be drawn for ADP and P_i than for ATP and H_2O. This puts the products ADP and P_i at a lower energy level than the starting materials, ATP and H_2O.

Phosphorylated Compounds	$\Delta G^{\circ\prime}$ (kJ/mol)[a]	Phosphoryl Group Transfer Potential
Phosphoenolpyruvate	− 61.9	Highest
1,3-Biphosphoglycerate	− 49.3	
Phosphocreatine	− 43.0	
ATP	− 30.5	
ADP	− 30.5	↓
Glucose 1-phosphate	− 20.9	
Glucose 6-phosphate	− 13.8	
Glycerol 1-phosphate	− 9.2	Lowest

[a]These values are for hydrolysis reactions (the transfer potential of the phosphoryl group to H_2O).

Table 14.6

$\Delta G^{\circ\prime}$ values and relative ranking for phosphoryl group transfer potential of important phosphorylated biochemicals

where

$$\Delta G^{\circ\prime} = -31.3 \text{ kJ/mol}$$

$$\text{ATP } + \text{ glucose } \underset{}{\overset{\text{hexokinase}}{\rightleftharpoons}} \text{ glucose 6-phosphate } + \text{ ADP}$$

where

$$\Delta G^{\circ\prime} = -16.7 \text{ kJ/mol}$$

The first reaction is an energy-releasing step in glycolysis. Energy from glucose degradation is captured in ATP; the ATP is then used in the second reaction to "activate" another glucose molecule for entry into glycolysis. Is the cell capable of direct phosphoryl transfer from PEP to glucose, thereby bypassing ATP formation? There is no enzyme that catalyzes this reaction:

$$\text{PEP } + \text{ glucose } \xrightarrow{\times} \text{ pyruvate } + \text{ glucose 6-phosphate}$$

With no ATP being produced, there would be no means for energy transfer to biosynthetic reactions or other energy-requiring processes. The recycling of ATP (see Figure 14.4) serves as a pipeline to transfer energy from catabolism to anabolism. Another class of energy-rich molecules is discussed in Window on Biochemistry 14.2.

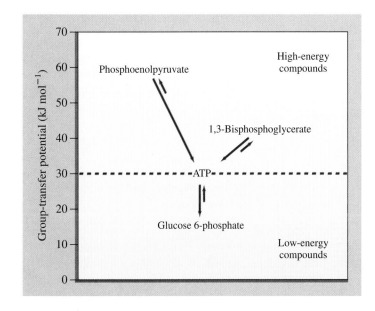

Figure 14.14

Transfer of phosphoryl groups. Compounds with high phosphoryl group transfer potential, such as phosphoenolpyruvate and 1,3-bisphosphoglycerate, may react spontaneously with ADP to produce ATP. Likewise, ATP may transfer a phosphoryl group to glucose to make glucose 6-phosphate. The reaction arrows indicate the favorable direction for each reaction.

Summary

■ Metabolism is the study of the chemistry, regulation, and energetics of the thousands of reactions occurring in a biological cell. The process of metabolism takes place via sequences of consecutive, enzyme-catalyzed reactions.

■ Organisms are of two metabolic types: (1) **Autotrophs** use atmospheric CO_2 as the sole source of carbon and energy from the sun to make other biomolecules; (2) **heterotrophs** obtain energy by ingesting complex carbon-containing compounds usually found in the autotrophs.

■ Metabolic processes are grouped into two categories: (1) **Catabolism** is the degradative phase that releases energy for the production of ATP; (2) **anabolism** is the synthetic phase, which uses the energy in ATP to make complex biomolecules.

■ In terms of chemistry, metabolic processes can be grouped into six categories: oxidation–reduction; group-transfer; hydrolysis; nonhydrolytic cleavage; isomerization; bond formation with energy from ATP.

■ ATP is the universal molecular carrier of free energy in organisms. The energy is released when a phosphoryl group in ATP is transferrred to another molecule such as water ($ATP + H_2O \rightleftharpoons ADP + HPO_4^{2-}$).

Study Exercises

Understanding Terms

14.1 Define the following terms in 25 words or less.

  a. Catabolism
  b. Anabolism
  c. Anaerobic organisms
  d. Metabolic pathway
  e. ATP
  f. Convergence of pathways
  g. Dehydrogenases
  h. $\Delta G^{\circ\prime}$
  i. Phosphoryl group transfer potential
  j. PEP
  k. Thioesters
  l. Differential centrifugation
  m. Chromatography
  n. Electrophoresis
  o. Acetyl CoA

Reviewing Concepts

14.2 Classify each of the following reactions into one of the six types of reactions defined in Table 14.2.

  a.

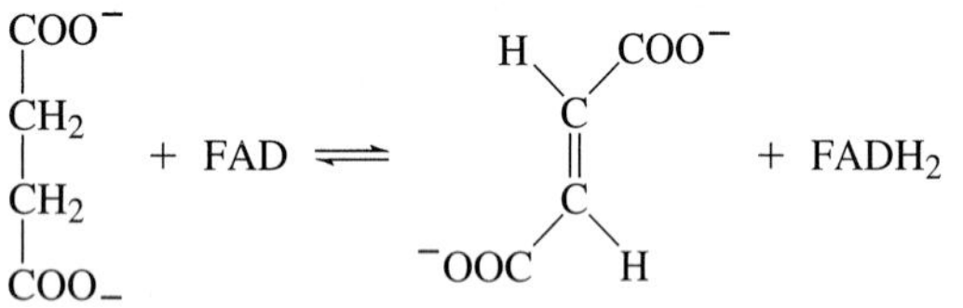

  b.

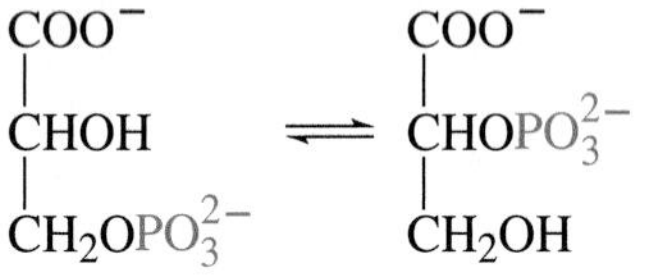

  c.

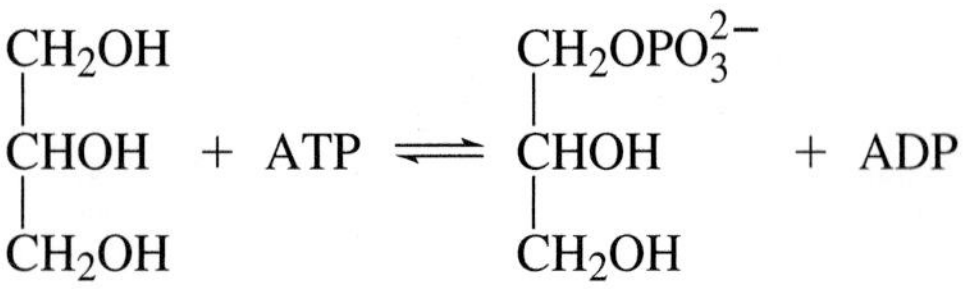

  d.

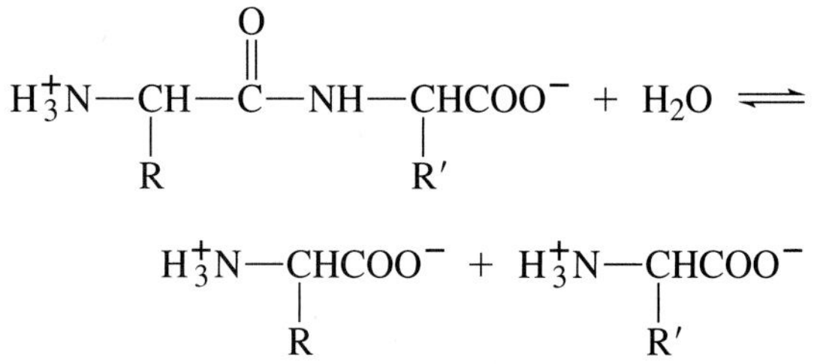

  e.

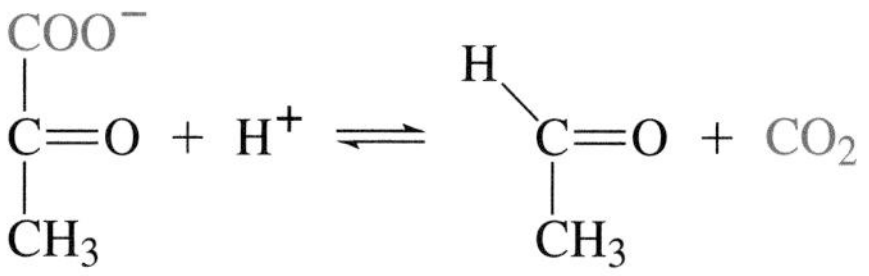

  f.

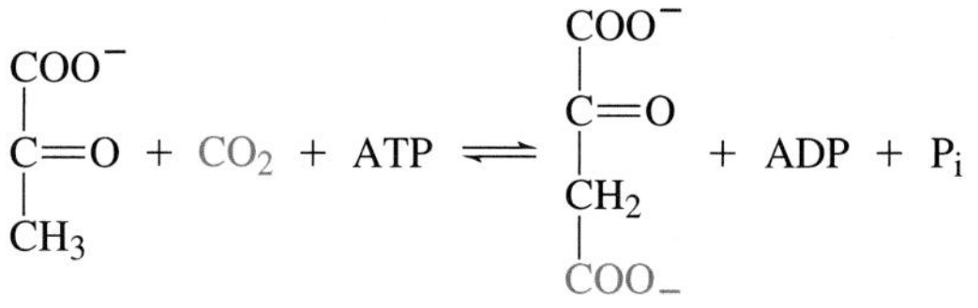

14.3 Name the general class of enzyme that catalyzes each reaction in Problem 14.2.

14.4 Several statements about stage III of catabolism are written below. Circle those that are correct and explain why the incorrect statements are wrong.

  a. The reactions in stage III are common to all fuel molecules.
  b. The overall $\Delta G^{\circ\prime}$ for stage III is a positive value.
  c. Stage III involves hydrolysis of proteins, starch, and triacylglycerols to their respective monomeric components.
  d. Stage III produces the majority of ATP generated in the cell.

14.5 The following terms can be used to characterize catabolism or anabolism. Assign each term to the proper direction of metabolism.

　　Reduction reactions
　　Energy input necessary
　　Energy release
　　Convergence of reactions
　　Cleavage of ATP
　　Biosynthesis
　　Oxidation reactions
　　Production of ATP
　　Divergence of reactions
　　Degradation

14.6 Consider the following reaction:

$$\text{pyruvate} + \text{NADH} + \text{H}^+ \rightleftharpoons \text{lactate} + \text{NAD}^+$$

The K'_{eq} is 1.7×10^{11}. Calculate $\Delta G^{\circ\prime}$.

14.7 How many phosphoanhydride bonds are present in each of the following biomolecules?

　a. ATP
　b. ADP
　c. AMP
　d. 1,3-Bisphosphoglycerate
　e. Phosphoenolpyruvate
　f. Glucose 1-phosphate
　g. Glyceraldehyde 3-phosphate
　h. Fructose 1,6-bisphosphate
　i. Pyrophosphate

14.8 Answer the following questions about ATP.

　a. How many phosphoanhydride bonds are present?
　b. What kind of chemical linkage is present between the ribose and the triphosphate group?
　c. How are the negative charges on ATP usually neutralized in the cell?
　d. What kind of chemical bond links adenine and ribose?

14.9 Consider the following reaction catalyzed by alcohol dehydrogenase. Answer the following questions about the reaction proceeding from left to right.

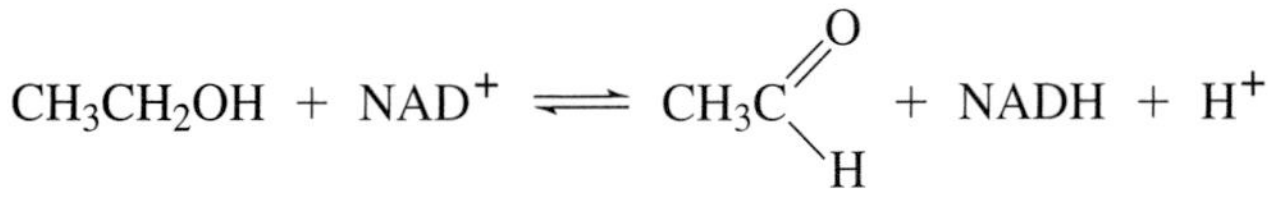

$$\text{CH}_3\text{CH}_2\text{OH} + \text{NAD}^+ \rightleftharpoons \text{CH}_3\text{C}\!\!\underset{\text{H}}{\overset{\text{O}}{<}} + \text{NADH} + \text{H}^+$$

　a. What is the oxidizing agent?
　b. What is the reducing agent?

14.10 Look again at the reaction in Problem 14.9 and answer the following questions about the reaction proceeding from right to left.

　a. What is the oxidizing agent?
　b. What is the reducing agent?

14.11 Draw as many resonance structures as you can for HPO_4^{2-} (P_i).

HINT: Begin with

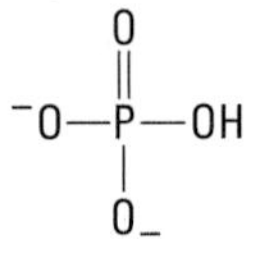

$$^-\text{O}-\!\!\underset{\text{O}_-}{\overset{\text{O}}{\underset{|}{\overset{\|}{\text{P}}}}}\!\!-\text{OH}$$

14.12 What is the common structural feature in NAD^+, FAD, ATP, and CoASH? Do they have any common biological functions?

14.13 In this chapter it is stated that "standard conditions do not normally exist in living cells." Compare and contrast standard conditions and normal cellular conditions.

14.14 Discuss the metabolic differences between autotrophs and heterotrophs.

14.15 What is the difference between ΔG° and $\Delta G^{\circ\prime}$?

14.16 Draw as many resonance structures as you can for pyrophosphate, PP_i.

14.17 Draw as many resonance structures as you can for ATP.

Working Problems

14.18 Arrange the following compounds in order of increasing oxidation level beginning with the least oxidized (or most reduced).

　a. 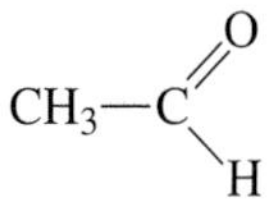 $\text{CH}_3-\text{C}\!\!\underset{\text{H}}{\overset{\text{O}}{<}}$

　b. $\text{CH}_2=\text{CH}_2$

　c.  $\underset{|}{\overset{\text{COOH}}{\text{COOH}}}$

　d. CH_3COOH
　e. $\text{CH}_3\text{CH}_2\text{OH}$
　f. CH_3-CH_3
　g. CO_2

HINT: Begin series with (f) ethane.

14.19 Consider the following reaction:

$$\text{glucose 1-phosphate} \rightleftharpoons \text{glucose 6-phosphate}$$

After 1 M reactant and product were mixed and allowed to reach equilibrium at 25°C, the concentration of each compound was measured:

$$[\text{glucose 1-phosphate}]_{eq} = 0.1\ M$$
$$[\text{glucose 6-phosphate}]_{eq} = 1.9\ M$$

Calculate K'_{eq} and $\Delta G^{\circ\prime}$ for the reaction. What is the favorable direction for this reaction?

14.20 Arrange the following phosphorylated compounds in the order of increasing phosphoryl group transfer potential.

a.

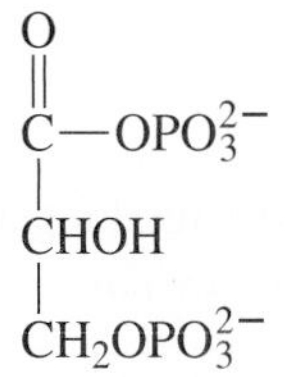

$$
\begin{array}{c}
O \\
\parallel \\
C\text{—}OPO_3^{2-} \\
\mid \\
CHOH \\
\mid \\
CH_2OPO_3^{2-}
\end{array}
$$

1,3-Bisphosphoglycerate (1,3-BPG)

b. ATP

c. AMP

d.

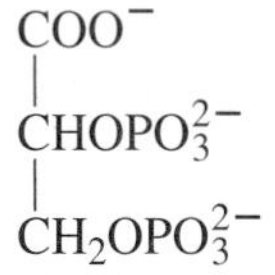

$$
\begin{array}{c}
COO^- \\
\mid \\
CHOPO_3^{2-} \\
\mid \\
CH_2OPO_3^{2-}
\end{array}
$$

2,3-Bisphosphoglycerate (2,3-BPG)

e. PEP
f. ADP

14.21 What is the spontaneous direction of each of the following reactions under standard conditions?

a. PEP + ADP $\rightleftharpoons$ pyruvate + ATP
b. ATP + H_2O $\rightleftharpoons$ ADP + P_i
c. ADP + H_2O $\rightleftharpoons$ AMP + P_i
d. Glucose 6-phosphate + ADP $\rightleftharpoons$ glucose + ATP
e. Glucose 6-phosphate $\rightleftharpoons$ glucose 1-phosphate

HINT: Use the data in Table 14.6 to answer this question.

14.22 Consider the following general reaction:

$$A \rightleftharpoons B$$

After A and B were mixed in 1 M concentrations and allowed to come to equilibrium at 25°C, the concentration of each was measured and found to be equal ([A] = [B]). Calculate K'_{eq} and $\Delta G^{\circ\prime}$ for the reaction. What is the spontaneous direction for this reaction? How much energy is released in this reaction under standard conditions?

14.23 Why is it an advantage to the cell for the mainstream catabolic reactions to converge to the common intermediate acetyl CoA?

14.24 Consider the following phosphoryl group transfer reaction which is the first step in glucose catabolism by glycolysis:

glucose + ATP $\rightleftharpoons$ glucose 6-phosphate + ADP

The $\Delta G^{\circ\prime}$ for the reaction is -16.7 kJ/mol.

a. Calculate K'_{eq} for the reaction.
b. Define K'_{eq} in terms of reactant and product concentrations.

c. What is the ratio of [glucose 6-phosphate] to [glucose] if the ratio of [ADP] to [ATP] is 10?

14.25 Write the products for the following hydrolysis reactions.

a.

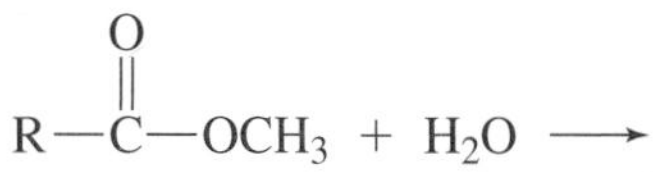

$$
\begin{array}{c}
O \\
\parallel \\
R\text{—}C\text{—}OCH_3 + H_2O \longrightarrow
\end{array}
$$

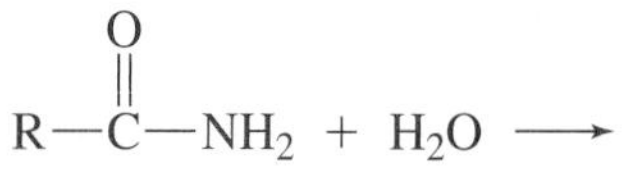

b.

$$
\begin{array}{c}
O \\
\parallel \\
R\text{—}C\text{—}NH_2 + H_2O \longrightarrow
\end{array}
$$

c.

$$
\begin{array}{c}
O \\
\parallel \\
R\text{—}C\text{—}N\text{—}CHCOO^- + H_2O \longrightarrow \\
\mid \quad\mid \\
H \quad R'
\end{array}
$$

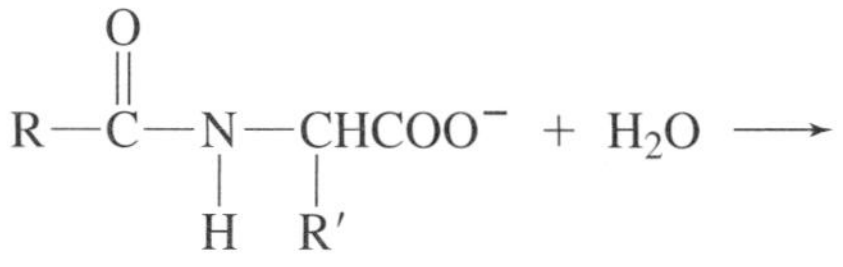

d.

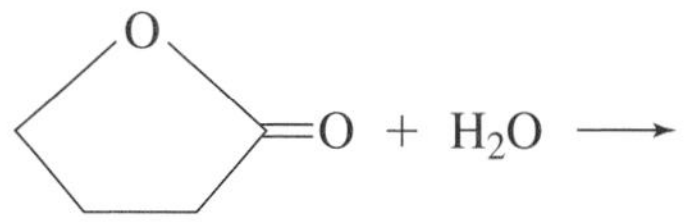

e.

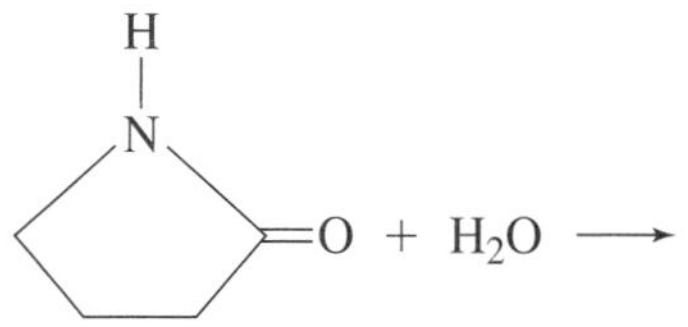

14.26 It has been said that in order for life to reproduce, develop, and thrive, there are three basic requirements: (1) a blueprint (directions), (2) materials, and (3) energy. What specific biomolecules or biological processes provide each of these needs?

14.27 Consider the following reaction of glucose 6-phosphate.

glucose 6-phosphate + H_2O $\rightleftharpoons$ glucose + P_i

a. Define the type of chemistry. Choose from redox, isomerization, and phosphoryl group transfer (hydrolysis).
b. The $\Delta G^{\circ\prime}$ for the uncatalyzed reaction is -13.8 kJ/mol. What is the spontaneous direction for the reaction under standard conditions?
c. An enzyme was discovered that catalyzes the reaction. What is the $\Delta G^{\circ\prime}$ for the reaction when the enzyme is present?
  1. Greater than -13.8 kJ/mol
  2. Less than -13.8 kJ/mol
  3. -13.8 kJ/mol

14.28 In the process of photosynthesis, the carbon atoms of CO_2 are used to make glucose. Study the structure of glucose in Figure 8.8 and determine if the overall process is one of oxidation or reduction. Explain.

14.29 Phosphocreatine is a compound in our muscles that has a high phosphoryl group transfer potential. It can be used to make ATP from ADP. Complete the following reaction by drawing a structure for all missing products.

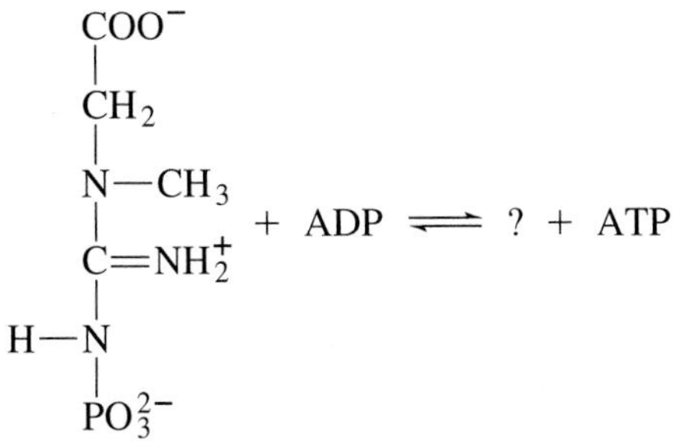

$$
\begin{array}{c}
COO^- \\
| \\
CH_2 \\
| \\
N-CH_3 \\
| \\
C=NH_2^+ \\
| \\
H-N \\
| \\
PO_3^{2-}
\end{array}
+ \ ADP \ \rightleftharpoons \ ? \ + \ ATP
$$

14.30 In animal metabolism, glucose is converted to CO_2 and H_2O. Is this overall process of glucose breakdown one of oxidation or reduction? Explain.

14.31 The $\Delta G^{\circ\prime}$ for the transfer of a phosphoryl group from ATP to H_2O is -30.5 kJ/mol. Calculate the K'_{eq} under standard conditions.

14.32 Complete the following reactions and predict if at the equilibrium point, the ratio of [products]/[reactants] is greater or less than one. Assume that an appropriate catalyst is present.

a.
$$
\begin{array}{c}
O \\
\parallel \\
R-C-S-CH_3 \ + \ H_2O \ \rightleftharpoons
\end{array}
$$

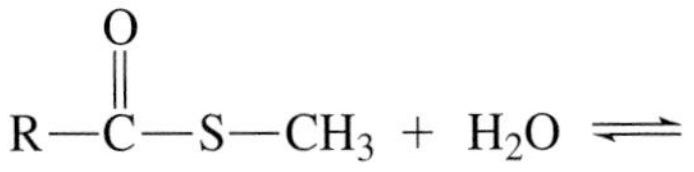

b. $ATP + H_2O \rightleftharpoons$

c. $Glucose + ATP \rightleftharpoons$

Writing Biochemistry

14.33 Prepare a list of the three concepts in this chapter you found most confusing. Write the concepts in the form of a question and then write an answer for each question.

14.34 Write a 100-word essay describing the differences between hydrolysis reactions and nonhydrolytic cleavage reactions.

Applying Biochemistry

14.35 What does it mean when we say that *"In metabolism, ATP is the universal carrier of biochemical energy"?* Discuss your answer in at least 100 words using appropriate reactions.

14.36 How is thermodynamics like riding the bus home after classes? For a hint, review the CD/ROM Concept Review unit on "Thermodynamics."

Further Reading

Banfalvi, G., 1994. The metabolic clockwork. *Biochem. Educ.* 22:137–139.

Bentley, R., 2000. From reactive C_2 units to acetyl coenzyme A. *Trends Biochem. Sci.* 25:302–305.

Brosemer, R., 1989. Metabolism: where to begin? *Biochem. Educ.* 17:31–33.

Carusi, E., 1992. It's time we replaced 'high-energy phosphate group' with 'phosphoryl group'. *Biochem. Educ.* 20:145–147.

Copley, S., 2000. Evolution of a metabolic pathway for degradation of a toxic xenobiotic. *Trends Biochem. Sci.* 25:261–265.

Elliott, K., et al., 1999. Teaching metabolism for the future—Concepts not pathways. *The Biochemist* 21(1):9–39.

Fell, D., 1997. *Understanding the Control of Metabolism.* London: Portland Press.

Frayn, K., 1996. *Metabolic Regulation: A Human Perspective.* London: Portland Press.

Hanson, R., 1990. Oxidation states of carbon as aids to understanding oxidative pathways in metabolism. *Biochem. Educ.* 18:194–196.

Manchester, K., 2000. Arthur Harden: An unwitting pioneer of metabolic control analysis. *Trends Biochem. Sci.* 25:89–92.

Punekar, N., 2000. A simple setup to illustrate metabolic pathway dynamics. *Biochem. Mol. Biol. Educ.* 28:248–250.

Rasmussen, N., 1996. Cell fractionation biochemistry and the origins of 'cell biology.' *Trends Biochem. Sci.* 21:319–321.

Srinivasen, 1999. The role of mutants in metabolic regulation. *Biochem. Educ.* 27:215–216.

Chapter 15

Metabolism of Carbohydrates

All organisms obtain energy from the oxidative breakdown of glucose and other carbohydrates. In some cells and organisms, such as brain, erythrocytes, and many bacteria, this is the sole or major source of energy. Biochemists have long pursued knowledge of the mechanisms of carbohydrate metabolism. The practical uses of carbohydrate metabolism in the processes of beer and wine making, leavening bread, and fermentation of milk products are described in the earliest written historical records. A lack of scientific understanding of fermentation processes did not limit the profits of the brewers, bakers, and cheesemakers or prohibit the enjoyment of their products by consumers. Scientific studies of carbohydrate metabolism were initiated in the middle of the nineteenth century. The most prominent researcher of this time, Louis Pasteur, studied the fermentation process as carried out by living bacterial and yeast cells. Pasteur held firmly to his belief in the vital-force theory, that only whole, living organisms are capable of carrying out metabolism and other biochemical processes. However, the vital-force theory was put to rest when Hans and Eduard Büchner demonstrated in the 1890s that cell-free extracts of yeast (made by breaking open yeast cells and using their contents) could ferment glucose, sucrose, and other carbohydrates to ethanol. This discovery made it possible to isolate enzymes from cell extracts and to study the reactions and products of individual metabolic steps.

The research of biochemists in England, Germany, and the United States provided the detailed information necessary to outline the important pathways of carbohydrate metabolism by the 1940s and 1950s. See Figure 15.1 for a preview. The most important pathway, called **glycolysis,** transforms glucose to pyruvate in anaerobic metabolism with production of a small amount of energy in the form of ATP and NADH. Pyruvate still has much potential energy, which can be extracted only by aerobic metabolism. The **phosphogluconate**

Figure 15.1

Major pathways of glucose metabolism in plants and animals. Glucose is stored in the form of glycogen (animals) and starch (plants). Free glucose may be oxidized to pyruvate (glycolysis), which has several possible fates depending on conditions and type of cell. Glucose in some cells is converted to ribose 5-phosphate. The major route for energy metabolism is glycolysis, the oxidation of glucose to pyruvate. Like glucose, pyruvate is a major metabolic junction. The most important pathway for aerobic energy metabolism is oxidation of pyruvate to acetyl CoA and continuation of catabolism by the citric acid cycle and oxidative phosphorylation. Glucose may be synthesized from pyruvate (gluconeogenesis) and polymerized into glycogen (animals) or starch (plants).

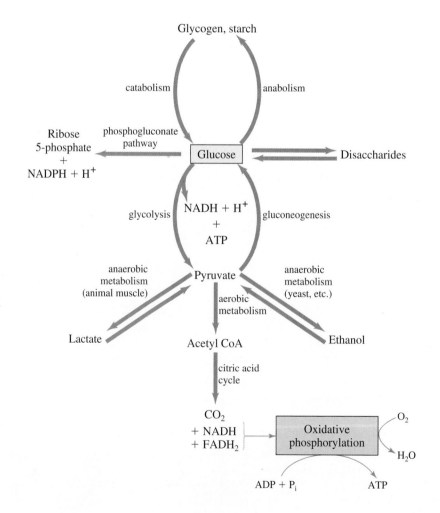

Breads and vegetables are abundant sources of dietary carbohydrates including amylose and amylopectin.

pathway **(pentose phosphate pathway),** an auxiliary route for glucose oxidation in animals, produces the pentose ribose 5-phosphate and the reduced cofactor NADPH (see Chapter 16). Glucose present in excess of catabolic needs is stored in starch (plants) or glycogen (animals) or, in some organisms, is used to make fats. Free glucose can be released from the stored forms if energy is required by an organism. Glucose concentrations can also be increased by synthesis from pyruvate or lactate by **gluconeogenesis.** In addition to its role in energy metabolism, glucose also serves as a precursor for the synthesis of structural polysaccharides, such as cellulose, disaccharides, such as sucrose and lactose, and other monosaccharides such as galactose and fructose. In this chapter, we explore the detailed pathways of carbohydrate metabolism, emphasizing their biological functions, coordination, and regulation.

15.1 The Energy Metabolism of Glucose

Glycolysis (*glyco* = "sweet substances"; *lysis* = "splitting") is the earliest discovered and most important process of carbohydrate metabolism. It is believed that the pathway is universal for extraction of the energy available in carbohydrates by all plants and animals. Because molecular oxygen is not required in any of the reaction steps, the glycolytic pathway is used by anaerobic as well as aerobic organisms. Glycolysis in aerobic organisms serves as an initial phase to prepare glucose for additional energy production. Some energy in the form of ATP and NADH is generated during stage II of carbohydrate metabolism (see Figure 14.6), but considerably more is harvested in the further oxidation of pyruvate to acetyl CoA, which enters the citric acid cycle and cellular respiration (stage III). To anaerobic organisms, ATP and NADH produced from glycolysis is the *only* significant energy available from carbohydrate metabolism. The major sugar entering glycolysis is glucose, but other monosaccharides, such as fructose and galactose, may also be degraded to pyruvate.

As we introduce each metabolic pathway, we will be especially interested in its intracellular location. The enzymes for most pathways we encounter will reside in the soluble cytoplasm, in the mitochondria (animal and plant cells), in chloroplasts (plant cells), in the nucleus, or in other organelles. The enzymes responsible for glycolysis are present in the cytoplasm of cells. The pathway was probably used in primitive organisms even before the appearance of oxygen in the atmosphere or the evolutionary development of eukaryotic cells with organelles. The glycolytic enzymes from many different organisms have been isolated and studied and our understanding of their mechanism and regulation is greater than for any other metabolic pathway.

The First Five Reactions of Glycolysis

The glycolytic pathway consists of ten enzyme-catalyzed reactions that begin with a hexose substrate and split it into two molecules of the α-ketoacid, pyruvate. The ten reactions, using glucose as initial substrate, are shown in Figure 15.2. The enzyme name for

Interactive Animation: Glycolysis

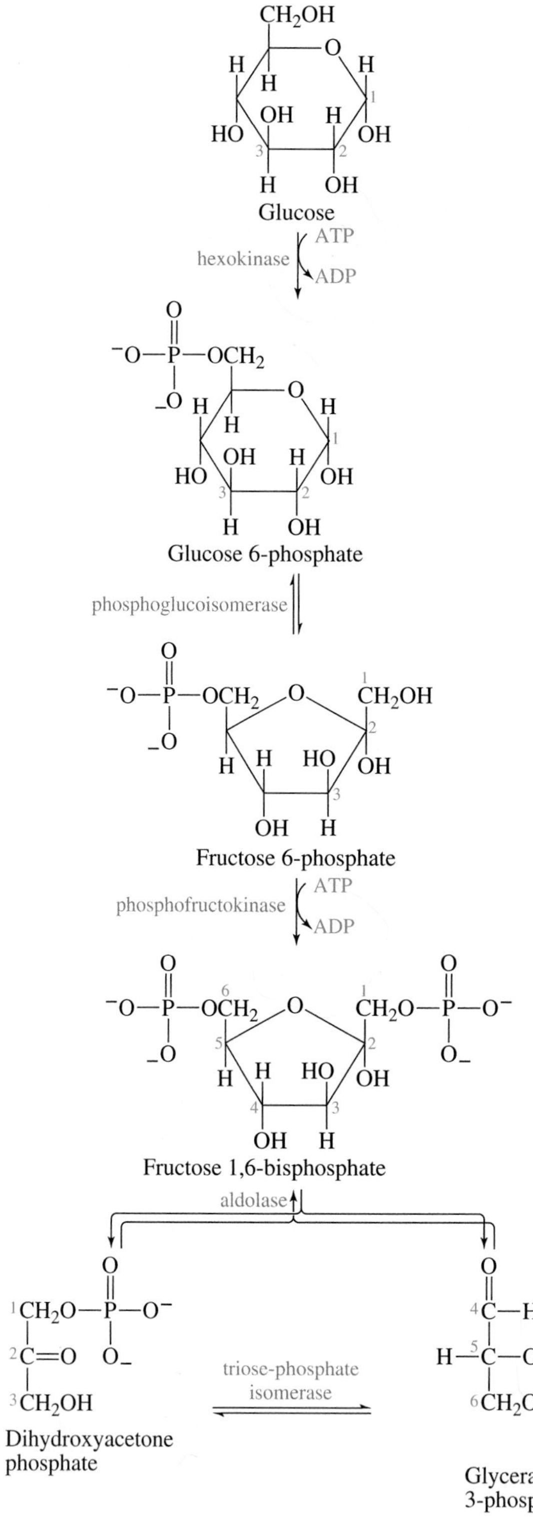

Figure 15.2

The reactions of glycolysis. α-D-Glucose enters as the substrate to be degraded. In the first five steps, the glucose is phosphorylated and cleaved to two molecules of glyceraldehyde 3-phosphate. In the second five steps, glyceraldehyde 3-phosphate is transformed to pyruvate. The complete reaction sequence leads to the formation of two ATP and two NADH for each glucose that enters.

Glucose

ATP
hexokinase
ADP

Glucose 6-phosphate

phosphoglucoisomerase

Fructose 6-phosphate

ATP
phosphofructokinase
ADP

Fructose 1,6-bisphosphate

aldolase

$^1CH_2O-P-O^-$
$^2C=O$
$^3CH_2OH$
Dihydroxyacetone phosphate

triose-phosphate isomerase

^4C-H
$H-^5C-OH$
$^6CH_2O-P-O^-$
Glyceraldehyde 3-phosphate

(*continued*)

each step and the type of reaction are described in Table 15.1. All reactions can be included in one of the six categories introduced in Chapter 14, Section 14.2. From an energy standpoint, we can define two stages of glycolysis. The first half of the reactions (reactions 1–5) are considered an "investment stage." Catabolism is a degradative phase that is characterized by oxidation (formation of NADH) and of energy release and capture to form ATP from ADP. However, none of the first five steps in glycolysis is oxida-

Figure 15.2—*continued*

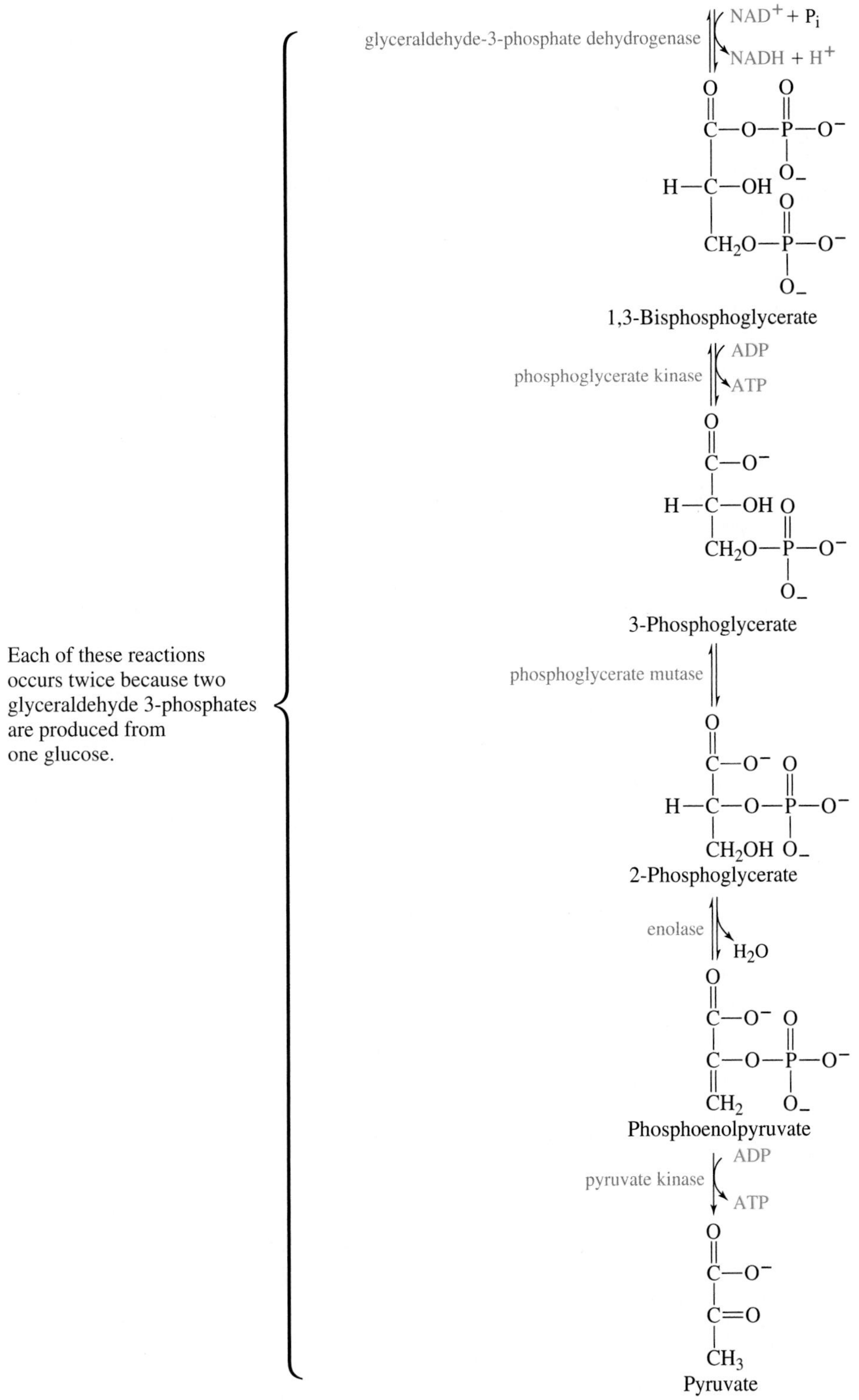

glyceraldehyde-3-phosphate dehydrogenase $NAD^+ + P_i$
                                           $NADH + H^+$

1,3-Bisphosphoglycerate

phosphoglycerate kinase   ADP
                          ATP

3-Phosphoglycerate

phosphoglycerate mutase

2-Phosphoglycerate

enolase H_2O

Phosphoenolpyruvate

pyruvate kinase   ADP
                  ATP

Pyruvate

Each of these reactions occurs twice because two glyceraldehyde 3-phosphates are produced from one glucose.

tive, and an expenditure of energy (ATP) is required. These steps have the purpose of "activating" the entering glucose; hence we call them investment steps. The input of energy from ATP increases the potential energy of the glucose so that chemical reactions can more readily occur. The outcome of the first five reactions is best understood by studying the **net reaction** of this initial phase. A net reaction describes the input on the left side and the output on the right side of the equation. It is obtained by adding all

Table 15.1

The reactions of glycolysis with common enzyme names and reaction type

Reaction Number	Reaction	Enzyme[a]	Reaction Type[b]
1	Glucose + ATP ⟶ glucose 6-phosphate + ADP + H$^+$	Hexokinase	2
2	Glucose 6-phosphate ⇌ fructose 6-phosphate	Phosphoglucoisomerase	5
3	Fructose 6-phosphate + ATP ⟶ fructose 1,6-bisphosphate + ADP + H$^+$	Phosphofructokinase	2
4	Fructose 1,6-bisphosphate ⇌ dihydroxyacetone phosphate + glyceraldehyde 3-phosphate	Aldolase	4
5	Dihydroxyacetone phosphate ⇌ glyceraldehyde 3-phosphate	Triose phosphate isomerase	5
6	Glyceraldehyde 3-phosphate + P$_i$ + NAD$^+$ ⇌ 1,3-bisphosphoglycerate + NADH + H$^+$	Glyceraldehyde 3-phosphate dehydrogenase	1,2
7	1,3-Bisphosphoglycerate + ADP ⇌ 3-phosphoglycerate + ATP	Phosphoglycerate kinase	2
8	3-Phosphoglycerate ⇌ 2-phosphoglycerate	Phosphoglycerate mutase	5
9	2-Phosphoglycerate ⇌ phosphoenolpyruvate + H$_2$O	Enolase	4
10	Phosphoenolpyruvate + ADP + H$^+$ ⟶ pyruvate + ATP	Pyruvate kinase	2

[a]Enzymes are listed by common names.
[b]Reaction type: (1) oxidation–reduction, (2) phosphoryl group transfer, (3) hydrolysis, (4) nonhydrolytic cleavage (addition or elimination), (5) isomerization–rearrangement, and (6) bond formation coupled to ATP cleavage.

pertinent reactions and canceling metabolites that appear on opposite sides of the equations. The net reaction for the first five steps of glycolysis is derived as follows:

1. glucose + ATP ⇌ glucose 6-phosphate + ADP
2. glucose 6-phosphate ⇌ fructose 6-phosphate
3. fructose 6-phosphate + ATP ⇌ fructose 1,6-bisphosphate + ADP
4. fructose 1,6-bisphosphate ⇌ glyceraldehyde 3-phosphate + dihydroxyacetone phosphate
5. dihydroxyacetone phosphate ⇌ glyceraldehyde 3-phosphate

Sum: glucose + 2 ATP ⟶ 2 glyceraldehyde 3-phosphate + 2 ADP

A study of the net reaction shows that in the investment stage of glycolysis, a six-carbon substrate is split into two molecules of a three-carbon metabolite. For each glucose entering the pathway, two ATP molecules are consumed for activation purposes.

The net reaction is a useful tool for observing the overall outcome of a reaction series, but it does not provide information about detailed metabolic intermediates. It summarizes the salient features of a pathway:

1. A net reaction shows the fate of each carbon atom. Note that the number of carbon atoms on one side must equal the number on the other side. The entering glucose has six carbons; the two product molecules of glyceraldehyde 3-phosphate each have three carbon atoms.
2. A net reaction shows the input and output of ATP/ADP and NAD$^+$/NADH. It is important to know about NADH and ATP production. NADH, a source of reductive energy, is used in stage III of aerobic metabolism to make ATP by oxidative phosphorylation of ADP. The number of ATPs formed in a net reaction indicates the amount of energy available from the chemical transformation of a substrate.

In summary, a net reaction provides a convenient tally of the carbon input and output, ATP yield, and NADH yield, all important aspects of a metabolic pathway.

As shown in Table 15.1, the first five reactions of glycolysis utilize only three types of reaction chemistry. Phosphoryl group transfers and isomerizations account for four of the reactions. Perhaps the reaction of most interest is reaction 4, the aldolase-catalyzed cleavage step from which glycolysis gets part of its name ("lysis"). In Chapter 14 and in

Structure Tutorials:
Hexokinase;
Phosphofructokinase

Table 15.1, reaction 4 is described as nonhydrolytic cleavage or addition/elimination of a functional group. You may also recognize the reverse of reaction 4 as an **aldol condensation.**

The Second Five Reactions of Glycolysis

The second half of glycolysis (see reactions 6–10), the "dividend stage," transforms each glyceraldehyde 3-phosphate to another three-carbon metabolite, pyruvate. Here we recover our initial investment of two ATPs plus a dividend of two more ATPs. In addition, two NADH are generated. The net reaction for the second half is

$$2 \text{ glyceraldehyde 3-phosphate} + 2 \text{ P}_i + 4 \text{ ADP} + 2 \text{ NAD}^+ \longrightarrow$$
$$2 \text{ pyruvate} + 4 \text{ ATP} + 2 \text{ NADH} + 2 \text{ H}^+ + 2 \text{ H}_2\text{O}$$

Several reactions in the dividend stage are of special interest. Reaction 6 displays two types of chemistry, oxidation of an aldehyde to a carboxylic acid (with NADH formation) and phosphoryl group transfer to the carboxylic acid, making the reactive compound 1,3-bisphosphoglycerate (1,3-BPG). 1,3-BPG has sufficient potential energy for ATP synthesis (reaction 7). A second ATP-generating step occurs in the last reaction (reaction 10). The ATP and NADH balance sheet for the reactions of glycolysis is shown in Table 15.2. The overall, net reaction for the ten steps of glycolysis is

$$\text{glucose} + 2 \text{ P}_i + 2 \text{ ADP} + 2 \text{ NAD}^+ \longrightarrow$$
$$2 \text{ pyruvate} + 2 \text{ ATP} + 2 \text{ NADH} + 2 \text{ H}^+ + 2 \text{ H}_2\text{O}$$

In summary, two molecules of ATP and NADH are generated for each glucose split into two molecules of pyruvate. Our earlier preview of glycolysis as an oxidation process releasing energy for ATP production has now come to fruition. Two carbon centers in glucose (carbons 3 and 4; $-\overset{|}{\underset{\text{OH}}{\text{CH}}}-\overset{|}{\underset{\text{OH}}{\text{CH}}}-$) are oxidized to the carboxylic acid level in a pyruvate. Carbons 3 and 4 in glucose 6-phosphate each become C1 in pyruvate, $\overset{3,2,1}{\text{CH}_3\overset{|}{\underset{\text{O}}{\text{C}}}\text{COO}^-}$. Two phosphorylated metabolites, 1,3-bisphosphoglycerate and phosphoenolpyruvate (1,3-BPG and PEP), have sufficient phosphoryl group transfer energy for ATP production. (What individual reaction is the source of the two water molecules on the right side of the net reaction?)

The pathway of glycolysis is rigorously regulated according to the cell's needs for ATP and NADH. The major control point in glycolysis is reaction 3, the phosphorylation of fructose 6-phosphate catalyzed by phosphofructokinase (PFK). The enzyme is allosteric with activity regulated by the concentrations of various effectors (see Section 15.5).

Summarizing, in the process of glycolysis, energy has been extracted from the oxidation of glucose, which has resulted in the production of ATP and NADH. The remnants of the glucose carbon skeleton in the form of two pyruvate molecules become further oxidized in aerobic metabolism to CO_2 and H_2O with even greater production of ATP and NADH. In anaerobic metabolism, the pyruvate continues in fermentation processes.

Table 15.2
The ATP and NADH balance sheet for glycolysis

Number[a]	Reaction per Glucose	ATP Change per Glucose[b]	NADH Change per Glucose[b]
1	Glucose ⟶ glucose 6-phosphate	−1	0
3	Fructose 6-phosphate ⟶ fructose 1,6-bisphosphate	−1	0
6	2 Glyceraldehyde 3-phosphate ⇌ 2 1,3-bisphosphoglycerate	0	+2
7	2 1,3-Bisphosphoglycerate ⇌ 2 3-phosphoglycerate	+2	0
10	2 Phosphoenolpyruvate ⟶ 2 pyruvate	+2	0
Total change		+2	+2

[a]The number corresponds to the reaction number in Table 15.1.
[b]A minus sign indicates loss of ATP by cleavage of a phosphoanhydride bond; a plus sign indicates formation of ATP (from ADP) or NADH (from NAD^+).

15.2 Entry of Other Carbohydrates into Glycolysis

In terms of abundance, glucose is the most important carbohydrate of energy metabolism in plants and animals. (Even plants need energy at night, which they obtain by oxidation of the glucose synthesized during daylight.) Other carbohydrates are present in our diets and can serve as energy sources. Figure 15.3 outlines the entry of important carbohydrates into glycolysis. The ports of entry for other carbohydrates include glucose 6-phosphate, fructose 6-phosphate, dihydroxyacetone phosphate, and glyceraldehyde 3-phosphate. As we study Figure 15.3, an important concept becomes evident. Only one to two reactions are required for the entry of other carbohydrates into glycolysis. An important exception is galactose, which must pass through a complex series of reactions.

Dietary Carbohydrates

Carbohydrates in our diets are present in three chemical forms: polysaccharides, disaccharides, and monosaccharides. We ingest polysaccharides of two major types, **starch** (amylose

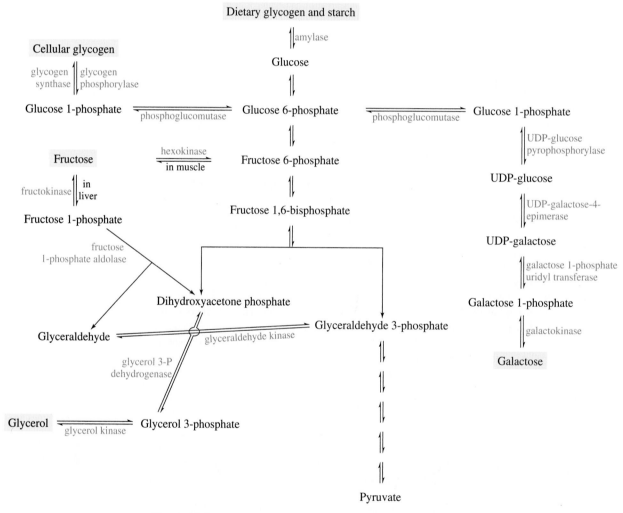

Figure 15.3

Input of other carbohydrates into the glycolytic pathway. The figure shows the entry of glucose units from dietary starch or cellular glycogen. Disaccharides in the diet enter glycolysis after hydrolysis to glucose and fructose or galactose. Fructose enters glycolysis in two possible routes depending on the type of cell. Galactose, produced from dietary lactose, is converted to glucose 6-phosphate. Glycerol, a product from triacylglycerol hydrolysis, enters in the form of dihydroxyacetone phosphate. All carbohydrates in our diet can readily be converted to a glycolytic intermediate.

and amylopectin) from vegetables and cereal grains, and **glycogen** from animal tissue. (Cellulose is also a polysaccharide in our diets, but it serves as bulk fiber without undergoing metabolism; Chapter 8, Section 8.4). Dietary polysaccharides (except cellulose) are hydrolyzed in the mouth in a reaction catalyzed by a salivary enzyme called amylase. The acidic pH of the stomach decreases the rate of enzyme-catalyzed hydrolysis but increases the extent of H^+-catalyzed hydrolysis. Glucose, which is the primary product of starch and glycogen hydrolysis, is absorbed through the intestinal walls to the blood, where it is transported to peripheral tissue, including skeletal muscle, brain, and heart. Approximately one-third of dietary glucose goes to skeletal and heart muscle for energy production and storage; about one-third goes to the brain, which relies solely on glucose oxidation for energy; and the final one-third goes to the liver for storage in the form of glycogen.

Major disaccharides in our diets include maltose, sucrose (table sugar), and lactose (milk sugar). Maltose is produced during the breakdown of dietary polysaccharides by amylases. Sucrose is a relatively abundant dietary component. In addition to its use as a sweetening agent, it also is present in most fruits and vegetables. For example, each 100 g of raw carrot contains over 3 g of sucrose; a banana contains 6.5 g of sucrose per 100 g of weight. Milk and milk products are our major source of lactose. About 5% of the composition of bovine milk is lactose. The disaccharides are hydrolyzed by glycosidases, enzymes present in the small intestines:

Maple syrup is a rich source of carbohydrates including glucose and fructose.

$$\text{maltose} + \text{H}_2\text{O} \xrightleftharpoons{\text{maltase}} 2 \text{ glucose}$$

$$\text{sucrose} + \text{H}_2\text{O} \xrightleftharpoons{\text{sucrase}} \text{fructose} + \text{glucose}$$

$$\text{lactose} + \text{H}_2\text{O} \xrightleftharpoons{\text{lactase}} \text{glucose} + \text{galactose}$$

The monosaccharide products are transported through the intestinal wall and into the blood.

Dietary monosaccharides include glucose and fructose. These are the components of common table sugar (sucrose) used to sweeten beverages and foods (see Chapter 8, Section 8.3). Fructose is also found in honey, in many fruits and vegetables, and corn syrup used to sweeten soft drinks. A 100 g portion of honey is composed of 42 g of fructose and 34 g of glucose. A 100 g portion of an apple contains 7.6 g of fructose, whereas a 100 g portion of a carrot contains 1 g of fructose. By the time dietary carbohydrates reach our internal cells, they are all in the form of the monosaccharides, primarily glucose, fructose, and galactose.

Fructose, Galactose, and Glycerol

Fructose enters into glycolysis by two different pathways depending on the type of tissue (see Figure 15.3). In skeletal muscle, fructose is accepted as a substrate by the normal glycolytic enzyme, hexokinase, but its binding is only 1/20 the affinity of glucose. In just one phosphoryl transfer step, fructose can enter the mainstream of glycolysis as fructose 6-phosphate. Liver cells contain another enzyme, fructokinase, that has two differences from hexokinase: (1) It has a higher affinity for fructose than other hexoses and (2) it catalyzes phosphoryl group transfer from ATP to the one position to produce fructose 1-phosphate. Before entry into the mainstream of glycolysis, an aldolase-type cleavage and additional phosphorylation must occur. The net reaction for liver fructose entry into glycolysis (fructose + 2 ATP $\rightleftharpoons$ 2 glyceraldehyde 3-phosphate + 2 ADP) is the same as for glucose, even though different enzymes are used.

Galactose requires five reactions to transform it to glucose 6-phosphate. Galactokinase catalyzes the initial phosphorylation at the expense of ATP. The product galactose 1-phosphate is then modified with a uridyl group to UDP-galactose (Figure 15.4). This unusual carbohydrate derivative marks the galactose for special use in metabolism (see Section 15.4). Eventually, glucose 1-phosphate is formed from the original carbon skeleton of galactose. A key isomerization step is required for entry of glucose 1-phosphate into glycolysis:

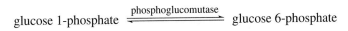

$$\text{glucose 1-phosphate} \xrightleftharpoons{\text{phosphoglucomutase}} \text{glucose 6-phosphate}$$

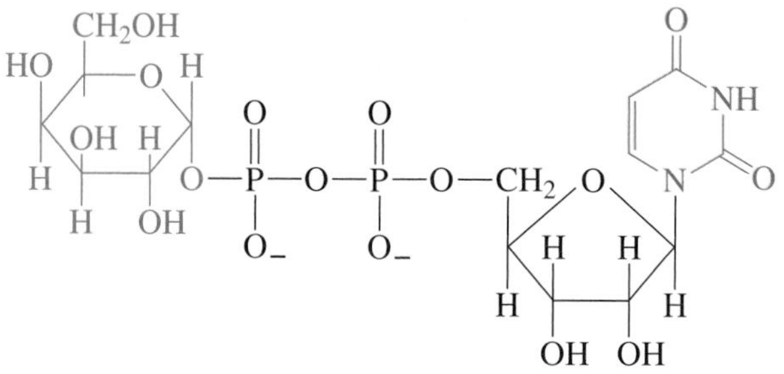

Figure 15.4

Structure of uridine diphosphate-galactose (UDP-galactose). UDP attached to galactose marks the monosaccharide for specialized metabolism.

UDP-galactose

A malfunction in the series of reactions preparing galactose for entry into glycolysis leads to the disease **galactosemia,** which is inherited as an autosomal recessive trait. Most cases are caused by a deficiency of the enzyme galactose-1-phosphate uridyl transferase. Without the normal pathway for galactose metabolism open, the monosaccharide accumulates and is transformed to a toxic product, galactitol. Symptoms of the disease, which appear soon after birth, include a failure to thrive, enlargement of the liver, jaundice, cataracts of the eye, and eventually mental retardation. Vomiting and diarrhea are also experienced by patients after ingestion of dairy products or other foods containing galactose. Newborn infants are routinely screened for galactosemia by measuring galactose concentrations in the blood. Affected infants are fed milk formula in which sucrose or other carbohydrates are substituted for galactose and lactose. As children mature, they develop another enzyme that metabolizes galactose.

The trihydroxyl compound, glycerol, is released during the degradation of fats (triacylglycerols and glycerophospholipids see Chapter 9, Section 9.2). Although glycerol is not a carbohydrate, with just two, simple reactions (phosphoryl transfer and oxidation), its carbon skeleton can enter into glycolysis as dihydroxyacetone phosphate (see Figure 15.3).

Glucose from Intracellular Glycogen and Starch

Dietary glycogen and starch are metabolized by hydrolysis in the mouth, stomach, and intestines and distributed to cells in monosaccharide form. Plants and animals also synthesize polysaccharides in cells during times of plenty, when glucose is present in excess. (The average content of glycogen in liver and skeletal muscle is about 10% and 1%, respectively, of the total wet weight.) Glycogen stored in animal cells and starch reserves in plant cells are used during times of energy need, when intracellular levels of glucose are low. The reaction to release glucose from stored fuel is not simple hydrolysis, catalyzed by amylases as with dietary polysaccharides, but cleavage by inorganic phosphate, P_i (called **phosphorolytic cleavage**). The reaction differs in another way from hydrolysis. The glucose residues are released as the derivative, glucose 1-phosphate:

$$(\text{glucose})_n + P_i \rightleftharpoons (\text{glucose})_{n-1} + \text{glucose 1-phosphate}$$

Glycogen or starch is represented as $(\text{glucose})_n$ where n is the number of glucose residues. The cleavage takes place only at a nonreducing end of the polysaccharide (Figure 15.5) and is catalyzed by glycogen phosphorylase (animals) or starch phosphorylase (plants). Glucose units (in the form of glucose 1-phosphate) are removed sequentially by cleavage of the $\alpha(1 \longrightarrow 4)$ glycosidic bonds. The release of each glucosyl residue uncovers yet another nonreducing end on which the enzyme can act. The amylose molecule contains only one nonreducing end for attack, but amylopectin and glycogen contain many nonreducing ends that can be acted upon by phosphorylase to release large amounts of glucose 1-phosphate in a short period of time. (Review the structures of glycogen and starch in Figures 8.19 and 8.20.) The phosphorylase reaction is of significance because it balances the level of stored cellular fuel (polysaccharides) and directly usable fuel molecules (monosaccharides). We sometimes refer to the phosphorolytic cleavage as a **mobilization reaction** because it releases a mobile or usable molecule (sugar phosphate) from an

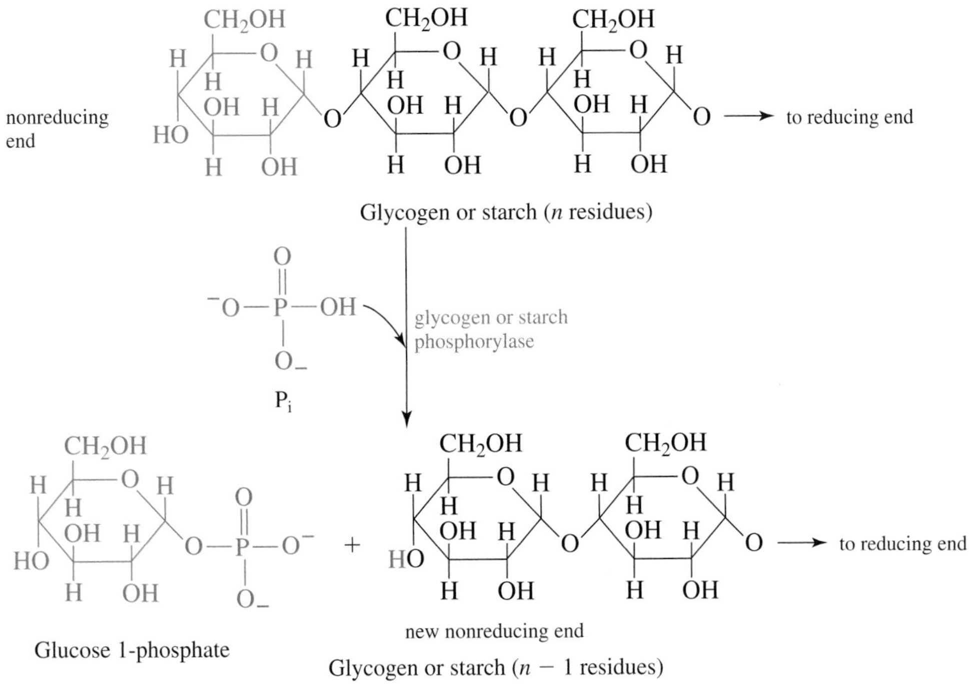

Figure 15.5
The action of glycogen or starch phosphorylase on the nonreducing end(s) of glycogen or starch. The enzyme catalyzes the cleavage of $\alpha(1 \rightarrow 4)$ glycosidic bonds by o-phosphoric acid (P_i). Glucose is produced as glucose 1-phosphate.

immobile fuel reservoir molecule (glycogen or starch). The phosphorylase can act only on $\alpha(1 \rightarrow 4)$ glycosidic bonds, not on $\beta(1 \rightarrow 4)$ or $\alpha(1 \rightarrow 6)$ bonds. The glucose 1-phosphate from cleavage of glycogen or starch can readily enter glycolysis and energy metabolism by isomerization to the mainstream intermediate, glucose 6-phosphate, a reaction catalyzed by phosphoglucomutase.

Energetically, the process of phosphorolytic cleavage catalyzed by phosphorylase is more favorable for the cell than hydrolytic cleavage catalyzed by amylase. The reason for this can be seen by comparing the reactions necessary to prepare the glucose for glycolysis:

Hydrolytic cleavage catalyzed by amylase:

$$(glucose)_n + H_2O \rightleftharpoons glucose \overset{ATP}{\rightleftharpoons} glucose\ 6\text{-}phosphate$$

Phosphorolytic cleavage catalyzed by phosphorylase:

$$(glucose)_n + P_i \rightleftharpoons glucose\ 1\text{-}phosphate \rightleftharpoons glucose\ 6\text{-}phosphate$$

Note that no expenditure of energy in the form of ATP is required for phosphorolytic cleavage. Because of its strategic position in metabolism at the boundary separating fuel reserves for future needs (polysaccharides) from readily usable glucose 6-phosphate, we can expect the phosphorylase to play a major role in the regulation of glucose metabolism.

15.3 Pyruvate Metabolism

In the previous section we discovered that the carbon skeleton of hexoses entering glycolysis is split into two molecules of pyruvate. This is a general pathway that is present in all kinds of organisms and cells, both aerobic and anaerobic. The product, pyruvate, occupies a key position in cellular metabolism as an important metabolic junction. Pyruvate is presented with several metabolic choices depending on the specific organism, the type of cell, and the intracellular conditions in a cell. The most important determining factor in pyruvate's fate is the availability of oxygen. In aerobic cells with plentiful oxygen, pyruvate continues in a mode of oxidative degradation through the citric acid cycle and oxidative phosphorylation (stage III) until each carbon is oxidized to CO_2 and all potential energy is collected in ATP. Electrons removed from oxidized substrates are collected in NADH and $FADH_2$ and transferred to the terminal electron acceptor, oxygen.

Metabolism of pyruvate in anaerobic organisms, or in aerobic cells when oxygen concentrations are lower, is very different from aerobic metabolism. Because oxygen is less available, other molecules must serve as electron acceptors to oxidize NADH to NAD^+ so metabolism can continue. The extraction of energy from carbohydrates and other organic substrates without using O_2 as an electron acceptor is called **fermentation.** The various kinds of fermentation are named by using the major end product of each pathway. We shall explore the details of lactate fermentation and ethanol fermentation.

Lactate Fermentation

The most common fermentation process, the transformation of glucose to lactate, occurs in a variety of anaerobic microorganisms and in animal muscle during periods of strenuous activity (anaerobic metabolism). Lactate fermentation consists of just one reaction beyond pyruvate, the formation of the three-carbon compound, lactate. This oxidation–reduction reaction, catalyzed by lactate dehydrogenase, is comprised of a two-electron transfer from the coenzyme NADH to the keto group of pyruvate to form lactate:

$$CH_3\underset{\underset{\text{Pyruvate}}{\overset{\|}{O}}}{C}COO^- + NADH + H^+ \rightleftharpoons CH_3\underset{\underset{\text{Lactate}}{\overset{|}{OH}}}{C}HCOO^- + NAD^+$$

The net reaction of lactate fermentation (glucose $\longrightarrow$ 2 lactate) is

$$\text{glucose} + 2\,ADP + 2\,P_i \longrightarrow 2\,\text{lactate} + 2\,ATP + 2\,H_2O$$

Why is there no appearance of NADH and NAD^+ in the net reaction? The two molecules of NADH generated during glycolysis are used to reduce the two molecules of pyruvate in the formation of lactate. The NADH is therefore oxidized to NAD^+ and they both drop out of the net reaction for fermentation. The recycling of NAD^+ and NADH sustains the process of glycolysis in anaerobic cells. This is best illustrated by Figure 15.6, which couples the two enzyme reactions using NAD^+/NADH, glyceraldehyde-3-phosphate dehydrogenase (E_1), and lactate dehydrogenase (E_2).

You have probably had first-hand experience with lactate fermentation in your skeletal muscle. During strenuous exercise, your muscles are using oxygen for respiration faster than your blood supplies it to the muscles. Your muscle cells are in an "oxygen-debt" condition. Low oxygen concentrations reduce the importance of aerobic metabolism of pyruvate. Pyruvate is now shunted toward lactate formation. The cramps and soreness you experience the next day are due to an increased concentration of H^+ from lactic acid in muscle cells. Lactate may appear to be a "dead-end metabolite"; however, it can be used to synthesize new glucose by return to pyruvate. Lactate dehydrogenase, which catalyzes the lactate–pyruvate interchange, has many isoenzymic forms, which are distributed throughout peripheral muscle and have different levels of activity (see Chapter 7, Section 7.3).

Figure 15.6

The coupling of glycolysis with lactate fermentation. Reduction of pyruvate recycles NADH to NAD^+. E_1 = glyceraldehyde-3-phosphate dehydrogenase, E_2 = lactate dehydrogenase. The net yield of NADH formation in the conversion of glucose to lactate is zero.

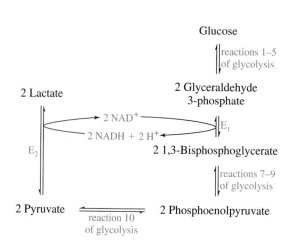

Many varieties of anaerobic bacteria carry out lactate fermentation during growth on the disaccharide lactose in milk. The products of lactate fermentation—cheese, yogurt, buttermilk, and sour cream—are very important commercially. Buttermilk is produced by inoculating milk with *Streptococcus lactis,* which causes curdling and coagulation of proteins. This is caused by the formation of lactic acid, which denatures the proteins by pH change. Cheeses are produced by treating milk with *Lactobacillus* cultures, rennet (a protease), and other enzymes. *Lactobacillus* is also used to ferment shredded cabbage leaves to make sauerkraut. Yogurts are prepared by incubating milk with cultures of *Lactobacillus bulgaricus* and *Streptococcus thermophilus.*

Ethanol Fermentation and Metabolism

Another economically important fermentation process is the production of ethanol by strains of yeast and other microorganisms. Ethanol fermentation occurs by a two reaction sequence:

$$\underset{\text{Pyruvate}}{CH_3CCOO^-} \underset{1}{\overset{\text{pyruvate}\atop\text{decarboxylase}}{\rightleftharpoons}} \underset{\text{Acetaldehyde}}{CH_3CH} \underset{2}{\overset{\text{alcohol dehydrogenase}}{\rightleftharpoons}} \underset{\text{Ethanol}}{CH_3CH_2}$$

CO$_2$ + NADH + H$^+$ → NAD$^+$

Policeman administering a breathalyzer test for ingestion of ethyl alcohol.

Reaction 1, a nonhydrolytic cleavage step, is catalyzed by pyruvate decarboxylase, which contains thiamine pyrophosphate (vitamin B$_1$) as a coenzyme (see Chapter 7, Section 7.1). Thiamine pyrophosphate is associated with many decarboxylases that use α-ketoacids as substrates. The second reaction, reduction of the carbonyl in acetaldehyde, results in the formation of ethanol. Acetaldehyde serves as the final electron acceptor in ethanol fermentation. (Draw a sequence of reactions for ethanol fermentation showing the recycling of NADH as in Figure 15.6.)

The commercial importance of ethanol fermentation should be obvious to all. Wine, which has an ethanol content in the range of 10 to 13%, is made using cultures of *Saccharomyces ellipsoideus.* Concentrations of alcohol higher than 13% are growth inhibitory to the yeast. Beverages with a higher ethanol content are obtained by distillation. Cultures of *Saccharomyces cerevisiae* are used for brewing beer. The baker also relies on the yeast *S. cerevisiae.* Bread dough rises because of the CO$_2$ produced during yeast fermentation of sucrose. Ethanol and other oxidation products expelled during baking lead to the characteristic aroma of freshly baked bread.

Human consumption of ethanol in beverages has major biochemical, clinical, and social implications. Ingested ethanol is first oxidized in the liver by cytoplasmic alcohol dehydrogenase:

$$\underset{\text{Ethanol}}{CH_3CH_2OH} + NAD^+ \rightleftharpoons \underset{\text{Acetaldehyde}}{CH_3CHO} + NADH + H^+$$

A second oxidation step is catalyzed by aldehyde dehydrogenase:

$$\underset{\text{Acetaldehyde}}{CH_3CHO} + NAD^+ \rightleftharpoons \underset{\text{Acetate}}{CH_3COO^-} + NADH + H^+$$

Acetate is further oxidized to CO$_2$ and H$_2$O by the citric acid cycle and respiration (see Chapters 16 and 17). Up to 95% of ingested ethanol is oxidized by these processes. The balance is excreted in urine or exhaled. The rate-controlling step in ethanol metabolism in humans is probably the alcohol dehydrogenase step. The average rate of ethanol metabolism is approximately 100 mg/kg of body weight per hour. The phrase "holding one's liquor" is actually based on two factors, weight of an individual and genetic variation in alcohol dehydrogenase activity caused by the presence of isoenzymes. The most significant metabolic change brought about by ethanol metabolism is the excessive production of reduced cofactor, NADH. This impedes the normal functioning of metabolic pathways such as the citric acid cycle and gluconeogenesis, which rely on the availability of NAD$^+$.

Cutting Edge: Alcohol abuse

"Gasohol" is prepared by mixing gasoline with ethanol produced by fermentation of corn carbohydrates.

🔵 **Cutting Edge: Methanol as fuel**

Ethanol fermentation is also of significant commercial importance in producing the gasoline substitute "gasohol." Carbohydrates from sugarcane and corn plants are fermented to ethanol, which is mixed with fossil fuels to reduce the consumption of petroleum products. Gasohol consists of 90% gasoline and 10% ethanol.

To conclude, we explore the unusual energetics of fermentation processes. The net reactions of lactate and ethanol fermentation show the production of only two ATPs per glucose. The simple anaerobic organism carrying out only fermentation must be content with this low level of energy, which represents about 5% of the potential energy in glucose. Our skeletal muscle cells have the advantage that they can perform both anaerobic and aerobic metabolism of pyruvate.

15.4   Biosynthesis of Carbohydrates

We have now seen that all cells are dependent on glucose and other monosaccharides for energy metabolism and biosynthesis. These are supplied by diet or by the phosphorolytic cleavage of storage glycogen. For some cells, this source of glucose may not fulfill every metabolic need. Between meals or during periods of fasting, blood glucose concentrations may become too low for appropriate levels of metabolism, especially in brain cells. (The human brain needs a continuous supply of glucose that averages more than 100 g per day.) To fill this need, organisms must synthesize glucose from smaller precursor molecules and distribute it to peripheral tissue. In addition to energy needs, cells require glucose derivatives for the synthesis of glycolipids, glycoproteins, and structural polysaccharides. Glycogen stored in liver is mobilized between meals to maintain proper blood glucose levels, but it only lasts 12 to 24 hours.

Synthesis of Glucose

🔵 **Interactive Animations: Gluconeogenesis**

The starting point for the synthesis of new glucose can be pyruvate, lactate, glycerol, many of the amino acids, or other small biomolecules. We call the synthesis of glucose from non-carbohydrate precursors **gluconeogenesis** ("the beginning of new sugar"). Like glycolysis, the synthesis of glucose is a universal pathway, present in plants, animals, fungi, and microorganisms. The liver is the major site of glucose synthesis in higher animals, where the important precursors are pyruvate, lactate, glycerol, and some amino acids (Figure 15.7). In microorganisms, glucose is synthesized from small precursors, such as acetate and propionate. Plants synthesize glucose and sucrose from CO_2 using energy from the sun (see Chapter 17). The glyoxylate cycle discussed in Chapter 16 allows plants and some microorganisms to produce sugars from amino acids and fatty acids. Since animals do not have this cycle, they are not able to carry out net synthesis of glucose from fatty acids.

Although gluconeogenesis is sometimes called "reverse glycolysis," this is a misnomer because the glycolytic pathway is irreversible. However, it was not necessary for an entirely new pathway of glucose synthesis to evolve (discussed in Chapter 14, Section 14.1). Eco-

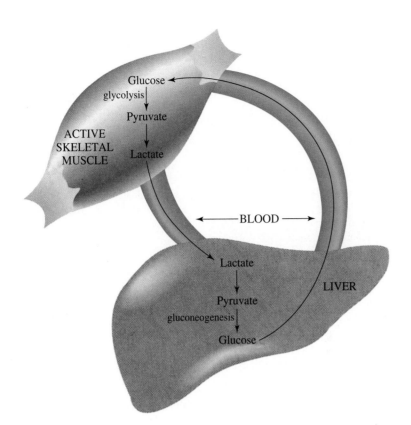

Figure 15.7
Glucose is synthesized primarily in the liver and transported via blood to muscle. Active skeletal muscle degrades the glucose by glycolysis and lactate fermentation. Lactate is transported to the liver where it is used in gluconeogenesis. This circular process is called the Cori cycle.

nomical and efficient processes to degrade and synthesize glucose evolved by the sharing of common reactions. Seven of the ten glycolytic reactions can be used in reverse. Only three of the ten reactions in glycolysis are thermodynamically irreversible; these must be bypassed in gluconeogenesis. The three glycolysis reactions that must be replaced are shown along with their respective $\Delta G^{\circ\prime}$ values in Table 15.3. We will next look at the molecular details of each of the three detours around these energetically unfavorable reactions.

Bypass I: Pyruvate → Phosphoenolpyruvate

The transformation of pyruvate to PEP has the highest energy barrier of any reaction in the gluconeogenic pathway. The sequence of reactions is shown in Table 15.4 (reactions 1 and 2) and Figure 15.8. The process in higher animals begins with pyruvate in the mitochondrial matrix. Some pyruvate is generated here from amino acids, but pyruvate also is transported from the cytoplasm. Pyruvate is carboxylated to oxaloacetate by pyruvate carboxylase, which is present only in mitochondria. This reaction type is category 6, synthesis of a carbon–carbon bond using energy from ATP (see Table 14.2). The essential cofactor of pyruvate carboxylase is biotin, which serves as a carrier of "activated CO_2" (see Table 7.1 for structure and function). The product oxaloacetate is reduced to malate by the enzyme malate dehydrogenase in conjunction with NADH serving as electron donor. Malate is transported from mitochondria to the cytoplasm, where it is reoxidized to oxaloacetate. This

Table 15.3
The irreversible reactions of glycolysis that are bypassed in gluconeogenesis

Number[a]	Reaction	Enzyme	$\Delta G^{\circ\prime}$ (kJ/mol)
1	Glucose + ATP → glucose 6-phosphate + ADP	Hexokinase	−16.7
3	Fructose 6-phosphate + ATP → fructose 1,6-bisphosphate + ADP	Phosphofructokinase	−14.2
10	PEP + ADP → pyruvate + ATP	Pyruvate kinase	−31.4

[a]The number corresponds to the reaction number in Table 15.1.

Table 15.4

The reactions of gluconeogenesis beginning with pyruvate

Number	Reaction
1	Pyruvate + CO_2 + ATP $\longrightarrow$ oxaloacetate + ADP + P_i + H^+
2	Oxaloacetate + GTP $\rightleftharpoons$ phosphoenolpyruvate + CO_2 + GDP
3	Phosphoenolpyruvate + H_2O $\rightleftharpoons$ 2-phosphoglycerate
4	2-Phosphoglycerate $\rightleftharpoons$ 3-phosphoglycerate
5	3-Phosphoglycerate + ATP $\rightleftharpoons$ 1,3-bisphosphoglycerate + ADP + H^+
6	1,3-Bisphosphoglycerate + NADH + H^+ $\rightleftharpoons$ glyceraldehyde 3-phosphate + NAD^+ + P_i
7	Glyceraldehyde 3-phosphate $\rightleftharpoons$ dihydroxyacetone phosphate
8	Glyceraldehyde 3-phosphate + dihydroxyacetone phosphate $\rightleftharpoons$ fructose 1,6-bisphosphate
9	Fructose 1,6-bisphosphate + H_2O $\longrightarrow$ fructose 6-phosphate + P_i
10	Fructose 6-phosphate $\rightleftharpoons$ glucose 6-phosphate
11	Glucose 6-phosphate + H_2O $\rightleftharpoons$ glucose + P_i

roundabout process of oxidation–reduction could be avoided if oxaloacetate could pass through the mitochondrial membrane system. However, oxaloacetate is too polar for simple diffusion and there are no transport proteins for it. Continuing with PEP formation, cytoplasmic oxaloacetate is decarboxylated to phosphoenolpyruvate by **phosphoenolpyruvate carboxykinase.** The source of the phosphoryl group is GTP rather than the more common ATP. The overall, net reaction for bypass I is

$$\text{pyruvate} + \text{ATP} + \text{GTP} \longrightarrow \text{phosphoenolpyruvate} + \text{ADP} + \text{GDP} + P_i$$

The energy released from the transfer of two phosphoryl groups (from ATP and GTP) is used to lift pyruvate to the energy level of PEP.

The pathway from lactate to glucose is also an important anabolic process. Here a shortened version of bypass I can be used (Figure 15.8). When pyruvate is the precursor for glucose, the only source of NADH is from the oxidation of malate to oxaloacetate. However, when synthesis begins with lactate, NADH is generated by lactate dehydrogenase. The abbreviated pathway for lactate to PEP is shown in Figure 15.8. Note that the same amount of energy from ATP and GTP is required for either pyruvate or lactate use. Lactate may also be used to synthesize glucose via cytoplasmic phosphoenolpyruvate carboxykinase.

Figure 15.8

Reactions of gluconeogenesis, bypass I. Cytoplasmic pyruvate or lactate is used to synthesize phosphoenolpyruvate by enzymes in the cytoplasm and mitochondria.

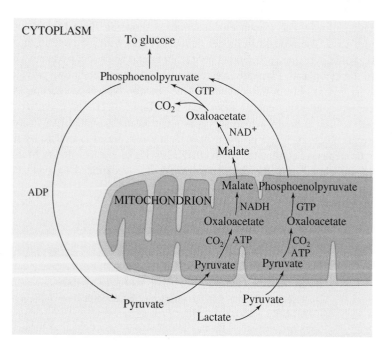

Bypass II: Fructose 1,6-Bisphosphate ⟶ Fructose 6-Phosphate

Phosphofructokinase is the major regulatory enzyme in glycolysis catalyzing the irreversible phosphoryl transfer from ATP to fructose 6-phosphate. Reversal of this reaction for gluconeogenesis would require the synthesis of an ATP. Since an insufficient amount of energy is present in fructose 1,6-bisphosphate to do this, we must bypass this step (Figure 15.9). In gluconeogenesis, the phosphoryl group is removed by hydrolysis catalyzed by fructose 1,6-bisphosphatase:

$$\text{fructose 1,6-bisphosphate} + H_2O \rightleftharpoons \text{fructose 6-phosphate} + P_i$$

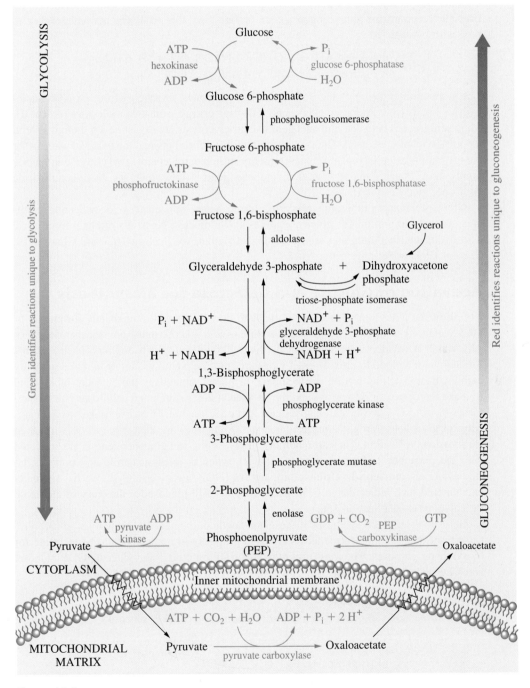

Figure 15.9
Glucose metabolism showing reactions of glycolysis and gluconeogenesis.

Bypass III: Glucose 6-phosphate → Glucose

The third and final detour followed by gluconeogenesis is the removal of the phosphoryl group from glucose 6-phosphate (Figure 15.9):

$$\text{glucose 6-phosphate} + H_2O \rightleftharpoons \text{glucose} + P_i$$

The enzyme glucose 6-phosphatase catalyzes the hydrolysis of the phosphoryl group in a reaction similar to bypass II.

Summary of Gluconeogenesis

We have now completed the pathway for glucose synthesis beginning with pyruvate or lactate. The reactions for gluconeogenesis are outlined in Table 15.4. The net reaction beginning with pyruvate is

$$2 \text{ pyruvate} + 4 \text{ ATP} + 2 \text{ GTP} + 2 \text{ NADH} + 2 \text{ H}^+ + 6 \text{ H}_2O \rightarrow$$
$$\text{glucose} + 2 \text{ NAD}^+ + 4 \text{ ADP} + 2 \text{ GDP} + 6 \text{ P}_i$$

The production of a single glucose requires two pyruvates, the energy from six phosphoanhydride bonds (four ATP, two GTP), and reductive energy from two molecules of NADH. Major differences between gluconeogenesis and glycolysis are evident even though several identical enzymes are used. The most important difference between the two pathways is the energy requirement. Glycolysis releases energy sufficient to generate two phosphoanhydride bonds in ATP per glucose degraded; gluconeogenesis consumes six phosphoanhydride bonds per glucose synthesized.

Other precursors, including glycerol, amino acids, and citric acid cycle intermediates, are also important for glucose synthesis. We discuss glucose synthesis from citric acid cycle intermediates and amino acids in Chapter 16, Section 16.3 and Chapter 19, Section 19.3.

Activation of Glucose and Galactose for Biosynthesis

Not all of the glucose available to an organism may be needed immediately for energy or other metabolic purposes. Higher animals store excess glucose in the polysaccharide glycogen, which is usually accumulated in the liver. Upon need, glucose is mobilized and distributed via the blood to provide energy resources to peripheral tissues or to make specialty carbohydrates such as lactose (milk sugar). In plants, glucose is the building block used to make the disaccharide sucrose and the polysaccharides starch and cellulose. The biosyntheses of lactose, sucrose, starch, glycogen, and cellulose are, of course, linked to the pathways of glycolysis and gluconeogenesis, but the reactions are distinctive in their mode of carbohydrate activation. The carbohydrates in glycolysis and gluconeogenesis are activated with the phosphoryl group (sugar phosphates), whereas in disaccharide and polysaccharide syntheses, **nucleoside diphosphate sugars** (NDP sugars) are used. We have already encountered such molecules in UDP-galactose and UDP-glucose for the entry of galactose into glycolysis (see Section 15.2).

Here we discuss the synthesis of these activated compounds and their roles in carbohydrate reactions. The nucleotide labels are used primarily to mark sugars to be set aside for biosynthetic purposes. The most important are UDP-glucose, ADP-glucose, GDP-glucose, and UDP-galactose. Structures are shown in Figures 15.4 and 15.10.

Formation of NDP-glucose

A general reaction can be written to show the formation of NDP-glucose:

$$\text{NTP} + \text{glucose 1-phosphate} \rightleftharpoons \text{NDP-glucose} + \text{PP}_i$$

NTP = nucleoside triphosphate, ATP or UTP or GTP
NDP-glucose = nucleoside diphosphate glucose or ADP-glucose or UDP-glucose or
    GDP-glucose

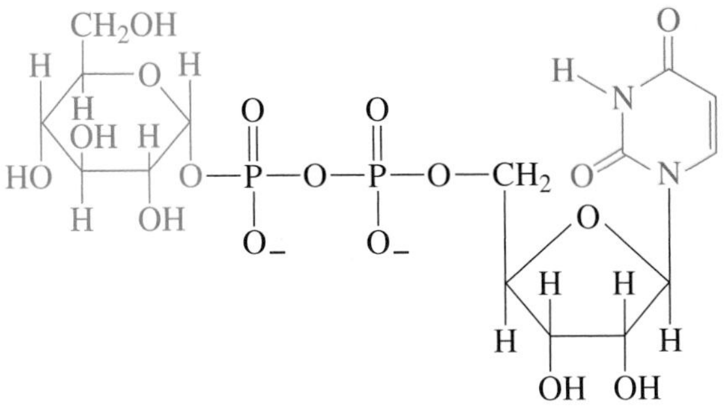

Figure 15.10
Structures of nucleotide sugars:
(a) UDP-glucose, (b) ADP-
glucose, and (c) GDP-glucose.
These activated forms of glucose
are used for biosynthesis.

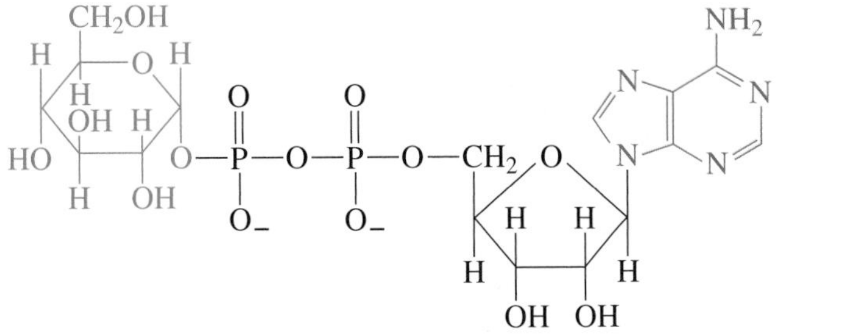

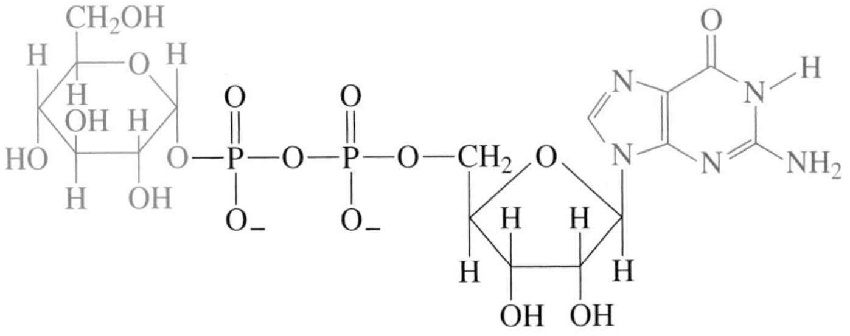

(a) UDP-glucose

(b) ADP-glucose

(c) GDP-glucose

PP_i = pyrophosphate,

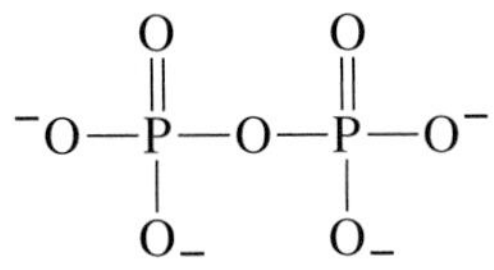

Enzyme = ADP-glucose pyrophosphorylase or
   UDP-glucose pyrophosphorylase or
   GDP-glucose pyrophosphorylase

The K'_{eq} for this reaction is approximately 1, which at equilibrium results in about equal concentrations of reactants and products. The enzyme inorganic pyrophosphatase catalyzes the hydrolysis of product PP_i, thus shifting the equilibrium toward formation of NDP-glucose:

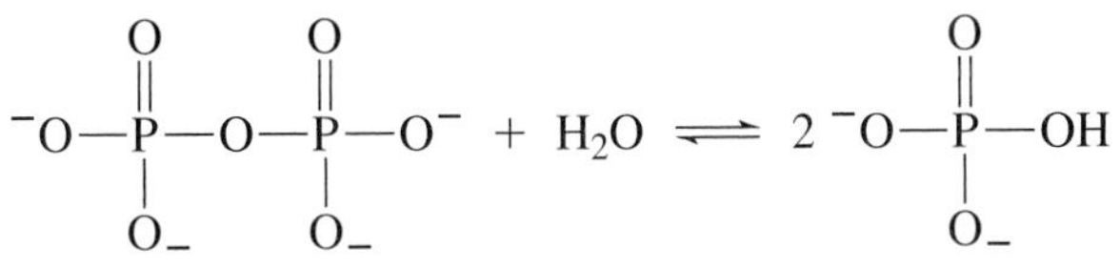

Formation of UDP-galactose

The synthesis of UDP-galactose has two possible routes:

1. Isomerization of UDP-glucose

$$\text{UDP-glucose} \rightleftharpoons \text{UDP-galactose}$$

Enzyme = UDP-glucose-4-epimerase

2. Exchange of UDP

galactose 1-phosphate + UDP-glucose $\rightleftharpoons$ UDP-galactose + glucose 1-phosphate

Enzyme = galactose-1-phosphate uridyl transferase

Both of these reactions are used to feed galactose into glycolysis (see Section 15.2). The presence of a defective gene for uridyl transferase, which is a cause of galactosemia, prohibits the entry of galactose into mainstream metabolism.

UDP-glucose, UDP-galactose, GDP-glucose, or ADP-glucose are the activated forms of monosaccharides used for the synthesis of glycogen, starch, lactose, sucrose, and cellulose.

Synthesis of Polysaccharides

Glycogen

Excess glucose in animals is stored in the polysaccharide glycogen:

$$\text{UDP-glucose} + (\text{glucose})_n \rightleftharpoons \text{UDP} + (\text{glucose})_{n+1}$$

Preexisting glycogen — Lengthened glycogen

Glucose, activated and tagged by attachment to UDP, is added to the nonreducing ends of an existing glycogen molecule (Figure 15.11). The enzyme glycogen synthase catalyzes the formation of a new $\alpha(1 \rightarrow 4)$ glycosidic linkage between the incoming glucose and existing glycogen. Each of the nonreducing ends of glycogen may act as a site for attach-

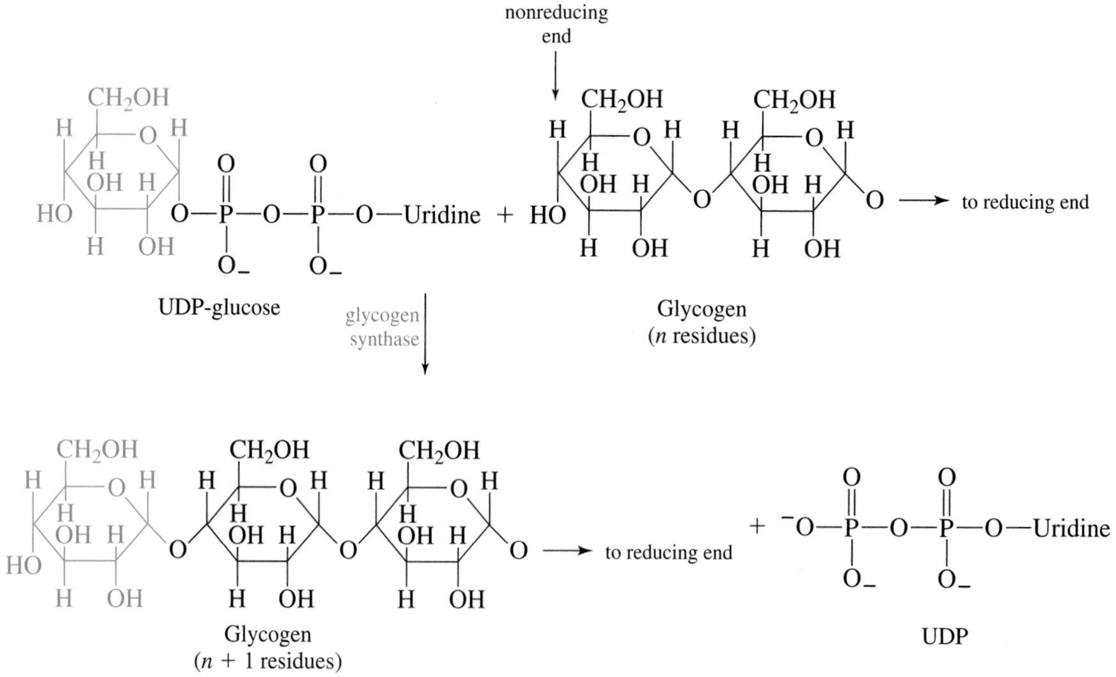

Figure 15.11

The action of glycogen synthase. A glucose residue activated by attachment to UDP is added to a nonreducing end of existing glycogen. Glucose units may be added to all nonreducing ends of a glycogen molecule.

ment of new glucose units for storage. Note that this is the same location at which glycogen phosphorylase acts (see Section 15.2). The net reaction for incorporating one glucose 6-phosphate residue into existing glycogen is

$$\text{glucose 6-phosphate} + \text{UTP} + (\text{glucose})_n + H_2O \longrightarrow (\text{glucose})_{n+1} + \text{UDP} + 2\,P_i$$

Only one phosphoanhydride bond is consumed in the storage of a glucose molecule. This is a very economical way to store glucose since, when mobilized, the glucose can yield over 30 phosphoanhydride bonds when oxidized under aerobic conditions.

Starch

The storage of glucose in plants follows a pathway similar to that in animals except that the glucose is activated by ADP, not UDP, and is accumulated in starch, $(\text{glucose})_n$:

$$\text{ADP-glucose} + (\text{glucose})_n \rightleftharpoons \text{ADP} + (\text{glucose})_{n+1}$$

Preexisting   Lengthened
starch   starch

Starch synthase catalyzes the addition of new glucose residues to nonreducing ends of an existing starch molecule by formation of $\alpha(1 \longrightarrow 4)$ glycosidic bonds.

Cellulose

Cellulose, the major structural polysaccharide in the cell walls of plants and some bacteria, is composed of glucose residues linked in $\beta(1 \longrightarrow 4)$ glycosidic bonds (see Chapter 8, Section 8.4). The synthetic route to cellulose is similar to that of starch. The activated form of glucose is UDP-glucose in some organisms, but GDP-glucose in others.

$$\text{UDP-glucose or GDP-glucose} + (\text{glucose})_n \rightleftharpoons \text{UDP or GDP} + (\text{glucose})_{n+1}$$

Preexisting   Lengthened
cellulose   cellulose

Synthesis of Disaccharides

Lactose

The disaccharide lactose is actively synthesized in the mammary gland of humans and other animals following the birth of young. Lactose is formed by combining activated galactose with glucose:

$$\text{UDP-galactose} + \text{glucose} \rightleftharpoons \text{UDP} + \text{lactose}$$

The enzyme system **lactose synthase** catalyzes the formation of a $\beta(1 \longrightarrow 4)$ glycosidic bond between the two monosaccharides (see Window on Biochemistry 15.1).

Sucrose

The disaccharide sucrose is present in most fruits and vegetables, but especially sugarcane and sugar beets. Sucrose, a storage form of monosaccharides for energy and biosynthetic processes, is transported throughout a plant via the phloem. The disaccharide is synthesized in two steps from UDP-activated glucose and phosphate-activated fructose:

$$\text{UDP-glucose} + \text{fructose 6-phosphate} \xrightleftharpoons{\text{sucrose-6-phosphate synthase}} \text{sucrose 6-phosphate} + \text{UDP}$$

$$\text{sucrose 6-phosphate} + H_2O \xrightleftharpoons{\text{phosphatase}} \text{sucrose} + P_i$$

15.5   Regulation of Carbohydrate Metabolism

We have now explored in detail the pathways of glycolysis and gluconeogenesis. At this point you probably have one of two possible feelings:

1. You are in awe of the many intricacies involved in carbohydrate metabolism.
2. You are overwhelmed by the maze of reactions.

If the latter is more correct, perhaps this section will bring some order to what you may perceive as chaos. Here we discuss the control of carbohydrate metabolism to learn how cells and organisms balance energy and biosynthetic needs. Regulation of pathways is important in order to achieve constant levels of ATP. It is not necessarily important for ATP concentrations to be at a high level, but the ATP recycling rate must be balanced; that is, ATP should be made as rapidly as it is used. The most effective way to do that is to control the relative levels of free glucose and stored glucose. If at a particular moment energy is needed by a cell, then glycolysis must operate and glycogen must be degraded, releasing glucose 1-phosphate for use in glycolysis. If, on the other hand, energy is not needed, the glycolysis pathway should be closed and the glucose stored for future use by increasing glycogen

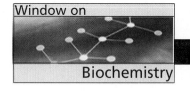

Window on
Biochemistry

15.1 Synthesis of Milk Sugar—α-Lactalbumin Rules

The biosynthesis of milk sugar, lactose, is catalyzed by the enzyme system, lactose synthase (see text). Lactose synthase was shown several years ago to be comprised of two distinct proteins, initially called Protein A and Protein B. The A protein was discovered to be an enzyme that catalyzes the synthesis of amino sugars needed for antibody production:

$$\text{UDP-galactose} + N\text{-acetylglucosamine} \xrightleftharpoons{\text{protein A}} \text{UDP} + N\text{-acetyllactosamine}$$

Protein A was characterized as **galactosyl transferase,** which catalyzes the formation of a β (1 → 4) glycosidic bond between galactose and another monosaccharide. If Protein B is added to the reaction mixture, the synthesis of N-acetyllactosamine is inhibited and the synthesis of lactose is stimulated:

$$\text{UDP-galactose} + \text{glucose} \xrightleftharpoons{\text{proteins A and B}} \text{UDP} + \text{lactose}$$

Protein B changes the substrate specificity of galactosyl transferase (A) to accept glucose rather than N-acetylglucosamine as a substrate.

Protein B has been identified as **α-lactalbumin,** a protein first isolated from milk in 1899. α-Lactalbumin is a relatively small protein with 129 amino acid residues and a molecular weight of 14,200 daltons. It is a metalloprotein with one calcium bound per protein molecule. Because its concentration is rather high in milk (1 mg/mL), the protein was at first thought to serve a nutritional role. We now know that it functions as a major regulatory protein in the synthesis of lactose. Galactosyl transferase and the regulatory protein, α-lactalbumin bind together to form the active complex, lactose synthase.

During pregnancy, the protein hormone, **prolactin,** is present in the mammary gland where it induces lactation and stimulates the synthesis of α-lactalbumin. However, the hormone progesterone, which is also present during pregnancy, counteracts the action of prolactin. During pregnancy then, only low concentrations of α-lactalbumin are present so galactosyl transferase actively catalyzes the formation of N-acetyllactosamine for antibody production (see figure). Progesterone concentrations decline at the onset of lactation and after parturition allowing the prolactin to actively stimulate α-lactalbumin synthesis. Now with both components of the lactose synthase system present (galactosyl transferase and α-lactalbumin), lactose is produced (see figure).

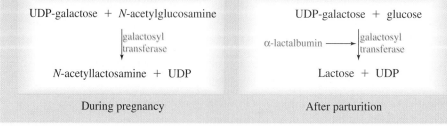

During pregnancy   After parturition

synthesis. This is all accomplished by having enzymes at strategic locations that can "turn on" and "turn off" reactions. Think of the metabolic pathways as networks of pipelines in which the flow of material may be in either direction. Then, think of enzymes as valves that can be opened or closed to direct the flow in the proper direction and at the proper rate. In Chapter 7, Sections 7.2 and 7.3, we discuss the various ways that enzyme action can be regulated. These included allosteric actions, covalent modification, proteolytic cleavage, and isoenzymic forms. At this point it is recommended that you review these sections before you proceed further. The above forms of regulation act at the level of a specific reaction. Metabolic rates are also regulated at the gene level by controlling the synthesis rate of an enzyme. Recall that enzymes may be constitutive or induced (see Chapter 12, Section 12.4). Enzymes are also influenced by hormones. These regulators rarely interact directly with enzyme molecules, but they send their messages via molecules, such as cyclic AMP (see Chapter 2, Section 2.5). If the enzymes are metabolic valves on pipelines, then the controllers of these valves are the effector molecules that turn the enzymes "on" or "off."

Those enzymes that play important regulatory roles in carbohydrate metabolism are listed in Table 15.5 and their metabolic location is highlighted in Figure 15.12. We will consider each in detail (see Window on Biochemistry 15.2).

Glycogen Phosphorylase and Glycogen Synthase

Glycogen phosphorylase and glycogen synthase are present in most animal cells and tissues, but they are of particular significance in skeletal muscle and liver. They are regulated differently in liver and muscle because the substrate glycogen plays slightly different roles in the two tissues. The liver serves as a storehouse of glycogen for the whole organism. Excess glucose in the blood is taken up by the liver, where it is used for glycogen synthesis. When glucose is needed, liver glycogen phosphorylase mobilizes glucose for distribution to all tissues via blood. The low levels of glycogen in muscle are degraded by phosphorolytic cleavage and glycolysis to produce ATP locally for muscle contraction.

Glycogen synthase is a dimeric protein that exists in two interconvertible forms, the (a) form and the (b) form (Figure 15.13). The (a) form, when present, is active under all cellular conditions. The activity of the (b) form, when present, depends on the positive modulator, glucose 6-phosphate. The (a) form is converted to the (b) form by transfer of two phosphoryl groups from ATP to two serine hydroxyl groups in glycogen synthase. Cyclic AMP–dependent protein kinase or protein kinase A catalyzes the reaction. The phosphoryl groups on (b) are removed by hydrolysis catalyzed by phosphoprotein phosphatase. This cycle of reactions interconverts the (a) and (b) forms. In Chapter 7, Section

Table 15.5
Regulatory enzymes in carbohydrate metabolism

Enzyme Name	⊕Effector[a]	⊖Effector[a]	Comments
Glycogen phosphorylase			
Form (a)	—	—	Active under all conditions.
Form (b)	AMP	G6P[b]	The concentration is regulated by hormones.
Glycogen synthase			
Form (a)	—	—	Active under all conditions.
Form (b)	G6P[b]	—	The concentration is regulated by hormones.
Hexokinase	—	G6P[b]	Feedback inhibition.
Phosphofructokinase	AMP, F 2,6-BP[c]	ATP, citrate	Major control point in glycolysis.
Fructose-1,6-bisphosphatase	—	AMP, F 2,6-BP[c]	Regulatory step in glucose synthesis.
Pyruvate kinase	—	ATP, acetyl CoA	Also regulated by isoenzymic forms.
Pyruvate carboxylase	Acetyl CoA	—	Helps sustain levels of glucose and oxaloacetate.

[a]+ Effectors stimulate the enzyme; − effectors inhibit the enzyme.
[b]G6P = glucose 6-phosphate.
[c]F 2,6-BP = fructose 2,6-bisphosphate.

Figure 15.12
Strategic locations for regulation in carbohydrate metabolism. Regulatory enzymes are listed in green.

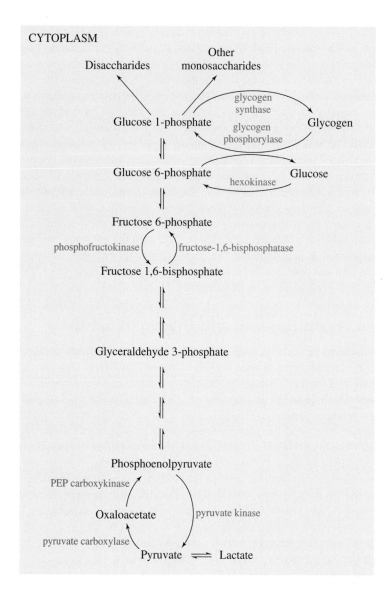

7.3, we explained the same type of chemistry occurring with glycogen phosphorylase, except the result was the opposite. Phosphorylation of two serine residues on the phosphorylase by protein kinase converts the less active (b) form into the fully active (a) form. This mode of regulation ensures that the two enzymes (glycogen synthase and glycogen phosphorylase) are never fully active at the same time. In resting or mildly active muscle, the only forms present are the (a) glycogen synthase (active) and the (b) glycogen phosphorylase, whose rates of activity depend on metabolite concentration.

The interconversions of the two enzymic forms of glycogen phosphorylase and glycogen synthase are controlled by the hormones epinephrine and glucagon, which activate protein kinases. These hormones have different sites of action. Epinephrine acts on muscle cells, where it stimulates glycogen breakdown and deactivates glycogen synthesis. Glucagon acts on liver cells, where the results are the same as for epinephrine. These two hormones act by controlling the level of cyclic AMP, which stimulates protein kinases in the various tissues.

Hexokinase

The entry of free glucose into the glycolytic pathway is regulated by the allosteric enzyme hexokinase. The enzyme is inhibited by its product, glucose 6-phosphate. This mode of metabolic control, called **feedback inhibition,** regulates the intracellular concentration of glucose 6-phosphate.

Chinese stamp recognizing the analysis of the insulin structure by X-ray crystallography. Insulin is an important regulator of glucose metabolism.

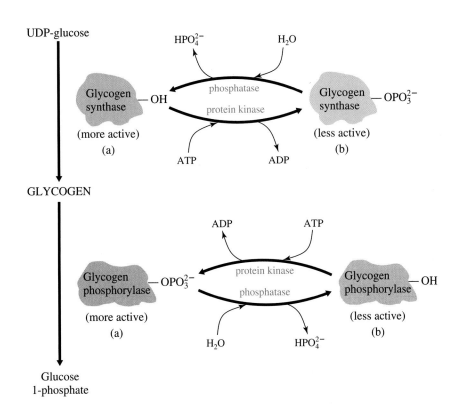

Figure 15.13
Coordinated regulation of glycogen synthase and glycogen phosphorylase.

Phosphofructokinase

Once glucose is phosphorylated by hexokinase, it readily continues through glycolysis to produce ATP. If there is no need to produce ATP, the glycolytic pathway is closed by the key regulatory enzyme, phosphofructokinase (PFK), which catalyzes the phosphorylation of fructose 6-phosphate. PFK is allosterically inhibited by high levels of ATP. The effector ATP binds to the enzyme at two sites, the active site (as a substrate) and a regulatory site (as a modulator). When bound at the regulatory site, it lowers the affinity of the enzyme for the substrate fructose 6-phosphate.

PFK is also inhibited by citrate, the physiological form of citric acid. This compound is synthesized by the citric acid cycle in later aerobic metabolism of pyruvate (see Chapter 16). If cellular levels of citrate are sufficient, then glycolysis is slowed to reduce the rate of production of pyruvate and citrate.

The compound fructose 2,6-bisphosphate (F 2,6-BP) acts to stimulate PFK in liver cells (Figure 15.14). Although its mode of action is not yet entirely clear, it is thought to bind to PFK thus increasing the enzyme's affinity for the normal substrate fructose 6-phosphate (Figure 15.15).

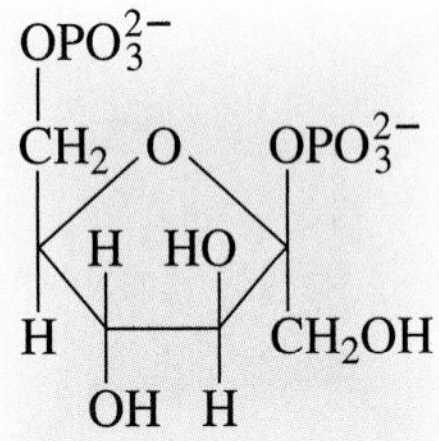

Structure Tutorial: Hexokinase; Phosphofructokinase

Fructose 1,6-Bisphosphatase

Fructose 1,6-bisphosphatase is inhibited by fructose 2,6-bisphosphate and AMP. Recall that F 2,6-BP has the opposite effect on PFK. Thus, F 2,6-BP is now recognized as a key factor in the balance of glycolysis and gluconeogenesis rates in the liver. The concentration of F 2,6-BP is regulated by the hormone glucagon. Glucagon lowers the intracellular levels of F 2,6-BP, thus allowing the inhibition of glycolysis and enhancing gluconeogenesis. Regulation at the strategic position of PFK and fructose 1,6-bisphosphatase (FBP) is extremely important. Let's review the two reactions at this metabolic step:

$$\text{ATP} + \text{fructose 6-phosphate} \xrightleftharpoons{\text{PFK}} \text{ADP} + \text{fructose 1,6-bisphosphate}$$

$$\text{fructose 1,6-bisphosphate} + \text{H}_2\text{O} \xrightleftharpoons{\text{FBP}} \text{fructose 6-phosphate} + \text{P}_\text{i}$$

Fructose 2,6-bisphosphate

Figure 15.14

The structure of fructose 2,6-bisphosphate, an allosteric modulator that stimulates the action of phosphofructokinase.

Figure 15.15

Reciprocal regulation of phosphofructokinase and fructose 1,6-bisphosphatase.

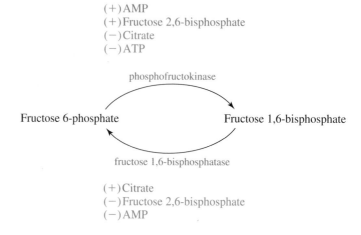

(+)AMP
(+)Fructose 2,6-bisphosphate
(−)Citrate
(−)ATP

phosphofructokinase

Fructose 6-phosphate ⟶ Fructose 1,6-bisphosphate

fructose 1,6-bisphosphatase

(+)Citrate
(−)Fructose 2,6-bisphosphate
(−)AMP

If both enzymes were to be active at the same time, then the net result is obtained by adding the two reactions:

$$ATP + H_2O \rightleftharpoons ADP + P_i + heat$$

This reaction represents apparent "wasteful hydrolysis" of ATP because the energy in the phosphoanhydride bond is not used for synthesis or work. This sequence of reactions

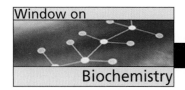

Window on

Biochemistry

15.2 Diabetes Mellitus—Type I and Type II

Diabetes mellitus has caused the death of millions of people throughout the world. It is the sixth leading cause of death by disease in the United States. The current estimate is that in the United States about 16 million people, or about 6% of the population, have the disease. Unfortunately, perhaps 5 million of these individuals have not yet been diagnosed with the disease and are not receiving proper treatment. The total annual cost for the care and treatment of diabetics in the United States is about $100 billion per year.

Diabetes is characterized by abnormally high concentrations of glucose in the blood. **Type I diabetes** (insulin-dependent), which represents only about 5–10% of the total number of cases, is due to inadequate secretion of insulin by the islet of Langerhans in the pancreas. Treatment usually involves administration of insulin. **Type II diabetes** (insulin-resistant) represents about 90–95% of the cases. Here, insulin is produced normally, but the hormone is less effective perhaps because there is a shortage of insulin receptors in cells. Treatments for the type II form include drugs that enhance the body's use of insulin. Current drugs are Avandia, a thiazolidine-dione, and the forthcoming oral hypoglycemic agent Starlix, a derivative of phenyl-alanine. In both disease types, glucose is abundant, but transport of the important energy molecule to the interior of cells is greatly reduced. Individuals with type I diabetes produce increased levels of **ketone bodies** (see Window on Biochemistry 20.1).

New research on the type II disease has found that genetically it is a multifactorial disorder with several genes playing a role. Genetic diseases with these complex origins are extremely difficult to study. Of the several genes involved in the type II disease, one susceptible to mutations encodes the protein **calpain-10,** a member of the protein family of calcium-activated neutral proteases, which are regulatory proteins found in all human cells. Their functions are not yet understood, but they may play a role in nerve damage in the brain. Sequence variations in the gene for calpain-10 lead to amino acid changes. Calpain-10 with amino acid substitutions have been shown to affect the rate of insulin-stimulated glycolysis. New approaches to prevention, diagnosis, and treatment of type II diabetes are expected to result from these new biochemical and genetic studies.

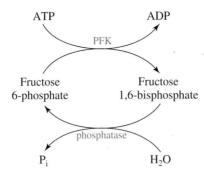

Figure 15.16
Coupling of two reactions to form a substrate cycle. This set of reactions appears to result in the "wasteful hydrolysis" of ATP as indicated in the sum of the two reactions. Some organisms use this sequence to generate body heat.

Sum: ATP + $H_2O \longrightarrow$ ADP + P_i + heat

is called a **substrate cycle** (some biochemists call them futile cycles) (Figure 15.16 and Chapter 20). Proper regulation by F 2,6-BP of the enzymes probably prohibits such cycles, so under normal conditions they are unlikely to occur. However, in some cells and organisms, substrate cycles may serve to generate necessary heat to maintain body temperature. For example, the honeybee uses substrate cycles to generate heat to activate flight muscles and is thus able to fly in cold weather. Bumblebees are unable to fly in cold weather because they lack the necessary enzymes to complete substrate cycles.

Pyruvate Kinase and Pyruvate Carboxylase

Pyruvate kinase (PK) is a tetrameric protein that requires a monovalent ion, Na^+ or K^+, for activity. ATP acts to inhibit pyruvate kinase and, hence, slow pyruvate formation.

Pyruvate carboxylase (PC) is an important regulatory site for gluconeogenesis. PC catalyzes the formation of oxaloacetate by addition of CO_2 to pyruvate. PC is active only in the presence of the metabolite **acetyl CoA.** The metabolic logic of this control step will not be entirely clear until we explore the aerobic metabolism of pyruvate to acetyl CoA and its entry into the citrate cycle in Chapter 16. Briefly, acetyl CoA needs the metabolite oxaloacetate in order to be oxidized by the citrate cycle.

Summary

- All organisms obtain energy from the oxidative degradation of glucose and other carbohydrates. The most active pathway for carbohydrate metabolism in plants and animals is **glycolysis.** The glycolytic pathway consists of ten enzyme-catalyzed reactions that begin with a hexose substrate, usually glucose, and oxidatively cleave it into two molecules of pyruvate.

- Plants and animals store energy in the form of starch and glycogen, respectively.

- Pyruvate, the product of glycolysis, is at an important metabolic junction. In aerobic metabolism, pyruvate continues in oxidative decarboxylation to acetyl CoA. In anaerobic metabolism, pyruvate is fermented to lactate in humans.

- If sufficient levels of glucose are not present, it may be synthesized from pyruvate, lactate, glycerol, and some amino acids, by the process of **gluconeogenesis.**

- Monosaccharides used for the synthesis of di- and polysaccharides are first activated to nucleoside diphosphate derivatives.

- Carbohydrate metabolism is regulated to maintain constant levels of ATP. This is done by controlling the activity of several enzymes including glycogen phosphorylase, glycogen synthase, hexokinase, phosphofructokinase, pyruvate kinase, and pyruvate carboxylase.

Study Exercises

Understanding Terms

15.1 Define the following terms in 25 words or less.

a. Glycolysis	**j.** Substrate cycle
b. Net reaction	**k.** Hexokinase
c. Ethanol fermentation	**l.** Nucleoside diphos-
d. Galactosemia	phate sugars
e. Lactate fermentation	**m.** Lactose synthase
f. Gluconeogenesis	**n.** α-Lactalbumin
g. UDP-galactose	**o.** Phosphofructokinase
h. Glycogen	**p.** Pyruvate kinase
i. Feedback inhibition	

Reviewing Concepts

15.2 Define each of the following metabolic steps according to the types of chemistry used in Table 14.2.

a. Pyruvate → acetaldehyde + CO_2

b. Pyruvate + CO_2 + ATP →
$$\text{oxaloacetate} + \text{ADP} + \text{P}_i$$

c. Glucose 6-phosphate + H_2O → glucose + P_i

d. Fructose 1,6-bisphosphate → glyceraldehyde 3-phosphate + dihydroxyacetone phosphate

15.3 Lactate appears to be a dead-end waste product of the muscle cell that cannot be used for energy. Outline how it can be used to make glucose.

15.4 Write out reactions using names of intermediates for the synthesis of sucrose beginning with two molecules of glucose.

15.5 Avidin, a protein found in egg white, is a specific inhibitor of biotin-containing enzymes. Which of the following metabolic processes would be blocked by avidin?

a. Glucose → 2 glyceraldehyde 3-phosphate

b. Glucose → 2 lactate

c. 2 Lactate → glucose

d. Pyruvate → ethanol + CO_2

e. Pyruvate → oxaloacetate

f. Pyruvate → phosphoenolpyruvate

g. Fructose → 2 pyruvate

15.6 Using structures of substrates and products, write a reaction catalyzed by each of the following enzymes:

a. Phosphofructokinase

b. Lactose synthase

c. Amylase

d. Glycerol kinase

e. Glycerol-3-phosphate dehydrogenase

f. Galactosyl transferase + α-lactalbumin

15.7 The term (glucose)$_n$ is used in this chapter to represent the polysaccharides starch, glycogen, and cellulose. Is it correct to represent all of these polymers with a sin-

gle structure? Compare and contrast the structural differences among these three glucose polymers.

15.8 The carbohydrates listed below are all substrates of the enzyme hexokinase. Write the structure of the product formed in each reaction.

a.

Glucose

b.
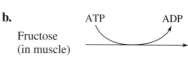
Fructose (in muscle)

c.
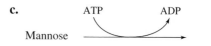
Mannose

15.9 Write the name of a negative modulator for each of the following regulatory enzymes.

a. Glycogen phosphorylase

b. Hexokinase

c. Phosphofructokinase

15.10 ATP acts as an important modulator of metabolic regulation. What is the effect of increasing cellular levels of ATP on each of the following enzymes?

a. Pyruvate kinase

b. Phosphofructokinase

c. Hexokinase

15.11 Compare and contrast glycolysis with gluconeogenesis in terms of each of the following characteristics.

a. Starting substrate(s) for each

b. Energy requirements in terms of ATP and GTP

c. Use of NAD^+ and NADH

15.12 Write the three reactions of glycolysis that are not used in gluconeogenesis. What do the three reactions have in common?

HINT: Check Table 15.3.

15.13 The reactions below define how organisms obtain free glucose or modified glucose from polysaccharides. Complete the reactions by drawing the form of glucose produced.

a. Cellulose + H_2O $\xrightleftharpoons{\text{cellulase}}$

b. Glycogen + H_2O $\xrightleftharpoons{\text{amylase}}$

c. Glycogen + P_i $\xrightleftharpoons{\text{phosphorylase}}$

d. Amylose + H_2O $\xrightleftharpoons{\text{amylase}}$

15.14 Compare the anaerobic metabolism of pyruvate in yeast with that in human muscle. Describe chemical reactions and results.

15.15 In untreated diabetics the concentration of blood sugar increases to dangerous levels. Draw the structure for "blood sugar."

Working Problems

15.16 Derive the net reaction for each of the overall metabolic processes:

   a. Glucose 1-phosphate $\longrightarrow$ 2 pyruvate
   b. Glyceraldehyde 3-phosphate $\longrightarrow$ lactate
   c. Glycerol $\longrightarrow$ pyruvate
   d. Pyruvate $\longrightarrow$ ethanol

15.17 In this chapter two reactions were coupled to produce a substrate cycle. The two reactions were the phosphoryl transfer to fructose 6-phosphate and the hydrolysis of a phosphoryl group from fructose 1,6-bisphosphate. Can you find two other reactions from carbohydrate metabolism that form a substrate cycle?

15.18 Using the names of all metabolic intermediates, outline the process of ethanol fermentation carried out in yeast cells growing on sucrose.

HINT: sucrose + H_2O $\xrightleftharpoons{\text{invertase}}$ glucose + fructose

15.19 Write the net reaction for the complete degradation of maltose to pyruvate.

15.20 Outline the metabolic reactions in *Lactobacillis* bacteria growing on the lactose in milk.

HINT: lactose + H_2O $\xrightleftharpoons{\text{lactase}}$ glucose + galactose

15.21 Compounds labeled with radioactive ^{14}C have been used to study the glycolytic pathway. Show how the product pyruvate would be labeled if glycolysis were started with each of the following compounds.

   a. Glucose, ^{14}C at C1
   b. Glyceraldehyde 3-phosphate, ^{14}C at C1
   c. Glycerol, ^{14}C at C2

15.22 Assume that carbons 3 and 4 in glucose are labeled with radioactive ^{14}C. Show how the product pyruvate would be labeled were that labeled glucose fed to bacteria.

15.23 In a lab experiment, you are using a preparation of isolated liver cells to study gluconeogenesis. You add lactate and radiolabeled CO_2 ($^{14}CO_2$) to the extract and incubate 15 min. What atoms of glucose product would be labeled with ^{14}C? Assume that the incubation mixture has appropriate concentrations of all necessary cofactors including NAD^+, ATP, and GTP.

15.24 How many phosphoanhydride bonds are required to store one free glucose in glycogen?

HINT: Derive the net reaction for glucose $\longrightarrow$ glycogen. How does this compare to the reaction glucose 6-phosphate $\longrightarrow$ glycogen shown in Section 15.4?

15.25 Write the net reaction for the following processes.

   glucose $\longrightarrow$ 2 pyruvate (glycolysis)
   glucose $\longrightarrow$ 2 lactate (fermentation)

Compare the processes in terms of these characteristics:
   a. Starting carbohydrate
   b. Final carbon product
   c. Yield of ATP
   d. Yield of NADH

15.26 Outline the pathway for synthesis of glucose from two glycerol molecules.

15.27 Study the structure of 1,3-bisphosphoglycerate, an intermediate in glycolysis. Describe the chemical bond linkage for each phosphoryl group in the molecule. Why is the phosphoryl group linked to the carboxyl group more reactive than the phosphoryl group linked to the hydroxyl group?

15.28 Derive the net reaction for the formation of PEP from lactate and compare it with the net reaction pyruvate $\longrightarrow$ PEP.

15.29 Calculate $\Delta G^{\circ\prime}$ for the first step of glycolysis.

   glucose + ATP $\rightleftharpoons$ glucose 6-phosphate + ADP

15.30 You have just discovered a new enzyme that catalyzes the following reaction.

   glucose + ATP $\rightleftharpoons$ glucose 3-phosphate + ADP

   a. Write the structures for glucose and glucose 3-phosphate.
   b. Would you expect the $\Delta G^{\circ\prime}$ for transfer of the phosphoryl group from glucose 3-phosphate to H_2O to be more, less, or about the same as for glucose 6-phosphate? Explain.

Writing Biochemistry

15.31 Immediately after the concepts in this chapter have been presented or discussed in class, spend a few minutes to respond in writing to three questions: (a) What were the three most important concepts from the chapter? (b) What concepts were a review for you? (c) What concepts did you find confusing?

15.32 Select one of the concepts from this chapter that you listed as confusing in Exercise 15.31(c). Phrase the concept in the form of a question. Use your text or other biochemistry books to gather information to answer the question. Write a 100- to 200-word paragraph that answers the question to your satisfaction.

15.33 Your professor has assigned each student to give a 15-minute presentation in class on some concept from this chapter. Select one of the articles listed in "Further Reading." Read the article and prepare an outline of the contents that you can use for your class presentation.

Applying Biochemistry

15.34 One of the first enzymes whose catalytic action was described by the induced fit model was hexokinase, the first enzyme in the glycolytic pathway. Describe in your own words the "induced fit model" for the catalytic action of hexokinase. Explain how it defines enzyme action better than the "lock and key model" proposed by Emil Fischer.

15.35 Explain why the Cori cycle of reactions (see Figure 15.7) is not a futile cycle of energy waste.

15.36 Write out the reactions to show how glycerol can be used as a precursor for glucose synthesis via gluconeogenesis.

HINT: See Figure 15.9.

Further Reading

Baugh, M., 1991. Aerobic evolution—A fascinating world. *Educ. Chem.* 20:20–22.

Bullock, C., 1990. The biochemistry of alcohol metabolism—A brief review. *Biochem. Educ.* 18:62–66.

Christoffersen, J., Christofferson, M., and Hunding, A., 1998. Comments on optimization of glycolysis. *Biochem. Educ.* 26:290–291.

Darvey, I., 2000. Gluconeogenesis. *Biochem. Educ.* 28:80–82.

Elliott, K., et al., 1999. Teaching metabolism for the future—Concepts not pathways. *The Biochemist* 21(1):9–39.

Fell, D., 1997. *Understanding the Control of Metabolism.* London: Portland Press.

Filippusson, H., 1998. Exercise biochemistry as an approach to the study of intermediary metabolism. *Bicohem. Educ.* 26:119–120.

Frayn, K., 1996. *Metabolic Regulation: A Human Perspective.* London: Portland Press.

Leinhard G., Slot, J., and Mueckler, M., 1992. How cells absorb glucose. *Sci. Am.* 266(1):86–91.

Manchester, K., 2000. Arthur Harden: an unwitting pioneer of metabolic control analysis. *Trends Biochem. Sci.* 25:89–92.

Manchester, K., 2000. Optimization of energy coupling. *Biochem. Educ.* 28:18–19.

McCoy, M., 1998. Biomass ethanol inches forward. *Chem. Eng. News* December 7:29–32.

Mulimani, V., 1995. Teaching metabolic pathways to understand its regulation. *Biochem. Educ.* 23:32–34.

Waddell, T., et al., 1999. Optimization of glycolysis. *Biochem. Educ.* 27:12–13.

Wang, J., 1998. The insulin connection: Dorothy Hodgkin and the Beijing Insulin Group. *Trends Biochem. Sci.* 23:497–500.

Watford, M., 1988. What is the metabolic fate of dietary glucose? *Trends Biochem. Sci.* 13:329–330.

Chapter 16

Production of NADH and NADPH: The Citric Acid Cycle, the Glyoxylate Cycle, and the Phosphogluconate Pathway

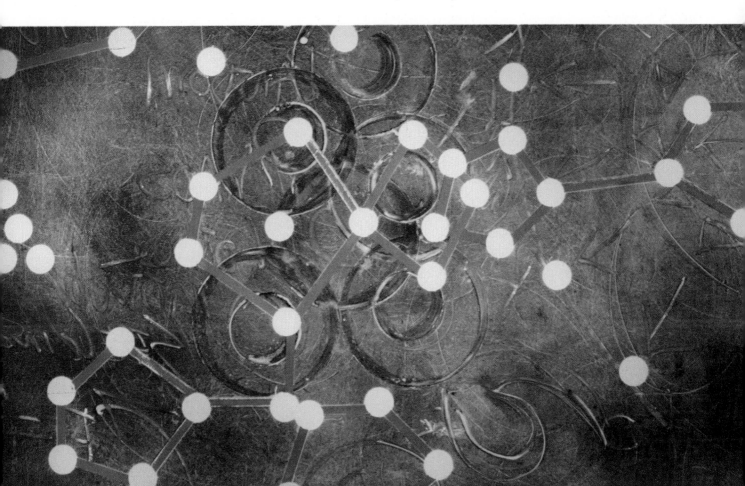

n the previous chapter, we explored the universal pathway of glycolysis: the oxidative degradation of glucose and other monosaccharides to pyruvate. The net reaction derived for the process indicated an output of two pyruvates, two ATPs, and two NADHs for each glucose entering the reaction sequence. The energy collected in the phosphoanhydride bonds of two ATPs represents only about 7% of the total potential energy present in glucose. However, some organisms, albeit relatively simple anaerobic ones, can live nicely on this low yield. Their only concern is the oxidation of NADH, which they achieve by fermentation so that glycolysis can be sustained. During the early stages of evolution, before the introduction of oxygen into the atmosphere, all organisms were subject to this energy limitation. But anaerobic metabolism is energetically inefficient because it leaves over 90% of the energy in carbohydrates untapped and it does not allow for more complex biochemical development in an organism. The evolutionary development of cell machinery to use O_2 opened new metabolic pathways that allowed more efficient use of the energy in food.

Glycolysis has also been described as a preliminary phase, preparing glucose for entry into aerobic metabolism. When O_2 is present as the final electron acceptor, the glycolysis product

Figure 16.1

The positions of the pyruvate dehydrogenase complex and the citric acid cycle in intermediary metabolism. The pyruvate dehydrogenase complex provides the link or bridge between the first and third stages of glucose metabolism. The product of glycolysis, pyruvate, is converted under aerobic conditions to acetyl CoA, which enters the citric acid cycle. Important products of the citric acid cycle include ATP (or GTP) and reduced cofactors, NADH and $FADH_2$.

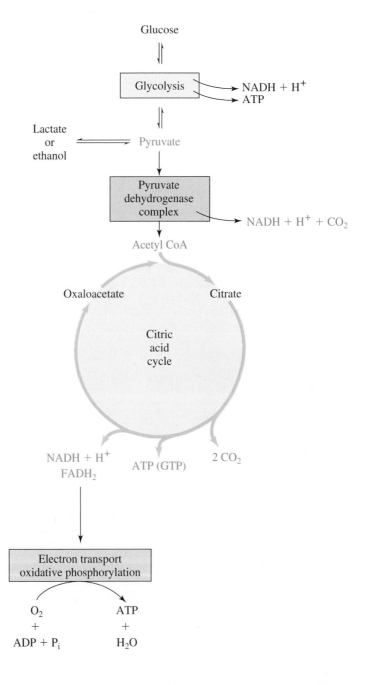

pyruvate is oxidized completely to CO_2 and water by entry into stage III of metabolism (see Figures 14.6 and 16.1). Chapter 16 focuses on the pathways available for aerobic oxidation of the carbon skeleton of pyruvate to CO_2. Aerobic metabolism usually occurs in the mitochondria. These pathways begin with the **pyruvate dehydrogenase complex,** which directs the conversion of pyruvate to acetyl CoA and thus serves as the bridge between glycolysis and aerobic metabolism. Next comes the central and universal pathway of aerobic metabolism—the **citric acid cycle**—which oxidizes acetyl CoA to CO_2 with the production of energy-rich NADH and $FADH_2$. In Chapter 17, reduced cofactors made by the cycle are seen to be used to make ATP by electron transport to O_2 and oxidative phosphorylation. Although the citric acid cycle plays a major role in catabolism, its intermediates also serve as important precursors for anabolic processes, such as the synthesis of amino acids, nucleotide bases, and porphyrins. In our discussion, we shall emphasize the strict coordination and regulation of metabolic processes to ensure cell economy, the proper flow of metabolites in the reaction pathways, and the correct balance between anabolism and catabolism.

The citric acid cycle is present in all forms of aerobic life, but in some organisms (plants and some microorganisms) it is modified by the addition of a shunt called the **glyoxylate cycle** that allows these organisms to utilize acetate as a source of energy and carbons for biosynthesis. The glyoxylate cycle provides plants and some microorganisms with the ability to use acetate to make carbohydrates, a metabolic transformation unavailable to animals.

The chapter concludes with an introduction to the **phosphogluconate pathway, (pentose phosphate pathway),** an auxiliary route for glucose oxidation in animals. The biological role of this specialized pathway is to produce the reduced and reactive cofactor, NADPH, and the important synthetic intermediate, ribose 5-phosphate.

16.1 The Pyruvate Dehydrogenase Complex

Pyruvate and Mitochondria

Pyruvate, the product of glycolysis, has been described as a molecule at the crossroads; it has several metabolic options (see Figure 16.1). Two of those options, lactate fermentation and ethanol fermentation, were discussed in detail in Chapter 15 and a third, entry into aerobic metabolism, is briefly mentioned. Here we describe the chemical details of pyruvate oxidation in aerobic cells. Before we do this we must describe a change of scenery for pyruvate and aerobic metabolism. Most of the enzymes for glycolysis and gluconeogenesis are present in the cellular cytoplasm. Energy production by continued oxidation of pyruvate is localized in **mitochondria,** cell compartments specialized in aerobic metabolism (Figure 16.2).

Mitochondria, organelles that are present in virtually all eukaryotic cells, are often called the "powerhouses of the cell." Mitochondria are also present in phototropic cells (wherein light is the source of energy), where they produce ATP during catabolism. We first encountered the mitochondria in our discussion of cell structure (see Figure 1.11). With average diameters between 0.5 and 1 μm and lengths of 10 μm, mitochondria are smaller than cell nuclei and chloroplasts. Some cells have more than 1000 mitochondria each. In higher animals, muscle cells, which have high demands for ATP, are rich in mitochondria. Mitochondrial structure and function have been studied by biochemists and biologists for about 150 years. The role of mitochondria in oxidative reactions and oxygen metabolism was first explored in crude extracts of tissues and cells by Otto Warburg in Germany in 1913. Intact mitochondria, separated from other organelles, were first studied by Albert Lehninger and Eugene Kennedy in the United States in 1948; they showed that mitochondria contained the biochemical apparatus for the citric acid cycle and respiration (uptake of O_2, electron transport, and ATP synthesis). In morphological terms, mitochondria are characterized by the presence of internal compartments resulting from two membrane systems. The outer membrane is permeable to most important small metabolites, but the inner membrane has specific transport properties regulated by integral proteins. The region between the two membranes is called the intermembrane space. The highly folded inner membrane forms the boundary to a fluid-filled internal region called the matrix. This region, which has a gelatinous consistency, contains enzymes for further metabolism of pyruvate, including

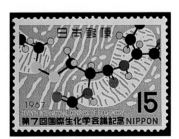

Japanese stamp with mitochondria in the background to recognize the 7th International Congress of Biochemistry.

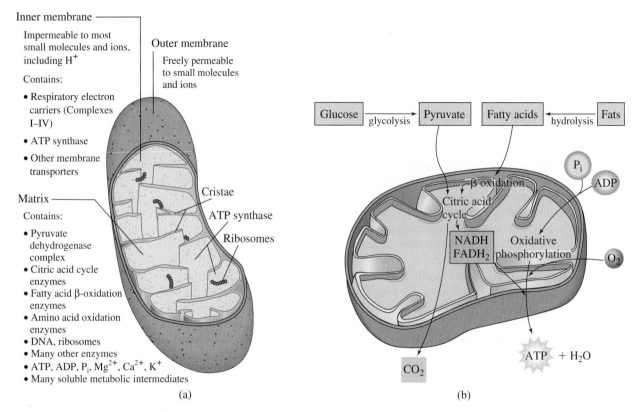

Figure 16.2

Structures of a mitochondrion showing the double membranes, which have differing transport properties. The reaction processes of aerobic metabolism are localized in the inner membrane and the mitochondrial matrix: (a) important bioprocesses and their location in the mitochondrion, (b) how the mitochondrion is involved in energy metabolism. Fats and carbohydrates (glucose) are converted to $CO_2 + H_2O$ with generation of energy in the form of ATP. Oxidative phosphorylation consists of electron transport and ATP synthesis.

the pyruvate dehydrogenase complex, the citric acid cycle, and the electron-transport chain (see Chapter 17). The functional proteins responsible for aerobic metabolism are either solubilized in the gellike matrix or bound to the inner membrane. For pyruvate and other cytoplasmic metabolites to continue in aerobic metabolism, they must be moved from cytoplasm through the two mitochondrial membranes to the matrix. Pyruvate, a polar, ionic species, is able to diffuse through the relatively polar outer membrane but unable to pass through the inner membrane by simple diffusion. It is transported by pyruvate translocase, a permease that functions by exchanging pyruvate and hydroxide ion in an antiport manner (see Chapter 9, Section 9.6). To balance the electrical charge on each side of the membrane, a hydroxide ion is moved out for each pyruvate translocated in.

Oxidation of Pyruvate

Now that pyruvate has taken this journey through the mitochondrial membranes, it is present in the matrix, where it can continue in aerobic metabolism. Before pyruvate can enter into the citric acid cycle, the central pathway of aerobic metabolism, the carbon skeleton must first undergo three chemical transformations:

1. Decarboxylation (loss of CO_2).
2. Oxidation of the keto group on C2 to a carboxyl group.
3. Activation by linkage to coenzyme A through a thioester bond.

These changes are complicated and require three enzymes, five coenzymes, and five distinct chemical reactions. The complete enzyme package responsible for pyruvate oxidation to

Enzyme	Abbreviation	Coenzyme
Pyruvate dehydrogenase	E_1	Thiamine pyrophosphate (TPP)
Dihydrolipoyl transacetylase	E_2	Lipoamide, coenzyme A (CoASH)
Dihydrolipoyl dehydrogenase	E_3	Flavin adenine dinucleotide (FAD), nicotinamide adenine dinucleotide (NAD)

Table 16.1
Enzymes and coenzymes of the pyruvate dehydrogenase complex

acetyl CoA is called the **pyruvate dehydrogenase complex** and is a classic example of a **multienzyme complex.** The overall reaction catalyzed by the complex can be summarized as follows:

$$\underset{\text{Pyruvate}}{\begin{array}{c}COO^- \\ | \\ C=O \\ | \\ CH_3 \end{array}} \quad \overset{\begin{array}{c}\text{TPP,} \\ \text{lipoamide,} \\ \text{FAD}\end{array}}{\underset{\begin{array}{c}\text{pyruvate dehydrogenase complex} \\ (E_1 + E_2 + E_3)\end{array}}{\xrightarrow{\hspace{2cm}}}} \quad \underset{\text{Acetyl CoA}}{\begin{array}{c}SCoA \\ | \\ C=O \\ | \\ CH_3 \end{array}} + CO_2$$

CoASH   NAD⁺ → NADH + H⁺

The formal names of the three enzymes in the complex, designated as E_1, E_2, and E_3, and their associated cofactors are given in Table 16.1. Three of the cofactors are present as prosthetic groups; that is, they are covalently attached to their enzymes.

Interactive Animation: Pyruvate dehydrogenase complex

The Pyruvate Dehydrogenase Complex in Action

Figure 16.3 outlines the mode of action of the pyruvate dehydrogenase complex. The combined steps function as a catalytic cycle. Entering the cycle are pyruvate, NAD⁺, and **coenzyme A** (CoASH). Exiting are acetyl CoA, NADH, and CO_2. The three enzymes with their associated prosthetic groups function as a package or complex. It is useful to view the intermediates produced from pyruvate passing from one enzyme directly to another, with each enzyme carrying out its catalytic duty. Step 1, decarboxylation, is brought about by

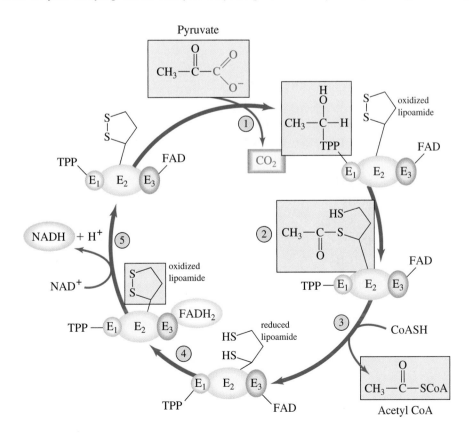

Figure 16.3

Steps in the oxidation of pyruvate to acetyl CoA and CO_2 by the pyruvate dehydrogenase complex. The enzymes (E_1, E_2, and E_3) are shown linked to their prosthetic groups. E_1 = pyruvate dehydrogenase; E_2 = dihydrolipoyl transacetylase; E_3 = dihydrolipoyl dehydrogenase.

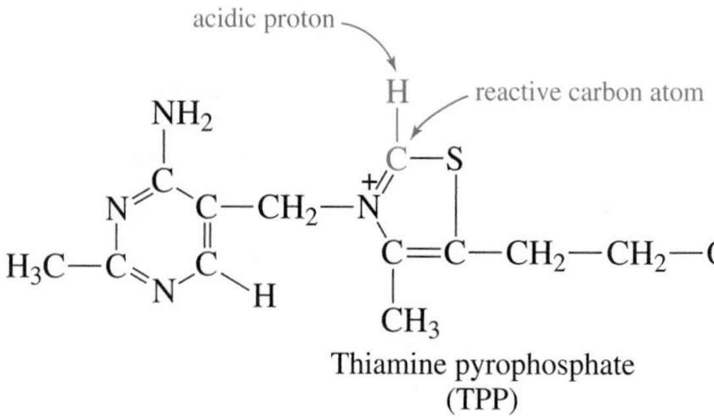

Figure 16.4

Structure of the cofactor of decarboxylation reactions, thiamine pyrophosphate (TPP). The hydroxyethyl group from pyruvate is bound to the reactive carbon atom. A reactive carbanion is generated by dissociation of the acidic proton.

Thiamine pyrophosphate
(TPP)

pyruvate dehydrogenase (E_1) with assistance by thiamine pyrophosphate (TPP) (Figure 16.4). The vitamin thiamine is used as a building block for the synthesis of TPP (see Table 7.1). In general, TPP works with enzymes that are responsible for decarboxylation from α-keto acids. We were introduced to TPP as a cofactor in a similar reaction, the decarboxylation of pyruvate in ethanol fermentation (see Chapter 15, Section 15.3). In that case, free acetaldehyde was generated, whereas with pyruvate dehydrogenase the C_2 product remains bound to E_1 as a hydroxyethyl derivative (see Figure 16.3). In step 2, the C_2 unit is oxidized and transferred as the acetyl group to **lipoamide,** the prosthetic group associated with E_2, dihydrolipoyl transacetylase. Lipoamide, which is derived from the vitamin lipoic acid, has a disulfide bond (in the oxidized state) that undergoes reversible reduction to thiol groups (Figure 16.5). Thus, lipoamide as a prosthetic group is involved in reactions of acyl group transfer coupled with oxidation–reduction. The C_2 unit, acetyl, is linked to one of the — SH groups of lipoamide in a thioester bond. The final product, **acetyl CoA,** is formed in step 3 by transfer of the acetyl group from E_2–lipoamide to the — SH group of **coenzyme A,** again forming a thioester bond (see Window on Biochemistry 16.1).

We should pause at this point to explain the importance of this process. **Coenzyme A** (also called CoASH; the — SH denotes the reactive sulfhydryl group) is a complex molecule containing the vitamin pantothenic acid (see Table 7.1 and Figure 16.6). CoASH has an important role as a carrier of acetyl or other acyl groups, which are activated by linkage to

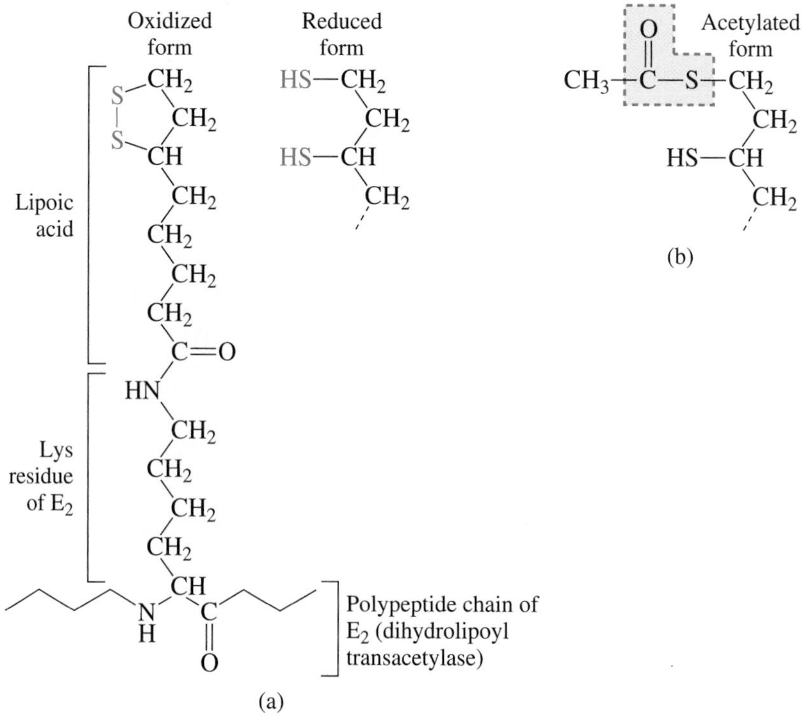

Figure 16.5

(a) Structure of the E_2–lipoamide complex showing oxidation–reduction of the sulfur atoms of lipoamide. (b) A thioester bond is shown linking the acetyl group to lipoamide.

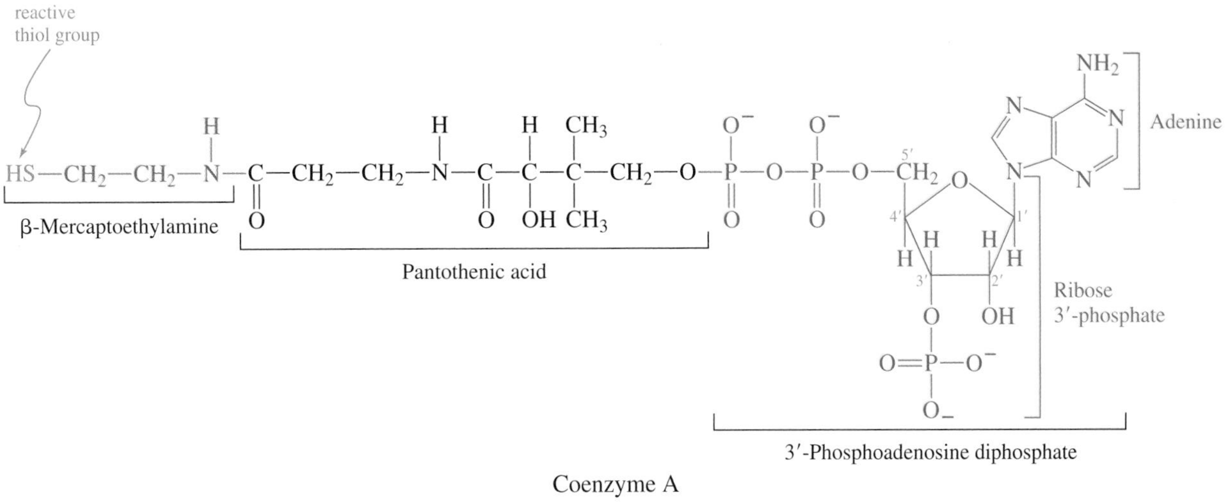

Figure 16.6

Structure of coenzyme A (CoASH) showing the pantothenic acid moiety and the reactive thiol group (—SH) for activation of acyl groups (*red arrow*).

the —SH group to form a thioester bond ($-\text{s}-\overset{\text{O}}{\overset{\|}{\text{C}}}_{\text{R}}$). The thioester is of high energy and some of the energy released from pyruvate oxidation is stored in that bond. (See Window on Biochemistry 14.2.) The energy may be released by transfer of the acetyl group, for example, to H_2O:

$$\text{CoAS}-\overset{\text{O}}{\overset{\|}{\text{C}}}-\text{CH}_3 + H_2O \rightleftharpoons \text{CoASH} + \text{CH}_3\text{COOH} \quad \Delta G^{\circ\prime} = -31.4 \text{ kJ/mol}$$

This is energy available to drive a thermodynamically unfavorable reaction. We will encounter the use of CoASH as an acyl group carrier in a number of metabolic processes, including fatty acid oxidation and amino acid and cholesterol biosynthesis. The acyl groups most commonly activated by attachment to CoASH are acetyl ($-\overset{\text{O}}{\overset{\|}{\text{C}}}_{\text{CH}_3}$), malonyl ($-\overset{\text{O}}{\overset{\|}{\text{C}}}_{\text{CH}_2\text{CO}_2^-}$), succinyl ($-\overset{\text{O}}{\overset{\|}{\text{C}}}_{\text{CH}_2\text{CH}_2\text{CO}_2^-}$), or acyl ($-\overset{\text{O}}{\overset{\|}{\text{C}}}_{\text{R}}$), where R represents hydrocarbon chains of varying lengths.

Returning to the details of the pyruvate oxidation reaction, we see that the last reaction (see step 3, Figure 16.3) has expelled acetyl CoA formed by the decarboxylation of pyruvate and linkage to CoASH. We have accomplished the transformation of pyruvate to acetyl CoA, which can now enter the citric acid cycle, but the pyruvate dehydrogenase complex cycle is not complete. The multienzyme complex must be returned to its original chemical form so that more pyruvate can be oxidized. The primary action required is the oxidation of the thiol groups of lipoamide. In two oxidation–reduction steps (see steps 4 and 5, Figure 16.3 and Figure 16.5), two electrons and two protons are transferred from reduced E_2–lipoamide to E_3–FAD and finally to NAD^+. Dihydrolipoyl dehydrogenase (E_3) catalyzes these oxidation–reduction steps. Its prosthetic group, FAD (flavin adenine dinucleotide), is similar in biological function to NAD^+. FAD is derived from the vitamin riboflavin, whereas NAD^+ is derived from the vitamin niacin (see Table 7.1). Both FAD and NAD^+ are cofactors for oxidation–reduction reactions; that is, they exist in oxidized and reduced forms. In the pyruvate dehydrogenase complex, the FAD serves as a conduit to channel electrons from lipoamide to NAD^+. The structures for NAD^+, NADH, FAD, and $FADH_2$ are shown in Figure 16.7. Although their general biochemical functions are similar, they have distinctive characteristics in their actions. Both NAD^+ and FAD accept or donate two electrons and hydrogens in oxidation–reduction reactions, but the mechanism for each is different. NAD^+

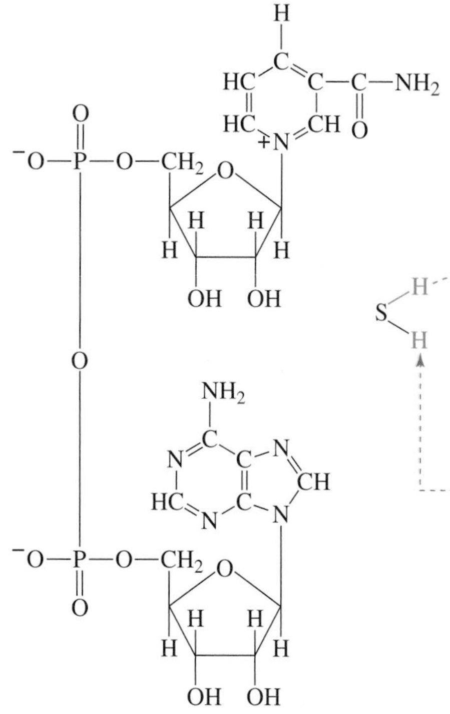

Nicotinamide adenine dinucleotide (NAD$^+$)

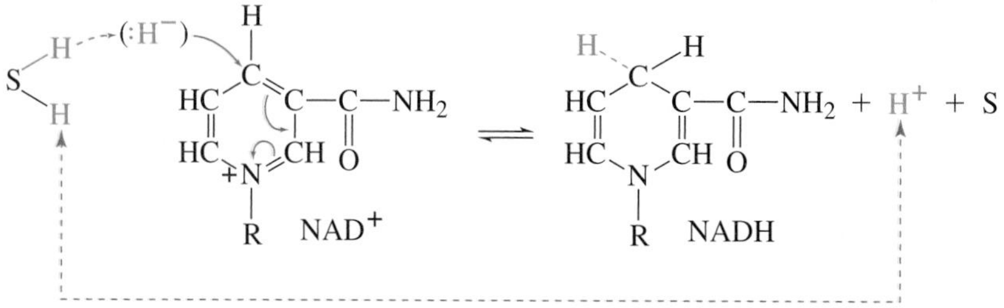

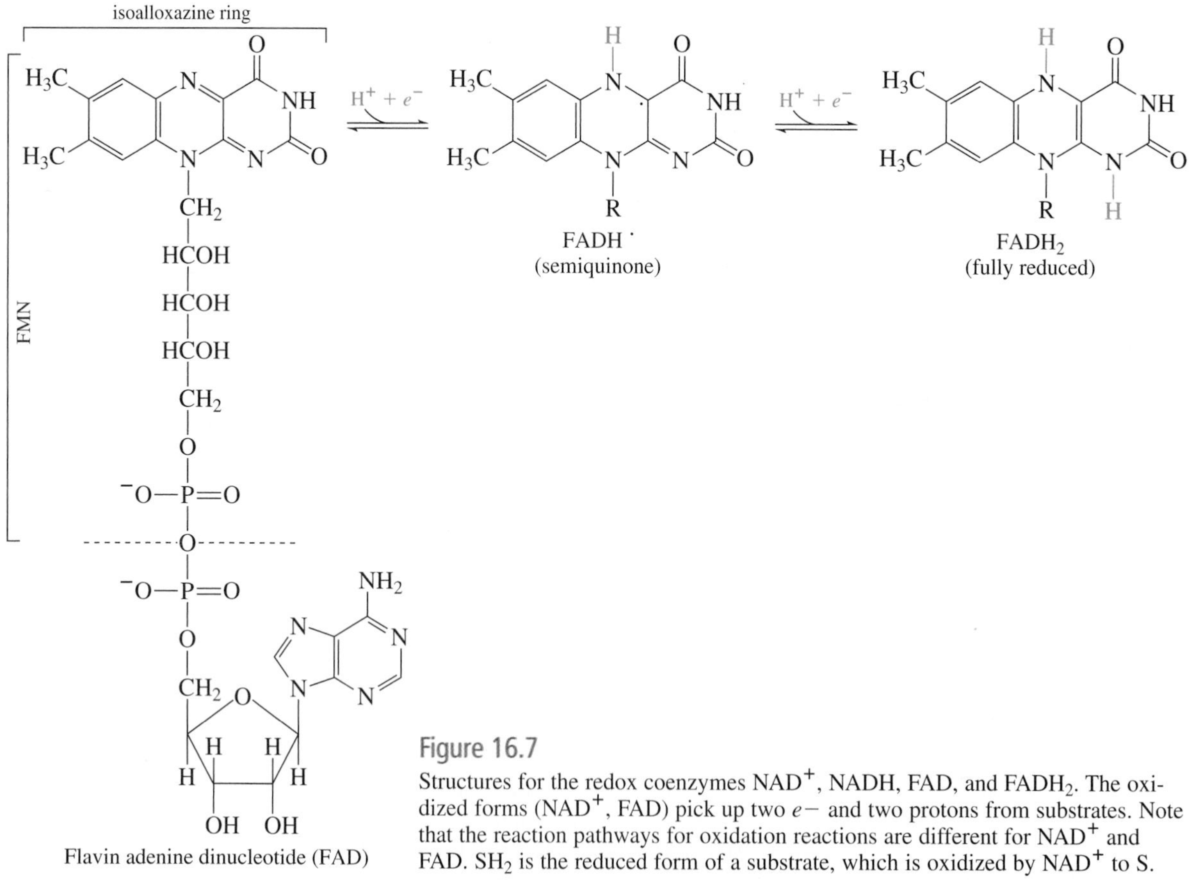

isoalloxazine ring

FMN

Flavin adenine dinucleotide (FAD)

FADH$^·$
(semiquinone)

FADH$_2$
(fully reduced)

Figure 16.7

Structures for the redox coenzymes NAD$^+$, NADH, FAD, and FADH$_2$. The oxidized forms (NAD$^+$, FAD) pick up two $e-$ and two protons from substrates. Note that the reaction pathways for oxidation reactions are different for NAD$^+$ and FAD. SH$_2$ is the reduced form of a substrate, which is oxidized by NAD$^+$ to S.

exists in two forms, the oxidized form, NAD^+, and the reduced form, with two additional electrons and a hydrogen, NADH. The oxidized form picks up electrons from the substrate of a dehydrogenase:

$$NAD^+ + RCH_2OH \xrightleftharpoons[\text{dehydrogenase}]{\text{alcohol}} NADH + H^+ + RC \overset{O}{\underset{H}{\diagup}}$$

The formal electron exchange of the two electrons from the alcohol to NAD^+ probably involves transfer of a hydride ion, H^-, and a proton, H^+. The NAD^+ is reduced to NADH. The proton released from the substrate joins the solvent. We write the reduced product as $NADH + H^+$ to indicate both the H^- and the H^+. The use of NADH as a reducing agent works in reverse. The hydride is transferred from NADH to the carbonyl group of a substrate. A proton is pulled from solvent to neutralize the oxyanion formed. NAD^+ is usually limited to redox reactions with alcohols and ketones (or aldehydes). Substrates we have encountered or will encounter for NAD^+-linked dehydrogenases are glyceraldehyde 3-phosphate, pyruvate, lactate, ethanol, isocitrate, α-ketoglutarate, and malate. As with NAD^+, FAD exists in two forms, an oxidized form (FAD) and a reduced form with two additional electrons and hydrogens ($FADH_2$). FAD picks up two electrons and two hydrogens from the substrate:

$$FAD + R—CH_2—CH_2—R' \xrightleftharpoons[]{\text{dehydrogenase}} \underset{H}{\overset{R}{\diagdown}} C = C \underset{R'}{\overset{H}{\diagup}} + FADH_2$$

In contrast to the reduction of NAD^+, both hydrogens are transferred to the cofactor FAD and $FADH_2$ is produced. FAD can also function in an alternate mode by accepting only one electron and hydrogen. This produces an FAD free radical called the semiquinone form.

Summary of the Pyruvate Dehydrogenase Complex

After the five steps in Figure 16.3, the pyruvate dehydrogenase complex is now in a catalytically active form and ready to accept another pyruvate for a second round of oxidative decarboxylation. The five reactions are coordinated by the multienzyme cluster. The physical placement of each of the three enzymes and its cofactor in the complex is such that the reaction intermediates can literally be handed from one enzyme active site to another. The intermediates probably remain bound to the surface of the complex throughout the entire process and are not free to roam in solution. Multienzyme complexes are often used to make difficult chemical transformations economical, efficient, rapid, and without unwanted side reactions.

The net reaction for the oxidation of pyruvate by the pyruvate dehyhydrogenase complex is relatively simple considering the complicated chain of events:

$$\text{pyruvate} + \text{CoASH} + NAD^+ \longrightarrow \text{acetyl CoA} + NADH + H^+ + CO_2$$

Our goal of transforming pyruvate to an activated form for direct entry into aerobic metabolism has been accomplished. In the process, we have generated an energized acetyl group (acetyl CoA) and reducing power in the form of NADH.

It is difficult to envision the prospect of aerobic life without all the components necessary for the pyruvate dehydrogenase complex. Indeed, if thiamine (a component of TPP) is deficient in the diets of humans, the disease **beriberi** develops (see Chapter 7, Section 7.1). Symptoms of this nutritional disease include loss of neural and cardiovascular function. The disease occurs most often in countries where rice is a staple food. Rice, especially white rice, has a relatively low content of thiamine; polished rice (hulls removed) has essentially none. Thiamine is abundant in whole, unmilled cereals and grains. For children in the United States of a century ago, beriberi was a possible fact of life because milled wheat used in bread products contained little thiamine. All commercial bakeries in the United States must now, by law, use thiamine-enriched flour.

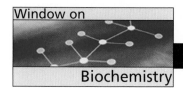

Window on

Biochemistry

16.1   Lipoic Acid—The Universal Antioxidant

The coenzyme **α-lipoic acid** (sometimes called thioctic acid), which functions as the prosthetic group, **lipoamide,** is involved in reactions of acyl group transfer coupled with oxidation–reduction (i.e., the pyruvate dehydrogenase complex). Lipoic acid was first isolated and characterized in 1951. In these early studies, it required 10 tons of liver to isolate just 30 mg of purified lipoic acid. Although lipoic acid was initially considered a vitamin, new studies have shown that our cells do synthesize it, but in very small quantities. The new studies also uncovered possible additional biochemical functions for the compound. Two new roles include (1) serving as an **antioxidant,** and (2) chelating metal ions. Lipoic acid is sometimes called the "universal antioxidant" because it is more powerful in blocking oxidation processes destructive to the cell than either of the more well-known antioxidants, vitamins C and E. In addition, lipoic acid regenerates the active species of both of these vitamins, a process we call **redox cycling:**

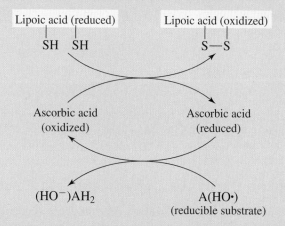

Another advantage of lipoic acid as an antioxidant is its solubility in both fatty tissue and in water-based tissue. Vitamin C is soluble only in water-based tissue and vitamin E only in fat tissue (i.e., nerve tissue). Antioxidants function by preventing oxidative damage to cellular molecules. By using oxidation–reduction chemistry, antioxidants neutralize the especially reactive radicals that are generated from oxygen and hydrogen peroxide under normal physiological conditions. Some of these radicals, called **reactive oxygen species (ROS),** are superoxide ($O_2^{\cdot-}$), hydroxyl (HO·), hydroperoxy (HOO·), and peroxynitrite (OONO$^-$, formed from the reaction of nitric oxide, NO, with superoxide). The ROS are especially involved in the process of lipid peroxidation whereby the unsaturated fatty acids of membranes are chemically modified (see page 215).

Lipoic acid may also serve a useful purpose as a chelator of metals. It is able to bind copper, iron, zinc, cadmium, and mercury ions. During oxidative stress, metal ions, especially iron, which may cause oxidative damage, are released from proteins and perhaps detoxified by redox reactions of lipoic acid.

Some of the possible benefits of lipoic acid are currently undergoing therapeutic trials. Lipoic acid is being examined for the treatment of diabetes, cataracts, glaucoma, and heavy metal poisoning. In addition, new research has shown that it inhibits the HIV reverse transcriptase. Human cells probably make enough lipoic acid for its metabolic roles, but some nutritionists recommend that dietary supplements may be needed for it to function as an antioxidant and metal chelator. Natural lipoic acid is present in spinach, liver, and brewers yeast.

16.2 The Citric Acid Cycle

🔘 Interactive Animation: Citric acid cycle

In 1937, the German biochemist Hans Krebs postulated that a cycle of eight reactions comprised the central pathway of aerobic metabolism. The sequence of reactions was proposed by Krebs after bringing together years of intense biochemical research on pyruvate metabolism carried out by himself, Albert Szent-Gyorgi, and Franz Knoop. Several names are now used in reference to this cyclic metabolic pathway: **the citric acid (citrate) cycle,** the Krebs cycle, and the tricarboxylic acid (TCA) cycle. Briefly, this pathway is a catalytic cycle that has two purposes:

1. The degradation of the C_2 unit of acetyl CoA (from pyruvate and other sources, including fatty acid oxidation) to CO_2, yielding energy that is collected in the form of ATP or GTP and reducing power in the form of NADH and $FADH_2$.
2. Supplying precursors for the biosynthesis of amino acids, porphyrins, and pyrimidine or purine bases for nucleotides.

It is instructive at this point to emphasize the fundamental differences between glycolysis, a pathway we are now familiar with, and the citric acid cycle. Glycolysis is a linear pathway, whereas the citric acid cycle functions in a *cyclic* manner. Further, the enzymes of glycolysis are located in the cytoplasm of the cell; citrate cycle enzymes are present in the mitochondrial matrix, as discovered by Eugene Kennedy and Albert Lehninger in 1948.

Reactions of the Citric Acid Cycle

With the general characteristics of the citric acid cycle now provided, we turn to look at the chemical details of this metabolic pathway. Figure 16.8 shows the reactions with structural intermediates, enzymes, and cofactors. As we travel around the cycle, we should make comparisons with carbohydrate metabolism, especially glycolysis. Most intermediates in carbohydrate metabolism are activated by phosphorylation. Few intermediates in the citric acid cycle are activated, but when necessary they are linked via a thioester bond to CoASH (acetyl CoA, succinyl CoA). We will see a variety of reaction types in the citric acid cycle, just as in carbohydrate metabolism. One-half of the reactions involve substrate oxidation indicating the importance of the cycle as a general catabolic pathway. All of the intermediates in the cycle have two or three carboxyl groups that exist as carboxylate ions. All intermediates except succinyl CoA and succinate have another functional group (in addition to carboxyl groups), such as a hydroxyl group, a double bond, or a keto group. Each step in the cycle involves chemical changes at these other functional groups. The carboxyl groups, with the exception of two decarboxylation steps, are left intact.

The reactions of the citric acid cycle are summarized in the following eight steps. Enzymes, prosthetic groups, and reaction types are listed in Table 16.2.

Step 1

The cycle begins with the combination of acetyl CoA and oxaloacetate. Formally, the reaction is described as the addition of the C_2 unit (acetyl) to the keto double bond of the C_4 acid, oxaloacetate, to produce the C_6 compound, citrate. The common name for the enzyme is **citrate synthase.** (The name **synthase** is used for enzymes catalyzing addition to a double bond, or elimination to form a double bond, in which no energy from cleavage of a phosphoanhydride bond is required. **Synthetase** is the name used for addition–elimination reactions *with* energy donated by ATP, GTP, etc.) Water plays an important role in hydrolyzing the reaction intermediate, citroyl CoA.

Step 2

Step 2 could certainly be considered an isomerization–rearrangement; however, because its mechanism involves elimination of water to form a double bond and addition of water to a double bond, it is placed in category 4 and the formal enzyme class is a **lyase.** The enzyme, commonly called **aconitase** in reference to the reaction intermediate aconitate, has a unique prosthetic group, an iron–sulfur cluster. The purpose of step 2 in the overall scheme of the

Citric acid, a tricarboxylic acid, is present in citrus fruits.

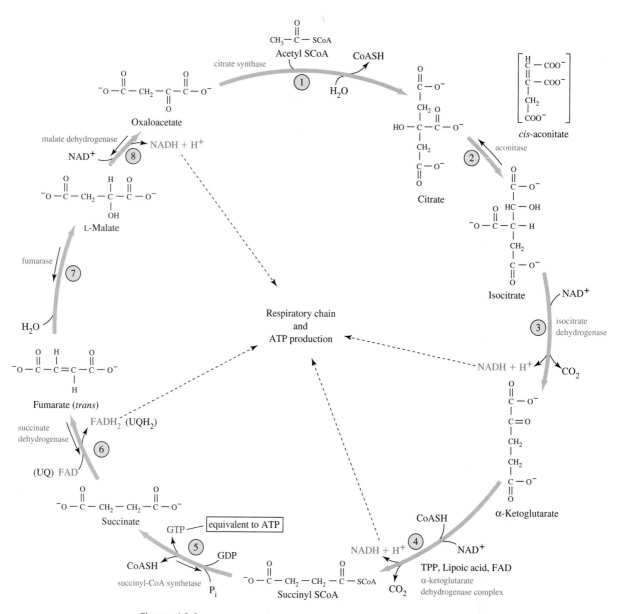

Figure 16.8

The steps of the citric acid cycle, which correspond to the reactions in Table 16.2. Acetyl CoA enters the cycle by combining with oxaloacetate to produce citrate. The two carbons of acetyl CoA provide the top two carbons of citrate and eventually are oxidized and released as CO_2. *cis*-Aconitate is an intermediate formed in the aconitase reaction. Electrons removed from cycle intermediates are used to form NADH and $FADH_2$.

cycle may not be completely clear. Our study of this offers a chance to review organic chemistry. The citric acid cycle is oxidative (as is all of catabolism), but citrate, with a tertiary alcohol functional group, cannot be oxidized without carbon–carbon bond cleavage. By the isomerization of citrate to isocitrate (which contains a secondary alcohol functional group), the path is opened for further oxidation of the carbon skeleton, which occurs in step 3.

Step 3

In this step, isocitrate from rearrangement of citrate is oxidized. The oxidation of the hydroxyl group of isocitrate results in an unstable intermediate that decarboxylates to the C_5 acid, α-ketoglutarate. The enzyme, **isocitrate dehydrogenase,** functions with NAD^+ as electron acceptor. NAD^+ and NADH are most often the cofactors for the interconversion of hydroxyl and keto groups in substrates.

Table 16.2
The reactions of the citric acid cycle

Reaction Number[a]	Reaction	Enzyme	Prosthetic Group	Reaction Type[b]
1	Acetyl CoA + oxaloacetate + H_2O $\rightleftharpoons$ citrate + CoA + H^+	Citrate synthase		3, 4
2	Citrate $\rightleftharpoons$ *cis*-aconitate $\rightleftharpoons$ isocitrate	Aconitase	Fe–S	4
3	Isocitrate + NAD^+ $\rightleftharpoons$ α-ketoglutarate + CO_2 + NADH + H^+	Isocitrate dehydrogenase		1, 4
4	α-Ketoglutarate + NAD^+ + CoA $\rightleftharpoons$ succinyl CoA + CO_2 + NADH + H^+	α-Ketoglutarate dehydrogenase complex	Lipoamide, FAD, TPP	1, 4
5	Succinyl CoA + P_i + ADP or GDP $\rightleftharpoons$ succinate + ATP or GTP + CoA	Succinyl-CoA synthetase		2
6	Succinate + FAD (enzyme bound) $\rightleftharpoons$ fumarate + $FADH_2$ (enzyme bound)	Succinate dehydrogenase	FAD, Fe–S	1
7	Fumarate + H_2O $\rightleftharpoons$ L-malate	Fumarase		4
8	L-Malate + NAD^+ $\rightleftharpoons$ oxaloacetate + NADH + H^+	Malate dehydrogenase		1

[a]The reaction numbers correspond to the steps in Figure 16.8.
[b]Reaction type: 1, oxidation–reduction; 2, phosphoryl group transfer; 3, hydrolysis; 4, nonhydrolytic cleavage (addition or elimination); 5, isomerization–rearrangement; 6, bond formation coupled to ATP cleavage.

Step 4

The fourth step, the oxidation of α-ketoglutarate, is more complex than step 3, but we have already met the chemical reactions used here. The **α-ketoglutarate dehydrogenase complex** displays the same chemistry as the pyruvate dehydrogenase complex, but the acyl group activated by CoASH is the C_4 unit, succinyl, rather than acetyl. The oxidative decarboxylation of α-ketoglutarate requires five chemical steps, three enzymes, and the same cofactors as for the pyruvate dehydrogenase complex. The purpose of this step is to collect the energy from α-ketoglutarate decarboxylation in the high-energy compound succinyl CoA.

Step 5

Succinyl coenzyme A (succinyl CoA) is a thioester compound with a free energy level greater than that of a phosphoanhydride bond ($\Delta G^{\circ\prime}$ for hydrolysis of succinyl CoA is -36 kJ/mol). In step 5 of the cycle, the energy is transformed from a thioester bond to the form of ATP or GTP. Higher animals have two isoenzymic forms of **succinyl-CoA synthetase,** one that prefers ADP as a phosphoryl group acceptor and one that prefers GDP. Plants and microorganisms make only ATP in this reaction.

This step illustrates the generation of ATP (or GTP) by **substrate-level phosphorylation.** Most ATP in the cell is produced by **oxidative phosphorylation:** the oxidation of reduced cofactors (NADH, $FADH_2$) with transfer of electrons to the final electron acceptor of metabolism, O_2, and release of energy to drive the reaction ADP + P_i $\rightleftharpoons$ ATP. This latter process is *indirectly* linked to substrate oxidation and occurs as the last stage of aerobic metabolism (see Figure 14.6) and after electrons have been collected in NADH and $FADH_2$. Substrate-level phosphorylation of ADP (or GDP) is *directly* coupled with a substrate or metabolite reaction.

Step 6

The succinate product from step 5 is oxidized to fumarate in a reaction catalyzed by **succinate dehydrogenase,** an enzyme with an FAD prosthetic group.[1] It also contains an iron–sulfur cluster similar to that of aconitase. This is our first encounter with the formation

[1]Some biochemists argue that ubiquinone rather than FAD should be shown as the final election acceptor. See study exercise 16.32.

of a carbon–carbon double bond by oxidation. The cofactor for such redox reactions is FAD, whereas NAD^+ usually functions in redox reactions interconverting hydroxyl and carbonyl functional groups. The compound malonate, although similar in structure to succinate, is very toxic to organisms (Figure 16.9). It acts as a competitive inhibitor of succinate dehydrogenase, thereby blocking the citric acid cycle and aerobic metabolism.

Step 7

The unsaturated dicarboxylic acid fumarate becomes a substrate for the addition of water. The enzyme **fumarate hydratase,** more commonly called **fumarase,** catalyzes the stereospecific addition of water to the double bond, forming only the L stereoisomer of malate.

Step 8

Oxaloacetate is formed in the last step by oxidation of the hydroxyl group of L-malate. The enzyme, **malate dehydrogenase** is linked to the coenzyme NAD^+.

Adding these eight steps results in the net reaction for the citric acid cycle:

$$\text{Acetyl CoA} + 3\,NAD^+ + FAD + \text{ADP or GDP} + P_i + 2\,H_2O \longrightarrow$$
$$2\,CO_2 + 3\,NADH + 3\,H^+ + FADH_2 + \text{ATP or GTP} + CoA$$

Summary of the Citric Acid Cycle

To summarize the accomplishments of the cycle:

1. Acetate, C_2, enters as acetyl CoA and two carbons leave the cycle in two separate reactions as CO_2.
2. Three molecules of NAD^+ are reduced by dehydrogenase-catalyzed reactions to NADH.
3. One molecule of FAD is reduced to $FADH_2$.
4. A phosphoanhydride bond in ATP or GTP is generated from energy stored in a CoA thioester.

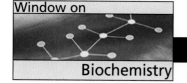

Window on

Biochemistry

16.2   Fluoroacetate—Blocking the Citric Acid Cycle

The enzyme aconitase is one of the target sites for the toxic action of **fluoroacetate (FA).** FA is a natural product found in many South American, African, and Australian plants and is a minor component in locoweed (see photo) found in the western United States. The salt form of the chemical, sodium fluoroacetate sold as "Compound 1080," is used by ranchers throughout the world to control coyotes, rabbits, and fox. Most mammals are fatally poisoned by less than 2 mg per kilogram of body weight. Poisoned animals initially show signs resembling alcohol intoxication, but they soon experience convulsions that usually result in death. Use of the commercial biocide, Compound 1080, and ingestion of toxic plants also causes the death of domestic animals like cattle and horses as well as eagles and other endangered species.

Fluoroacetate is not itself toxic to cells; however, enzymes treat it like acetate and convert it by normal pathways to the substance **fluorocitrate,** which is highly toxic. Because of this lethal modification of fluoroacetate, it is sometimes called a **suicide substrate.**

Acetyl-CoA
synthase   Citrate
                                            synthase

COO⁻
| CoASH
CH₂F
Fluoroacetate

O
‖
C—SCoA
|
CH₂F
Fluoroacetyl CoA

CoASH
oxaloacetate

CH₂COO⁻
|
HO—C—COO⁻
|
CHF—COO⁻
Fluorocitrate

$$
\begin{array}{c}
COO^- \\
| \\
CH_2 \\
| \\
CH_2 \\
| \\
COO_- \\
\text{Succinate}
\end{array}
\qquad
\begin{array}{c}
COO^- \\
| \\
CH_2 \\
| \\
COO_- \\
\text{Malonate} \\
\text{(competitive inhibitor of} \\
\text{succinate dehydrogenase)}
\end{array}
$$

Figure 16.9
Structures of succinate, the normal substrate for succinate dehydrogenase, and a competitive inhibitor of succinate dehydrogenase, malonate. Malonate is toxic because it blocks a reaction in the citric acid cycle. For other inhibitors, see Window on Biochemistry 16.2.

We have discussed only the oxidation of pyruvate as a source of acetyl CoA for the citric acid cycle. The catabolism of fatty acids by β oxidation generates large quantities of acetyl CoA. The catabolism of the amino acids leucine, isoleucine, lysine, phenylalanine, tryptophan, and tyrosine also forms acetyl CoA.

The citric acid cycle, like glycolysis, produces substrate-level ATP (or GTP) by collecting energy produced by oxidation reactions. Both glycolysis and the citric acid cycle, however, produce low levels of ATP. Table 16.3 summarizes the production of ATP, NADH, and $FADH_2$ by the combination of the pyruvate dehydrogenase complex and the citric acid cycle. Much more ATP is produced during the recycling (oxidation) of NADH and $FADH_2$ by electron transport and oxidative phosphorylation.

We have viewed the citric acid cycle as a catalytic unit whose purpose is to degrade acetyl CoA. During the functioning of the cycle in acetyl CoA degradation, there is no net loss of oxaloacetate (or other intermediate, for that matter). Oxaloacetate used to bring in an acetyl CoA is always regenerated and can assist the entry of another acetyl CoA. The catalytic cycle continues as long as there is a supply of acetyl CoA, NAD^+, FAD, and ADP (or GDP). Citric acid cycle molecules also are used in anabolic processes. Some intermediates, especially citrate, α-ketoglutarate, succinyl CoA, and oxaloacetate, are withdrawn from the cycle to be used as synthetic precursors.

Fluorocitrate displays two physiological actions as a toxin: (1) It is a potent inhibitor of the citric acid cycle enzyme, aconitase, and (2) it blocks citrate transport in and out of the mitochondria. Fluoroacetate/fluorocitrate intoxication is characterized by elevated citrate levels in the heart, kidney, and brain.

© Farrell Grehan/Photo Researchers

Table 16.3

The ATP, NADH, and $FADH_2$ balance sheet for the pyruvate dehydrogenase complex and the citric acid cycle

Reaction Number[a]	Reaction	ATP (GTP) Change[b]	NADH Change[b]	$FADH_2$ Change[b]
		(per pyruvate)		
	Pyruvate oxidation			
	Pyruvate dehydrogenase complex	0	+1	0
	Pyruvate oxidation total	0	+1	0
	Citric acid cycle			
3	Isocitrate + NAD^+ $\rightleftharpoons$ α-ketoglutarate + CO_2 + NADH + H^+	0	+1	0
4	α-Ketoglutarate + NAD^+ + CoA $\rightleftharpoons$ succinyl CoA + CO_2 + NADH + H^+	0	+1	0
5	Succinyl CoA + GDP or ADP + P_i $\rightleftharpoons$ succinate + GTP or ATP + CoASH	+1	0	0
6	Succinate + FAD $\rightleftharpoons$ fumarate + $FADH_2$	0	0	+1
8	L-Malate + NAD^+ $\rightleftharpoons$ oxaloacetate + NADH + H^+	0	+1	0
	Citric acid cycle total	+1	+3	+1
	Grand Total	+1	+4	+1

[a] The reaction numbers correspond to the steps in Figure 16.8.

[b] A + number indicates a production of ATP, NADH, and $FADH_2$. For example, for each pyruvate that is converted to acetyl CoA by the pyruvate dehydrogenase complex, 1 NADH is formed.

16.3   The Citric Acid Cycle in Regulation and Biosynthesis

Regulation of Aerobic Pyruvate Metabolism

We have seen that the pathways of carbohydrate metabolism are carefully coordinated and regulated. This guarantees that metabolic products are available when needed and it avoids excessive energy use or overproduction of metabolites. In a similar fashion, the entry of pyruvate and acetyl CoA into aerobic metabolism is under strict control. To study regulation at this stage of metabolism, it is helpful to consider that glycolysis, the pyruvate dehydrogenase complex, and the citric acid cycle are a continuous sequence of reactions. In Chapter 15, we noted that pathways were usually regulated by compounds that acted as indicators of the energy state of the cell or by intermediates or final products of a pathway. Enzymes at strategic locations are sensitive to AMP, ADP, ATP, NAD^+, and NADH concentrations as well as to products of the reactions, such as acetyl CoA, citrate, and glucose 6-phosphate. The pathways introduced in this chapter are controlled by allosteric enzymes at four locations. The primary regulatory points are the pyruvate dehydrogenase complex, citrate synthase, isocitrate dehydrogenase, and the α-ketoglutarate dehydrogenase complex (Figure 16.10). The regulatory properties of the enzymes are summarized in Table 16.4.

Because of its strategic position as the gatekeeper to aerobic metabolism, we expect to see strict regulation of the pyruvate dehydrogenase complex enzyme system. Compounds that inhibit activity are primarily the energy products of aerobic metabolism (NADH and ATP). Acetyl CoA also acts as an inhibitor in an example of feedback inhibition. The action of fatty acids as reaction inhibitors will not become completely clear until Chapter 18. Fatty acids produce acetyl CoA during catabolism by β oxidation. If fatty acids are abundant, there is less need for acetyl CoA from pyruvate oxidation. The pyruvate dehydrogenase complex is stimulated by AMP, NAD^+, and CoASH.

Three enzymes in the citric acid cycle have regulatory properties:

Citrate synthase: The enzyme is allosterically inhibited by products or intermediates of the cycle, NADH, ATP, succinyl CoA, and citrate. This is a form of feedback inhibition.

Isocitrate dehydrogenase: ATP and ADP act as allosteric modulators to maintain the appropriate rate of the cycle.

α-Ketoglutarate dehydrogenase complex: As with other regulatory enzymes, we note inhibition by succinyl CoA and NADH, both products of the cycle.

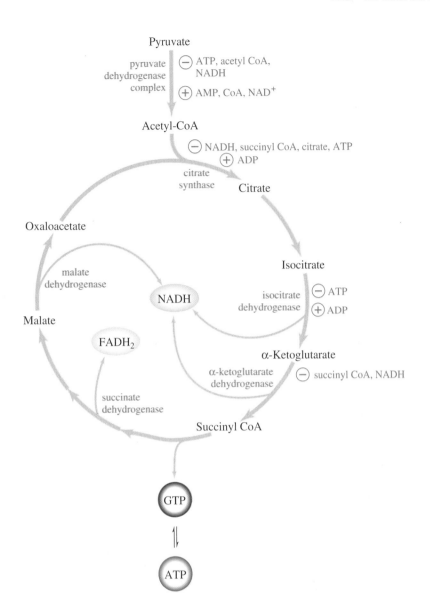

Figure 16.10

Metabolic pathways of pyruvate and acetyl CoA showing key regulatory enzymes. $\ominus$ indicates negative modulators; $\oplus$ indicates positive modulators.

In summary, we see a common thread in metabolic regulation. Catabolic reactions are slowed when the cell already contains sufficient quantities of metabolic products and high-energy compounds.

Anabolic Roles of the Citric Acid Cycle

We have focused our attention on the catabolic roles of the citric acid cycle: to degrade acetyl CoA (from pyruvate, amino acids, and fatty acids) and produce ATP, NADH, and

Table 16.4
The important regulatory enzymes of pyruvate and acetyl CoA metabolism

Enzyme Name	$\oplus$ Modulators	$\ominus$ Modulators	Comments
Pyruvate dehydrogenase complex	AMP, NAD$^+$, CoA	ATP, acetyl CoA, NADH	Also regulated by covalent modification.
Citrate synthase	ADP	NADH, succinyl CoA, citrate, ATP	Activity depends on metabolite concentrations.
Isocitrate dehydrogenase	ADP	ATP	Activity depends on metabolite concentrations.
α-Ketoglutarate dehydrogenase complex	—	Succinyl CoA, NADH	Activity depends on metabolite concentrations.

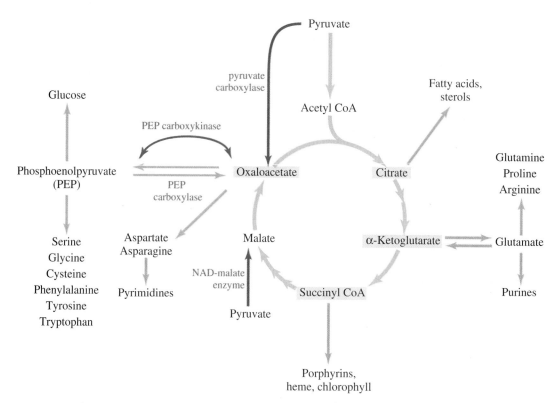

Figure 16.11

Use of citric acid cycle intermediates for synthesis of biomolecules. Molecules removed from the cycle include citrate, α-ketoglutarate, succinyl CoA, and oxaloacetate. Biomolecules synthesized from intermediates diverted from the citrate cycle include phosphoenolpyruvate (PEP), aspartate, glutamate, pyrimidines, purines, and porphyrins. Anaplerotic reactions to replenish citric acid cycle intermediates are shown with blue arrows.

FADH$_2$. Biochemical studies have uncovered other important and diverse functions for the citric acid cycle. Most significantly, the pathway has been shown to be **amphibolic;** that is, it functions in catabolism *and* anabolism. The cycle is an important source of biosynthetic precursors. Intermediates of the cycle are removed and their carbon skeletons used as starting points for the synthesis of important biomolecules, such as amino acids, nucleotides, and porphyrins (Figure 16.11). You may already have recognized that the carbon skeletons of the α-keto acids oxaloacetate and α-ketoglutarate are the same as those of the amino acids aspartate (and asparagine) and glutamate (and glutamine), respectively. The intermediate succinyl CoA is the starting point for the synthesis of protoporphyrins for use in the important pigment molecules heme and chlorophyll. We have already discussed the role of oxaloacetate in the gluconeogenesis process (see Chapter 15, Section 15.4).

The use of citric acid cycle intermediates for biosynthesis raises an important question. If these intermediates are drained away, what effect does this have on the efficiency of the citric acid cycle in catabolism? Clearly, if fewer molecules of the intermediates are present,

Table 16.5

Anaplerotic reactions to replenish the citric acid cycle intermediates oxaloacetate and malate

Enzyme	Reaction	Comments
Pyruvate carboxylase	Pyruvate + CO$_2$ + ATP + H$_2$O ⇌ oxaloacetate + ADP + P$_i$	Also the starting point for gluconeogenesis.
PEP carboxykinase	Phosphoenolpyruvate + CO$_2$ + GDP ⇌ oxaloacetate + GTP	The reverse reaction is important in gluconeogenesis.
NAD–malate enzyme	Pyruvate + CO$_2$ + NADH + H$^+$ ⇌ malate + NAD$^+$	Found in plants and microorganisms.

the cycle has a reduced capacity to accept and oxidize acetyl CoA. This problem has been solved by the evolution of reactions to replenish missing intermediates. Biochemists have discovered several reactions that resupply key intermediates. We call these reactions **anaplerotic** ("to fill in"). Important anaplerotic reactions are listed in Table 16.5. The pyruvate carboxylase reaction is present primarily in liver and kidney cells, where it is used to begin gluconeogenesis or to replenish the citric acid cycle. Two forms of phosphoenolpyruvate (PEP) carboxykinase are found in most cells, one in the cytoplasm and one in mitochondria. The anaplerotic NAD–malate enzyme, which is one of the few known to replenish malate, is especially important in higher plants and microorganisms.

16.4   The Glyoxylate Cycle

Higher plants and some microbes (*Escherichia coli, Pseudomonas,* algae) can utilize the compound acetate, CH_3COO^-, as a source of energy and as a biosynthetic precursor. They do this by making use of an auxiliary reaction pathway not present in higher animals: a citric acid cycle modified by a two-reaction shunt (the **glyoxylate cycle**) that forms succinate, a C_4 dicarboxylic acid, from two molecules of acetyl CoA. This process becomes important when acetate must serve as the sole or major source of carbon for an organism.

Reaction Steps of the Glyoxylate Cycle

Acetate in plants and some microbes must first be activated before entry into mainstream catabolic and anabolic pathways. In a reaction catalyzed by acetyl CoA synthetase, energy from ATP is used to produce the thioester bond in acetyl CoA:

$$acetate + CoA + ATP \rightleftharpoons acetyl\ CoA + AMP + PP_i$$

Again, as in protein synthesis (see Chapter 12, Section 12.2), ATP undergoes abnormal cleavage to produce AMP and pyrophosphate, PP_i. This mode of phosphoanhydride bond cleavage is used when an extra burst of energy is required to drive a reaction to completion. The extra energy comes from hydrolysis of pyrophosphate, PP_i.

$$PP_i + H_2O \xrightarrow{\text{pyrophosphatase}} 2\ P_i$$

The steps of the glyoxylate cycle and its linkage to the citric acid cycle are shown in Figure 16.12. Table 16.6 shows each reaction of the glyoxylate cycle along with a description of the reaction type. The modified cycle begins in the same way as the normal citric acid cycle (reaction 1). Acetyl CoA is accepted into the cycle by reaction with oxaloacetate. The product citrate is isomerized by dehydration and hydration to isocitrate (reaction 2). The branch to the glyoxylate cycle begins at this point. Instead of an oxidation–reduction reaction on isocitrate (as is the case for the citric acid cycle), it undergoes a cleavage reaction (reaction 3).

Isocitrate, a C_6 carbon skeleton, is split into succinate (C_4) and glyoxylate (C_2). The enzyme catalyzing this category 4 reaction (nonhydrolytic cleavage) is **isocitrate lyase.** Formally, this reaction involves the addition or elimination of functional groups to remove or form a double bond. The unusual product, **glyoxylate,** has an aldehyde and a carboxylate functional group. Succinate is expelled from the cycle and used by the plant or microbe for biosynthetic purposes. Glyoxylate continues in the cycle by combining with acetyl CoA to produce malate and CoA (reaction 4). The nonhydrolytic cleavage step (remember that an enzyme catalyzes a reaction in both directions) is catalyzed by **malate synthase.**

The net reaction for the glyoxylate cycle is

$$2\ acetyl\ CoA + NAD^+ + 2\ H_2O \longrightarrow succinate + 2\ CoA + NADH + H^+$$

It is instructive to compare this net reaction with that of the citric acid cycle. In the glyoxylate cycle, two acetyl CoAs enter, bringing in four carbons that are combined into succinate. Only one acetyl CoA enters the citric acid cycle per turn and is eventually oxidized to two CO_2. The glyoxylate cycle produces reducing power in NADH, but just one per cycle. The citric acid cycle yields three NADH and one $FADH_2$ per turn, and that with just one acetyl CoA. Finally,

Figure 16.12
Reactions of the glyoxylate cycle. The glyoxylate cycle in germinating seeds converts acetyl CoA to carbohydrates. The enzymes for the cycle are present in plant glyoxysomes, but products are transported to the mitochondrial matrix and the cytoplasm.

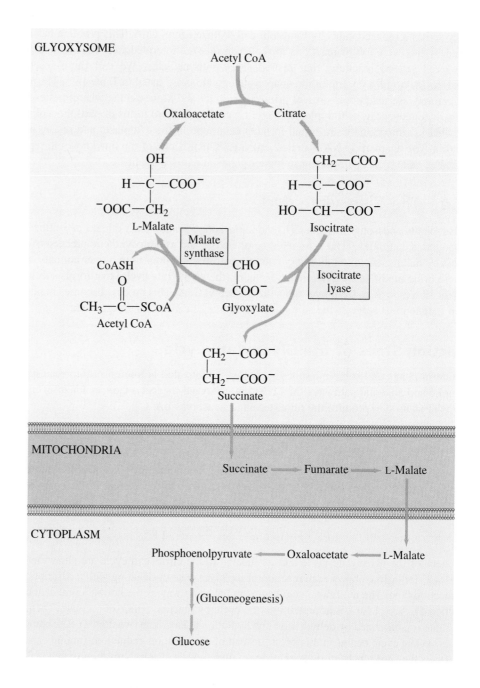

Table 16.6
The reactions of the glyoxylate cycle

Reaction Number	Reaction	Enzyme	Reaction type[a]
1	Acetyl CoA + oxaloacetate + H_2O ⇌ citrate + CoA	Citrate synthase	3, 4
2	Citrate ⇌ isocitrate	Aconitase	4
3	Isocitrate ⇌ succinate + glyoxylate	Isocitrate lyase	4
4	Glyoxylate + acetyl CoA + H_2O ⇌ malate + CoA	Malate synthase	4
5	Malate + NAD^+ ⇌ oxaloacetate + NADH + H^+	Malate dehydrogenase	1

[a]Reaction type: 1, oxidation–reduction; 2, phosphoryl group transfer; 3, hydrolysis; 4, nonhydrolytic cleavage (addition–elimination); 5, isomerization–rearrangement; 6, bond formation coupled to ATP cleavage.

the glyoxylate cycle generates no energy in the form of phosphoanhydride bonds; the citric acid cycle yields one ATP or GTP per turn. Clearly the two pathways have different characteristics and serve different biochemical functions. In organisms possessing both the glyoxylate cycle and the citric acid cycle, isocitrate is at a metabolic junction. This intermediate can undergo oxidative decarboxylation and proceed through the reactions of the citric acid cycle.

The glyoxylate cycle is especially important in the seeds of higher plants. This set of reactions allows them to use energy stored in seed oil (triacylglycerols) to make carbohydrates. Plant seeds have specialized cell organelles called **glyoxysomes** that contain the enzymes isocitrate lyase and malate synthase in addition to the usual citric acid cycle enzymes. Thus, the plant seed is able to convert acetyl CoA (from fatty acid degradation) to carbohydrates, something that is impossible for animals. Succinate produced by the glyoxylate cycle leaves the glyoxysome and enters the mitochondrial matrix, where it is converted to oxaloacetate (by citric acid cycle enzymes); this can then be used for gluconeogenesis (see Chapter 15, Section 15.4). Note that these processes require the transfer of metabolites across compartment membranes (see Figure 16.12).

Sunflower seeds use the glyoxylate cycle to make carbohydrates from fats.

16.5 The Phosphogluconate Pathway

The glycolytic pathway accounts for the great majority of glucose oxidized by animals. The products of glycolysis are energy in the form of ATP and NADH and the three-carbon compound pyruvate, which continues in aerobic metabolism via continued oxidation, the citric acid cycle, and cellular respiration. Some animal cells possess a secondary pathway for glucose metabolism that leads to specialized products. The **phosphogluconate pathway** (also called the pentose phosphate pathway) oxidizes glucose to produce the pentose ribose 5-phosphate and reducing power in the form of NADPH. Ribose 5-phosphate is a precursor for synthesis of nucleotides, nucleic acids, and several enzyme cofactors. NADPH, which has a structure similar to NADH (Figure 16.13), is required for reductive reactions in biosynthetic processes in tissues responsible for fatty acid and steroid formation. These tissues include adipose tissue (fat cells), mammary gland, the liver, and

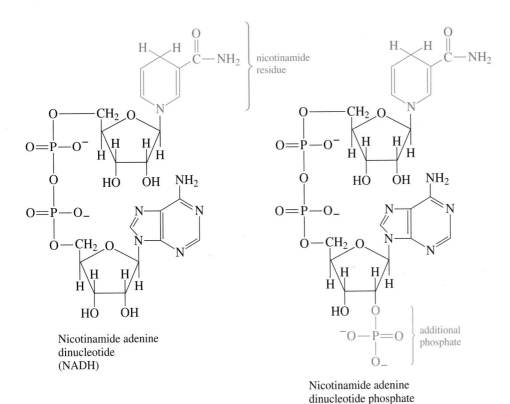

Nicotinamide adenine dinucleotide (NADH)

Nicotinamide adenine dinucleotide phosphate (NADPH)

Figure 16.13

Structure for the dehydrogenase cofactors, NADH and NADPH.

the adrenal cortex. This secondary pathway of glucose metabolism is also important in erythrocytes, which rely on NADPH for proper function. The phosphogluconate pathway is not present in skeletal muscle.

The transformation of glucose by the phosphogluconate pathway to ribose 5-phosphate takes place by a five-reaction oxidative process. The structural intermediates for each step are shown in Figure 16.14. Table 16.7 contains a list of the reactions, the enzyme names, and the type of chemistry for each step. The pathway is similar to glycolysis in that it is characterized by oxidation–reduction processes; however, the outcome is very different from glycolysis. The net reaction for the phosphogluconate pathway beginning with glucose is

$$\text{glucose} + \text{ATP} + 2\ \text{NADP}^+ + \text{H}_2\text{O} \longrightarrow$$
$$\text{ribose 5-phosphate} + \text{CO}_2 + 2\ \text{NADPH} + 2\ \text{H}^+ + \text{ADP}$$

Only one ATP is required for glucose activation, and all carbons in glucose are accounted for by ribose 5-phosphate and CO_2. The purpose of the pathway is to produce the specialized products, ribose 5-phosphate and two molecules of NADPH per glucose that enters.

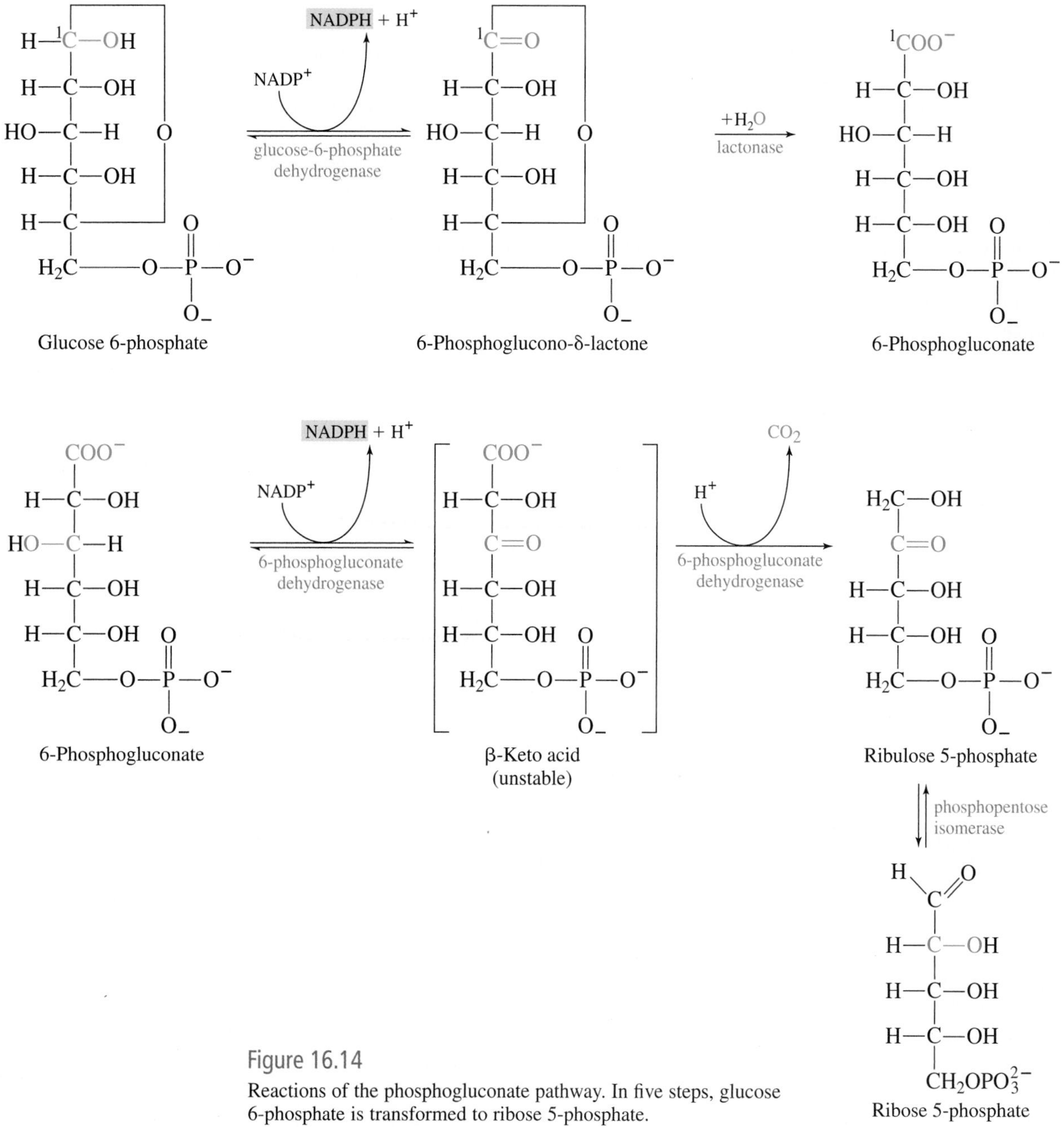

Figure 16.14

Reactions of the phosphogluconate pathway. In five steps, glucose 6-phosphate is transformed to ribose 5-phosphate.

Table 16.7

The reactions of the phosphogluconate pathway

Reaction Number	Reaction	Enzyme[a]	Reaction Type[b]
1	Glucose + ATP $\rightleftharpoons$ glucose 6-phosphate + ADP	Hexokinase	2
2	Glucose 6-phosphate + NADP$^+$ $\rightleftharpoons$ 6-phosphoglucono-δ-lactone + NADPH + H$^+$	Glucose-6-phosphate dehydrogenase	1
3	6-Phosphoglucono-δ-lactone + H$_2$O $\rightleftharpoons$ 6-phosphogluconate	Lactonase	3
4	6-Phosphogluconate + NADP$^+$ $\rightleftharpoons$ D-ribulose 5-phosphate + NADPH + H$^+$ + CO$_2$	6-Phosphogluconate dehydrogenase	1, 4
5	D-Ribulose 5-phosphate $\rightleftharpoons$ ribose 5-phosphate	Phosphopentose isomerase	5

[a]Enzymes are listed by common names.
[b]Reaction type: 1, oxidation–reduction; 2, phosphoryl transfer; 3, hydrolysis; 4, nonhydrolytic cleavage (addition or elimination); 5, isomerization–rearrangement; 6, bond formation coupled to ATP cleavage.

The phosphogluconate pathway ends with these five reactions in some tissue, but in others it continues in a nonoxidative mode to make fructose 6-phosphate, glucose 6-phosphate, and glyceraldehyde 3-phosphate. These latter reactions link the phosphogluconate pathway to glycolysis. Regulation of the phosphogluconate pathway is simply by product demand; removal of NADPH regenerates NADP$^+$ and thus pulls the pathway.

Approximately 10% of the individuals of African or Mediterranean descent have a genetic deficiency of the phosphogluconate pathway enzyme, glucose-6-phosphate dehydrogenase (reaction 2 in Table 16.7 and Figure 16.14). Erythrocytes are especially dependent on NADPH from the pathway to maintain the tripeptide glutathione in a reduced state:

$$\gamma\text{-Glu-Cys-Gly} + \text{NADPH} + \text{H}^+ \rightleftharpoons 2\,\gamma\text{-Glu-Cys-Gly} + \text{NADP}^+$$

(oxidized glutathione with S–S bridge between two γ-Glu-Cys-Gly chains; reduced glutathione with SH)

Glutathione (oxidized)   Glutathione (reduced)

Reduced glutathione maintains hemoglobin in the reduced Fe(II) state necessary for oxygen binding (see Chapter 5, Section 5.5). If the erythrocytes in these affected individuals with low glucose 6-phosphate dehydrogenase are stressed with drugs, such as the antimalarial primaquine, they undergo massive destruction (hemolytic anemia). The genetic deficiency of glucose 6-phosphate dehydrogenase does have a positive side: It provides a resistance to malaria for the affected individuals.

Summary

■ Energy released from the degradation of carbohydrates and other substrates is stored in ATP and in reduced cofactors. The cofactors NAD$^+$, NADP$^+$, and FAD serve in redox reactions to assist in the oxidation of substrates. The reduced cofactors are recycled by the electron transport process linked to respiration.

■ NADH is also generated by the pyruvate dehydrogenase complex, which oxidizes pyruvate to acetyl CoA.

■ The central pathway of aerobic metabolism is the **citric acid cycle.** This pathway degrades the two carbons in acetyl CoA to CO$_2$ and produces NADH, FADH$_2$, and GTP.

■ The citric acid cycle also plays important anabolic roles. Intermediates such as oxaloacetate and alpha-ketoglutarate are removed and used to synthesize the amino acids aspartate and glutamate, respectively. Succinyl CoA is the starting point for the synthesis of protoporphyrin.

■ Some organisms, especially higher plants and microorganisms, have a modified citric acid cycle called the **glyoxylate cycle.**

■ Animals possess a specialized pathway for glucose metabolism—the **phosphogluconate pathway.** This process converts glucose to ribose 5-phosphate and produces NADPH for fatty acid synthesis.

Study Exercises

Understanding Terms

16.1 Define the following terms in 25 words or less:

a. Mitochondrion
b. Lipoamide
c. FAD
d. Beriberi
e. Amphibolic
f. Succinyl CoA
g. Substrate-level phosphorylation
h. Anaplerotic reactions
i. Fluorocitrate
j. Glyoxysomes
k. Lipoic acid
l. Acetyl CoA
m. Phosphogluconate pathway
n. Glyoxylate
o. Ribulose 5-phosphate
p. Thioester bond
q. Multienzyme complex
r. Ribose 5-phosphate

Reviewing Concepts

16.2 Match the correct biochemical function with each of the cofactors listed.

Cofactor	Biochemical Function
1. NAD$^+$	a. Carrier of activated acyl groups
2. FAD	b. Carrier of activated CO_2
3. CoASH	c. Oxidation of hydroxyl groups
4. Lipoamide	d. Reductive reactions for fatty acid synthesis
5. TPP	e. Oxidation to form carbon–carbon double bonds
6. Biotin	f. Carrier of activated acyl groups coupled with oxidation–reduction
7. NADH	g. Reduction of carbon–carbon double bonds
8. FADH$_2$	h. Reduction of carbonyl groups
9. NADPH	i. Decarboxylation of α-keto acids

16.3 Calculate the total amount of substrate-level ATP (or GTP), NADH, and FADH$_2$ produced from the complete degradation of glucose by glycolysis, the pyruvate dehydrogenase complex, and the citric acid cycle.
HINT: Use Tables 15.2 and 16.3.

16.4 Write out, using structures, the reaction catalyzed by NAD–malate enzyme. Describe the oxidation–reduction that takes place in the reaction.

$$\text{pyruvate} + CO_2 + \text{NADH} + H^+ \rightleftharpoons$$
$$\text{malate} + \text{NAD}^+$$

16.5 Name the intermediate from the citric acid cycle that is the starting precursor for each of the compounds listed below. Choose from the following citric acid cycle intermediates: oxaloacetate, α-ketoglutarate, succinyl CoA.

a. Aspartate
b. Porphyrin for heme
c. Asparagine
d. Glutamine
e. Glucose
f. Glutamate
g. Phosphoenolpyruvate
h. Uracil

16.6 Match each of the enzymes listed with its regulatory properties.

Enzyme	Regulation
1. Pyruvate dehydrogenase complex	a. Allosteric activation by ADP; inhibited by NADH, succinyl CoA
2. Citrate synthase	b. Inhibited by succinyl CoA, ATP
3. Isocitrate dehydrogenase	c. Stimulated by AMP, CoA
4. α-Ketoglutarate dehydrogenase complex	d. Stimulated by ADP, inhibited by ATP

16.7 Prepare a balance sheet for the reactions of the glyoxylate cycle.
HINT: Use Table 16.3 as a model.

16.8 Study the following reaction, the hydrolysis of pyrophosphate:

$$PP_i + H_2O \rightarrow 2\ P_i$$

a. Write out the reaction using structures.
b. What type of bond is cleaved?
c. When is this reaction important in metabolism?

16.9 What is the cellular location of the enzymes of the citric acid cycle?

a. Nucleus
b. Cytoplasm
c. Mitochondria
d. Ribosomes

16.10 Discuss the similarities and differences of the pyruvate dehydrogenase complex and the α-ketoglutarate dehydrogenase complex.

16.11 Prepare an energy balance table (like that in Table 15.2) for the phosphogluconate pathway beginning with glucose.

16.12 List the major differences and similarities of glycolysis (glucose $\rightarrow$ pyruvate) and the phosphogluconate pathway (glucose $\rightarrow$ ribose 5-phosphate).

16.13 Define each of the following metabolic steps according to the types of chemistry used in Table 14.2.

a. Ribulose 5-phosphate $\rightarrow$ ribose 5-phosphate
b. 6-Phosphogluconate + NADP$^+$ $\rightarrow$ ribulose 5-phosphate + NADPH + H$^+$ + CO$_2$

16.14 Describe the structural difference between NADH and NADPH.

16.15 In what part of the cell does each of the following metabolic processes take place? Select from cytoplasm or mitochondrial matrix.

a. Glycolysis _____
b. Citric acid cycle _____
c. Phosphogluconate pathway _____
d. Gluconeogenesis; oxaloacetate $\rightarrow$ glucose _____
e. Gluconeogenesis; pyruvate $\rightarrow$ malate _____

16.16 Describe the functional differences between NADH and NADPH.

16.17 List the enzymes in the citric acid cycle that are in the class of oxidoreductases.

16.18 List the enzymes in the phosphogluconate pathway that are oxidoreductases.

Working Problems

16.19 Explain the advantage to the cell of each of the following regulatory actions:

   a. The pyruvate dehydrogenase complex is inhibited by ATP.
   b. Isocitrate dehydrogenase is stimulated by ADP.
   c. Pyruvate carboxylase is stimulated by acetyl CoA.
   d. Citrate synthase is inhibited by NADH.
   e. The pyruvate dehydrogenase complex is inhibited by fatty acids.

16.20 Derive the net reaction for each pathway:

   a. Pyruvate $\rightarrow$ 3 CO_2
   b. Acetyl CoA $\rightarrow$ 2 CO_2
   c. Acetyl CoA $\rightarrow$ succinate (glyoxylate cycle)
   d. Glucose 6-phosphate $\rightarrow$
   $$\text{ribose 5-phosphate} + CO_2$$

HINT: For (a), pyruvate dehydrogenase complex plus the citric acid cycle.

16.21 Outline a pathway showing how some microorganisms can use ethanol as a sole source of carbon.

16.22 Calculate the total yield of substrate-level ATP (or GTP), NADH, and $FADH_2$ produced from the complete degradation of pyruvate by the pyruvate dehydrogenase complex and the citric acid cycle.

16.23 What enzymes and metabolic reactions would be slowed in an individual with beriberi?

16.24 Draw the structures for two other compounds besides malonate that may be competitive inhibitors of succinate dehydrogenase.

16.25 Complete the following oxidation–reduction reactions. Assume that the appropriate enzyme is present in each case.

   a. $CH_3CH_2OH + NAD^+ \rightleftharpoons$
   b. $FADH_2 + NAD^+ \rightleftharpoons$
   c.
   $$\text{lipoamide} \overset{SH}{\underset{SH}{\diagdown}} + FAD \rightleftharpoons$$
   d. $^-OOCCH_2CH_2COO^- + FAD \rightleftharpoons$

16.26 Identify two reactions in glycolysis that are examples of substrate-level phosphorylation of ADP.

16.27 Consider the following reaction from the glyoxylate cycle:

$$
\begin{array}{c}
{}^*COO^- \\
| \\
{}^*C \\
{\diagup}\,{\diagdown} \\
H \quad O
\end{array}
+
\begin{array}{c}
CH_3 \\
| \\
C=O \\
| \\
SCoA
\end{array}
+ H_2O \rightleftharpoons
\begin{array}{c}
COO^- \\
| \\
CHOH \\
| \\
CH_2 \\
| \\
COO_-
\end{array}
+ CoASH
$$

Assume that the two carbons of glyoxylate are labeled with radioactive $^{14}C^*$. What carbon(s) of product malate would be labeled with ^{14}C?

16.28 The net reaction for the citric acid cycle shows the entry of two water molecules. In what specific reactions of the cycle are these used and what kind of chemistry occurs in each step?

16.29 Ribulose 5-phosphate and ribose 5-phosphate are important molecules in the phosphogluconate pathway. What is the structural relationship between these compounds?

   a. Enantiomers
   b. Epimers
   c. Anomers
   d. Structural isomers
   e. No relationship

16.30 NADPH is an inhibitor of the phosphogluconate pathway. What is the metabolic logic behind this regulatory process?

Writing Biochemistry

16.31 Fluoroacetate poisoning is characterized by elevated cellular levels of citrate. Write a 100-word paragraph explaining how the higher concentrations occur.

16.32 Some biochemists argue that the succinate dehydrogenase reaction should be written using ubiquinone (UQ, oxidized form) and ubiquinol (UQH_2, reduced form), rather than FAD and $FADH_2$. [See reference by S. Brown, *Biochem. Educ.* 28:52–54 (2000).]

$$\text{Succinate} + UQ \rightleftharpoons \text{Fumarate} + UQH_2$$

Read the reference and write a 200 to 300-word summary of the main points of the article.

Applying Biochemistry

16.33 Propose some possible regulatory processes for control of the glyoxylate cycle. For help, review the regulation of the citric acid cycle in this chapter and the CD/ROM unit on the cycle.

16.34 Outline a sequence of reactions that illustrate the anaplerotic synthesis of citrate from pyruvate.

HINT: Begin with the enzyme pyruvate carboxylase.

16.35 Your instructor has assigned you the task of preparing a 15-minute class presentation on the biochemistry of alcohol abuse. Review the CD/ROM unit on the topic and prepare an outline for your talk.

Further Reading

Behal, R., Buxton, D., Robertson, V., and Olson, M., 1993. Regulation of the pyruvate dehydrogenase multienzyme complex. *Annu. Rev. Nutr.* 13: 497–520.

Bentley, R., 2000. From reactive C_2 units to acetyl coenzyme A. *Trends Biochem. Sci.* 25:302–305.

Brown, S., 2000. Does succinate oxidation yield $FADH_2$ or ubiquinol? *Biochem. Educ.* 28:52–54.

Capaldi, R., 2000. The changing face of mitochondrial research. *Trends Biochem. Sci.* 25(5):212–214.

Elliott, K., 1999. Teaching metabolism for the future—Concepts not pathways. *The Biochemist* 21(1):9–39.

Fell, D., 1996. *Understanding the Control of Metabolism.* London: Portland Press.

Frayn, K., 1996. *Metabolic Regulation: A Human Perspective.* London: Portland Press.

Frey, T., 2000. The internal structure of mitochondria. *Trends Biochem. Sci.* 25(7):319–324.

Halliwell, B., 1999. Vitamin C: poison, prophylactic or panacea? *Trends Biochem. Sci.* 24:255–259.

Manchester, K., 1998. Albert Szent Gyorgyi and the unraveling of biological oxidation. *Trends Biochem. Sci.* 23:37–40.

Raha, S., and Robinson, B., 2000. Mitochondria, oxygen free radicals, disease, and ageing. *Trends Biochem. Sci.* 25:502–508.

Tariq, V., Stefani, L., and Butcher, A., 1995. Citric acid production. *Biochem. Educ.* 23:145–148.

Vulliamy, T., Mason, P., and Luzzatto, L., 1992. The molecular basis of glucose 6-phosphate deficiency. *Trends Genet.* 8:138–143.

Chapter 17

ATP Formation by Electron-Transport Chains

This German stamp honors the biochemist Otto Warburg who developed methods for studying biological oxidation processes.

In the previous three chapters we have studied fundamentals of the early stages of metabolism. The oxidation steps in catabolism have been defined as stepwise processes designed to remove electrons from substrates and collect them in the reduced cofactors, NADH, NADPH, and $FADH_2$. Another important event is the substrate-level formation of ATP. During the study of each metabolic pathway we have kept a tally of these important metabolic products (see Tables 15.2 and 16.3). For example, the oxidation of glucose by glycolysis, pyruvate oxidation, and the citric acid cycle generates a total of ten NADH, two $FADH_2$, and four substrate-level phosphoanhydride bonds (two ATP and two GTP) while each of the six carbon atoms of glucose is oxidized to CO_2. We will see in Chapter 18 that fatty acids travel the β-oxidation pathway leading to acetyl CoA, which enters the citric acid cycle to oxidize each carbon atom to CO_2 with production of NADH, $FADH_2$, and substrate-level ATP (or GTP).

Only a small amount of energy in the form of reactive phosphoanhydride bonds (ATP or GTP) is extracted from fuel molecules during the transfer of electrons from substrates to cofactors in stages I and II and the citric acid cycle. The remainder of the oxidative energy from food resides in the form of electrons with high transfer potential in the reduced cofactors, NADH and $FADH_2$. The energy in the reduced cofactors is recovered by employing molecular oxygen as a terminal electron acceptor (oxidizing agent). The participation of O_2 allows more complete oxidation of substrates than is possible in anaerobic metabolism. The electrons from NADH and $FADH_2$ are not passed directly to O_2 but are transferred through a series of electron carriers that undergo reversible reduction and oxidation. The series of carriers located in the inner mitochondrial membrane of plant and animal cells is called an **electron-transport chain.** Electron transport results in the availability of large amounts of free energy. Part of the energy released by electron transport is used to pump protons through the inner mitochondrial membrane, thus generating a transmembrane proton gradient. Energy from the gradient drives the synthesis of ATP ($ADP + P_i \rightleftharpoons ATP + H_2O$). The chemical reaction of phosphorylation is catalyzed by the enzyme complex **ATP synthase.** The combined processes of electron transport and ATP synthesis (called **oxidative phosphorylation**) are the final stages of aerobic metabolism (Figure 17.1).

Electron-transport chains are universal metabolic features found in all aerobic organisms, including plants. Oxidative phosphorylation occurs in the mitochondria of plants for catabolic metabolism; however, plants use another type of electron-transport chain to provide energy for photosynthesis. Photosynthetic organisms use light-absorbing pigments to collect energy from the sun and specialized electron-transport chains to drive electrons to a higher energy level. A goal of photosynthesis is to generate ATP and reducing power as NADPH in order to make glucose by fixing CO_2. The molecular system for photosynthesis is present in plant organelles called chloroplasts. In bacteria, ATP is synthesized by energy coupling of electron-transport chains in the cell membrane.

17.1   Mitochondrial Electron Transport

Our primary objective in previous chapters of Part IV was to outline the catabolism of carbohydrates. After preliminary reactions in stages I and II, the catabolic pathways of these primary nutrients converge to stage III, the citric acid cycle, electron transport, and oxidative phosphorylation. The important metabolic processes of stage III occur in mitochondria (see Figure 16.2). The components of the biochemical apparatus for respiration (uptake of O_2, electron transport, and ATP synthesis) are present as integral proteins in the folded inner membrane or as peripheral proteins on the membrane. The extensive folding of the membrane provides a large surface area on the matrix side for placement of many molecular systems for high levels of ATP production. Components of the electron-transport chain are arranged in the membrane in packages called **respiratory assemblies.** The inner membrane also possesses knoblike spheres projecting from its surface into the matrix. These protrusions, called F_1 particles, contain the components of the ATP synthase complex and are the sites for coupling electron transport and ATP production. Mitochondria are complete biochemical factories that

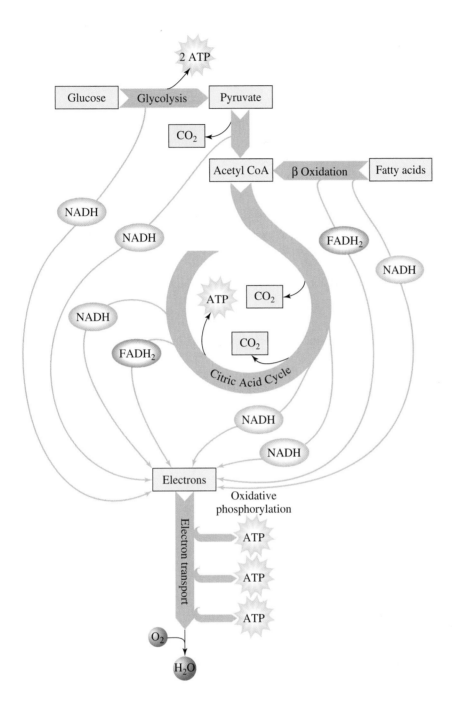

Figure 17.1

Convergence of preliminary metabolic pathways into aerobic metabolism. Pyruvate from glycolysis is decarboxylated to acetyl CoA, which enters the citric acid cycle. Acetyl CoA is also produced by the oxidation of fatty acids. NADH and $FADH_2$, generated in the pathways, bring electrons to the electron-transport chain. Here their electrons are used to reduce O_2 to water. The oxidized cofactors NAD^+ and FAD return to be reduced by nutrient oxidation. ATP is generated during electron transport. All entering nutrients including glucose and fatty acids are oxidized to CO_2 and H_2O.

can independently carry out respiration if provided with O_2, ADP, P_i, reducible cofactors (NAD^+ and FAD), and oxidizable substrates such as pyruvate, fatty acids, and citric acid cycle intermediates.

Electron Transport and Oxidative Phosphorylation

Now that we have described the cellular location of aerobic metabolism, we turn to a discussion of the biochemical components and characteristics of electron transport and oxidative phosphorylation. Electrons removed from nutrients and metabolic intermediates during oxidative reactions are transferred by dehydrogenases to the cofactors NAD^+ and FAD. The general reaction displaying this oxidation–reduction process is

$$AH_2 \;+\; NAD^+ \;\xrightleftharpoons{\text{dehydrogenase}}\; A \;+\; NADH \;+\; H^+$$

Table 17.1

Reactions catalyzed by NAD- and FAD-linked dehydrogenases

NAD Linked

Glyceraldehyde 3-phosphate + P_i + NAD^+ $\rightleftharpoons$ 1,3-bisphosphoglycerate + NADH + H^+

Pyruvate + CoA + NAD^+ $\rightleftharpoons$ acetyl CoA + CO_2 + NADH + H^+

Isocitrate + NAD^+ $\rightleftharpoons$ α-ketoglutarate + NADH + H^+ + CO_2

α-Ketoglutarate + CoA + NAD^+ $\rightleftharpoons$ succinyl CoA + CO_2 + NADH + H^+

Malate + NAD^+ $\rightleftharpoons$ oxaloacetate + NADH + H^+

FAD Linked

Succinate + FAD $\rightleftharpoons$ fumarate + $FADH_2$

where AH_2 and A represent reduced and oxidized metabolites from glycolysis, pyruvate dehydrogenation, and the citric acid cycle. A similar process occurs with FAD, where BH_2 and B represent metabolites from the citric acid cycle and other pathways:

$$BH_2 + FAD \xrightleftharpoons{\text{dehydrogenase}} B + FADH_2$$

Important dehydrogenase-catalyzed reactions from various pathways are reviewed in Table 17.1.

Thus, electrons from substrates are collected in NADH and $FADH_2$. Cells contain only a limited supply of the oxidized cofactors NAD^+ and FAD. Their reduced products must be recycled if metabolism is to be continuous. Recycling is accomplished by oxidation brought about by transfer of electrons from NADH and $FADH_2$ to the terminal electron acceptor, O_2:

$$NADH + H^+ + \tfrac{1}{2}O_2 \xrightarrow{\quad\overset{ADP + P_i \quad ATP}{\frown}\quad} NAD^+ + H_2O$$

$$FADH_2 + \tfrac{1}{2}O_2 \xrightarrow{\quad\underset{ADP + P_i \quad ATP}{\frown}\quad} FAD + H_2O$$

There are two important consequences shown in these reactions:

1. The oxidized cofactors that are reformed (NAD^+ and FAD) are thus available for continued oxidative metabolism.
2. Energy released during electron transport is coupled to ATP synthesis.

The Electron-Transport Chain

Although the above reactions for cofactor oxidation are rather simplistic, they do show how the two processes of electron transport and ATP synthesis are combined. The reactions are written as single steps, but each represents a net reaction obtained by addition of several reactions. In fact, electrons are shuttled from NADH and $FADH_2$ to O_2 through a series of carriers called the respiratory electron-transport chain (Figure 17.2). The chain is composed of four large protein complexes called NADH-coenzyme Q reductase (complex I), succinate-coenzyme Q reductase (complex II), cytochrome *c* reductase (complex III), and cytochrome *c* oxidase (complex IV). Most of the carriers within each complex are integral membrane proteins with prosthetic groups that can accept and donate electrons. In other words, the carriers can undergo reduction and oxidation reactions.

Two characteristics of the electron-transport chain are of special importance:

1. The order of the electron carriers in the chain
2. The quantity of energy available upon oxidation of NADH and $FADH_2$

The carriers are arranged in order of increasing electron affinity; hence, electrons can flow spontaneously from one carrier to the next. (The electrons flow "downhill" in terms of energy.) Electrons in NADH are at the highest energy level of any among the carriers. Stated in different terms, NADH is a strong reducing agent (in fact, the strongest of the carriers). Electrons from NADH pass to complex I (NADH-CoQ reductase), which is composed of

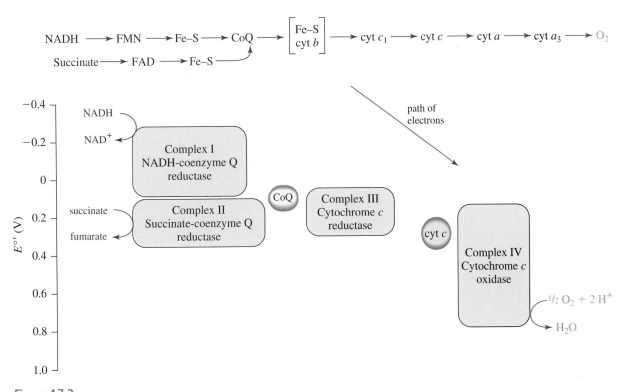

Figure 17.2
The four complex carriers in the mitochondrial electron-transport chain. The relative energy level of each is noted on the scale of $E^{\circ\prime}$ to the left. $E^{\circ\prime}$ refers to the standard reduction potential measured in volts. The four complexes transfer electrons from reduced cofactors NADH and FADH$_2$ to molecular O$_2$. The flow of electrons is shown in the reaction sequence at the top. CoQ refers to coenzyme Q, which exists in two forms: ubiquinone, the oxidized form, and ubiquinol, the reduced form.

proteins with FMN and iron–sulfur (Fe–S) clusters. FADH$_2$ is oxidized by Fe–S clusters that are part of succinate-CoQ reductase (complex II) and also acyl CoA dehydrogenase. Thus electrons from the two cofactors enter the electron-transport chain through separate branches that converge at CoQ. Electrons from reduced CoQ (CoQH$_2$) are transported by complex III to cytochrome *c* (via cytochrome *c* reductase) and finally through cytochrome *c* oxidase (complex IV) to O$_2$. (The order of carriers also represents their physical proximity to one another in the inner membrane of the mitochondria.)

The relative energy levels of the electron chain carriers are quantified by using **standard reduction potentials,** $E^{\circ\prime}$. The reduction potential is a measure of how easily a compound can be reduced (how easily it can accept electrons). $E^{\circ\prime}$ is measured under standard temperature and pressure conditions and at a pH of 7, just as we did for $\Delta G^{\circ\prime}$ (see Chapter 14, Section 14.3). All compounds are compared to the redox potential of H$^+$/H$_2$ which is assigned an $E^{\circ\prime}$ of 0.0 V. Compounds with positive $E^{\circ\prime}$ values are better electron acceptors than H$^+$. The more positive the value of $E^{\circ\prime}$ for the oxidized form of a carrier, the better it functions as an electron acceptor. The final and best electron acceptor of the mitochondrial electron chain is O$_2$, which has an $E^{\circ\prime}$ of +0.82 V. The more negative the $E^{\circ\prime}$ of a carrier, the weaker it acts as an oxidizing agent and the better it acts as a reducing agent (electron donor). The $E^{\circ\prime}$ for NADH, the best reducing agent in the chain, is -0.32 V. The $E^{\circ\prime}$ for each carrier in the transport chain is given in Table 17.2. Note that the $E^{\circ\prime}$ values for respiratory carriers range between the limits of -0.32 V (NAD$^+$/NADH) and +0.82 V (O$_2$/H$_2$O).

A second important characteristic of the electron transport chain is the *amount* of energy released upon electron transfer from one carrier to another. The energy available when electrons are passed from carrier to carrier can be calculated in terms of the familiar constant, the standard free energy change, $\Delta G^{\circ\prime}$.

$$\Delta G^{\circ\prime} = -nF\Delta E^{\circ\prime}$$

Concept Review:
Redox reactions

Table 17.2
Standard reduction potentials, $E^{\circ\prime}$ for mitochondrial electron carriers

Redox Reaction[a]	$E^{\circ\prime}$ (V)
$NAD^+ + 2\ H^+ + 2\ e^- \longrightarrow NADH + H^+$	−0.32
$FMN + 2\ H^+ + 2\ e^- \longrightarrow FMNH_2$	−0.30
$FAD + 2\ H^+ + 2\ e^- \longrightarrow FADH_2$	−0.18
$2\ H^+ + 2\ e^- \longrightarrow H_2$	0.00
Ubiquinone $+ 2\ H^+ + 2\ e^- \longrightarrow$ ubiquinol	0.05
Cytochrome $b_{(ox)} + e^- \longrightarrow$ cytochrome $b_{(red)}$	0.08
Cytochrome $c_{1(ox)} + e^- \longrightarrow$ cytochrome $c_{1(red)}$	0.22
Cytochrome $c_{(ox)} + e^- \longrightarrow$ cytochrome $c_{(red)}$	0.25
Cytochrome $a_{(ox)} + e^- \longrightarrow$ cytochrome $a_{(red)}$	0.29
Cytochrome $a_{3(ox)} + e^- \longrightarrow$ cytochrome $a_{3(red)}$	0.55
$\frac{1}{2}\ O_2 + 2\ H^+ + 2\ e^- \longrightarrow H_2O$	0.82

[a](ox), oxidized; (red), reduced.

where

n = number of electrons transferred from one carrier to another (usually one or two)

F = the faraday constant, 96.5 kJ/volt·mole. This is a unit conversion factor to change units of volts to kJ/mol

$\Delta E^{\circ\prime}$ = the difference in reduction potential between the two carriers, in volts = $(E^{\circ\prime}_{acceptor} - E^{\circ\prime}_{donor})$

We use this equation to calculate the amount of energy available when electrons are transferred from NADH through the chain to O_2. Normally two electrons pass from an NADH ($E^{\circ\prime}_{donor} = -0.32$ V) to O_2 ($E^{\circ\prime}_{acceptor} = +0.82$ V). The total energy released during this transfer is calculated as follows:

$$NADH + H^+ + \frac{1}{2}\ O_2 \longrightarrow NAD^+ + H_2O$$

$$\Delta G^{\circ\prime} = -(2)\left(96.5\ \frac{kJ}{\ \ \ \ \ \ }\right)[+0.82 - (-0.32)]\ V$$

$$\Delta G^{\circ\prime} = -(2)(96.5)(1.14)\ kJ/mol$$

$$\Delta G^{\circ\prime} = -220\ kJ/mol$$

If a similar calculation is completed for electron flow from $FADH_2$ to O_2, $\Delta G^{\circ\prime} = -152$ kJ/mol. Clearly, a large amount of energy is available for ATP synthesis when electrons travel the electron-transport chain. (The reaction ADP + $P_i \rightleftharpoons$ ATP + H_2O has a $\Delta G^{\circ\prime}$ of +31 kJ/mol.)

It is important to note that all the energy is released not in a *single* step of electron transfer but in incremental amounts at each redox step. The quantity of energy released can be calculated between adjacent carriers or over any interval of the chain. The energy released at three specific steps in the chain is collected in the form of a transmembrane proton gradient and used to drive the synthesis of ATP (ADP + $P_i \rightleftharpoons$ ATP + H_2O).

17.2 Components of the Electron-Transport Chain

Interactive Animation: Oxidative phosphorylation

The electron-transport chain is composed of several serially ordered components that participate in electron flow. Each component is capable of undergoing reduction and oxidation; that is, it can accept electrons from the preceding carrier and pass them to the next carrier in line. The flow of electrons is spontaneous and thermodynamically favorable because the next carrier has a greater affinity for electrons than the previous one. To aid our chemical study of these carriers, we divide the chain into four segments referred to in the previous section: NADH-CoQ reductase (complex I), succinate-CoQ reductase (complex II), cytochrome c reductase (complex III), and cytochrome c oxidase (complex IV). There are four molecular types of redox centers in the complexes (in addi-

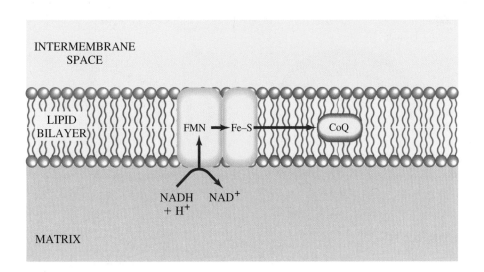

Figure 17.3

Components of NADH-CoQ reductase, complex I. The proteins in complex I have the prosthetic groups, FMN and Fe–S. Electrons enter the complex from NADH and flow to CoQ.

tion to NADH and FADH$_2$): flavin mononucleotide (FMN), iron–sulfur clusters (Fe–S), ubiquinone (CoQ), and **cytochromes.**

Complex I

The carriers in complex I are shown in Figure 17.3. Electrons flow from NADH to the first carrier in the complex, **flavin mononucleotide** (Figure 17.4). FMN, a relative of FAD, is reduced by accepting two electrons from NADH. Electrons from the product, FMNH$_2$, pass next to a prosthetic group composed of an **iron–sulfur cluster** (Figure 17.5). The most common forms have equal numbers of iron and sulfur (Fe$_2$S$_2$ and Fe$_4$S$_4$). Sulfur atoms either are from those in cysteine residues in proteins or are from inorganic sulfur. The iron atoms in these complexes cycle between the oxidized ferric state (Fe^{3+}) and the reduced ferrous state (Fe^{2+}). Complex I terminates at the carrier, **ubiquinone,** also called **coenzyme Q** or **CoQ.** (The name is derived from its ubiquitous presence in nature.) The structure of the CoQ found commonly in mammalian systems is shown in Figure 17.6. CoQ is the smallest and most hydrophobic of all the carriers. The long hydrocarbon side chain of CoQ acts as an anchor to hold it in the nonpolar inner mitochondrial membrane. When ubiquinone (CoQ) accepts two electrons and two protons it is reduced to ubiquinol (CoQH$_2$; Figure 17.6). A semiquinone intermediate is formed after transfer of one electron.

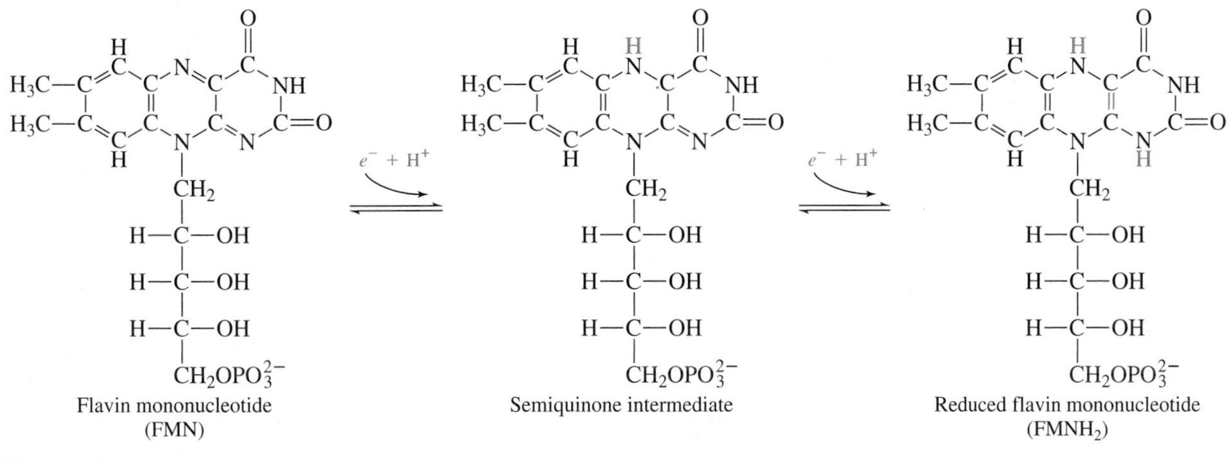

Flavin mononucleotide (FMN)   Semiquinone intermediate   Reduced flavin mononucleotide (FMNH$_2$)

Figure 17.4

Structures of the redox cofactors, FMN and FMNH$_2$. FMN accepts one electron and a proton to form a semiquinone intermediate. A second electron and proton produce FMNH$_2$.

Figure 17.5

The structures of iron–sulfur complexes in electron-transport proteins: (a) Fe_2S_2 and (b) Fe_4S_4. In addition to inorganic sulfide ions (S^{2-}), sulfur atoms from cysteine residues in the proteins also bind to iron. The iron atoms can undergo reversible oxidation–reduction, $Fe^{3+} + e^- \rightleftharpoons Fe^{2+}$.

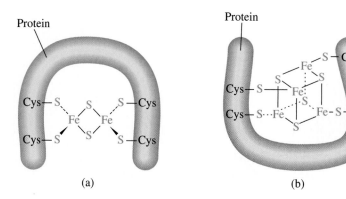

(a)   (b)

Complex II

Coenzyme Q also serves as the entry point for electrons from the cofactor $FADH_2$. At least two forms of complex II are present in mitochondrial membrane. The enzymes succinate dehydrogenase (citric acid cycle) and acyl-CoA dehydrogenase (β oxidation; see Chapter 18) direct the transfer of electrons from their substrate molecules (succinate and acyl CoA) to CoQ via $FADH_2$. Both of these enzymes have iron–sulfur clusters as prosthetic groups and are present as integral proteins in the inner mitochondrial membrane (Figure 17.7). All electrons that enter the transport chain from NADH and $FADH_2$ must pass through the ubiquinone/ubiquinol pair.

Complex III

In the next segment of the chain, complex III, electrons are transferred from ubiquinol to cytochrome *c* (Figure 17.8). The electron carriers in complex III are iron–sulfur clusters and cytochromes. The iron–sulfur clusters are similar to those already described. **Cytochromes** are electron-transferring proteins that contain a heme prosthetic group (Figure 17.9). Note that the cytochromes resemble the oxygen-carrying proteins myoglobin and hemoglobin. Cytochromes are highly colored because of the presence of the heme prosthetic group. The red-brown color in animal muscle is due to cytochromes (and myoglobin). Wild game birds have dark breast meat because of a high concentration of mitochondrial cytochromes in active flight muscle. Domestic birds (chickens, turkeys, etc.) have white breast meat because of a lower concentration of cytochromes in less active flight muscle. Each cytochrome molecule, with its heme group, is able to accept and donate only one electron per redox cycle. The iron atom in the heme cycles through the reduced (Fe^{2+}) and oxidized (Fe^{3+}) forms. The heme structures in their different protein environments lead to slightly different reduction potentials for the cytochromes (see Table 17.2).

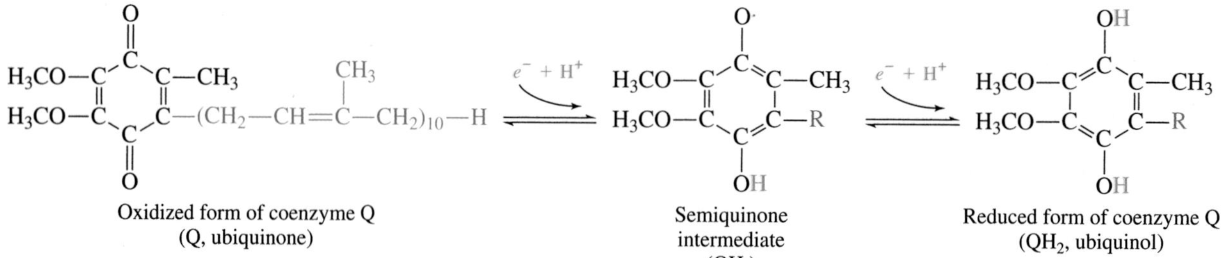

Oxidized form of coenzyme Q
(Q, ubiquinone)

Semiquinone intermediate
(QH·)

Reduced form of coenzyme Q
(QH_2, ubiquinol)

Figure 17.6

The structures and redox action of the two forms of coenzyme Q, ubiquinone and ubiquinol. The oxidized form, CoQ, accepts two electrons and two protons to produce $CoQH_2$. A semiquinone intermediate is produced after transfer of one electron and one proton.

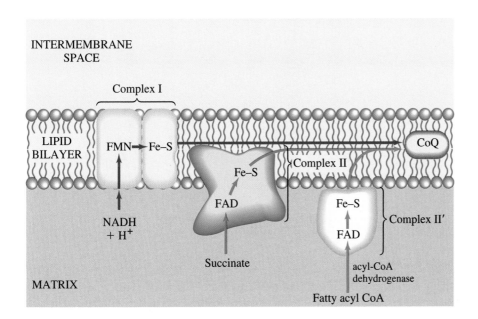

Figure 17.7

Entry of FADH$_2$ electrons into electron transport (complexes II and II$'$). Electrons from succinate and fatty acyl CoA substrates are transported by specific FAD-linked dehydrogenases to CoQ.

The transfer of electrons from reduced CoQ (CoQH$_2$) is not as direct as indicated in the text or figures. Complications in electron transfer arise because CoQH$_2$ is a two-electron donor and the cytochromes are one-electron acceptors. Electrons entering complex III via CoQH$_2$ separate into interlinking paths that result in the **Q cycle** (see Figure 17.10) Some electrons from CoQH$_2$ are used to reduce cytochrome c (via cytochrome c_1). In the figure, two CoQH$_2$ molecules each transfer a single electron to a cytochrome c. The two CoQH$\cdot$ semiquinone products then each transfer an electron to a cytochrome b. The two electrons from the cytochrome b are used to generate a new CoQH$_2$. The net result is that two CoQH$_2$ molecules, bringing in four electrons, are oxidized. Two electrons are used to reduce two cytochrome c molecules and two electrons are used to regenerate a CoQH$_2$:

$$2 \text{ CoQH}_2 + 2 \text{ cyt } c_{(\text{oxid})} + \text{CoQ} \rightleftharpoons 2 \text{ CoQ} + 2 \text{ cyt } c_{(\text{red})} + \text{CoQH}_2 + 2 \text{ H}^+$$

The Q cycle is thought to assist in maintaining the proton gradient necessary for ATP synthesis.

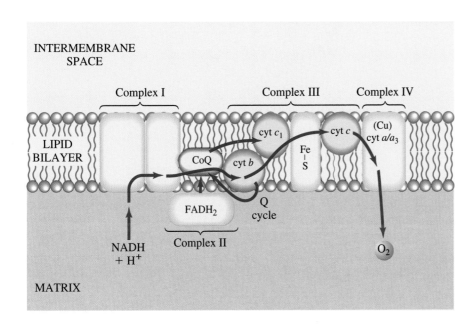

Figure 17.8

Complex III transfers electrons from CoQH$_2$ to cytochrome c (cyt c). The electron carriers in Complex III are iron–sulfur clusters and cytochromes.

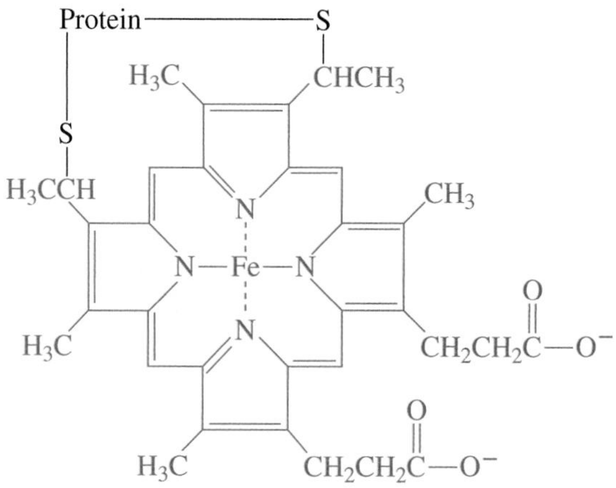

Figure 17.9

The structure of heme C, the kind found in cytochrome c.

Complex IV

The final complex in the electron-transport chain is composed of cytochromes a and a_3, combined into a carrier called **cytochrome c oxidase.** It consists of ten protein subunits with two types of prosthetic groups, two hemes (a and a_3), and two copper atoms (see Figure 17.8). Electrons passing from cytochrome c to O_2 flow through the iron atoms in the two hemes and the copper ions. Copper cycles through cuprous (Cu^+) and cupric (Cu^{2+}) ions. Of all the carriers in the chain, cytochromes a and a_3 are the only ones capable of *direct transfer* of electrons to O_2. A total of four electrons is necessary to reduce O_2 to H_2O:

$$O_2 + 4\,e^- + 4\,H^+ \rightleftharpoons 2\,H_2O$$

The flow of electrons from NADH (or $FADH_2$) through the various complexes to O_2 drives the movement of protons across the inner membrane (from matrix to intermembrane space). A proton gradient provides energy for ATP formation.

Elucidating the actual sequence of respiratory electron carriers has been a popular subject for biochemical study. Several lines of experimental evidence support the order as we have outlined here:

1. Complex studies on the unique absorption of light by the oxidized and reduced forms of each carrier. The ultraviolet–visible absorption spectrum of a carrier varies with redox state.
2. Experimental measurements of the standard reduction potentials, $E^{\circ\prime}$, for each component of the chain. This required the tedious task of isolating, purifying, and measuring each carrier.

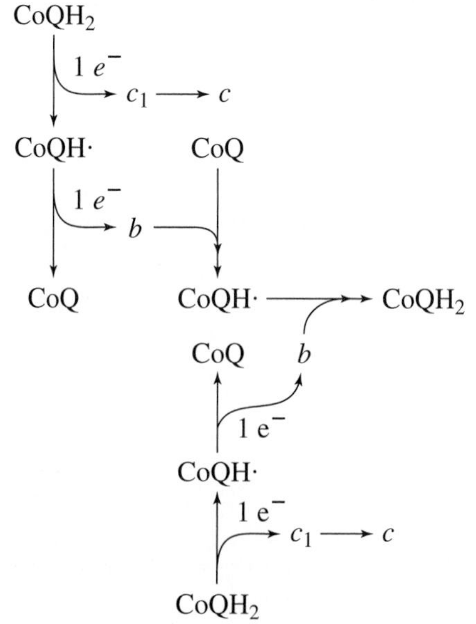

Figure 17.10

The Q cycle of electron transport in complex III.

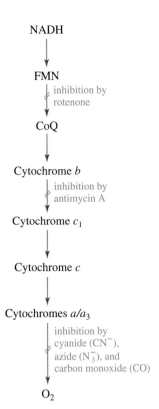

Figure 17.11
Inhibition of mitochondrial electron transport. Several chemicals are shown that block electron transport at specific locations. Rotenone and antimycin A have little effect on human mitochondrial electron transport except at very high concentrations; however, CN^-, N_3^-, and CO are extremely toxic to humans and other animals.

3. The use of chemicals that inhibit the flow of electrons at specific points in the chain. In Figure 17.11 are shown three types of electron flow inhibitors and their sites of action. Rotenone, a fish poison and insecticide, inhibits electron transfer between NADH and CoQ (probably between the iron–sulfur cluster and CoQ in complex I). The antibiotic antimycin A inhibits flow of electrons from cytochrome b to cytochrome c_1 (in complex III) in bacterial electron-transport chains. Cyanide (CN^-), azide (N_3^-), and carbon monoxide (CO) block electron flow from cytochromes a and a_3 to O_2. The CN^-, N_3^-, and CO may act by complexing with copper atoms in cytochrome oxidase, thereby interfering with the binding of O_2 to complex IV.

17.3 Oxidative Phosphorylation

The process of oxidative phosphorylation is the combination of two distinct activities, which can be summarized in reaction form:

1. The flow of electrons from NADH (or $FADH_2$) to molecular oxygen via the electron-transport chain:

$$NADH + H^+ + \tfrac{1}{2} O_2 \longrightarrow H_2O + NAD^+ \qquad \Delta G^{\circ\prime} = -220 \text{ kJ/mol}$$

The downhill flow of electrons from a high energy level in NADH and $FADH_2$ to O_2 releases large amounts of energy, measured in terms of $\Delta G^{\circ\prime}$, to drive the second process:

2. The phosphorylation of ADP by inorganic phosphate synthesizing ATP:

$$ADP + P_i \rightleftharpoons ATP + H_2O \qquad \Delta G^{\circ\prime} = +31 \text{ kJ/mol}$$

Step 2 is catalyzed by the inner mitochondrial membrane enzyme, **ATP synthase.** The energy from electron transport is collected and converted into a form appropriate to drive the reaction of ATP synthesis. In the intact mitochondrion, the two processes of electron transport and ATP synthesis are linked together or coupled. Each process is dependent on the other. Electrons do not flow from nutrients to O_2 unless there is a need for ATP synthesis (high concentrations of ADP must be present). ATP cannot be generated by oxidative phosphorylation unless there is energy from electron transport.

Interactive Animation:
Oxidative phosphorylation

The stoichiometry of oxidative phosphorylation may be studied by using isolated, intact mitochondria that are supplied with O_2, oxidizable substrates such as pyruvate or succinate, ADP, P_i, and NAD^+ or FAD. Each molecule of pyruvate oxidized by the pyruvate dehydrogenase complex causes the formation of one molecule of NADH (two electrons from pyruvate are transferred to NAD^+). When the two electrons in the reduced cofactor pass through the electron-transport chain, about three molecules of ATP are made for each oxygen atom reduced. (Because of proton leakage back into the matrix, the exact number of ATP is between two and three, probably about 2.5 ATP. For ease of later calculations and for consistency, we will use 3 ATP.) Thus, the net overall reaction for mitochondrial NADH oxidation is

$$NADH + H^+ + 3\,ADP + 3\,P_i + \tfrac{1}{2}\,O_2 \longrightarrow NAD^+ + 4\,H_2O + 3\,ATP$$

Each molecule of succinate, upon oxidation by succinate dehydrogenase, leads to the formation of one $FADH_2$. Two electrons from $FADH_2$ enter the electron-transport chain at CoQ and lead to the production of about two ATPs upon reduction of O_2. (Write the net overall reaction for the oxidation of the $FADH_2$ by O_2.)

Coupling of Electron Transport with ATP Synthesis

The mechanistic study of oxidative phosphorylation has been an active and controversial area of biochemical research. Beginning in the early 1940s, several American and British biochemists, including S. Ochoa, A. Lehninger, D. Green, E. Racker, P. Mitchell, and P. Boyer, attempted to describe the coupling details of electron transport and ATP synthesis. One of the first experimental findings was the relationship between the number of moles of ATP generated per mole of oxygen reduced to H_2O. This is often expressed as the P/O ratio; that is, the number of molecules of ATP formed (ADP + P_i) per pair of electrons that moves through the chain to oxygen. For NADH oxidation, P/O is about 3. For $FADH_2$, P/O is about 2. (Experimental measurements show actual ratios of 2.5 and 1.5, respectively.) These results can be explained by assuming there are *three* discrete coupling sites along the electron chain between the entry of NADH and oxygen (Figure 17.12). Electrons from $FADH_2$ enter the chain at a point beyond the first coupling site. Passage of two electrons through each site leads to the formation of an ATP at that site. After years of intense study we now know that there is one coupling site in each of three main enzyme complexes: Site 1 is in NADH-CoQ reductase (complex I), site 2 is in cytochrome *c* reductase (complex III), and site 3 is in cytochrome *c* oxidase (complex IV).

Figure 17.12

Locations of the three energy-coupling sites in the mitochondrial electron-transport chain. Site 1 is between the NADH and CoQ; site 2 is in the cytochrome *c* reductase region, and site 3 is in cytochromes a/a_3. As two electrons pass through each site, one ATP is produced by phosphorylation of ADP.

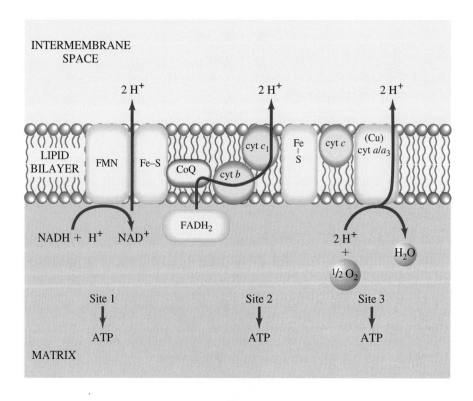

After the coupling sites and their location were identified, the next challenge for bio-chemists was to define the mechanism of coupling: How is energy from electron transport transduced (changed) to a form of energy that can be used to phosphorylate ADP? This requires a change from electrochemical (redox) energy to chemical (phosphoanhydride bond) energy. Several mechanisms of coupling have been proposed over the years:

1. It was first suggested that electron transport caused the formation of high-energy covalent intermediates that transferred energy to ATP generation. No experimental evidence for such energy-rich intermediates has yet been found.

2. A second hypothesis was that electron transport energy was stored in high-energy pro-tein conformational forms. As with the previous suggestion, no one has obtained experi-mental evidence for conformational intermediates.

3. A very different mechanism was proposed by Peter Mitchell in 1961. His hypothesis, the **chemiosmotic coupling mechanism,** suggests that *electron transport through the car-riers in the inner membrane causes the unidirectional pumping of protons from the inner mitochondrial matrix to the other side of the membrane* (into the intermembrane space; Fig-ure 17.13). Mitchell and others, over a period of several years, obtained important experi-mental results that supported this idea. Proton pumping occurs at each of the three coupling

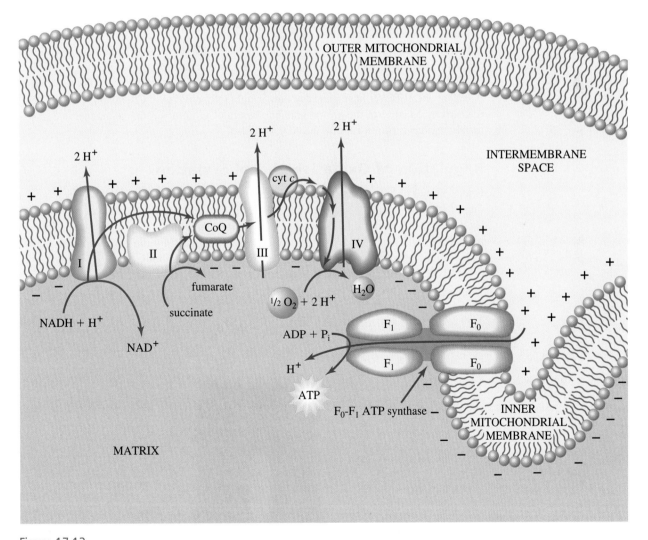

Figure 17.13

Mechanism of chemiosmotic coupling. As electrons flow through the three coupling sites, protons are pumped through the inner membrane. A proton differential or gradient is built up because proton concentration becomes higher in the intermembrane space relative to the matrix. Upon collapse, this high-energy situation provides the driving force for ATP synthesis. The F_0-F_1 ATP synthase is asso-ciated with the inner membrane. I, II, III, and IV refer to the four electron-transport complexes.

sites as a pair of electrons passes the site. The concentration of H^+ becomes higher in the intermembrane space than in the matrix; that is, a proton gradient is established. In carefully designed experiments, Mitchell was able to measure a difference of about 1.4 pH units between the two sides of the membrane (a higher concentration of protons or lower pH in the intermembrane space). The proton gradient can be established and maintained because protons cannot diffuse through the inner membrane back into the matrix. With a greater positive charge on one side of the membrane, an electrochemical potential difference exists across the membrane. In terms of thermodynamics, the proton gradient is an unstable, highly energetic system. Collapse of the gradient releases free energy that is channeled to the synthesis of ATP.

Components of ATP Synthase

We turn now to the process of ATP synthesis by phosphorylation of ADP using energy from collapse of the proton gradient. This is catalyzed by the enzyme complex **ATP synthase.** The ATP synthase system is present as spheres projecting from the matrix side of the inner membrane. ATP synthase consists of two units (or factors), F_1 and F_0 (Figure 17.14). The knobs, comprised of F_1 particles, can be removed from the membrane by mechanical shaking. F_1 contains a catalytic site for ATP synthesis where ADP and P_i are brought together for combination. The F_1 component is a peripheral protein that is held in place on the membrane by interaction with an integral membrane component, F_0. F_0 serves as the transmembrane channel for proton flow. The F_0-F_1 complex serves as the molecular apparatus for coupling proton movement to ATP synthesis. In spite of extensive research, the detailed mechanism of ATP formation by the complex is still not entirely clear, but recent results reported by J. Walker and P. Boyer have provided new insight (Nobel Prize in Chemistry, 1997). Energy released by collapse of the proton gradient (return of protons through a channel in F_0 to the matrix region) is transmitted to the ATP synthase.

Regulation of Oxidative Phosphorylation

Under normal physiological conditions electron transport is tightly coupled to phosphorylation of ADP. Recall that a major function of energy metabolism is to maintain constant ATP levels. Electrons do not flow down the electron-transport chain to O_2 unless ADP is present for phosphorylation. Thus, the flow of electrons is regulated by the concentration of ADP. This form of regulation is similar to the control of other metabolic reactions. If high levels of ADP are present in cells ATP must have been consumed to increase ADP levels). The

Figure 17.14

The structure of the ATP synthase (F_0-F_1 complex). F_0 acts as a channel for flow of protons back to the matrix, which results in collapse of the gradient. When protons flow through the ATPase (F_1 component) it is activated to catalyze the phosphorylation of ADP. α, β, γ, δ, and ϵ label protein subunits in F_1.

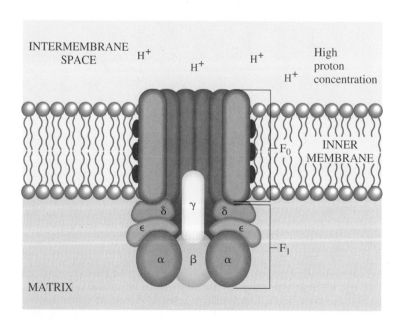

increased level of ADP triggers an increase in the rate of catabolic reactions on nutrients. The enzymes glycogen phosphorylase, phosphofructokinase, citrate synthase, and others are stimulated to enhance the rate of nutrient oxidation. This leads in turn to the production of substrate-level ATP and GTP, but more importantly, to NADH and $FADH_2$; these can be oxidized by the electron-transport chain, thereby producing ATP. Again, we see how all metabolic processes are coordinated to maintain a proper balance of ATP and ADP (see Chapter 15, Section 15.5).

Under some specialized physiological conditions, the tight coupling of electron transport and ADP phosphorylation is disrupted (the two processes are uncoupled). The rate of electron transport is no longer controlled by the ADP concentrations but by thermodynamics. If the two processes are unlinked, electrons speed down the electron transport chain at an uncontrolled rate. Large amounts of O_2 and nutrient are consumed but no ATP is synthesized. The energy from electron transport normally used to make ATP is released in the form of heat. The strategy of uncoupling is used by some newborn animals and hibernating mammals such as bears to maintain body heat. Some animals (including human infants) have large amounts of a specialized type of adipose tissue called **brown fat.** The color is the result of high concentrations of mitochondria containing highly colored cytochromes. In brown fat mitochondria, an integral inner membrane protein called uncoupling protein or thermogenin provides a channel for the return of pumped protons from the intermembrane space back into the matrix. This route of proton flow bypasses the F_0-F_1 complex; thus, the energy is released as heat instead of ATP. For a brief period in the 1940s, the chemical 2,4-dinitrophenol was prescribed as a dieting aid. Individuals using this drug experienced heavy breathing (excessive consumption of O_2), loss of body weight (uncontrolled rate of breakdown of nutrients, fats, and carbohydrates), and excessive perspiring and fever (an increase in body temperature). Studies showed that 2,4-dinitrophenol was an uncoupling agent. Although it was effective in burning off fat, its use in diet pills was halted because of a very high level of toxicity.

Hibernating bears keep warm by uncoupling electron transport in brown fat mitochondria.

17.4 Recycling of Cytoplasmic NADH

We now have the information needed to estimate the energy yield (in terms of ATP) upon complete oxidation of carbohydrates. We have tallied the production of ATP (or GTP), NADH, and $FADH_2$ during the preliminary stages of catabolism. The amount of substrate-level ATP is quite low for glucose; however, large amounts of reduced cofactors are produced, which are recycled by respiration. The complete oxidation of glucose yields 36 or 38 ATP (assuming a P/O of 3), depending upon the cellular location of its catabolism (Table 17.3). The reaction for the complete degradation of glucose in skeletal muscle is as follows:

$$\text{glucose} + 36\ \text{ADP} + 36\ \text{P}_i + 36\ \text{H}^+ + 6\ \text{O}_2 \longrightarrow 6\ \text{CO}_2 + 36\ \text{ATP} + 42\ \text{H}_2\text{O}$$

A brief outline of glucose oxidation and the energy yield are displayed in Table 17.3. NADH is generated by glycolysis (two molecules), pyruvate dehydrogenase (two molecules), and citric acid cycle (six molecules). $FADH_2$ is produced only by the citric acid cycle (two $FADH_2$ per glucose). Substrate-level phosphorylation leads to a total of four ATP (in some organisms two ATP and two GTP). All of the $FADH_2$ and the NADH, except that produced in glycolysis (cytoplasm), is made in the mitochondrial matrix and can be oxidized directly by the electron-transport chain. NADH generated outside the mitochondria cannot be transported through the inner mitochondrial membrane into the matrix region where respiratory assemblies are located. Cytoplasmic NADH must be recycled by electron shuttle systems, which carry electrons through the membrane in the form of reduced substrates. The **glycerol 3-phosphate shuttle** functions in skeletal muscle and brain (Figure 17.15). Electrons from NADH are used to make glycerol 3-phosphate. The dehydrogenase catalyzing the reduction of dihydroxyacetone phosphate on the cytoplasmic side is NAD linked. Glycerol 3-phosphate is oxidized by a membrane-bound, FAD-linked dehydrogenase, leading to $FADH_2$. The electrons originally in NADH are now in $FADH_2$, which enters electron transport at CoQ (via complex II), leading to two ATP. We can make the general statement that

Table 17.3
Energy yield from the complete oxidation of glucose

Metabolic Stage	NADH[a]	FADH$_2$	Substrate-level Phosphorylation
Glycolysis	2 (cyto)	0	2 (ATP)
Pyruvate dehydrogenation	2 (mito)	0	0
Citric acid cycle	6 (mito)	2	2 (GTP)
Subtotal	2 (cyto) 8 (mito)	2	4

Oxidative Phosphorylation

Glycerol 3-Phosphate Shuttle

2 cyto NADH × 2 ATP	=	4 ATP
8 mito NADH × 3 ATP	=	24 ATP
2 FADH$_2$ × 2 ATP	=	4 ATP
Substrate level	=	4 ATP (or GTP)
Grand total		36 ATP

Malate–Aspartate Shuttle

2 cyto NADH × 3 ATP	=	6 ATP
8 mito NADH × 3 ATP	=	24 ATP
2 FADH$_2$ × 2 ATP	=	4 ATP
Substrate level	=	4 ATP (or GTP)
Grand total		38 ATP

[a]cyto, cytoplasmic; mito, mitochondrial.

cytoplasmic NADH produced in skeletal muscle and brain is oxidized by the glycerol 3-phosphate shuttle, resulting in only two ATP produced by oxidative phosphorylation.

On the other hand, the **malate–aspartate shuttle,** which is present in heart and liver, leads to three ATP for each cytoplasmic NADH. Electrons from cytoplasmic NADH are carried through the inner membrane by the substrate malate, which is oxidized in the matrix by mitochondrial malate dehydrogenase, an NAD-linked enzyme (Figure 17.16). Thus, in heart and liver, cytoplasmic NADH oxidized by the malate–aspartate shuttle results in three ATP

Figure 17.15

Reactions of the glycerol 3-phosphate shuttle. Cytoplasmic NADH is oxidized during the reduction of dihydroxyacetone phosphate to glycerol 3-phosphate. This reduced substrate is oxidized to dihydroxyacetone phosphate using a membrane-bound, FAD-linked dehydrogenase as catalyst. The electrons in FADH$_2$ enter the electron-transport chain at CoQ via complex II, bypassing complex I and coupling site 1.

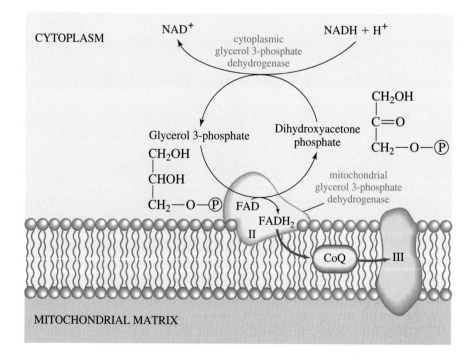

CYTOPLASM

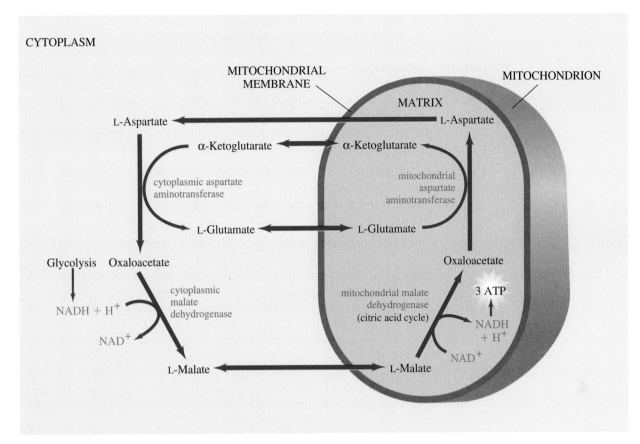

Figure 17.16

Reactions of the malate–aspartate shuttle. Cytoplasmic NADH is oxidized during the reduction of oxaloacetate to malate. Malate diffuses into the matrix, where it is oxidized to oxaloacetate using an NAD-linked dehydrogenase as catalyst. Thus, three ATP molecules are produced upon oxidation of the matrix-generated NADH. Oxaloacetate is transformed to aspartate, which diffuses to the cytoplasm where it is converted back to oxaloacetate.

in oxidative phosphorylation. In summary, each glucose oxidized in skeletal muscle and brain produces 36 ATP; glucose oxidized by heart and liver cells yields a total of 38 ATP. Of the 36 ATP produced from glucose oxidation in skeletal muscle, only 4 come from substrate-level phosphorylation. Thirty-two ATPs are generated by oxidative phosphorylation. Therefore $32/36 \times 100$ or 89% of the ATP generated from glucose is produced during oxidative phosphorylation. (For another use of oxygen, see Window on Biochemistry 17.1.)

17.5 Photosynthetic Electron Transport

Photosynthesis

Organisms that have survived over millions of years have done so by using one or both of two options to meet cellular energy needs. **Heterotrophs,** also called chemotrophs, use the chemical option. They ingest other plants and animals and oxidize organic compounds (carbohydrates and fats). The final stage of energy metabolism, oxidative phosphorylation, generates ATP using energy from electron transport. The other way to generate cellular energy is the photosynthetic option, used by **phototrophs.** They absorb energy from solar radiation and divert the energy through electron-transport chains to synthesize ATP and generate reducing power in the form of NADPH. These energetic products of photosynthesis (ATP, NADPH) are used to make carbohydrates from the simple molecules CO_2 and H_2O. The phototrophs include both prokaryotic and eukaryotic organisms, such as higher plants, algae, and photosynthetic bacteria.

Interactive Animation: Photosynthesis

In chemical terms, respiration, as described in Sections 17.1 through 17.4, is the production of energy by **oxidative decarboxylation** of nutrients. Pyruvate, for example, is ultimately oxidized to CO_2 and H_2O. In contrast, photosynthesis can be defined as **reductive carboxylation** of organic substrates. The chemical processes of photosynthesis involve covalent combining ("fixing") of CO_2 with acceptor molecules and reduction of a carboxylic acid level carbon to that of a carbohydrate (aldehyde and alcohol). The energy for carbon–carbon bond formation and reduction comes from solar radiation. Thus the photosynthetic process is composed of two coupled phases as shown in Figure 17.17:

1. Absorption of light energy by chlorophyll and other pigments (the "photo" phase, sometimes called the "light" reactions).
2. Carbon metabolism to make glucose, sucrose, and starch (the "synthesis" phase, sometimes called the "dark" reactions because light is not directly needed).

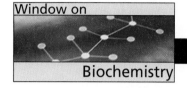
Window on
Biochemistry

17.1　Cytochromes P450—Enzymes That Use Oxygen to Detoxify Chemicals

It is estimated that 90% of the molecular oxygen consumed in cells is used for the process of oxidative phosphorylation. The remaining 10% of the oxygen is consumed by enzymes that catalyze specialized metabolic reactions.

One of the largest groups of oxygen-metabolizing enzymes are the **oxygenases**—enzymes that incorporate oxygen atoms into substrates. Within this group, the most important and the largest family of enzymes are the **cytochromes P450.** Several hundred of these enzymes have been isolated and characterized. The type of reaction most often observed for the P450 enzymes is **hydroxylation** shown in the reaction below where R—H represents a relatively unreactive carbon hydrogen bond, and R—OH represents the hydroxylated product:

$$R—H + NADPH + H^+ + O_2 \rightleftharpoons R—OH + NADP^+ + H_2O$$

Note that the overall type of reaction is oxidation–reduction. (What is oxidized and what is reduced?) Because molecular oxygen is a relatively unreactive substrate, these enzymes require a cofactor, usually a metal ion. (Remember that cytochrome oxidase requires metal ions to enhance the reduction of oxygen in electron transport.) The cytochromes P450 are characterized by the presence of a heme cofactor that binds and activates the oxygen substrate. The reactive species of bound oxygen is the perferryl, FeO_3^+. P450 enzymes, in their reduced form, bind CO and display a characteristic and strong absorption band at 450 nm. [The name of this group of enzymes is derived from their intense color (P stands for pigment) and from the unique wavelength of the spectral absorption, 450 nm.] Reactions catalyzed by cytochromes P450 are somewhat common in biochemistry as is evidenced by the many hydroxylation reactions in steroid hormone biosynthesis (see pp. 219–222; 510–514). But what is especially unusual is the action of these enzymes toward **xenobiotics,** chemicals foreign to an organism. These foreign compounds can be drugs like phenobarbital and acetaminophen, polychlorinated biphenyls (PCBs), and environmental carcinogens like benzpyrene. The catalytic action of the cytochromes P450 helps achieve an important result, namely, the detoxification of the xenobiotics. Hydroxylation of xenobiotics (which are usually nonpolar) makes them more water-soluble and much easier to eliminate from the body. Occasionally a substrate is made more toxic upon hydroxylation. For example, the polycyclic aromatic hydrocarbon benzpyrene is hydroxylated to products that are carcinogenic.

Another interesting and important characteristic of the cytochromes P450 is that their synthesis is induced, usually in the liver, by xenobiotics. Chemicals foreign to the body stimulate the synthesis of enzymes that use the xenobiotics as substrates. This process, of course, enhances the efficiency of detoxification and elimination of xenobiotics.

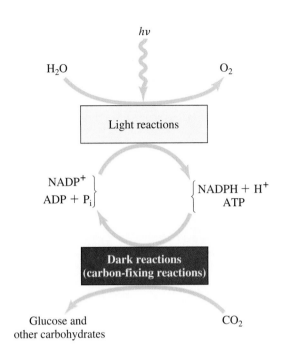

Figure 17.17
Two phases of photosynthesis linked together in a cycle. During the light reactions, light energy, shown as a photon, hv, is used to make NADPH and ATP. These products are used to provide energy for the synthesis of glucose and other "dark" reactions.

These two reaction phases can be combined into a single oxidation–reduction reaction:

$$CO_2 + 2\,H_2A \xrightarrow{\text{light}} [CH_2O] + 2\,A + H_2O$$

where

$$H_2A = \text{electron donor}$$
$$A = \text{oxidized form of } H_2A$$
$$[CH_2O] = \text{organic molecules in the form of carbohydrates}$$

The identity of the electron donor, H_2A, depends on the photosynthetic organism. Photosynthetic bacteria use inorganic molecules, such as hydrogen sulfide (H_2S), hydrogen gas (H_2), or ammonia (NH_3), or organic compounds, such as lactate or isopropanol. For higher plants and algae, the electron donor, H_2A, is water, which is oxidized to O_2 (product A in the reaction above). The following discussion on photosynthesis focuses on the oxygenic phototrophs, organisms that use water as electron donor (H_2A).

Chloroplasts

The biochemical assemblies for aerobic energy metabolism, including electron transport and ATP synthesis, are compartmentalized in the mitochondria of eukaryotic cells. The apparatus for light absorption and carbon-fixing reactions in eukaryotic photosynthetic cells, such as green plants, is localized in the **chloroplasts** (Figure 17.18). These organelles can be studied in an intact form by electron microscopy or isolated from crude extracts of plants by gradient centrifugation. Plant leaves have about 20 to 50 chloroplasts per cell. Green plant chloroplasts are usually globular or discoid and have lengths in the range of 5 to 10 μm, making them as much as 50 times larger than average mitochondria. Like mitochondria, chloroplasts have a dual membrane system. A continuous outer membrane is permeable to small molecules and ions. A highly folded inner membrane system serves as a boundary to separate an internal compartment. The **stroma,** a gel-like, unstructured matrix within the inner compartment, contains most of the enzymes for the dark reactions. Also within the stroma is an extensive internal network of membranes folded into flattened sacs, called **thylakoids.** The thylakoids are arranged in stacks called grana. Photosynthetic prokaryotes such as bacteria and blue-green algae have no chloroplasts but have their photosynthetic apparatus bound to the plasma membrane.

Biochemical activities are localized in the chloroplasts in much the same way they are in the mitochondria. The stroma contains the soluble enzymes for carbon-fixing reactions that

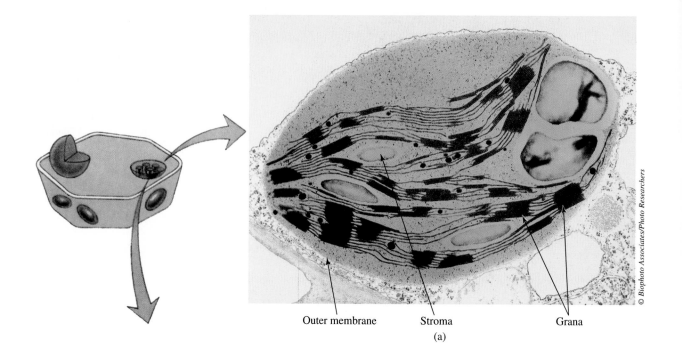

Outer membrane   Stroma   Grana

(a)

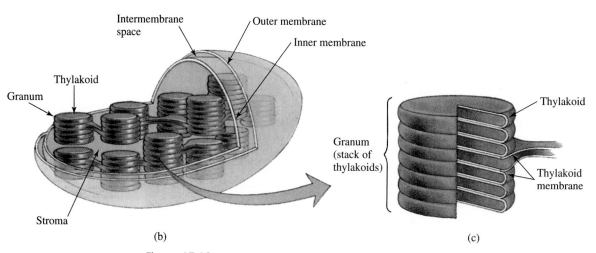

(b)   (c)

Figure 17.18

Structural and biochemical features of the chloroplast: (a) electron micrograph, (b) schematic diagram, and (c) detailed schematic of thylakoid.

form carbohydrates (dark reactions). The thylakoid membranes are the sites of the light-receiving pigments, including chlorophyll, the carriers for electron transport, and the components required for ATP and NADPH synthesis. It is important to understand the major distinctions between ADP phosphorylation (ATP synthesis) that occurs in mitochondria and the contrasting process in chloroplasts:

1. In mitochondrial respiration, electrons flow from reducing agents (NADH and $FADH_2$) to O_2, producing H_2O and free energy. The electrons flow "downhill" in the electron-transport chain, yielding energy that is coupled to ATP synthesis. Oxygen, the final electron acceptor, is consumed by reduction to H_2O.

2. In chloroplasts, electrons flow from H_2O to an electron acceptor, $NADP^+$. This is the reverse of the analogous mitochondrial process and is an energy-requiring process. Solar energy excites electrons in H_2O, pushing them "uphill" through a series of carriers to $NADP^+$. Oxygen is evolved as a product of water oxidation, and reduced cofactor, NADPH, is generated.

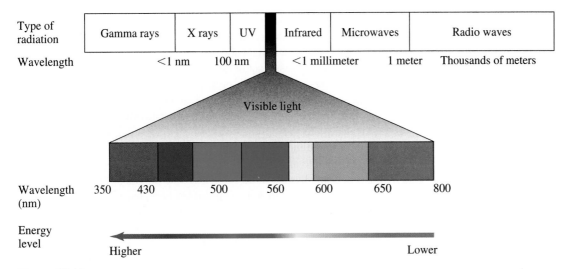

Type of radiation	Gamma rays	X rays	UV		Infrared	Microwaves	Radio waves
Wavelength	<1 nm	100 nm			<1 millimeter	1 meter	Thousands of meters

Visible light

| Wavelength (nm) | 350 | 430 | 500 | 560 | 600 | 650 | 800 |

Energy level   Higher ←———————————————————————— Lower

Figure 17.19

The electromagnetic spectrum, which is composed of a continuum of wavelengths. Visible light with wavelengths from 350 to 800 nm drives photosynthesis. The energy of the light decreases with longer wavelengths. The color associated with each wavelength range is also shown.

Biomolecules and Light

To explore the mechanism by which light energy transfers electrons from a poor reducing agent (H_2O) to make a strong electron donor (NADPH), we first need to review the physical properties of light and the interaction of light with molecules. The electromagnetic spectrum as shown in Figure 17.19 is composed of a continuum of waves with different energy levels. The region of primary importance in photosynthesis is visible light, with wavelengths in the range of 350 to 800 nm. Note that the visible region makes up only a very small portion of the entire spectrum. Light quantities are defined in terms of particles called **photons** represented by hv. The least energetic photons are represented by radio waves, microwaves, and infrared waves. On the opposite end of the spectrum are the very energetic photons from nuclear radiation, X rays, and the ultraviolet region, which are capable of doing extensive

A rainbow has all the wavelengths of the visible light spectrum.

biological damage by breaking covalent bonds in proteins, nucleic acids, and other biomolecules. Light in the visible region has sufficient energy to excite the valence electrons of some molecules. When photons strike and are absorbed by molecules, electrons are excited to higher energy levels. Molecules possess a set of quantized energy levels, as shown in Figure 17.20. Although several states are possible, only two electronic states are shown, a ground-state, G, and the first excited state, S_1. These two states differ in the distribution of valence electrons. When electrons are promoted from a ground state orbital in G to an orbital of higher energy in S_1, an electronic transition is said to occur. The energy associated with visible light is sufficient to promote molecules from one electronic state to another, that is, to move electrons from one orbital to another. Within each electronic energy level is a set of vibrational levels. These represent changes in the stretching and bending of covalent bonds. The importance of these energy levels will not be discussed here, but transitions between these levels are the basis of infrared spectroscopy. The electronic transition for a molecule from G to S_1, represented by the vertical arrow in Figure 17.20, has a high probability of occurring if the energy of the photon corresponds to the energy necessary to promote an electron from energy level E_1 to energy level E_2. Molecules in their excited states are usually unstable; upon removal of the light source the electrons may return to their ground-state

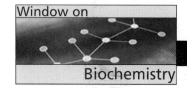

Window on
Biochemistry

17.2   Dragon Fish Use Chlorophylls to See Red

The dragon fish, *Malocosteus niger,* which lives in the pitch dark of the ocean's depths, uses a red searchlight to find food, to scare away predators, and perhaps even to communicate with its own species. *M. niger* emits red light from a tear-dropped organ just below its eye and emits a blue-green light from a spot further back (see picture). This, of course, is an example of **bioluminescence,** the emission of light by a living organism. Bioluminescence is a common occurrence in the oceans—perhaps as many as 80% of marine organisms emit light. What is unusual about *M. niger* is the color of the light: They are the only known organism that emits the color red. Most known bioluminescence is blue-green, and the majority of marine organisms probably cannot even see red.

Visual pigments, which are colored biomolecules found in the light receptors of the eye, are comprised of the protein **opsin** and a chromophore, usually **retinal,** a derivative of vitamin A. Even the primary visual pigments in the eye

The dragon fish, *M. niger,* emits far-red light from a red tear-dropped shape organ just under its eye. It also emits blue-green light from the smaller spot further back. Its jaw, partway open here, can distend wide enough to eat organisms bigger than itself.

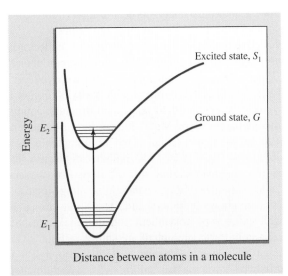

Figure 17.20
Quantized energy levels in molecules, showing the ground (G) state and the first excited state (S_1). The arrow represents the transition of a molecule from the ground state energy level to the S_1 excited state level. This transition readily occurs when the wavelength of light has energy equivalent to the energy difference between energy levels E_1 and E_2.

of *M. niger* are not sensitive enough to see its own red light. However, the dragon fish has accessory visual pigments in the retina that act as "photosensitizers" to detect red. When the accessory pigment molecules absorb red light, they transmit the visual signal to the blue-green receiving pigments by a currently unknown mechanism. These red-sensitive pigment molecules have been isolated from the retina of the dragonfish and found to be derivatives of a special chlorophyll (see structure). Analysis of the new pigments showed them to be derived from bacteriochlorophylls *c* and *d*. Because the dragon fish, like all animals, are unable to synthesize chlorophylls, the organism must obtain the pigments in the diet. The dragon fish eats mostly copecods, which are small crustaceans. Copecods eat protozoans that feed on bacteria containing the specific chlorophylls. The red-sensitive forms of the chlorophylls are modified in their passage through the food chain because they no longer contain the magnesium ion or the long hydrocarbon side chain.

The red beacon of dragon fish appears to be a private channel of communication. They can illuminate prey without the prey knowing, and perhaps the dragon fish use the light to talk to each other.

R^1 = ethyl, propyl, isobutyl, or neopentyl
R^2 = methyl or ethyl
R^3 = hydrogen or methyl

Carotenoids in these aspen leaves provide the brilliant colors.

levels, with emission of the additional energy as heat or light (fluorescence), or excited electrons in some molecules may be transferred to other molecules.

The initial step in photosynthesis is absorption of light by collector molecules in photosynthetic cells. Photosynthetic cells contain light-absorbing compounds, called pigments, that are physically located in thylakoid membranes. All photosynthetic organisms possess one or more types of the class of green pigments, **chlorophylls** (Figure 17.21). The distinctive characteristic of the chlorophylls is their polycyclic structure, comprised of four substituted pyrroles coordinating the metal ion Mg^{2+}. Except for the metal ion, chlorophylls have a striking resemblance to the heme group found in myoglobin, hemoglobin, and the cytochromes, which bind iron (see Figure 17.9). Another distinctive feature of the chlorophylls is the presence of a long hydrocarbon side chain, which enhances their solubility in membranes. The extended conjugated bond system (alternating single and double bonds) of the chlorophyll ring system allows these molecules to display strong absorption of visible light, making them well suited as photosynthetic collector molecules. Structures of chlorophyll *a*, chlorophyll *b*, and bacteriochlorophyll are shown in Figure 17.21. Chlorophylls *a* and *b* are found in most green plants, whereas bacteriochlorophyll is found in photosynthetic bacteria. Experimental evidence that chlorophylls play the major role in photosynthetic light absorption comes from the construction of a **photochemical action spectrum** (Figure 17.22 and Window on Biochemistry 17.2).

In addition to chlorophylls, photosynthetic cells contain **secondary pigments** or accessory pigments, which absorb light in wavelength ranges where chlorophyll is not as efficient; thus, they supplement the photochemical action spectrum of the chlorophylls. Accessory pigments are of two types, the carotenoids and the phycobilins (Figure 17.23). The carotenoids are represented by β-carotene (a precursor of vitamin A) and xanthophyll. All carotenoids contain 40 carbon atoms and display extensive conjugation. Most carotenoids absorb in the 400 to 500 nm range and thus are colored yellow, red, or orange. Xanthophyll's color is a bright yellow. Many of the bright colors of flower petals, fruits, and vegetables are due to the presence of carotenoids. The yellow and red colors of fall leaves result from the preferential destruction of the green chlorophylls, revealing the colors of the carotenoids. Phycobilins, which are linear tetrapyrroles, absorb in the 550- to 630-nm range. (Chlorophylls are *cyclic* tetrapyrroles.) A common red phycobilin is phycoerythrobilin. The variety of colors among plant species is due to differences in the relative abundance of chlorophylls and the accessory pigments. Green colors are

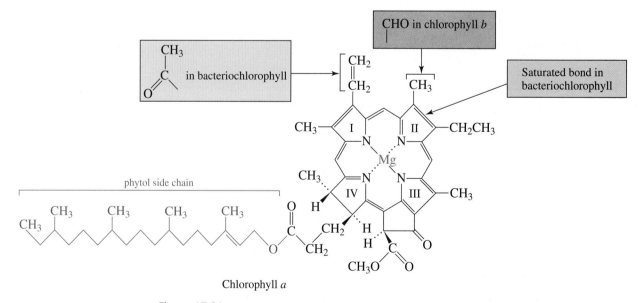

Chlorophyll *a*

Figure 17.21
The molecular structures of chlorophylls found in photosynthetic organisms. The three types shown are *a*, *b*, and bacteriochlorophyll.

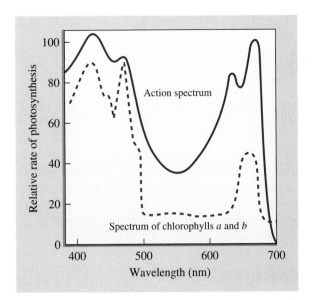

Figure 17.22
A photochemical action spectrum for a typical green plant. The solid line represents the biological action spectrum. This is a graph of the efficiency of each wavelength in sustaining plant growth as measured by CO_2 uptake and O_2 evolution (relative rate of photosynthesis). The dotted line is the combined absorption spectrum of chlorophylls *a* and *b*. These two lines are similar in shape, indicating the importance of chlorophyll *a* and *b* molecules in supporting photosynthesis.

caused primarily by a greater proportion of chlorophylls, whereas other colors such as red, green, and purple are due to an abundance of accessory pigments.

Photosynthetic Light Reactions

How does light drive the synthesis of the chemically energetic compounds ATP and NADPH, which are used in the dark reactions? In thermodynamic terms, this requires the transduction of electromagnetic energy (light) into chemical bond energy. In 1937, Robert Hill, using leaf extracts containing chloroplasts, discovered experimental evidence for electron-transport (oxidation–reduction) processes. When he illuminated chloroplasts

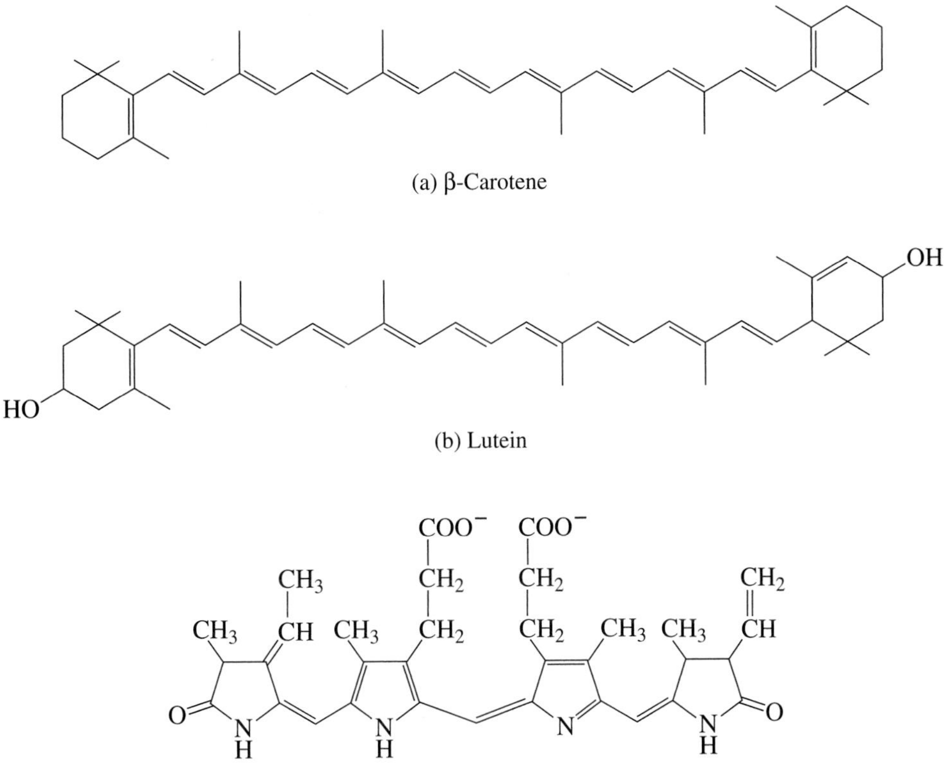

(a) β-Carotene

(b) Lutein

(c) Phycoerythrobilin

Figure 17.23
The molecular structure of accessory pigments: (a) β-carotene; (b) lutein, a xanthophyll; and (c) the red phycobilin, phycoerythrobilin. These biomolecules assist chlorophyll *a* and *b* in absorbing light for photosynthesis.

in the presence of various synthetic electron acceptors, Hill observed two simultaneous processes, neither of which occurred in the dark:

1. Reduction of the added electron acceptor.
2. Oxidation of H_2O with evolution of O_2.

Hill was able to show that electrons flowed (apparently through an electron-transport chain) from H_2O (an oxidizing agent) to the electron acceptor. It was later shown that the physiological electron acceptor in green plant chloroplasts was $NADP^+$ and the reaction occurring was

$$2\ H_2O + 2\ NADP^+ \xrightarrow{\text{light}} 2\ NADPH + 2\ H^+ + O_2$$

The distinctions between electron flow in chloroplasts and that in mitochondria now become evident. The above reaction is the reverse of the process in mitochondria; electrons flow from the electron donor H_2O to the acceptor $NADP^+$. Because the direction of electron flow is thermodynamically uphill, light energy is required.

Photosystems I and II

Light is absorbed in photosynthetic organisms by functional units called photosystems or light-harvesting complexes present in thylakoid membranes. These are groupings of chlorophyll and accessory pigments that vary in content and define the absorbing properties of various cells. Two specific types of photosystems have been found. Each contains one primary acceptor molecule (the photochemical reaction center), usually chlorophyll, and a set of accessory pigments (other chlorophylls, carotenoids, etc.) that act as antenna molecules to funnel additional light to the primary acceptor (Figure 17.24). **Photosystem I** (also called P700) consists of the primary acceptor chlorophyll *a* and accessory pigments that absorb light in the range of 600 to 700 nm. **Photosystem II** (also called P680) contains chlorophylls

Figure 17.24

Schematic of a photosystem, showing absorption of light by accessory pigments, chlorophyll (chl) and β-carotene (car). Light is directed to a chlorophyll *a* reaction center. *Source:* Based on Wolfe, 1993.

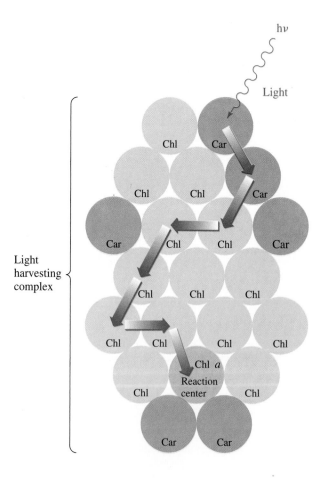

a and *b* plus accessory pigments and absorbs light primarily at 680 nm. All photosynthetic cells have photosystem I. Both photosystems I and II are found in O_2-evolving organisms such as higher plants, algae, and cyanobacteria. Photosynthetic bacteria, which do not evolve O_2, have only photosystem I.

As shown in Figure 17.25, the two photosystems in green plants are linked together by an electron-transport chain. The coordinated action of the two photosystems begins with absorption of light by photosystem I. Light absorbed by chlorophylls and accessory pigments is transmitted to the reaction center designated P700. Upon excitation by light, an electron in a P700 molecule is moved to a higher energy level, generating an excited molecule, P700*. The activated electron, now at a high energy level, is passed through a chain of electron carriers beginning with A_0, a form of chlorophyll, and ending with $NADP^+$. In reaction form the first electron-transfer step is

$$P700^* + A_0 \longrightarrow P700^+ + A_0^-$$

In words, electron transfer from P700* results in a deactivated, electron-deficient reaction center (P700$^+$) and a reduced electron acceptor, designated A_0^-. (P700$^+$ is a strong oxidizing agent and quickly accepts an electron from photosystem II.) The order of electron carriers in the chain beginning with A_0 allows the spontaneous ("downhill") flow of electrons. This is possible because of the input of light to raise the electron to a high energy level. A_0 transfers its electron to phylloquinone (A_1), which forwards an electron to an iron–sulfur protein (Fe–S). Next in line to accept an electron from reduced Fe–S is another iron–sulfur protein, ferredoxin (Fd). The best known ferredoxin is that of spinach chloroplasts, which has a molecular weight of 10,700 and a Fe_2S_2 redox center. Finally, electrons are transferred from reduced ferredoxin to $NADP^+$ by a flavoprotein catalyst, ferredoxin-$NADP^+$ oxidoreductase:

$$2\ Fd_{reduced} + NADP^+ + 2\ H^+ \longrightarrow 2\ Fd_{oxidized} + NADPH + H^+$$

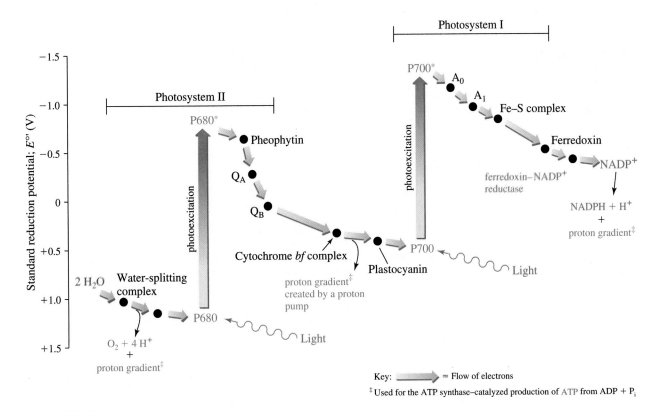

Figure 17.25

Linkage of photosystems I and II into the **Z** scheme for electron transport. This describes the light-driven flow of electrons from water to $NADP^+$. *Source:* Based on Moran et al., 1994.

The electron activated in P700 by light has been used to reduce $NADP^+$ to NADPH, which also has activated electrons but at a lower energy level than P700*. An electron must pass from each of two P700* molecules to completely reduce $NADP^+$.

Excitation and transfer of an electron from P700 leaves an unstable, electron-deficient species, $P700^+$. The "electron hole" is refilled by another electron-transport chain that is driven by photosystem II. Illumination of photosystem II with a chlorophyll reaction center, P680, produces an activated form, P680*. The energetic electron in P680* is quickly passed on to pheophytin (Ph), a chlorophyll-like electron acceptor:

$$P680* + Ph \xrightarrow{\text{light}} P680^+ + Ph^-$$

Reduced pheophytin, Ph^-, transfers an electron to plastoquinone Q_A, a protein-bound species, and Q_B, a free form. Q_A and Q_B have structures similar to coenzyme Q in mitochondria. The next component of the electron-transport chain linking photosystems I and II is composed of a cytochrome *bf* complex. This is an assembly of several integral membrane proteins that use heme groups and iron–sulfur centers to transfer electrons from reduced Q_B to a blue copper protein, plastocyanin (PC). The reduced copper center in PC transfers an electron to $P700^+$, the electron-deficient form of P700.

The electron-deficient $P680^+$ formed in the above reaction is filled with electrons from H_2O. In this process, water, which does not readily give up electrons (it is a poor reducing agent), is oxidized by the metalloprotein, **water-splitting complex.** Water is oxidized to O_2 as follows:

$$2\,H_2O \longrightarrow 4\,H^+ + 4\,e^- + O_2$$

Water-splitting complex, which contains a cluster of four manganese ions, transfers one electron at a time from water to $P680^+$ molecules. The mechanism of the water-splitting complex is not completely understood, but experimental evidence supports a role for the manganese ions in the electron-transfer reactions.

The electron flow from H_2O to $NADP^+$ is described by the **Z scheme,** so-called because of its shape. The Z scheme links together photosystems I and II. The net reaction for the photochemically driven oxidation of H_2O and reduction of $NADP^+$ is

$$2\,H_2O + 2\,NADP^+ \xrightarrow{8\,h\nu} O_2 + 2\,NADPH + 2\,H^+$$

Eight photons (four delivered to each photosystem) are required to transfer the four electrons necessary to split each H_2O. Those four electrons end up in NADPH, two electrons per molecule of NADPH.

Photophosphorylation

In the 1950s, Daniel Arnon at Berkeley discovered that photoinduced electron flow in spinach chloroplasts as described by the Z scheme is accompanied by the phosphorylation of ADP. This process, called **photophosphorylation,** converts light energy into chemical bond energy. Photosynthetic organisms use the product ATP for energy-requiring metabolic processes. Experimental measurements made on illuminated chloroplasts provide evidence that photophosphorylation occurs by a mechanism very similar to oxidative phosphorylation. Photoinduced electron transfer from H_2O to $NADP^+$ pumps protons through the thylakoid membrane from the stromal side (outside) to the inner compartment (Figure 17.26). On the outer surface of thylakoid membranes are protein complexes, CF_0 and CF_1, that together comprise the ATP synthase of chloroplasts. Their biochemical roles are similar to the analogous proteins, F_0 and F_1, in the mitochondrial membrane (see Figure 17.14). CF_0 is a protein that spans the thylakoid membrane and acts as a proton channel. CF_1 is a peripheral membrane protein possessing binding and catalytic sites for combining ADP and P_i. The yield of ATP in photophosphorylation is difficult to measure accurately but the best estimates are one to two ATP formed per pair of electrons transferred from H_2O to $NADP^+$. This compares to approximately three ATP formed per pair of electrons transferred from NADH to O_2 in mitochondrial respiration.

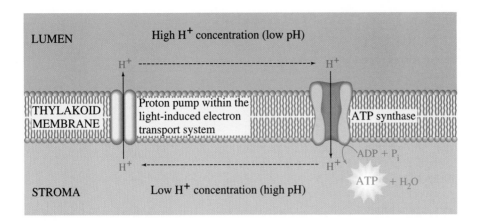

Figure 17.26
Pumping of protons through the thylakoid membrane during photoinduced electron transport. Electron transfer through the series of electron carriers pumps protons from the stromal side to the inner compartment (lumen). The resulting proton concentration difference provides energy for ADP phosphorylation. ATP synthase of chloroplasts couples the energy from the collapse of the proton concentration difference to the synthesis of ATP. The ATP synthase is comprised of CF_0 and CF_1 protein complexes.

The Z-shaped scheme for electron flow shown in Figure 17.25 is *noncyclic;* that is, there is a beginning point (H_2O) and an end point (NADPH). The products NADPH and O_2 continue to accumulate as long as light, H_2O, and $NADP^+$ are available for reaction. An alternative pathway of electron flow and photophosphorylation that is *cyclic* is also present in green plant chloroplasts. Cyclic electron flow results in the production of ATP but no NADPH or O_2 evolution. This alternate path is limited to the electron-transport chain associated with photosystem I. Electrons excited in P700* are passed through some of the initial carriers in the electron chain but they never reach $NADP^+$ (Figures 17.25 and 17.27). They are shunted from ferredoxin to the cytochrome *bf* complex, where they flow through plastocyanin back to the electron-deficient $P700^+$. Continuous illumination of photosystem I results in a continuous cycle of electron flow. Since there is no reduction of $NADP^+$, there is no need for the oxidation (splitting) of water and no production of O_2. There is, however,

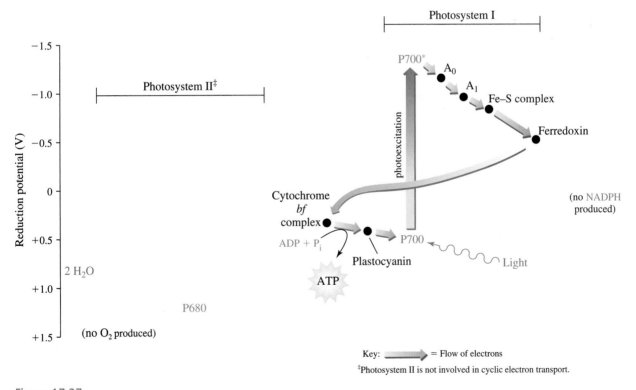

Figure 17.27
Pathway of cyclic electron flow in green plant chloroplasts. Electron flow from the cytochrome *bf* complex to plastocyanin generates a proton gradient, which is used to synthesize ATP. No NADPH or O_2 are produced. *Source:* Based on Moran et al., 1994.

proton pumping across the thylakoid membrane and, hence, ATP synthesis. In summary, cyclic electron flow has the following characteristics:

1. No NADPH is produced.
2. No water is oxidized.
3. No O_2 evolution occurs.
4. But ADP is phosphorylated to ATP.

Cyclic electron flow and accompanying cyclic photophosphorylation are thought to be important in regulating the cellular concentrations of ATP and NADPH. Metabolic logic would predict the cyclic pathway to be allowed when plant cells have sufficient amounts of NADPH but may need ATP.

17.6   Synthesis of Carbohydrates by the Calvin Cycle

We have seen how light energy is trapped by pigments in the chloroplasts of photosynthetic cells and transduced to energy in the form of phosphoanhydride bonds (ATP) and reducing power (NADPH). Both of these reactive products of the light phase of photosynthesis are available for biosynthesis or other energy-requiring processes in the cells as described in Chapters 14 to 19. In this section we describe how green plants use the ATP and NADPH to incorporate CO_2, an inorganic form of carbon, into simple organic molecules that are used as building blocks for glucose, sucrose, starch, and other carbohydrates (Figure 17.28). These metabolic reactions are called "dark reactions" because they do not require direct involvement of light, just the products of the photosynthetic process (ATP and NADPH). Specific reaction types include the formation of new carbon–carbon bonds (fixation of carbon) using CO_2 and the reduction of the incorporated CO_2 to the level of carbohydrates (aldehyde and alcohol). Some of the reactions we encounter in photosynthetic carbohydrate synthesis are new; however, others have been introduced in Chapter 15. The dark reactions are presented as individual reactions, but they all fit into a broader picture of a metabolic pathway called the **Calvin cycle.** Division of the overall process into four stages assists in learning the reactions and their importance.

Stage I: Addition of CO_2 to an Acceptor Molecule (Carbon Fixation)

The key reaction in photosynthetic carbon fixation was discovered in the laboratory of Melvin Calvin at Berkeley in the 1940s. Researchers in his laboratory illuminated algae in the presence of radiolabeled carbon dioxide ($^{14}CO_2$) and used paper chromatography to identify organic chemical products containing ^{14}C. One of the first metabolites to be identified was 3-phosphoglycerate (3-PG), with ^{14}C concentrated in the carboxyl carbon atom.

Figure 17.28

Coupling of dark and light reactions in photosynthesis. The products of the light reactions of photosynthesis (ATP and NADPH) are used to synthesize carbohydrates. The oxidation (recycling of NADPH) and cleavage of ATP are completed by reactions of the Calvin cycle, which incorporate CO_2 into organic acceptor molecules (called dark reactions). The carbohydrate product from the Calvin cycle is represented as glucose, $C_6H_{12}O_6$.

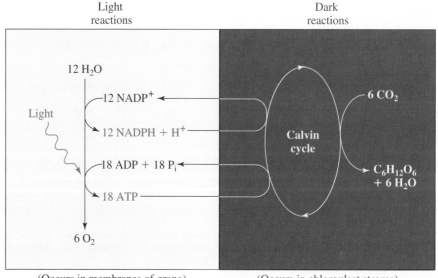

Further experiments showed the metabolite to be produced in a reaction catalyzed by the enzyme **ribulose 1,5-bisphosphate carboxylase/oxygenase,** called rubisco for short (Figure 17.29). The enzyme catalyzes the addition of inorganic carbon (CO_2) to ribulose 1,5-bisphosphate to form a six-carbon intermediate, followed by cleavage of the intermediate. This results in two molecules of the three-carbon compound 3-phosphoglycerate. (The enzyme has dual catalytic activities, carboxylation and oxygen insertion.) Rubisco from plants is a large protein with a molecular weight of 560,000 and is located in the chloroplast stroma, where in most plants it comprises at least 15% of the total chloroplast protein. The high concentration of rubisco in chloroplasts indicates a significant role for the enzyme in plant metabolism. It is estimated that 4×10^{13} g of the enzyme are made on the planet each year, which is equivalent to a million grams produced per second.

Cutting Edge: Methanol as fuel

Stage II: Entry of 3-Phosphoglycerate into Mainstream Metabolism

This is not our first encounter with the metabolite 3-phosphoglycerate. In our earlier introduction to 3-PG, we found it was produced in the cytoplasm during glycolysis (see Chapter 15, Section 15.3) and had two possible fates: catabolism to pyruvate for energy production or anabolism to glucose and storage in glycogen. 3-Phosphoglycerate synthesized in chloroplast stroma is used for the synthesis of glucose by a pathway similar to the reverse of glycolysis. Phosphorylation of the carboxyl group (C1) of 3-PG produces the important glycolytic intermediate, 1,3-bisphosphoglycerate (1,3-BPG):

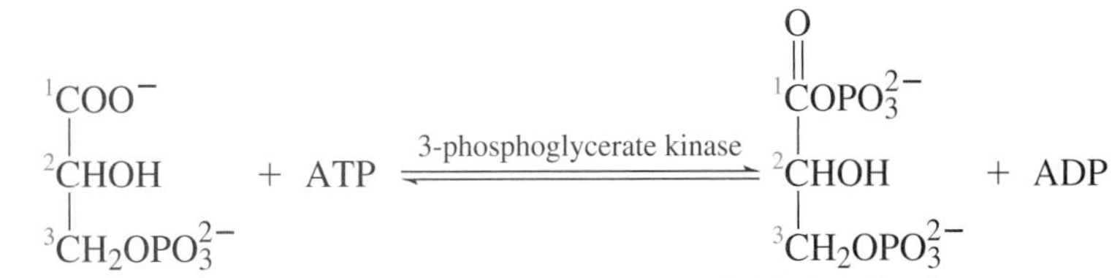

3-Phosphoglycerate 1,3-Bisphosphoglycerate

Stromal 1,3-BPG is reduced in a reaction requiring the enzyme glyceraldehyde 3-phosphate dehydrogenase and NADPH as the nucleotide cofactor:

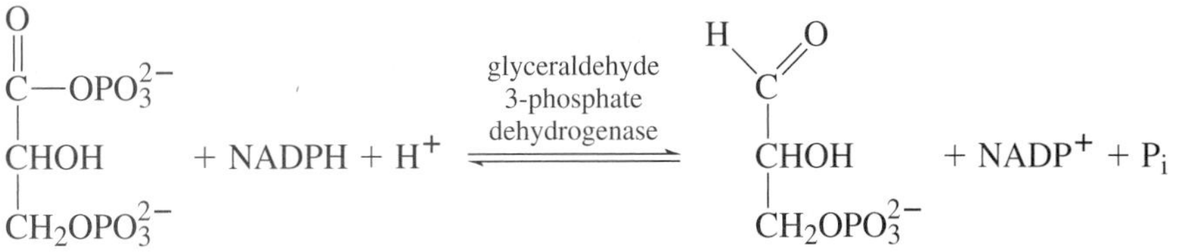

Recall that NADH is the cofactor for the analogous reaction for cytoplasmic glucose synthesis (see Chapter 15, Section 15.4).

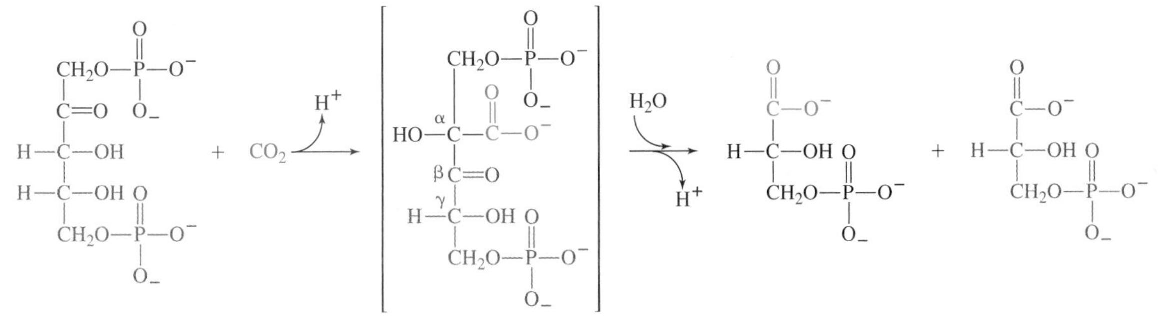

Ribulose 1,5-bisphosphate β-Keto acid intermediate 3-Phosphoglycerate 3-Phosphoglycerate

Figure 17.29

Carbon-fixing reaction catalyzed by ribulose-1,5-bisphosphate carboxylase. This represents stage I of the dark reactions, addition of CO_2 to an acceptor molecule, ribulose 1,5-bisphosphate.

Stage III: Synthesis of Carbohydrates from Glyceraldehyde 3-Phosphate

The formation of carbohydrates from glyceraldehyde 3-phosphate (and its isomerized relative, dihydroxyacetone phosphate, DHAP) follows pathways in gluconeogenesis we have already explored (Table 17.4; also see Chapter 15, Section 15.4). The primary carbohydrates synthesized by plants are monosaccharides, such as glucose and fructose, the disaccharide sucrose, and polysaccharide starch. Glucose and fructose are often represented as photosynthetic end products, but in reality they are produced in a phosphorylated form (glucose 1-phosphate, glucose 6-phosphate, or fructose 6-phosphate) and used for sucrose and starch synthesis. Sucrose is synthesized as previously described from UDP-glucose and fructose 6-phosphate (Table 17.5). Most sucrose is synthesized in the plant cytoplasm from DHAP transported from stroma. The polysaccharide starch is formed by adding activated glucose (ADP-glucose) to an existing starch chain (Table 17.6). Stages I, II, and III of the Calvin cycle are outlined in Figure 17.30.

Stage IV: Completion of the Calvin Cycle by Regeneration of Ribulose 1,5-Bisphosphate

At this stage in our discussion on carbohydrate synthesis we have fixed a single CO_2 onto the C_5 carbon sugar (ribulose 1,5-bisphosphate) and generated two C_3 units (3-phosphoglycerate). In addition, we have described how two C_3 units are used to build monosaccharides, which can be used to construct disaccharides and polysaccharides. We complete the Calvin cycle by showing how ribulose 1,5-bisphosphate is regenerated. To account for the six carbons used to make glucose, six Calvin cycles must be completed with entry of six molecules of CO_2. This stoichiometry can be shown in reaction form:

$$6 \text{ ribulose 1,5-bisphosphate} + 6\,CO_2 + 12\,NADPH + 12\,H^+ + 12\,ATP + 6\,H_2O \longrightarrow$$
$$12 \text{ glyceraldehyde 3-phosphate} + 12\,NADP^+ + 12\,ADP + 12\,P_i$$

As shown in Figure 17.30, only 2 of the 12 glyceraldehyde 3-phosphates are used for the synthesis of glucose (see Table 17.4). The balance of 10 molecules of glyceraldehyde 3-phosphate (C_3) are necessary to complete the Calvin cycle. The 10 C_3 units are rearranged to generate 6 C_5 molecules. The sequence of reactions outlining this transformation requires some new chemistry, which uses intermediates with three, four, five, six, and seven carbons. Included in the set of reactions is a new chemical process, transketolation, the transfer of a C_2 unit from a ketose to an aldose. This transfer produces a new aldose (aldose′) and a new ketose (ketose′):

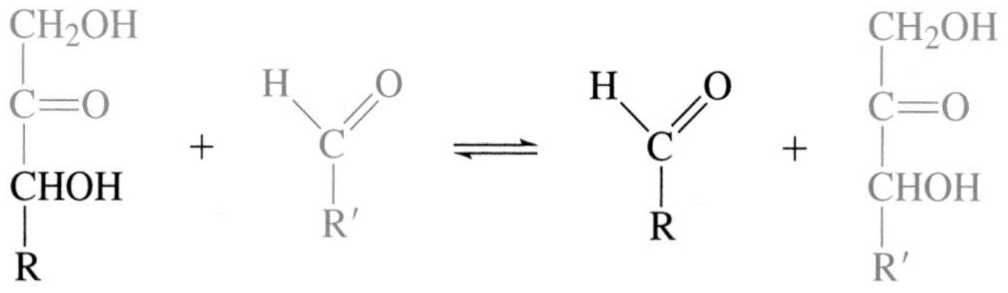

Table 17.4
Synthesis of phosphorylated glucose in green plants

Glyceraldehyde 3-phosphate $\xrightleftharpoons{\text{isomerase}}$ dihydroxyacetone phosphate	
DHAP + glyceraldehyde 3-phosphate $\xrightleftharpoons{\text{aldolase}}$ fructose 1,6-bisphosphate	
Fructose 1,6-bisphosphate + H_2O $\xrightleftharpoons{\text{phosphatase}}$ fructose 6-phosphate + P_i	
Fructose 6-phosphate $\xrightleftharpoons{\text{isomerase}}$ glucose 6-phosphate	
Glucose 6-phosphate $\xrightleftharpoons{\text{phosphoglucomutase}}$ glucose 1-phosphate	

Glucose 1-phosphate + UTP $\rightleftharpoons$ UDP-glucose + PP$_i$
UDP-glucose + fructose 6-phosphate $\rightleftharpoons$ sucrose 6-phosphate + UDP
Sucrose 6-phosphate + H$_2$O $\rightleftharpoons$ sucrose + P$_i$

Table 17.5
Synthesis of sucrose in green plants beginning with glucose 1-phosphate

Five trioses (three glyceraldehyde 3-phosphates, two dihydroxyacetone phosphates) enter and three pentoses (two xylulose 5-phosphates, one ribose 5-phosphate) exit. The C$_5$ sugars are isomerized to ribulose 5-phosphate. Before the ribulose 5-phosphate molecules can reenter the cycle, they must be phosphorylated. If each of these reactions is carried out twice, we can account for the recycling of ten triose molecules.

All of the reactions for synthesis of glucose (C$_6$H$_{12}$O$_6$) from CO$_2$ can be combined into a single net reaction summarizing the Calvin cycle (dark reactions):

$$6\ CO_2 + 12\ NADPH + 12\ H^+ + 18\ ATP + 12\ H_2O \rightarrow$$
$$C_6H_{12}O_6 + 12\ NADP^+ + 18\ ADP + 18\ P_i$$

Notice the need for large numbers of the reactive molecules, NADPH and ATP, which are generated by the light reactions. For each CO$_2$ incorporated into carbohydrates, two molecules of NADPH and three molecules of ATP are required.

Photorespiration

In our earlier discussion of rubisco, we focused on its action as a carboxylase (fixation of CO$_2$). As indicated by the enzyme name, it can act as an oxygenase by substituting O$_2$ in place of CO$_2$ as a substrate:

$$\text{Ribulose 1,5-bisphosphate} + O_2 \xrightarrow{\text{rubisco}} \text{Phosphoglycolate} + \text{3-Phosphoglycerate}$$

The substrate ribulose 1,5-bisphosphate is oxidized and cleaved into two parts, a C$_2$ unit, phosphoglycolate, and a C$_3$ unit, 3-phosphoglycerate. The action of rubisco as an oxygenase appears to be counterproductive to photosynthesis because it leads to O$_2$ *consumption* (characteristic of animal metabolism) and, in addition, reduces carbohydrate production by generating less 3-phosphoglycerate. It is estimated that the carboxylase to oxygenase activities ratio in the chloroplast is 3:1. The process of O$_2$ consumption observed in plants is called **photorespiration.**

Some plants have adapted to the apparently wasteful process of photorespiration by developing an optional pathway for CO$_2$ fixation. This pathway was discovered in tropical plants such as sugarcane, corn, and sorghum by two Australian biochemists, Marshall Hatch and Rodger Slack. The **Hatch–Slack pathway** begins with the incorporation of CO$_2$ into phosphoenolpyruvate instead of ribulose 1,5-bisphosphate (Figure 17.31). Because of the presence of four-carbon intermediates (oxaloacetate and malate) in the Hatch–Slack pathway, it is sometimes called the C$_4$ pathway and plants that use this pathway, C$_4$ plants. (Those that use only the Calvin cycle are C$_3$ plants, so named because the first metabolite

Courtesy of Agricultural Research Center, USDA

Sugarcane plants use the Hatch–Slack pathway to fix CO$_2$.

Glucose 1-phosphate + ATP $\rightleftharpoons$ ADP-glucose + PP$_i$
ADP-glucose + (glucose)$_n$ $\rightleftharpoons$ (glucose)$_{n+1}$ + ADP
preexisting   extended
starch   starch

Table 17.6
Synthesis of starch in green plants beginning with glucose 1-phosphate

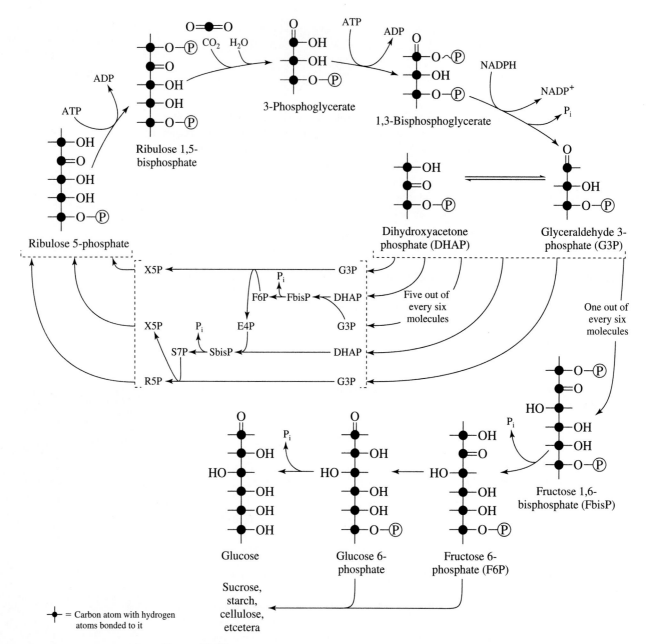

Figure 17.30

Reactions of stages I, II, and III of the Calvin cycle. E4P = erythrose 4-phosphate; SbisP = sedoheptulose 1,7-bisphosphate; S7P = sedoheptulose 7-phosphate; R5P = ribose 5-phosphate; X5P = xylulose 5-phosphate.

from CO_2 fixation, 3-phosphoglycerate, contains three carbons. C_3 plants include wheat and most others grown in the temperate parts of the world.) In C_4 plants the Hatch–Slack pathway is linked to the Calvin cycle. The CO_2 produced from malate decarboxylation is refixed by rubisco into ribulose 1,5-bisphosphate. The linkage of the Hatch–Slack pathway with the Calvin cycle is shown in Figure 17.31. Note that the combined pathways occur in two different cells: mesophyll cells and bundle sheath cells. Mesophyll cells produce malate, which is transported to bundle sheath cells where it is decarboxylated. The CO_2 product is utilized by the Calvin cycle. Intermediates from the C_4 pathway effectively carry CO_2 to those cells that catalyze CO_2 reduction and assimilation.

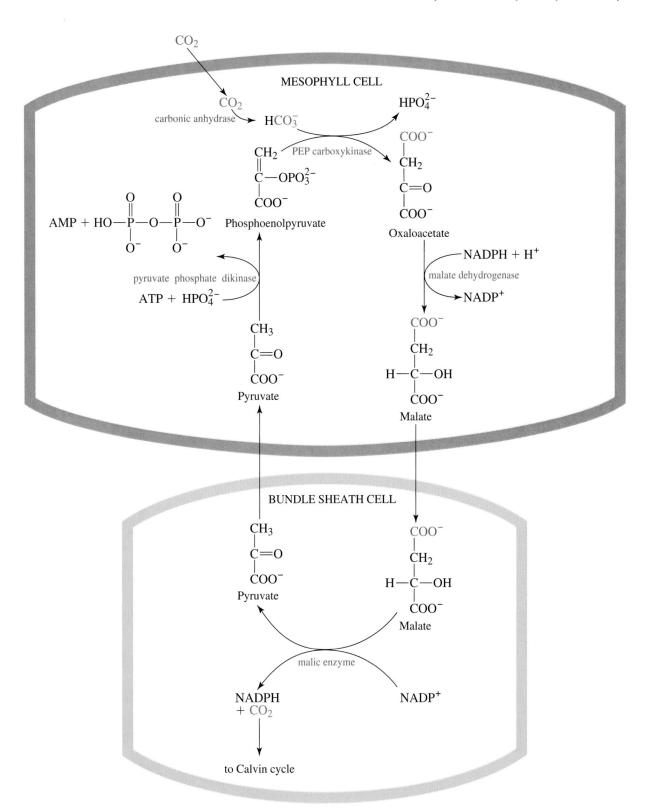

Figure 17.31
Combination of the Hatch–Slack pathway and the Calvin cycle in C_4 plants. The linkage requires reactions in mesophyll cells and bundle sheath cells.

Summary

■ Most of the energy available in fuel metabolism is released during electron transport and ATP synthesis, Stage III of metabolism. The combined processes are called **oxidative phosphorylation.**

■ The cellular sites of electron transport and ATP synthesis are the mitochondria, organelles characterized by two membrane systems and specialized for aerobic metabolism.

■ During catabolism, electrons removed from nutrients are transferred by dehydrogenases to the cofactors NAD^+ and FAD to produce NADH and $FADH_2$. These reduced cofactors are oxidized by transfer of electrons, through the electron transport chain, to O_2.

■ During electron transport, unidirectional pumping of protons generates a high energy gradient that collapses, coupling the energy to ATP synthesis.

■ Photosynthetic organisms absorb energy from solar energy radiation and direct the energy through electron transport chains to synthesize ATP and NADPH. These high-energy products are used for making carbohydrates from CO_2 and H_2O. The process occurs in the chloroplasts of plants.

■ The initial step in photosynthesis is absorption of light by chlorophyll and other accessory pigments. Electrons are then transferred through two photosystems to produce NADPH and ATP.

■ In the dark reactions, ATP and NADPH made by photoabsorption and electron transport are used as a source of energy to incorporate CO_2 into sugars. This reaction is initiated by ribulose 1,5-bisphosphate carboxylase/oxygenase. The reaction converts the six-carbon metabolite ribulose 1,5-bisphosphate into two molecules of 3-phosphoglycerate.

Study Exercises
Understanding Terms

17.1 Define the following terms in 25 words or less.

a. Electron transport
b. Mitochondrial matrix
c. Dehydrogenase
d. Standard reduction potential, $E°'$
e. Cytochrome
f. Iron–sulfur clusters
g. Chemiosmotic coupling
h. Thermogenin
i. Thylakoids
j. Accessory pigments
k. Ferredoxin
l. Photophosphorylation
m. 3-Phosphoglycerate
n. Photorespiration
o. ATP synthase
p. Ubiquinone
q. Chlorophyll
r. Malate-aspartate shuttle
s. Carbon fixation
t. Chloroplasts
u. Photosystem I
v. Rubisco

Reviewing Concepts

17.2 Which of the following statements about mitochondrial respiration are true? Rewrite false statements to make them true.

a. The electron carriers involved in electron transport are present in the cytoplasm of the cell.
b. NADH is a more powerful reducing agent than $FMNH_2$.
c. The final electron acceptor is O_2.
d. The electron carriers are arranged in order of increasing electron affinity.
e. Fe^{3+} atoms in cytochromes and iron–sulfur proteins act as electron acceptors.
f. $FADH_2$ can transfer one or two electrons to an acceptor.
g. During electron transport, protons are pumped from the inner mitochondrial matrix to the other side of the membrane.

HINT: For part b, check Table 17.2

17.3 Calculate $\Delta G°'$ for several electron-transfer steps listed below. Is enough energy released for ATP synthesis?

a. NADH → CoQ
b. $FADH_2$ → CoQ
c. Cyt c → cyt a (one electron)
d. $CoQH_2$ → 2 cyt c
e. $FMNH_2$ → CoQ

17.4 For each pair of electron carriers listed below, identify the better reducing agent (electron donor).

a. NADH:$FADH_2$
b. $FADH_2$:$CoQH_2$
c. Cyt a/a_3 (reduced):cyt c (reduced)
d. Cyt b (reduced):$CoQH_2$
e. H_2O:NADH
f. Succinate:$FADH_2$

17.5 For each pair of electron carriers listed below, identify the better oxidizing agent (electron acceptor).

a. CoQ:FAD
b. NAD^+:O_2
c. Cyt c (oxidized):cyt a/a_3 (oxidized)
d. Cyt b (oxidized):cyt c (oxidized)

e. NAD^+:FAD

f. Cyt c (oxidized):O_2

17.6 Describe the metabolic differences between heterotrophic and phototrophic organisms.

17.7 What is the role of each of the following in mitochondrial oxidative phosphorylation?

a. Cyt c **d.** F_0

b. F_1 **e.** CoQ

c. Cyt a/a_3 **f.** Iron–sulfur proteins

17.8 Which of the following statements about photophosphorylation are true? Rewrite each false statement so it is true.

a. Electrons flow from NADPH to O_2.

b. A proton concentration gradient is produced across the thylakoid membrane.

c. The major photoreceptor molecule in green plants is chlorophyll a.

d. Noncyclic photophosphorylation results in production of NADPH, oxidation of H_2O, evolution of O_2, and phosphorylation of ADP.

e. Cyclic photophosphorylation results in production of ATP, oxidation of water, and evolution of O_2.

f. Electrons will flow spontaneously from reduced ferredoxin, FdH_2, to $NADP^+$.

17.9 Define a role for the accessory pigments (carotenoids, etc.) found in chloroplasts.

17.10 Describe the biochemical differences between cyclic and noncyclic photophosphorylation.

17.11 What is the role of each of the following in chloroplast photophosphorylation?

a. Chlorophyll a

b. β-Carotene

c. P700

d. Ferredoxin

e. Plastocyanin

f. Manganese water-splitting complex

g. CF_0

h. CF_1

17.12 What electron carrier groups (cofactors or prosthetic groups) are associated with the following respiratory enzyme complexes? Choose from the list of electron-carrying groups below.

Enzyme complexes	Electron-carrying groups
a. Succinate-CoQ reductase	1. FMN
b. NADH-CoQ reductase	2. Fe–S
c. Cytochrome c reductase	3. FAD
d. Cytochrome c oxidase	4. Heme
	5. Cu
	6. NAD^+

17.13 Match each of the biomolecules or complexes listed with its primary function taken from the second list.

Biomolecules	Biochemical functions
a. Carotenoids	1. Light acceptor for photosystem I
b. Chlorophylls	
c. P700	2. Light acceptor for photosystem II
d. P680	
e. Water-splitting complex	3. Secondary pigments
	4. Primary pigments
f. Calvin cycle	5. Synthesis of carbohydrates

$$6.\ 2\,H_2O + 2\,A \xrightarrow{h\nu} 2\,AH_2 + O_2$$

17.14 Discuss the similarities and differences between the two chemical activities of ribulose 1,5-bisphosphate carboxylase/oxygenase.

17.15 Identify the cellular location of each of the reactions in Table 17.1.

17.16 Describe the major differences and similarities between electron transport in mitochondria and in chloroplasts.

17.17 What metal ions are important in mitochondrial electron transport?

17.18 What metal ions are important in chloroplast electron transport?

Solving Problems

17.19 A rat injected with 2,4-dinitrophenol shows an increase in body temperature, an increase in breathing rate, and a decrease in body weight. Explain the effect of 2,4-dinitrophenol.

17.20 The uncoupling agent 2,4-dinitrophenol acts by making the inner mitochondrial membrane "leaky." Protons can readily diffuse through such a leaky membrane. Explain how this disrupts oxidative phosphorylation.

17.21 What is the yield of phosphoanhydride bonds produced from the complete catabolism of each of the following compounds?

a. Fructose (in heart muscle)

b. Acetyl CoA

c. Maltose (in heart muscle)

d. Pyruvate

e. Glucose 6-phosphate (in skeletal muscle)

HINT: Derive the net equation for degradation to CO_2 and H_2O.

17.22 Describe the biochemistry that explains color changes of tree leaves in autumn.

HINT: Chlorophylls are green; carotenoids and phycobilins are red and yellow.

17.23 Compare the synthesis of glucose by the Calvin cycle with gluconeogenesis in animal cells.

HINT: Derive a net reaction for each of the processes.

17.24 Describe the biochemical similarities and differences of mitochondria and chloroplasts.

17.25 Explain the differences between the synthesis of ATP by substrate-level phosphorylation and oxidative phosphorylation.

17.26 Outline the steps for sucrose synthesis beginning with four molecules of glyceraldehyde 3-phosphate.

17.27 What amino acid residue plays an important role in binding iron at the redox site in the iron–sulfur proteins?

17.28 The following reaction represents a method of electron flow in chloroplasts. Complete the reaction.

$$Fe^{2+} + Cu^{2+} \longrightarrow$$

17.29 Describe the kind of chemistry in each of the following reactions.
  a. $AH_2 + NAD^+ \rightleftharpoons A + NADH + H^+$
  b. $ADP + P_i \rightleftharpoons ATP + H_2O$
  c. Glyceraldehyde 3-phosphate $\rightleftharpoons$ dihydroxyacetone phosphate
  d. Ribulose 1,5-bisphosphate $+ CO_2 \rightleftharpoons$ 2 3-phosphoglycerate

17.30 What chemicals cause the yellow color in autumn leaves?

17.31 Review the general reaction catalyzed by the enzyme cytochrome P450 in Window on Biochemistry 17.1. The type of reaction is oxidation–reduction. List all reactants that are oxidized in the reaction. List all reactants that are reduced in the reaction.

Writing Biochemistry

17.32 Read the reference in the "Further Reading" section by Govindjee and Coleman, "How plants make oxygen", *Sci. Am.* 262(2):50–58 (1990). Write a 200-word summary of the article.

17.33 You need to prepare a quick written review of this chapter to study for an exam. Write a 200- to 300-word summary of the first half of this chapter (Sections 17.1–17.4.)

17.34 Look again at exercise 16.32 on the subject of cofactors for the succinate dehydrogenase reaction. Now that you have studied the concepts of electron transport, answer the question again. How do you think the reaction catalyzed by succinate dehydrogenase should have been presented in Chapter 16, using $FADH_2$ or ubiquinone?

Applying Biochemistry

17.35 The uncoupling agent, dinitrophenol, was used many years ago as a dieting agent. It was effective in this regard, but its use was halted because it was extremely toxic and several individuals died. Describe the biochemical action of dinitrophenol as a weight-reducing agent.

17.36 Your biochemistry instructor has announced a class quiz covering the topic of photorespiration. To study for the quiz, read the appropriate section in this text and in an advanced biochemistry book. Write a 100-word summary of your reading.

Further Reading

Barber, J., and Anderson, B., 1994. Revealing the blueprint of photosynthesis. *Nature* 370:31–34.

Beinert, H., Holm, R., and Munck, E., 1997. Iron–sulfur clusters. *Science* 277:653–659.

Boyer, P., 1997. The ATP synthase—A splendid molecular machine. *Annu. Rev. Biochem.* 66:717–750.

Boyer, R., 2000. Isolation and characterization of plant pigments. In *Modern Experimental Biochemistry,* 3rd ed., pp. 333–343. San Francisco: Benjamin/Cummings.

Brown, S., 2000. Does succinate oxidation yield $FADH_2$ or ubiquinol? *Biochem. Educ.* 28:52–54.

Brunori, M., 2001. Nitric oxide, cytochrome *c* oxidase, and myoglobin. *Trends Biochem. Sci.* 26:21–23.

Capaldi, R., 2000. The changing face of mitochondrial research. *Trends Biochem. Sci.* 25:212–214.

Cornely, K., 1999. The electron transport game. *Biochem. Educ.* 27:74–76.

Douglas, R., et al., 1998. Dragon fish see red using chlorophyll. *Nature* 393:423–424.

Elliott, K., 1999. Teaching metabolism for the future: Concepts not pathways. *The Biochemist* 21:9–39.

Gibson, F., 2000. The introduction of *Escherichia coli* and biochemical genetics to the study of oxidative phosphorylation. *Trends Biochem. Sci.* 25:342–344.

Gonzalez, G., 1995. An easy approach to the Calvin cycle. *Biochem. Educ.* 23:132–133.

Govindjee and Coleman, W., 1990. How plants make oxygen. *Sci. Am.* 262(2):50–58.

Holliday, R., 2001. Aging and the biochemistry of life. *Trends Biochem. Sci.* 26:68–71.

Junge, W., Lill, H., and Englebrecht, S., 1997. ATP synthase. *Trends Biochem. Sci.* 22:420–423.

Kiberstis, P., 1999. Mitochondria make a comeback. *Science* 283:1475.

Konigsberg, M., 1999. A simple model to facilitate students' understanding of the mitochondrial respiratory chain. *Biochem. Educ.* 27:9–11.

Nagyvary, J., and Bechert, J., 1999. New insights into ATP synthesis. *Biochem. Educ.* 27:193–199.

Chapter 18

Metabolism of Fatty Acids and Lipids

Cheese and other dairy products melt in your mouth because they contain low melting triacylglycerols.

The last three chapters have described how glucose and other carbohydrates are degraded to pyruvate by glycolysis followed by aerobic oxidation to CO_2 via acetyl CoA and the citric acid cycle. Energy released by these metabolic processes is collected in substrate-level ATP and in the reduced cofactors, NADH and $FADH_2$. Reduced cofactors are recycled by transfer of electrons via the electron-transport chain to O_2 with concomitant production of ATP. Although glucose is an important energy source, humans and other animals are not able to store large quantities of this carbohydrate. It is estimated that in an average adult, glycogen reserves are depleted in 12 to 15 hours after eating. This time frame is shortened if the person is conducting moderate or active exercise. Glycogen is useful only as a short-term fuel molecule. For long-term energy, organisms turn to the utilization of stored fat. A 70-kg human has fuel reserves of about 350,000 to 400,000 kJ in total body fat, compared to approximately 7500 kJ of available energy in total glycogen and glucose. Experiments on humans have shown that greater than 50% of the energy needs in liver, kidney, and skeletal muscle come from fat breakdown. Heart muscle may get up to 95% of its energy from fat oxidation. Under some conditions, fats become the predominant fuel molecules. For example, in fasting animals and in untreated diabetes, fats are the major or sole source of energy in muscle cells. Hibernating animals and migrating birds also do not eat food for long periods but rely on stored fats for energy. In germinating seedlings of many plants such as the sunflower, stored fats are the primary fuel molecules.

Fuel molecules in the fat category are primarily the fatty acids, which are stored in cells as triacylglycerols. The fatty acids are especially appropriate as long-term energy molecules because they provide much more energy per unit weight than carbohydrates. Fatty acids release approximately 38 kJ/g, compared to only 16 kJ/g for carbohydrates. Fatty acids are highly reduced and must proceed through several oxidation steps before complete degradation to CO_2. An additional advantage of fatty acids as fuel molecules comes from their mode of storage. Glucose is stored in the hydrophilic molecules glycogen (animals) or starch (plants). Each gram of these polysaccharides is associated with 2 to 3 g of water. This means that only one-quarter to one-third of the total weight of stored carbohydrate is available for energy metabolism. Triacylglycerols, nonpolar molecules that avoid water, are stored in an anhydrous form in adipose tissue.

To use the stored energy, fatty acids must be released from the triacylglycerols and transported to the mitochondria of peripheral tissue, where they are degraded. Catabolism of fatty acids is brought about by **β oxidation,** a four-step process yielding acetyl CoA, which feeds into aerobic metabolism at the citric acid cycle. When fuel molecules are available in amounts greater than energy needs, fatty acids are synthesized, linked to glycerol, and stored in adipose tissue. Fatty acid synthesis and β oxidation display similar chemistries; however, different enzyme systems and even different cell compartments are used for the two processes. Degradation by β oxidation occurs in mitochondria, whereas fatty acid synthesis occurs in cytoplasm.

Here we explore the absorption, transport, metabolism, and nutritional implications of fatty acids. We also take a look at one of the most important non–fatty acid lipids, **cholesterol.** Its synthesis from acetyl CoA and metabolism to important, specialized biomolecules is emphasized. The transport of the very water-insoluble cholesterol throughout organisms is aided by protein molecules. The medical link between the transport of cholesterol and cardiovascular health is also explored.

18.1   Metabolism of Dietary Triacylglycerols

To be utilized for energy, the fatty acids must be delivered to cells where β oxidation occurs (liver, heart, skeletal muscle). There are three primary sources of fatty acids for energy metabolism in humans and other animals:

1. Dietary triacylglycerols
2. Triacylglycerols synthesized in the liver
3. Triacylglycerols stored in adipocytes (fat cells) as lipid droplets

The digestion, absorption, and transport of dietary fats in higher animals are complicated by their lack of water solubility. Several steps of preparation are required before triacylglyc-

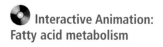

Interactive Animation: Fatty acid metabolism

erols are ready for storage or energy use. Those steps, from emulsification in the small intestines to storage in adipocytes or catabolism in muscle, are outlined in Figure 18.1. Bacteria obtain fats by consuming other organisms and storing the triacylglycerols as oil microdroplets in their cytoplasm. In higher plants, fats are of metabolic significance only during periods of seed germination. (See Window on Biochemistry 18.1).

Initial Digestion of Fats

Little happens to dietary triacylglycerols until they enter the small intestines. Here they are dispersed into microscopic micelles after mixing with bile salts released from the gallbladder. The bile salts, such as cholate and glycocholate (see Figure 9.11), act as detergent molecules to assist in the solubilization of dietary fats. The action of the bile salts makes the fats more accessible to water-soluble enzymes such as pancreatic lipase, which degrades the triacylglycerols to a mixture of free fatty acids, glycerol, monoacylglycerols, and diacylglycerols. The lipase catalyzes hydrolysis of ester linkages between the glycerol skeleton and fatty acids. The products are then taken up into the intestinal mucosa, where they are resynthesized into triacylglycerols. Before the newly formed triacylglycerols are released into the bloodstream, they combine with dietary cholesterol and specialized proteins into molecular aggregates called **lipoproteins.** Several different types of lipoproteins are found in blood. The most important ones for triacylglycerol transport are called **chylomicrons.**

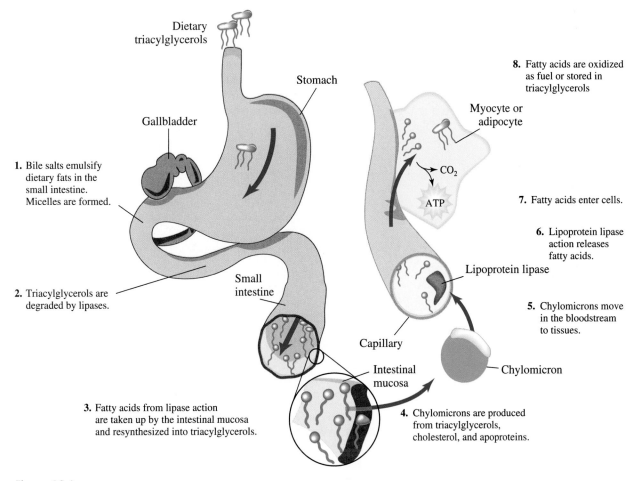

8. Fatty acids are oxidized as fuel or stored in triacylglycerols

7. Fatty acids enter cells.

6. Lipoprotein lipase action releases fatty acids.

5. Chylomicrons move in the bloodstream to tissues.

4. Chylomicrons are produced from triacylglycerols, cholesterol, and apoproteins.

3. Fatty acids from lipase action are taken up by the intestinal mucosa and resynthesized into triacylglycerols.

2. Triacylglycerols are degraded by lipases.

1. Bile salts emulsify dietary fats in the small intestine. Micelles are formed.

Dietary triacylglycerols

Stomach

Gallbladder

Small intestine

Capillary

Intestinal mucosa

Chylomicron

Lipoprotein lipase

Myocyte or adipocyte

CO_2

ATP

Figure 18.1

Digestion, mobilization, and transport of dietary triacylglycerols. Fats in the diet do not begin the digestion process until they are in the small intestines, where they are emulsified with bile salts (step 1). Steps 1 to 8 prepare the fats for transport to muscle and fat cells, where they are degraded by β oxidation or stored.

These contain a central core of triacylglycerols coated with cholesterol, proteins, and phospholipids to enhance water solubility (Figure 18.2). Dietary triacylglycerols make up between 80% and 90% of the contents of the chylomicrons. Triacylglycerols synthesized in the liver are combined into other plasma lipoprotein packages, including very low density and low density lipoproteins (50% and 10% triacylglycerols, respectively). In the blood, the triacylglycerol containing chylomicrons and other water-soluble lipoproteins are transported to peripheral tissues, including adipose, where a cell surface enzyme, **lipoprotein lipase,** promotes the release of free fatty acids by hydrolysis (see Figure 18.1). The fatty acids are taken up by cells for energy metabolism (muscle) or storage (adipocytes). The lipoproteins of blood plasma are quite efficient at maintaining lipids in a solubilized form. A human blood sample of 100 mL contains on average 480 mg of lipid, which is comprised of about 200 mg of cholesterol, 120 mg of triacylglycerol, and 160 mg of phospholipid. All of the lipids in the blood are packaged in lipoproteins for ease of transport in the water-based medium.

The mobilization (release) of fatty acids from storage in adipocytes is regulated by the hormones epinephrine and glucagon, which are released into the bloodstream in response to low glucose concentrations in blood. The hormone molecules present in blood bind to receptor sites on the surface of adipocytes. Here they initiate a process that generates an intracellular second messenger, **cyclic AMP** (see Chapter 2, Section 2.5). Cyclic AMP, acting indirectly through a protein kinase, activates hormone-sensitive **triacylglycerol lipase** (Figure 18.3). The lipase catalyzes the hydrolysis of stored triacylglycerols to yield free fatty acids and glycerol. This process is similar to the hormone-regulated mobilization of glucose from stored glycogen except that glucose is released in an activated form (glucose 1-phosphate). The relatively nonpolar fatty acids generated in the cytoplasm of adipocyte cells diffuse through the membrane into the bloodstream. Here they bind to a blood protein, serum albumin, and are transported to peripheral tissue for energy metabolism. In heart and skeletal muscle, the fatty acids are released from albumin and diffuse into the cells. The glycerol, monoacylglycerol, and diacylglycerol from adipocytes may be used to rebuild tri-

Figure 18.2

Schematic diagram of a chylomicron showing the central core of triacylglycerols and cholesteryl esters coated with a shell of proteins and lipids. The outside shell is composed of cholesterol, phospholipids, and apolipoproteins. Source: From Devlin, 1922.

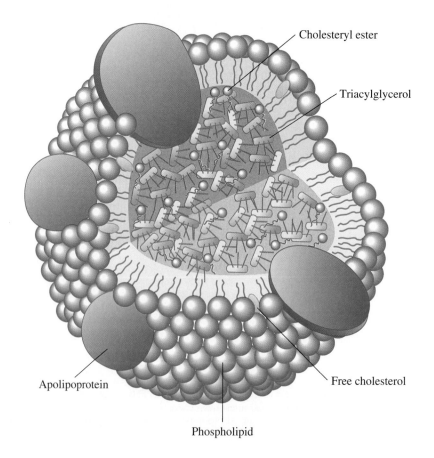

Cholesteryl ester

Triacylglycerol

Apolipoprotein

Free cholesterol

Phospholipid

acylglycerols in the liver. Some free glycerol may enter mainstream metabolism by conversion to dihydroxyacetone phosphate (see Chapter 15, Section 15.2).

Fatty Acids in Muscle Cells

Fatty acids that have been delivered to muscle tissue via blood capillaries diffuse through cell plasma membranes into the cytoplasm. All pathways of aerobic metabolism, including the β oxidation of fatty acids, occur in the mitochondria. Fatty acids that enter into the cytoplasm of muscle cells must be transported through the double membrane system of mitochondria before oxidative degradation can occur in the matrix region. Immediately upon entry into muscle cells, fatty acids are activated and at the same time labeled for energy degradation. This is similar to the action taken when glucose enters the cytoplasm. Glucose is activated by phosphoryl group transfer to produce glucose 6-phosphate (hexokinase). A hint about the type of activation for fatty acids comes from the metabolism of pyruvate and the citric acid cycle. When it is necessary to activate acids, they are linked as thioesters to CoASH (for example, acetyl CoA, succinyl CoA). The activation of fatty acids in muscle cytoplasm plays a broader role in that energizing the acids is coupled to their transport through the mitochondrial membranes. The enzyme **acyl CoA synthetase** catalyzes the activation process:

$$RCO^- + CoASH + ATP \rightleftharpoons CoASCR + AMP + PP_i$$
$$\quad \| \qquad\qquad\qquad\qquad\qquad\quad \|$$
$$\quad O \qquad\qquad\qquad\qquad\qquad\quad O$$

The K'_{eq} for this reaction is approximately 1, producing a reaction mixture of about 50% fatty acid CoA thioester and 50% free fatty acid. The reaction is driven to completion by

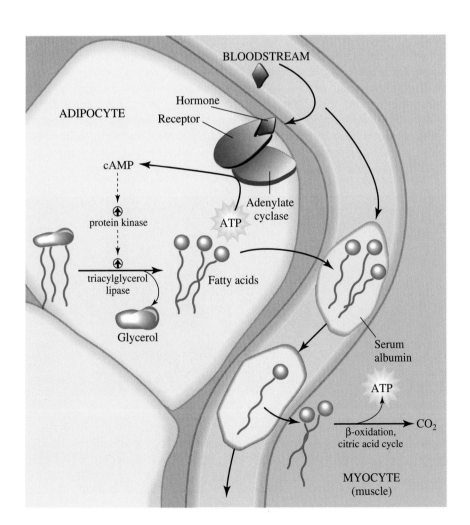

Figure 18.3

Hormone-activated mobilization of fatty acids stored in adipocytes. Triacylglycerol lipase in adipocytes is activated by protein kinase, which is stimulated by the second messenger, cyclic AMP. The cyclic AMP is produced from ATP in response to the binding of epinephrine or glucagon to a cell receptor site. Adenylate cyclase catalyzes the formation of cyclic AMP in the cell. Fatty acids mobilized in adipocytes diffuse into the bloodstream, where they are carried by albumin. Upon arrival in muscle cells, fatty acids are oxidized for energy.

removal of the product PP$_i$ by pyrophosphatase-catalyzed hydrolysis. We have encountered this process, which releases an extra burst of energy, at least twice before: in the activation of amino acids for protein synthesis (see Chapter 12, Section 12.2) and for the activation of acetate (see Chapter 16, Section 16.4). The activation step involves an unusual enzyme-bound intermediate, acyl adenylate or acyl-AMP. This reactive intermediate is commonly used throughout metabolism to couple ATP cleavage to thermodynamically unfavorable

Window on

Biochemistry

18.1   Olestra—A Modified Sugar Becomes a Fat Substitute

Nutritionists recommend that fats should provide no more than 30% to 35% of the total calories in our diets. Because of a demand for reduced-fat food, many food-manufacturing companies are searching for noncaloric fat substitutes. Fats bring special, desirable qualities to food: the creamy texture of ice cream, the smooth melting of chocolate, and the crispness of potato chips. The Food and Drug Administration (FDA) of the United States has recently approved the fat substitute, Proctor and Gamble's **Olestra,** for use in snack foods like potato chips, corn chips, and crackers. Olestra is a synthetic mixture of hexa-, hepta-, and octaesters formed between the hydroxyl groups of common table sugar, sucrose, and long-chain fatty acids (see structure). Only certain fatty acids may be used: Unsaturated fatty acids cannot be less then 25%, the content of C-12 and C-14 fatty acids cannot be greater than 1% each, C-20 and longer fatty acids should not be greater than 20%, and C-16 and C-18 fatty acids should comprise at least 78% of the fatty acids.

Olestra molecules have structures related to the natural fats, triacylglycerols, and give food the same rich, creamy texture and palatability. However, sucrose polyesters are different enough so they cannot become substrates for the digestive lipases. The bulkiness of the Olestra molecules does not allow them to be bound by the lipases; therefore, the ester bonds cannot be hydrolyzed. Olestra is not metabolized like natural fats and passes directly through the digestive tract unchanged. Therefore, the sucrose polyesters provide no calories. On the downside, Olestra molecules are hydrophobic and fat-soluble vitamins (A, D, E, K) and plant carotenoids in the diet bind to the fake fat and therefore are depleted from the body. Foods containing Olestra are supplemented with the fat-soluble vitamins to replenish those lost. Olestra also causes gastrointestinal distress and diarrhea in a small percentage of individuals. This problem has been remedied to some extent by increasing the percentage of the C-16 and C-18 fatty acids.

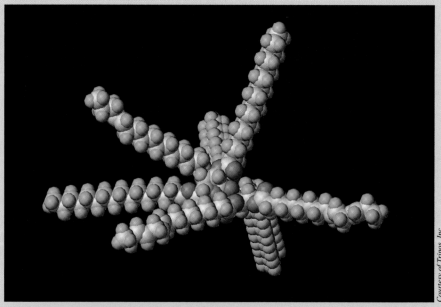

Courtesy of Tripos, Inc.

reactions. Recall that thioesters such as acetyl CoA, succinyl CoA, and fatty acyl CoA are reactive compounds. (See Window on Biochemistry 14.2.) The energy released from the cleavage of both phosphoanhydride bonds in ATP is coupled to the formation of CoA esters (Figure 18.4). This is shown in the overall, net reaction for fatty acid activation:

$$\underset{\displaystyle \overset{\|}{O}}{RC}O^- + CoASH + ATP + H_2O \longrightarrow \underset{\displaystyle \overset{\|}{O}}{RC}SCoA + AMP + 2\,P_i$$

This brings us another step closer to β oxidation, but the fatty acid CoA ester must still be transported from the cytoplasm to the mitochondrial matrix where degradation occurs. The CoA esters of fatty acids are able to diffuse through the outer mitochondrial membrane into the intermembrane space between the inner and outer membranes. The CoA esters, however, do not readily diffuse through the inner membrane. Transport through the inner membrane is aided by transfer of the long-chain acyl group to the hydroxyl group of the small molecule, **carnitine** (Figure 18.5). The enzyme carnitine acyltransferase I is present in the intermembrane space to promote formation of the ester linkage between carnitine and the acyl group. The inner membrane integral protein, acyl carnitine/carnitine translocase, provides a channel for transport of the chemically modified fatty acid through the membrane. Once inside, the acyl carnitine is cleaved to free carnitine and the fatty acyl group is transferred to matrix CoASH (carnitine acyltransferase II; Figure 18.6). Carnitine is returned to the intermembrane space by the translocase in a form ready to carry another fatty acyl group through the inner membrane.

18.2 Catabolism of Fatty Acids

Beta Oxidation

The study of fatty acid degradation has a long and interesting history. The general pathway was elucidated in the early 1900s from experimental data obtained by the German biochemist, Franz Knoop. He fed dogs a series of straight-chain fatty acids that were labeled with a phenyl group substituted on the terminal methyl group (Figure 18.7). When fatty acids with an even number of chain carbons were used, Knoop always isolated a phenylacetate derivative from the dogs' urine. Fatty acids with an odd number of carbon atoms led to a derivative of benzoate. From these creative experiments, Knoop concluded that fatty acids were degraded in a stepwise fashion by removal of a C_2 unit at each step. He proposed that an initial oxidation process occurred on the β carbon of the fatty acid (carbon number 3) followed by cleavage of the bond between carbons α and β (carbons number 2 and 3). Knoop called the pathway **β oxidation,** a name that has been retained. The experiments completed by Knoop were of great significance for what they revealed about fatty acid metabolism and because they illustrated the first use of tracers or markers to follow nutrients through metabolism. Knoop made an excellent choice for the labeling group because the phenyl group is quite unreactive under physiological conditions. If we were to repeat Knoop's experiments today, we would use radiolabeled fatty acids, but these were not available until the 1940s.

Further studies on fatty acid metabolism were reported in the late 1940s and 1950s. Eugene Kennedy and Albert Lehninger in the United States discovered that the enzymes for β oxidation were present in the mitochondrial matrix with other enzymes for aerobic metabolism and that ATP was required for fatty acid activation. Details of the acyl-CoA synthetase activation step linking fatty acids to CoASH were discovered and studied by Feodor Lynen (Germany) and Paul Berg (United States).

The Steps of Beta Oxidation

Beta oxidation of saturated fatty acids begins after their entry into the mitochondrial matrix. Carnitine acyl transferase II catalyzes the final step in the transport by linking the fatty acid to mitochondrial CoASH. Beta oxidation of the fatty acyl CoA takes place by a recurring series of four steps: (1) oxidation of a carbon–carbon single bond by FAD to form a carbon–carbon

Cutting Edge: Fat blockers

Interactive Animation: Fatty acid metabolism

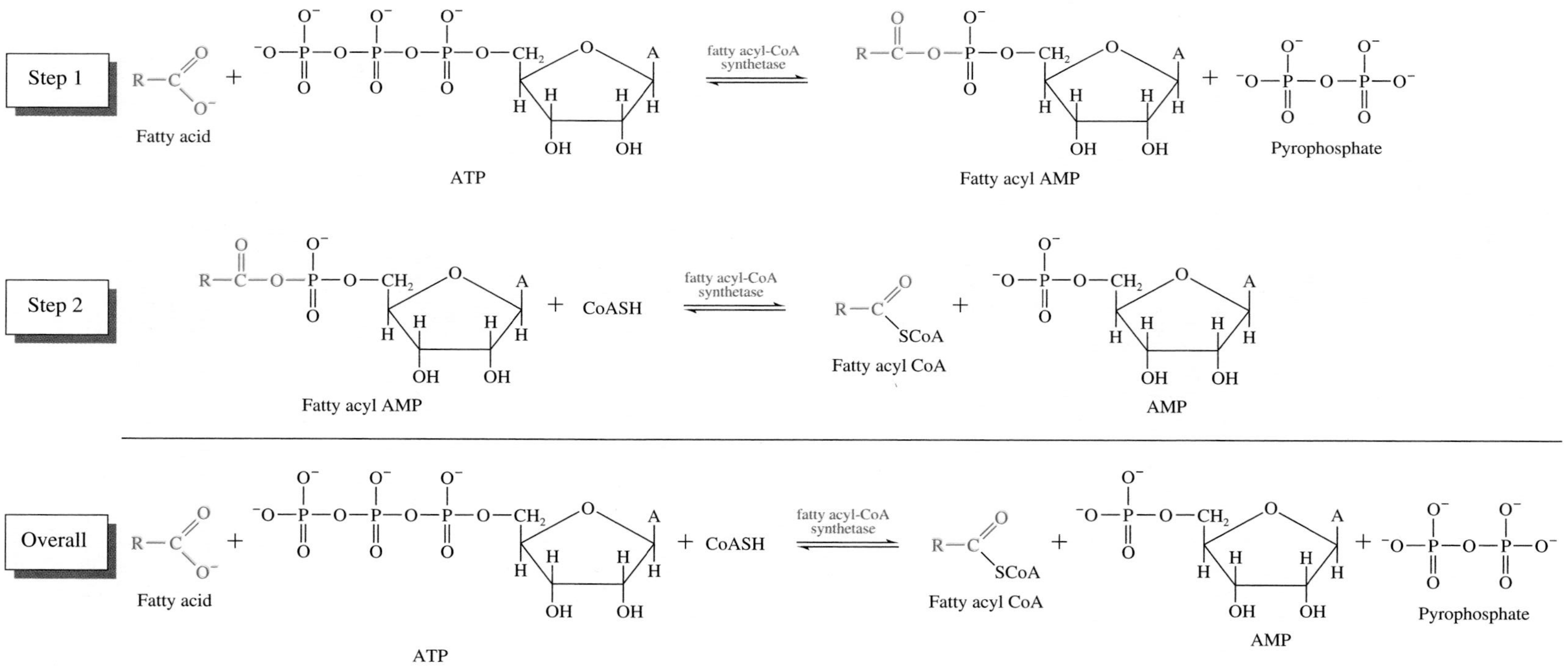

Figure 18.4
Mechanism of acyl-CoA synthetase. Step 1: The fatty acid is activated by reaction with ATP. Pyrophosphate cleavage of ATP leads to formation of the fatty acyl AMP, an enzyme-bound intermediate. Step 2: The fatty acyl group is transferred from AMP to CoASH. Extra energy to drive the reaction comes from the hydrolysis of pyrophosphate. The overall reaction is derived by adding Steps 1 and 2.

CH₃
|
CH₃—⁺N—CH₃
|
CH₂
|
HO—C—H
|
CH₂
|
COO⁻

Carnitine

CH₃
|
CH₃—⁺N—CH₃
|
O   CH₂
‖ |
R—C—O—C—H
|
CH₂
|
COO⁻

Acylcarnitine

Figure 18.5

The structures of carnitine and acyl carnitine. Carnitine assists in the transport of fatty acids through the inner mitochondrial membrane into the matrix, where energy metabolism of fatty acids occurs. The acyl group of a fatty acid is linked to carnitine by an ester bond.

double bond; (2) addition of H_2O to the double bond, with formation of a hydroxyl group on one carbon; (3) oxidation of the hydroxyl group by NAD^+ to produce a keto group; and (4) carbon–carbon bond cleavage, releasing acetyl CoA. Although the pathway is often defined as a cycle, it is probably more accurate to describe it as a spiral. Each complete turn of the spiral consists of the four enzyme-catalyzed steps, resulting in release of one molecule of acetyl CoA and the fatty acid chain shortened by two carbons (Figure 18.8). As shown in the figure, the degradation of palmitoyl CoA (C_{16}) requires seven trips around the spiral. The number of rounds does not correspond directly to the number of acetyl CoA products (seven rounds, eight acetyl CoA) because two acetyl CoAs are formed during the last spiral.

The four steps with the required enzymes and type of reaction are outlined in Table 18.1. The steps for fatty acid activation and transport (reactions 1–4) are included with the β-oxidation steps (reactions 5–8). The catabolism of fatty acids consists of primarily oxidation steps (two of the four steps). The other two steps are addition–elimination (or nonhydrolytic cleavage).

The first three steps of β oxidation (reactions 1, 2, and 3 in Figure 18.8; reactions 5, 6, and 7 in Table 18.1) display the same kind of chemistry as three continuous steps from the citric acid cycle:

$$\text{succinate} \xrightarrow{\text{oxidation}} \text{fumarate} \xrightarrow{\text{addition of water}} \text{malate} \xrightarrow{\text{oxidation}} \text{oxaloacetate}$$

$$\text{acyl CoA} \longrightarrow \text{unsaturated acyl CoA} \longrightarrow \text{hydroxyacyl CoA} \longrightarrow \text{ketoacyl CoA}$$

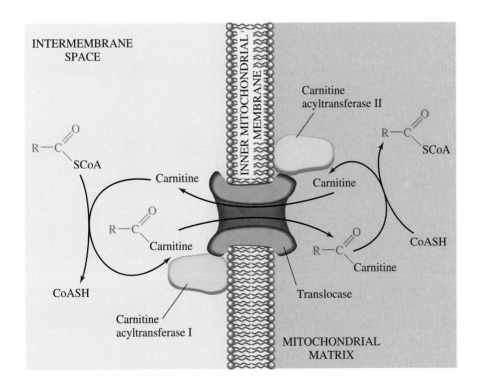

Figure 18.6

The role of carnitine in the delivery of fatty acids into the mitochondrial matrix. The acyl group of fatty acid CoA esters is transferred to carnitine in a reaction catalyzed by carnitine acyltransferase I. This reaction occurs in the intermembrane space (between the outer and inner membranes of the mitochondria). Acyl carnitine moves through the inner membrane with assistance by a translocase. In the mitochondrial matrix the acyl group is transferred back to CoASH in a reaction catalyzed by carnitine acyltransferase II.

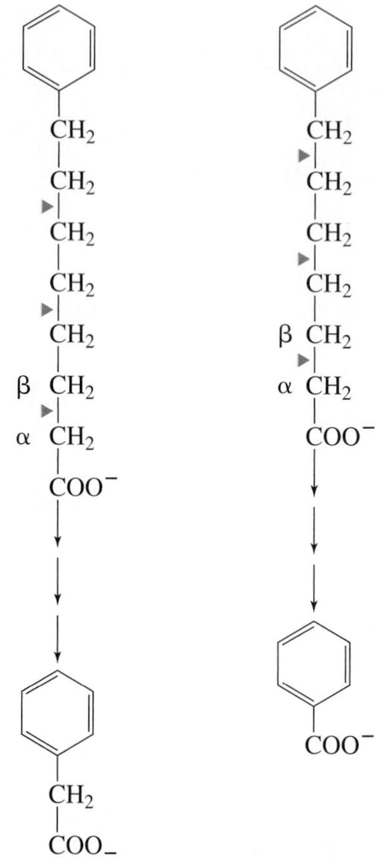

Figure 18.7

Labeling experiments carried out by Franz Knoop to determine the pathway of fatty acid catabolism. Phenyl-substituted fatty acids were fed to dogs and derivatives were isolated from urine. Fatty acids with even numbers of carbons in their chains led to phenylacetate formation; odd-numbered chains led to benzoate. The triangles indicate carbon–carbon bonds cleaved by β oxidation.

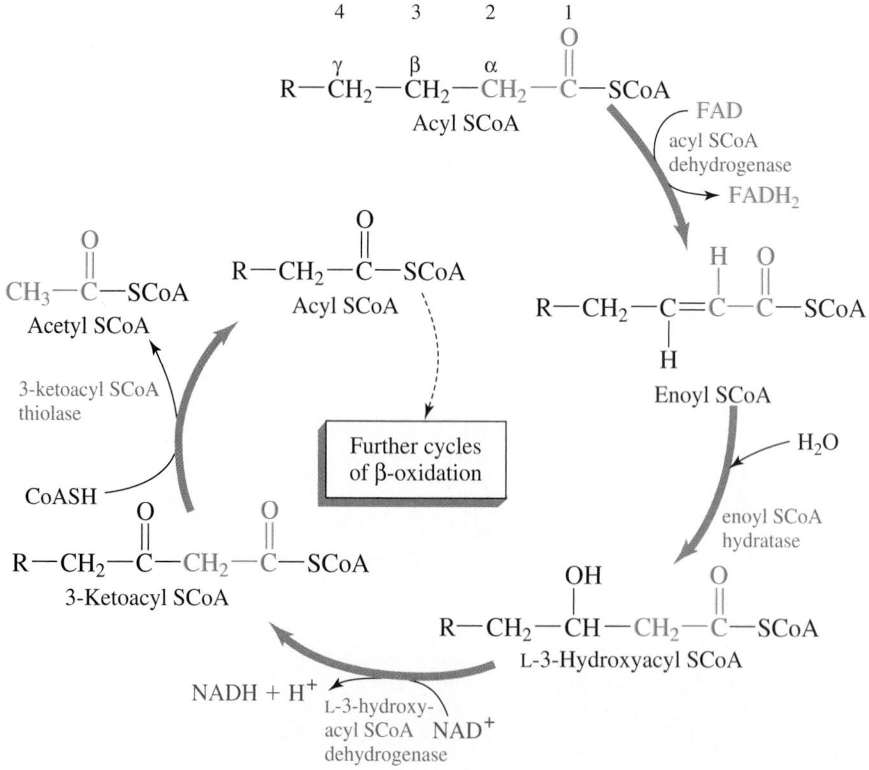

Phenylacetate
(from even-numbered carbon chain)

Benzoate
(from odd-numbered carbon chain)

Figure 18.8

The spiral pathway of fatty acid β oxidation. The fatty acyl CoA ester undergoes a four-step sequence of dehydrogenation (oxidation), hydration, oxidation, and carbon–carbon bond cleavage. Products of the first cycle include acetyl CoA and a fatty acyl CoA ester, shortened by two carbons. This CoA ester goes through the four-step reaction sequence until it is completely degraded to acetyl CoA.

Table 18.1
Reactions for fatty acid activation, transport, and the β-oxidation spiral

Reaction Number	Reaction	Enzyme	Reaction[a] Type
1	Fatty acid + CoASH + ATP $\rightleftharpoons$ acyl SCoA + AMP + PP$_i$	Acyl-SCoA synthetase	6
2	PPi + H$_2$O $\rightleftharpoons$ 2 P$_i$	Pyrophosphatase	3
3	Carnitine + acyl SCoA $\rightleftharpoons$ acyl carnitine + CoASH (intermembrane space)	Carnitine acyltransferase I	2
4	Acyl carnitine + CoASH $\rightleftharpoons$ acyl SCoA + carnitine (mitochondria)	Carnitine acyltransferase II	2
5	Acyl SCoA + E-FAD $\rightleftharpoons$ trans-Δ^2-enoyl SCoA + E-FADH$_2$[b]	Acyl-SCoA dehydrogenase	1
6	trans-Δ^2-Enoyl SCoA + H$_2$O $\rightleftharpoons$ L-3-hydroxyacyl SCoA	Enoyl-SCoA hydratase	4
7	L-3-Hydroxyacyl SCoA + NAD$^+$ $\rightleftharpoons$ 3-ketoacyl SCoA + NADH + H$^+$	Hydroxyacyl-SCoA dehydrogenase	1
8	3-Ketoacyl SCoA + CoASH $\rightleftharpoons$ acetyl SCoA + acyl SCoA[c]	β-Ketothiolase	4

[a]Reaction type: 1, oxidation–reduction; 2, group transfer; 3, hydrolysis; 4, nonhydrolytic cleavage (addition–elimination); 5, isomerization–rearrangement; 6, bond formation coupled to ATP cleavage.
[b]E-FAD and E-FADH$_2$ refer to the cofactor flavin adenine dinucleotide covalently linked to the enzyme.
[c]Acyl SCoA product is shortened by a C$_2$ unit.

The similarities of these two processes should assist in learning the reactions of β oxidation. In the final step of β oxidation (reaction 8), a C$_2$ unit in the form of acetyl CoA is cleaved from the β ketoacyl CoA ester and a shortened acyl CoA is prepared for another round of β oxidation.

The common, even-numbered, saturated fatty acids all follow the same pathway in β oxidation. They differ only in the number of spirals required to oxidize the hydrocarbon chain to acetyl CoA. The net reaction for the process of β oxidation of a saturated fatty acid, palmitoyl CoA (the CoA ester of palmitic acid), is as follows:

$$\text{palmitoyl SCoA} + 7\ \text{FAD} + 7\ \text{NAD}^+ + 7\ \text{CoASH} + 7\ \text{H}_2\text{O} \rightarrow$$
$$8\ \text{acetyl SCoA} + 7\ \text{FADH}_2 + 7\ \text{NADH} + 7\ \text{H}^+$$

The complete degradation of palmitoyl CoA requires seven turns of β oxidation and each round requires the entry of an FAD, NAD$^+$, H$_2$O, and CoASH.

Animals use the heat produced from fatty acid catabolism and oxidative phosphorylation to keep warm.

© K.D. McGraw/Rainbow

Figure 18.9

Combination of fatty acid activation, transport into the mitochondrial matrix, and β oxidation. Acetyl CoA derived from β oxidation is shown entering the citric acid cycle with production of NADH and FADH$_2$, which are oxidized by the respiratory chain.

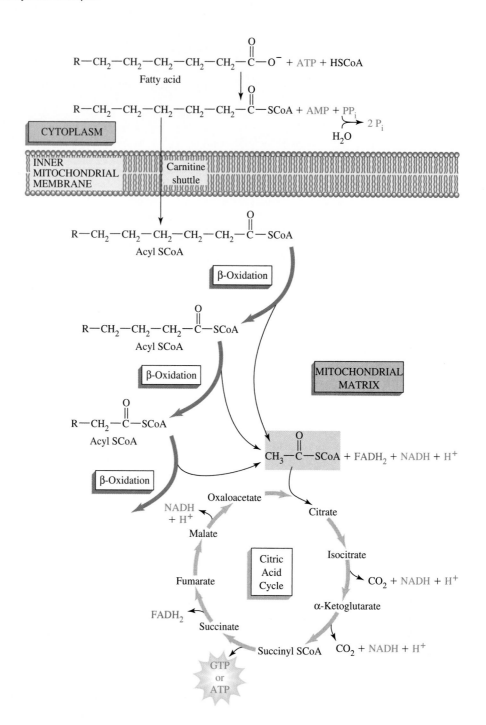

Significance of Beta Oxidation

Beta oxidation yields products that are of great metabolic importance. The net reaction for the *combined* activation and β oxidation of palmitic acid is the following:

$$\text{palmitic acid} + \text{ATP} + 7\ \text{FAD} + 7\ \text{NAD}^+ + 8\ \text{CoASH} + 8\ \text{H}_2\text{O} \longrightarrow$$
$$8\ \text{acetyl SCoA} + \text{AMP} + 2\ \text{P}_i + 7\ \text{FADH}_2 + 7\ \text{NADH} + 7\ \text{H}^+$$

The acetyl CoA is ready for further oxidation by the citric acid cycle. Recall that β-oxidation enzymes are present in the mitochondrial matrix, the same location as the citric acid cycle enzymes. The reduced products of β oxidation, NADH and FADH$_2$, are recycled by the electron-transport chain and ATP is generated by oxidative phosphorylation. The energy balance sheet for β oxidation of palmitate is outlined in Figure 18.9 and Table 18.2. These show

Metabolic Stage	NADH	FADH$_2$	Substrate-Level Phosphorylation
CoA activation	0	0	−2
β oxidation (seven cycles)	7 (mitochondria)	7	0
Citric acid cycle (eight cycles)	24 (mitochondria)	8	8
Subtotal	31	15	6

Oxidative Phosphorylation

31 NADH × 3 ATP   =   93 ATP
15 FADH$_2$ × 2 ATP   =   30 ATP

Substrate-Level Phosphorylation =   6 ATP

Grand total   129 ATP

Table 18.2
Energy yield from the complete oxidation of palmitate

the result of fatty acid activation and oxidation in a slightly different format from the net reaction. Table 18.2 is of value in calculating the total energy yield from the complete oxidation of fatty acids to CO_2 and H_2O with inclusion of the process of oxidative phosphorylation.

Palmitate, when catabolized by β oxidation and the citric acid cycle, yields 31 NADH, 15 FADH$_2$, and 6 substrate-level ATP (or GTP). Because all NADH from fatty acid oxidation is mitochondrial in origin, there is no need for an electron shuttle system as for cytoplasmic NADH. The complete oxidation of palmitate to CO_2 and H_2O yields 129 ATP.

Beta Oxidation of Unsaturated Fatty Acids

Many triacylglycerols in our diets contain unsaturated fatty acids, especially if we follow modern nutritional recommendations. Some cooking oils have an unsaturated fatty acid content of up to 94% (see Table 9.3). Can these fatty acids be completely degraded by the four reactions of β oxidation? Our first answer might be yes, because a fatty acid with a double bond is actually produced as an intermediate in β oxidation. However, the double bond in the intermediate enoyl CoA is trans, whereas the double bonds in naturally occurring fatty acids are cis. Also, the locations of the double bonds are usually not the same. To degrade unsaturated fatty acids completely, we need the four enzymes of β oxidation plus two auxiliary enzymes. To illustrate the use of all six enzymes, we shall outline the complete β oxidation of the fatty acid, 9,12-hexadecadienoic acid (16:2$^{\Delta 9,12}$) in Figure 18.10.

Unsaturated fatty acids are absorbed, transported, and activated in the same way as saturated fatty acids. We begin with the acid in the activated form located in the mitochondrial matrix. As shown in Figure 18.10, the fatty acyl CoA ester is able to undergo three normal β-oxidation rounds. The double bonds at C9 to C10 and C12 to C13 do not interfere with this process. The products at this point are three acetyl CoA molecules, reduced cofactors, and a shortened acyl CoA ester, *cis,cis*-$\Delta^{3,6}$-decadienoyl CoA (a C10 unit). The next normal step, addition of water, cannot be catalyzed by the hydratase because the double bond is cis rather than trans and the double bond is in the wrong location, Δ^3 rather than Δ^2. An auxiliary enzyme, **enoyl-CoA isomerase,** catalyzes the rearrangement of the *cis*-Δ^3 double bond to a *trans*-Δ^2 double bond (Figure 18.11a). The intermediate can now continue through one round of β oxidation. (Note that the double bond is already in place so the first dehydrogenation step can be bypassed.) The fatty acyl CoA ester product is now a *cis*-Δ^4-acyl CoA with eight carbons. This substrate is dehydrogenated in the normal way to the *trans*-Δ^2, *cis*-Δ^4 intermediate. Because of its rigidity, this diene cannot be acted upon by the next β-oxidation enzyme (hydratase). Instead, the second auxiliary enzyme, **2,4-dienoyl CoA reductase,** acting with the coenzyme NADPH, converts the intermediate to a *trans*-Δ^3-enoyl CoA (see Figures 18.10 and 18.11b). One of the double bonds is reduced in the reaction and the other moved. This intermediate must be rearranged by the enoyl-CoA isomerase before final β oxidation. Three more complete rounds of normal β oxidation are required to degrade the diunsaturated acid to a total of eight molecules of acetyl CoA.

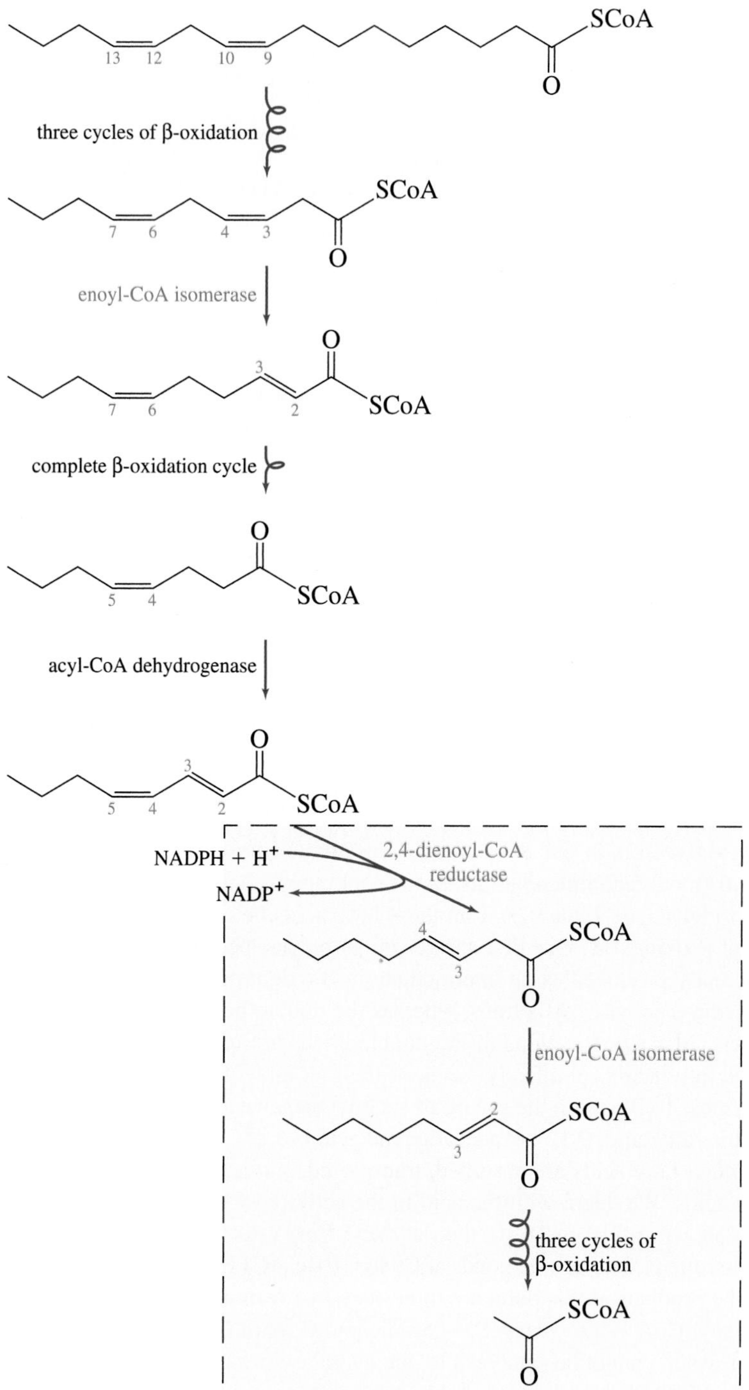

Figure 18.10

The metabolism of the unsaturated fatty acid 16:2$^{\Delta 9,12}$, showing the use of the four reactions of β oxidation plus the auxiliary enzymes enoyl-CoA isomerase and 2,4-dienoyl-CoA reductase.

Beta Oxidation of Fatty Acids with Odd Numbers of Carbons

Fatty acids with an odd number of carbons are minor components in nature, but some are present in significant amounts in plants and marine organisms. These are catabolized by normal β oxidation. The final cleavage step, however, yields an acetyl CoA *and* a C$_3$ unit, **propionyl CoA** (Figure 18.12). Propionyl CoA does not enter the citric acid cycle as acetyl CoA does but must be transformed to succinyl CoA. Three reactions are required for this conversion: (1) carboxylation to D-methylmalonyl CoA, (2) isomerization to L-methylmalonyl CoA, and (3) rearrangement to succinyl CoA. The enzyme, methyl-malonyl CoA mutase (reaction 3), is of special interest because its cofactor is the unusual

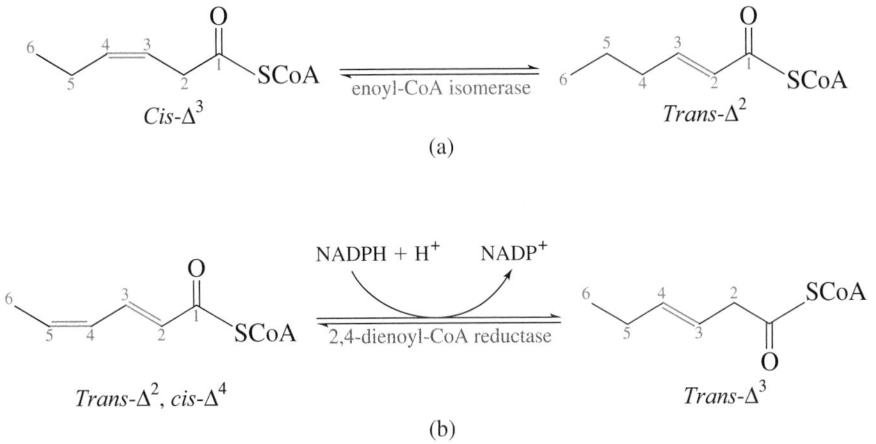

Figure 18.11

The reaction catalyzed by (a) enoyl-CoA isomerase and (b) 2,4-dienoyl-CoA reductase. The enzymes are auxiliary enzymes used for oxidation of unsaturated fatty acids.

molecule coenzyme B_{12} (deoxyadenosylcobalamin). This coenzyme, which is derived from vitamin B_{12}, is associated with enzymes catalyzing carbon skeleton rearrangements and with enzymes involved in purine and thymidine synthesis. Ribonucleotide reductase (see Chapter 19, Section 19.5), an enzyme found in many prokaryotic species, uses a vitamin B_{12} cofactor. Neither plants nor animals are able to synthesize vitamin B_{12}; it can be synthesized only by a few species of bacteria. (The structure and biological activity of the vitamin are given in Table 7.1.) Carnivorous animals obtain sufficient amounts of B_{12} from meat in their diet; however, herbivorous animals depend on intestinal bacteria for synthesis of B_{12}. Only very small amounts (about 3 μg/day) are required by adult humans so nutritional deficiency is rare. Symptoms of deficiency are observed in individuals with **pernicious anemia** and occasionally in strict vegetarians. It was discovered in 1926 that pernicious anemia could be prevented by eating large quantities of liver. The active compound in liver was identified in 1948 and called vitamin B_{12} or cobalamin. Pernicious anemia is caused not by a deficiency of B_{12}, but by impaired absorption of the vitamin in the intestines. Large quantities of B_{12} obtained in the diet (or by injection) permit an affected individual to absorb sufficient amounts. Symptoms of pernicious anemia are sometimes observed in vegetarians, but are slow in developing because the normal adult liver stores enough vitamin B_{12} for 3 to 5 years.

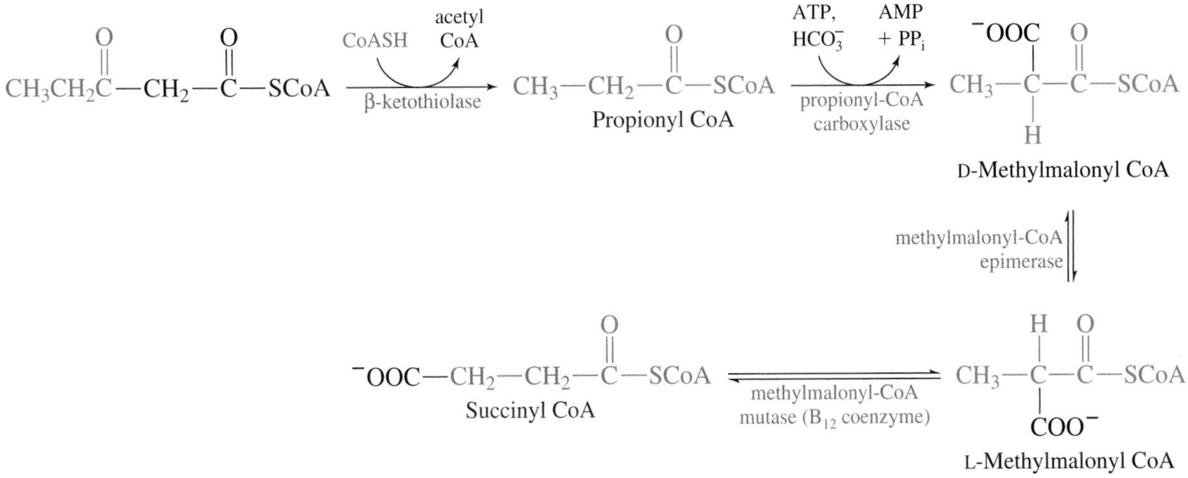

Figure 18.12

The final cleavage step in the β oxidation of an odd-numbered fatty acid and the transformation of the product, propionyl CoA, to succinyl CoA for entry into the citric acid cycle. The propionyl CoA undergoes carboxylation to a C_4 metabolite, D-methylmalonyl CoA. After two rearrangement steps, succinyl CoA emerges.

18.3   Biosynthesis of Fatty Acids

When fatty acids are ingested in the diet in amounts greater than needed, they are stored as triacylglycerols in adipocytes. The amount of fat that can be stored in higher animals, including humans, is unlimited. To add to the problem, glucose and other carbohydrates ingested in excess of immediate energy needs are also converted to fat (Figure 18.13). Initially, excess glucose is stored in glycogen, but only limited amounts of the hydrophilic polysaccharide can be accumulated. When the limits of glycogen storage have been reached, the glucose is degraded to acetyl CoA (by glycolysis and the pyruvate dehydrogenase complex), which is then used to synthesize fatty acids for storage.

Experimental studies on fatty acid synthesis began soon after the announcement of Knoop's β-oxidation results. It was widely predicted that the reactions of fatty acid synthesis would be the reverse of those for β oxidation. Some of the same types of chemistry are used in the two processes, but in other ways they are very different. Since we are now familiar with the process of fatty acid degradation, we can introduce fatty acid synthesis by comparing the differences (Table 18.3). Note that the enzymes for fatty acid synthesis are in the cytoplasm. Perhaps the most novel aspect of fatty acid synthesis is the use of a C_3 precursor, malonyl CoA, to add carbon units to a growing fatty acid chain.

The Steps of Fatty Acid Synthesis

Fatty acid synthesis from acetyl CoA occurs in the cytoplasm. All pathways of acetyl CoA production that we have studied (pyruvate dehydrogenase complex and β oxidation) occur in the mitochondrial matrix. Mitochondrial acetyl CoA is unable to diffuse through the inner membrane into the cytoplasm. A shuttle system transports the carbons of acetyl CoA disguised as citrate through the membranes (Figure 18.14). In the mitochondria, acetyl CoA reacts with oxaloacetate to produce citrate (citrate synthase). Citrate is transported by the integral membrane protein, tricarboxylate translocase, into cytoplasm, where it is cleaved to

Figure 18.13

The linkage between carbohydrate and fatty acid metabolism in animals. Acetyl CoA is produced from β oxidation of dietary fatty acids and from glycolysis plus pyruvate oxidation. Excess acetyl CoA may be used to make fatty acids stored in triacylglycerols. Note that animals are unable to convert fatty acids or acetyl CoA into glucose. Plants use the glyoxylate cycle to convert acetyl CoA to carbohydrates.

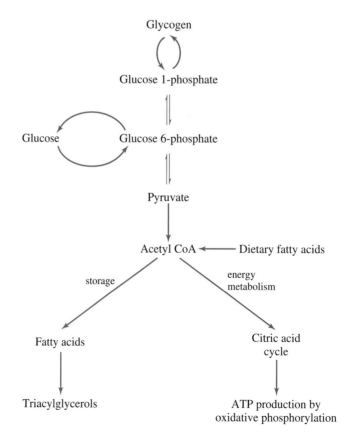

Characteristic	β Oxidation	Biosynthesis
Cellular location	Mitochondrial matrix	Cytoplasm
Activation and labeling of intermediates	CoA thioesters	Thioesters of acyl carrier protein (ACP)
Enzymes	Four distinct, nonassociated proteins	Fatty acid synthase, a multi-enzyme complex in mammals
Process	Two-carbon fragments removed as acetyl CoA	Two-carbon elongation using malonyl ACP
Fatty acid size	All sizes are degraded	Only palmitate is made
Redox reaction cofactors	$NAD^+/NADH$ and $FAD/FADH_2$	$NADP^+/NADPH$

Table 18.3
Differences between β oxidation of fatty acids and fatty acid synthesis

acetyl CoA and oxaloacetate (citrate lyase). Oxaloacetate is eventually returned to the mitochondria, but acetyl CoA in the cytoplasm is now available for fatty acid synthesis. The final product of the **fatty acid synthase** complex is palmitate.

Palmitate synthesis begins at the terminal methyl end and proceeds toward the carboxylate (Figure 18.15). The first two carbons of the palmitate chain (carbons 15 and 16) come directly from acetyl CoA. All other carbons come originally from acetyl CoA but must be activated as malonyl CoA. Acetyl-CoA carboxylase, a biotin-requiring enzyme, catalyzes the formation of malonyl CoA:

$$CH_3\overset{\overset{O}{\|}}{C}SCoA + CO_2 + ATP \rightleftharpoons \underset{\underset{COO_-}{|}}{CH_2}\overset{\overset{O}{\|}}{C}SCoA + ADP + P_i$$

Acetyl CoA   Malonyl CoA

This reaction, which can be considered one of the first steps of fatty acid synthesis, is the rate-controlling step. A positive modulator for the allosteric enzyme is citrate. The enzyme is inhibited by palmitoyl CoA, the end product of the fatty acid synthesis system.

From this point on, all intermediates are linked to a small molecular weight protein called **acyl carrier protein** (ACP-SH), which is a component of fatty acid synthase. As we might predict, ACP has properties similar to those of CoASH (Figure 18.16). The vitamin pantothenic acid is a component of both activating groups. The intermediates for fatty acid synthesis are linked via a thioester bond to the —SH group at one end of the ACP.

To prepare for fatty acid synthesis, one acetyl CoA and one malonyl CoA must be brought into the synthase complex. They are attached to —SH groups, one on an enzyme, β-ketoacyl-ACP synthase (K-SH), and one on ACP (malonyl CoA). These reactions, as well

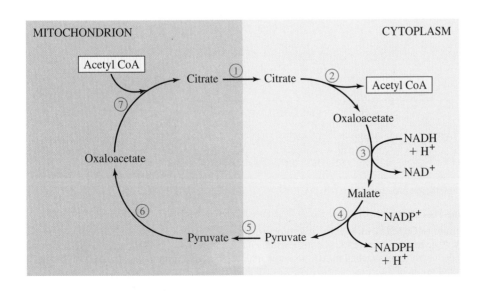

Figure 18.14
Transport of citrate from mitochondria to cytoplasm by tricarboxylate translocase (step 1). Citrate in the cytoplasm is cleaved to acetyl CoA and oxaloacetate by the enzyme citrate lyase (step 2). Cytoplasmic acetyl CoA is used for fatty acid synthesis. Oxaloacetate is converted by malate dehydrogenase (step 3) and malic enzyme (step 4) to pyruvate, which may be transported to mitochondria by pyruvate translocase (step 5). In the mitochondria, pyruvate is transformed to oxaloacetate and citrate by pyruvate carboxylase (step 6) and citrate synthase (step 7).

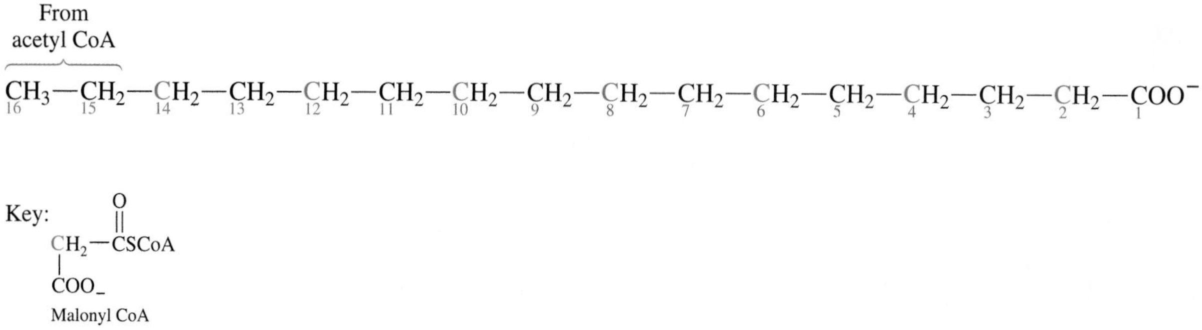

From
acetyl CoA

$$\overbrace{\underset{16}{CH_3}-\underset{15}{CH_2}}-\underset{14}{CH_2}-\underset{13}{CH_2}-\underset{12}{CH_2}-\underset{11}{CH_2}-\underset{10}{CH_2}-\underset{9}{CH_2}-\underset{8}{CH_2}-\underset{7}{CH_2}-\underset{6}{CH_2}-\underset{5}{CH_2}-\underset{4}{CH_2}-\underset{3}{CH_2}-\underset{2}{CH_2}-\underset{1}{COO^-}$$

Key:

$$\underset{\underset{COO_-}{|}}{CH_2}-\overset{\overset{O}{||}}{C}SCoA$$

Malonyl CoA

Figure 18.15

Origin of the carbons in palmitate synthesized in cytoplasm. Carbons 15 and 16 are derived directly from acetyl CoA. All other carbons come from malonyl CoA. The palmitate shown is the product when precursor malonyl CoA is labeled at C2 as shown. Carbons in green indicate carbons coming from the middle carbon of malonyl CoA.

as all the reactions of fatty acid synthesis, are listed in Table 18.4 and shown in Figure 18.17. The enzyme catalyzing the transfer of the acetyl group from CoASH via ACP-SH to K-SH is acetyl-CoA-ACP transacetylase (reaction 2). The free —SH group on ACP can now accept a malonyl unit (reaction 3). Malonyl-CoA-ACP transferase catalyzes the entry of the malonyl group into the fatty acid synthase. The two starting precursors for fatty acid synthesis are now activated and the synthesis of the fatty acid chain can begin. A sequence of four chemical steps is carried out with the acetyl and malonyl groups (reactions 4–7). The reactions consist of a spiral that is the chemical reverse of β oxidation: (1) formation of a carbon–carbon single bond, (2) reduction of a keto group, (3) dehydration to form a carbon–carbon double bond, and (4) reduction of the double bond to form the saturated fatty acid chain.

One round through the fatty acid synthase spiral has produced a four-carbon carboxyl group (butyryl) linked to ACP. In the second round, two more carbons are added from malonyl ACP to the butyryl unit to make a C_6 β-keto intermediate that goes through reduction,

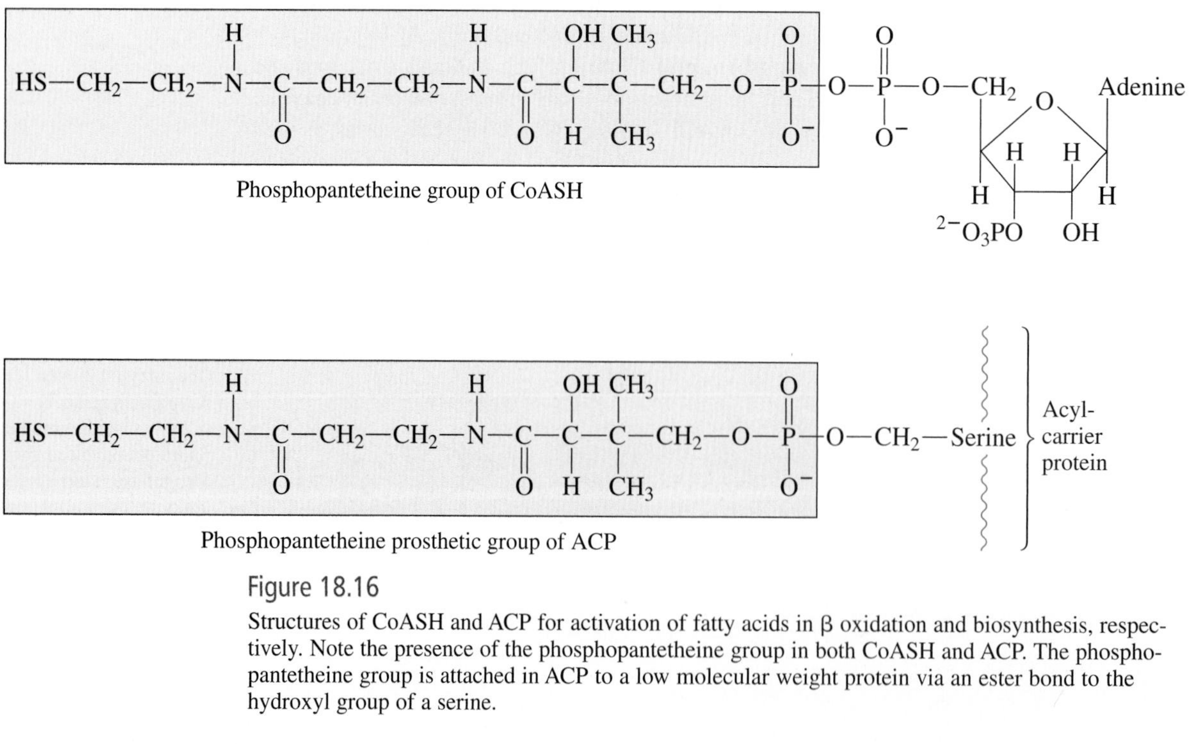

Phosphopantetheine group of CoASH

Phosphopantetheine prosthetic group of ACP

Figure 18.16

Structures of CoASH and ACP for activation of fatty acids in β oxidation and biosynthesis, respectively. Note the presence of the phosphopantetheine group in both CoASH and ACP. The phosphopantetheine group is attached in ACP to a low molecular weight protein via an ester bond to the hydroxyl group of a serine.

Table 18.4
The reactions of fatty acid synthesis

Reaction Number	Reaction	Enzyme	Reaction Type[a]
1	Acetyl SCoA + CO_2 + ATP $\rightleftharpoons$ malonyl SCoA + ADP + P_i + H^+	Acetyl-CoA carboxylase	6
2	Acetyl SCoA + ACP-SH $\rightleftharpoons$ acetyl-S-ACP + CoASH	Acetyl-CoA-ACP transacetylase	2
3	Malonyl SCoA + ACP-SH $\rightleftharpoons$ malonyl-S-ACP + CoA	Malonyl-CoA transferase	2
4	Acetyl-S-K + malonyl-S-ACP $\rightleftharpoons$ acetoacetyl-S-ACP + K-SH + CO_2	β-Ketoacyl-ACP synthase	4
5	Acetoacetyl-S-ACP + NADPH + H^+ $\rightleftharpoons$ D-3-hydroxybutyryl-S-ACP + $NADP^+$	β-Ketoacyl-ACP reductase	1
6	D-3-Hydroxybutyryl-S-ACP $\rightleftharpoons$ crotonyl-S-ACP + H_2O	3-Hydroxyacyl-ACP dehydratase	4
7	Crotonyl-S-ACP + NADPH + H^+ $\rightleftharpoons$ butyryl-S-ACP + $NADP^+$	Enoyl-ACP reductase	1

[a]Reaction type: 1, oxidation–reduction; 2, group transfer; 3, hydrolysis; 4, nonhydrolytic cleavage (addition–elimination); 5, isomerization–rearrangement; 6, bond formation coupled to ATP cleavage.

dehydration, and reduction. The two new carbons become numbers 11 and 12 in the palmitate product. A total of seven turns of this set of four reactions results in palmitoyl-ACP. At this point, the C_{16} unit is hydrolyzed from the ACP, yielding free palmitate. The hydrolytic enzyme palmitoyl thioesterase is part of the fatty acid synthase complex. The net reaction for fatty acid synthesis, showing the initial acetyl CoA starter and the entry of carbon atoms as malonyl units, is as follows:

$$\text{acetyl CoA} + 7 \text{ malonyl CoA} + 14 \text{ NADPH} + 14 \text{ H}^+ \rightarrow$$
$$CH_3(CH_2)_{14}COOH + 7 CO_2 + 8 \text{ CoA} + 14 \text{ NADP}^+ + 6 H_2O$$

Fatty acids longer than palmitate are synthesized by an elongation enzyme system in the endoplasmic reticulum. The reactions are similar to the ones seen with fatty acid synthase, with new carbons added in the form of malonyl CoA. All intermediates, however, are activated by attachment to CoASH rather than ACP.

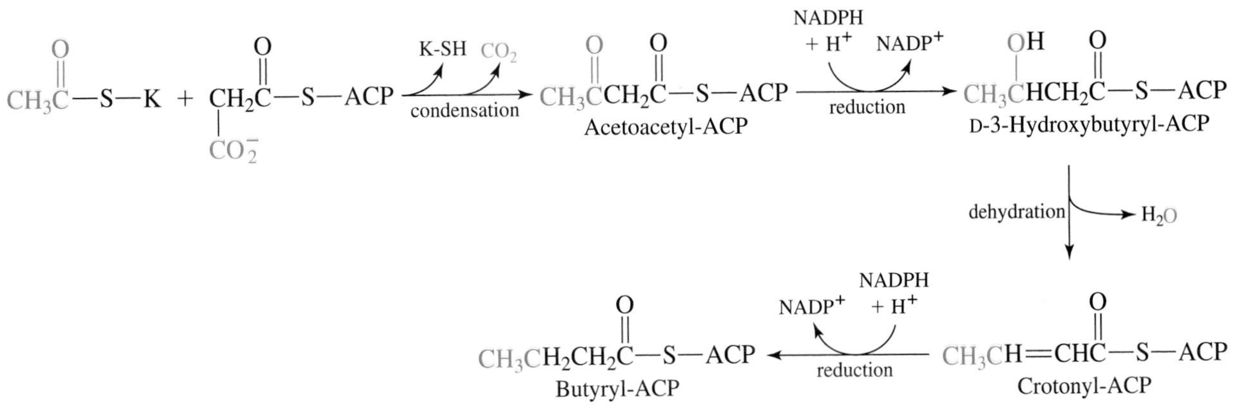

Figure 18.17
Reactions for the synthesis of fatty acids. To begin a fatty acid, the acetyl group linked to β-keto-ACP synthase and malonyl-ACP combine to produce a C_4 intermediate that is further changed to the completely saturated butyryl-ACP. Note that this sequence is the reverse of β oxidation. K-SH = β-ketoacyl-ACP synthase.

Biosynthesis of Unsaturated Fatty Acids

Unsaturated fatty acids are synthesized in the endoplasmic reticulum by enzymes called **fatty acyl-CoA desaturases.** These enzymes catalyze unique oxidation–reduction reactions. The dehydrogenation of stearic acid (18:0) to oleic acid ($18:1^{\Delta 9}$) is an example:

$$\text{stearoyl CoA} + \text{NADPH} + \text{H}^+ + \text{O}_2 \longrightarrow \text{oleoyl CoA} + \text{NADP}^+ + 2\,\text{H}_2\text{O}$$

Molecular oxygen as a substrate accepts four electrons in this reaction, two from the stearoyl CoA and two from NADPH.

Two unsaturated fatty acids, linoleate ($18:2^{\Delta 9,12}$) and linolenate ($18:3^{\Delta 9,12,15}$), which are essential for proper growth and development, cannot be synthesized by fatty acyl-CoA desaturases in mammals. We must obtain them in the diet from plant sources, so they are called **essential fatty acids.** A scaly dermatitis develops if they are excluded in the diets of humans and rats. Linoleate is an important constituent of skin sphingolipids and a precursor of the fatty acid arachidonate ($20:4^{\Delta 5,8,11,14}$). Arachidonate is used for the synthesis of leukotrienes, prostaglandins, and thromboxanes (see Chapter 9, Section 9.4).

Regulation of Fatty Acid Metabolism

The two major processes of β oxidation and synthesis must be coordinated so they do not occur simultaneously. If both processes were to occur at the same time, a wasteful substrate

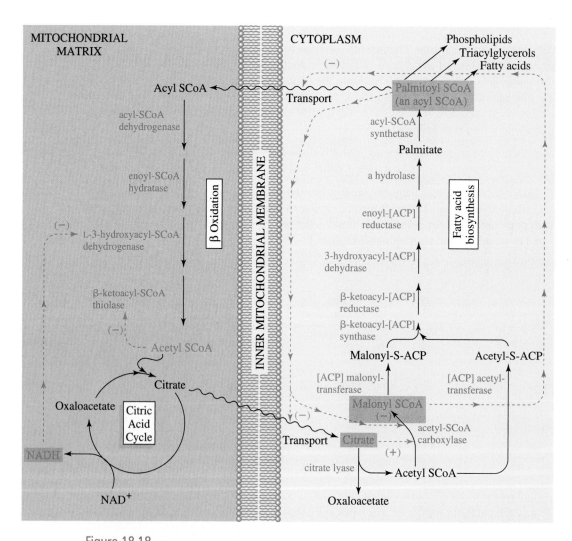

Figure 18.18

Regulation of fatty acid metabolism. Positive effectors are highlighted in green and negative effectors in red.

cycle would develop. The metabolism of fatty acids is controlled by the availability of the nutrients, carbohydrates and fatty acids (Figure 18.18). If glucose is abundant, high levels of citrate are produced. Citrate is a positive modulator for acetyl-CoA carboxylase, which catalyzes the formation of malonyl CoA, the rate-controlling step in fatty acid synthesis. With increasing citrate, fatty acid synthesis is stimulated. Beta oxidation of fatty acids is inhibited by the malonyl CoA produced. It blocks the action of carnitine acyltransferase I, an enzyme responsible for transport of fatty acid CoA esters into the mitochondrial matrix, where they are oxidized. The rate of fatty acid synthesis is also regulated by the end product, palmitoyl CoA, which inhibits acetyl-CoA carboxylase. In summary, with abundant glucose, fatty acid oxidation is turned off and fatty acid synthesis is active, but regulated by levels of palmitoyl CoA. During fasting or when glucose and glycogen levels become low, the hormones epinephrine and glucagon stimulate hormone-sensitive lipase to release free fatty acids from storage. Lowered glucose levels result in less citrate; therefore, stimulation of acetyl-CoA carboxylase is removed. Less malonyl CoA to inhibit carnitine acyltransferase I means that fatty acids can flow into the mitochondrial matrix for β oxidation. The rate-controlling step for β oxidation is the entry of fatty acids into mitochondria. The increase of fatty acids in the mitochondria enhances β oxidation under conditions of low blood glucose.

18.4 Cholesterol

Cholesterol may be one of the most dreaded chemicals in nature. We read and hear of it daily in biology and nutrition classes and even in the popular press. Mostly, we are reminded of the strong correlation between the increased incidence of cardiovascular diseases and high levels of serum cholesterol (Figure 18.19). Scientific evidence certainly supports a "bad side" to cholesterol. However, we hear little about its "good side." This molecule is essential to the life of all animals; indeed, it is a molecule we must live with. Cholesterol plays several roles in biochemistry: (1) It is an essential component of animal membranes, where it regulates fluidity, and (2) it is a precursor of many important biomolecules, including the steroid hormones, bile salts, and vitamin D. Health problems related to cholesterol may be caused not directly by the presence of the molecule, but by an inefficiency in its physiological transport.

Interactive Animation: Cholesterol

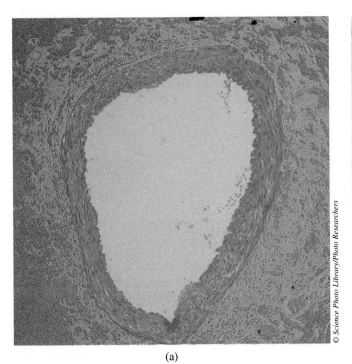
(a)

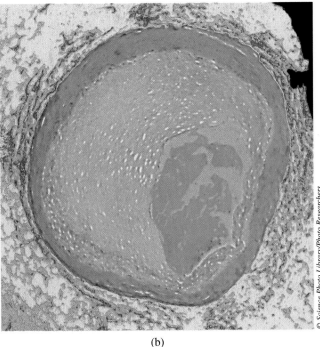
(b)

© Science Photo Library/Photo Researchers

Figure 18.19
Cross section of a normal artery (a) and an artery blocked with plaque consisting of cholesterol (b).

Cholesterol, a water-insoluble molecule, is very difficult for organisms to dissolve and transport. However, this very property supports its role in membrane structure. Our livers produce 800 to 1000 mg of cholesterol daily. In addition, our diets may add hundreds of milligrams daily. It is fortunate that dietary cholesterol reduces the rate of in vivo cholesterol synthesis. In the next two sections, we explore the structure and function of this notorious and biochemically complex molecule. Studies in the chemistry, biochemistry, and physiology of cholesterol have been at the forefront of science for many years. Thirteen Nobel prizes have been awarded to workers who have studied this molecule and its physiological functions.

Cholesterol Biosynthesis

Cholesterol was first isolated from natural sources (gallstones) in 1784, but its complicated structure was not elucidated until the 1930s. Most of the steps in cholesterol biosynthesis were discovered by Konrad Bloch and his colleagues at Harvard University in the 1940s. By using radiolabeled carbon substrates, they were able to show that all 27 carbons of cholesterol come from the simple C_2 compound, acetate. As we study the steps of cholesterol biosynthesis, we will also gain understanding of the diverse ways complex biomolecules are synthesized. The synthesis of cholesterol will be studied in four stages.

Stage 1

In this sequence, acetyl CoA is used to synthesize **mevalonate,** a C_6 compound. Three molecules of acetyl CoA combine in two successive reactions to form 3-hydroxy-3-methylglutaryl CoA (HMG CoA) (Figure 18.20). The reactions to initiate mevalonate synthesis take place in

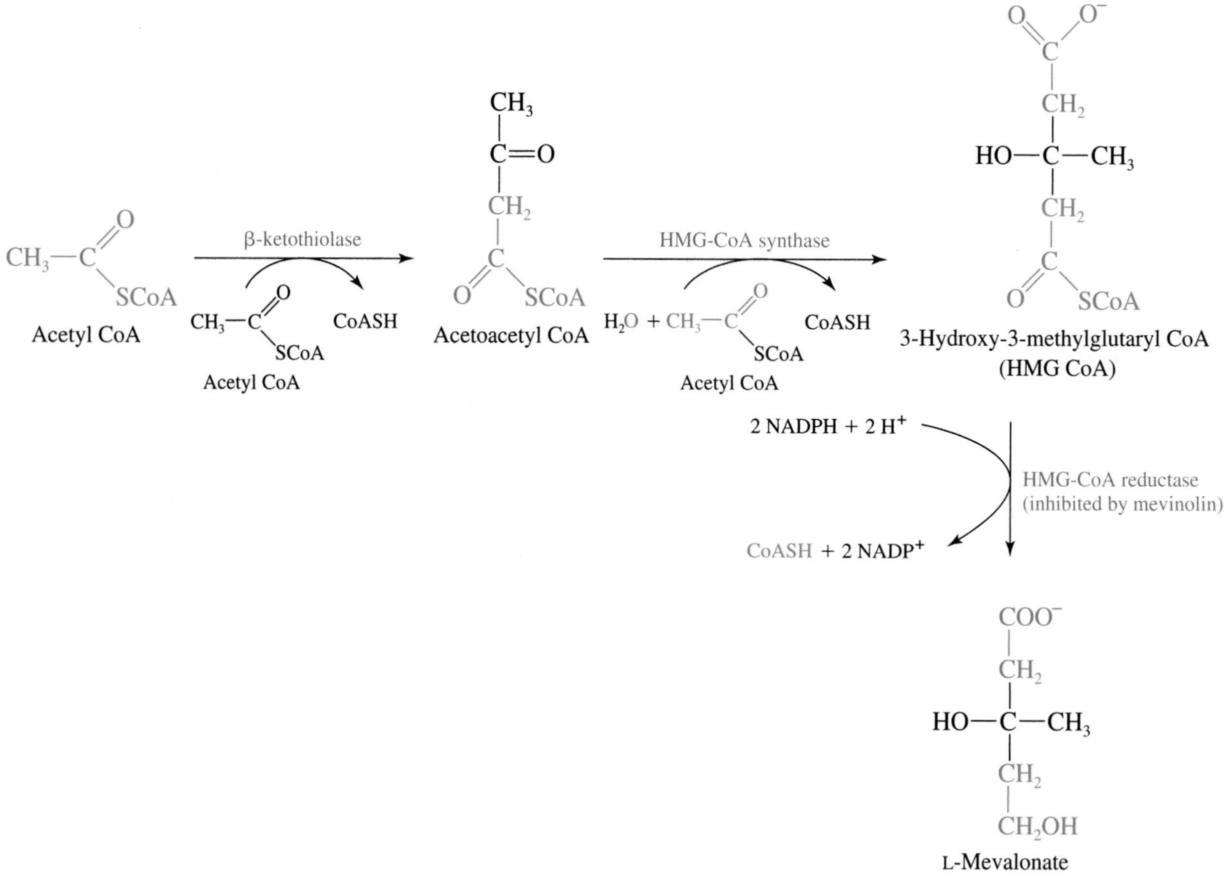

Figure 18.20

Stage 1 in the biosynthesis of cholesterol, acetyl CoA ⟶ mevalonate.

the cytoplasm of liver cells. The enzymes are part of the endoplasmic reticulum membrane, but the products are generated in the cytoplasm. HMG CoA in mitochondrial matrix is cleaved by HMG-CoA lyase to give acetoacetate (a ketone body) and acetyl CoA. In cytoplasm, HMG CoA has a different fate. The enzyme **HMG-CoA reductase** catalyzes the formation of mevalonate. The reductase has a preference for NADPH as coenzyme. This reaction is the major regulatory step in cholesterol synthesis. The enzyme is allosterically controlled by dietary cholesterol or some cholesterol product. Abundant cholesterol in the diet reduces the activity of HMG-CoA reductase. Drugs have been designed that inhibit HMG-CoA reductase, thereby lowering blood cholesterol levels.

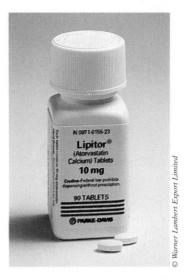

The cholesterol-lowering drug Lipitor® acts by inhibiting the enzyme HMG-CoA reductase.

Stage 2

The next four steps convert mevalonate, a C_6 intermediate, to **activated isoprenes,** C_5 intermediates. Although the loss of a single carbon might appear to be a simple, straight-forward process, this was one of the last stages of cholesterol synthesis to be understood. In three successive steps, mevalonate accepts three phosphoryl groups, each from an ATP. The product, 3-phospho-5-pyrophosphomevalonate, is decarboxylated to isopentenyl pyrophosphate (Figure 18.21). This latter intermediate is one of two activated compounds called **isoprenes.** Isopentenyl pyrophosphate is in equilibrium with another isoprene form, dimethylallyl pyrophosphate. Discovery of the elusive activated isoprenes opened the door to a greater understanding of the biosynthesis of many important terpene molecules, including cholesterol, limonene, lycopene, natural rubber, and vitamins A, D, E, and K.

Stage 3

Squalene, a C_{30} hydrocarbon, is synthesized by successive condensation of several isoprene units. Two isoprenes condense to form geranyl phosphate (a C_{10} compound). Addition of another isoprene produces the C_{15} compound, farnesyl pyrophosphate. The condensation of

Figure 18.21

Stage 2 in the biosynthesis of cholesterol, mevalonate ⟶ isoprenes. The isoprene, isopentenyl pyrophosphate is later isomerized to the isoprene dimethylallyl pyrophosphate.

two of the C_{15} intermediates results in the highly unsaturated hydrocarbon, squalene (Figure 18.22). Many of the intermediates of this condensation process have been isolated as natural oils in plants and animals.

Stage 4

The cyclization of squalene and its conversion to cholesterol are some of the most interesting and remarkable reactions in biochemistry. Many chemists and biochemists spent years of their lives studying the cyclization of squalene to lanosterol and further reactions leading to **cholesterol** (Figure 18.23). The key intermediate initiating the cyclization process is squalene 2,3-epoxide, a reactive compound formed using an oxygen atom from molecular oxygen. The enzyme squalene monooxygenase catalyzes this redox

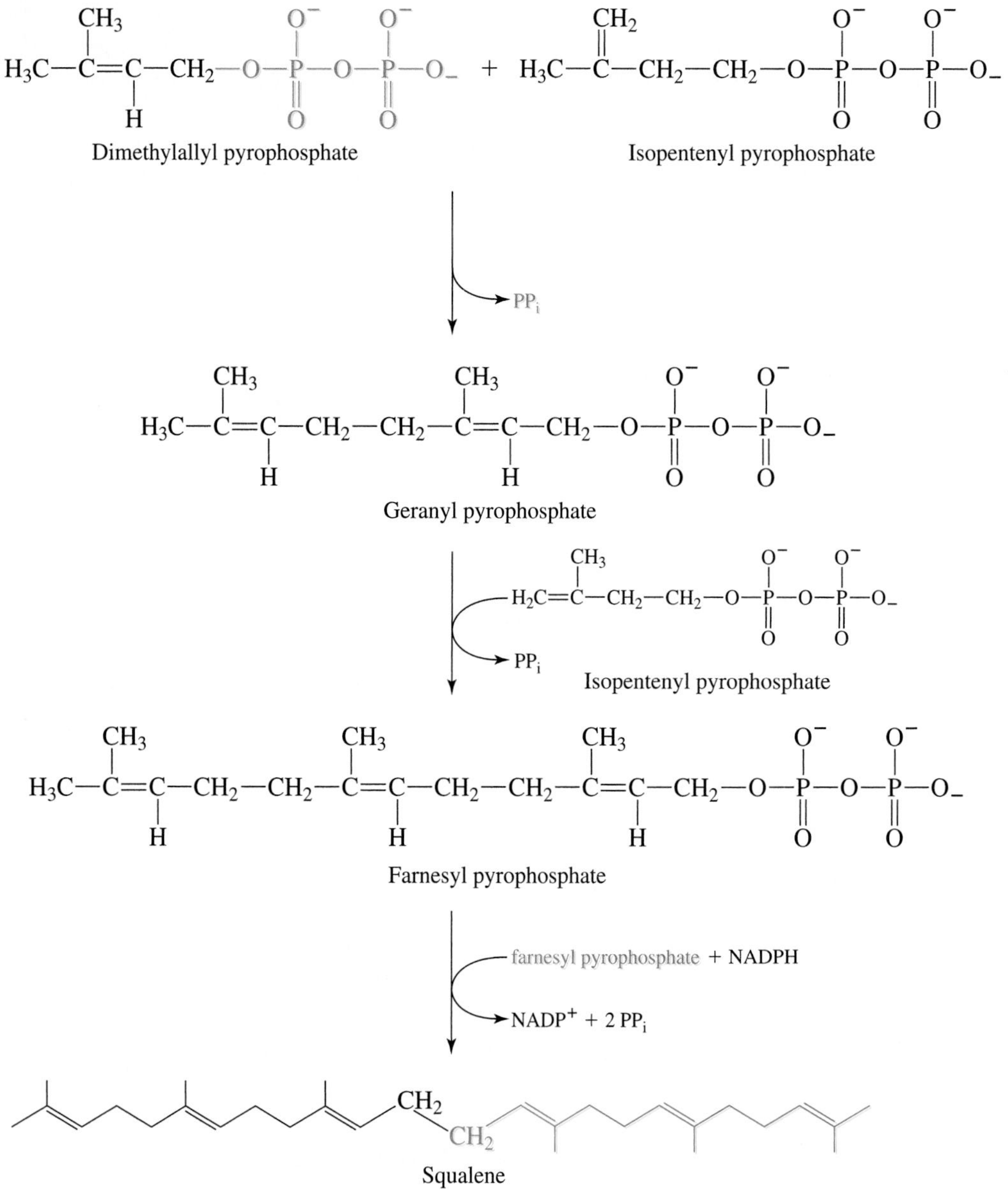

Figure 18.22

Stage 3 in the biosynthesis of cholesterol, isoprenes ⟶ squalene. Six isoprenes are condensed to squalene, a C_{30} hydrocarbon.

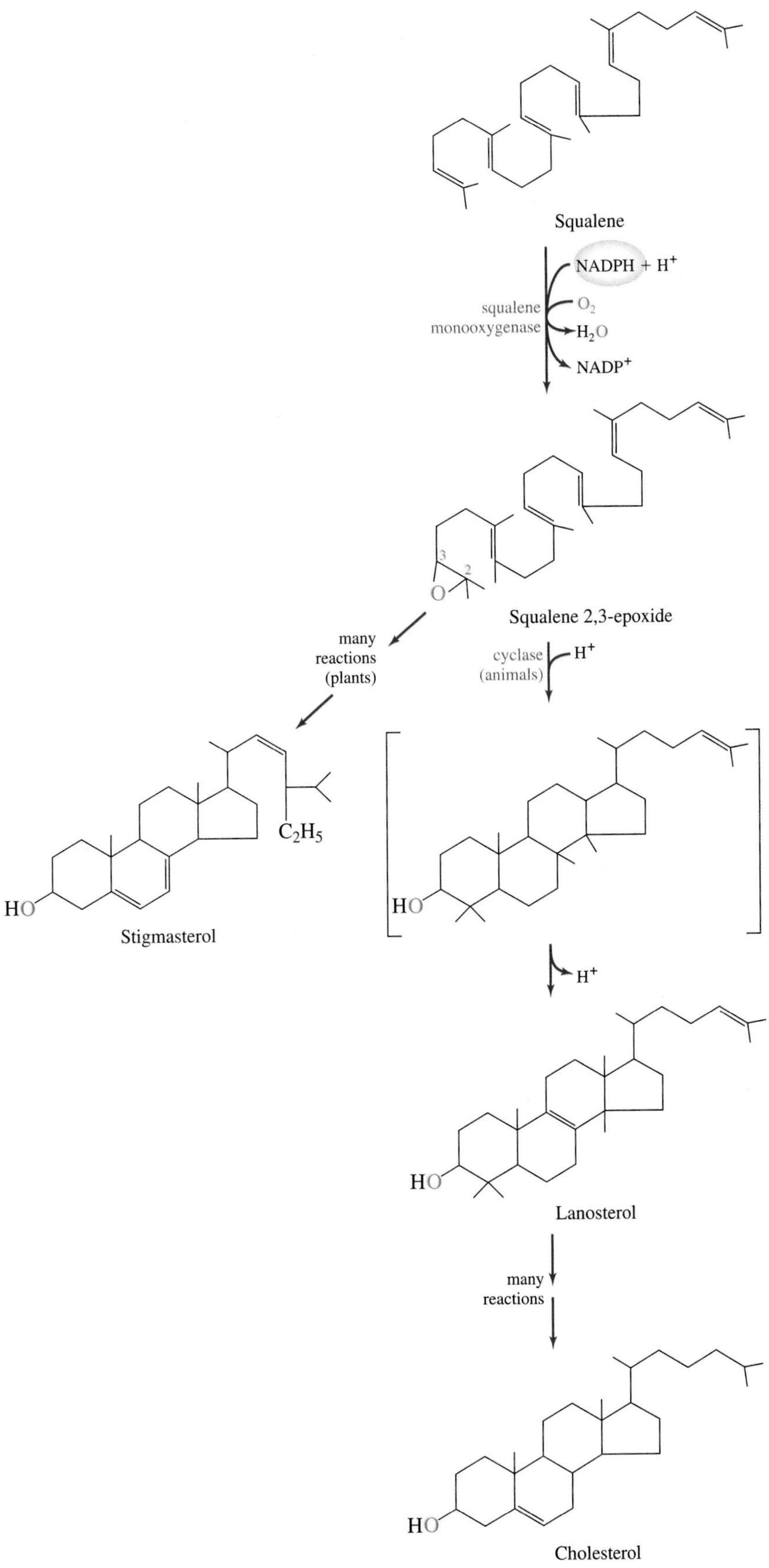

Figure 18.23
Stage 4 in the biosynthesis of cholesterol, squalene ⟶ cholesterol.
Squalene is cyclized after formation of an epoxide intermediate.

Squalene

squalene
monooxygenase

NADPH + H$^+$

O$_2$

H$_2$O

NADP$^+$

Squalene 2,3-epoxide

many
reactions
(plants)

cyclase
(animals)   H$^+$

Stigmasterol

C$_2$H$_5$

HO

HO

H$^+$

Lanosterol

HO

many
reactions

Cholesterol

HO

This stamp issued by the United States honors Percy Julian for his synthesis of cortisone and other important steroids.

reaction. The first intermediate in cholesterol synthesis that has an intact steroid ring system is lanosterol. At least 20 enzyme-catalyzed steps involving removal of methyl groups and rearrangements of double bonds are required to transform lanosterol to cholesterol. Squalene is also produced in plants but is cyclized to stigmasterol rather than cholesterol.

Cholesterol as a Precursor for Other Steroids

Cholesterol is the starting point for synthesis of many other important biomolecules (Figure 18.24). **Bile salts** are polar, steroid derivatives synthesized in the liver, stored in the gallbladder, and secreted into the intestines to assist in the solubilization of dietary fats. Their biosynthesis includes several hydroxylation steps catalyzed by oxygenases. Cholate and glycocholate are two of the most common bile salts (see Figure 9.11).

There are five major classes of steroid hormones. All are derived from the carbon skeleton of cholesterol. Progesterone regulates physiological changes in pregnancy; the glucocorticoids influence metabolism by promoting gluconeogenesis, glycogen formation, and fat breakdown; the mineralocorticoids regulate the absorption of Na^+, Cl^-, and HCO_3^- in the kidney; the androgens promote development of male sex characteristics and estrogens promote development of female sex characteristics. The general biosynthetic routes and structures for several important steroid hormones are shown in Figure 18.25.

The cholesterol skeleton is also used for the synthesis of vitamin D_3, which regulates calcium and phosphorus metabolism. The synthetic route begins with transformation of cholesterol to 7-dehydrocholesterol. This intermediate in skin cells undergoes rearrangement initiated by UV radiation from the sun (Figure 18.26). The product, cholecalciferol, is hydroxylated by liver enzymes to the most active form of the vitamin, 1,25-dihydroxycholecalciferol. Human deficiency of vitamin D_3 results in rickets, a nutritional disease characterized by bone malformation resulting from improper calcium and phosphorus metabolism. Adults deficient in vitamin D_3 develop **osteomalacia,** a weakening of bones. People living in areas with relatively low amounts of sunshine are most susceptible to the disease. Today many foods are fortified with vitamin D_3 or diets are supplemented with cod-liver oil, which is a good source of the vitamin.

Figure 18.24

Important biomolecules synthesized from cholesterol.

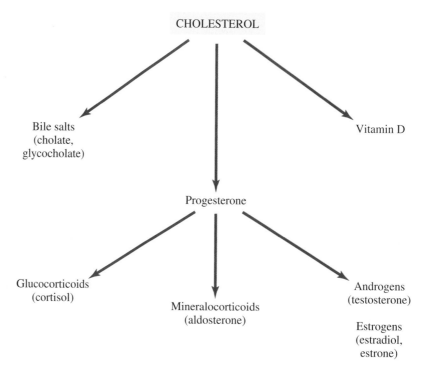

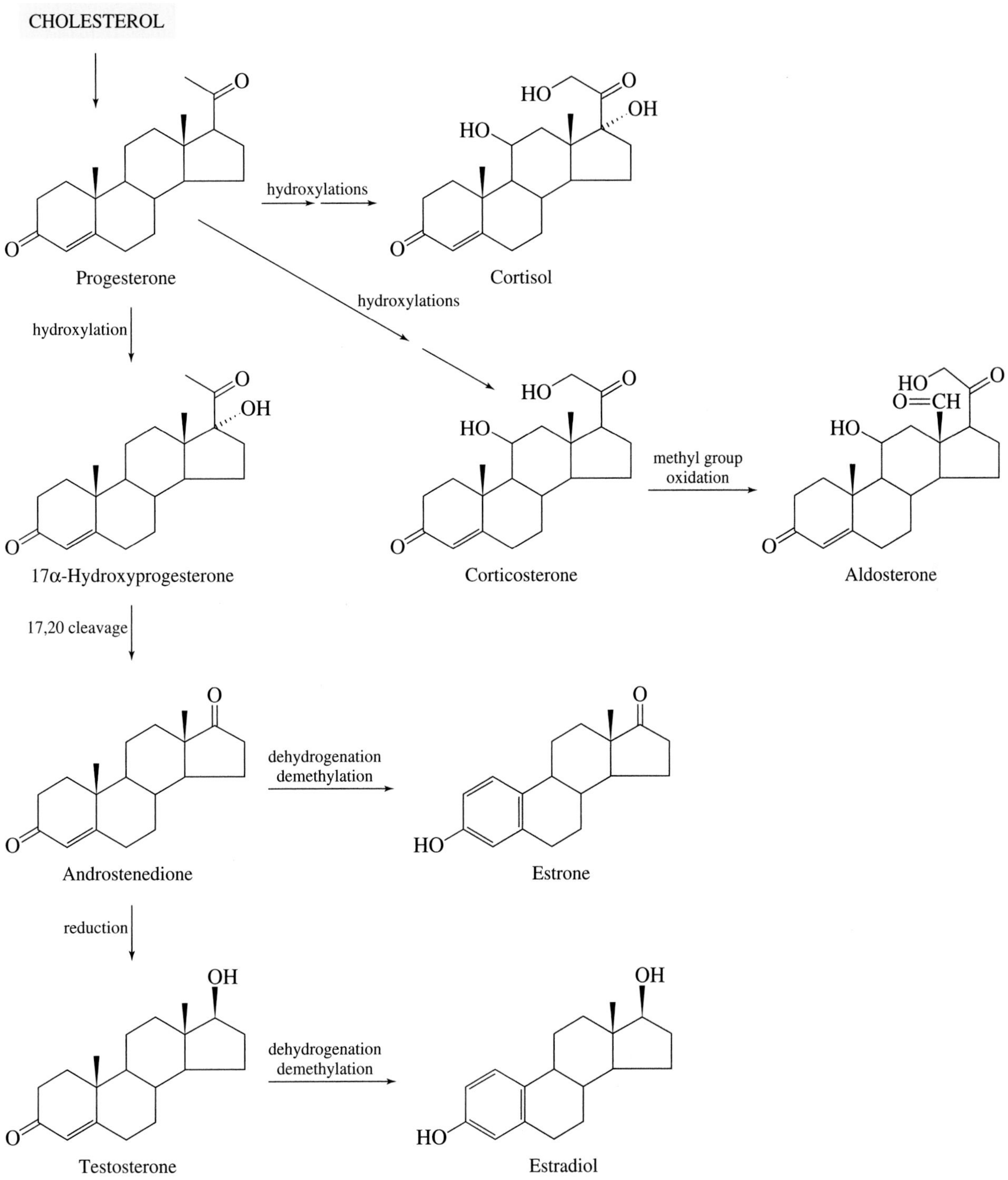

CHOLESTEROL

Figure 18.25
Biosynthetic routes from cholesterol to the structures of important steroid hormones.

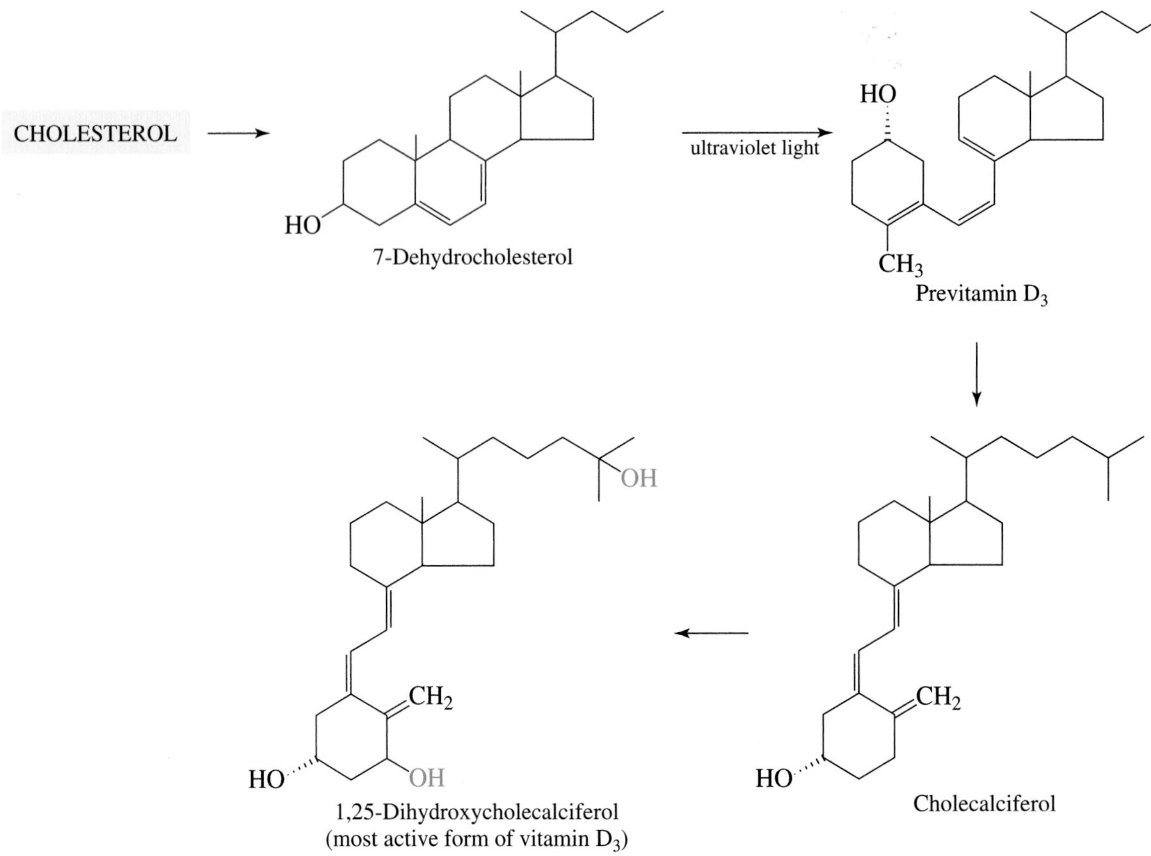

Figure 18.26
The synthesis of vitamin D_3 from cholesterol. A key step is a rearrangement process initiated by sunlight.

18.5　Transport of Lipids in Blood

Lipoproteins

Most of the dietary lipids and those synthesized in the cell are nonpolar molecules, displaying very low water solubility. Cholesterol and the triacylglycerols are essentially insoluble in water. However, they are needed by cells throughout higher organisms for construction of membranes, for use as fuel molecules, or for biosynthetic purposes. The most important lipids for transport are triacylglycerols and two forms of cholesterol, free cholesterol and cholesteryl esters. The esters are formed by transfer of fatty acyl groups (C_{16} and C_{18}) from CoASH to the hydroxyl group of cholesterol (acyl CoA-cholesterol acyl transferase). Nonpolar lipids are carried in the blood as lipoprotein complexes. These are serum particles comprised of specific proteins called apolipoproteins and diverse combinations of triacylglycerols, cholesterol, and cholesteryl esters. The lipids and proteins are noncovalently associated into water-soluble complexes. Chylomicrons are formed by the association of dietary triacylglycerols and cholesterol with specific proteins to assist in lipid digestion and transport. Several classes of lipoproteins have been characterized that differ in the specific types of apolipoprotein and in the combinations of lipids. They are experimentally separated by centrifugation and classified according to density. Since lipids are lighter than proteins, the higher the lipid to protein ratio, the less dense the lipoproteins. Table 18.5 lists the four main types of lipoproteins along with their densities and compositions. Even though the compositions of the lipoproteins differ, the structures are similar. They are composed of a central core of neutral, nonpolar lipid surrounded by a shell of phospholipids and proteins (see Figure 18.2). The surface of the aggregate is coated with

Table 18.5
Names, physical properties, and composition of major lipoproteins

Lipoprotein	Density (g/mL)	Diameter (Å)	Composition (wt %)			
			Protein	Cholesterol[a]	Phospholipids	Triacylglycerols
Chylomicrons	< 0.95	800–5000	2	4	9	85
Very low density	0.95–1.006	300–800	10	20	20	50
Low density	1.006–1.063	180–280	25	45	20	10
High density	1.063–1.2	50–120	55	17	24	4

[a]Includes free cholesterol + cholesteryl esters.

hydrophilic groups of polar lipids and proteins to interact with water for solubilization. The shell is similar to a membrane except it is a *single* layer. At least nine types of apolipoproteins have been identified. Their molecular weights range from 7000 to over 500,000. They have a relatively high content of hydrophobic amino acids to interact with the lipid interface of the core of each lipoprotein. Here we discuss the production and function of the major lipoproteins: the **chylomicrons,** the **very low density lipoproteins,** the **low density lipoproteins,** and the **high density lipoproteins.**

Chylomicrons: These are the least dense lipoproteins and consist of 98 to 99% lipid material. The lipid content is primarily triacylglycerols, and the chylomicrons can be considered droplets of fat with a surface layer of protein and polar lipids. They are assembled in the intestines from dietary lipids and absorbed into the bloodstream, where they are transported to peripheral tissue. There the enzyme lipoprotein lipase releases fatty acids from the triacylglycerols. The lipoprotein after loss of most triacylglycerols becomes a chylomicron remnant, which is cholesterol rich.

Very Low Density Lipoproteins (VLDL): These molecular aggregates are produced in the liver from triacylglycerols synthesized in the liver. Their function is to deliver synthesized lipids to adipose and other peripheral tissue for storage or energy use. The fatty acids are released in the same manner as are chylomicrons, resulting in the formation of low density lipoproteins.

Low Density Lipoproteins (LDL): The LDL particles are the major carriers of cholesterol in the blood. They move cholesterol and cholesteryl esters from the liver, where they are synthesized, to peripheral tissues. The predominant lipids are cholesteryl esters containing the polyunsaturated fatty acid, linoleate ($18:2^{\Delta 9,12}$). The role of the LDLs in the regulation of cholesterol metabolism is discussed later in this section.

High Density Lipoproteins (HDL): The HDLs have the highest content of protein of any of the lipoproteins (55% protein, 45% lipid); therefore, they are the most dense. The primary lipids in the core are cholesterol and cholesteryl esters. Like the LDLs, HDLs transport cholesterol and its esters; but unlike the LDLs, they move cholesterol in the *opposite* direction, from peripheral tissue to the liver. HDLs are synthesized in the liver but in a rather incomplete form; they have little lipid at this point. They move through the bloodstream, where they collect excess cholesterol and transport it to the liver. This process is called **reverse cholesterol transport.** The HDLs are often called "good" transport forms of cholesterol because their action reduces plasma cholesterol levels.

Cholesterol and Health

Cholesterol, although required for proper cell growth and development, is a potentially dangerous molecule. Therefore, its transport by the lipoproteins and uptake into cells must be very carefully regulated. Too much cholesterol in blood and tissue leads to its accumulation, especially in the arteries, causing blockage (**atherosclerosis**). Cholesterol and cholesteryl esters are taken up by cells using endocytosis, or engulfment. The process begins with the binding of LDL-containing cholesterol to specific binding sites (protein receptors on the plasma membrane) of cells (Figure 18.27). The binding sites on the cell membrane are called coated pits. The LDL invaginates into the plasma membrane and fuses with part of the membrane to form an endocytotic vesicle. The vesicle, now inside the cell, fuses with organelles called **lysosomes.** Degradative enzymes present in the lysosomes catalyze the hydrolysis of fatty acid esters, thereby releasing free cholesterol for membrane construction.

The importance of the endocytotic process and the presence of LDL receptors has recently been evaluated from a clinical standpoint. One of the main causes of hypercholesterolemia (abnormally high levels of blood cholesterol) are genetic defects in the protein components required for cholesterol transport and uptake. Individuals who completely lack functional LDL receptors develop **familial hypercholesterolemia,** a genetic disease that causes cholesterol levels to approach 700 mg/100 mL of blood (150–200 mg/100 mL is normal). Cholesterol is deposited in arterial plaques, which results in **atherosclerosis** (hardening of the arteries). Many of these individuals die from heart attacks during childhood.

When excessive cholesterol is present in normal individuals it can begin to accumulate in blood vessels, leading to obstruction of blood flow and heart failure. This form of atherosclerosis is a leading cause of death in the Western world. The incidence of atherosclerosis and other cardiovascular diseases shows a positive correlation with the level of blood cholesterol and lipoprotein concentrations. Individuals with high serum levels of LDL cholesterol (in relation to HDL levels) show an increased incidence of heart disease. High levels of HDL cholesterol are inversely related to coronary diseases. We are strongly encouraged to reduce our dietary cholesterol uptake, but there are other factors, controllable and uncon-

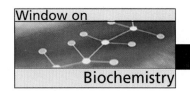

Window on
Biochemistry

18.2 Boosting "Good Cholesterol" Using an HDL Apoprotein A-I Mutant

Several blockbuster, cholesterol-reducing drugs have been introduced during the last decade: Mevacor, Zocor, Pravachol, and Lipitor. These drugs have total sales of $5 billion per year. All of these current pharmaceuticals are classified as **statins,** agents that inhibit the rate-controlling enzyme in cholesterol synthesis, 3-hydroxy-3-methylglutaryl CoA (HMG CoA) reductase. The statins function in the treatment of heart disease by lowering serum LDL cholesterol ("bad cholesterol"), but only 30% of users experience a reduction in heart attacks and strokes. Because so little protection is achieved from the statins, the search is on for new drugs. The cholesterol drugs of the next decade are likely to be those that increase the serum concentration of HDL or "good cholesterol."

One of the major proteins in the outer shell of the HDL complexes is **apoprotein A-1,** which has 243 amino acid residues and a molecular weight of approximately 29,000 daltons. A natural variant of this protein that appears to protect individuals from coronary disease has been found in a family living in the village of Limone in northern Italy. Members of this family have high serum triacylglycerol and cholesterol levels, but few clinical signs of atherosclerosis. The mutated apoprotein A-1 is characterized by a single arginine to cysteine substitution. The protein was discovered by physicians at the University of Milano using gel electrophoresis, hence the protein is called **apoprotein A-1 Milano.**

Studies on humans and other animals with heart disease have shown that the injection of recombinant (genetically engineered) apoprotein A-1 Milano improves blood flow to the arteries of the heart, brain, and body. Recent published results of human clinical studies reported that when the mutated protein was infused into people with high cholesterol, they showed increased elimination of cholesterol from the body by the process of **reverse cholesterol transport.** One disadvantage of the apoprotein A-1 Milano treatment is that therapeutic proteins cannot be taken orally, but must be injected. It is possible that injections can be infrequent if the apoprotein A-1 treatment is combined with a prescribed statin. No side effects have been identified in the use of the modified apoprotein. Further clinical trials need to be completed before it can be proven that the protein does indeed protect against heart disease.

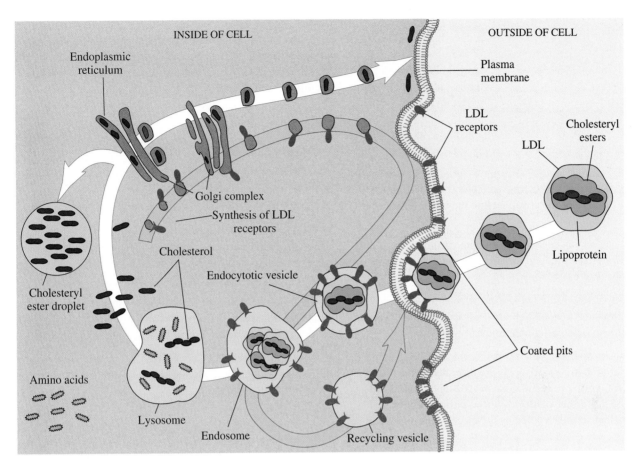

Figure 18.27
The uptake of cholesterol into cells by endocytosis. The LDL receptor is synthesized in the endoplasmic reticulum and migrates to the cell surface. The lipoprotein brought into the cell is degraded to release cholesterol. Amino acids are the result of lipoprotein degradation.

trollable, that can also lessen the risk of atherosclerotic development. Perhaps the most important factor is our genetic makeup, something we cannot change. Also, gender is important: females show lower levels of serum cholesterol, higher levels of HDL, and lower incidence of cardiovascular disease than males. There are, however, many factors that influence cholesterol plasma level that can be altered: smoking, consumption of alcohol, high-fat diet, lack of exercise, and high stress all tend to reduce HDL cholesterol levels and/or increase LDL cholesterol levels. (See Window of Biochemistry 18.2.)

For individuals whose blood cholesterol level is not reduced by altering lifestyle, drugs are available that lower cholesterol levels up to 30%. One drug currently on the market is

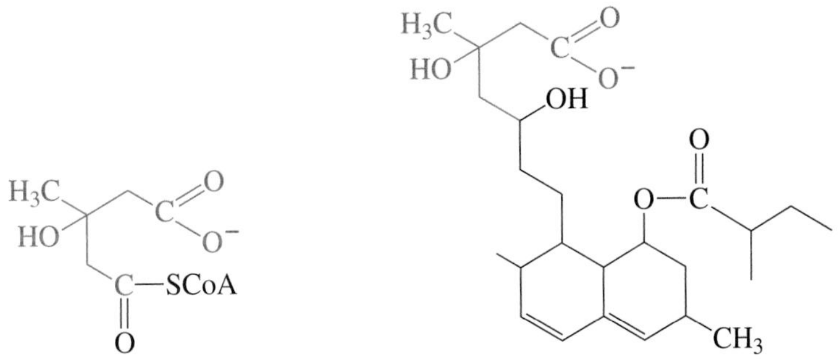

3-Hydroxy-3-methylglutaryl CoA
(HMG CoA)

Mevinolin

Figure 18.28
The structure of mevinolin, a drug used to treat hypercholesterolemia. Mevinolin is a competitive inhibitor of HMG-CoA reductase. The structure of the normal substrate, HMG CoA, is shown for comparison.

mevinolin (Figure 18.28). It is produced by various strains of fungi. This drug and others reduce blood cholesterol even in patients with familial hypercholesterolemia. The mode of action of mevinolin is well understood. It is a potent competitive inhibitor of the enzyme HMG-CoA reductase, the catalyst for the rate-controlling step of in vivo cholesterol synthesis. Figure 18.28 compares the structure of the normal substrate HMG CoA with mevinolin.

Summary

- Dietary fats, triacylglycerols, are solubilized in the intestinal tract by bile salts and absorbed into the blood for transport to peripheral tissue.

- When needed for energy metabolism, cellular fatty acids are mobilized from triacylglycerols by lipase, which catalyzes the hydrolysis of ester bonds.

- Free fatty acids in the cytoplasm of cells are activated by thioester linkage to CoASH and transported by carnitine through the mitochondrial membrane.

- Catabolism of fatty acids by beta oxidation occurs in the mitochondrial matrix. Fatty acids are degraded to the C_2 unit, acetyl CoA with production of NADH and $FADH_2$. The acetyl CoA is further oxidized by the citric acid cycle. The reduced cofactors are recycled by the electron transport chain.

- Fatty acids are synthesized from malonyl CoA, which is produced by the carboxylation of acetyl CoA.

- Cholesterol is synthesized in the cytoplasm by combining acetyl CoA. The process involves the intermediates mevalonate, isoprene, and squalene. The rate-controlling step is catalyzed by HMG-CoA reductase.

- Cholesterol, triacylglycerols, and other water-insoluble lipids are transported through the blood of animals as lipoprotein complexes, chylomicrons, VLDL, LDL, and HDL.

- Although cholesterol is required for proper cell growth, defects in its transport are the cause of artery blockage and other heart-related conditions.

Study Exercises
Understanding Terms

18.1 Define the following terms in 25 words or less.
- **a.** Bile salts
- **b.** Lipoproteins
- **c.** β oxidation
- **d.** Cholesterol
- **e.** Carnitine
- **f.** Acyl carrier protein
- **g.** HMG CoA
- **h.** Malonyl CoA
- **i.** Fatty acyl-CoA desaturases
- **j.** Essential fatty acids
- **k.** Isoprene
- **l.** Chylomicrons
- **m.** Olestra
- **n.** Mevalonate
- **o.** Isoprenes
- **p.** Low density lipoproteins
- **q.** High density lipoproteins
- **r.** Apoprotein A-1
- **s.** Reverse cholesterol transport

Reviewing Concepts

18.2 Explain the role of bile salts in the digestion of fats.

18.3 Listed below are regulatory enzymes and their allosteric modulators. Briefly explain the reasons for this kind of regulation.
- **a.** HMG-CoA reductase: inhibited by cholesterol
- **b.** Triacylglycerol lipase: stimulated by adrenaline
- **c.** Acetyl-CoA carboxylase: inhibited by palmitoyl CoA

18.4 Identify each of the lipoproteins described below. Choose from chylomicrons, VLDL, LDL, and HDL.
- **a.** Which lipoprotein has the lowest density? Why?
- **b.** Which lipoprotein carries the highest percentage of cholesterol and cholesteryl esters?
- **c.** Which lipoprotein has the highest percentage of protein?
- **d.** Which lipoprotein has the highest percentage of triacylglycerol?
- **e.** Which lipoprotein removes cholesterol from circulation?
- **f.** Which lipoprotein has the highest density? Why?

18.5 Draw the structure of the ester formed by the following reaction.

$$\text{Cholesterol + oleoyl CoA} \xrightarrow{\text{acyltransferase}}$$

18.6 What is the function of the phospholipids in lipoproteins?

18.7 Write the structures for the products of the following reactions.
- **a.** HMG CoA ⇌ \
enzyme: HMG-CoA lyase
- **b.** *trans*-Δ^2-butenoyl-ACP + NADPH + H^+ ⇌ \
enzyme: enoyl-ACP reductase

c. HMG CoA + 2 NADPH + 2 H$^+$ $\rightleftharpoons$
enzyme: HMG-CoA reductase

d. Cholesteryl linoleate + H$_2$O $\rightleftharpoons$
enzyme: cholesteryl esterase

e. Squalene + O$_2$ + NADPH + H$^+$ $\rightleftharpoons$
enzyme: squalene monooxygenase

f. CH$_3$CH$_2$CCH$_2$CSCoA + CoASH $\rightleftharpoons$
$\quad\quad\quad\quad$‖$\quad\quad$‖
$\quad\quad\quad\quad$O$\quad\quad$O

enzyme: β-ketothiolase

g. Citrate $\rightleftharpoons$
enzyme: citrate lyase

h. Palmitoyl ACP + H$_2$O $\rightleftharpoons$
enzyme: palmitoyl thioesterase

i. 2 Acetyl CoA $\rightleftharpoons$
enzyme: β-ketothiolase

j. Trimyristin + 3H$_2$O $\rightleftharpoons$
enzyme: lipase

18.8 The list below includes metabolic reactions and processes for the catabolism of fatty acids. Arrange them in the correct order

a. Thiolytic cleavage
b. Formation of acyl carnitine
c. FAD-linked dehydrogenation
d. Fatty acid in cytoplasm
e. Linkage of fatty acid to CoASH
f. Hydration of double bond
g. NAD$^+$-linked dehydrogenation
h. Transport of fatty acid to mitochondria

HINT: Begin with d.

18.9 From the list below, identify those reactants or characteristics that describe β oxidation and fatty acid synthesis.

a. Enzymes located in cytoplasm
b. Requires NADPH + H$^+$
c. Acyl CoA
d. Malonyl CoA
e. Enzymes located in mitochondria
f. Requires FAD
g. Acyl carrier protein (ACP)

18.10 Which of the following compounds or metabolic processes are related to cholesterol synthesis or degradation?

a. Squalene$\quad\quad$**f.** Cholic acid
b. β oxidation$\quad$**g.** Carnitine transport
c. Acetyl CoA$\quad$**h.** Action of lipase
d. Mevalonate$\quad$**i.** Chylomicrons
e. Palmitate$\quad\quad$**j.** HMG CoA

18.11 The drug mevinolin inhibits the enzyme hydroxymethylglutaryl-CoA reductase. What effect will this drug have on the rate of cholesterol biosynthesis?

18.12 Below is a list of enzymes discussed in this chapter. Classify each of the enzymes into one of the six standard enzyme groups.

Enzyme	Classification
a. Lipase	_____
b. Acyl-CoA synthetase	_____
c. Acyl-CoA dehydrogenase	_____
d. β-ketothiolase	_____
e. Acetyl-CoA carboxylase	_____
f. Squalene monooxygenase	_____

HINT: See Table 14.2.

18.13 Several enzymes used in this chapter are listed below. On the line provided, write the primary cellular location for each enzyme. Choose between cytoplasm and mitochondria.

Enzyme	Cell location
a. Carnitine acyltransferase II	_____
b. Enoyl-CoA hydratase	_____
c. HMG-CoA lyase	_____
d. Acetyl-CoA carboxylase	_____
e. HMG-CoA reductase	_____

Solving Problems

18.14 Each of the following enzymes requires the assistance of a cofactor. For each enzyme, list the most likely cofactor.

a. Acyl-CoA dehydrogenase
b. Hydroxyacyl-CoA dehydrogenase
c. Methylmalonyl-CoA mutase
d. β-Hydroxybutyrate dehydrogenase
e. Acetyl-CoA carboxylase

HINT: Select from NAD, FAD, biotin, and coenzyme B12.

18.15 Why is it not possible for humans to convert fatty acids into glucose?

18.16 In a laboratory experiment you are studying the incorporation of the malonyl group into fatty acids. You are using malonyl CoA that is labeled with ^{14}C, radioactive carbon. To complete the experiment you mix acetyl CoA, labeled malonyl CoA, and the fatty acid synthase complex. What carbons in the palmitate product are labeled if the malonyl CoA is labeled as follows?

$$^-O_2{}^{14}CCH_2CSCoA$$
$$\quad\quad\quad\quad\quad‖$$
$$\quad\quad\quad\quad\quad O$$

18.17 Show the pathway for complete β oxidation of each of the following fatty acids.

a.

b.

18.18 Derive the net reaction for the complete β oxidation of stearic acid and prepare an energy balance sheet as in Table 18.2.

18.19 Explain how a diet low in carbohydrate and fat helps reduce body fat.

18.20 Why is it necessary for the apolipoproteins to have two domains, a hydrophobic domain and a hydrophilic domain?

18.21 Describe the structure and chemical contents of chylomicron remnants and compare to the structure and chemical contents of chylomicrons.

18.22 Using names of intermediates and enzymes, show how the carbons in glucose can be used to synthesize fatty acids.

18.23 The metabolic process of β oxidation is inhibited by NADH. Explain the metabolic reasoning behind this mode of regulation.

18.24 The use of ATP in energy metabolism sometimes results in the formation of AMP and PP_i rather than the more common ADP + P_i. Give an example from this chapter of the pyrophosphate cleavage of ATP and describe advantages of this cleavage, if any.

18.25 Explain the metabolic reasons behind the following statement: "Carbohydrate molecules provide only about one-half the caloric energy of a fatty acid molecule of similar molecular weight."

18.26 Individuals who have low concentrations of carnitine in their muscles have increased levels of triacylglycerols. Explain.

HINT: What is the function of carnitine in fatty acid catabolism?

18.27 Beta oxidation is inhibited by acetyl CoA. Explain the metabolic reasoning behind this mode of regulation.

18.28 In your laboratory experiment for this week you are studying the synthesis of cholesterol by using radiolabeled (^{14}C) acetyl CoA. The methyl group of acetyl CoA is labeled with ^{14}C. The labeled acetyl CoA is incubated with the appropriate enzymes to convert it to mevalonate. Draw the structure for the mevalonate produced and draw circles around the carbons labeled with ^{14}C.

18.29 What enzymes, in addition to the standard β-oxidation enzymes, are needed to degrade each fatty acid below to acetyl CoA?
  a. Oleate
  b. Arachidonate

Writing Biochemistry

18.30 Your biochemistry professor has asked each student to give a 15-minute presentation in class on some topic in Chapter 18. You have selected to present a paper on one of the current drugs for treatment of high serum cholesterol (Lipitor, Zocor, Mevachor, etc.). Use biochemistry textbooks, journals, and the Internet to learn all you can about your topic. Write out your presentation in 300 to 500 words. The paper may be in the style of a "Window on Biochemistry."

18.31 Study the reference by Darvey listed in "Further Reading": *Biochem. Educ.* 27:209–210 (1999). Use 100 to 200 words to answer the question, what factors are responsible for the greater yield of ATP per carbon atom when fatty acids are completely oxidized to carbon dioxide and water compared with complete oxidation of glucose?

Applying Biochemistry

18.32 Your grandfather, who recently had a heart attack, is taking the cholesterol-lowering drug Lipitor. He is not a scientist, but he knows you are studying biochemistry. How would you answer his question, "How does this drug work?" Write out your explanation in about 100 words.

18.33 Biochemists at the Laboratory for Universal Life have discovered a new bacterium in a soil sample brought back from Mars. While studying fatty acid metabolism in the bacterium, they discovered a new pathway which they called **gamma oxidation.** The reactions are identical to beta oxidation except that the gamma carbon is oxidized. Write out reactions for this new pathway beginning with lauric acid.

Further Reading

Banfalvi, G., 1999. Estimating the energy content of metabolites. *Biochem. Educ.* 27:79–82.

Bentley, R., 2000. From "reactive C_2 units" to acetyl coenzyme A. *Trends Biochem. Sci.* 25:302–305.

D'Andrea, G., 1994. Beta-oxidation of polyunsaturated fatty acids. *Biochem. Educ.* 22:89–91.

Darvey, I., 1998. How does the ratio of ATP yield from the complete oxidation of palmitic acid to that of glucose compare with the relative energy contents of fat and carbohydrate? *Biochem. Educ.* 26:22–23.

Darvey, I., 1999. What factors are responsible for the greater yield of ATP per carbon atom when fatty acids are completely oxidized to carbon dioxide and water compared with glucose? *Biochem. Educ.* 27:209–210.

Doyle, E., 1997. Olestra? The jury's still out. *J. Chem. Educ.* 74:370–372.

Elliott, K., 1999. Teaching metabolism for the future: Concepts not pathways. *The Biochemist* 21(1):9–39.

Fell, D., 1997. *Understanding the Control of Metabolism.* London: Portland Press.

Filippusson, H., 1998. Exercise biochemistry as an approach to the study of intermediary metabolism. *Biochem. Educ.* 26:119–120.

Frayn, K., 1996. *Metabolic Regulation: A Human Perspective.* London: Portland Press.

Garber, K., 1999. Boosting good cholesterol. *Modern Drug Discovery* September/October: 67–75.

Hajjar, D., and Nicholson, A., 1995. Atherosclerosis. *Am. Scientist* 83: 460–467.

Jandacek, R., 1991. The development of Olestra, a noncaloric substitute for dietary fat. *J. Chem. Educ.* 68:476–479.

Lawn, R., 1992. Lipoprotein (a) in heart disease. *Sci. Am.* 266(6):54–60.

Prince, D., and Welschenbach, M., 1998. Olestra: A new food additive. *J. Am. Diet. Assoc.* 98:565–569.

Ritter, S., 1999. Anabolic steroids. *Chem. Engin. News,* Sept. 6:42–52.

Wang, J., 1998. The insulin connection: Dorothy Hodgkin and the Beijing Insulin Group: *Trends Biochem. Sci.* 23:497–500.

Wilson, J., 2000. The biological basis for atherosclerosis. *The Scientist,* Oct. 30:20.

Chapter 19

Metabolism of Amino Acids and Other Nitrogenous Compounds

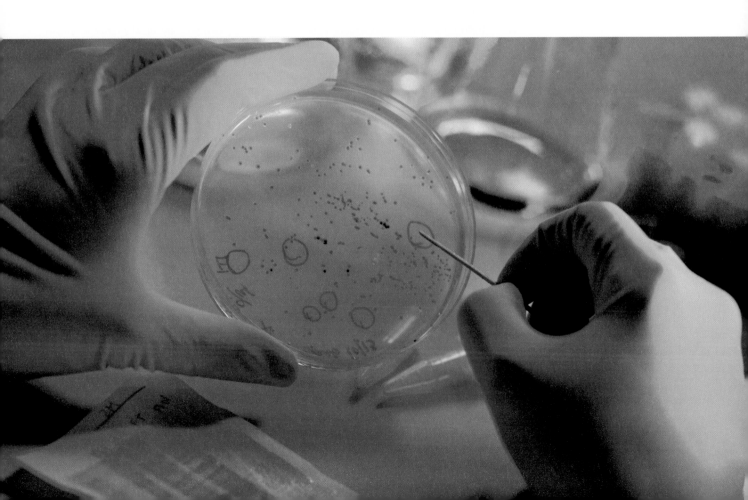

Red meat is an important dietary source of protein nitrogen.

© Mathew Klein/Photo Researchers

U p to this point our metabolic studies have focused on molecules that contain only the three most abundant elements in organisms: carbon, hydrogen, and oxygen. Their metabolism is complete after oxidation to CO_2 and H_2O, with a release of energy that is used to form ATP by electron transport and oxidative phosphorylation. We turn now to biomolecules that contain, in addition to carbon, hydrogen, and oxygen, the fourth most abundant element: nitrogen. Nitrogen is naturally present in both inorganic and organic forms, but in life the most abundant forms are organic: proteins, amino acids, purines, pyrimidines, porphyrins, catecholamines, vitamins, and many other small molecules.

This chapter begins by exploring the entry of nitrogen atoms into living organisms. The most important process is **nitrogen fixation,** the conversion of an unreactive molecule, nitrogen ($N\equiv N$), to a more biologically accessible molecule, ammonia (NH_3). After we show how ammonia is assimilated into amino acids, we explore the metabolism of amino acids. Twenty different structures for the amino acids result in 20 different catabolic routes and an equal number of anabolic pathways. All amino acids are catabolized to carbon skeletons that are oxidized by the citric acid cycle. The released nitrogen is either taken up for biosynthetic processes or eliminated in the form of urea, uric acid, or ammonia, depending on the organism.

This chapter introduces biochemical concepts not yet encountered, including nitrogen fixation; new chemistry involving nitrogen functional groups, such as transfer of amino groups; a new metabolic cycle (the urea cycle); biosynthesis and catabolism of amino acids; and the use of amino acids as precursors for important biomolecules (purines, pyrimidines, porphyrins, and others). The many topics covered here may at first appear to be unrelated; however, the primary unifying thread is the presence of nitrogen atoms in the compounds.

19.1 The Nitrogen Cycle

Nitrogen atoms appear in many different types of inorganic and organic compounds in the atmosphere and biosphere. The most common inorganic forms are listed with their oxidation numbers in Table 19.1. Note that nitrogen atoms exist in a wide range of oxidation states from nitrate, the most highly oxidized form, to ammonia, the most highly reduced form. By far the most abundant natural form of nitrogen is nitrogen gas, N_2, which makes up almost 80% of the atmosphere. This gas serves as a reservoir from which the element can be removed for use by organisms in the biosphere. All nitrogen atoms in life-forms originate from this source. However, only a few organisms—blue-green algae and soil bacteria—can tap into this huge supply by converting the very unreactive N_2 gas into a form usable by most other organisms. The flow of nitrogen atoms between the atmosphere and biosphere is defined by the **nitrogen cycle** (Figure 19.1). To summarize the cycle, N_2 gas is reduced to ammonia by nitrogen-fixing bacteria. Ammonia in the presence of water becomes the ammonium ion, NH_4^+. The ammonia generated in the soil by N_2 fixation can be used by plants and higher animals, but almost all is rapidly oxidized to nitrites by soil bacteria (genus *Nitrosomonas*):

$$2\,NH_3 + 3\,O_2 \longrightarrow 2\,NO_2^- + 2\,H_2O + 2\,H^+$$

Another group of soil bacteria, genus *Nitrobacter,* continues the oxidation to nitrates:

$$2\,NO_2^- + O_2 \longrightarrow 2\,NO_3^-$$

Table 19.1

Forms of inorganic nitrogen found in the atmosphere and biosphere

Name	Structure	Oxidation Number of Nitrogen Atom
Nitrate ion	NO_3^-	+5 (most oxidized)
Nitrite ion	NO_2^-	+3
Hyponitrite ion	$N_2O_2^{2-}$	+1
Nitrogen gas	N_2	0
Hydroxylamine	NH_2OH	−1
Ammonia	NH_3	−3 (least oxidized)

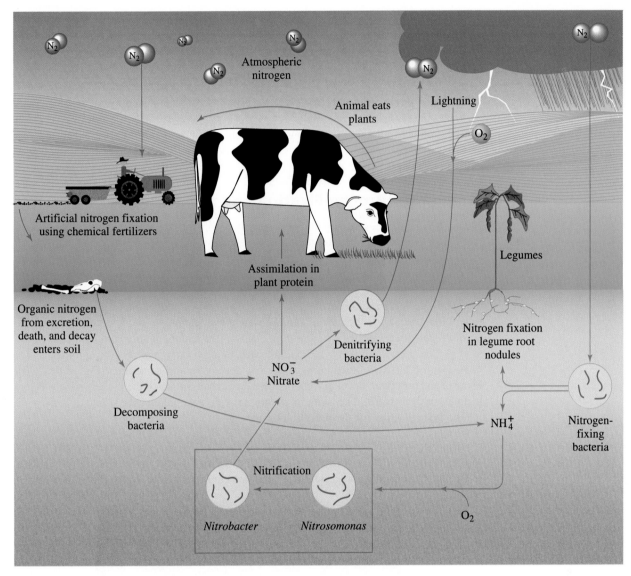

Figure 19.1

The nitrogen cycle, defining the flow of nitrogen atoms between the atmosphere and organisms.

These two sequential processes, called **nitrification,** provide most of the nitrite and nitrate needed by higher plants. Nitrogen can also enter the soil during lightning storms. High-voltage discharge catalyzes the oxidation of N_2:

$$N_2 \xrightarrow{O_2} 2\,NO \xrightarrow{O_2} 2\,NO_2^- \xrightarrow{O_2} 2\,NO_3^-$$

Oxides of nitrogen are washed from the air into the soil by water vapor and rain. Hydration of NO_3^- to nitric acid (HNO_3) and NO_2^- to nitrous acid (HNO_2) produces two important components of acid rain. Some bacteria are capable of denitrification, that is, the production of N_2 from nitrate and the return of the N_2 gas to the atmospheric reservoir.

Soil bacteria and plants can reverse the nitrification process by the action of nitrate and nitrite reductases:

$$NO_3^- + NADPH + H^+ \xrightleftharpoons[]{\text{nitrate reductase}} NO_2^- + NADP^+ + H_2O$$

$$NO_2^- + 3\,NADPH + 4\,H^+ \xrightleftharpoons[]{\text{nitrite reductase}} NH_4^+ + 3\,NADP^+ + H_2O + OH^-$$

Ammonium ion formed in the soil by these reduction processes and by nitrogen fixation is used by plants to synthesize amino acids. Although animals can use NH_4^+ directly in metabolic reactions, they obtain all nitrogen through the food chain by ingesting proteins of plants and other animals.

Nitrogen from plants and animals is recycled to the soil by two processes:

1. Excretion of nitrogen in the form of urea or uric acid, which is converted to ammonium ion by microorganisms.
2. The proteins and other nitrogenous components from dead and decaying plants and animals are hydrolyzed to amino acids and other compounds, which release ammonia upon microbial degradation.

Biological Nitrogen Fixation and the Nitrogenase

For many organisms, especially animals, the only usable form of inorganic nitrogen is ammonia. This is produced in the nitrogen cycle by reduction of N_2, a process carried out by a relatively few species of bacteria. The nitrogen-fixing bacteria can be classified into two groups: (1) nonsymbiotic microorganisms that live independently and (2) microorganisms that form specific symbiotic relationships with some higher plants. Group 1 microorganisms include the free-living soil bacteria *Klebsiella, Azotobacter,* and *Clostridia,* as well as the cyanobacteria (blue-green algae) found in soil and water. The symbiotic microorganisms (group 2) develop species-specific associations with plants of the legume family (pea, soybean, clover, alfalfa). The most important symbiotic nitrogen-fixing bacteria are the *Rhizobia* genus. The bacteria infect the roots of legumes, where they develop tumorlike nodules (Figure 19.2). The bacteria and host plant develop a cooperative association in which the plant produces carbohydrate for the bacteria. The bacteria, in turn, provide ammonia for the plant.

The basic reaction for the reduction of nitrogen (nitrogen fixation) is

$$N \equiv N + 3\,H_2 \longrightarrow 2\,NH_3$$

The triple bond of N_2 is extremely unreactive, so high levels of energy are required for reduction. The chemical process, carried out on an industrial scale for the production of crop

Figure 19.2

Root nodules formed in a symbiotic relationship between nitrogen-fixing bacteria *(Rhizobia)* and a legume (alfalfa).

© C. P. Vance/Visuals Unlimited

fertilizer, requires conditions of 200 atm of pressure, 450°C, and an iron catalyst. In contrast, biological nitrogen fixation, which is catalyzed by specialized enzymes in the **nitrogenase complex,** occurs at ambient temperatures and 1 atm of pressure (0.8 atm of N_2 substrate) and requires the presence of abundant ATP and a source of energetic electrons.

The nitrogenase complex has been studied extensively in several organisms. Components of the complex and mechanistic aspects of the process appear to be very similar in all nitrogen-fixing organisms, both symbiotic and nonsymbiotic. The fundamental requirements for nitrogen fixation are as follows:

1. A source of electrons—oxidative or photosynthetic.
2. An electron-transfer protein.
3. A source of energy—ATP.
4. Nitrogenase, the enzyme complex.
5. A final electron acceptor as substrate, usually N_2.

The original source of electrons varies with the organism and cellular conditions. It may be an oxidizable metabolite like NADH, pyruvate, or H_2. In photosynthetic organisms, light energy associated with an electron transport chain provides energetic electrons. The electrons are transferred from the original source to the nitrogenase via a redox protein that varies from species to species. In *Clostridia* and *Rhizobia* it is the protein **ferredoxin** (contains iron and sulfur), and in *Azotabacter* it is a flavoprotein called **flavodoxin.** The nitrogenase enzyme complex is comprised of two metalloproteins: the **iron–sulfur protein (Fe–S protein),** which binds ATP, and the **molybdenum–iron protein (Mo–Fe protein),** which is thought to contain the binding site for the nitrogen substrate.

The flow of electrons from the original source to the substrate N_2 is shown in Figure 19.3. Electrons pass from reduced ferredoxin (or flavodoxin) to the Fe–S protein of the nitrogenase. ATP binding and hydrolysis provide energy for the electron transfer processes. Electrons are finally transferred to the Mo–Fe protein with bound N_2 to produce NH_4^+. Although intermediates in the reduction of N_2 have not been isolated, it is presumed the process involves imine, $HN=NH$, and hydrazine, H_2N-NH_2. A curious aspect of the nitrogenase is its lack of substrate specificity. Acetylene, $HC \equiv CH$, and cyanic acid, $HC \equiv N$, are two additional substrates that can bind and be reduced.

In the overall reduction of N_2, a total of eight electrons must be transferred through this series of proteins; six electrons reduce N_2 to two NH_4^+ and two electrons are used to produce the product H_2 from two H^+. The overall, net reaction for the nitrogenase complex is as follows:

$$N_2 + 10\,H^+ + 8\,e^- + 16\,ATP \longrightarrow 2\,NH_4^+ + 16\,ADP + 16\,P_i + H_2$$

The equation is written in a balanced form; however, the stoichiometry, especially the exact number of ATP molecules, is not known. The large amount of ATP required gives an indication of the high levels of energy necessary for the reduction of N_2.

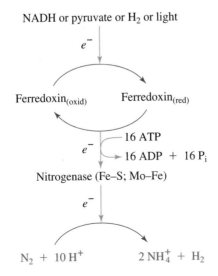

Figure 19.3

Flow of electrons in the nitrogen-fixation process.

One characteristic of the nitrogenase complex that has made it troublesome to study is its extreme sensitivity to O_2. The enzyme system is completely inactivated in the presence of O_2. This makes the protein complex very difficult to isolate, purify, and characterize. All experimental work must be done in the absence of oxygen. Nitrogen-fixing bacteria have varying strategies to avoid O_2 inactivation of their enzymes. Most nonsymbiotic bacteria live under anaerobic conditions. The nitrogenase complex in symbiotic organisms is protected by an oxygen-binding protein called leghemoglobin. This protein, which interacts with O_2 in a fashion similar to hemoglobin, is produced by the host plant.

Research on the nitrogenase complex continues at an intense pace. There is still much to learn about the structures of the protein components and detailed mechanisms of electron transfer, substrate binding, and the role of ATP. Another very active area of research is the attempt to transfer, by DNA recombinant techniques, the nitrogenase complex and related genes [nitrogen fixation (*nif*) cluster] to nonfixing bacteria.

19.2 Biosynthesis of Amino Acids

Use of Nitrogen

We have now explored the unusual reactions of inorganic nitrogen compounds carried out by plants and microorganisms, but animal cells are unable to perform any of these reactions. All organisms, including animals, are able to use the product of nitrogen fixation, NH_4^+, to incorporate nitrogen into organic molecules. High levels of ammonium ion are toxic to cells, so they must detoxify it by incorporation into organic forms. Reactions leading to three compounds, glutamate, glutamine, and carbamoyl phosphate, play key roles in assimilating ammonium. The two α-amino acids, glutamate and glutamine, also participate in the synthesis of other amino acids and nitrogen-containing compounds. Carbamoyl phosphate is used for the synthesis of urea, pyrimidines, and other compounds (see Section 19.5).

Glutamate is synthesized by the reductive amination of α-ketoglutarate:

$$\begin{array}{c}
\text{COO}^- \\
|\\
\text{C}=\text{O} \\
|\\
\text{CH}_2 \quad + \text{ NH}_4^+ + \text{ NADPH } + \text{ H}^+ \xrightarrow[\text{dehydrogenase}]{\text{glutamate}} \\
|\\
\text{CH}_2 \\
|\\
\text{COO}_-
\end{array}
\qquad
\begin{array}{c}
\text{COO}^- \\
|\\
\text{H}_3^+\text{NCH} \\
|\\
\text{CH}_2 \quad + \text{ NADP}^+ + \text{ H}_2\text{O} \\
|\\
\text{CH}_2 \\
|\\
\text{COO}_-
\end{array}$$

α-Ketoglutarate $\qquad\qquad\qquad\qquad\qquad\qquad\qquad$ Glutamate

This reaction, which occurs in all life-forms, uses an intermediate from the citric acid cycle (α-ketoglutarate) as an acceptor of NH_4^+. The reverse of the reaction is likely more important in the catabolism of amino acids and as an anaplerotic reaction to replenish α-ketoglutarate in the citric cycle (see Section 19.3). Glutamate is converted to glutamine by reaction with another ammonium ion:

$$\begin{array}{c}
\text{COO}^- \\
|\\
\text{H}_3^+\text{NCH} \\
|\\
\text{CH}_2 \quad + \text{ NH}_4^+ + \text{ ATP} \xrightarrow[\text{synthetase}]{\text{glutamine}} \\
|\\
\text{CH}_2 \\
|\\
\text{COO}_-
\end{array}
\qquad
\begin{array}{c}
\text{COO}^- \\
|\\
\text{H}_3^+\text{NCH} \\
|\\
\text{CH}_2 \\
|\\
\text{CH}_2 \quad + \text{ ADP } + \text{ P}_i \\
|\\
\text{C}-\text{NH}_2 \\
\|\\
\text{O}
\end{array}$$

Glutamate $\qquad\qquad\qquad\qquad\qquad\qquad\qquad$ Glutamine

Energy released in the cleavage of a phosphoanhydride bond (in ATP) is needed to incorporate the NH_4^+ into an amide linkage. This reaction is a regulatory step in bacterial nitrogen metabolism. The enzyme glutamine synthetase is feedback inhibited by several compounds that are end products of glutamine metabolism (histidine, tryptophan, AMP, CTP, and others). All organisms have both enzymes, glutamine synthetase and glutamate dehydrogenase; however, the quantitative role of each in ammonia assimilation and glutamate turnover is unknown.

Essential and Nonessential Amino Acids

All species of life are able to synthesize at least some of the 20 amino acids in proteins. Many species of plants and bacteria can synthesize all 20 amino acids. Humans and other animals can make only 9 of the amino acids. Amino acids that cannot be synthesized in sufficient quantities and must be supplied in the diet of humans are called **essential amino acids** (Table 19.2). Dietary proteins from plants and other animals that eat plants are sources of these amino acids for humans. The other 11 amino acids are synthesized by humans and are therefore **nonessential amino acids** in the diet.

Each of the 20 amino acids has its own unique pathway of biosynthesis. It is fortunate for the biochemistry student that there are several common features among these reactions:

1. There are six biosynthetic families of amino acids based on common precursors (Figure 19.4).
2. All amino acids obtain their carbon skeletons from a glycolysis intermediate, a citric acid cycle intermediate, or a phosphogluconate pathway intermediate.
3. The amino group of each amino acid usually comes from glutamate.

Synthesis of the Nonessential Amino Acids

The reactions for biosynthesis of nonessential amino acids are relatively straightforward and are identical in all organisms. We shall look at some highlights of amino acid biosynthesis here. Alanine, for example, is made in one step from pyruvate:

$$\text{pyruvate} + \text{glutamate} \underset{\text{aminotransferase}}{\overset{\text{alanine}}{\rightleftharpoons}} \text{alanine} + \alpha\text{-ketoglutarate}$$

In a reaction called transamination (see Section 19.3) the amino group of glutamate is transferred to pyruvate. Aspartate is made in a similar fashion:

$$\text{oxaloacetate} + \text{glutamate} \underset{\text{aminotransferase}}{\overset{\text{aspartate}}{\rightleftharpoons}} \text{aspartate} + \alpha\text{-ketoglutarate}$$

The biosynthesis of serine begins with the carbon skeleton of 3-phosphoglycerate, an intermediate in glycolysis. The major steps include oxidation–reduction, transamination, and hydrolysis (Figure 19.5). The synthesis of glycine from serine introduces an unusual biochemical process, the transfer of one-carbon units. The vitamin **tetrahydrofolate** (FH$_4$) is an essential cofactor for this activity (Figure 19.6). Humans are unable to synthesize the

Cutting Edge:
Homocysteine

Essential	Nonessential
Histidine	Alanine
Isoleucine	Arginine
Leucine	Asparagine
Lysine	Aspartate
Methionine	Cysteine
Phenylalanine	Glutamate
Threonine	Glutamine
Tryptophan	Glycine
Valine	Proline
	Serine
	Tyrosine

Table 19.2
Essential and nonessential amino acids in humans

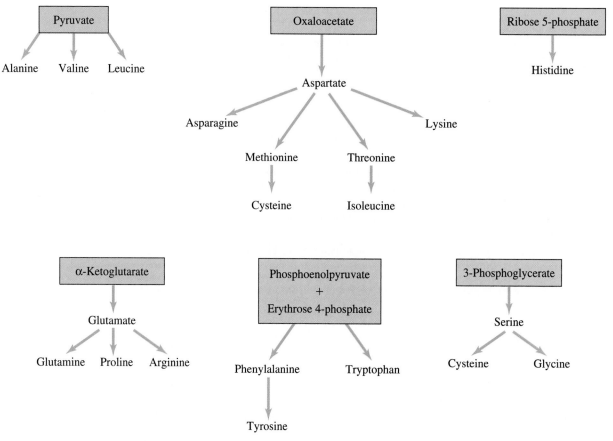

Figure 19.4

The six biosynthetic families of amino acids in plants and bacteria. The important precursors are pyruvate, oxaloacetate, ribose 5-phosphate, α-ketoglutarate, phosphoenolpyruvate + erythrose 4-phosphate, and 3-phosphoglycerate. The nonessential amino acids are synthesized by identical pathways in humans.

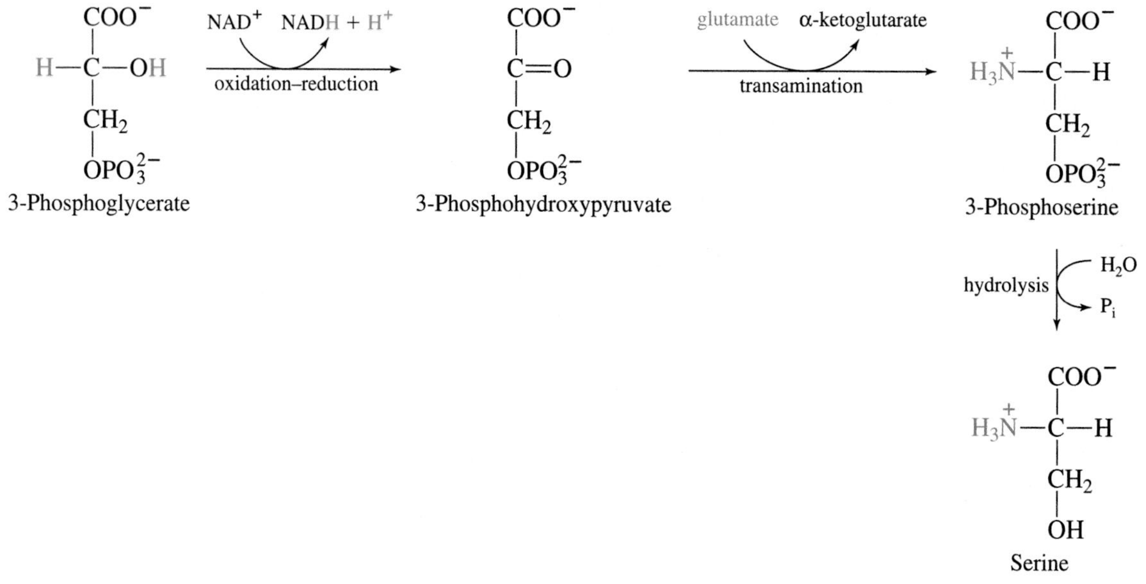

Figure 19.5

The synthesis of serine from 3-phosphoglycerate, an intermediate in glycolysis. The biochemical steps include oxidation–reduction, transamination, and hydrolysis.

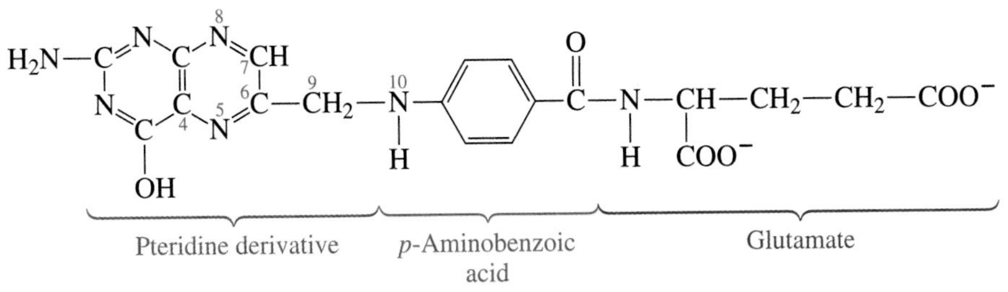

Figure 19.6
Structure of the cofactor tetrahydrofolate (FH_4), which acts as a one-carbon unit carrier.

One-carbon Unit	Atom and Position[a]	Relative Oxidation State
—CH$_3$, methyl	N^5	Most reduced
—CH$_2$—, methylene	N^5, N^{10}	Intermediate
—C(=O)H, formyl	N^5 or N^{10}	Most oxidized

[a]Indicates the nitrogen atom(s) that bond the one-carbon unit as numbered in Figure 19.6.

Table 19.3
One-carbon units carried by the tetrahydrofolate cofactor (FH_4)

complex pteridine ring system of FH_4, so we must obtain it in our diets or from intestinal microorganisms. The vitamin functions as a carrier of activated one-carbon units in varying oxidation states (Table 19.3). In the conversion of serine to glycine, a single carbon unit from serine is transferred to tetrahydrofolate (Figure 19.7). The N^5,N^{10}-methylene-FH_4 can be reduced to N^5-methyl-FH_4, which is used in the synthesis of methionine:

$$\text{homocysteine} + N^5\text{-methyl-}FH_4 \rightleftharpoons \text{methionine} + FH_4$$

Another important one-carbon unit carrier in metabolism is **S-adenosylmethionine** (SAM) as shown in Figure 19.8. The reactive compound SAM is produced by cleavage of both phosphoanhydride bonds and the phosphoester bond of ATP. The methyl group on the sulfur atom is highly activated and can be transferred to other compounds. Table 19.4 lists potential methyl group acceptors and their methylated products.

Synthesis of the Essential Amino Acids

The biosynthesis of the essential amino acids in bacteria and plants requires longer and more complex pathways than biosynthesis of nonessential amino acids. They cannot be synthesized in sufficient quantities in humans because we are deficient in at least one of the enzymes from each pathway in plants and microorganisms. We shall not examine all of these pathways in detail, but the biosynthesis of the aromatic amino acids phenylalanine, tyrosine, and tryptophan in plants and bacteria presents interesting biochemical reactions. (Tyrosine is not an essential amino acid in humans because it is normally synthesized from phenylalanine.) The carbons for these amino acids are derived from erythrose 4-phosphate (a product of the phosphogluconate pathway) and phosphoenolpyruvate (a product of glycolysis). All three amino acids share a single biosynthetic pathway to the metabolite chorismic acid (chorismate) (Figure 19.9). **Chorismate** is at a metabolic junction; one branch leads through anthranilic acid

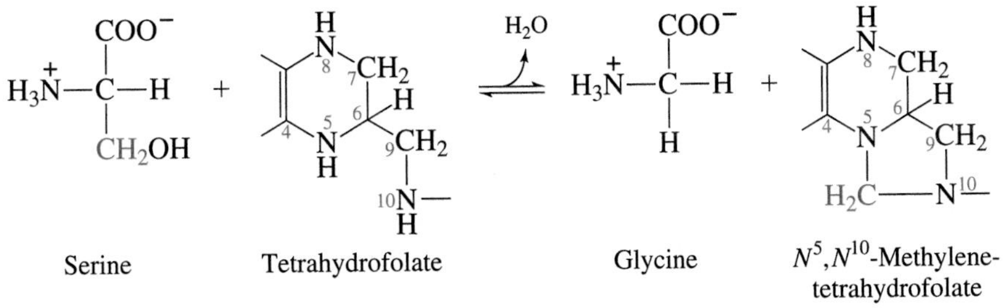

Figure 19.7
The role of tetrahydrofolate in carrying one-carbon units. The one-carbon unit in this example is the methylene group bonded between N^5 and N^{10}.

Serine   Tetrahydrofolate   Glycine N^5,N^{10}-Methylene-tetrahydrofolate

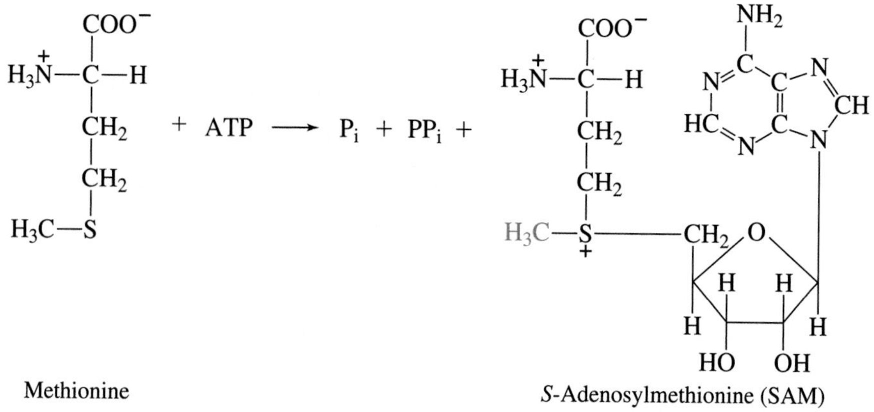

Figure 19.8

The structure and synthesis of the cofactor *S*-adenosylmethionine (SAM). The methyl group on sulfur of SAM (in *red*) is reactive and can be transferred to another molecule.

$$\text{Methionine} + \text{ATP} \longrightarrow P_i + PP_i + \textit{S}\text{-Adenosylmethionine (SAM)}$$

Table 19.4

The use of *S*-adenosylmethionine (SAM) in the synthesis of methylated biomolecules

Methyl Group Acceptor	Methylated Product
Protein amino acid residues	Methylated lysine, methylated arginine, methylated aspartate, methylated glutamate
DNA nucleotide bases	*N*-Methyladenine, 5-methylcytosine
t-RNA	Methylated bases
Norepinephrine	Epinephrine

(anthranilate) to tryptophan, the other through prephenic acid (prephenate) to tyrosine and phenylalanine.

Amino acids that are synthesized or ingested are available for all the needs of an organism. The most important uses include the synthesis of proteins and nucleotide bases for nucleic acids (see Section 19.5). Although many factors are important, the rate of amino acid biosynthesis is controlled primarily by **feedback inhibition.** In most cases, the first step of a biosynthetic pathway for an amino acid is inhibited by the final product. This allows an organism to adjust the level of amino acids depending on specific needs.

19.3 Catabolism of Amino Acids

The catabolism of amino acids has one major difference when compared to the catabolism of carbohydrates and fats: amino acids require the removal of the amino group, which other nutrients do not have. Under normal dietary circumstances, amino acids are not important fuel molecules for major tissues. They do, however, provide significant metabolic energy under some conditions (Chapter 20):

1. When dietary amino acids are in excess of those needed for synthesis of proteins and other molecules. Organisms, except for plants, are unable to store amino acids.
2. When the normal process of protein turnover (recycling) releases free amino acids that may be available for catabolism.
3. During times of starvation or untreated diabetes, structural and catalytic proteins are degraded and the amino acids are used for energy metabolism.

Each amino acid has its own unique pathway of catabolism. The burden of learning the steps is eased somewhat because the degradative pathway for each amino acid is essentially the same in all organisms. In our study of amino acid catabolism we will again observe a general metabolic principle: *Biomolecules are converted in as few steps as possible to a mainstream metabolite.* In other words, the major existing catabolic pathways are used for amino acid degradation. To this end, the initial process that occurs for most amino acids is removal of the amino group (deamination). This step results in two paths

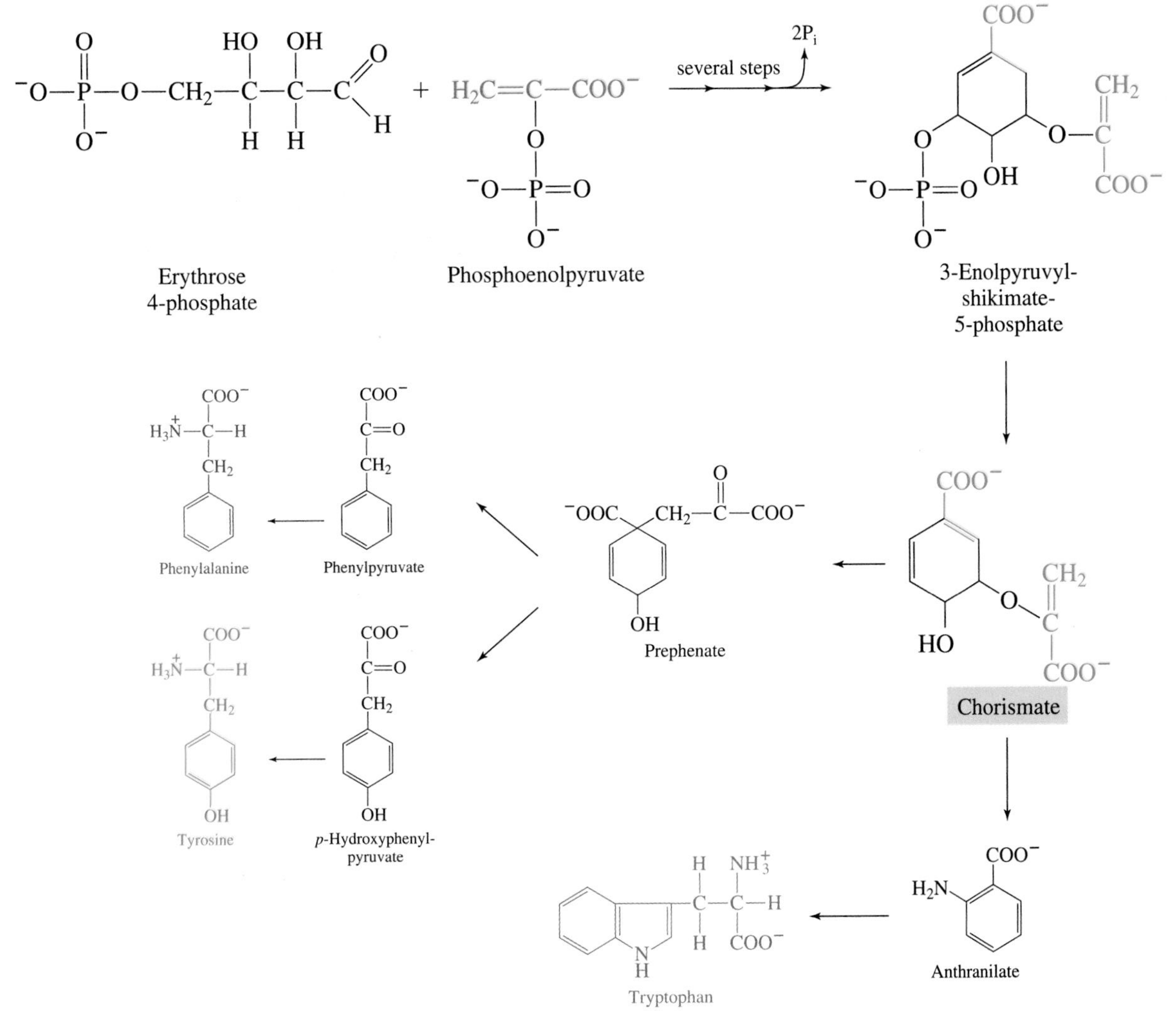

Figure 19.9

General pathway for the synthesis of the aromatic amino acids phenylalanine, tyrosine, and trypto-phan. A common route is followed from erythrose 4-phosphate and phosphoenolpyruvate to choris-mate. Chorismate is transformed to phenylalanine, tyrosine, and tryptophan in separate pathways. Phosphoenolpyruvate is labeled in red to show origin of atoms in chorismate.

for amino acid catabolism, one for the amino group and one for the remaining carbon skeleton (Figure 19.10). For essentially all amino acids, the carbon skeleton after removal of the NH_3 emerges as an intermediate that feeds directly into the citric acid cycle. The amino group, upon removal, is transformed to NH_4^+. The fate of this toxic substance depends on the species of organism and the cellular conditions (see Section 19.4).

Amino acids in our diet come from ingested proteins. Degradation of dietary protein begins in the stomach with enzymatic hydrolysis. The primary proteolytic enzyme in the stomach is pepsin, which cleaves peptide bonds on the amino side of the amino acid residues phenylalanine, tyrosine, and tryptophan. This results in shortened proteins, which pass from the stomach into the small intestines. Here the shorter proteins encounter several peptidases, including trypsin, chymotrypsin, aminopeptidase, and carboxypeptidase. Many of these enzymes are produced in an inactive or zymogen form that is activated by cleavage (see Chapter 7, Section 7.3). The specific action of these peptidases is described in Table 19.5. The combined action of the enzymes results in a mixture of free amino acids, which are transported across the epithelial cell lining of small intestines into the blood. The amino acids are distributed to peripheral tissue for biosynthetic use and to the liver for catabolism.

Figure 19.10
General pathway showing the
stages of amino acid catabolism.

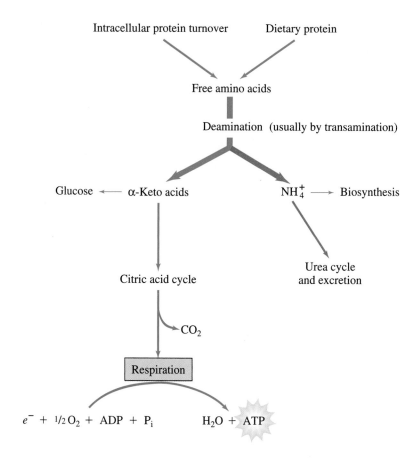

Figure 19.10
General pathway showing the
stages of amino acid catabolism.

Transamination

The stages of amino acid catabolism are outlined in Figure 19.10. The initial phase for most amino acids is the removal of the amino group by a process common to the amino acids. In **transamination** the amino group is transferred to an α-keto acid, usually α-ketoglutarate:

$$
\begin{array}{c}
\text{COO}^- \\
|\\
\text{C}=\text{O} \\
|\\
\text{CH}_2 \\
|\\
\text{CH}_2 \\
|\\
\text{COO}_-
\end{array}
\;+\;
\begin{array}{c}
\text{COO}^- \\
|\\
\text{H}_3^+\text{N}-\text{CH} \\
|\\
\text{R}
\end{array}
\;\rightleftharpoons\;
\begin{array}{c}
\text{COO}^- \\
|\\
\text{H}_3^+\text{N}-\text{CH} \\
|\\
\text{CH}_2 \\
|\\
\text{CH}_2 \\
|\\
\text{COO}_-
\end{array}
\;+\;
\begin{array}{c}
\text{COO}^- \\
|\\
\text{C}=\text{O} \\
|\\
\text{R}
\end{array}
$$

α-Ketoglutarate   Amino acid   Glutamate   α-Keto acid

The general name for the enzyme responsible for this reaction is **aminotransferase.** The acceptor of the amino group (α-ketoglutarate) becomes an amino acid and the carbon

Table 19.5
Peptidases that function in the
digestion of proteins in the
stomach and small intestines

Peptidase	Sites of Catalytic Hydrolysis
Pepsin	Amino side of Phe, Tyr, Trp
Trypsin	Carboxyl side of Lys and Arg
Chymotrypsin	Carboxyl side of Phe, Tyr, Trp
Carboxypeptidase	Sequential removal of amino acids beginning at C-terminus
Aminopeptidase	Sequential removal of amino acids beginning at N-terminus

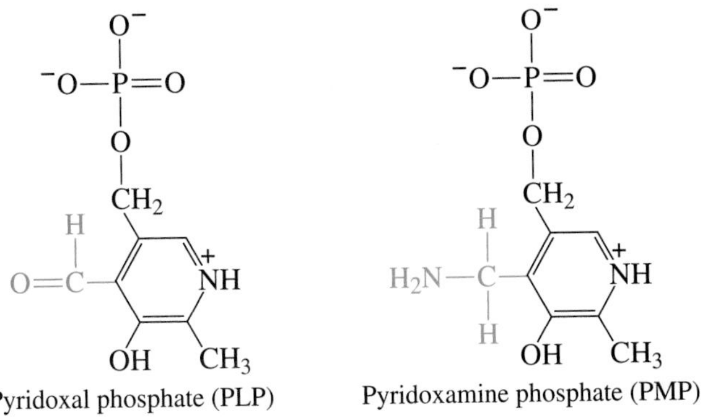

Figure 19.11
Structure of two forms of the prosthetic group pyridoxal phosphate (PLP and PMP). PLP is the acceptor of amino groups transferred from amino acids during transamination. PMP is the amino group carrier.

Pyridoxal phosphate (PLP)   Pyridoxamine phosphate (PMP)

skeleton of the amino group donor (amino acid) becomes an α-keto acid. As a specific example, consider the transamination of the amino acid alanine. In a single step, alanine is degraded to pyruvate, a mainstream metabolite.

$$\text{α-ketoglutarate } + \text{ alanine} \underset{}{\overset{\substack{\text{alanine}\\\text{aminotransferase}}}{\rightleftharpoons}} \text{ glutamate } + \text{ pyruvate}$$

The purpose of transamination is to remove amino groups from the various α-amino acids and collect them in a single type of molecule, glutamate. Glutamate then acts as a single source of amino groups for continued nitrogen metabolism (biosynthesis or excretion). Many different aminotransferases have been isolated and studied. Most are present in the cytoplasm. All aminotransferases have the same type of prosthetic group: pyridoxal phosphate (PLP). PLP is derived from the vitamin pyridoxine (B_6). Its structure and function were introduced in Table 7.1. When bound to an aminotransferase, PLP acts as an intermediate carrier of the transferred amino group. It accepts the amino group from the amino acid and donates it to the α-keto acid. PLP cycles between two forms, a form with an aldehyde group, **pyridoxal phosphate** (PLP), and a form with the amino group, **pyridoxamine phosphate** (PMP) as shown in Figure 19.11. Transfer of the amino group from PMP to α-ketoglutarate regenerates PLP. During its action, PLP is covalently bound to a lysine residue at the active site of the aminotransferase.

Catabolism of Carbon Skeletons

We turn now to the fate of the amino acid carbon skeleton remaining after removal of the amino group. The different amino acids are degraded to one or two of seven possible intermediates, all eventually accepted by the citric acid cycle (Figure 19.12). Amino acids sometimes are divided into two groups based on their mode of degradation. Those amino acids degraded to acetyl CoA or acetoacetyl CoA are called **ketogenic amino acids** because they can produce ketone bodies (see Chapter 20, Section 20.2). Amino acids degraded to pyruvate, α-ketoglutarate, succinyl CoA, fumarate, or oxaloacetate are termed **glucogenic amino acids** because their carbon skeletons may be used for the synthesis of glucose. We shall not discuss catabolic details of all 20 amino acids; only a few with clinical and health significance are considered here.

The branched-chain amino acids isoleucine, leucine, and valine share some steps in their catabolism. All three begin the process with transamination catalyzed by a branched-chain aminotransferase (Figure 19.13). The α-keto acid products are then decarboxylated in a reaction similar to the ones catalyzed by the pyruvate dehydrogenase complex and the α-ketoglutarate dehydrogenase complex. The three CoA esters then proceed in catabolism in three separate branches.

Phenylalanine catabolism is unusual in that transamination does not occur in the first step. Instead, phenylalanine is first hydroxylated to form tyrosine (Figure 19.14). The oxygen

Figure 19.12

The fate of each carbon skeleton in amino acid catabolism. Most metabolites eventually enter the citric acid cycle.

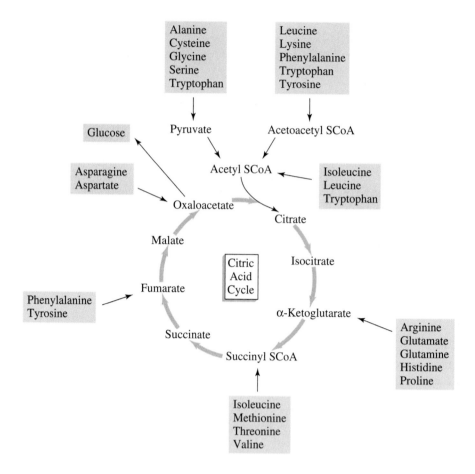

atom of the hydroxyl group substituted onto the ring comes originally from O_2 (phenylalanine hydroxylase). In the next step tyrosine is transaminated to *p*-hydroxyphenylpyruvate, which is oxidized to homogentisate. The next step, resulting in the cleavage of the ring, is catalyzed by homogentisate oxidase. The pathway culminates with the formation of fumarate and acetoacetate, which are oxidized by the citric acid cycle.

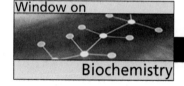

Window on
Biochemistry

19.1 Genetic Defects in Amino Acid Metabolism

The idea that the deficiency of an enzyme can cause disease was first recognized about 100 years ago. In the early 1900s, Archibald Garrod proposed that some diseases are caused by an individual being unable to produce an enzyme in an amount or form adequate for normal function. The diseases are the result of mutations in the genes that encode the proteins. Garrod also recognized that these genetic disorders could be diagnosed because some metabolic intermediates accumulated in tissues and blood serum and were eventually eliminated in the patient's urine. Such diseases with a genetic origin are now called **"inborn errors of metabolism."**

Amino acid metabolism is particularly susceptible to genetic disorders; in fact, the first metabolic disorder recognized by Garrod was **alcaptonuria**, which is caused by a deficiency of the enzyme **homogentisate oxidase** (see Figure 19.14). The metabolic intermediate, homogentisate, accumulates in blood and is excreted in the urine. Upon exposure to air, homogentisate polymerizes to a black pigment, thus darkening the urine. Fortunately, this condition, which occurs in about one out of every 200,000 infants born in the United States, is relatively benign, except for the embarrassment of black urine.

The deficiency of **phenylalanine hydroxylase** results in the genetic disorder **phenylketonuria (PKU)**. This disease is much more common (one in every 15,000

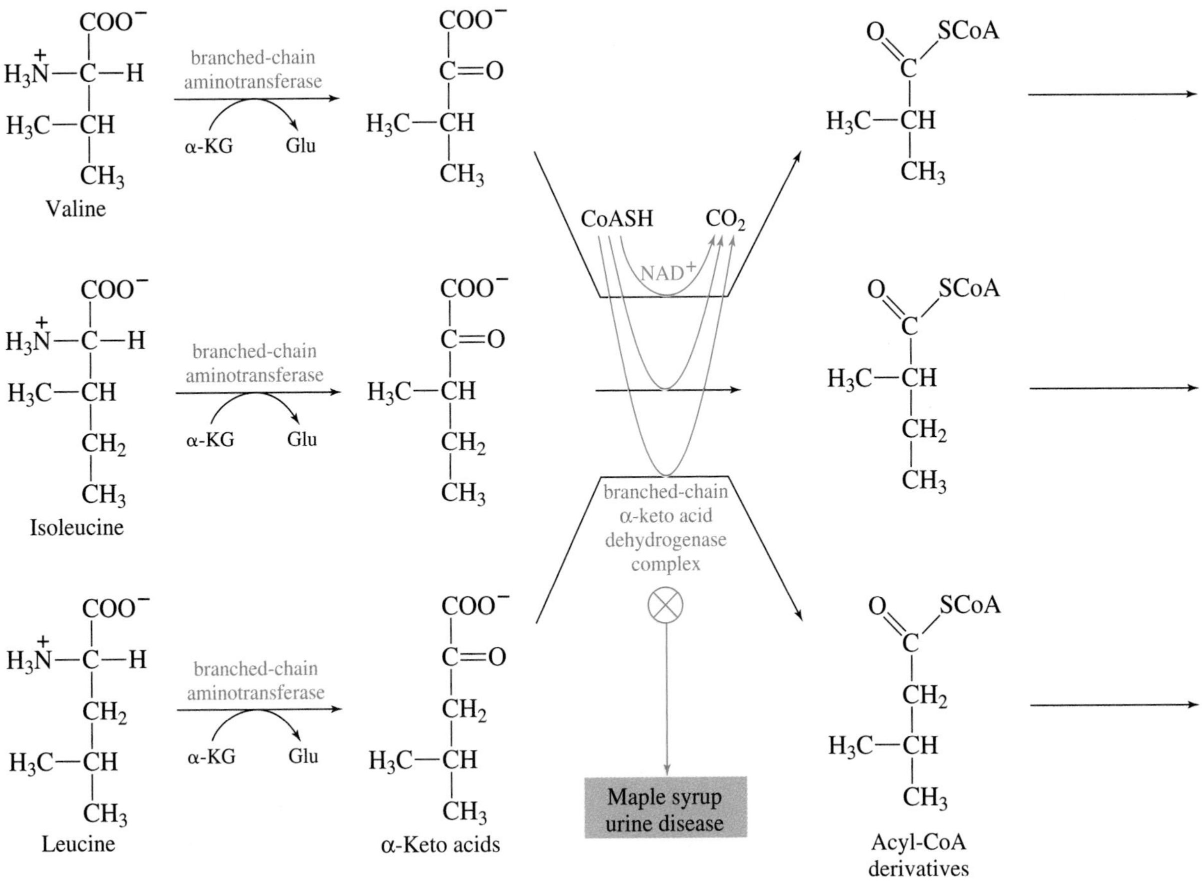

Figure 19.13
Pathway for catabolism of the branched-chain amino acids valine, isoleucine, and leucine. After two common steps, transamination and oxidative decarboxylation, each amino acid is converted to an acyl CoA derivative. Each CoA ester is degraded by a separate pathway. A genetic deficiency of the dehydrogenase complex, which acts on all three α-keto acids, is the cause of maple syrup urine disease.

infants) and more serious than alcaptonuria. In PKU, the high concentrations of phenylalanine are directed into an alternate, abnormal degradative pathway. Instead of hydroxylation to tyrosine (Figure 19.14), phenylalanine is transaminated directly by nonspecific aminotransferases to phenylpyruvate. This product affects developing brain cells by inhibiting mitochondrial pyruvate translocase. If untreated, PKU eventually leads to mental retardation and early death. Most hospitals do a routine test on all newborns in order to detect the disease. Individuals diagnosed with PKU must restrict their dietary intake of phenylalanine, including products sweetened with aspartame (Nutrasweet), a dipeptide of aspartate and phenylalanine methyl ester.

Maple syrup urine disease is a genetic disorder caused by a deficiency of an enzyme involved in the degradation of the aliphatic amino acids valine, isoleucine, and leucine. The deficient enzyme is **branched-chain alpha-keto acid dehydrogenase complex** (see Figure 19.13). The alpha-keto acids, because they cannot be oxidized, accumulate in the blood and are excreted in the urine. The urine has the smell of maple syrup. Patients with this disease must restrict their dietary intake of the three branched-chain amino acids. If untreated, affected individuals experience mental retardation and early death.

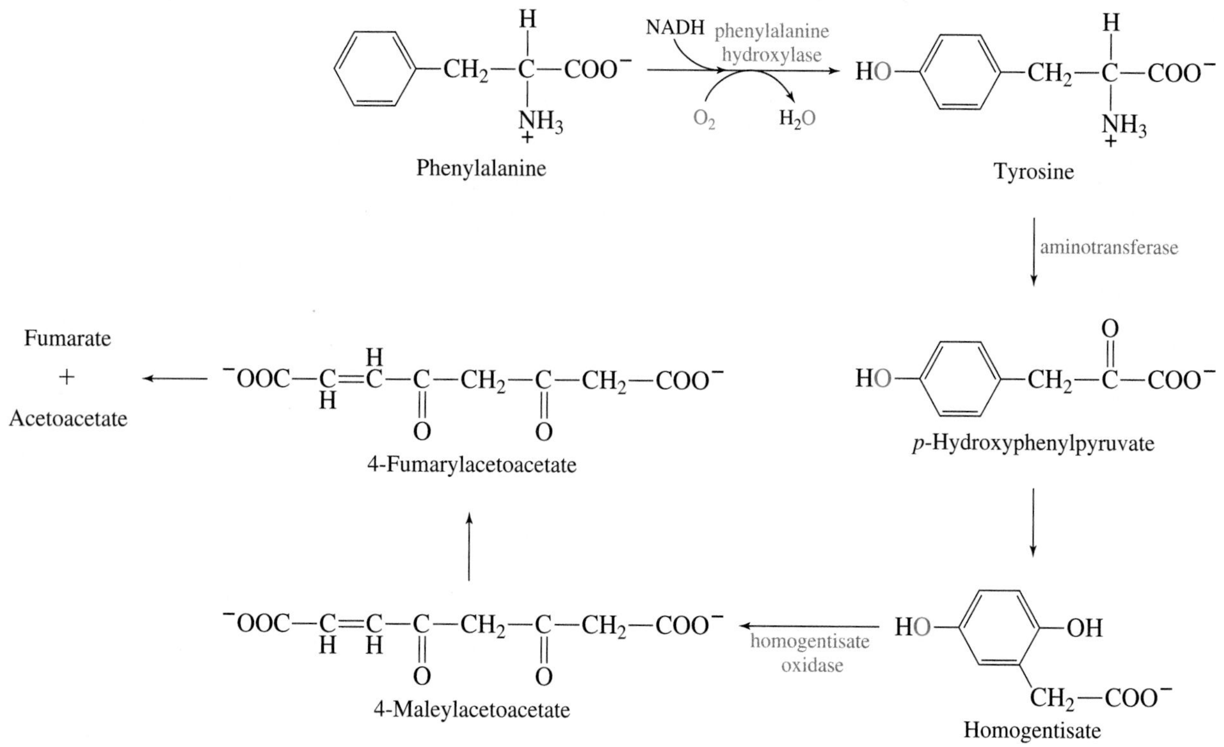

Phenylalanine → (NADH, phenylalanine hydroxylase, O_2, H_2O) → Tyrosine

Tyrosine → (aminotransferase) → p-Hydroxyphenylpyruvate

p-Hydroxyphenylpyruvate → Homogentisate

Homogentisate → (homogentisate oxidase) → 4-Maleylacetoacetate

4-Maleylacetoacetate → 4-Fumarylacetoacetate

4-Fumarylacetoacetate → Fumarate + Acetoacetate

Figure 19.14

Pathway for catabolism of phenylalanine and tyrosine. A deficiency of phenylalanine hydroxylase is the cause of phenylketonuria (PKU). The deficiency of homogentisate oxidase is the cause of alcaptonuria. (See Window on Biochemistry 19.1.)

19.4 Elimination of NH_4^+

The amino groups removed during the catabolism of amino acids are collected in glutamate by the transamination process. Accumulation of the nitrogen in a single type of molecule is efficient and economical because it requires just one set of reactions to recover the nitrogen. The combination of transamination with glutamate deamination leads to a coupled catalytic process of amino group removal from amino acids:

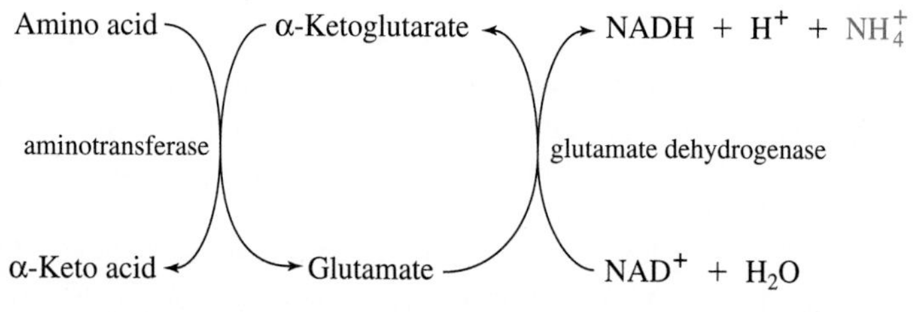

Amino acid / α-Keto acid — (aminotransferase) — α-Ketoglutarate / Glutamate — (glutamate dehydrogenase) — NADH + H⁺ + NH_4^+ / NAD⁺ + H_2O

The amino group removed by this combined process is released as NH_4^+, a form of nitrogen toxic to cells except at low concentrations. The toxicity of NH_4^+ is caused in large part by its effect on the enzyme glutamate dehydrogenase:

$$\alpha\text{-ketoglutarate} + NADH + H^+ + NH_4^+ \rightleftharpoons \text{glutamate} + NAD^+ + H_2O$$

High concentrations of NH_4^+ shift the equilibrium of this reaction to the right, thus lowering the concentration of α-ketoglutarate available for the citric acid cycle. This is especially detrimental to developing brain cells, which depend on aerobic metabolism of glucose.

Ammonium ion above physiological needs is a waste product and is excreted in different chemical forms depending on the species. Terrestrial vertebrates, including mammals, excrete excess ammonium ion in the form of urea. Birds, insects, and reptiles convert excess ammonium ion to uric acid for disposal. Marine invertebrates excrete NH_4^+ directly (Figure 19.15).

(a) Urea

(b) Uric acid

NH_4^+
(c) Ammonium ion

Figure 19.15

Chemicals used by organisms for excretion of excess nitrogen.

The Urea Cycle

We shall consider the details of NH$_4^+$ elimination only in animals. All vertebrates have a set of enzymes that catalyzes the **urea cycle.** This series of reactions was first reported by Hans Krebs in 1932, 5 years before his publication of the citric acid cycle. The transformation of NH$_4^+$ into urea requires a five-step cyclic pathway (Figure 19.16 and Table 19.6). The entry

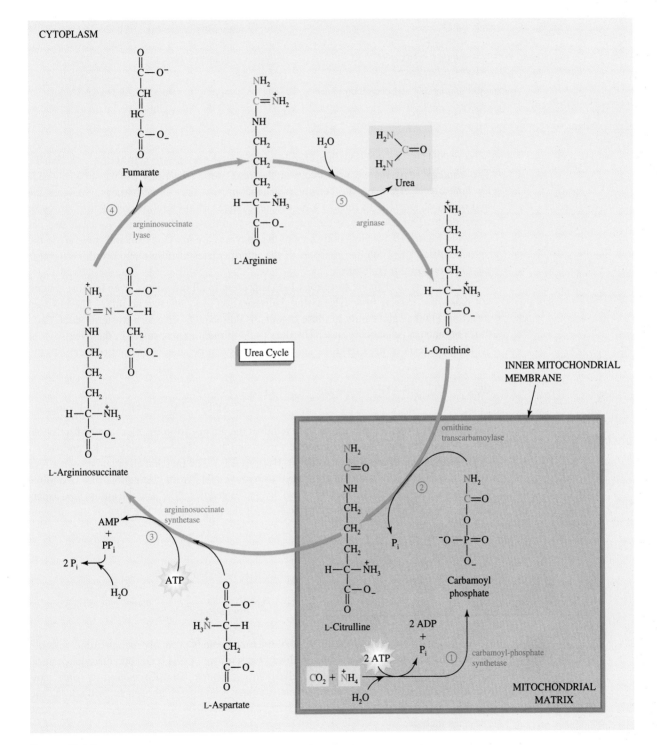

Figure 19.16

The reactions of the urea cycle, showing structures of intermediates and cellular location of enzymes. Note that some reactions are in cytoplasm whereas others are in the mitochondrial matrix. The numbered reactions correspond to Table 19.6.

Table 19.6
Reactions of the urea cycle

Reaction Number	Reaction	Enzyme	Reaction Type[a]
1	$CO_2 + NH_4^+ + 2\,ATP + H_2O \rightleftharpoons$ carbamoyl phosphate $+ 2\,ADP + P_i$	Carbamoyl phosphate synthetase	6
2	Carbamoyl phosphate + ornithine $\rightleftharpoons$ citrulline + P_i	Ornithine transcarbamoylase	2
3	Citrulline + aspartate + ATP $\rightleftharpoons$ argininosuccinate + AMP + PP_i	Argininosuccinate synthetase	6
4	Argininosuccinate $\rightleftharpoons$ arginine + fumarate	Argininosuccinate lyase	4
5	Arginine + $H_2O \rightleftharpoons$ urea + ornithine	Arginase	3

[a]Reaction type: 1, oxidation–reduction; 2, group transfer; 3, hydrolysis; 4, nonhydrolytic cleavage (addition–elimination); 5, isomerization–rearrangement; 6, bond formation coupled to ATP cleavage.

point of NH_4^+ is the reaction catalyzed by **carbamoyl phosphate synthetase** (reaction 1). The strategy used by the urea cycle in the next steps (reactions 2–4) is to synthesize the amino acid arginine. The enzyme **arginase,** present only in vertebrates, catalyzes the hydrolysis of arginine to release urea and regenerate ornithine, thus completing the cycle (reaction 5). The structures of the intermediates and the cellular location for each step are shown in Figure 19.16. Note that some of the enzymes are in cytoplasm and some are in the mitochondria. This requires transport of the intermediates citrulline and ornithine through the mitochondrial membranes.

The pathway to arginine (reactions 1–4) is present in all organisms, even in those that do not make and excrete urea. This is the pathway used to synthesize arginine. The last step (reaction 5) takes place only in those organisms that excrete urea as a waste product. Only those organisms possess arginase. The urea cycle in mammals is solely in the liver.

The synthesis of urea requires energy input, as can be seen in the net reaction for the urea cycle:

$$CO_2 + NH_4^+ + 3\,ATP + aspartate + 2\,H_2O \rightleftharpoons$$
$$urea + 2\,ADP + 2\,P_i + AMP + PP_i + fumarate$$

The formation of one molecule of urea requires the energy from the cleavage of four phosphoanhydride bonds. This is accounted for in step 1 ($2\,ATP \rightarrow 2\,ADP + P_i$) and step 3 ($ATP \rightarrow AMP + PP_i$; followed by $PP_i + H_2O \rightarrow 2\,P_i$). The fact that an organism can afford this much energy indicates the great need to remove NH_4^+ from circulation. The two nitrogen atoms in urea come directly from two diverse molecules: NH_4^+ and the amino acid aspartate. However, the original source of both nitrogens is glutamate. The flow of the nitrogen can be shown by these reactions:

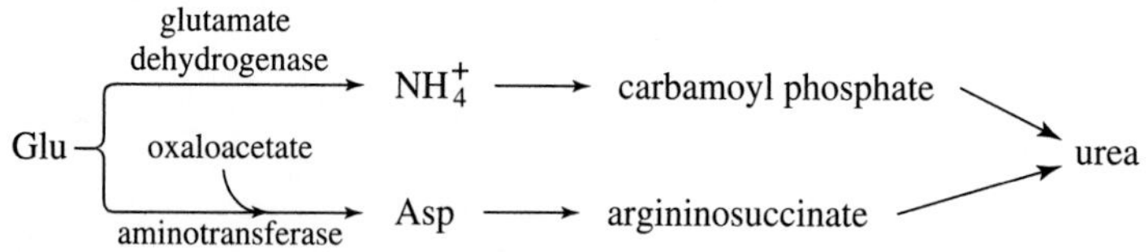

A complete block of any of the steps of the urea cycle is probably incompatible with life because there is no known alternate pathway for removal of NH_4^+. Genetic disorders caused by a partial blockage of reactions in the urea cycle have been discovered in some individuals. Some affected enzymes include arginase, carbamoyl phosphate synthetase, and ornithine transcarbamoylase. Clinical symptoms caused by a partial deficiency of these enzymes include:

1. Elevated levels of NH_4^+ in blood and urine (hyperammonemia)
2. Nausea and illness after ingestion of proteins
3. Gradual mental retardation if not treated

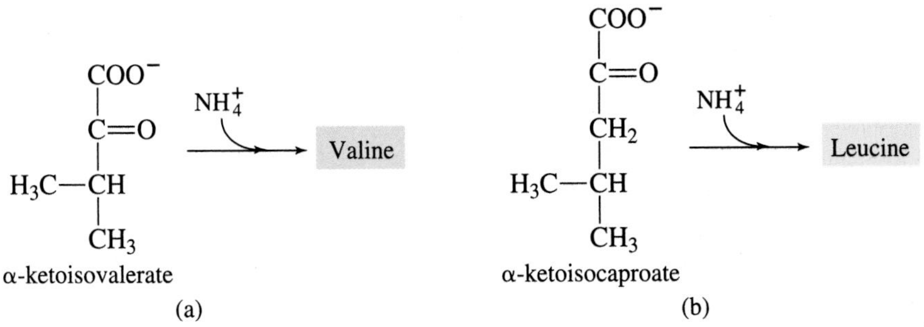

Figure 19.17
Keto acids that are used in the treatment of urea cycle genetic defects: (a) α-Ketoisovalerate is transaminated to valine and (b) α-ketoisocaproate is transaminated to leucine.

Individuals with these symptoms are treated by dietary changes. Patients are fed low-protein diets supplemented with mixtures of α-keto acids. The purpose of the α-keto acids is twofold:

1. The α-keto acids pick up excess NH_4^+ by combined reactions of glutamate dehydrogenase and transamination.
2. If the proper α-keto acids are chosen, they can be converted to essential amino acids that may be lacking in the low-protein diet. For example, the α-keto acids in Figure 19.17 would be transaminated to valine and leucine.

19.5 Amino Acids As Precursors of Other Biomolecules

When we think of functional roles for the amino acids, protein construction comes first to mind. The amino acids may be catabolized for energy purposes, although in a balanced diet, they are used less often than carbohydrates and fats. This section explores the role of amino acids in specialized metabolic pathways that lead to important natural products. There are numerous examples given in Table 19.7, but we shall consider only the synthesis of porphyrins, nucleotide bases, biogenic amines, and nitric oxide.

Porphyrins

Throughout this book we have encountered several proteins that have **heme** as a cofactor. Hemoproteins include hemoglobin, myoglobin, the cytochromes, and the enzyme catalase. Heme is a combination of an iron atom with a porphyrin ring. If a modified porphyrin ring is

Table 19.7
Specialized nitrogen-containing natural products

Bioactive Product	Biological Function	Amino Acid Precursor(s)
Alkaloids	Nitrogen bases in plants	Ornithine, Asp, Lys, Tyr, Trp, Phe, His
γ-Aminobutyric acid (GABA)	Inhibitory neurotransmitter	Glu
Auxin	Plant growth hormone	Trp
Catecholamines	Neurotransmitters, hormones	Tyr, Phe
Glutathione	Redox tripeptide	Gly, Glu, Cys
Histamine	Allergic response, stomach acid secretion	His
Melanin	Skin pigments	Tyr, Phe
Melatonin	Regulates sleep cycles	Trp
Nitric oxide	Cell messenger	Arg
Phosphocreatine	Energy molecule in muscle	Gly, Arg, Met
Porphyrin	Heme and chlorophyll	Gly
Purine bases	RNA, DNA, cofactors	Asp, Gly, Gln
Pyrimidine bases	RNA, DNA, cofactors	Asp
Serotonin	Neurotransmitter (hormone)	Trp
Spermine, spermidine	DNA packaging	Met, ornithine
Thyroxine	Hormone	Tyr

combined with magnesium ion, the result is chlorophyll, the plant pigment and acceptor of light energy for photosynthesis. Porphyrin synthesis, indeed, is important in plants, bacteria, and animals. In an average adult human, up to 900 trillion molecules of hemoglobin are synthesized per second. Each molecule of hemoglobin requires four molecules of heme. In contrast to its complex structure and multiple uses in the cell, the porphyrin ring has very humble beginnings: glycine and succinyl CoA. The outline for porphyrin synthesis in animals is shown in Figure 19.18. The condensation of glycine and succinyl CoA, which results in **δ-aminolevulinate (ALA),** is catalyzed by δ-aminolevulinate synthase. Regulation of porphyrin synthesis occurs at this step. The rate of the reaction is controlled by the concentration of heme, which acts as a feedback inhibitor of the synthase, but heme also may act at the gene level as a signal molecule to repress the production of the synthase. The pathway to porphyrin continues with condensation of four ALA molecules until the final form of the ring, protoporphyrin IX, is produced. This form of porphyrin is the most common one found in hemo-

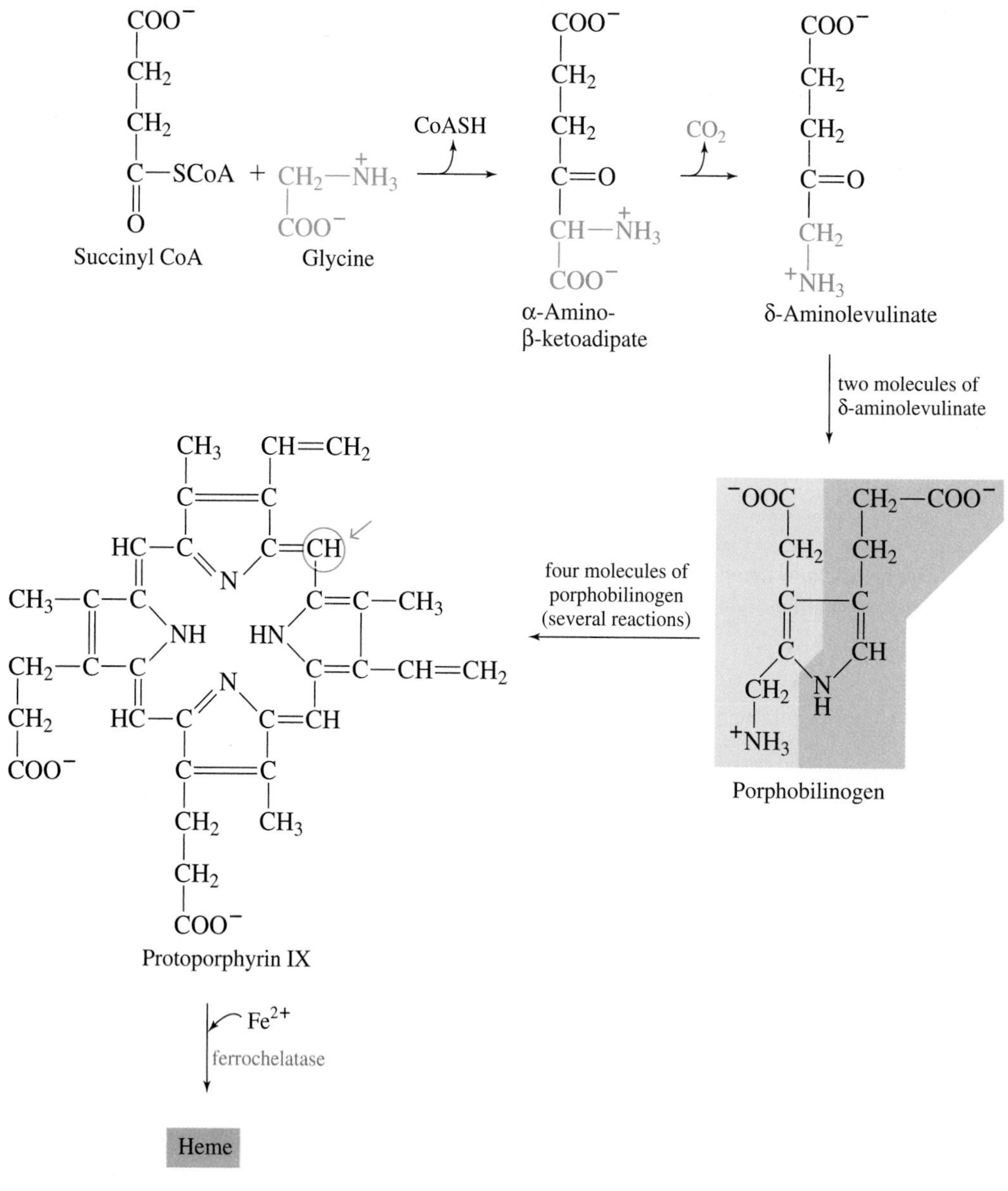

Figure 19.18
Outline for synthesis of protoporphyrin IX and heme from glycine and succinyl CoA. The red circle and arrow indicate the carbon atom in protoporphyrin IX oxidized to CO during catabolism of heme by heme oxygenase.

proteins. The incorporation of iron into the ring is catalyzed by the enzyme ferrochelatase. The route of porphyrin synthesis in plants is the same as animals from ALA to the final product, but the ALA is synthesized from glutamate rather than glycine and succinyl CoA.

The enzymatic degradation of hemoglobin in animals represents an important metabolic process because it releases iron to be reutilized and recycles protein. The average human red blood cell has a life span of about 120 days. Aged cells are removed from circulation and degraded in the spleen. Heme released in this process is degraded by ring opening. The enzyme heme oxygenase catalyzes this cleavage, which releases a carbon atom as CO and leads ultimately to the product **bilirubin,** a linear tetrapyrrole derivative (Figure 19.19). The site of ring cleavage is marked by the arrow in Figure 19.18. To ease its continued metabolism and excretion in urine, bilirubin is made more water soluble by attachment of carbohydrate molecules to the propionate side chains. The presence of yellow bilirubin gives urine its characteristic color.

Several metabolic disorders are caused by defects in heme metabolism. Some of these diseases result in the accumulation of heme breakdown products. In general, the term **porphyrias** is used for these disorders. A common symptom of all porphyrias is the condition **jaundice.** The outward, noticeable signs of jaundice are a yellowing of skin and whites of the eyes caused by an excessive accumulation of bilirubin. Jaundice is frequently encountered in premature newborn children. If not corrected, serum bilirubin levels become excessively high and bilirubin sometimes crystallizes in brain cells. Deafness, mental retardation, and neurologic damage may result. The accumulation of bilirubin in newborn infants apparently is caused by a relative deficiency of the enzyme that catalyzes linkage of carbohydrates to bilirubin. The enzyme is not formed in the liver in amounts sufficient until the normal time of birth. Affected infants are treated by exposure to a special fluorescent light (bilirubin-reducing lamp). The lamp rays photochemically degrade bilirubin to products that are less toxic and can be metabolized and excreted by the infant.

Biogenic Amines and Other Products

Specialized metabolic pathways are used to synthesize important biologically active products from amino acids (see Table 19.7). These reactions often take place in specific cellular locations and the products function in that locale. Here we consider the synthesis of catecholamine neurotransmitters, hormones, biogenic amines, and nitric oxide. Many of these products are the result of decarboxylation reactions carried out on amino acids. The decarboxylation of histidine in animals forms **histamine** (Figure 19.20a). This product, which is

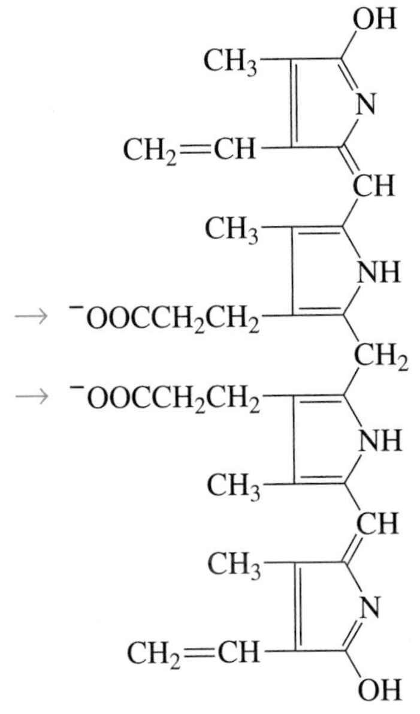

Bilirubin

Figure 19.19

The structure of bilirubin, the degradation product of heme. The red arrows indicate the site of attachment of carbohydrate groups to enhance solubility.

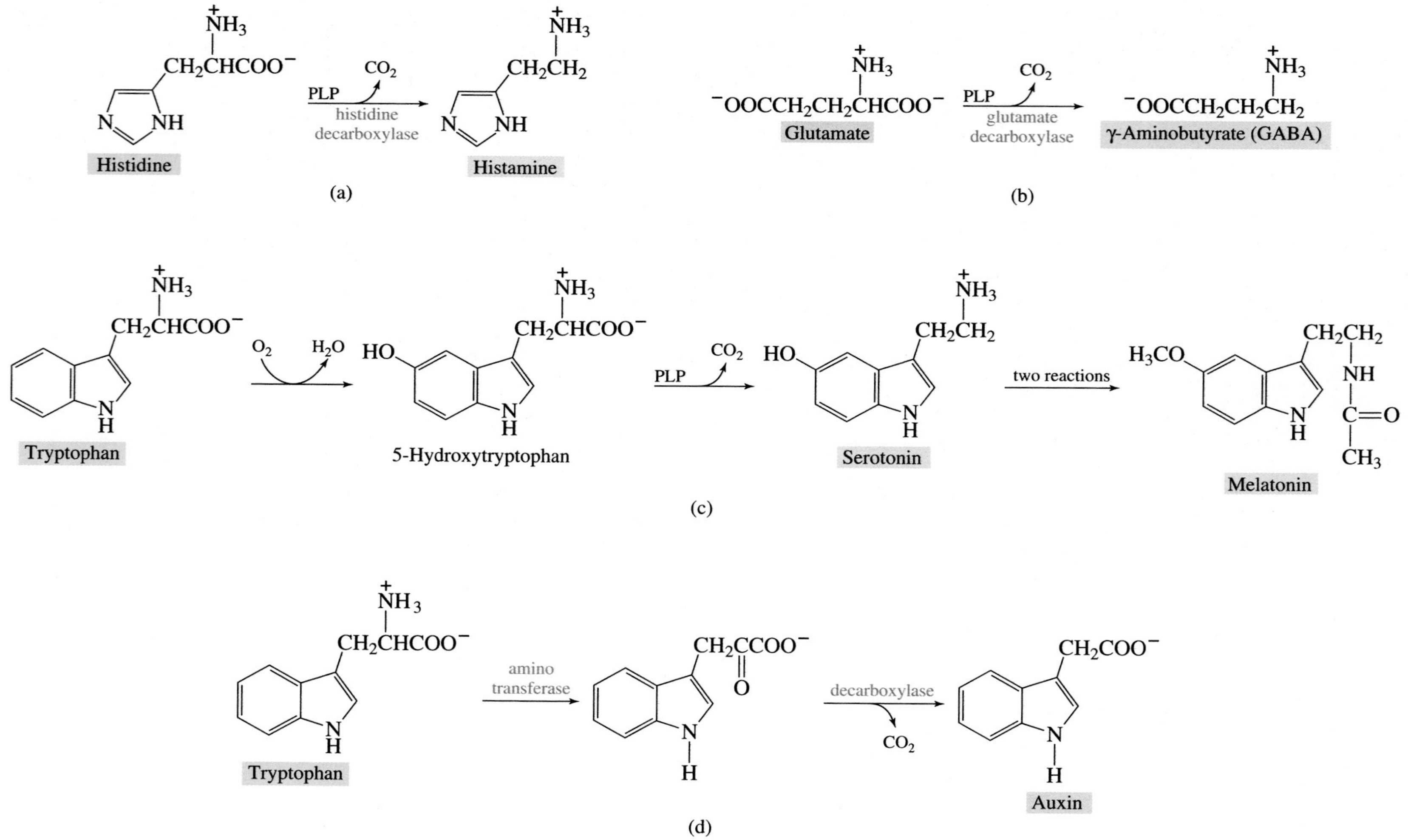

Figure 19.20

Structures of specialized products of amino acid reactions: (a) the synthesis of histamine from histidine; (b) the synthesis of the inhibitory neurotransmitter γ-aminobutyrate from glutamate; (c) the synthesis of serotonin and melatonin from tryptophan; and (d) the synthesis of the plant growth regulator, auxin, from tryptophan.

released in excessive amounts during allergic response to allergens, is a powerful vasodilator and causes much suffering in humans. Pharmaceutical companies are attempting to design drugs that relieve the symptoms of allergic reactions. The strategy is twofold: to make either drugs that inhibit histidine decarboxylase or drugs that interfere with the binding of histamine to its receptor protein, which initiates the allergic response. Drugs that block the binding of histamine to receptors in the gastric mucosa (stomach) can reduce HCl production. The decarboxylation of glutamate produces the inhibitory neurotransmitter **γ-aminobutyrate (GABA)** (Figure 19.20b).

The aromatic amino acids phenylalanine, tyrosine, and tryptophan are among the most active in specialized metabolism. The hydroxylation and decarboxylation of tryptophan result in the neurotransmitter **serotonin,** which is a precursor for melatonin (Figure 19.20c), the regulator of light–dark cycles. Tryptophan in plants is the precursor of the important growth hormone **indole 3-acetate (auxin)** (Figure 19.20d). Specialized metabolism of tyrosine leads to important neurotransmitters, hormones, and skin pigments. Tyrosine is used for the synthesis of dihydroxyphenylalanine (dopa), **dopamine, norepinephrine,** and **epinephrine** (Figure 19.21). The carbon skeleton of tyrosine is also used to construct the thyroid hormone **thyroxine.** In the pigment-producing cells of skin (melanocytes), tyrosine is the building block for synthesis of the skin-darkening pigment **melanin** (Figure 19.22). The enzyme tyrosinase, an oxygenase, is activated by sunlight, which results in greater

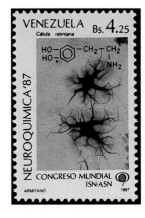

This Venezuelan stamp with the structure of dopamine and retinal neurons commemorates the World Neurochemistry Congress.

Figure 19.21

Tyrosine is the precursor for several biogenic amines. (a) The synthesis of the important neurotransmitters dopamine, norepinephrine, and epinephrine from tyrosine (SAM = S-adenosylmethionine; SAH = S-adenosylhomocysteine). (b) Structure of thyroxine, a hormone synthesized from two molecules of tyrosine.

Figure 19.22
The synthesis of skin pigment melanins from tyrosine.

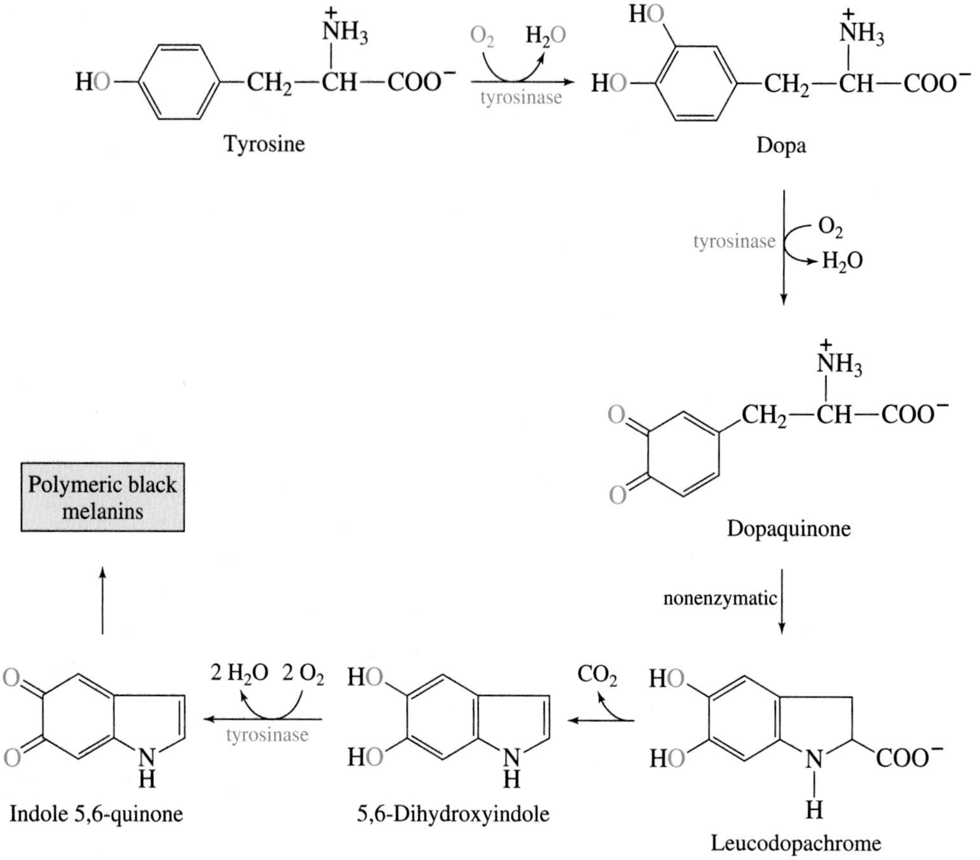

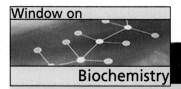

Tyrosine
O_2 H_2O
tyrosinase
Dopa
tyrosinase
O_2
H_2O
Dopaquinone
nonenzymatic
Leucodopachrome
CO_2
5,6-Dihydroxyindole
2 H_2O 2 O_2
tyrosinase
Indole 5,6-quinone
Polymeric black melanins

Window on Biochemistry

19.2 Treating Depression—Neurotransmitters and Monoamine Oxidase Inhibitors

The aromatic amino acids phenylalanine, tyrosine, and tryptophan are especially active as precursors for the synthesis of biologically significant amines. Here we will explore the characteristics of several **neurotransmitters (NT)** that are synthesized from these amino acids. The catecholamine neurotransmitters **dopamine, norepinephrine,** and **epinephrine** are derived from phe and tyr (see Figure 19.21). Tryptophan is the starting point for **serotonin** (see Figure 19.20). Neurotransmitters are chemicals that play essential roles in the transmission of the nerve impulse from one nerve cell (neuron) to another (see page 236 and CD/ROM unit on Membrane Transport). The chemical molecules are released from the end of one neuron and migrate through the synaptic cleft to a receptor site on an adjoining neuron. This process triggers the impulse in the adjoining neuron. After a neurotransmitter has served its purpose, it must be inactivated so the nerve does not continue firing. The most important NT termination processes include uptake of the NT and metabolic transformation. We focus here on the action of the **monoamine oxidases (MAOs)** in degrading the NT. The following reaction shows the chemical changes brought about by an MAO on norephinephrine:

MAO–FAD
MAO–$FADH_2$
+ H_2O
+ NH_3

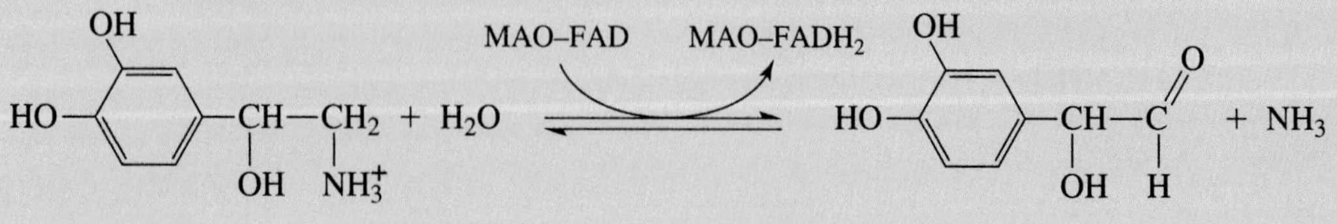

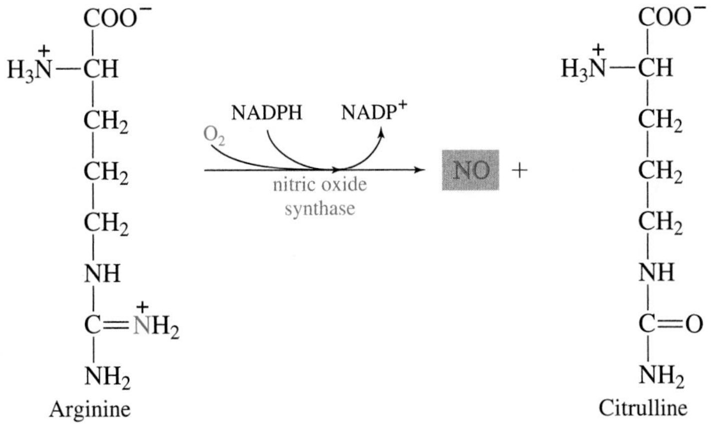

Figure 19.23
Arginine as the precursor of the
cell messenger nitric oxide. The
reaction is catalyzed by nitric
oxide synthase. The colored nitrogen atom in arginine is the origin
of the nitrogen in NO.

production of the melanins and darkening of the skin. Individuals with a genetic deficiency of tyrosinase have no skin pigmentation (albinos). The browning of apples, bananas, and mushrooms when exposed to oxygen is due to uncontrolled oxidation and polymerization of plant diphenols catalyzed by tyrosinase.

A most unusual biosynthetic use of an amino acid has very recently been discovered. The important metabolic regulator and local hormone, **nitric oxide** (NO), is derived from arginine (Figure 19.23). The enzyme nitric oxide synthase is an oxygenase that depends on the calcium ion, Ca^{2+}, for activity. NO serves as a messenger in the signal transduction processes.

Cutting Edge: Nitric
oxide and Viagra

Purine and Pyrimidine Nucleotides

The nucleotides are composed of a nitrogenous base (purine or pyrimidine), a carbohydrate (ribose or deoxyribose), and phosphate (see Figure 10.1). The nucleotides are components in many important biomolecules we have already encountered. First and foremost, they are

MAOs are present in the brain, nerve cells, liver, intestines, and kidney. The most common form of the enzyme has an FAD cofactor and is attached to the outer membrane of mitochondria.

It is estimated that between 5% and 20% of individuals will suffer from major depression at some time during their life. One of the most likely causes for the various forms of clinical depression is the improper processing (synthesis, degradation, uptake) of certain neurotransmitters, especially norepinephrine and serotonin. Among the many antidepressant drugs are the **MAO inhibitors (MAOIs).** These drugs act by slowing the degradation of the amine neurotransmitters and thus increasing their concentrations. The most commonly used MAOIs are tranylcypromine (Parnate) and phenelzine (Nardil). The side effects of these drugs include fluctuations in blood pressure, daytime sleepiness, dry mouth, weight gain, and sexual dysfunction. A dangerous aspect in the use of the MAOIs is due to the fact that the MAOs not only function in brain chemistry, but also are essential in the degradation of dietary amines. Users of the MAOIs must avoid certain foods containing the chemical tyramine (the decarboxylated form of tyrosine). If the MAOs in the intestines are inhibited, tyramine reaches dangerous levels and may cause blood pressure fluctuations. Foods that contain tyramine include aged cheese, hard sausages, chocolate, and red wine. Individuals using MAOIs must also avoid pseudoephidrine (Sudafed), diet pills containing phenolpropanolamine, and many other medications. Many over-the-counter and prescription medications are amines, and their concentrations could reach toxic levels if they cannot be properly degraded by the MAO.

Albinos have a genetic deficiency of the enzyme tyrosinase.

the building blocks of DNA and RNA. They are also found in cofactors including FAD, NAD^+, $NADP^+$, CoASH, and SAM. In this section we explore the metabolism and clinical aspects of the nucleotides.

Most organisms have two types of biosynthetic pathways for the nucleotides: **de novo synthesis of nucleotide bases** (synthesis from new precursors) and **salvage synthesis of nucleotide bases** (recycling). The de novo synthesis of pyrimidine nucleotides begins with construction of the heterocyclic ring, followed by the attachment to ribose phosphate. The origin of each ring atom is shown in Figure 19.24. A general outline for biosynthesis of pyrimidines is shown in Figure 19.25. The final complete nucleotides from this pathway are uridine 5′-triphosphate (UTP) and cytidine 5′-triphosphate (CTP). The first step, catalyzed

Figure 19.24

The origin of each atom in the biosynthesis of the (a) pyrimidine and (b) purine rings.

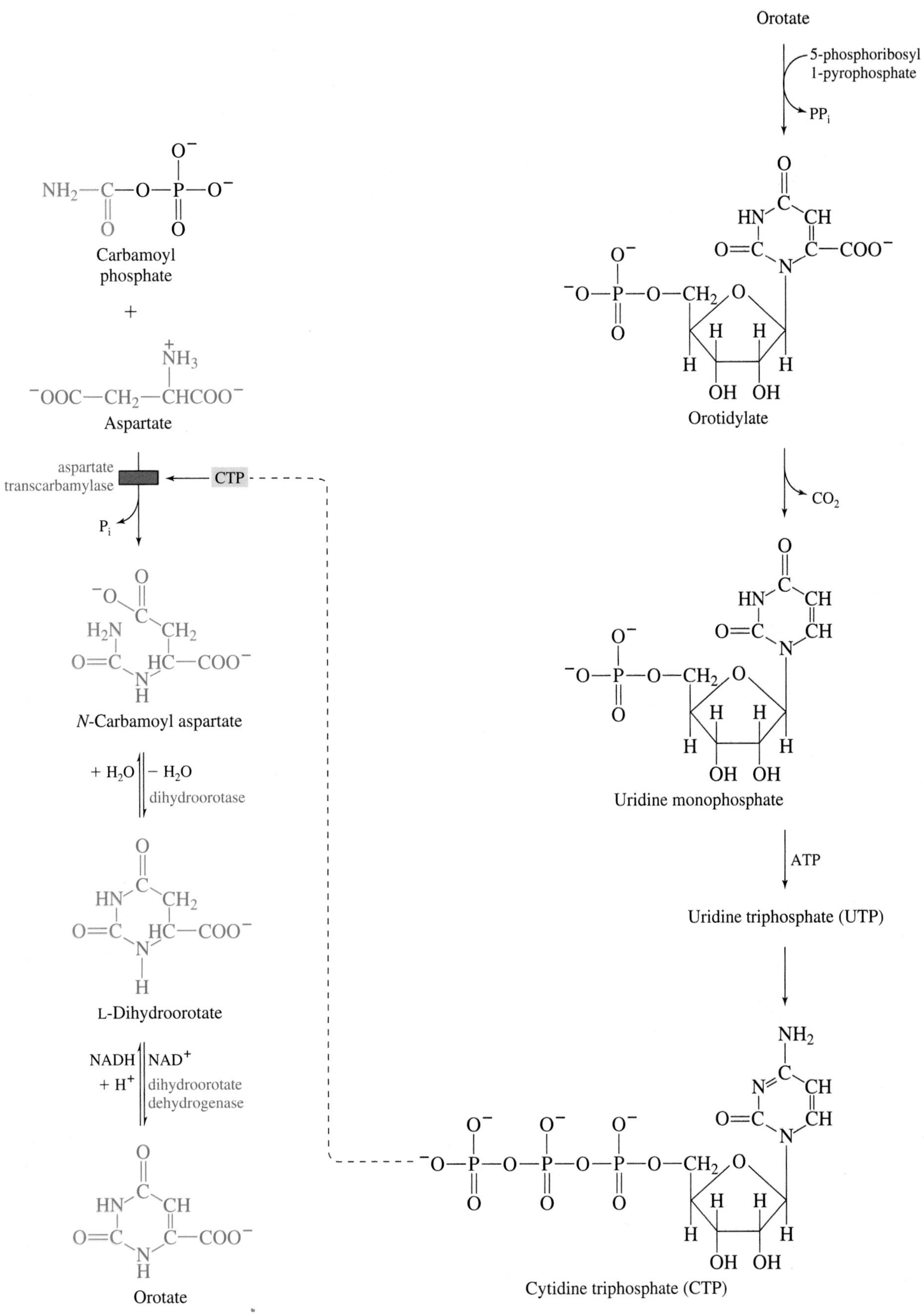

Figure 19.25

A general outline for the synthesis of pyrimidine nucleotides UTP and CTP. The pyrimidine ring begins with carbamoyl phosphate and aspartate. The final product, CTP, is a negative modulator for aspartate transcarbamoylase.

by the allosteric enzyme aspartate transcarbamoylase (ATCase), is the rate-limiting step. CTP is a negative modulator for this enzyme.

In contrast to pyrimidine nucleotide biosynthesis, purine nucleotides begin with an activated ribose phosphate derivative as a foundation, and atoms of the heterocyclic ring are added as shown in Figures 19.24 and 19.26. The first intermediate with a completed purine ring is inosine monophosphate (IMP), which branches to the nucleotides GMP and AMP. These products act as feedback inhibitors at several points in the biosynthetic scheme.

Another important source of purine and pyrimidine nucleotides is the salvage pathway.

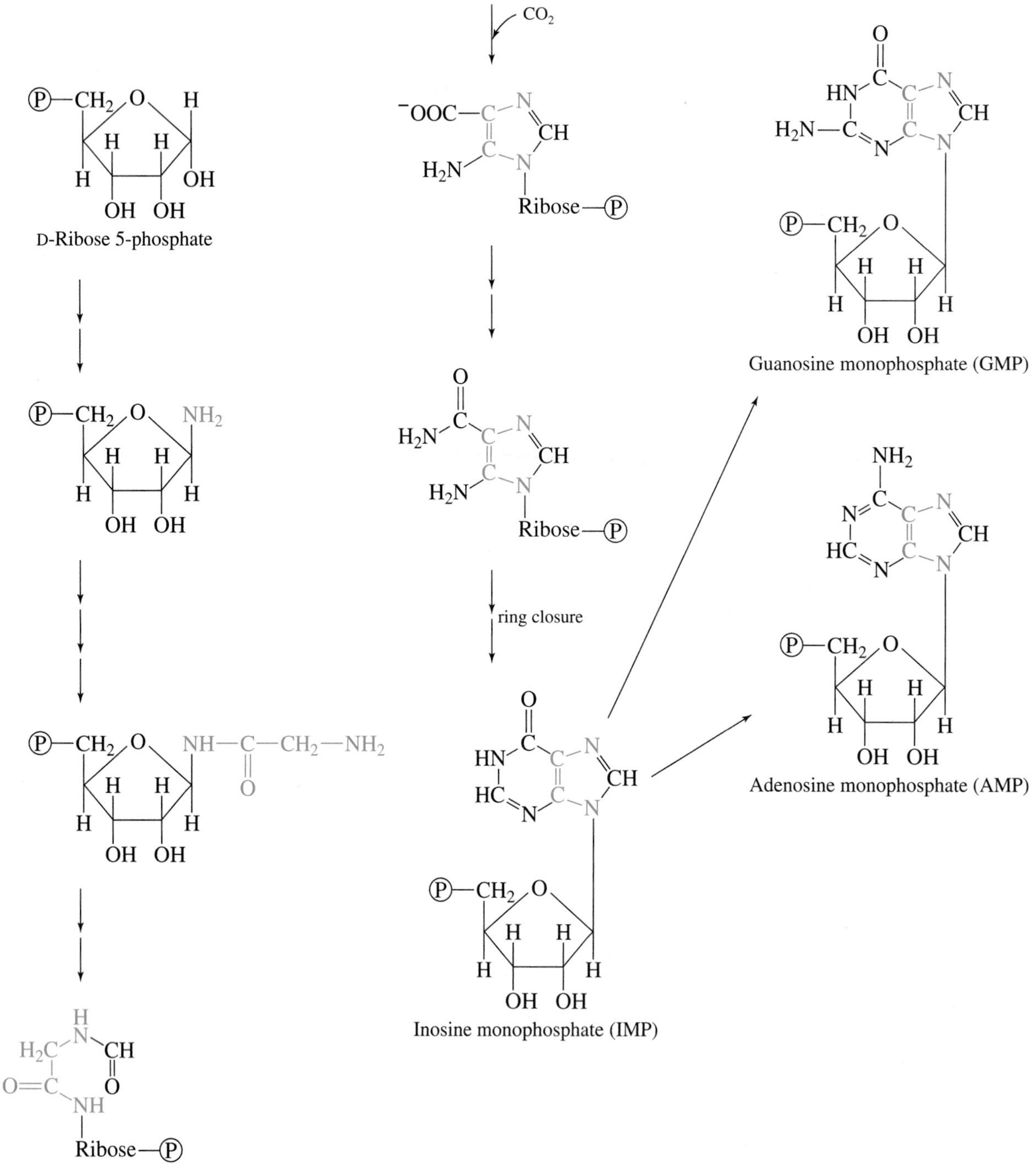

Figure 19.26

A general outline for the synthesis of purine nucleotides beginning with ribose 5-phosphate. The pathway leads to guanosine monophosphate and adenosine monophosphate. ℗ indicates a phosphate group.

Up to 90% of the nitrogen bases we ingest in food or synthesize are recycled. Significant amounts of energy in the form of ATP are required for de novo synthesis, so the salvage pathways provide an alternate, economical route of nucleotide synthesis. Two important salvage enzymes are adenine phosphoribosyltransferase and hypoxanthine-guanine phosphoribosyltransferase. Both of these enzymes bring together the proper base with activated ribose in the form phosphoribosyl pyrophosphate (PRPP). The following reactions are catalyzed:

$$\text{adenine} + \text{PRPP} \rightleftharpoons \text{AMP} + \text{PP}_i$$
$$\text{hypoxanthine} + \text{PRPP} \rightleftharpoons \text{IMP} + \text{PP}_i$$
$$\text{guanine} + \text{PRPP} \rightleftharpoons \text{GMP} + \text{PP}_i$$

The purine and pyrimidine nucleotides we have discussed to this point all have had ribose as their carbohydrate component; that is, they are ribonucleotides. These compounds are, of course, used in cofactor synthesis and in RNA construction. The nucleotides containing deoxyribose are of importance primarily for their incorporation in DNA. All four deoxyribonucleotides are synthesized from the ribonucleotides by a reduction process. The overall reaction shows the necessary reactants:

$$\text{ribonucleoside diphosphate} + \text{NADPH} + \text{H}^+ \rightleftharpoons$$
$$\text{deoxyribonucleoside diphosphate} + \text{NADP}^+ + \text{H}_2\text{O}$$

The necessary enzyme is **ribonucleotide reductase.** The source of electrons for the reduction is NADPH, and the active ribonucleotide substrate forms are the diphosphates ADP, GDP, UDP, and CDP. The change in structure of the substrates upon reduction is shown in Figure 19.27. The 2′-hydroxy group is substituted by a hydrogen. The final products are dADP, dGDP, dUDP, and dCDP.

DNA contains the pyrimidine base thymine rather than uracil. The uracil ring is methylated to the thymine base in preparation for DNA synthesis:

$$\text{dUMP} + N^5,N^{10}\text{-methylene-FH}_4 \underset{}{\overset{\text{thymidylate synthetase}}{\rightleftharpoons}} \text{dTMP} + \text{FH}_4$$

The enzyme thymidylate synthetase transfers a methyl group from methylene tetrahydrofolate. The form of the substrate is the monophosphate. Rapidly growing tumor cells must maintain a high level of DNA synthesis. Inhibition of thymidylate synthetase would slow down growth of such cells by limiting the amount of dTMP available. This role is played by the widely used chemotherapeutic agents fluorouracil and methotrexate.

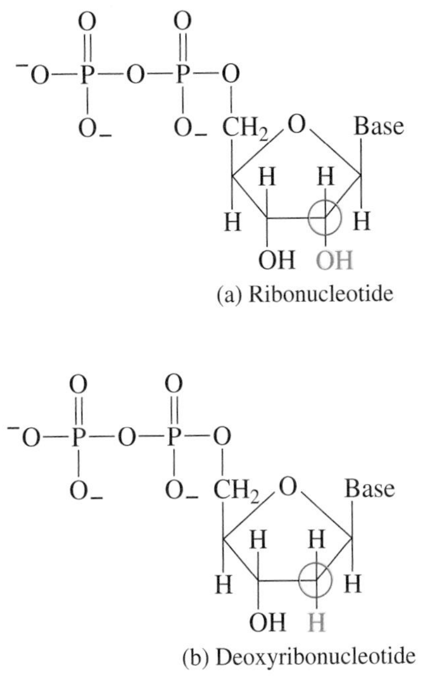

(a) Ribonucleotide

(b) Deoxyribonucleotide

Figure 19.27

The structures of a (a) ribonucleotide and (b) a deoxyribonucleotide. The circled carbon atom undergoes an oxidation change.

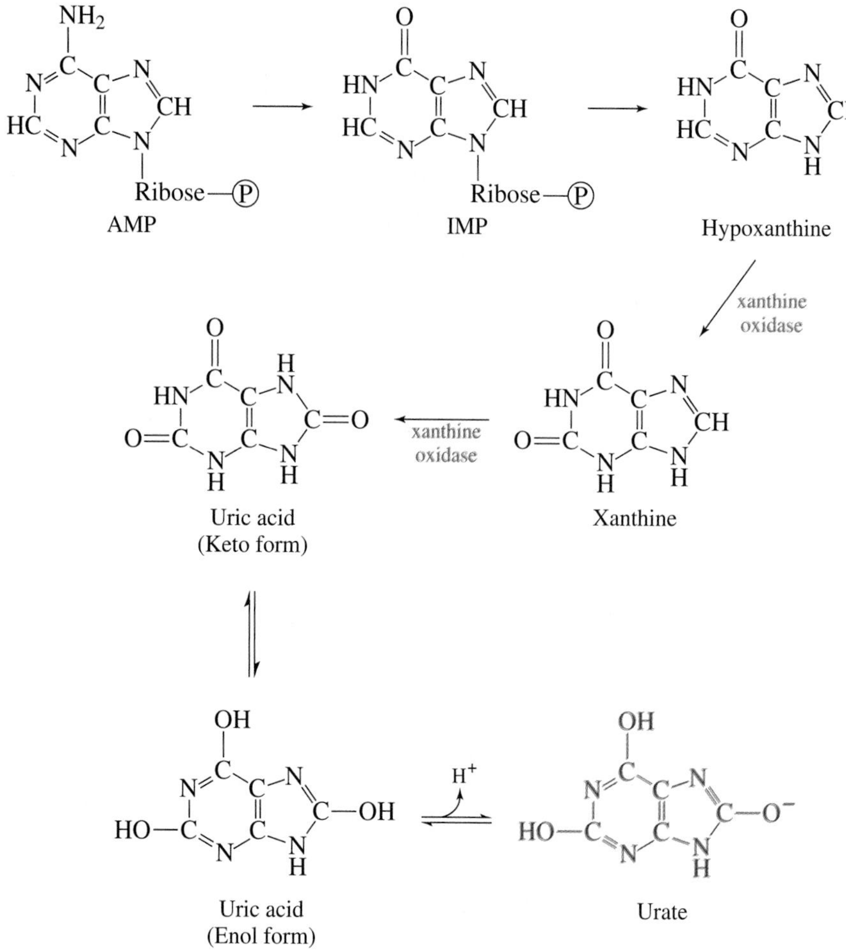

Figure 19.28

The degradation of the purine nucleotide AMP to uric acid and urate. A key enzyme is xanthine oxidase. (P) indicates a phosphate group.

AMP

IMP

Hypoxanthine

xanthine oxidase

Xanthine

xanthine oxidase

Uric acid
(Keto form)

Uric acid
(Enol form)

Urate

Allopurinol

Figure 19.29

The structure of the drug allopurinol used for treatment of gout.

Metabolic disorders for example—gout and Lesch-Nyhan syndrome—affect nucleotide metabolism. All purine rings are catabolized to uric acid, which is transformed to the anionic form, urate, under physiological conditions (Figure 19.28).

Gout: The major biochemical feature of gout is an elevated level of serum urate. It causes inflammation of the joints by precipitation as sodium urate crystals. The excessive production of urate by purine degradation may have many causes, including diet and disease. The treatment of gout includes the use of the drug allopurinol (Figure 19.29). This drug has a structure similar to xanthine and acts by inhibiting the enzyme xanthine oxidase.

Lesch–Nyhan Syndrome: This genetic disorder is caused by a deficiency of the salvage enzyme hypoxanthine-guanine phosphoribosyltransferase. Children with this deficiency display compulsive self-destructive behavior. They bite their fingers and lips and are aggressive toward others. They also show signs of mental retardation. Because there is less salvaging of bases, more uric acid is produced, causing gout. Allopurinol relieves symptoms of gout, but it has no effect on other Lesch–Nyhan symptoms.

Summary

■ Nitrogen is present in nature in both inorganic and organic biomolecules, but the most abundant natural form is N_2 gas. Before this form can be utilized by most organisms, it must be "fixed" or reduced to ammonia by the nitrogenase complex in nitrogen-fixing bacteria.

■ All species of life are able to synthesize at least some of the 20 amino acids for proteins. Humans and other higher animals can make only nine of the amino acids in quantities sufficient for normal growth. The others must be supplied in the diet.

■ Dietary proteins are degraded in the stomach and intestinal tract by hydrolysis catalyzed by proteolytic enzymes. The free amino acids are distributed to peripheral tissue for protein synthesis. Under some cellular conditions, amino acids are degraded for energy.

■ The initial phase of amino acid catabolism is usually transamination in which the amino group is transferred to an α-keto acid, most often α-ketoglutarate. The reactions are catalyzed by aminotransferases with pyridoxal phosphate cofactors. The remaining carbon skeletons are degraded by the citric acid cycle.

■ Excess nitrogen in humans is usually present in the form of the ammonium ion, which must be removed from the cell. This is done by the five enzyme-catalyzed reactions of the urea cycle.

■ Specialized metabolic pathways use amino acids to synthesize important biomolecules such as histamine, GABA, serotonin, melatonin, indole 3-acetate, the neurotransmitters (dopa, dopamine, epinephrine), NO, and the nucleotide bases adenine, guanine, thymine, uracil, and cytosine.

Study Exercises

Understanding Terms

19.1 Define the following terms in 25 words or less.

    a. Essential amino acids
    b. Transamination
    c. Pyridoxamine phosphate
    d. Carbamoyl phosphate
    e. δ-Aminolevulinate
    f. Porphyrias
    g. Histamine
    h. Mo–Fe protein
    i. Pepsin
    j. Allopurinol
    k. Monoamine oxidase
    l. Catecholamines
    m. Phenylalanine hydroxylase
    n. Alcaptonuria
    o. Serotonin
    p. Norepinephrine

Reviewing Concepts

19.2 Each of the following metabolic disorders results in the accumulation of a metabolite. Match each disorder with a metabolite from the list.

Metabolic Disorder	Metabolite
1. Maple syrup urine disease	a. Phenylalanine
2. Alcaptonuria	b. NH_4^+
3. Phenylketonuria	c. Sodium urate
4. Hyperammonemia	d. α-Keto acids
5. Porphyria	e. Homogentisate
6. Jaundice	f. Bilirubin
7. Gout	g. Porphyrin
8. Lesch–Nyhan syndrome	

19.3 For each of the metabolic disorders listed in Problem 19.2, list the deficient enzyme, if any.

19.4 Briefly define the biochemical function for each of the following cofactors in amino acid metabolism. Write a reaction showing the use of each cofactor.

    a. Pyridoxal phosphate
    b. $NAD^+/NADH + H^+$ or $NADP^+/NADPH + H^+$
    c. Tetrahydrofolate
    d. S-Adenosylmethionine
    e. CoASH

19.5 Listed below are the essential reactants, product, components, and processes of the nitrogenase complex found in nitrogen-fixing organisms. Arrange each item in the correct order of electron transfer and nitrogen fixation.

    a. N_2
    b. Source of energized electrons
    c. Nitrogenase reductase (Mo–Fe protein)
    d. Reduction of ferredoxin
    e. NH_3
    f. Nitrogenase (Fe–S protein)

HINT: Begin with b.

19.6 Match each of the components of the nitrogenase complex system with its appropriate cofactor.

Component	Cofactor
1. Ferredoxin	a. Fe–S
2. Flavodoxin	b. Mo–Fe
3. Nitrogenase reductase	c. FAD
4. Nitrogenase	

19.7 Study Figure 19.9 and answer the following questions. Select your answers from the list of intermediates below.

    a. What intermediates are present in the biosynthetic pathway for tryptophan?
    b. What intermediates are present in the biosynthetic pathway for phenylalanine?
    c. What intermediates are in the pathway for biosynthesis of both phenylalanine and tryptophan?
    d. What intermediates are present in the biosynthetic pathway for tyrosine?

e. What intermediates are present in the biosynthetic pathway for both phenylalanine and tyrosine?

Intermediates
1. Chorismate
2. Shikimate
3. Anthranilate
4. Prephenate
5. Phenylpyruvate
6. *p*-Hydroxyphenyl pyruvate

19.8 What are the starting compounds for the de novo synthesis of pyrimidine nucleotides? Select correct answers from the following list.

a. Aspartate
b. Phosphoribosyl pyrophosphate
c. NH_4^+
d. Xanthine
e. CO_2

19.9 Using the names of intermediates, outline the pathway for the synthesis of the amino acid arginine in humans, beginning with carbamoyl phosphate and ornithine.

19.10 The anabolic pathway of a biomolecule is often determined using radiolabeled molecules. Show how the smaller labeled molecules below are incorporated into the larger molecules.

a. $[^{15}N]$ aspartate $\longrightarrow$ AMP
b. $[^{14}C]$ glycine (carbon 2) $\longrightarrow$ AMP

19.11 The enzymes below are from this chapter. Classify each according to Table 14.2.

Enzyme	Classification
a. Nitrate reductase	_____
b. Alanine aminotransferase	_____
c. Arginase	_____
d. Phenylalanine hydroxylase	_____

19.12 What metal ions are present in the nitrogenase complex?

19.13 Describe a clinical treatment for phenylketonuria.

Working Problems

19.14 Write structures for all the products of nitrogenase complex action on each of the following substrates. What structural feature do the substrates have in common?

a. N≡N
b. HC≡CH
c. HC≡N

19.15 Individuals with the protein deficiency disease kwashiorkor suffer loss of skin pigmentation. Suggest a biochemical basis for this observation.

19.16 Write the structure of the product from the direct transamination of each compound below.

a. Aspartate
b. Alanine
c. Valine
d. Phenylalanine
e. Ornithine

19.17 The following α-keto acids are used to treat individuals with hyperammonemia. What amino acids are formed from the α-keto acids under physiological conditions?

a. $CH_3\overset{\overset{O}{\|}}{C}COO^-$

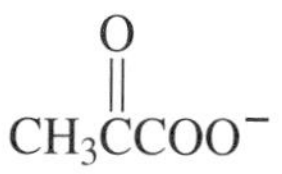

b. $CH_3CH(CH_3)-\overset{\overset{O}{\|}}{C}COO^-$

c. $CH_2CH_2CH_2CH_2\overset{\overset{O}{\|}}{C}-COO^-$ with $\overset{+}{N}H_3$

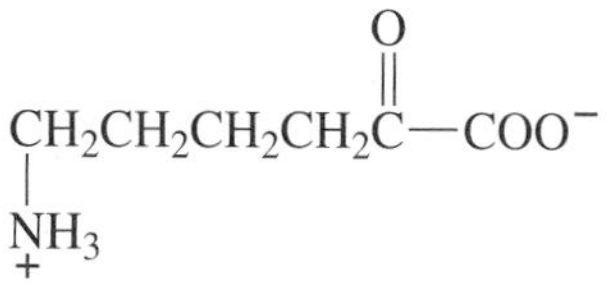

19.18 The natural products drawn below are synthesized from amino acids. For each natural product, name the amino acid precursor.

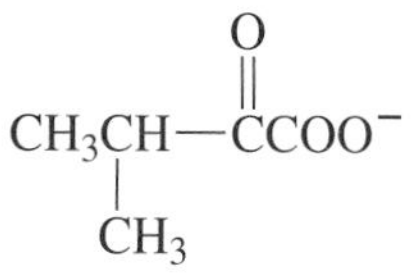

a. Psilocybin (hallucinogenic substance from plants)

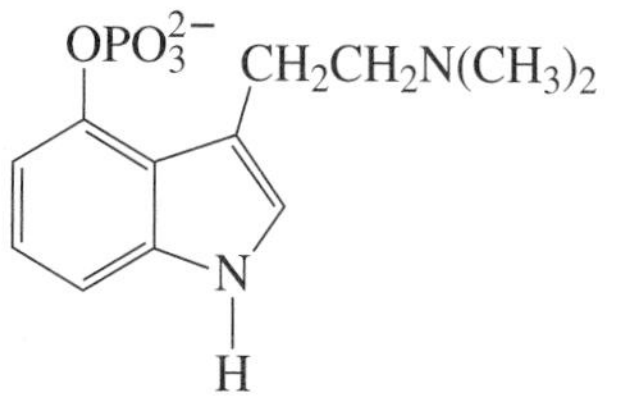

b. Dopamine

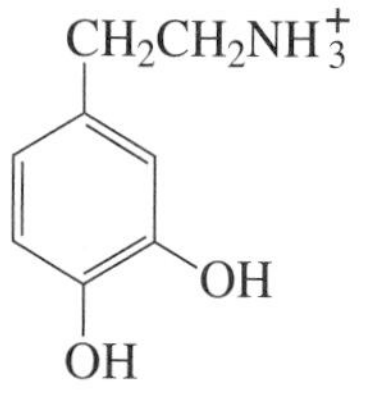

c. $CH_2CH_2CH_2COO^-$ with $\overset{+}{N}H_3$
γ-Aminobutyric acid

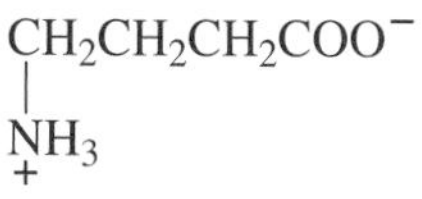

d. $CH_2CH_2CH_2CH_2NH\overset{\overset{H}{|}\overset{\|}{N}}{C}NH_2$ with $\overset{+}{N}H_3$
Agmatine (formed in bacteria)

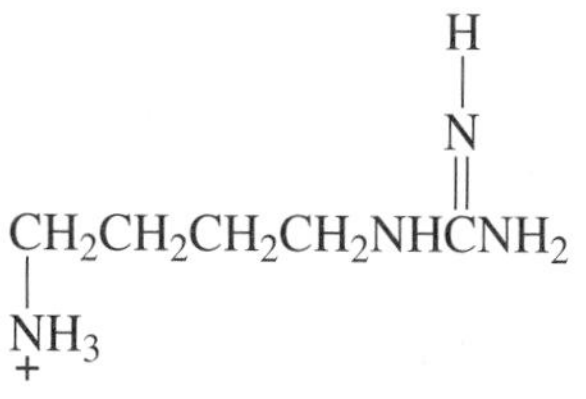

e.

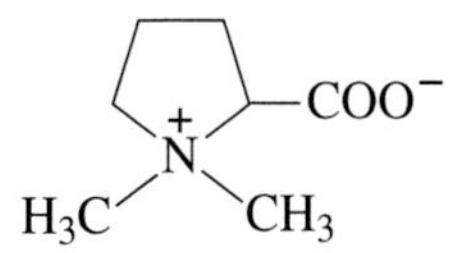

A bacterial product

19.19 Define the type of chemistry that has occurred in the following transformations. Outline the steps for each overall reaction.

a.

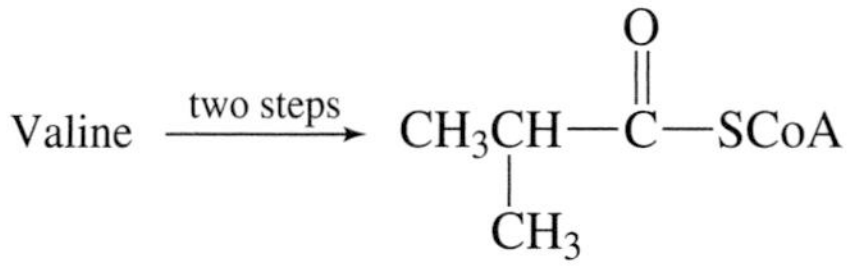

$$\text{Valine} \xrightarrow{\text{two steps}} \text{CH}_3\text{CH}-\overset{\displaystyle O}{\overset{\|}{C}}-\text{SCoA}$$
$$\underset{\text{CH}_3}{|}$$

b.

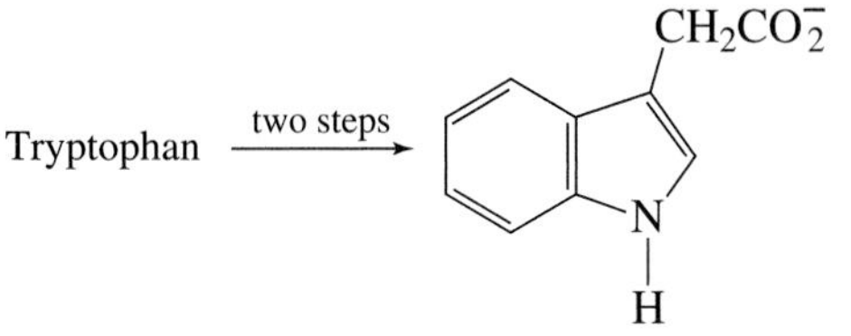

Tryptophan $\xrightarrow{\text{two steps}}$

Auxin

c.

Tyrosine $\xrightarrow{\text{four steps}}$

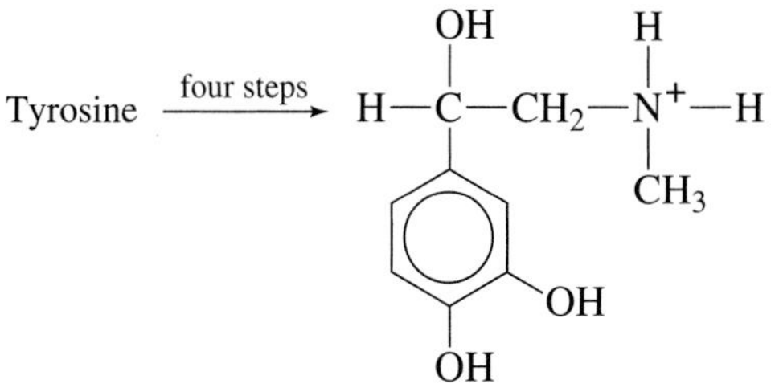

Epinephrine

19.20 Using the names of intermediates, outline the complete catabolism of the amino acid alanine to CO_2, H_2O, and NH_4^+. You will need to use pathways from earlier chapters.
HINT: Begin with alanine → pyruvate.

19.21 Using the names of intermediates, outline the conversion of NH_4^+ produced in Problem 19.20 to urea.

19.22 Each of the following amino acids may be catabolized to a citric acid cycle intermediate. Identify the citric acid cycle intermediate for each.

a. Glutamate **c.** Asparagine
b. Aspartate   **d.** Glutamine

19.23 Which of the following compounds are excreted by organisms in order to remove excess ammonia?

a. Urea   **c.** Uracil   **e.** Uric acid
b. NH_4^+ **d.** Uridine   **f.** Ornithine

19.24 Why is NH_4^+ toxic in higher animals?

19.25 Urate is the final catabolite of the purine ring. Which of the following would yield urate in complete metabolic degradation?

a. ATP   **f.** UDP-glucose
b. Xanthine **g.** FAD
c. NAD^+ **h.** DNA
d. CMP   **i.** RNA
e. CoASH   **j.** Tryptophan

19.26 Discuss and explain the following statement: Phenylalanine is an essential amino acid in humans but tyrosine is not.

19.27 Recall from your introductory chemistry class that ammonia, NH_3, is a gas. Why does it not diffuse out of cells?
HINT: The pK_a for the following reaction is about 9.5.

$$\text{NH}_3 + \text{H}^+ \rightleftharpoons \text{NH}_4^+$$

19.28 The carbohydrates and fatty acids are important fuel molecules because they yield metabolic energy in the form of ATP when they are catabolized. Are the purines and pyrimidines important fuel molecules? Explain.

19.29 Do you expect birds to have the enzyme arginase?

19.30 Write reactions showing the action of monoamine oxidases on serotonin and dopamine.

Writing Biochemistry

19.31 Several metabolic diseases are discussed in this chapter. Select one and read about it in this text, in other biochemistry textbooks, and in the literature. Write a 500-word paper describing characteristics of the disease. Include biochemical details, genetic aspects, clinical symptoms, diagnosis, and treatment of the disease.

19.32 The editor of your college newspaper has asked you to write an article on the drug Viagra. Review the CD/ROM unit on "Nitric oxide and Viagra" and write a 100-word article for the paper. Remember that most readers of the paper are not science majors.

Applying Biochemistry

19.33 Using chemical structures, show reactions for the synthesis of homocysteine in humans. For references, you should review the CD/ROM unit on "Homocysteine" and also study an advanced biochemistry book. Begin your reaction sequence with methionine.

Further Reading

Anand, U., and Anand, C., 1997. Teaching the urea cycle. *Biochem. Educ.* 25:20–21.
Anand, U., and Anand, C., 1999. Connecting links between the urea cycle and the TCA cycle: A tutorial exercise. *Biochem. Educ.* 27:153–154.

Battersby, A., 1994. How nature builds the pigments of life: The conquest of B_{12}. *Science* 264:1551–1557.

D'Andrea, G., 2000. Classifying amino acids as gluco(glyco)genic, ketogenic, or both. *Biochem. Educ.* 28:27–28.

Elliott, K., 1999. Teaching metabolism for the future: concepts not pathways. *The Biochemist* 21(1):9–39.

Gibson, F., 1999. The elusive branch-point compound of aromatic amino acid biosynthesis. *Trends Biochem. Sci.* 24:36–38.

Leigh, G., 1998. Fixing nitrogen any which way. *Science* 279:506–507.

Manchester, K., 1995. Glutamate dehydrogenase: A reappraisal. *Biochem. Educ.* 23:131–133.

Mayer, B., and Hemmens, B., 1997. Biosynthesis and action of nitric oxide in mammalian cells. *Trends Biochem. Sci.* 22:477–481.

Smil, V., 1997. Global population and the nitrogen cycle. *Sci. Am.* 227(1): 76–81.

Snyder, S., and Bredt, D., 1992. Biological roles of nitric oxide. *Sci. Am.* 266(5):68–77.

Torchinsky, Y., 1987. Transamination: Its discovery, biological and chemical aspects. *Trends Biochem. Sci.* 12:115–117.

Warren, M., et al., 1996. The maddening business of King George III and porphyria. *Trends Biochem. Sci.* 21:229–234.

Chapter 20

Integration, Coordination, and Specialization in Metabolism

The major pathways for metabolism of carbohydrates, fats, and amino acids have been detailed in the previous five chapters. The individual steps of anabolism and catabolism were outlined along with information on enzymes, cofactors, chemical reactions, and regulation. Emphasis was placed on the reactions and pathways that involved use of the reduction–oxidation (redox) cofactors (NAD^+, $NADP^+$, FAD) and energy-rich, phosphorylated metabolites (ATP, ADP, PEP, etc.). The bioenergetics of reactions and pathways were quantified by keeping tally of the redox cofactors and ATP (or GTP) produced or utilized. **Net reactions** that clearly define the input and output for a process were derived for each pathway (Section 15.1). In most instances, the predominant cell or tissue type and the intracellular location for the pathways were identified. The enzymes for most metabolic pathways were found to be localized in the cytoplasm or the mitochondria of the respective cells. However, what was often missing from the previous details was a description of the coordinated metabolism for the whole organism. Each pathway and tissue type was studied independently, and the importance of connections between and among pathways was not obvious.

All cells do not have the biochemical machinery to express all of the biochemical pathways. They may express different parts at different times, or they may rely on other cells to provide pathways. The myriad of reactions must be regulated and coordinated so that organisms are able to maintain resources for routine biochemical processes, but organisms must also be able to adapt to changes in the internal and external conditions. It would be wasteful in terms of nutrients and energy for all of the enzymatic reactions of an organism to proceed at a maximum rate at all times. The coordination of metabolic processes is important in prokaryotes, the simplest single-celled organisms; however, it becomes essential in multicellular organisms such as humans and other animals where different organs and tissues have special roles to play in the overall well-being of the organism.

In this chapter the larger picture of how metabolic pathways in a whole organism are interlinked and coordinated is explored with emphasis on the biochemical requirements and responsibilities of each major cell type. Examples will be described to show how organisms are able to respond to varying conditions by changing the importance of metabolic pathways and by opening alternate routes. The chapter will combine the metabolism of the important energy biomolecules, carbohydrates, fats, and amino acids, to define the metabolic roles of each. The perspective will be that of a whole metabolic system and will focus on animal, especially human, biochemistry.

20.1 Overall Strategies of Metabolism

Review of Metabolism

In the previous chapters we grouped metabolism into two paths or directions depending on the biochemical purposes. Catabolism was defined as a degradative pathway, whereby biomolecules such as fats, carbohydrates, and amino acids are oxidized to CO_2 and H_2O with production of the energy-rich molecules NADH, NADPH, $FADH_2$, and ATP. In anabolism, small precursor molecules are combined to synthesize larger, sometimes complex, biomolecules. Energy for synthesis is supplied primarily by ATP, NADH, NADPH, and $FADH_2$.

The major pathways of energy metabolism in animals are reviewed below and combined in Figure 20.1. These are the pathways primarily involved in the energy metabolism of carbohydrates, fats, and amino acids.

1. Glycolysis. This pathway degrades a molecule of glucose to two molecules of pyruvate and in the process produces a net yield of two molecules of ATP and two molecules of NADH. Glycolysis accepts glucose from intracellular storage (glycogen) and from dietary mono-, di-, and polysaccharides. The reactions of glycolysis display anaerobic metabolism; they do not require oxygen, but they prepare the carbons from carbohydrates for entry into aerobic metabolism (pyruvate dehydrogenase complex, citric acid cycle, and oxidative phosphorylation). During anaerobic metabolism in the cell, pyruvate from glucose oxidation is reduced to lactate. The enzymes of the glycolytic pathway are present in the cytoplasm of the cell (Section 15.1) and are highly concentrated in brain cells and skeletal muscle.

Figure 20.1
Integration of the major pathways
of fuel metabolism in animals.

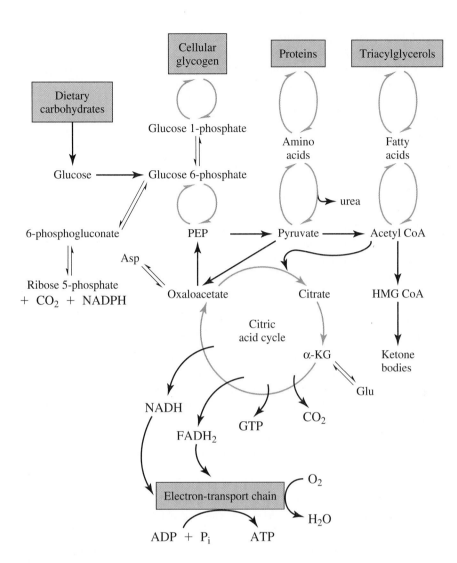

2. Gluconeogenesis. Precursors that contain three carbons, such as pyruvate, lactate, glycerol, and some amino acids, are used to make glucose. Many of the reactions of glycolysis proceed in reverse for gluconeogenesis, but the complete reactions for the two processes are not identical (Section 15.4). Most gluconeogenesis occurs in the liver. Glucose in excess of immediate needs is stored in glycogen (Section 15.5).

3. The phosphogluconate pathway (ribose phosphate pathway). This is a specialized pathway of carbohydrate metabolism that oxidizes glucose to ribose 5-phosphate with production of NADPH. The enzymes for the pathway are present in the cytoplasm of liver cells and adipose tissue, where they generate NADPH needed for fatty acid synthesis (Section 16.5), antioxidant action (see Window on Biochemistry 16.1), and drug detoxification (see Window on Biochemistry 17.1).

4. Beta oxidation and fatty acid synthesis. Fatty acids released from dietary fats and from stored triacylglycerols are degraded to acetyl CoA with generation of abundant NADH and FADH$_2$ (Section 18.2). The enzymes responsible for beta oxidation are located in the cell mitochondria of skeletal muscle, heart, and adipose tissue. Fatty acid synthesis, a pathway present in the cytoplasm of liver and adipose tissue, begins with acetyl CoA which first must be activated by carboxylation to malonyl CoA (Section18.3). Fatty acids are carried in the blood stream by serum albumin and by the VLDL.

5. Amino acid degradation and synthesis. Amino acids that are not needed for protein synthesis or for specialized metabolism (Section 19.5) are degraded to carbon skeletons that may enter the mainstream of fuel metabolism. The process of transamination converts amino acids to α-keto acids that eventually are degraded by the citric acid cycle (Section 19.3). There are several entry points into mainstream metabolism for the amino acid carbon skeletons including pyruvate, acetyl CoA, α-ketoglutarate, succinyl CoA, fumarate, and oxaloacetate (Figure 19.10). The amino groups, which are collected primarily in glutamate, are transformed into urea by the urea cycle for elimination from the body (Section 19.4).

6. The citric acid cycle. This central pathway collects acetyl CoA from the degradation of carbohydrates (via the pyruvate dehydrogenase complex), from the beta oxidation of fatty acids, and from the oxidation of some amino acids (Sections 16.2 and 19.3). The sequence requiring eight enzyme steps oxidizes acetyl CoA to CO_2 and H_2O. Each acetyl CoA that enters the catalytic cycle generates three NADH, one $FADH_2$ (or $CoQH_2$), and one ATP (or GTP). The citric acid cycle may also serve as a source of small building blocks for use in anabolism. For example, α-ketoglutarate is transformed to glutamate, and oxaloacetate may be converted either to aspartate or to phosphoenolpyruvate (Section 16.3).

7. Oxidative phosphorylation. This complex process occurring in the mitochondria is actually a coupling of two activities: (a) electron transport from NADH and $FADH_2$ (produced in the catabolic pathways above) to oxygen and (b) the phosphorylation of ADP to generate ATP (Section 17.3). This process allows for the recycling of the redox cofactors ($NADH/NAD^+$; $FADH_2/FAD$), and produces much ATP for energy.

Intermediates That Connect Pathways

The integration of the major metabolic pathways as displayed in Figure 20.1 is instructive in that it shows how the various pathways interconnect. Even though the pathways involve different substrates, different intermediates, different enzymes, different chemistries, different regulation, and different biochemical purposes, they are linked by the sharing of some common intermediates. Those metabolites that link pathways are also called **"branch points"** because many reactions converge or diverge at these intersections. The most important metabolites that lie at these strategic positions are glucose 6-phosphate, pyruvate, acetyl CoA, and oxaloacetate.

1. Glucose 6-phosphate. This is a major metabolic junction for carbohydrate metabolism, especially in the liver. It links the pathways of glycolysis, glycogen metabolism, and the specialized 6-phosphogluconate pathway. The connection to glycogen is of particular interest because glucose 6-phosphate lies at the crossroads of glucose storage (glycogen synthesis) and glucose utilization (glycolysis). Glucose, after phosphorylation catalyzed by hexokinase (muscle and liver) or glucokinase (in liver only, but low glucose affinity), has the option of either being stored in glycogen for later use or being used for immediate fuel. The direction and flux for the glucose 6-phosphate will depend on cell and environmental conditions. Glucose 6-phosphate in liver or adipose tissue may also be shunted to the 6-phosphogluconate pathway to generate NADPH.

2. Pyruvate. This metabolite is part of the bridge that links anaerobic carbohydrate metabolism (glycolysis) with aerobic metabolism—the pyruvate dehydrogenase complex and the citric acid cycle. It is also the entry point for utilization of muscle lactate and the carbon skeletons of some glucogenic amino acids. In terms of anabolism, pyruvate serves as the starting point (via oxaloacetate and phosphoenolpyruvate) for gluconeogenesis and for the anaplerotic reactions to replenish citric acid cycle intermediates.

3. Acetyl CoA. This metabolite can be considered the common product from Stage II metabolism (Figure 14.6) of all three types of fuel molecules: carbohydrates, fatty acids, and some amino acids. Acetyl CoA is further degraded to CO_2 and H_2O by entry into the citric acid cycle. In anabolic terms, acetyl CoA is the starting point for fatty acid and steroid synthesis. Under certain metabolic conditions where acetyl CoA entry into the citric acid cycle is reduced because of limiting oxaloacetate, excess acetyl CoA may be shunted toward the production of ketone bodies (see Section 20.2 and Window on Biochemistry 20.1).

4. Oxaloacetate. Although an intermediate in the citric acid cycle, this metabolite has connections to amino acid metabolism, to glucose metabolism, and to pyruvate metabolism. Oxaloacetate is the entry point for metabolism of the carbon skeletons of the amino acids aspartate and asparagine. Oxaloacetate may also be used for gluconeogenesis (through phosphoenolpyruvate and pyruvate).

20.2   Metabolic Specialization and Integration

Metabolic Profiles of Organs

In a differentiated organism containing many cell types, only the liver is able to carry out most of the pathways reviewed in Section 20.1. Each major organ has its own metabolic profile and specific biochemical functions that are determined by a unique set of enzymes. What we observe, then, is a division of labor among the various types of cells. Each organ is specialized in terms of the type of metabolic fuel used, the type of fuel stored (if any), and the metabolic fuel available for transport to other organs (Table 20.1). This sharing of resources requires continuous metabolite exchange by the organs that is brought about by the bloodstream. Blood carries metabolites including glucose, fatty acids and other lipids, amino acids, and oxygen, as well as hormones that signal changes in metabolism. Blood also removes metabolic waste products including urea and CO_2 from cells. Figure 20.2 displays the metabolic interactions among fuel-metabolizing organs.

In this section we will describe the metabolic profiles of five types of tissue including brain, skeletal muscle, heart muscle, liver, and adipose tissue (see Table 20.1). Later in the chapter there will be a discussion of how physiological and environmental disturbances can change conditions and thus modify the metabolic relationships among the organs.

1. Brain. With an average weight of only 1.2 to 1.3 kg, the brain makes up a small portion of an average 60- to 70-kg human. However, brain cells have voracious appetites accounting for at least 20% of the total energy demands of the body. Glucose, degraded by aerobic metabolism, is the preferred fuel of brain cells. The monosaccharide is degraded by the common and combined processes of glycolysis, the pyruvate dehydrogenase complex, the citric acid cycle, electron transport to oxygen, and phosphorylation of ADP. The brain consumes about 100 to 120 g of glucose per day and uses 15% to 20% of the total oxygen consumed by the body. Much of the energy is needed to produce ATP that is essential to maintain membrane potentials for nerve impulse transmission (Na^+–K^+ ATPase pump, Section 9.7). The brain's needs for glucose and oxygen are constant whether the body is at rest or in motion. Energy needs in the brain also do not change during periods of wakefulness or sleep. Brain cells do not store glycogen, so there must be a constant supply of glucose from the bloodstream. Even though the metabolic requirements are strict, brain cells can readily adapt to metabolic adversity such as starvation or disease. Under these conditions, some brain cells switch to metabolism of ketone bodies that are made from acetyl CoA in the liver and distributed via the bloodstream (see Sections 20.2 and 20.4 and Window on Biochemistry 20.1).

2. Skeletal muscle. Metabolism in muscle tissue differs from brain in three important ways: (a) Muscle can use a variety of fuels—glucose, fatty acids, and sometimes, amino acids; (b) energy demands of muscle can undergo wide variations in the level of activity; and (c) muscle is able to store glucose in the form of glycogen. At rest, muscle cells prefer fatty acids that are supplied by the bloodstream for fuel, but during muscle contraction, glucose, which is mobilized from glycogen, becomes the preferred fuel. Glycogen is attacked by

Interactive Animation:
Metabolic process location

Tissue	Fuel Stored	Fuel Used	Fuel Molecules Exported
Brain	None	Glucose, ketone bodies	None
Skeletal muscle (at rest)	Glycogen	Fatty acids, glucose	None
Skeletal muscle (active)	None	Glucose	Lactate, alanine
Heart muscle	None	Fatty acids, glucose, lactate, ketone bodies	None
Liver	Glycogen, triacylglycerols	Glucose, fatty acids, lactate, amino acids, glycerol	Fatty acids, glucose, ketone bodies
Adipose tissue	Triacylglycerols	Fatty acids, glucose	Fatty acids, glycerol

Table 20.1
Metabolic profiles of major organs

Figure 20.2

Metabolism in the major organs. Arrows show the fuel metabolites imported and exported by each organ.

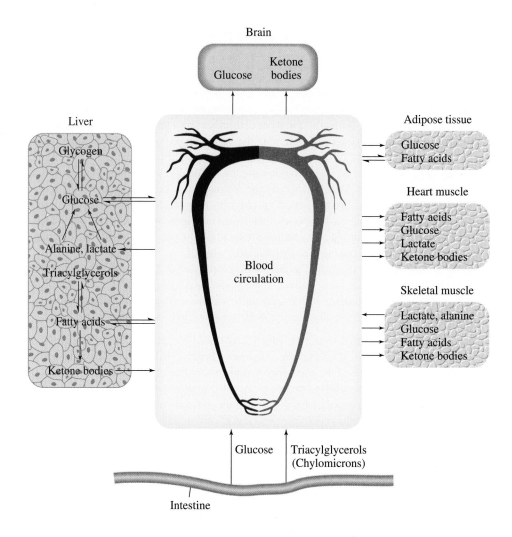

glycogen phosphorylase to produce glucose 1-phosphate, which is isomerized to glucose 6-phosphate. Since muscle cells contain no glucose 6-phosphatase, an enzyme that catalyzes the hydrolytic removal of the phosphoryl group, they cannot share glucose with other cells (Section 15.4). During lower levels of exertion, when fatty acids are degraded to acetyl CoA, muscle cells function aerobically and account for up to 30% of the total oxygen consumption of the body. During periods of moderate and high muscular activity, anaerobic metabolism of glucose produces elevated levels of lactate.

In times of maximum exertion or in times of emergency energy need, muscle cells can also utilize the energy in **phosphocreatine**, but the concentrations of the energy-rich molecule are low and would last less than one minute:

$$\text{phosphocreatine} + \text{ADP} \xrightleftharpoons[]{\overset{\text{phosphocreatine}}{\text{kinase}}} \text{creatine} + \text{ATP}$$

Skeletal muscle cells may use amino acids produced from protein breakdown as metabolic fuel; however, extensive loss of protein, especially enzymes, may cause damage to muscle function (see Section 20.4).

3. Heart muscle. Heart muscle in terms of function and metabolism differs in several ways from skeletal muscle: (a) The possible range of activity levels of the heart is much less than that of muscle; (b) heart muscle stores little or no glycogen; and (c) heart muscle functions primarily aerobically. When the body is at rest, the heart relies on aerobic metabolism of fatty acids for energy but may turn to glucose, lactate, pyruvate, or ketone bodies upon higher energy demand. These fuel molecules are supplied to heart muscle by transport through the blood.

4. Liver. Whereas most other tissues emphasize catabolic degradation, the liver plays both catabolic and anabolic roles. The liver functions to synthesize fatty acids, glucose, amino acids, and ketone bodies for distribution via the bloodstream to other tissues. The liver is also equipped enzymatically for catabolism of a wide variety of fuel molecules including glucose, lactate, fatty acids, and amino acids. Liver cells contain the only storage of glycogen that can be utilized for glucose release into the bloodstream.

The liver plays an important role in controlling the level of blood glucose. This is possible through the action of **glucokinase,** an enzyme unique to the liver. When blood glucose levels become high, especially after eating, the liver can take up the glucose, phosphorylate it, and store it in glycogen. When blood glucose levels are low, glycogen breakdown catalyzed by glycogen phosphorylase is initiated. The glucose 1-phosphate produced is isomerized to glucose 6-phosphate. Transfer of the phosphoryl group to water (hydrolysis) catalyzed by glucose 6-phosphatase provides free glucose for distribution by blood to other tissue.

5. Adipose tissue. Adipose tissue, composed of fat cells (also called adipocytes), represents the most important organ for the long-term storage of fuel molecules. It is estimated that in the average human of 70 kg, the triacylglycerols stored in adipose tissue represent about 50% to 75% of the total energy stored in the body. Assuming moderate exertion levels, this would provide energy for one to two months in the absence of other caloric intake. The adipocyte is metabolically designed for the synthesis and degradation of triacylglycerols. Release of stored fatty acids by hydrolysis is catalyzed by hormone-sensitive lipase. The fatty acids are released into the bloodstream for distribution and use by other cells. When fatty acids are not being used as a fuel source, they may be stored in the form of triacylglycerols. The phosphogluconate pathway is also important in adipose tissue for the production of NADPH needed for fatty acid synthesis.

Integrated Metabolic Pathways

A prominent metabolic feature of the liver is to supply other tissues with molecules for energy metabolism. The liver supports skeletal muscle metabolism by providing glucose via the bloodstream. The liver is also a source of fatty acids needed by heart and adipose tissue and a source of ketone bodies, which are important fuel molecules for skeletal muscle, brain, and heart. Here we describe pathways that provide energy-rich metabolites for distribution to needy cells.

1. The Cori cycle. Skeletal muscle, especially during active contraction, obtains most of its energy from glycolysis, the oxidation of glucose to pyruvate. During anaerobic metabolism in very active muscle, the pyruvate, rather than being decarboxylated to acetyl CoA, is reduced to lactate. This is favored by a high level of NADH in active muscle. Excess lactate is released from muscle by the bloodstream and is taken up by the liver, where it is used for gluconeogenesis (Figure 20.3). The starting point for the synthesis of glucose may be lactate, pyruvate, glycerol, or some amino acids, although lactate is quantitatively the most important. Lactate transported to the liver from muscle is

Interactive Animation: Cori cycle

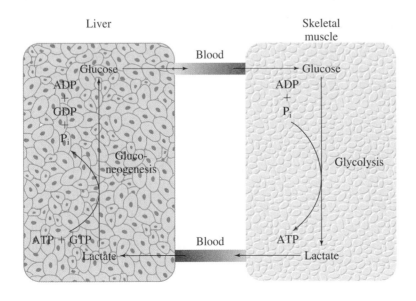

Figure 20.3
The Cori cycle (see also Figure 15.7).

Figure 20.4
The glucose–alanine cycle.

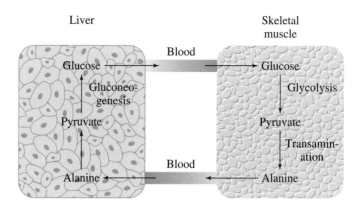

oxidized to pyruvate, which is then used for glucose synthesis. This process involving the sharing of metabolites between skeletal muscle and the liver is called the **Cori cycle**.

2. The glucose–alanine cycle. Another pathway that links the metabolism of the liver with skeletal muscle is the **glucose–alanine cycle** (Figure 20.4). This pathway is analogous to the Cori cycle except that the amino acid alanine rather than lactate is transferred from muscle to the liver. In muscle, pyruvate produced from glycolysis becomes a substrate for transamination and is converted to alanine (Section 19.3):

$$\text{pyruvate } + \text{ glutamate } \underset{}{\overset{\text{aminotransferase}}{\rightleftharpoons}} \text{ alanine } + \text{ alpha-ketoglutarate}$$

Alanine transported via the bloodstream to the liver undergoes transamination to generate pyruvate, which is used for gluconeogenesis. The resulting glucose is returned to skeletal muscle for energy metabolism. Glutamate produced in the liver is deaminated by the action of glutamate dehydrogenase, and the released ammonia is converted to urea (Section 19.4).

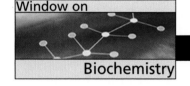

20.1   Dieting and Ketosis

Ketosis is a medical condition characterized by elevated levels of the ketone bodies in the patient's blood and urine. The ketone bodies—acetoacetate, β-hydroxybutyrate, and acetone—are produced in the body when there are excessive amounts of acetyl CoA due to high rates of fatty acid degradation by beta oxidation. Two of the ketone bodies, acetoacetate and hydroxybutyrate, are relatively strong acids, so ketosis can cause a lowering of blood pH, or **acidosis.** Untreated acidosis leads to coma and death. Individuals on certain popular low-carbohydrate, low-fat diets are indeed in a mild ketotic state. Such diets may be effective because metabolism is adjusted to use body fat for energy. Dieters are able to purchase "keto-stix" from a pharmacy to test if they are ketotic. These test strips are dipped into urine to see if ketone bodies are being eliminated. Many commercial strips contain sodium nitroprusside, which changes color upon the oxidation of acetoacetate. Although a mild state of ketosis is probably not dangerous (most nondieting individuals have a normal ketone body level in the blood of about 20 micromolar compared to blood levels as high as 7000 micromolar in severe cases of ketosis), physiological damage may occur. Dieting individuals may experience dehydration (caused by excessive urination to rid the body of acids), electrolyte imbalance (loss of Na^+ and K^+ with acids in urine), kidney stress (caused by excessive urination), difficulty in thought concentration, and bad breath (acetone). In addition, there may be indiscriminate degradation of protein in muscle tissue. The amino acids are used to replenish glucose supplies. Nutritionists recommend that individuals on low-carbohydrate diets should eat high levels of protein, low carbohydrate, and low fat; consume large quantities of water; and supplement their diet with vitamins and salt. It is most important that the diet be supervised by a physician or nutritionist.

Production and Distribution of Ketone Bodies

The beta oxidation of fatty acids, under normal conditions in healthy individuals, yields acetyl CoA, most of which enters the citric acid cycle for continued oxidation. Normal conditions are defined as the proper balance of carbohydrate and fatty acid degradation where carbohydrates provide at least half of the energy. Under conditions of fasting, starvation, untreated diabetes mellitus, or a low-carbohydrate diet, acetyl CoA is produced in abnormally large amounts because of excessive breakdown of fatty acids. The common factor in all of these conditions is the lack of carbohydrates or impaired utilization of carbohydrates, especially glucose. The unbalanced metabolism of fats and carbohydrates changes the flow of nutrients in the various pathways. Without the availability of carbohydrates, fatty acids become the fuel molecule of choice for heart, skeletal muscle, and liver. Brain cells cannot utilize fatty acids for energy but require a high and continuous source of glucose, which must be synthesized by gluconeogenesis (see Chapter 15, Section 15.4):

$$\text{oxaloacetate} \longrightarrow \text{PEP} \longrightarrow \text{glucose 6-phosphate}$$

Let's summarize the cellular conditions present when fatty acid catabolism by beta oxidation is predominant:

1. Elevated levels of acetyl CoA are produced from the breakdown of fatty acids.
2. Oxaloacetate levels are reduced because the molecule is being drained from the citric acid cycle to make glucose.

The result is an abundance of acetyl CoA, but entry into the citric acid cycle is impaired because of oxaloacetate deficiency. Excess acetyl CoA is diverted to the formation of **ketone bodies:**

$$\text{excess acetyl CoA} \longrightarrow \begin{array}{c} \text{acetoacetate,} \\ \text{D-}\beta\text{-hydroxybutyrate,} \\ \text{acetone} \end{array}$$

Ketone bodies

The generation of ketone bodies occurs predominantly in the liver, but the products are distributed throughout the body via the blood. Heart and skeletal muscle and even brain cells can use acetoacetate and D-β-hydroxybutyrate as fuel molecules. The biochemistry of ketone body formation is shown in Figure 20.5. In two successive reactions, three

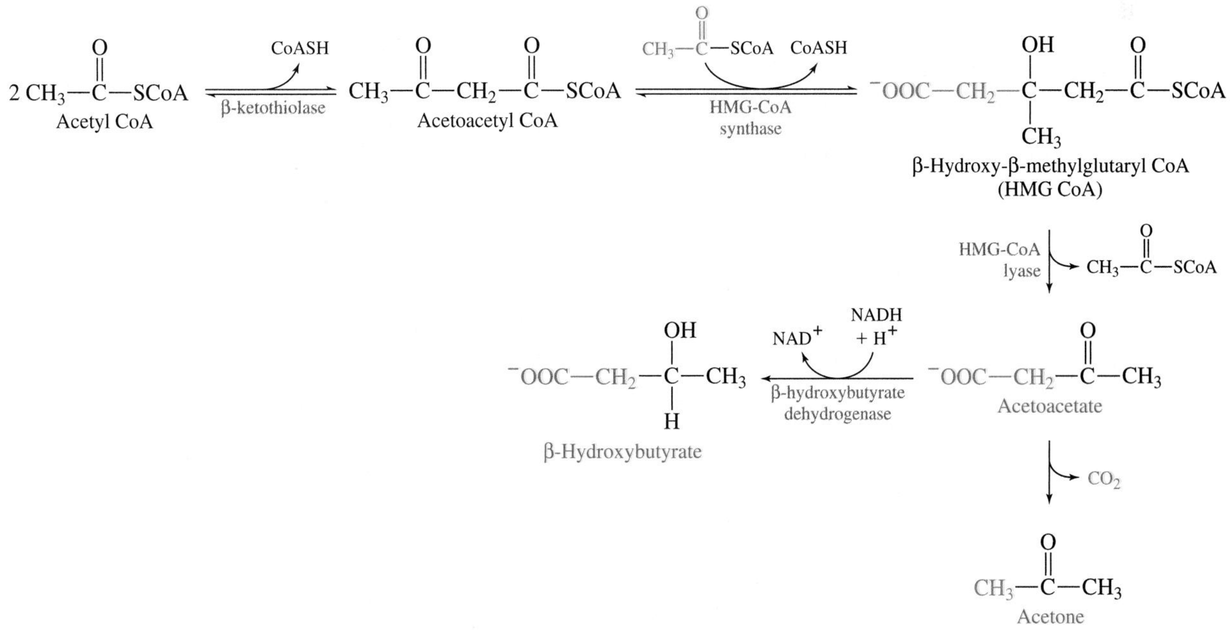

Figure 20.5

The pathway of ketone body formation from acetyl CoA. The ketone bodies—acetoacetate, acetone, and β-hydroxybutyrate—are produced by the condensation of three molecules of acetyl CoA.

molecules of acetyl CoA combine to form β-hydroxy-β-methylglutaryl CoA (HMG CoA). The first reaction, catalyzed by β-ketothiolase, is just the reverse of the last step of beta oxidation. Acetoacetyl CoA combines with another acetyl CoA in a reaction promoted by HMG-CoA synthase. The product, HMG CoA, is cleaved by HMG-CoA lyase to acetoacetate and acetyl CoA. The product, acetoacetate, is at a junction leading to two reactions: (1) decarboxylation to acetone or (2) reduction to D-β-hydroxybutyrate. Note that only two of the three ketone body products are actually ketones. Acetone, which is volatile, is expelled from the body during breathing and often can be detected on the breath of fasting individuals or patients with untreated diabetes. Individuals with elevated levels of ketone bodies in the blood are in a medical condition called **ketosis.** (See Window on Biochemistry 20.1.)

20.3   Metabolic Control by Hormones

The many metabolic activities occurring in a cell and organism require exact coordination and control. This is especially important for highly developed organisms that are comprised of various organs and types of tissue as described in Section 20.1. Metabolic activities are altered by the action of extracellular chemical signals that are secreted by specific cells and distributed throughout the organism via the blood. Each chemical messenger, whatever its origin, has a unique chemical structure and biological effect. Target cells using receptors on their surfaces recognize only those signal molecules intended for them. Here we will focus on the hormones, those signaling molecules synthesized and secreted by endocrine glands and transported to their site of action via the bloodstream. Our understanding of how these and other chemical messengers elicit their effects has increased recently. We originally believed that all hormones interacted directly, inside the cell, with proteins, enzymes, and nucleic acids whose activities they altered. The concepts of **signal transduction** and **second messengers** have been developed to explain the mechanism of action of the hormones and other chemical signals (see Section 2.5).

The hormones we will explore here are those that control fuel metabolism in animals—**insulin, glucagon,** and **epinephrine** (Table 20.2). Blood glucose levels must be maintained within a rather narrow range. In a resting individual several hours after a meal, blood glucose concentrations average 4.5 mM (about 80 mg/100 mL of blood). Soon after a meal containing a normal amount of carbohydrate (90 g), protein (30 g), and fat (20 g), blood glucose levels might reach a high of 6–7 mM (about 100–125 mg/100 mL of blood). At the other extreme, blood glucose levels after a few hours of moderate exercise could be reduced to about 3.5 mM (about 60 mg/100 mL of blood). Even after weeks of starvation, blood glucose levels do not drop below the 3 mM range. The hormones insulin and glucagon work in reciprocal ways to maintain blood glucose levels at the concentration of about 4.5 mM (Figure 20.6).

Insulin

Insulin is a polypeptide with 51 amino acids and a molecular weight of 5800 daltons (Figure 4.11). It is synthesized in the beta cells of the pancreas and secreted into the bloodstream in response to increased levels of blood glucose. Some of the biochemical actions displayed after secretion include (see Table 20.2):

- Increase in the permeability of muscle and adipose cells to glucose uptake.
- Stimulation of glycogen breakdown and glycolysis in the liver.
- Inhibition of triacylglycerol hydrolysis in adipocytes.
- Stimulation of triacylglycerol synthesis in the liver and adipose tissue.
- Increase in the activity of pyruvate dehydrogenase complex.

It is often said that insulin acts as a signal for the "fed state" in animals and functions to cap blood glucose levels.

This Belgium stamp shows the protein chain of insulin over a background of test tubes with solutions used to analyze the hormone.

Table 20.2
Biochemical actions of major hormones (see also Figure 20.6).

Hormone	Biochemical Actions Increased or Stimulated	Biochemical Actions Decreased or Inhibited
Insulin	Muscle and adipose tissue permeability to glucose	Gluconeogenesis
	Glycolysis	Triacylglycerol hydrolysis
	Glycogen synthesis	Glycogen breakdown
	Triacylglycerol synthesis	Blood glucose levels
Glucagon	Cyclic AMP in liver and adipose tissue	Glycogen synthesis
	Glycogen breakdown	Glycolysis
	Blood glucose levels	
Epinephrine	Cyclic AMP levels in skeletal muscle	Glycogen synthesis
	Triacylglycerol hydrolysis	
	Glycogen breakdown	
	Blood glucose levels	

Glucagon

Glucagon is a polypeptide with 29 amino acids and a molecular weight of 3500 daltons. It is synthesized and secreted by alpha cells of the islets of Langerhans in the pancreas. In contrast to insulin, glucagon acts as a signal for the "starved state" (see Table 20.2) and functions to increase blood glucose levels. Glucagon is released when blood sugar levels decline from the normal range. Its target organ is the liver, where it stimulates the cells to enhance glycogen breakdown (through glycogen phosphorylase) and release glucose into the bloodstream. The message from glucagon is sent into the cell by the process of signal transduction. The principal effect is to activate adenylate cyclase in order to increase cyclic AMP levels in liver cells:

$$\text{ATP} + \text{H}_2\text{O} \xrightleftharpoons{\text{adenylate cyclase}} \text{cyclic AMP} + \text{PP}_i$$

Epinephrine

Epinephrine (also known as adrenaline) is a catecholamine that functions in the regulation of carbohydrate and fatty acid metabolism. It is released from the adrenal medulla in response to environmental or metabolic stress; hence, it is often called the "fight or flight" hormone. Its intracellular action is elicited through signal transduction. Binding of epinephrine to specific receptors on muscle cells activates adenylate cyclase to produce cyclic AMP. Increased cyclic AMP levels in muscle cells activate lipase to release fatty acids from triacylglycerols and also increase glucose levels by stimulating glycogen phosphorylase and inhibiting glycogen synthase. The ultimate result is to provide elevated levels of glucose and

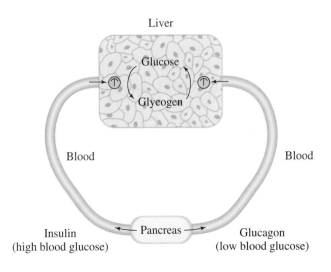

Figure 20.6

Control of blood glucose levels by the coordination of insulin and glucagon.

fatty acids for energy metabolism, especially in time of emergency need. Epinephrine action is enhanced by inhibiting the secretion of insulin and stimulating glucagon secretion. Epinephrine also acts on adipose tissue by increasing cyclic AMP levels, thereby stimulating hormone-sensitive lipase (Section 18.1).

20.4 Metabolic Responses to Stressful Conditions

The details of metabolic regulation, coordination, and specialization have been outlined in the previous sections, assuming that the metabolic conditions are the traditional three meals a day, eight hours of rest, and moderate activity. Humans do not always live under these "normal" or average conditions. They must deal with times of low caloric intake and/or extreme levels of activity requiring large amounts of energy. The flexibility of metabolic coordination provides an array of options available for modifying the metabolic system. Humans and other animals are able to respond quickly to neutralize conditions that place stresses on metabolism. Some examples of the kinds of disturbances that we encounter include fasting or starvation, exercising (sprinting or running a marathon), and diseases (diabetes). Here we explore the consequences of a "normal" lifestyle on metabolism and compare them to the adjustments required for dealing with stresses on metabolism.

Conventional Lifestyle

Before we describe disturbances that modify metabolism, we need to outline the biochemical events of the normal feeding cycle. The following sequence of events is typical after a carbohydrate-containing meal as previously defined:

- Blood glucose levels begin to rise soon after the meal. This is due to metabolism of carbohydrates in the mouth, stomach, and intestines. Monosaccharides, primarily glucose, are absorbed from the intestines into the bloodstream, where they are distributed to the brain, the liver, and skeletal muscle.
- An increase in blood glucose stimulates secretion of insulin and suppression of glucagon. These events promote uptake of glucose into the liver, where glycogen synthesis is enhanced and glycogen breakdown is inhibited.
- Fatty acid synthesis is stimulated in the liver with enhanced storage of triacylglycerols. Triacylglycerols are also distributed by VLDL in the bloodstream for storage in adipose tissue.
- Abundant glucose available in muscle is stored in glycogen.

    A few hours after the meal, blood glucose levels begin to fall and the following events occur:

- Insulin secretion is decreased and glucagon secretion increases.
- In order to maintain a constant level of blood glucose, glycogen is mobilized in the liver (activation of glycogen phosphorylase and inhibition of glycogen synthase).
- Lipase action in adipocytes is activated by removal of insulin inhibition.
- Decreasing levels of insulin slow glycolysis in muscle, liver, and adipocytes by reducing their permeability to glucose.
- Glucose produced in liver (by gluconeogenesis) is transported to brain.

Disturbances That Modify Metabolism

Starvation/Fasting

Now, let us assume that food intake is halted for many days. How does the body's metabolism compensate for this kind of stress? The first line of defense is the use of metabolic fuels stored for reserve. Table 20.3 lists the total amounts of metabolic energy stored as glycogen, triacylglycerols, and protein in an average human. The estimated times that a fuel would last are calculated assuming different levels of activity. Figure 20.7 is a plot of blood plasma concentration changes for glucose, fatty acids, and ketone bodies during prolonged starvation.

- The stores of triacylglycerols in an average human are sufficient to last for 1 to 2 months or more depending on the level of activity. Degradation of fatty acids generates increased quantities of acetyl CoA, thereby triggering production of ketone bodies.

Table 20.3
Storage of metabolic fuel in humans[a]

Tissue/fuel Type	Approximate Fuel Reserve (kJ)[b]	Estimated Time Period Fuel Would Last		
		Starvation[c] (days)	Walking[d] (days)	Marathon[e] (min)
Adipose/triacylglycerols	337,000	34	11	4012
Liver/glycogen	1500	0.15	0.05	18
Muscle/glycogen	6000	0.6	0.2	71
Free glucose (blood, etc.)	320	0.03	0.01	4
Total body protein	150,000	15	5	1786

[a]Adapted from *Biochemistry for the Medical Sciences* by E. Newsholme and A. Leech (1983), John Wiley & Sons, Chichester, UK, and *Human Nutrition and Dietetics* by Davidson et al. (1979), 7th edition, Churchill Livingstone, Edinburgh, UK.
[b]One nutritional Calorie (one kcal) is equal to 4.184 kJ.
[c]Assuming an energy expenditure of about 10,000 kJ per day.
[d]Assuming an energy expenditure of about 30,000 kJ per day.
[e]Assuming an energy expenditure of about 84 kJ per minute.

- The total muscle and liver glycogen (7500 kJ; about 1800 nutritional Calories) would last only 12 to 18 hours depending on the level of activity. Some tissues such as muscle, heart, and liver begin using fatty acids for fuel when glycogen and glucose levels decline. The brain, however, requires glucose for proper function. The liver cannot keep up with the brain's demand for glucose, but it can compensate by producing ketone bodies (see Section 20.2). Ketone body concentrations in the blood can range from 20 micromolar in fed individuals to as high as 7000 micromolar in times of high stress. More brain cells begin to use ketone bodies for energy metabolism; by the third day after the start of fasting, about one-third of the brain's energy comes from ketone body oxidation. After a few weeks, up to two-thirds of the brain's energy comes from metabolism of ketone bodies. The medical term describing the consequences of high blood concentrations of ketone bodies is **ketosis** (see Window on Biochemistry 20.1).

- Muscle protein with a potential for releasing up to 150,000 kJ appears to be a source of abundant energy in times of need. Amino acid oxidation can provide some relief; however, extensive muscle protein degradation can lead to loss of important enzymes and a

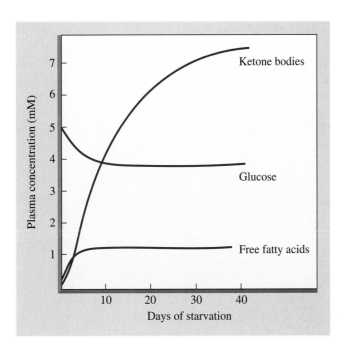

Figure 20.7

Changes in blood plasma concentrations of major metabolites during prolonged starvation.

wasting of muscle tissue. The rate of protein use changes greatly over time. At day three after fasting begins, about 150 g of protein per day are degraded. When muscle cells use ketone body oxidation for energy, protein use may drop to as low as 20 g per day.

Dieting

The human body will also experience the above sequence of events during dieting to lose excess weight. Perhaps the most widely used diet consists of high protein, low carbohydrates, and low fat (sometimes known as the Stillman or Atkins diet). The diet forces the body to proceed through the same biochemical adjustments as fasting/starvation but at a more moderate and controllable pace. Some dieting individuals show signs of mild ketosis after only 10 to 12 days. However, as described in Window on Biochemistry 20.1, dieting is not free of problems and it is best to attempt loss of weight only under the supervision of a physician or nutritionist.

Diabetes

Patients of the disease diabetes mellitus experience many of the same metabolic consequences as those who are fasting or dieting; however, the causes of the metabolic disturbances are quite different. In fasting or starving, there is an insufficient supply of glucose for normal metabolism so other energy pathways become important. In the case of diabetes, glucose is in abundance but cells cannot utilize it either because insulin is deficient (Type I) or because of an impairment in the insulin response mechanism (Type II; see Window on Biochemistry 15.2). With glucose metabolism essentially halted (except in the brain and liver), metabolic cycles adjust by using triacylglycerols for fuel. Pathways that become important are beta oxidation of fatty acids, gluconeogenesis, and ketone body metabolism. Untreated patients with type I diabetes soon develop ketosis. Ketone body concentrations in the blood may reach levels of 5000 to 7000 micromolar. Ketone body levels for normal "fed" individuals are in the range of 10 to 30 micromolar (see Window on Biochemistry 20.1).

Biochemistry of Exercise

Numerous biochemical and physiological studies have been conducted to explore the metabolic changes that occur during the many forms of exercising. Here we will limit our coverage to two extremes of exercise: sprinting and endurance (marathon) running.

Sprinting

A 100-m sprint (about 110 yards), which usually lasts about 10 seconds, has been defined as "exercising to exhaustion." The rate of energy expenditure for average sprinting is estimated at 200 kJ per minute (48 nutritional Calories per minute) or about 3.3 kJ per second (0.8 nutritional Calories per second). The sprinter achieves this high exertion rate for short periods by training to increase the amount of type II muscle fiber. These muscle fibers have a low blood supply, low concentrations of myoglobin, few mitochondria, and very low activities of the enzymes of aerobic metabolism: citric acid cycle enzymes and cytochrome oxidase of the electron transport chain. On the other hand, glycolytic enzymes and phosphocreatine kinase activities are high. The major source of energy in sprinters is anaerobic

Table 20.4
Biochemical characteristics of sprinting and endurance running

Characteristic	Sprint (100 m)	Marathon (42.2 km)
Rate of energy expenditure	200 kJ per minute or 3.3 kJ per second	84 kJ per minute
Total energy expended	33 kJ	12,000 kJ
Fuel	Anaerobic: phosphocreatine, glucose	Aerobic: glucose, fatty acids, glycogen, triacylglycerols
Cause of fatigue	Increase in intracellular proton concentration that lowers pH (lactic acid)	Depletion of muscle glycogen

Compound	Contents in Muscle (micromol per gram of muscle)		
	Resting	After Exercising 15 seconds	After Exercising 30 minutes
Glycogen	88	58	70
Lactate	1.1 (1.1)[b]	30.5 (1.2)	6.5 (1.3)
Phosphocreatine	17.1	3.7	18.8
ATP	4.6	3.4	4.0
H^+ (pH)	7.1	6.3	7.0

[a]The exercise activity consisted of three periods of sprinting for one minute separated by resting for one minute. Adapted from *Biochemistry for the Medical Sciences* by Newsholme and Leech (1983), John Wiley & Sons, Chichester, UK.
[b]The ( ) in the lactate line indicates concentrations in a marathon runner.

Table 20.5
Metabolic changes following a sprint.[a] Cell conditions are those after the activity period followed by 15 seconds and 30 minutes rest.

production of ATP from phosphocreatine and glucose/glycogen utilization (Table 20.4). It has been postulated that phosphocreatine is the primary source of energy for the first few seconds of a sprint, and then glycolysis takes over for the remainder of the run. Anaerobic metabolism leads to the production of high levels of lactate and intracellular protons (decrease in pH). Note the changes of these and other metabolites in Table 20.5. Perhaps most interesting are the small concentration changes for ATP. Recall that the function of metabolism is to maintain constant ATP levels. This is a perfect example to illustrate that this important principle of ATP function holds even under stressful conditions. The important biochemical characteristics for sprinting are summarized in Table 20.4 and Table 20.5. Fatigue in the sprinter is probably caused by the high concentration of intracellular protons.

Endurance Running (Marathon)

The standard marathon run is a distance of 42.2 km or 26 miles plus 385 yards. Trained runners complete the race in just over 2 hours and use energy at the average rate of 84 kJ per minute (20 nutritional Calories per minute). Examination of the muscle of trained endurance runners shows an abundance of type I fibers which are characterized by extensive blood supply, high levels of mitochondria, and high activities of the enzymes for the citric acid cycle and beta oxidation. In contrast to sprinters, the marathon runner's metabolism is based on aerobic processes. The major fuels for endurance runners are glucose and fatty acids from the blood. Also muscle glycogen and triacylglycerols are broken down to glucose and fatty acids, respectively. The changes in muscle lactate concentrations are small in 30 minutes of endurance running indicating little anaerobic metabolism (see Table 20.5). Fatigue in the marathon runner occurs when the carbohydrate stores of glycogen have been depleted. The biochemical characteristics of endurance running are outlined and compared to sprinting in Table 20.4.

20.5 Biochemical Factors in Obesity

Leptin Biochemistry

Obesity, the accumulation of excessive triacylglycerols in adipocytes, is clinically defined as having a body weight more than 20% over an ideal standard weight. In the United States, more than one-third of adults (over 50 million individuals) are obese. In the last 10 years, the number of obese Americans has increased from 25% to 34% of the adult population. Obesity is a major risk factor for diabetes, heart disease, high blood pressure, and stroke, as well as some cancers. In the past, obesity has been considered mainly as a consequence of overeating; however, now we have discovered biochemical control factors that make some individuals more prone to obesity than others. In 1995, researchers at The Rockefeller University identified a gene product in mice that is related to obesity control. The protein, **leptin**, appears to act as a hormone to regulate body weight by monitoring the amount of fat stored in the body of the mouse. The protein is a product of the *ob* gene, which is defective in some mice. Mice carrying two defective alleles of the *ob* gene do not produce leptin and are overweight. The defective mice are constantly hungry and they eat incessantly. When leptin is not made in mice, there is no signal to stop eating. Injection of leptin into these mice lowers feeding rates and also increases energy expenditure; thus the animals lose weight. After two weeks of leptin treatment, overweight mice lose as much as 30% of their weight. Leptin produced in normal mice is a 16-kilodalton protein that is synthesized in adipocytes and transferred via the bloodstream to the hypothalamus, the part of the brain that controls basic bodily functions such as food intake. Leptin in mice appears to function as a sensor to monitor the amount of fat stored in adipocytes. When fat stores are high, blood leptin levels are high. During starvation, leptin levels are low, which prompts eating and storage of fat.

The new knowledge of mouse leptin is very informative; however, it now appears that it cannot be directly transferred to studies on human obesity. In humans, leptin levels increase with the percentage of body fat. Investigators found that the greater the percent of fat and body mass, the higher the level of blood leptin. Researchers also discovered a direct correlation between human leptin levels and the body mass index (BMI; see Window on Biochemistry 20.2 to calculate your BMI). It appears that obesity in humans is not the result of defective leptin or insufficient levels of leptin, as in mice, but is instead caused by a malfunction of the leptin signaling mechanism. For example, some humans may have reduced sensitivity to leptin because they do not have enough receptor sites in the hypothalamus or the receptor sites may be defective. The human leptin receptor has been cloned, and it is

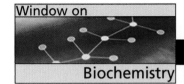

Window on

Biochemistry

Cutting Edge:
Fat blockers

20.2 Body Mass Index—Do I Have Too Much Body Fat?

Body mass index (BMI) is a scientific measurement that is used to determine a person's risk for weight-related health problems. Although the term may be new, the method has been used for many years by physicians and researchers who study obesity. You may be more familiar with the use of "weight-for-height tables" on the walls of hospitals and doctors' offices. There are many versions of these tables, but they are not especially effective because they do not take into account a person's frame size when considering an ideal weight. Although the BMI is not a perfect system, it provides a useful general guideline for judging a healthy weight and an appropriate amount of body fat. The calculation for BMI uses a mathematical formula that accounts for both a person's weight and body area (or "bulk") to indicate the extent of fat storage. The formula uses the "mks" system of units.

The BMI is a person's weight, in kilograms, divided by height, in meters squared:

$$BMI = \text{weight in kilograms}/(\text{height in meters})^2$$

The following conversion factors will be helpful in calculating your BMI:

1 inch = 2.54 cm   1 meter = 100 cm   1 lb = 0.454 kg

For example, the BMI for a person with a weight of 150 lb and a height of 5′ 10″ is calculated as follows:

$$\text{Weight (kg)} = 150 \text{ lb} \times 0.454 \text{ kg/lb} = 68.1 \text{ kg}$$
$$\text{Height (m)} = 70 \text{ in} \times 2.54 \text{ cm/in} \times 1 \text{ m}/100 \text{ cm} = 1.78 \text{ m}$$
$$BMI = 68.1/ (1.78)^2 = 21.5 \text{ kg/m}^2$$

The following guidelines may be used by individuals 34 years of age and younger to determine the extent of body fat:

- If your BMI is between 20 and 25, this indicates a relatively lean body and a healthy amount of body fat.
- If your BMI is between 25 and 30, you may consider your body to be "hefty" or perhaps even overweight and you may want to lower your health risk through dieting and/or exercising after discussions with a physician.
- If your BMI is over 30, it is an indication of moderate to severe obesity and a physician should be consulted.

expected that pharmaceutical companies will seek to design drugs for enhancing leptin response. There is also evidence that in humans other hormones like serotonin and neuropeptide Y may be involved in leptin control. We need to have a greater understanding of the signaling pathways involving leptin and other hormones before a rational approach to antiobesity drugs is possible.

Summary

- The important fuel molecules for most organisms are carbohydrates, fatty acids, and, sometimes, amino acids.

- The following anabolic and catabolic pathways are needed for complete degradation of the fuel molecules: glycolysis, gluconeogenesis, glycogen metabolism, phosphogluconate pathway, beta oxidation, fatty acid synthesis, amino acid oxidation, the citric acid cycle, and oxidative phosphorylation.

- Each type of cell has a metabolic profile that is defined by the molecules it makes, the molecules it uses for energy, and the molecules it exports to other tissue. This metabolic profile

is the result of the differential expression of enzymes which is regulated at the gene level.

- Energy metabolism is controlled by the three hormones insulin, glucagon, and epinephrine.

- Normal metabolism is modified by conditions of fasting, exercise, and disease. When glucose is not utilized, fatty acids are degraded, producing acetyl CoA which is converted to ketone bodies.

- One cause of obesity in humans may be the result of a defective leptin-signaling mechanism.

Study Exercises

Understanding Terms

20.1 Define the following terms in 25 words or less.

a. Catabolism
b. Anabolism
c. Glycolysis
d. Beta oxidation
e. Phosphocreatine kinase
f. Glucokinase
g. Cori cycle
h. Glucose–alanine cycle

i. Ketone bodies
j. Glucagon
k. Insulin
l. Epinephrine
m. Glycogen phosphorylase
n. Ketosis
o. Leptin
p. Acidosis
q. BMI
r. Lactate fermentation

Reviewing Concepts

20.2 Derive the net reaction for each of the metabolic processes involving the catabolism of carbohydrates. Include ATP, NAD, H_2O, P_i, substrates, and products.

a. Glucose $\rightleftharpoons$ 2 pyruvate
b. Glucose 1-phosphate $\rightleftharpoons$ 2 pyruvate
c. Glyceraldehyde 3-phosphate $\rightleftharpoons$ pyruvate
d. Glucose $\rightleftharpoons$ 2 lactate

20.3 Derive the net reaction for each of the metabolic processes involving the anabolism of carbohydrates.

a. 2 Pyruvate $\rightleftharpoons$ glucose
b. 2 PEP $\rightleftharpoons$ glucose
c. 2 Lactate $\rightleftharpoons$ glucose

20.4 Derive the net reaction for the complete beta oxidation of palmitic acid beginning with its activation by the CoA ester.

20.5 Write the structure of the product derived from the transamination of each amino acid below and describe where the product enters mainstream metabolism.

a. Aspartate
b. Alanine
c. Valine

20.6 Draw the structure of the amino acid that would be produced if each of the following was a reactant in a transamination reaction with glutamate.

a. Oxaloacetate
b. Pyruvate

20.7 Review the structures of the three ketone bodies and answer the following questions.

a. Which of the ketone bodies have ketone functional groups?
b. Which of the ketone bodies have carboxylic acid functional groups?
c. Which of the ketone bodies have hydroxyl functional groups?

20.8 Write the name of a negative modulator for each of the following enzymes.

a. Glycogen phosphorylase
b. Glycogen synthase
c. Hexokinase
d. Phosphofructokinase

20.9 Write the name of a positive modulator for each of the following enzymes.

a. Pyruvate carboxylase
b. Glycogen synthase
c. Adenylate cyclase
d. Glycogen phosphorylase
e. Hormone-sensitive lipase

20.10 Compare and contrast glycolysis with gluconeogenesis in terms of each of the following characteristics.

a. Starting substrate(s) for each
b. Yield of ATP
c. Final carbon product
d. Yield of NADH

20.11 The list below includes metabolic reactions and processes for the catabolism of fatty acids. Arrange the items in the order in which they occur in an organism.

a. Thiolytic cleavage
b. Formation of acyl carnitine
c. FAD-linked dehydrogenation
d. Fatty acid in cytoplasm
e. Linkage of fatty acid to CoASH
f. Hydration of double bond
g. NAD^+-linked dehydrogenation
h. Transport of fatty acid to mitochondria

HINT: Begin with d

20.12 Calculate the BMI for persons with the following characteristics. Would these individuals be considered about right or excessive in terms of body fat?

a. Height = 5′ 2″; weight = 120 lb
b. Height = 6′ 4″; weight = 240 lb

20.13 Convert the following energy values now in kJ to nutritional Calories.

a. 13,000 kJ
b. 20,000 kJ

20.14 Which of the following statements about the citric acid cycle are true?

a. It generates three NADH, one NADPH, and one GTP for each acetyl CoA.
b. It is the only metabolic process in aerobic organisms that produces ATP.
c. Its intermediates may be removed to make the amino acids glutamate and aspartate.
d. It is linked to the electron transport chain through succinate dehydrogenase.

20.15 Describe two biochemical effects of each of the following hormones.

a. Insulin
b. Glucagon
c. Epinephrine

20.16 Which of the following statements about mitochondrial respiration are true?

a. The electron carriers involved in electron transport are present in the cytoplasm of the cell.
b. The final electron acceptor is oxygen.
c. The electron carriers are arranged in order of increasing electron affinity.
d. During electron transport, protons are pumped from the inner mitochondrial matrix to the other side of the membrane.

20.17 Compare the biochemical differences between type I and type II muscle fibers. What is the advantage to sprinters to have a higher proportion of type II muscle fibers compared to endurance runners?

Solving Problems

20.18 How is pyruvate used to replenish intermediates in the citric acid cycle? Select the correct answer.

a. Pyruvate ⇌ lactate ⇌ succinate
b. Pyruvate ⇌ oxaloacetate
c. Pyruvate ⇌ PEP ⇌ oxaloacetate
d. Pyruvate ⇌ malate

20.19 Why is it an advantage to the cell for the mainstream catabolic reactions to converge to the common intermediate, acetyl CoA?

20.20 Lactate fermentation is the major source of energy for skeletal muscle during high-intensity exercises. Outline the pathway for the synthesis of glucose 6-phosphate from lactate that is left after exercise.

20.21 Explain in biochemical terms how a diet low in carbohydrates and fat may help reduce body fat.

20.22 The following is the sequence of reactions of gluconeogenesis from pyruvate to phosphoenolpyruvate (PEP):

$$\text{pyruvate} \underset{A}{\rightleftharpoons} \text{oxaloacetate} \underset{B}{\rightleftharpoons} \text{malate} \underset{C}{\rightleftharpoons} \text{oxaloacetate} \underset{D}{\rightleftharpoons} \text{PEP}$$

Match the capital letters indicating the reactions of the gluconeogenic pathway with the following statements. There may be more than one response to a statement.

a. Occurs in the mitochondria _____
b. Occurs in the cytoplasm _____
c. Produces CO_2 _____
d. Consumes CO_2 _____
e. Requires ATP _____
f. Requires GTP _____
g. Is regulated by acetyl CoA _____
h. Requires a biotin cofactor _____
i. Is also a reaction of the CAC _____
j. Is an anaplerotic reaction _____

20.23 Which of the following statements correctly describe what happens when acetyl CoA levels from fatty acid degradation are elevated?

a. Pyruvate carboxylase is activated.

b. Phosphoenolpyruvate carboxykinase is activated.

c. Phosphofructokinase is activated.

d. If ATP is needed, oxaloacetate is diverted to gluconeogenesis.

e. If ATP is not needed, oxaloacetate is diverted to gluconeogenesis.

f. Some acetyl CoA is directed to ketone bodies and fatty acid synthesis.

20.24 When glucose concentrations are low, brain cells may use ketone bodies for energy. Outline the general steps for energy production from degradation of acetoacetate and hydroxybutyrate.

20.25 Identify the location of each enzyme listed below. Choose from cytoplasm and mitochondria.

a. Glucose 6-phosphatase

b. Lactate dehydrogenase

c. Glycogen synthase

d. Pyruvate carboxylase

e. Malate dehydrogenase

20.26 Determine the net reaction for the Cori cycle.

20.27 Experienced marathon runners usually prepare for a race by "carbohydrate loading." The most effective carbohydrate for runners is starch not glucose or sucrose. Why?

20.28 Diabetics (Type I) who are not under proper treatment often have the smell of acetone on their breath. Why?

20.29 Suppose that a person begins a fast. Estimate how many days the person could live without additional food. Assume that Table 20.3 is representative of the fuel store and that the person requires 2400 nutritional Calories per day.

20.30 Which of the enzymes below are not needed for production of ketone bodies?

a. Acetoacetate decarboxylase

b. HMG-CoA lyase

c. HMG-CoA synthase

d. Acetone carboxylase

20.31 Upon your first study of the Cori cycle you might have thought that it functions like a substrate cycle with no useful purpose, just wasteful loss of ATP (see pp. 418–419). Explain why this is not the case.

20.32 Phosphocreatine is an energy-rich molecule found in skeletal muscle. The energy associated with the molecule may be released by transfer of the phosphoryl group to ADP catalyzed by phosphocreatine kinase. Calculate the $\Delta G°'$ for the reaction.

$$\text{phosphocreatine} + \text{ADP} \rightleftharpoons \text{creatine} + \text{ATP}$$

HINT: See Table 14.6.

Writing Biochemistry

20.33 Throughout this book, the important roles of ATP in energy metabolism have been emphasized. For example, the following statements have been made:

- "ATP serves as the universal carrier of biochemical energy" (page 369).
- "From an energy standpoint, the recycling of ATP is the central theme of metabolism" (page 369).
- "ATP should be made as rapidly as it is used" (page 414).
- The importance of maintaining a relatively constant concentration of ATP is shown in Table 20.5

Use these references to ATP and the general knowledge you have learned this term to write a 200- to 300-word essay on the role of ATP in energy metabolism.

20.34 Search for an article related to the topic "biochemistry of exercise." You may use the Internet, the World Wide Web, or a journal such as *Scientific American*, *Science*, or *Nature*. Write a 200-word summary of the article emphasizing biochemical concepts.

20.35 Section 20.4 covers the topic, "Metabolic responses to stressful conditions." It explores the metabolic adaptability of humans in neutralizing disturbances we encounter. Your instructor has just announced a quiz on this section. Write a 100-word outline or summary of the section to use as a study guide to prepare for the quiz.

 ## Applying Biochemistry

20.36 Based on what you learned in this chapter about exercising, what would be your last meal before participating in each of the following events? Explain the biochemical reasoning behind your answers.

a. A 100-m sprint

b. A 30-km endurance run

20.37 Contrast the major metabolic differences between skeletal muscle and heart muscle.

20.38 Determine the net reaction for the glucose–alanine cycle.

Further Reading

Banfalvi, G., 1999. Estimation of the energy content of metabolites. *Biochem. Educ.* 27:79–82.

Bronk, R., 1999. *Human Metabolism.* Essex: Addison Wesley Longman.

Campfield, L., Smith, F., and Burn, P., 1998. Strategies and potential molecular targets for obesity treatments. *Science* 280:1383–1387.

D'Andrea, G., 2000. Classifying amino acids as gluco(glyco)genic, ketogenic, or both. *Biochem. Educ.* 28:27–28.

Darvey, I., 1999. What factors are responsible for the greater yield of ATP per carbon atom when fatty acids are completely oxidized to carbon dioxide and water compared with glucose? *Biochem. Educ.* 27:209–210.

Elliott, K., 1999. Teaching metabolism for the future. *The Biochemist* 21(1):9–39.

Fell, D., 1997. *Understanding the Control of Metabolism.* London: Portland Press.

Filippusson, H., 1998. Exercise biochemistry as an approach to the study of intermediary metabolism. *Biochem. Educ.* 26:119–120.

Frayn, K., 1996. *Metabolic Regulation: A Human Perspective.* London: Portland Press.

Friedman, J., and Halaas, J., 1998. Leptin and the regulation of body weight in mammals. *Nature* 395:763–770.

Holliday, R., 2001. Aging and the biochemistry of life. *Trends Biochem. Sci.* 26:68–71.

Manchester, K., 2000. Arthur Harden: An unwitting pioneer of metabolic control analysis. *Trends Biochem. Sci.* 25:89–92.

Manchester, K., 2000. Optimization of energy coupling. *Biochem. Educ.* 28:18–19.

Mulimani, V., 1995. Teaching metabolic pathways to understand its regulation. *Biochem. Educ.* 23:32–34.

Newsholme, E., and Leech, A., 1983. *Biochemistry for the Medical Sciences.* Chichester, UK: John Wiley & Sons.

Rolland, F. et al., 2001. Glucose-sensing mechanisms in eukaryotic cells. *Trends Biochem. Sci.* 26:310–317.

Wang, J., 1998. The insulin connection: Dorothy Hodgkin and the Beijing Insulin Group. *Trends Biochem. Sci.* 23:497–500.

Watford, M., and Goodridge, A., 2000. Regulation of fuel utilization. In *Biochemical and Physiological Aspects of Human Nutrition.* Philadelphia: Saunders.

Weindruch, R., 1996. Caloric restriction and aging. *Sci. Am.* 274(1):46–52.

A

abzymes protein antibodies that catalyze chemical reactions like enzymes; catalytic antibodies

acetyl CoA an activated form of acetate formed by a thioester linkage between acetyl and coenzyme A (CoASH)

acidosis a medical condition characterized by an increase in $[H^+]$ of blood

acquired immune deficiency syndrome (AIDS) a medical condition associated with infection of the immune system cells by a retrovirus, HIV

activation energy the amount of energy required to convert molecules in a reacting system from the ground state to the transition state

activators regulatory proteins that bind to DNA and enhance the rate of gene transcription

active site the specific region on an enzyme where the substrate molecule binds

active transport movement of a biomolecule through a membrane with expenditure of energy

acyl carrier protein (ACP-SH) polypeptide containing the vitamin pantothenic acid, which activates acyl groups for fatty acid synthesis

adenosine triphosphate (ATP) a molecule that serves as a universal energy carrier and energy transfer agent in biochemical processes

S-**adenosylmethionine (SAM)** a biomolecule that serves as a carrier and transfer agent for the methyl group

adipocytes animal cells that are specialized for fat storage

aerobic the presence and use of oxygen

alcaptonuria a medical condition caused by a deficiency of the enzyme homogentisate oxidase

aldol condensation reactions that take place between carbanions and aldehydes or ketones leading to formation of a carbon–carbon single bond; catalyzed by lyases

aldose a carbohydrate that contains an aldehyde functional group

alkalosis a medical condition characterized by a decrease in $[H^+]$ of blood

allosteric interactions a change in conformational structure at one location of a multisubunit protein that causes a conformational change at another location on the protein

Ames test a laboratory procedure using bacterial cultures that determines whether or not a substance is a mutagen

amino acid any of a class of 20 molecules that are building blocks for proteins

aminoacyl-tRNA synthetases enzymes that catalyze the linkage of an amino acid to a specific tRNA to be used for protein synthesis

amino sugars carbohydrates that have one or more hydroxyl groups replaced by an amino group

amino terminus the amino acid end of a polypeptide chain that has an unreacted or derivatized α-amino group

aminotransferase a class of enzymes that catalyzes transamination reactions

amphibolic a metabolic pathway or metabolite that can function in catabolism and anabolism

amphiphilic a class of molecules that has a hydrophilic region and a hydrophobic region

anabolism the synthetic path of metabolism, which is characterized by the construction of larger biomolecules from simple precursors; usually requires input of energy

anaerobic refers to the absence of oxygen

anaplerotic reactions reactions that replenish metabolites that may be low in concentration (e.g., the reaction catalyzed by pyruvate carboxylase)

annealing reassociation of two free polynucleotide strands into the DNA double helix

anomers stereoisomers of cyclized carbohydrates that differ in stereochemistry only at the hemiacetal or hemiketal carbon center

antibody a protein in blood that selectively binds and neutralizes foreign substances

anticodon the sequence of three bases on tRNA that combines with the complementary triplet codon on mRNA

antigen a foreign substance that induces an organism to produce antibodies

antioxidant a chemical agent, such as vitamin C or E, that is easily oxidized, thereby preventing the oxidation of other substances

apoenzyme an enzyme in its polypeptide form without any necessary prosthetic groups or cofactors

archaebacteria a group of prokaryotes that are biochemically distinct from true bacteria and found mainly in extreme living conditions

atherosclerosis a medical condition characterized by deposition of cholesterol in arteries

ATP energy cycle the process by which ATP is generated by catabolism and used for anabolism and other energy-requiring processes

ATP synthase complex the enzyme complex in the inner mitochondrial membrane that couples electron transport to the synthesis of ATP from adenosine diphosphate (ADP)

autooxidation the reaction of O_2 with other molecules

autoradiography a laboratory procedure that uses X-ray film to detect radioactively labeled molecules; especially useful for visualization and characterization of DNA after gel electrophoresis

autotrophs organisms that can synthesize organic chemicals for structure and energy from CO_2 and other inorganic precursors

B

base pair two nitrogen bases (adenine–thymine or guanine–cytosine) held together by hydrogen bonds in DNA

bend an element of protein secondary structure; a reverse turn in protein conformation; sometimes called a loop

beriberi the medical condition in humans caused by a dietary deficiency of thiamine

bilayers two monolayers of polar lipids combined to form a membrane structure

bile salts oxidized derivatives of cholesterol that are amphiphilic and assist in solubilization of dietary lipids; they are made in the liver, stored in gallbladder, and secreted into the intestines

bilirubin a linear tetrapyrrole derivative produced by the catabolism of heme

bioenergetics the study of energy flow and transformations in living cells and organisms; ATP synthesis driven by oxidation–reduction and related processes

biogenic amines important biomolecules that contain the nitrogen atom in the form of an amino group

bioinformatics the application of computers to organizing the data of nucleic acid sequences and studying the relationships between protein sequence and structure

bioluminescence emission of light from a living organism

biotechnology the application of organisms, biological cells, cell components, and biological processes to practical operations and procedures

blotting transfer of molecules (especially recombinant DNA) from gels to paper for genetic analysis

Bohr effect binding of CO_2 and H^+ to hemoglobin decreases the affinity of O_2 binding to the hemoglobin

bond-forming reactions requiring energy from ATP reactions in which new bonds are formed using energy from ATP cleavage; catalyzed by ligases

buffer a solution comprised of a conjugate acid–base pair that resists changes in pH

C

Calvin cycle the reactions involved in the fixation of CO_2 into simple organic molecules to synthesize carbohydrates

carbohydrates a group of naturally occurring compounds that have aldehyde or ketone functional groups and multiple hydroxyl groups

carboxyl terminus the amino acid end of a polypeptide chain that has an unreacted or derivatized α-carboxyl group

carcinogen an agent that induces a mutation by causing damage to DNA that results in cancerous cells

carnitine a small organic molecule that assists in the transport of fatty acids across the mitochondrial membranes

catabolism the degradative path of metabolism in which complex organic molecules are oxidized to the simpler molecules CO_2, H_2O, and NH_3, usually resulting in the release of energy

catalytic RNA a form of RNA that catalyzes bioreactions, especially splicing of mRNA; ribozyme

catalytic site the location on an enzyme where substrates and cofactors bind; the active site

cell extracts prepared by breaking open cells and allowing the internal components to be suspended into a solution for further study; also called a crude extract or cell homogenate

cell membranes bilayer complexes of proteins and lipids that form the outer boundary of cells

cellulose polysaccharide composed of glucose; has a structural role in plants

central dogma of protein folding the primary structure of a protein determines the secondary and tertiary structures

centrifugation subjecting a biological sample to extreme gravitational forces by spinning at high rates, causing sedimentation of organelles and biomolecules

chain-termination sequencing method a laboratory method that uses dideoxyribonucleoside triphosphates to terminate synthesis; used for DNA sequencing

chaperones proteins that act as catalysts to guide the folding of other proteins

charge repulsion unfavorable interactions between like charges on a single molecular species; used to explain the energy transfer role for ATP

chemiosmotic coupling mechanism the synthesis of ATP is coupled to electron transport by formation and collapse of a high-energy proton gradient across the inner mitochondrial membrane

chiral molecules molecules that can exist in nonsuperimposable, mirror-image forms

chlorophylls green pigments composed of Mg^{2+} bound in a porphyrin polycyclic structure; absorb light for photosynthesis

chloroplasts double-membraned organelles in plants that are the sites of photosynthesis

cholesterol a lipid of the steroid class that is a component of membranes and a precursor for important biomolecules, especially hormones

chromatin complexes of DNA and proteins found in the eukaryotic nucleus

chromatography a laboratory technique that can separate molecules on the basis of charge, shape, size, or biological binding affinity

chromosomes packages of the functional units of DNA; in prokaryotes, the whole genome is on one circular chromosome; in eukaryotes, the genome consists of several chromosomes packaged with proteins in the nucleus

chylomicrons blood lipoproteins formed by the association of triacylglycerols with proteins to assist in the digestion and transport of dietary fats; consist of about 98% lipid and 2% protein

citric acid cycle a central metabolic pathway that oxidizes acetyl CoA to CO_2 with production of ATP, NADH + H^+, and $FADH_2$

cloning of DNA the introduction of a segment of DNA from one species into the DNA of another species, resulting in production of many copies of the hybrid DNA by replication

codon a sequence of three nucleotide bases on mRNA that interacts with the anticodon on tRNA; specifies the incorporation of a specific amino acid into a polypeptide

coenzyme a smaller organic or organometallic molecule that assists an enzyme in catalytic action; usually derived from a vitamin

coenzyme A (CoASH) a biomolecule containing pantothenic acid that serves as a carrier of acyl groups

coenzyme B_{12} a small molecule that is associated with enzymes catalyzing carbon skeleton rearrangements; also called deoxyadenosylcobalamin; derived from the vitamin B_{12}

cofactor a nonprotein, usually organic molecule that assists an enzyme in its catalytic action

competitive inhibitor a compound that blocks the activity of an enzyme by binding to its active site and competing with substrate binding

complementary base pairs combining by specific hydrogen bonding of adenine with thymine (or uracil) and guanine with cytosine in nucleic acids

condensation reaction a reaction process that covalently combines two reagents and results in the loss of a small molecule, usually water

conjugate acid the species HA produced by the reaction of H^+ with a base A^-

conjugate acid–base pair two species that differ in molecular structure by a proton (e.g., HA and A^-)

conjugate base the species A^- produced by the reaction of acid HA with OH^-

constitutive genes those that code for certain protein products that must be present at all times for general cell maintenance and metabolism

cooperative binding binding of a ligand molecule to one subunit of an oligomeric protein that enhances the binding of a ligand molecule to another subunit

cyclic AMP a form of adenosine monophosphate in which the 3′ and 5′ hydroxyl groups are linked in a cyclic phosphodiester; functions as a second messenger

cyclic photophosphorylation in photosynthesis, formation of ATP is linked to a cyclic flow of electrons within photosystem II; no $NADPH + H^+$ or O_2 is produced; opposite of noncyclic photophosphorylation

cytochromes heme-containing proteins present in the mitochondrial electron-transport chain that undergo redox reactions

cytochromes P450 family of heme enzymes that catalyze the hydroxylation of normally unreactive substrates

cytoplasm the gellike fluid medium located inside a cell but outside the organelles

cytoskeleton the three-dimensional fibrous matrix extended throughout the inside of the cell; the fibers are composed of protein

D

dalton (D) a mass unit used to define the size of a biomolecule; one dalton is equivalent to one atomic mass unit

dehydrogenases enzymes that catalyze oxidation–reduction reactions; also called oxidoreductases

denaturation the complete loss of secondary, tertiary, and quaternary structure in a protein; loss of secondary structure in DNA

de novo synthesis of nucleotide bases synthesis of purine and pyrimidine bases for nucleotides beginning with simple precursors

deoxyribonucleases enzymes that catalyze the hydrolytic cleavage of phosphoester bonds in DNA

diastereoisomers molecules that are stereoisomers but not enantiomers

disaccharide two monosaccharide molecules linked by a glycosidic bond

disulfide bond a covalent bond formed between the sulfhydryl groups of two cysteine residues

DNA fingerprinting a laboratory technique using restriction enzymes and electrophoresis to compare differences and similarities in the DNA of individuals

DNA ligase the enzyme that catalyzes the formation of the final phosphoester bond in DNA replication

DNA melting the thermal denaturation of double-stranded DNA by breaking complementary hydrogen bonding; unwinding of the two strands

DNA polymerases enzymes that catalyze the synthesis of DNA

domain a part of a polypeptide chain that folds into a compact unit and remains distinct even within the tertiary structure of the protein; a discrete portion of a protein with its own distinct function

double helix the structural arrangement of two polynucleotide chains to form double-stranded DNA

double reciprocal plot a graph of the reciprocal of the velocity of an enzyme-catalyzed reaction versus the reciprocal of the substrate concentration; *see also* Lineweaver–Burk equation

E

Edman degradation a laboratory method used to determine protein amino acid sequence

effectors small molecules that regulate the activity of allosteric enzymes; may stimulate or inhibit; also called modulators

eicosanoids a class of lipids characterized by their localized, hormonelike activities, very low cellular concentrations, and derivation from arachidonic acid; the three subclasses are prostaglandins, thromboxanes, and leukotrienes

electron-transport chain a series of molecular carriers that transfer electrons from a donor (usually NADH or $FADH_2$) to an acceptor (usually O_2)

electrophoresis a laboratory technique used to separate molecules by subjecting them to an electric field in a gel matrix

elongation factors proteins that guide the elongation (polymerization) phase of protein synthesis

enantiomers compounds that are mirror images of each other and not superimposable

endonuclease an enzyme that catalyzes the hydrolysis of internal phosphoester bonds in RNA or DNA

endoplasmic reticulum a highly folded, membrane-enclosed cellular compartment with many biological functions

energy coupling a system or reaction sequence in which energy from an energy-releasing process is used to drive an energy-requiring process

energy transduction the conversion of energy in one form to another form—for example, the conversion of energy from oxidation–reduction reactions to form a reactive chemical bond, as in ATP

enzyme a biomolecule, usually a protein, that acts as a biological catalyst to speed the rate of a biochemical reaction

enzyme specificity the ability of an enzyme to discriminate among possible substrate molecules

epimers stereoisomers that differ in configuration only at one chiral center

equilibrium constant (K_{eq}) the ratio of the two rate constants for a reversible reaction; the equilibrium ratio of the product concentrations to the reactant concentrations

ES complex molecular species consisting of a substrate (S) bound to the active site of an enzyme (E)

essential amino acids amino acids that cannot be synthesized by an organism and must be supplied in the diet for proper growth and development

essential fatty acids fatty acids that cannot be synthesized by an organism and must be supplied in the diet for proper growth and development

eukaryotes a class of organisms, including plants and animals, whose cells have a distinct membrane-enclosed nucleus and well-defined internal compartments (organelles)

exon the region in DNA that codes for transcription; as opposed to an intron

exonuclease an enzyme that catalyzes hydrolysis of nucleotide (phosphoester) bonds from free ends

F

facilitated diffusion protein-mediated transport of a biomolecule through a membrane

familial hypercholesterolemia a genetic medical condition caused by a lack of functional low density lipoprotein receptors; results in high plasma cholesterol levels

fatty acids biomolecules containing a carboxyl functional group connected to an aliphatic chain that is usually unbranched

fatty acid synthase complex the enzyme system that catalyzes synthesis of fatty acids from acetyl CoA and malonyl CoA

fatty acid synthesis the formation of long-chain fatty acids from acetyl CoA; catalyzed by fatty acid synthase complex

feedback inhibition a mode of metabolic regulation in which the final product of a metabolic pathway inhibits an enzyme in that pathway (usually the first)

fermentation the metabolism of carbohydrates to extract energy without using O_2 as an electron acceptor; ethanol or lactate fermentation

fibrous proteins water-insoluble proteins that usually have a structural function

Fischer projections figures used to represent three-dimensional stereoisomers in a two-dimensional format

flavin adenine dinucleotide (FAD) a cofactor associated with some dehydrogenase enzymes; derived from the vitamin riboflavin

fluid-mosaic model describes the arrangement of proteins and lipid bilayer in cell membranes

flux the flow of intermediates in metabolism; the rate at which substrates enter and exit a pathway

free energy change (ΔG) a measure of the energy from a reaction available to do useful work

free radical a molecule that has one or more unpaired electrons

G

G protein a membrane-bound protein that binds guanine nucleotides and thus signals adenylate cyclase to make cyclic AMP

galactosemia medical condition caused by a malfunction of galactose metabolism leading to toxic levels of galactose and metabolites

gel electrophoresis a laboratory technique that separates molecules according to size and charge by subjecting them to an electric field in a matrix

gene region of DNA that codes for a specific polypeptide; the functional unit of heredity

genetic code the rules by which the sequence of bases in mRNA specify the amino acid sequence of a polypeptide; read in triplets (codons)

genetic engineering see recombinant DNA

gene knockouts animals that have been genetically modified to either contain a defective gene or lack a gene

gene therapy medical procedures used to correct a genetic defect by inserting the normal gene into the cells of an organism

genome the total genetic information possessed by an organism in its chromosomal DNA

genomic library a collection of bacterial or yeast cells that have been transformed using recombinant vehicles with DNA inserts from a single species

globular proteins highly folded, water-soluble proteins that usually play a dynamic role in transport, immune protection, or catalysis

glucogenic amino acids amino acids that are degraded and converted to metabolites that can be used for the synthesis of glucose

gluconeogenesis the metabolic pathway for the synthesis of glucose from noncarbohydrate precursors

glycerol 3-phosphate shuttle a process for transferring electrons in cytoplasmic NADH to the mitochondrial matrix

glycerophospholipids polar lipid molecules that have the foundation molecule phosphatidic acid and fatty acids esterified to glycerol hydroxyl groups

glycogen the polysaccharide used by animals for the storage of glucose for energy metabolism

glycolysis the anaerobic metabolic pathway for conversion of glucose and other carbohydrates to pyruvate; generates ATP and $NADH + H^+$

glycoproteins proteins that contain covalently bonded carbohydrate units

glycoside the product (cyclic acetal) formed by the elimination of water from a hydroxyl group of a hemiacetal or hemiketal and another alcohol; type of covalent bond linking monosaccharides to polysaccharides

glyoxysomes single-membraned plant vesicles that contain oxidative enzymes

glyoxylate cycle modified version of the citric acid cycle found in plants and microorganisms that allows them to use acetate for carbohydrate synthesis

gout a disease characterized by elevated levels of serum uric acid and deposition of the acid in joints

group-transfer reactions the transfer of a chemical functional group from one molecule to another; catalyzed by transferases

H

Hatch–Slack pathway (C_4 pathway) an alternate pathway for CO_2 fixation found in some plants; CO_2 is incorporated into phosphoenolpyruvate rather than CO_2 into ribulose 1,5-bisphosphate

Haworth projection form a molecular representation for the cyclic form of sugars

helicases enzymes that catalyze the unwinding of double-stranded DNA

α-helix a rodlike structure formed by a tightly coiled polypeptide backbone; held together by intramolecular hydrogen bonding

helix–turn–helix motif a common DNA binding domain found in many prokaryotic regulatory proteins

heme a porphyrin ring with a bound iron atom; cofactor in hemoglobin and cytochromes

Henderson–Hasselbalch equation the mathematical relationship between the pK of an acid and the pH of a solution of that acid and conjugate base

heptoses carbohydrates with seven carbon atoms

heteropolymers macromolecules that are composed of chemically different monomeric units

heterotrophs organisms that cannot synthesize all needed organic compounds, but instead obtain complex carbon- and nitrogen-containing compounds in their diets

hexoses carbohydrates with six carbon atoms

high density lipoproteins blood lipoproteins formed by the association of phospholipids, cholesterol, and cholesteryl esters with protein to assist in the transport of water-insoluble lipids; consist of about 55% protein and 45% lipid

histones family of small proteins that contain large numbers of basic amino acid residues and bind to DNA

HMG-CoA reductase the enzyme that catalyzes the formation of mevalonate; the rate-limiting step in cholesterol synthesis

holoenzyme an enzyme in its complete form, including polypeptide(s) and cofactor

homopolymers macromolecules that are composed of chemically identical monomeric units

Human Genome Project a federally funded program with the goal to sequence the complete human genome; the first draft was announced in 2000

hybridization probe a DNA or RNA molecule that is complementary to a region in DNA; used to detect specific genes in DNA fingerprinting

hydrogen bonding a noncovalent attractive force between a hydrogen atom covalently bonded to an electronegative atom (O, S, N) and the lone pair of electrons on a second electronegative atom

hydrolases enzymes that use water to cleave bonds

hydrolysis reactions water is used to split a single molecule into two distinct molecules; catalyzed by hydrolases

hydrophilic characterizes a substance that dissolves in water

hydrophobic characterizes a substance that is nonpolar and insoluble in water

hydrophobic interactions the associations of nonpolar molecules or regions of nonpolar molecules in water that result in stability of the molecules

hyperchromic effect the heating of DNA solutions resulting in a significant increase in the absorption of ultraviolet light by the solution

I

induced-fit model used to describe conformational changes in an enzyme caused by the binding of a substrate molecule

inducible genes those genes that are activated by signal molecules to increase the level of a specific mRNA and protein

initial velocity the rate of an enzyme-catalyzed reaction immediately after mixing enzyme and substrate

initiation factors protein molecules that help start the protein synthesis process

integral proteins proteins that are embedded in a membrane

intercalating agents substances that insert between stacked base pairs in DNA, often leading to mutagenesis

intermediary metabolism the combined chemical reactions in a cell that carry out degradation and synthesis and form products (intermediates) at each reaction step

intron noncoding region in DNA; as opposed to an exon

ion-channels pores in integral membrane proteins that can be opened and closed like gates

ionic bonds formed by attractive forces between fully charged atoms or groups

iron–sulfur clusters protein prosthetic groups composed of iron atoms complexed to sulfur in the inorganic form or as cysteine; present in electron transport chains

irreversible inhibitor compounds that form covalent bonds or very strong noncovalent bonds with an enzyme, thus permanently damaging the enzyme

isoenzymes multiple forms of an enzyme that have similar but not identical amino acid sequences and reaction characteristics

isomerization reactions biochemical reactions that involve interconversion of molecules that are isomeric; catalyzed by isomerases

isoprene (2-methyl-1,3-butadiene) a C_5 molecule used as a building block for larger biomolecules, especially terpenes and steroids

J

jaundice a medical condition resulting in yellowing of the skin and the whites of eyes because of the accumulation of bilirubin

K

ketogenic amino acids amino acids that are converted to ketone bodies when degraded

ketone bodies the substances acetoacetate, β-hydroxybutyrate, and acetone; produced from excess fatty acid degradation

ketose a carbohydrate that contains a ketone functional group

ketosis a medical condition characterized by high concentrations of ketone bodies in blood

kilobase unit of length for DNA fragments; equal to 1000 nucleotides

kinases enzymes that catalyze the transfer of a phosphoryl group from ATP to an acceptor substrate; a subclass of the transferases

L

lactone a cyclic ester

lactose intolerance a medical condition caused by a deficiency in the enzyme lactase

lectins proteins that bind reversibly to specific carbohydrates; found primarily in plants

leghemoglobin a plant protein similar to hemoglobin; binds O_2 and protects the nitrogenase from O_2; produced by the host plant in a symbiotic relationship

Lesch–Nyhan syndrome a medical disorder caused by a deficiency in a purine salvage enzyme

leucine zipper motif a common binding domain that allows regulatory proteins to bind to each other (dimerization) or to other proteins

leukotrienes a subclass of eicosanoids; linear chains synthesized from arachidonate

ligases a class of enzymes that catalyzes bond-forming reactions using energy from ATP

Lineweaver–Burk equation obtained by taking the reciprocal of the Michaelis–Menten equation; a mathematical relationship between the rate of an enzyme-catalyzed reaction and the concentration of enzyme and substrate; graphed as the double-reciprocal plot.

lipases enzymes that catalyze the hydrolysis of lipids, especially triacylglycerols

lipids a class of biological compounds characterized by high solubility in organic solvents but low solubility in water

lipoproteins biomolecules composed of lipids and polypeptides that are used to transport insoluble molecules in blood; the four types are chylomicrons, very low density, low density, and high density

lock and key model used to describe the binding interaction between an enzyme and its substrate molecule

low density lipoproteins blood lipoproteins formed by the association of phospholipids, cholesterol, and cholesteryl esters with proteins to assist in the transport of water-insoluble lipids; consist of about 25% protein and 75% lipid

lyases enzymes that catalyze nonhydrolytic cleavage reactions

lysosomes cell organelles that contain hydrolytic enzymes in animals

M

malate–aspartate shuttle a process for transferring electrons in cytoplasmic NADH to the mitochondrial matrix

MAO inhibitors chemical agents that block the action of monoamine oxidases; used to treat depression

maple syrup urine disease a medical condition caused by a deficiency of the enzyme branched-chain α-keto acid dehydrogenase complex

Maxam–Gilbert sequencing procedure a laboratory technique that uses chemical cleavage and gel electrophoresis to sequence nucleic acid bases

maximum velocity (V_{max}) the initial velocity attained in an enzyme-catalyzed reaction when substrate concentration is extremely high

messenger RNA (mRNA) the product of DNA transcription; serves as a template for protein synthesis

metabolic pathway a sequence of biochemical reactions that has a specific purpose

metabolism the study of the biochemical reactions in an organism, including their coordination, regulation, and energy needs

metabolites compounds formed as intermediates during reactions in metabolic pathways

metal response elements a regulatory system that induces the synthesis of proteins that detoxify metals

mevalonate a C_6 compound synthesized from acetyl CoA; used for the synthesis of cholesterol

micelle an aggregate of molecules that has a polar region in contact with water and a nonpolar region that is water free

Michaelis constant (K_m) a numerical constant that quantifies the affinity of a substrate molecule for an enzyme active site

Michaelis–Menten equation a mathematical relationship between the rate of an enzyme-catalyzed reaction and the concentration of enzyme and substrate

micronutrient a substance, such as a metal ion, that is required in minute amounts by an organism for proper growth

mitochondria double-membraned organelles that contain many enzymes responsible for aerobic metabolism

mitochondrial matrix an interior, gelatinous region in the organelle that contains many enzymes for aerobic metabolism

mobilization reaction a reaction in metabolism that releases a small molecule from an immobile fuel reservoir molecule (glucose from glycogen)

molecular biology the study of the chemical structures and reactions that perform biological processes; especially of the molecular basis of genetics and protein synthesis

molecular cloning the process of splicing DNA from one species, inserting it into the DNA from another species, and making many copies of the hybrid DNA by replication

molecular recognition the ability of biomolecules to use noncovalent interactions to combine in a specific way

Monod–Wyman–Changeux (MWC) concerted model used to describe the action of regulatory molecules in modulating enzyme action

monomeric protein a protein that is composed of a single polypeptide chain (subunit) and thus has no quaternary structure

monosaccharide a carbohydrate that has a single carbonyl group and two or more hydroxyl groups

mucopolysaccharides polysaccharides with a viscous, jelly-like consistency that are found in connective tissue

multienzyme complex an aggregate of enzymes that catalyzes a specific reaction sequence

mutagen an agent that is capable of causing changes in DNA base sequence (mutations)

mutarotation the process of interconversion of anomers

mutases enzymes that catalyze intramolecular isomerizations; subclass of isomerases

mutations changes in the base sequence of DNA caused by spontaneous or induced events; inheritable change in DNA

N

native conformation the distinct, three-dimensional form of a protein as it exists under physiological conditions

net reaction an equation that represents the addition of two or more reactions occurring in sequence; displays the input on the left side of the equation and output on the right

nicotinamide adenine dinucleotide (NAD) a cofactor associated with some dehydrogenase enzymes; derived from niacin

N-glycosidic bond covalent linkage formed between an anomeric hydroxyl group (hemiacetal or hemiketal) and an amine; as in nucleosides and nucleotides

nitric oxide (NO) a molecule that is synthesized from arginine and serves as a metabolic regulator and local hormone

nitrogenase complex an enzyme complex that catalyzes reduction of N_2 to NH_3 (nitrogen fixation); found in only a few species of bacteria

nitrogen cycle a scheme that defines the flow of nitrogen atoms between the biosphere and the atmosphere

nitrogen fixation the process by which N_2 is reduced to NH_3, the more biologically accessible form of nitrogen; carried out by only a few types of microorganisms

noncompetitive inhibitor an agent that binds to an enzyme at a site other than the active site and reduces enzyme activity

noncovalent interactions the forces between atoms and molecules that do not involve sharing of electrons as in a covalent bond; includes ionic bonds, hydrophobic interactions, hydrogen bonding, and van der Waals forces

noncyclic photophosphorylation in photosynthesis, synthesis of ATP is linked to a one-way flow of electrons from H_2O through photosystems I and II to $NADP^+$, which produces NADPH + H^+; as opposed to cyclic photophosphorylation

nonessential amino acids amino acids that are synthesized by an organism and are not required in the diet

nonhydrolytic cleavage reactions reactions in which molecules are split without the action of water; usually involves cleavage of a carbon–carbon single bond; catalyzed by lyases

nucleases enzymes that catalyze the hydrolysis of phosphoester bonds in DNA or RNA

nucleic acids polymeric biomolecules composed of nucleotide units; active especially in storage and transfer of genetic information; RNA and DNA

nucleoproteins complexes composed of proteins and nucleic acids

nucleoside biomolecule composed of an organic nitrogenous base and a carbohydrate linked by an N-glycosidic bond

nucleotide diphosphate sugars activated compounds such as UDP-glucose, UDP-galactose, and ADP-glucose; used for metabolism of carbohydrates

nucleosomes complexes of histones and DNA; units of chromatin

nucleotides biomolecules composed of an organic nitrogenous base, a carbohydrate, and phosphate

nucleus a cellular organelle in eukaryotic cells that contains the chromosomal genetic material and associated components

O

O-glycosidic bond covalent linkage formed between an anomeric hydroxyl group and another alcohol; the linkage between monosaccharides in polysaccharides

Okazaki fragments short DNA molecules that are present during discontinuous replication

oligomeric protein a protein composed of more than one polypeptide subunit

oligonucleotide a short sequence of a nucleic acid containing ten or fewer nucleotides

oligosaccharide a polysaccharide with ten or fewer monosaccharides

oncogene a gene that, when mutated, can transform a normal cell to a cancerous one

operator the region on DNA where a repressor binds

operons units of related genes on chromosomes

organelles membrane-enclosed packages of organized biomolecules that perform specialized functions inside a eukaryotic cell

osteomalacia a medical condition characterized by a weakening of bones; caused by a deficiency of vitamin D_3

β oxidation the metabolic pathway for degradation of fatty acids to acetyl CoA

oxidative decarboxylation a metabolic sequence involving two chemical processes, transfer of electrons and loss of CO_2

oxidative phosphorylation the synthesis of ATP from ADP using energy from mitochondrial electron transfer from NADH + H^+ and $FADH_2$ to O_2

oxidoreductases a class of enzymes that catalyzes oxidation and reduction reactions; also called dehydrogenases

oxygenases enzymes that catalyze the incorporation of oxygen atoms into substrates

P

passive transport movement of a biomolecule through a membrane without expenditure of energy; simple or facilitated diffusion

patch clamp technique a laboratory technique used to isolate and study the action of membrane ion-channels

pentoses carbohydrates with five carbon atoms

peptidases enzymes that catalyze the hydrolysis of peptide (amide) bonds; a subclass of hydrolases

peptide bond a covalent bond between the α-amino group of one amino acid and the α-carboxyl group of another amino acid; also an amide bond

peripheral proteins proteins that are located on the surface of membranes

permease an integral protein that assists transport of a biomolecule through a membrane

pernicious anemia a medical condition caused by a deficiency of vitamin B_{12}

peroxisomes cell organelles that contain oxidative enzymes; found in animal cells

pH the negative logarithm of the hydrogen ion concentration; a measure of acidity

phage a virus for which the natural host is a bacterial cell

phenylketonuria (PKU) a medical condition caused by a deficiency of the enzyme phenylalanine hydroxylase

pheromones molecules produced by animals that serve as a behavioral stimulus to other individuals of the same species

phosphoanhydride bond a bond formed by elimination of a water molecule between two phosphate groups; as in ATP

phosphodiester bond the linkage between nucleotide units in RNA and DNA

phosphogluconate pathway an auxiliary route for glucose degradation that produces ribose 5-phosphate and NADPH + H$^+$

phosphorolytic cleavage the cleavage of a chemical bond by inorganic phosphate, HPO$_4^{2-}$, as in glycogen metabolism

phosphoryl group (-PO$_3^{2-}$) often transferred from ATP to an acceptor molecule

phosphoryl group transfer potential a measure of the relative tendency of a molecule to transfer its phosphoryl group to an acceptor molecule

photochemical action spectrum a graph that displays the efficiency of each wavelength of light to sustain plant growth

photons (hν) individual particles of light

photophosphorylation the process in which photoinduced electron flow in chloroplasts is accompanied by synthesis of ATP from ADP

photorespiration a light-dependent process in plants in which O$_2$ is consumed and CO$_2$ is released

photosynthesis the process in plants by which energy from light is used to drive carbohydrate synthesis

photosystem I (P700) system consisting of chlorophyll a and secondary pigments that absorb light at 600–700 nm

photosystem II (P680) system consisting of chlorophyll a and b plus secondary pigments that absorbs light at 680 nm

phytosterols plant biochemicals that have structures based on the steroid ring system (i.e., stigmasterol, sitosterol)

phototrophs organisms that absorb energy from solar radiation to make ATP and NADPH + H$^+$, which is then used to make carbohydrates

pK the negative logarithm of K, the dissociation constant for an ionization reaction

plasma lipoproteins biomolecules composed of lipids and proteins that function as carriers for water-insoluble lipids in blood; the four types are chylomicrons, very low density, low density, and high density

plasma membrane the boundary layer of proteins and lipids surrounding a cell

plasmids self-replicating, extrachromosomal DNA molecules found in bacteria; contain genetic information for the translation of proteins that confer a specialized characteristic; used as cloning vectors

polymerase chain reaction a laboratory method used to synthesize amplified quantities of specific nucleotide sequences of DNA from small amounts of DNA using heat-stable DNA polymerase and cycles of denaturation and replication

polypeptide a molecule with 10 to 100 amino acids linked by peptide (amide) bonds

polyribosome a cluster of ribosomes simultaneously translating an mRNA molecule, making several copies of the polypeptide product

polysaccharides several or many monosaccharides linked by glycosidic bonds, including cellulose and starch

polyunsaturated fatty acids fatty acids with two or more carbon–carbon double bonds

porphyrias medical disorders caused by defects in heme metabolism

porphyrin large, complex biomolecule composed of four pyrrole rings; often has a bound iron, as in heme

post-translational modification a set of reactions that changes the structure of newly synthesized polypeptides

primary structure the sequence of amino acids in a protein

primer a short strand of RNA used for initiation of replication in DNA synthesis

prions protein molecules that may undergo conformational changes and become infectious; they are thought to cause encephalopathies, including scrapie and mad cow disease

prokaryotes simple, unicellular organisms, mainly bacteria and blue-green algae, with neither a distinct cell nucleus nor internal cellular compartmentation

promoter a region on DNA where RNA polymerase binds to begin transcription

prostaglandins a subclass of the eicosanoids; contain a five-membered ring; made from arachidonate; have a wide range of hormonelike activities

prosthetic group a small organic molecule or metal ion associated with a protein usually by covalent or ionic bonds

protease inhibitors agents that block the action of enzymes that catalyze the cleavage of amide bonds in polypeptides

protein a biopolymer composed of amino acid monomer units

protein folding a polypeptide winding into a stable, three-dimensional arrangement (native conformation) that provides the maximum number of strong noncovalent interactions

protein kinase an enzyme that catalyzes a phosphoryl group transfer from ATP to an amino acid residue in a protein

protein synthesis process whereby amino acids are linked together by amide bonds; process is directed by ribosomes

protein targeting the sorting and transporting of proteins from the site of synthesis on ribosomes to compartments in the cell where they are needed

proteome the total number of proteins that are coded in all the genes of an organism

proteomics the broad field of study that investigates the structure and function of protein products made from the genome

pyridoxal phosphate (PLP) a cofactor for enzymes catalyzing transamination reactions

pyruvate dehydrogenase complex the enzyme system that catalyzes the conversion of pyruvate to acetyl CoA, thus linking glycolysis with aerobic metabolism

Q

quaternary structure the association of two or more polypeptide chains to form a multisubunit protein molecule

R

ras proteins a family of proteins from the tumor-causing rat sarcoma virus with amino acid sequences similar to G-proteins; a type of oncogene

reactive oxygen species intermediate and radical species generated from reactions of oxygen (i.e., superoxide, hydrogen peroxide, hydroxyl radical)

receptor protein a protein usually on the outer surface of a cell membrane that binds a signal molecule (hormone) and transmits a message to the cell's interior

recombinant DNA a form of DNA from one species that has been modified by incorporation of a DNA segment from another species; hybrid DNA

recombinant protein the polypeptide product made from transcribing a hybrid DNA and translating the mRNA

reducing sugar carbohydrates that contain a free aldehyde group and are capable of transferring electrons to (reducing) metal ions in solution

reductive carboxylation the chemical process of photosynthesis that involves covalent fixing of CO_2 and reduction of functional groups to produce carbohydrates

regulatory enzyme an enzyme in which the catalytic activity depends on the presence or absence of signal molecules called effectors; also known as allosteric enzyme

regulatory protein protein molecule that binds to DNA and influences the action of RNA polymerase, thus controlling the rate of protein synthesis; there are two types: activators and repressors

regulatory site specific locations on allosteric enzymes where effectors bind

release factors proteins that terminate the process of protein synthesis

replication the process by which existing DNA is used as a template for the synthesis of new DNA strands

replication fork the site at which incorporation of nucleotides occurs in DNA synthesis

repressors regulatory proteins that bind to DNA and block transcription

residue an amino acid after incorporation into a protein

resonance stabilization a decrease in the energy (stabilization) of a molecular species by the occurrence of resonance hybrid structures; an explanation given for the energy role of ATP

respiration the process by which cellular energy is generated by the oxidation of nutrients and the reduction of O_2

respiratory assemblies the series of carriers in the inner mitochondrial membrane that transports electrons from nutrient oxidation to O_2

restriction endonucleases hydrolytic enzymes that catalyze the cleavage of phosphoester bonds at specific nucleotide sequences in DNA

restriction enzyme map a figure drawn for a plasmid that displays the sites of cleavage by many restriction endonucleases and the number of fragments obtained

restriction fragment length polymorphisms (rflp) a method used for the analysis of DNA fingerprinting data; uses restriction endonucleases to cleave the DNA and electrophoresis to compare fragments from different individuals

reversible inhibitor compounds that readily associate with or dissociate from an enzyme; these compounds render enzymes inactive only when bound to them

ribonucleases enzymes that catalyze the hydrolytic cleavage of phosphoester bonds in RNA

ribosomal RNA (rRNA) RNA associated with ribosomes; functional in protein synthesis

ribosomes cell organelles composed of RNA and protein that are the site for protein synthesis

ribozyme a catalytic form of RNA that splices other forms of RNA during post-translational processing

ribulose 1,5-bisphosphate carboxylase/oxygenase (rubisco) the enzyme that catalyzes the addition of CO_2 to ribulose 1,5-bisphosphate, which is the first step of the Calvin cycle in photosynthesis

rickets a nutritional disease caused by improper calcium and phosphorus metabolism that results in bone malformation; caused by a deficiency of vitamin D_3

RNA polymerases enzymes that catalyze the synthesis of RNA

RNA replicase an enzyme that catalyzes RNA-directed RNA synthesis

S

salvage synthesis of nucleotide bases recycling of intact purine and pyrimidine bases to avoid energy-expensive synthesis from small molecules

Sanger sequencing procedure a laboratory technique that uses chain termination and gel electrophoresis to sequence nucleic acid bases

saponification base-catalyzed hydrolysis of the ester bonds in triacylglycerols

saturated fatty acids fatty acids in which all carbon–carbon bonds are single bonds

scissle bond the bond cleaved in the reactant during a chemical reaction

secondary pigments carotenoids, phycobilins, and other light-absorbing biomolecules that assist chlorophyll in accepting light energy for photosynthesis

secondary structure conformational arrangement of a polypeptide backbone in a regular, repeating pattern (i.e., α-helix, β-sheet)

second messenger an intracellular molecule that relays the signal from an extracellular molecule (hormone), such as cyclic AMP

sedimentation coefficient a number expressing the relative rate that a particle moves to the bottom of a spinning centrifuge tube

semiconservative replication method of DNA synthesis in which each new duplex is composed of an original strand and a newly synthesized strand

sequence homology the relationship between two proteins that display similar or identical amino acid sequence in some regions

sequential model designed by Koshland and used to explain the action of regulatory molecules in modulating the action of allosteric enzymes

β-sheet a secondary conformation in which the polypeptide chain is nearly fully extended and interacts with another polypeptide by hydrogen bonding

sickle cell anemia a genetic disease characterized by nonfunctional hemoglobin

signal transduction processes whereby the presence of a biomolecule exterior to a cell relays a biochemical command to the interior of the cell

simple diffusion the unassisted transport of a biomolecule through a membrane; requires no energy input

site-directed mutagenesis a genetic laboratory method used to modify the amino acid sequence of a protein

small nuclear ribonucleoproteins complexes of protein and RNA that affect RNA processing

sodium-potassium ATPase a membrane transport system that maintains proper intracellular and intercellular concentrations of Na^+ and K^+

sphingolipids polar lipid molecules that have the foundation molecule sphingosine

squalene a C_{30} noncyclic hydrocarbon that is an intermediate in cholesterol synthesis

standard free energy change ($\Delta G^{\circ\prime}$) the amount of energy from a reaction available to do useful work under standard conditions of 1 atm, 25°C, an initial concentration of reactants and products of 1 M, and pH of 7

standard reduction potential ($E^{\circ\prime}$) a measure of how easily a compound can be reduced under standard conditions

starch the polysaccharide used by plants for storage of glucose for energy metabolism

steady-state intermediate a reaction intermediate produced at the same rate that it is converted to a product

steroid hormones a class of compounds, derived from cholesterol, that have a wide range of physiological actions, especially metabolic regulation and sexual development

steroids lipids that have the characteristic structure of four fused rings (e.g., cholesterol)

structural motifs elements of secondary structure in polypeptides that are combined into native conformations (e.g., αα motif)

substrate cycle a net sequence of reactions that results in the cleavage of ATP with no apparent use of the energy released

substrate-level phosphorylation the synthesis of ATP from ADP using energy from the direct metabolism of a high-energy reactant; as opposed to oxidative phosphorylation

subunit individual parts of a large molecule; usually a polypeptide chain in an oligomeric protein

supercoiled DNA the twisting of the DNA double helix into new conformations; extra twisting of closed circular DNA

supramolecular assembly organized clusters of biomolecules that perform a specialized biological function (e.g., cell membranes, organelles)

synthase an enzyme that catalyzes the formation of a compound without need for energy from ATP; a subclass of ligases

synthetases enzymes that catalyze bond-forming reactions with use of energy from ATP cleavage; a subclass of ligases

T

telomerase an enzyme that catalyzes the formation of caps or specialized ends (telomeres) on DNA

telomeres specialized ends on DNA consisting of simple repeated sequences of nucleotides

template a molecule (usually DNA or RNA) that serves as a pattern for generation of a complementary molecule

termination the closing stage of protein synthesis, which is signaled by the stop codons and characterized by release of the new polypeptide from the 70S ribosome

terpenes biomolecules that are synthesized from the C_5 molecules, isoprenes

tertiary structure conformational arrangement of a polypeptide into a compact, globular form

tetrahydrofolate (FH_4) a vitamin that serves as a cofactor for reactions involving transfer of one-carbon units

thioester a compound formed by the covalent linkage of an acyl group to a thiol, $-SH$ (i.e. acetyl CoA)

thromboxanes a subclass of the eicosanoids; contain a six-membered, oxygen-containing ring; made from arachidonate; present in blood platelets

topoisomerases enzymes that control the extent of supercoiling in DNA

transamination a reaction process in which the amino group of an amino acid is transferred to an α-keto acid; catalyzed by aminotransferases

transferases enzymes that catalyze the transfer of a functional group from one molecule to another

transfer RNA (tRNA) the small form of RNA with covalently bound amino acid that reads the codon message on mRNA and incorporates the amino acid into the protein being synthesized

transcription the process by which the nucleotide sequence in DNA is converted to form complementary RNA; initial step in gene expression

transformation the introduction of hybrid DNA into a host organism, in which it can be replicated

transgenic an organism with a genome carrying one or more DNA sequences from a different species

transition state an energetic, unstable, short-lived species formed in a reaction; can be reversed to the reactant or transformed to the product

transition state analog a molecule that resembles the predicted transition state of a reaction; will bind tightly to the active site of an enzyme and serve as a competitive inhibitor

translation the protein synthesis process; the sequence message in mRNA is used to direct synthesis of proteins from amino acids

translocation the movement of the 70S ribosomal unit from one mRNA codon to the next during protein synthesis

triacylglycerols lipids formed by esterification of the three hydroxyl groups on glycerol with three fatty acids

turnover number the number of moles of substrate transformed to product per mole of enzyme in a defined time period

U

ubiquinone a redox center in the mitochondrial electron transport chain; a small, nonpolar molecule that becomes ubiquinol upon reduction; also called coenzyme Q

ubiquitin pathway a process for the labeling and degradation of defective or damaged proteins

uncompetitive inhibitor a compound that inhibits an enzyme by binding only to its ES complex

uncoupling protein a protein found in brown fat that allows the process of electron transport to continue, but disrupts ADP phosphorylation; also called thermogenin

unsaturated fatty acids fatty acids with one or more carbon–carbon double bonds

urea cycle a metabolic pathway that uses carbon in CO_2 and nitrogen present in glutamate and NH_3 to synthesize urea; a mechanism to detoxify NH_3

V

vacuoles membrane-enclosed sacs in cells used for storage of nutrients

van der Waals force attractive force between molecules with temporary dipoles induced by fluctuating electrons

vector an agent that serves as the carrier for foreign DNA in recombinant DNA technology; also called a vehicle

very low density lipoprotein blood lipoproteins formed by the association of phospholipids, cholesterol, cholesteryl esters, and triacylglycerols with proteins to assist in the transport of water-insoluble lipids; consist of 10% protein and 90% lipid

viruses infectious noncellular particles composed of DNA or RNA and wrapped in a protein package; can reproduce only in a host cell

vitalism a now defunct doctrine that living organisms have a vital force that distinguishes them from the inanimate world

vitamin an organic or organometallic molecule required for proper growth and development of organisms

W

western blot a technique used for identifying proteins based on interactions with specific antibodies; immunoblotting

X

xenobiotics organic compounds that are foreign to an organism; usually synthetic, unnatural

Y

yeast artificial chromosomes (yac) large fragments of DNA that are used as vehicles to prepare recombinant eukaryotic DNA

Z

zinc finger motif a common DNA binding domain found in eukaryotic regulatory proteins

Z scheme the direction of electron flow from H_2O to $NADP^+$ that links together photosystems I and II; cyclic photophosphorylation

zwitterion a molecule that has a positive and negative charge; thus, its net charge is zero

zymogen the inactive precursor for an enzyme that catalyzes the hydrolysis of peptide bonds

ANSWERS TO STUDY EXERCISES

Chapter 1

1.2 Lipids are relatively nonpolar molecules. Some like fatty acids in triacylglycerols are fuel molecules; the phosphoglycerol lipids are in membranes.

1.3 a.

$$CH_2C \overset{\displaystyle O}{\underset{\displaystyle OH}{\big<}}$$
$$\underset{NH_2}{|}$$
Glycine

b. H_2NCNH_2
      $\overset{\|}{O}$
Urea

c.
$$CH_3C \overset{\displaystyle O}{\underset{\displaystyle OH}{\big<}}$$
Acetic acid

d.
$$CH_3C \overset{\displaystyle O}{\underset{\displaystyle SH}{\big<}}$$
Thioacetic acid

e.
H   H
 \   /
  C = C
 /   \
N   N—H
 \   /
  C
  |
  H
Imidazole

f.
$$H_2N—CHC \overset{\displaystyle O}{\underset{\displaystyle OH}{\big<}}$$
$$\underset{CH_2}{|}$$
$$\underset{SH}{|}$$
Cysteine

g.
$$H_2CC \overset{\displaystyle O}{\underset{\displaystyle OH}{\big<}}$$
$$\underset{OH}{|}$$
Glycolic acid

h.
H
 \
  O—O
      \
       H
Hydrogen peroxide

i.
$$\overset{\displaystyle O}{\underset{\displaystyle OH}{HO—\overset{\|}{\underset{|}{P}}—OH}}$$
Phosphoric acid

All of the above compounds are naturally occurring.

1.4 C, H, O, N, S, P, and several others

1.5 N_2, O_2, CO_2, NO

1.6 Na^+, K^+, Fe^{3+}, Cu^{2+}, Mg^{2+}, and several others

1.7 Nucleic acids, proteins, carbohydrates, lipids, vitamins

1.8 See Figures 1.8 and 1.11.

1.9 Molecular biology emphasizes the study of biological processes and biomolecules that are involved in protein synthesis and genetic function.

1.10 a. Storage of genetic information
b. Transfer of genetic information
c. Building blocks for proteins
d. Components for construction of cell membranes
e. Nutrients for energy metabolism
f. Assist enzymes in catalytic action
g. Many function as enzymes to catalyze biochemical reactions
h. Acts as a reagent in hydrolytic reactions

1.11 Eukaryotic cells have a distinct membrane-enclosed nucleus and well-defined internal compartmentation. Prokaryotic cells do not have these. See Tables 1.1 and 1.2.

1.12 Cellulose, DNA, hemoglobin, proteins

1.13 a. Plants and animals   **e.** Plants
b. Plants and animals   **f.** Plants and animals
c. Animals   **g.** Plants
d. Animals

1.14 1. c   **2.** d or a   **3.** b   **4.** a   **5.** a

1.15 The cell homogenate should be centrifuged at $600g$ to sediment nuclei and cell debris. The supernatant is then centrifuged at $20,000g$. At this point the mitochondria would sediment into a pellet. See Window on Biochemistry 1.3.

1.16 a. Iron   **c.** Iron
b. Magnesium   **d.** Iron

1.17 Leaves are cut from a living bean plant, suspended in a buffer solution, and homogenized by grinding with sand in a mortar and pestle or with an electric homogenizer. All of these procedures should be done at ice temperatures so biomolecules are not thermally denatured.

1.18 $b > g > a > c > d > e > h > f$

1.19 DNA contains deoxyribose and exists in a double helix. RNA contains ribose and is usually single-stranded.

1.20 a. Peroxisomes   **d.** Ribosomes
b. Nucleus   **e.** Chloroplasts
c. Lysosomes   **f.** Mitochondria

1.21 Proceed through several centrifugation steps as shown in Window on Biochemistry 1.3. The enzyme would be present in the supernatant remaining after centrifugation at $100,000g$.

1.22 Mitochondria

1.23 This was the first laboratory synthesis of a molecule found in living cells using only inorganic or "lifeless" chemicals.

1.24 See Figure 1.2.

1.25 The pellet after centrifugation at $600g$.

1.26 Carbon can form multiple and stable covalent bonds with other carbon atoms and with other elements. This allows for a large and diverse collection of molecules.

1.27 Blueprint: DNA
Materials: foodstuffs (carbohydrates, fats, proteins)
Energy: metabolism (degradation of carbohydrates and fats)

1.31 a.

$$CH_3COOH + CH_3CH_2OH \longrightarrow CH_3\overset{\overset{\displaystyle O}{\|}}{C}OCH_2CH_3 + H_2O$$

b.

$$CH_3COOH + NH_3 \longrightarrow CH_3\overset{\overset{\displaystyle O}{\|}}{C}NH_2 + H_2O$$

c.

$$CH_3COOH + CH_3COOH \longrightarrow CH_3\overset{\overset{\displaystyle O\ \ O}{\|\ \ \|}}{C}OCCH_3 + H_2O$$

d.

$$CH_3COOH + CH_3SH \longrightarrow CH_3\overset{\overset{\displaystyle O}{\|}}{C}SCH_3 + H_2O$$

e.

$$CH_3COOH + \underset{\underset{H\ \ H\ \ H}{|\ \ |\ \ |}}{\underset{O\ \ O\ \ O}{CH_2CHCH_2}} \longrightarrow \underset{\underset{H\ \ H}{|\ \ |}}{\underset{O\ \ O\ \ O-CCH_3}{CH_2CHCH_2}}$$
$$+ H_2O$$

1.33 Because the concentration of water in a cell is relatively high, the equilibrium probably lies toward the left.

Chapter 2

2.2 DNA, RNA, and proteins carry instructions that tell cells how to complete biological processes. The information is present in the form of sequence of monomers in the biopolymers and in the form of native conformation (structure).

2.3 (a) True. (b) False; it is two strands in a double helix. (c) True. (d) False; A, T, G, C.

2.4 DNA contains deoxyribose; RNA contains ribose.
DNA is most often present as a double helix; RNA is most often present as a single polynucleotide strand.
DNA contains adenine, thymine, guanine, and cytosine; RNA contains adenine, uracil, guanine, and cytosine.

2.5 (a) 3′ TAAACTGG; (b) 3′ GATTCGGG

2.6 (a) 3′ AUGGC; (b) 3′ GGGAAA

2.7 3′ TTAGACCATG 5′   DNA
5′ AAUCUGGUAC 3′ RNA

2.8 5′ A-U-U-C-G-A 3′

2.9 3′ T-A-C-A-T-C 5′

2.10 3′ A-G-C-A-U-C 5′

2.11 mRNA = 3′ A-A-U-C-U-G-G-A-A 5′
protein = Lysine-Valine-stop

2.12
5′							3′	
T	A	A	C	A	G	T	T	(old)
A	T	T	G	T	C	A	A	(new)
3′							5′	

2.13 a. 5′ A-C-C-A-A-G-A-G-G-A-C-A-A-T-T-T-T-G-A-A-T-A-T-A-A-C-A 3′
b. 5′ U-G-U-U-A-U-A-U-U-C-A-A-A-A-U-U-G-U-C-C-U-C-U-U-G-G-U 3′

2.14 a. The synthesis of RNA using a DNA template.
b. The synthesis of DNA using an RNA template.
c. The formation of cyclic AMP from ATP.

2.15 Receptor protein in membrane; G protein; adenylate cyclase.

2.16 See Figure 2.10 and Section 2.5.

2.17 Cholera, whooping cough, cancer.

2.18 No, because there is no variation in the sequence of monomers.

2.19 DNA molecules are compacted by very extensive folding and packing.

2.20 A-T-G; T-A-G; G-A-T; T-G-A; A-G-T; G-T-A

2.21 DNA is chemically very stable and not very susceptible to breakdown under aqueous conditions in the cell.

2.22 The genetic message in mRNA is in the language of four nucleotide bases (A, U, G, C). Protein molecules are linear sequences of amino acids. Two different languages are involved in the transformation from RNA to proteins.

2.23 Miscommunication in the cell may cause production of cancer-causing proteins.

2.24 The biological process triggered to start by molecular recognition processes must also be stopped. Dissociation of molecules serves this purpose.

2.25 RNA may be the first functional biomolecule in the origin of life. It can serve as a template for replication processes (now performed by DNA) and as a catalyst for biological reactions (now performed primarily by protein enzymes).

2.26 Once a signal molecule triggers an action it is important for the molecule to be degraded so it does not continue to induce the action indefinitely.

Chapter 3

3.2 a. $1.6 \times 10^{-2}\ M$
b. $3.9 \times 10^{-8}\ M$
c. $2.5 \times 10^{-7}\ M$
d. $5.0 \times 10^{-5}\ M$
e. $1.0 \times 10^{-5}\ M$
f. $7.9 \times 10^{-8}\ M$

3.3 pH ranges: 1.4–3.4 and 8.8–10.8

3.4 a.

b.

3.5 a. $HCl \rightleftharpoons H^+ + Cl^-$
b. $CH_3COOH \rightleftharpoons CH_3COO^- + H^+$
c. $NH_4^+ \rightleftharpoons NH_3 + H^+$
d. $CH_3(CH_2)_{13}CH_2COOH \rightleftharpoons$
$\qquad CH_3(CH_2)_{13}CH_2COO^- + H^+$
e. $\underset{R}{H_3^+N-CHCOOH} \rightleftharpoons \underset{R}{H_3^+N-CHCOO^-} + H^+ \rightleftharpoons$
$\qquad \underset{R}{H_2N-CHCOO^-} + H^+$
f. $H_3PO_4 \rightleftharpoons H_2PO_4^- + H^+$
$\quad H_2PO_4^- \rightleftharpoons HPO_4^{2-} + H^+$
$\quad HPO_4^- \rightleftharpoons PO_4^{3-} + H^+$
g. $H_2O \rightleftharpoons H^+ + OH^-$
h. $H_2CO_3 \rightleftharpoons H^+ + HCO_3^-$
$\quad HCO_3^- \rightleftharpoons H^+ + CO_3^{2-}$

3.6 Gastric juice > cola > acid rain > coffee > blood.

3.7

Adenine   Thymine

3.8 a. True.
b. False; hydrogen bonding is a noncovalent attractive force between a hydrogen atom covalently bonded to an electronegative atom (O, S, N) and the lone pair of electrons on a second electronegative atom.
c. False; hydrophobic interactions are the association of nonpolar molecules or regions of nonpolar molecules in water that result in stability.
d. False; H^+ and OH^- form a covalent bond to produce water.
e. True.

3.9 a. Ion:ion
b. Ion:dipole
c. Ion:ion
d. Dipole:dipole
e. Ion:ion

3.10 a. H_2O
b. H_2O
c. CH_3CH_2OH
d. $\underset{OH\ OH}{CH_2CH_2}$
e. $\underset{O}{CH_3\overset{||}{C}NH_2}$
f. $\underset{NH_2}{CH_2CH_3}$
g. $\underset{O}{H_2N\overset{||}{C}NH_2}$

3.11 The number of moles of HOH in 1 L of water is:

$$\text{Moles of HOH} = \frac{\text{g of } H_2O}{\text{MW of } H_2O} = \frac{1000 \text{ g}}{18} = 55.5\ M$$

55.5 mol of HOH in 1 L; therefore the molarity is 55.5 M.

3.12 a. H_2O
b. H_2CO_3
c. $H_3^+NCH_2COOH$
d. CH_3COOH
e. H_3PO_4

3.13 a. CH_3CH_3
b. $\underset{OH}{CH_2CH_3}$
c. $H_3^+NCHCOO^-$ $\underset{H_3C\quad CH_3}{\overset{CH}{}}$
d. CH_3CH_2COOH
e. $CH_3CH_2CH_2COOH$

3.14 a. OH^-
b. $H_3^+NCH_2COO^-$
c. $CH_3(CH_2)_{10}COO^-$
d. CO_3^{2-}
e. $\underset{^+NH_3\qquad NH_2}{CH_2(CH_2)_{10}CH_2}$

3.15 (a) Ionic, (b) covalent, (c) hydrogen, (d) i. covalent, ii. hydrogen.

3.16 (a) True. (b) False; noncovalent bonds have bond strengths in the range of 1–30 kJ/mol. (c) True. (d) True.

3.17 See Table 3.2.

3.18 Ionic bonds, formed by attractive forces between unlike charged atoms and molecules, are usually much stronger than hydrogen bonds. See Table 3.2.

3.19 a, c, d, e, f

3.20 a, b, c

3.21 Lactic acid

3.22 $[HCO_3^-/H_2CO_3] = 10$.

3.23 Infusion of sodium bicarbonate solution into the blood.

3.24 pH = 9.6.

3.25 Normal rain: 0.16
Acid rain: 1.26×10^{-3}

3.26 a, c, e, h

3.27 $MgCO_3$ is a base that neutralizes aspirin, an acid, thus causing less discomfort.

3.28 a, c, d

3.29 a. Diprotic

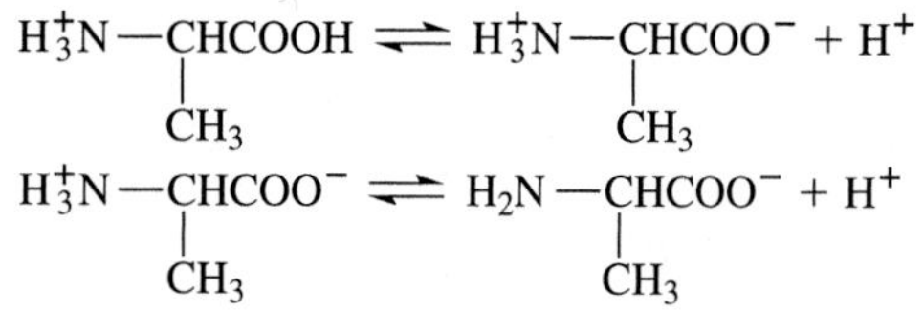

$$\overset{+}{H_3}N-CHCOOH \rightleftharpoons \overset{+}{H_3}N-CHCOO^- + H^+$$
$$\qquad\quad | \qquad\qquad\qquad\quad |$$
$$\qquad\quad CH_3 \qquad\qquad\qquad CH_3$$

$$\overset{+}{H_3}N-CHCOO^- \rightleftharpoons H_2N-CHCOO^- + H^+$$
$$\qquad\quad | \qquad\qquad\qquad\quad |$$
$$\qquad\quad CH_3 \qquad\qquad\qquad CH_3$$

b. Triprotic
c. Triprotic
d. Diprotic

3.30 b, d.

3.31

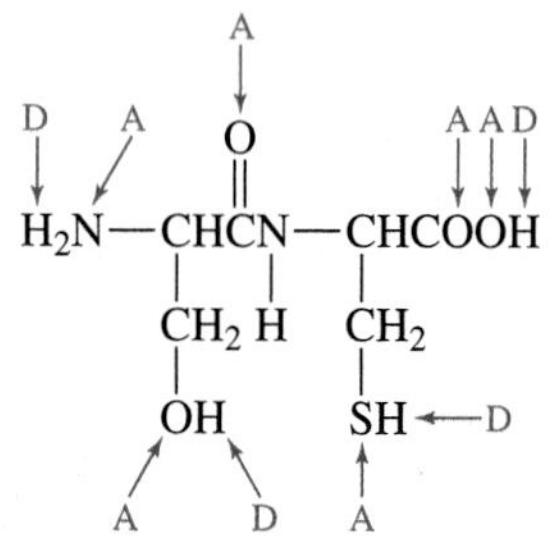

A = acceptor
D = donor

3.32 [lactate]/[lactic acid] = 14.13.

3.33 pH = 4.03.

3.34 0.19 mol of dibasic sodium phosphate; 0.31 mol of monobasic sodium phosphate

Chapter 4

4.2

Valine

Threonine

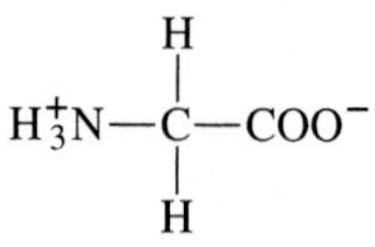

Glycine

4.3 The same shape as for alanine (Figure 4.4); $pK_1 = 2.4$; $pK_2 = 9.8$; $pH_I = 6.1$.
pH = 1; $\overset{+}{H_3}N-CH_2COOH$
pH = 2.3; $\overset{+}{H_3}N-CH_2COOH$ and $\overset{+}{H_3}N-CH_2COO^-$
pH 9.6; $\overset{+}{H_3}N-CH_2COO^-$ and $H_2N-CH_2COO^-$
pH 12.0; $H_2N-CH_2COO^-$

4.4 (a) pH = 3.1, (b) pH = 2.1, (c) pH = 2.4.

4.5 a. True   **e.** False
     b. False   **f.** False
     c. False   **g.** False
     d. True

4.6 (a) Ala, (b) Ala, (c) Val, (d) Phe, (e) Pro

4.7 (a) Val, (b) Phe, (c) Glu, (d) Ser, (e) Lys, (f) Asn

4.8

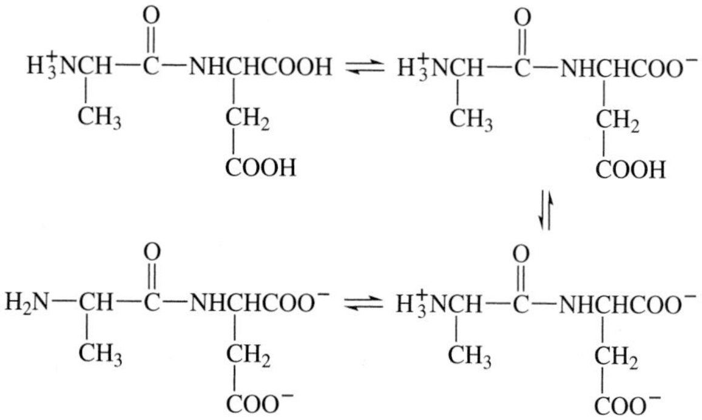

4.9

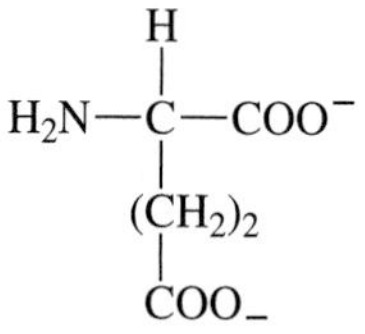

4.10 a. Identification of N-terminal amino acid in a protein.
   b. Identification of N-terminal amino acid in a protein.
   c. Cleavage of peptide bonds in a protein to yield free amino acids; used to determine amino acid composition.
   d. Determine amino acid sequence of a protein; Edman procedure.
   e. Catalyzes the hydrolysis of specific peptide bonds; used to assist protein sequencing.
   f. Detergent that denatures proteins; used to disrupt quaternary structure of multisubunit protein before electrophoresis.
   g. A reagent that reacts with amino acids for quantitative determination.
   h. Used to cleave a protein specifically on the C side of the amino acid methionine.

4.11

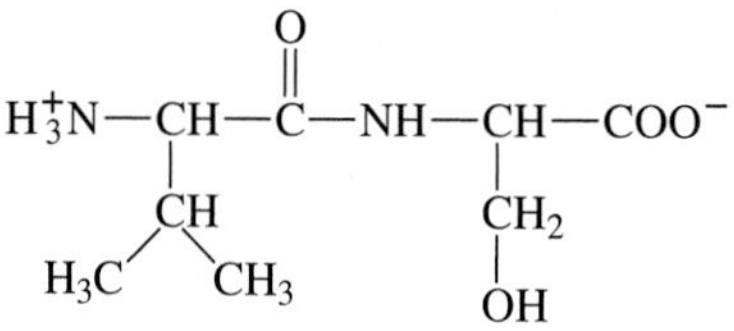

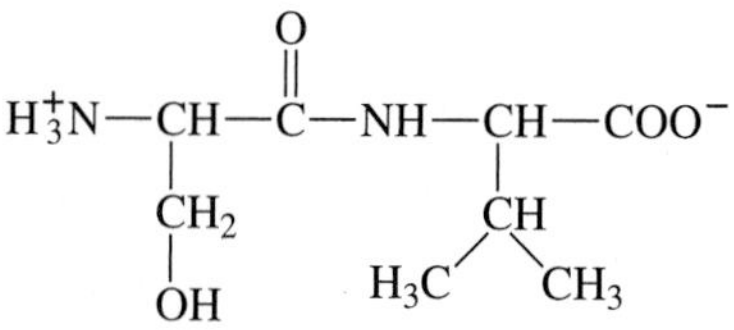

4.12 Globular: water soluble; play dynamic biological roles in transport, catalysis, and so on.
Fibrous: water insoluble; play structural roles in cells and organisms.

4.13 The average individual molecular weight for the 20 amino acids is 110.
  a. 5733/110 = 52 amino acids.
  b. 1318
  c. 2273

4.14 (a) 11,440, (b) 451,000

4.15 (a) Cys, (b) Asp, (c) Asn, (d) Thr, (e) His, (f) Gly

4.16 pH = 1 (predominant)

$$HC=C-CH_2-\overset{\overset{\displaystyle H}{|}}{C}-COOH$$

with ring: HN^+ ... NH, $\overset{+}{N}H_3$, ring carbon CH

pH = 5 (predominant)

$$HC=C-CH_2-CH-COO^-$$

with HN^+ ... NH, $\overset{+}{N}H_3$

pH = 11 (predominant)

$$HC=C-CH_2-CH-COO^-$$

with N ... NH, NH_2

4.17 Yes. The individual amino acids that make up these peptides are soluble so the peptides are expected to be soluble. In addition, the peptides have several charged or polar functional groups that can be solvated by water.

4.18 **a.** Cys-S — S-Cys
  b. Glu
  c. Gly + Ala

4.19 Ala, Val, Leu, Ile, Pro, Met, Phe, Trp

4.20 The message as represented by sequences of nucleotide bases in DNA is transcribed to the form of mRNA. Triplet codons in mRNA are read by the anticodon in tRNA to bring in the correct amino acid to be used for protein synthesis.

4.21 **a.** Separated by gel filtration
  b. Not separated by gel filtration
  c. Separated by gel filtration

4.22

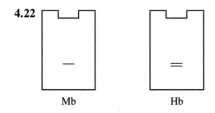

Mb   Hb

4.23 **a.** A and B chains are unchanged
  b. A: 2,4-dinitrophenyl-labeled Gly
     B: 2,4-dinitrophenyl-labeled Phe
  c. A: release of Asn
     B: release of Ala
  d. A: no change
     B: cleavage on the C side of Arg and Lys

4.24 **a.** Separation of molecules based on net charge.
  b. Separation of molecules based on their ability to bind to a specific support by noncovalent interactions.
  c. Separation of molecules based on their sizes.

4.25 **a.** Asp
  b. Phe
  c.

$$H_3^+N-\underset{\underset{\displaystyle COOH}{\overset{\displaystyle |}{\underset{\displaystyle |}{CH_2}}}}{CH}-\overset{\overset{\displaystyle O}{\|}}{C}-NH-\underset{\underset{\displaystyle C_6H_5}{\overset{\displaystyle |}{\underset{\displaystyle |}{CH_2}}}}{CH}-COOH$$

  d. −1

4.26
$$H_3^+N-Cys$$
$$|$$
$$S$$
$$|$$
$$S$$
$$|$$
$$^-OOC-Cys$$

4.27 Ser, Asn, Thr, His, Tyr, Cys

4.28 The Sanger reagent can be used only to identify the N-terminal amino acid of a polypeptide. During the analysis, the remainder of the protein is destroyed by hydrolysis of each peptide bond. Therefore, sequence analysis cannot be completed.

4.29 $Cys + Cys + CH_2CH_2-S-S-CH_2CH_2$
with SH, SH, OH, OH

4.30 Three

4.31 Glycine has such a small side chain (H) that it often acts like a Group II amino acid. For example, it is very water soluble. Histidine exists in two forms at pH 7.0. Which one is predominant?

4.32 Circular peptides where the amino terminus and carboxyl terminus react to form amide bonds. Thus there are no "ends" to the peptide.

4.33

$$HO-\underset{\underset{\displaystyle COO_-}{\overset{\displaystyle |}{\underset{\displaystyle |}{C}}}}{\overset{\overset{\displaystyle NH_3^+}{|}}{C}}-CH_2OH \rightleftharpoons HOH_2C-\underset{\underset{\displaystyle COO_-}{\overset{\displaystyle |}{\underset{\displaystyle |}{C}}}}{\overset{\overset{\displaystyle NH_3^+}{|}}{C}}-H$$

Chapter 5

5.2 About 1.5 amino acids per Å; therefore, about 13 amino acids in 20 Å.

5.3 a.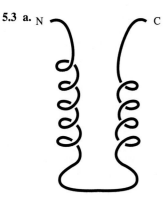

b.

c. See Figure 5.10.

d.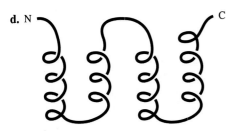

e.

5.4 See Figure 5.5.

5.5 Noncovalent Interactions: hydrogen bonding, ionic interactions, van der Waals forces, hydrophobic interactions; see Table 3.2.

5.6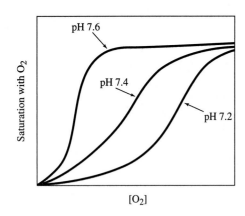

5.7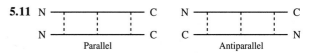

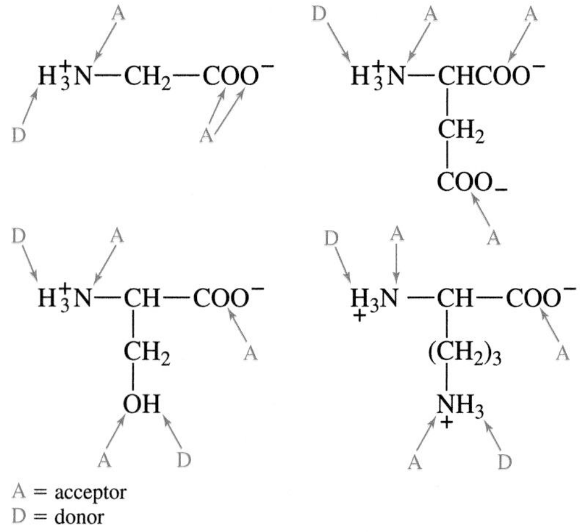

A = acceptor
D = donor

5.8 α-Helix is predominant; there are also bends and loops.

5.9 a. Covalent; amide bonds
b. Noncovalent; primarily hydrogen bonding
c. Noncovalent; hydrogen bonding, ionic, hydrophobic; covalent disulfide bonds
d. Noncovalent; hydrogen bonding, hydrophobic interactions, ionic

5.10 See Section 5.2.

5.11

N ——————— C   N ——————— C
N ——————— C   C ——————— N
      Parallel   Antiparallel

5.12 Proteins made by cells immediately after exposure to increases in temperature. These are now thought to be chaperones, proteins that assist other proteins in folding into their native conformation.

5.13 *Motif:* Combination of elements of secondary structure (α-helix, β-sheet, bends) into supersecondary structure; for example, the α-helix–loop–α-helix motif.
Domain: A stable, independently folded region in a protein that is usually made up of several motifs.

5.14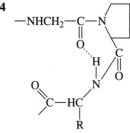

5.15 Inside: Phe, Val, Leu
Outside: Ser, Glu, Thr, His, Asn, Cys

5.16 a. With a high content of Gly and Ala, it would be expected to be primarily β-sheet, like silk.
b. Fibrous
c. Water insoluble
d. Probably plays a structural role.

5.17 a. A bend at proline would cause the polypeptide to form a U shape and fold into a β-sheet arrangement called a ββ motif.
  b. Random coil
  c. α-Helix
  d. A bend at proline to form a U shape and fold into an αα motif with the possibility of a disulfide linkage between two cysteine residues.
  e. A bend at proline to fold into an αα motif.
  f. Random coil

5.18 At pH 7, all of the lysine side chains are in the form $-NH_3^+$. The side chains repel each other and the polypeptide opens into a random coil. At pH 13, the side chains are $-NH_2$; therefore, the polypeptide chain can fold into an α-helix.

5.19 Proteins from thermophilic organisms would be expected to have a higher content of amino acids with nonpolar side chains so hydrophobic interactions could become more abundant.

5.20 Lane 1: myoglobin
Lane 2: bacteriorhodopsin
Lane 3: glucose oxidase
Lane 4: hemoglobin

5.21 Urea is a strong hydrogen bonding agent. It has both donor and acceptor atoms that can form hydrogen bonds with proteins and disrupt hydrogen bonds within the protein.

5.22 a,c

5.23 Hydrophobic interactions among the side chains can form within the region Val-Phe-Val-Leu-Phe. Two cysteine residues are responsible for the disulfide bond. The ionic bond is formed between an $-NH_3^+$ (perhaps lysine side chain) and $-COO^-$ (perhaps an aspartate side chain).

5.24 a. About 0.1–0.2 kP_a
  b. About 3 kP_a
  c. None; myoglobin has a higher affinity for O_2 at all O_2 pressures.

5.25 Hemoglobin has the ability to change its affinity for O_2 under different physiological conditions. It can load up with O_2 in some physiological regions and unload the O_2 in other physiological regions. See Section 5.5 and Figure 5.19.

5.26 Polyglutamate is a peptide made up only of glutamate. There will be many side chains with the carboxyl group. At pH 7, the carboxyl groups will be ionized to COO^-. Many negative charges will disallow the formation of an α-helix. At pH 3, the carboxyl groups are unionized, COOH. Therefore the polypeptide may fold into an α-helix.

5.27 a. Noncovalent interactions; especially hydrogen bonding
  b. Covalent bonds; especially amide bonds
  c. Noncovalent interactions; especially hydrophobic interactions and hydrogen bonding
  d. Noncovalent interactions; especially hydrophobic interactions and hydrogen bonding
  e. Covalent bonds; especially amide bonds
  f. Noncovalent interactions; especially hydrophobic interactions
  g. Noncovalent interactions; especially hydrogen bonding and hydrophobic interactions

  h. Covalent bonds; disulfide linkages
  i. Covalent bonds; amide (peptide) bonds

5.28 A folded protein has most of its nonpolar amino acids buried inside to avoid the solvation of water molecules on the surface. A denatured protein has nonpolar amino acids on the surface; therefore, the denatured protein will be less soluble.

5.29 (a) Globular, (b) globular, (c) globular, (d) fibrous, (e) globular, (f) fibrous

5.30 The regions that span the membrane must have a rather high percentage of nonpolar amino acids so they can interact with the nonpolar lipids and proteins in the membrane.

Chapter 6

6.2 $k_3 = 833 \text{ s}^{-1}$

6.3 Acetylcholine has greater affinity for enzyme. The longer nonpolar region of butyrylcholine probably interferes with its binding to acetylcholinesterase.

6.4 (a) Of the three noncovalent interactions, two are ionic: $-COO^-\ ^+H_3N-$, and one is hydrogen bonding. (b) Asp, Glu, Lys, Asn, Gln

6.5 *Competitive*
Binds to enzyme active site
Structure similar to substrate
Same V_{max}, different K_M
Uncompetitive
Binds to ES complex
Different V_{max}, different K_M
Noncompetitive
Binds to a site other than active site
Structures not similar to substrate
Different V_{max}, same K_M

6.6 a. $2H_2O_2 \rightarrow 2H_2O + O_2$
  b. Cellulose + $H_2O \rightarrow$ glucose
  c. Protein + $H_2O \rightarrow$ cleavage of peptide bonds on C side of Arg and Lys
  d.
$$CH_3COCH_2CH_2\overset{+}{N}(CH_3)_3 + H_2O \rightarrow$$
$$CH_3CO^- + CH_2CH_2\overset{+}{N}(CH_3)_3$$
  with O and OH below respectively
  e. Starch + $H_2O \rightarrow$ glucose

6.7 a. Increase of reaction rate
  b. Increase of reaction rate
  c. Decrease of reaction rate
  d. No effect
  e. Decrease for most enzymes because of denaturation
  f. Decrease of reaction rate because of thermal denaturation
  g. Increase of reaction rate

6.8 a, f, g

6.9 The stains from milk and blood are made primarily by proteins. Peptidases present in detergents degrade the proteins to smaller polypeptide units that are more water soluble than larger proteins.

6.10 Nucleophilic side chains contain atoms with lone pairs of electrons such as S, O, and N. For example, Cys, Ser, His, and Lys.

6.11 a. 2, Esterase
  b. 4, Peptidase
  c. 6, Cellulase
  d. 7, Phenylalanine hydroxylase
  e. 1, Amylase
  f. 3, Alcohol dehydrogenase
  g. 8, Phenylalanine decarboxylase

6.12 $1/v_0 = \dfrac{K_M + [S]}{V_{max}[S]}; \dfrac{1}{v_0} = \dfrac{K_M}{V_{max}[S]} + \dfrac{1}{V_{max}}$

6.13 Lys and Arg would have positive charges on their side chains at pH 7. Most His side chains would not be positively charged at pH 7.

6.14 Michaelis–Menten: $V_{max} = 122$ nmol/min
            $K_M = 39$ μM
    Lineweaver–Burk: $V_{max} = 125$ nmol/min
            $K_M = 38$ μM

6.15 Acetophenone is a competitive inhibitor.
    $V_{max} = 125$ nmol/min
    $K_M = 70$ μM

6.16

$$E{-}CH_2{-}SH \; + \; I{-}CH_2{-}\overset{\displaystyle O}{\overset{\|}{C}}NH_2 \longrightarrow$$

$$E{-}CH_2{-}S{-}CH_2{-}\overset{\displaystyle O}{\overset{\|}{C}}NH_2 \; + \; HI$$

6.17 Ethanol is oxidized by the same dehydrogenase, so it acts as a competitive inhibitor, blocking the enzyme from converting methanol to its toxic products.

6.18 Trypsin, pH = 7.9.
    Pepsin, pH = 1.8.
    In the stomach, where the pH is acidic, trypsin would have little or no activity. Pepsin would be more active under these conditions.

6.19 Most enzymes denature and become inactive at temperatures above 50°C.

6.20 The rate of the reaction could be measured by monitoring the disappearance of A or B. Also the appearance of A—B could be monitored.

6.21 (a) Curve 1, (b) curve 3, (c) curve 2

6.22 d

6.23 a. [S] = 0.001 M.
  b. [S] = 0.0005 M.

6.24 Pepsin and papain catalyze the hydrolysis of peptide bonds in proteins. Long fibrous proteins in meat are degraded to shorter polypeptide chains, making the meat tender.

6.25 Pb^{2+} reacts with cysteine sulfhydryl groups in proteins:
    $$E{-}SH + Pb^{2+} \rightleftharpoons E{-}S{-}Pb^{2+} + H^+$$

6.26 (a) Esterases would catalyze hydrolysis of the methyl ester. (b) Peptidases would catalyze hydrolysis of the amide bond.

6.27 a. $CH_3CH_2OH + CH_3COOH$
  b. $CH_3CH_2CCH_3,\; CH_3CH_2CH$
        (both with $\overset{\|}{O}$ ketone/aldehyde groups)

Both of these potential inhibitors have carbonyl groups but not a functional group that can be hydrolyzed.
  c. Diisopropylfluorophosphate

6.28 a. $C_6H_5COOH + CH_3OH$
  b. N.R.
  c. $CH_3CH_2COOH + NH(CH_3)_2$
  d. $\overset{\displaystyle H_3^+NCH_2}{\underset{CH_3}{|}} + CO_2$

  e. Phe + Gly

6.29 b. The K_M values are the same.
  c. The K_M values are different.

6.30 a. The V_{max} values are the same.
  b. The V_{max} values are different.
  c. The V_{max} values are different.

6.31 This drug would be expected to inhibit the synthesis of cholesterol, thus lowering blood cholesterol.

6.32 The V_{max} for the enzyme is about 70 μM/min; $\frac{1}{2} V_{max}$ is 35 μM/min. The [S] that yields v_0 of 35 is about 200 μM.

Chapter 7

7.2 a. Positive cooperativity, homotropic
  b. Negative cooperativity, heterotropic, negative effector

7.3 (a) NAD^+, (b) biotin, (c) coenzyme A, (d) NAD^+

7.4 a.

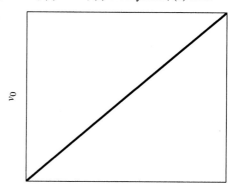

  b.

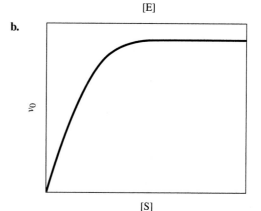

c.

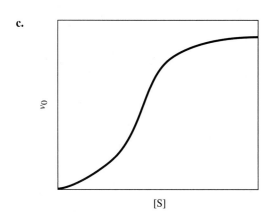

7.5 Catalytic sites or active sites are where substrate molecules bind to an allosteric enzyme. Regulatory sites are where regulatory molecules (effectors) bind.

7.6 *Allosterism:* The enzyme exists in multimeric form and has catalytic and regulatory binding sites. The enzyme is made more or less active by the reversible, noncovalent binding of effectors.
Covalent Modification: The catalytic activity of an enzyme is altered by reversible, covalent changes to specific amino acid side chains in the enzyme.
Proteolytic Cleavage: Some enzymes are synthesized initially in an inactive form, a zymogen, and require cleavage at one or a few specific peptide bonds to produce the active form.
Isoenzymes: Some enzymes are present in different isomeric forms. The forms have similar, but not identical amino acid sequences. All forms demonstrate enzymatic activity and catalyze the same biochemical reaction, but they may differ in kinetics.

7.7 a, d

7.8 a. Decarboxylation or loss of CO_2 from a substrate
   b. Transfer of an amino group from one molecule to another (transamination)
   c. Oxidation–reduction

7.9 c, e

7.10 b

7.11 a. Glycogen phosphorylase + 2 ATP ⇌
          (OH, OH)

   glycogen phosphorylase + 2 ADP
   (OPO_3^{2-}, OPO_3^{2-})

   b. Glycogen phosphorylase + 2 H_2O ⇌
   (OPO_3^{2-}, OPO_3^{2-})

   glycogen phosphorylase + 2 $H_2PO_4^-$
   (OH, OH)

7.12 *Coenzyme:* A small organic or organometallic molecule that assists an enzyme in catalytic action; usually derived from a vitamin.
Prosthetic Group: A small organic molecule or metal ion strongly associated with a protein, usually by covalent or ionic bonds.

7.13 a.
—NH—CHC— + ADP with O (double bond) and CH₂, OPO_3^{2-}

   b.
—NH—CH—C— + A with O (double bond), CH₂, SH, SH, CH₂ O, —NH—CH—C—

   c.
—NH—CHC— + ADP with O (double bond), CH₂, benzene ring, OPO_3^{2-}

   d.
—NHCHC— + HPO_4^{2-} with O (double bond), CH₂, OH

7.14 a. Riboflavin: FAD
   b. Vitamin B_6: Pyridoxal phosphate
   c. Niacin: NAD^+
   d. Folic acid: Tetrahydrofolate
   e. Vitamin B_1: Thiamine pyrophosphate

7.15 a. Alcohol dehydrogenase
   b. Dehydrogenase
   c. Carboxylase

7.16 Those metals essential in some form of life: Fe, Na, Mg, Cu, Mo, Ni, Co

7.17 For example, A → B → C → D → F. A metabolic sequence of reactions that results in a final product of importance (F) requires the formation of intermediate metabolites, B, C, and D. B, C, and D are perhaps of no biological use except to form F. To save on energy and materials, it is logical for the sequence to be regulated at A → B. The enzyme at step 1 is activated only when F is needed.

7.18 Proteolytic cleavage as is carried out on zymogens is an irreversible process and occurs only once in the lifetime of an enzyme molecule. In covalent modification the enzyme is reversibly altered by a set of regulatory enzymes, a process that may be repeated many times with the same enzyme molecule.

7.19 These enzymes would degrade important and essential proteins as soon as they were synthesized. During their storage in the pancreas they could degrade essential structural proteins and other enzymes.

7.20

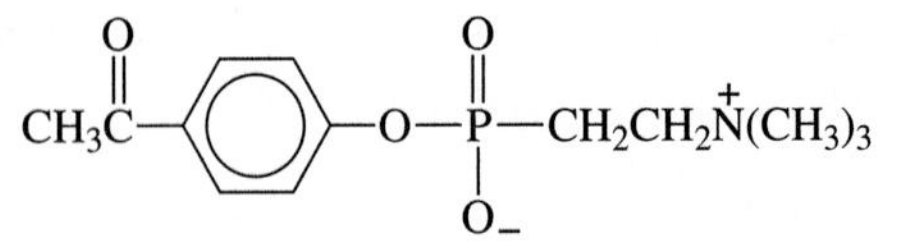

$$CH_3C{-}\underset{O}{\overset{O}{\|}}{-}\bigcirc{-}O{-}\underset{O_-}{\overset{O}{\underset{\|}{P}}}{-}CH_2CH_2\overset{+}{N}(CH_3)_3$$

7.21 The patient may have had a heart attack, causing damage to heart tissue. LDH molecules of the H_4 variety leaked out from the damaged heart tissue into the blood.

7.22 Urea is acting as a denaturing agent, changing the secondary and tertiary structure of ribonuclease P. Urea breaks the native hydrogen bonding in the ribozyme structure.

7.23 Binding of an effector at a secondary site can trigger conformational changes in the protein and change the structure at the active site so that the substrate can bind with more or less affinity.

7.24 No; the enzyme can take on specific secondary and tertiary structures (native conformation) that can bring together, into the same region, amino acid residues from various parts of the enzyme.

7.25 The RNA in ribonuclease P contains the active site for binding and catalytic breakdown of tRNA substrates.

7.26 The first step in a biosynthetic sequence is under regulation. HMG-CoA reductase (catalyzes reaction 1) is a regulatory enzyme.

7.27 a. $CH_3\underset{O}{\overset{\|}{C}}H + NADH + H^+$

b. $\underset{HOOC}{CH_2}\underset{COOH}{CH_2} + FAD$

c. $^-OOCCH_2\underset{O}{\overset{\|}{C}}COOH + ADP + P_i$

7.28 $—Ala—COO^- + H_3{}^+N—Arg—$

7.29 a. Oxidation–reduction
b. Oxidation–reduction
c. Carboxylation

7.30 a. Chlorophyll, ATP
b. Heme
c. Coenzyme B_{12}

Chapter 8

8.2

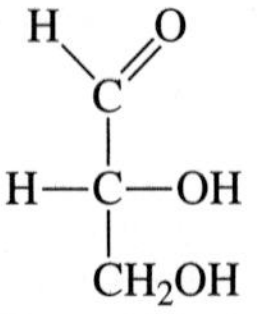

Glyceraldehyde (D, L pairs)   Dihydroxyacetone

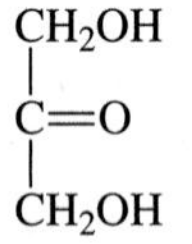

8.3 (a) O, (b) 3, (c) 1, (d) 4, (e) 3, (f) 4, (g) 2, (h) 4, (i) 4, (j) 3

8.4 a.

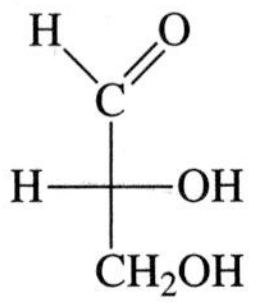

b.

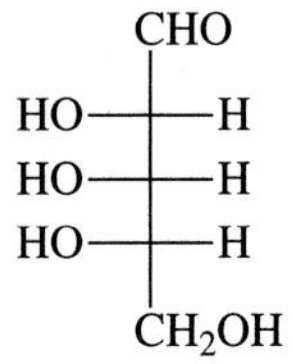

c.

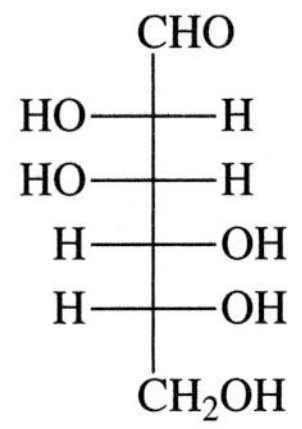

8.5

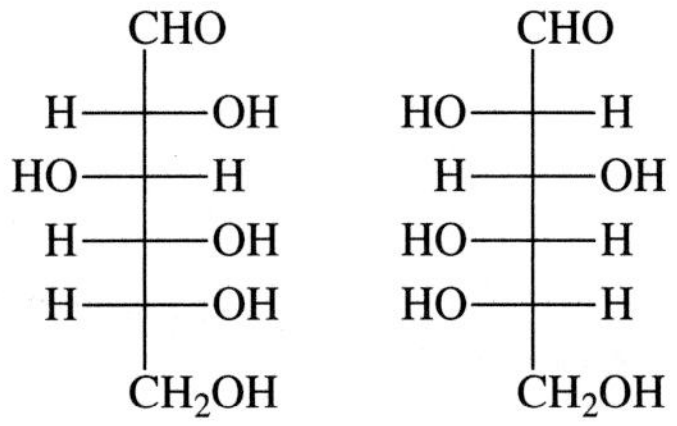

D-Glucose   L-Glucose

8.6 a, b, d, f

8.7 a. Aldehyde, hydroxyl groups
b. Aldehyde, hydroxyl groups
c. Acetal, hydroxyl groups
d. Amide, acetal, hydroxyl groups
e. Carboxylic acid, acetal, hydroxyl groups

8.8 (a) Lactose, (b) maltose, (c) glycogen branches; see Figures 8.16 and 8.20 for structures.

8.9 b, e, f, g

8.10 Hydroxyl group; amide group

8.11 (a) One, (b) one, (c) one, (d) two, (e) two

8.12 a. Hydroxyl group
b. Cyclic hemiketal
c. Hydroxyl group

8.13 (a) Ribose, (b) sucrose, (c) cellulose, (d) starch, (e) hyaluronic acid

8.14

	Amylose	Glycogen
a.	Plants	Animals
b.	Energy storage	Energy storage
c.	Polysaccharide	Polysaccharide
d.	Figure 8.19	Figure 8.20
e.	$\alpha(1 \rightarrow 4)$	$\alpha(1 \rightarrow 4)$
f.	None	$\alpha(1 \rightarrow 6)$

8.15 Glucose, mannose, galactose, N-acetylglucosamine, sialic acid

8.16 Immunological protection and cell–cell recognition events in animals

8.17 a. Aldose–ketose pairs   **e.** Epimers
b. Aldose–ketose pairs   **f.** Anomers
c. Epimers   **g.** Enantiomers
d. Epimers

8.18 Cyclic hemiacetal structures for glyceraldehyde and erythrose would require the formation of unstable three- and four-membered rings, respectively.

8.19 a.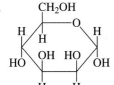

b.

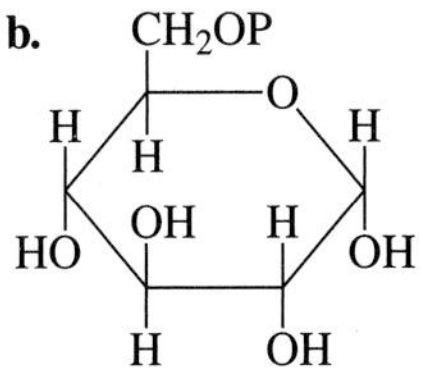

c.

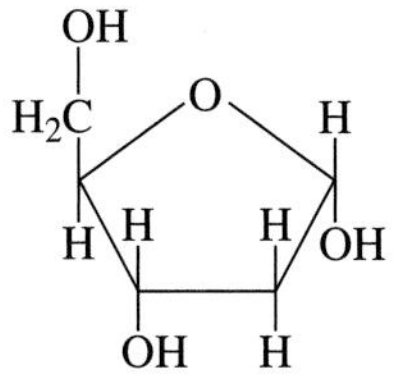

8.20

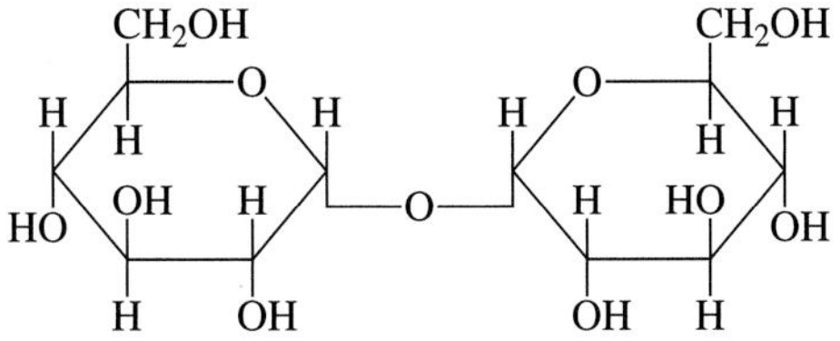

It is a nonreducing sugar.

8.21 a.

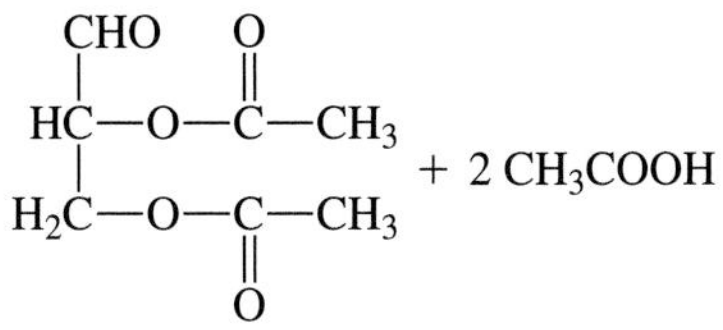

$+ 2\,CH_3COOH$

b. $CH_2-CH-CH_2$ with OH, OH, OH

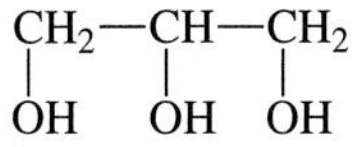

c. COOH
CHOH
CHOH
CH_2OH

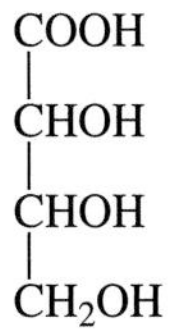

d.

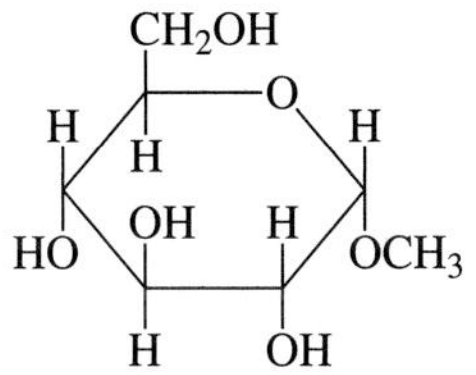

8.22 D-Glucose. The monosaccharide has many hydroxyl groups to hydrogen bond with water. 1-Hexanol only has one hydroxyl group.

8.23 Transfer of electrons with oxidation of a hydroxyl group to a carbonyl group and reduction of $NADP^+$ to $NADPH + H^+$

8.24 a. Cellulose $+ H_2O \rightleftharpoons$ free glucose
b. Starch $+ H_2O \rightleftharpoons$ glucose + maltose
c. Lactose $+ H_2O \rightleftharpoons$ galactose + glucose
d. $Fe^{2+} \xrightarrow{[O]} Fe^{3+}$

8.25 The many hydroxyl groups in saccharides can hydrogen bond with the solvent water.

8.26 a.

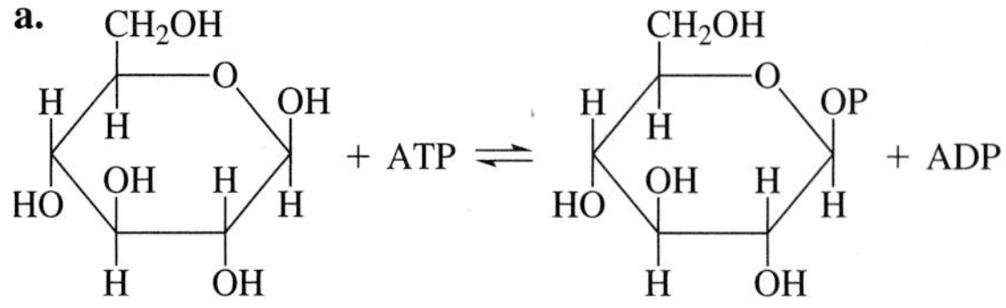

$+ ATP \rightleftharpoons$ ... $+ ADP$

b. See Figure 8.16.

c.

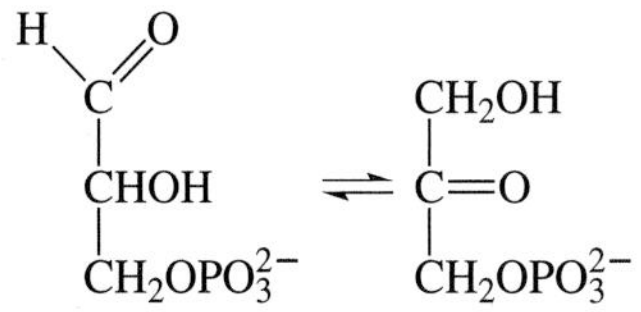

$\rightleftharpoons$

d. CHO
CHOH
CHOH
CHOH
CHOH
CH_2OH $+ NADH + H^+ \rightleftharpoons$ CH_2OH
CHOH
CHOH
CHOH
CHOH
CH_2OH $+ NAD^+$

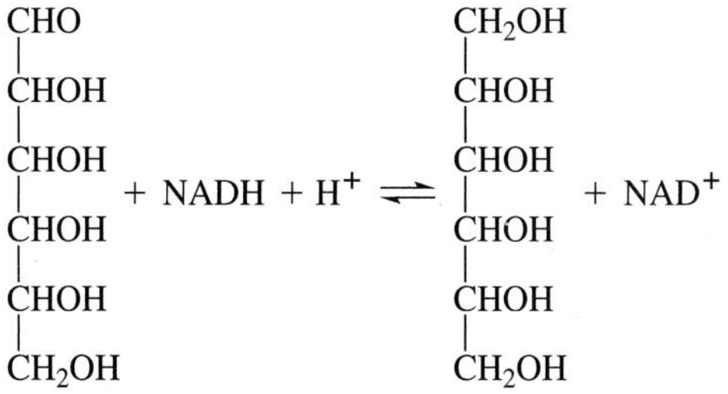

8.27 a. Phosphorylation
b. Hydrolysis
c. Isomerization
d. Oxidation–reduction

8.28 a. Kinase
b. Lactase
c. Isomerase
d. Dehydrogenase

8.29 Pretreat lactose-containing foods with β-galactosidase preparations. Avoid foods containing lactose.

8.30 *pH 1:*

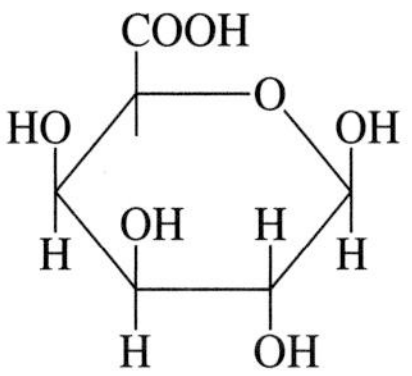

pH 7 and 12:

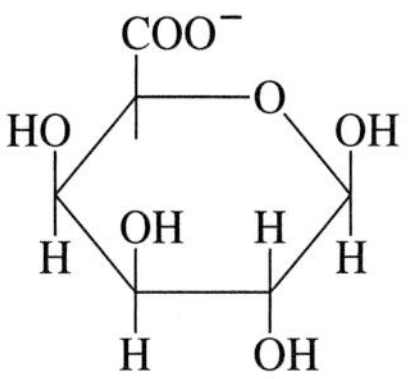

8.31.

8.32 For example, hydrogen bonding between the hydroxyl group on the side chain of serine and a hydroxyl group on glucose.

8.33

8.34

Chapter 9

9.2 a, d, e, h, k

9.3 **b.** $16:2^{\Delta 6,9}$
   c. $20:4^{\Delta 5,8,11,14}$

9.4 **a.** $CH_3(CH_2)_4CH = CHCH_2CH_2COOH$
   b. $CH_3(CH_2)_4CH = CHCH_2CH = CH(CH_2)_7COOH$
   c. $CH_3CH_2CH = CHCH_2CH = CHCH_2CH =$
$CH(CH_2)_7COOH$

9.5 **a.** $CH_3CH_2CH = CHCH_2CH = CHCH_2CH =$
$CHCH_2CH = CHCH_2CH =$
$CHCH_2CH_2CH_2COOH$
   b. $CH_3CH = CHCH_2CH = CHCH_2CH =$
$CHCH_2CH = CHCH_2CH = CHCH_2CH =$
$CH(CH_2)_3COOH$

9.6 **a.** $CH_3(CH_2)_5\underset{\underset{Br}{|}}{C}H-\underset{\underset{Br}{|}}{C}H(CH_2)_7COOH$
   b. $RCOO^-Na^+ + R'COO^-Na^+ + R''COO^-Na^+ + glycerol$
   c. $CH_3(CH_2)_{14}COO^-Na^+ + \underset{\underset{OH}{|}}{C}H_2(CH_2)_{28}CH_3$
   d.

   e. Addition of iodine to the double bonds in the unsaturated fatty acid chains.

9.7 Soap molecules are amphiphilic; that is, they have a nonpolar region and a polar region. In aqueous solution, the nonpolar regions can avoid water by aggregating into molecular arrangements with these regions held together by hydrophobic interactions.

9.8 Bile salts are amphiphilic molecules. Their nonpolar regions interact with nonpolar regions of dietary fats. Their polar regions form hydrogen bonds with water, thus making them somewhat soluble; they form emulsions in water.

9.9 **a.** Catalyzes the hydrolysis of ester bonds in triacylglycerols. Leads to the release of free fatty acids.
   b. Arachidonate $+ O_2 \rightleftharpoons$ various prostaglandins. See Figure 9.13.
   c. Permeases are proteins that enhance the transport of biomolecules through membranes.

9.10 K_M is the enzyme rate constant that defines the substrate concentration which yields a reaction rate of $\frac{1}{2}V_{max}$. K_t is the solute concentration that yields a transport rate of one-half the maximum transport rate.

9.11 **a.** pH 1: $CH_3CH_2CH_2COOH$
   b, c. pH 7 and 12: $CH_3CH_2CH_2COO^-$

9.12 $[CH_2 = CH(CH_2)_8COO^-]_2\ Zn^{2+}$
Most naturally occurring monounsaturated fatty acids have the carbon–carbon double bond between carbons 5 and 6 or 9 and 10.

9.13 Prostaglandins have a five-membered ring; leukotrienes do not. See Figure 9.13.

9.14 The lipid bilayer consists only of amphiphilic lipids associated by hydrophobic interactions. The fluid-mosaic model consists of lipids and proteins. The lipids are present in a lipid bilayer structure, but proteins are interspersed in the bilayer. See Section 9.5 for details.

9.15 See Chapter 9, Window on Biochemistry 9.4

9.16 The triacylglycerols in oils such as canola and olive oil, contain a greater amount of polyunsaturated fatty acids. Carbon–carbon double bonds are readily oxidized by O_2, leading to dark and strongly smelling products that cause rancidity.

9.17 1,2,3-Triacylglycerol < 1,3-diacylglycerol <
2-monoacylglycerol < glycerol.

9.18 $CH_3(CH_2)_6COOH$

9.19 **a.** $CH_3(CH_2)_{10}COO^-Na^+ + H_2O$
   b. $CH_3(CH_2)_{10}COO^-Na^+ + CH_3CH_2OH$
   c. $CH_3\underset{\underset{OH}{|}}{C}HCH_2COOH$
   d. $CH_3CH_2CH_2COOH$

9.20 **a.** Active transport
   b. Facilitated diffusion or active transport
   c. Simple or facilitated diffusion
   d. Simple diffusion
   e. Active transport
   f. Active transport

g. Facilitated diffusion or active transport
h. Facilitated diffusion or active transport

9.21

Polar head	Nonpolar Tail
a. —OH group	Hydrocarbon skeleton
b. $-O-\overset{\overset{\displaystyle O}{\|}}{\underset{\underset{\displaystyle O_-}{\|}}{P}}-O-CH_2CH_2\overset{+}{N}(CH_3)_3$	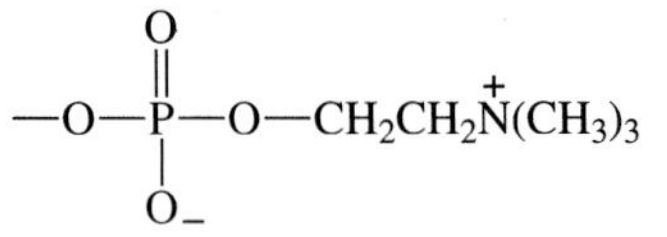 Remainder of molecule
c. $-O-\overset{\overset{\displaystyle O}{\|}}{\underset{\underset{\displaystyle O_-}{\|}}{P}}-O-CH_2CH_2\overset{+}{N}(CH_3)_3$	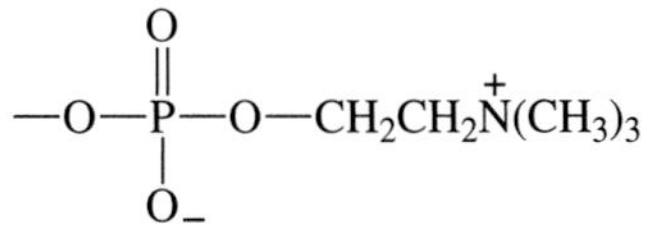 Remainder of molecule
d. Glucose or other carbohydrate	Remainder of molecule

9.22 $K_{transport}$ = the glucose concentration that yields a transport rate of $^1\!/_2\, V_{max}$; $K_t = 1.2$ mM.

9.23 a. H_2O is transported by simple diffusion because it is small and can fit between lipid and protein molecules in the membrane. The sodium ion is small but charged so it cannot diffuse through the nonpolar region of the membrane.
b. CO_2 is small and nonpolar; therefore, it can diffuse through a membrane. Bicarbonate is small but charged so it cannot be transported by simple diffusion.
c. NO is small and relatively nonpolar. Nitrite is charged.
d. Glucose 6-phosphate is highly charged and cannot be transported through membranes.

9.24 Aspirin causes a reduction in the cellular concentration of the prostaglandins. Products of prostaglandin synthase action on arachidonate cause fever, inflammation, and blood clotting.

9.25 All four lipids have the same foundation molecule, phosphatidic acid. See Figure 9.7a.

9.26 (a) Solid, (b) liquid; (c) liquid, (d) solid

9.27 Diethyl ether, chloroform, hexane, methanol

9.28 Because they are synthesized from a C_2 unit, acetate

9.29 The urine contains only water-soluble compounds.

9.30 Comparison of membrane transport and enzyme action:
Same: Both processes involve binding of a molecule, usually a smaller molecule (a substrate or ligand, L), to a protein (Pr). This forms a reversible complex:

$$Pr + L \rightleftharpoons Pr{:}L$$
$$E + S \rightleftharpoons ES$$

Saturation kinetics and ligand specificity are observed for both processes.
Different: Transport involves the movement of the ligand from one side of the membrane to the other with no chemical change in the structure of the ligand. Enzyme action involves the chemical transformation of the substrate to a product.

Chapter 10

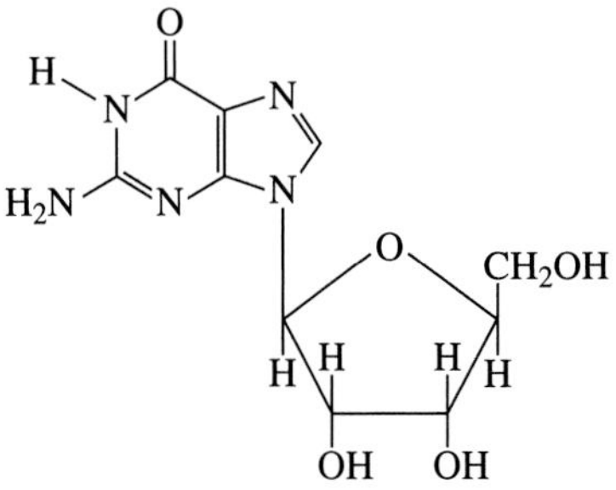

10.2 a.

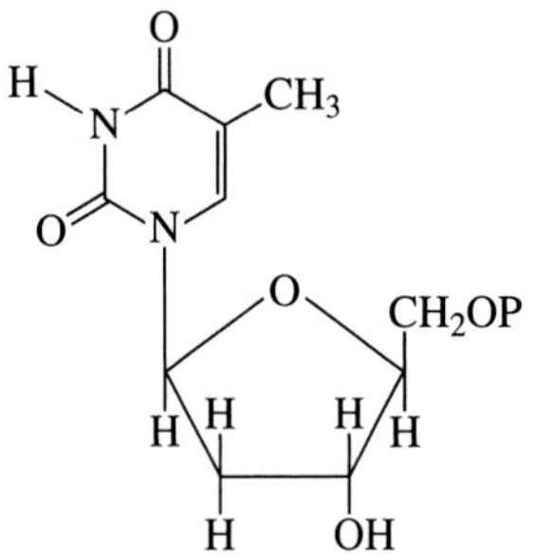

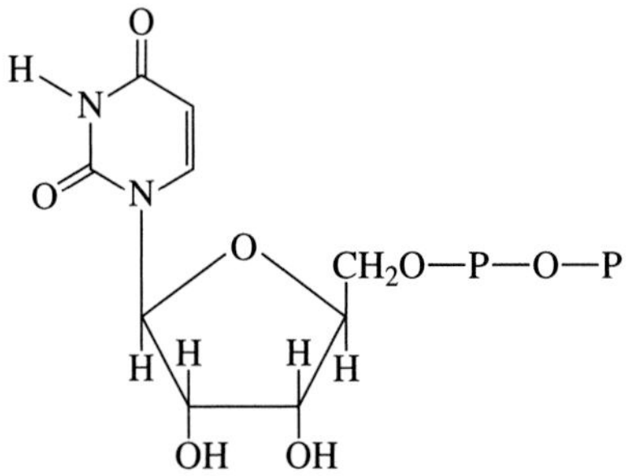

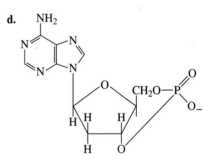

10.3 a. 2′-Deoxyadenosine
b. Adenosine
c. 2′-Deoxythymidine 5′-monophosphate
d. 2′-Deoxyguanosine 5′-diphosphate

10.4 a. 3′ TCGAATGCAGG
b. 3′ AUCCAUGAAGC

10.5 3′ UACAUGGCUAU

10.6

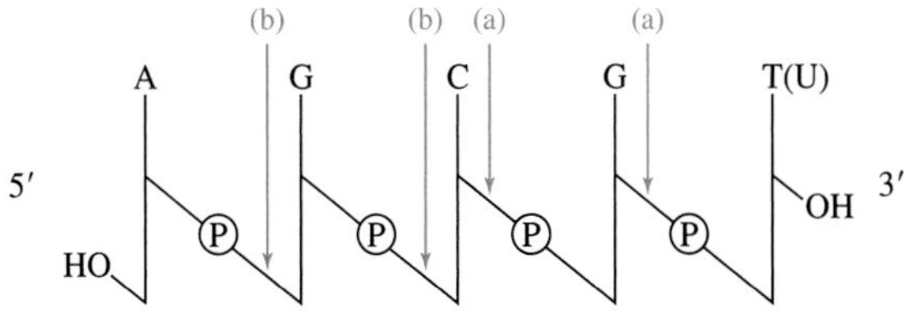

10.7

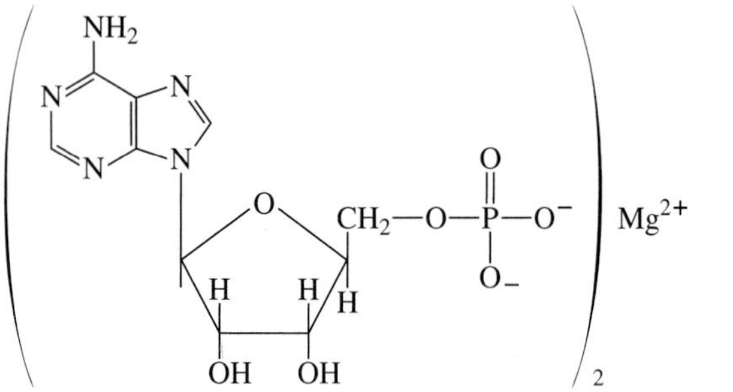

Mg^{2+} (with structure)

c. (structure)

10.8

Form	Helix Direction	Number of Base Pairs	Diameter
A	Right-handed	11	26 Å
B	Right-handed	10.5	20 Å
Z	Left-handed	12	18 Å

10.9 *Primary Structure:* DNA contains A, T, G, and C; RNA contains A, U, G, and C.
Secondary Structure: DNA exists primarily in the double helix; RNA is single stranded and the molecule can fold back onto itself to form short stretches of double-helix structure.
Tertiary Structure: The DNA double helix can fold into more complex structures called supercoiled DNA. RNA in its three forms can take on significant tertiary structure with hairpin turns, double helixes, internal loops, and bulges.

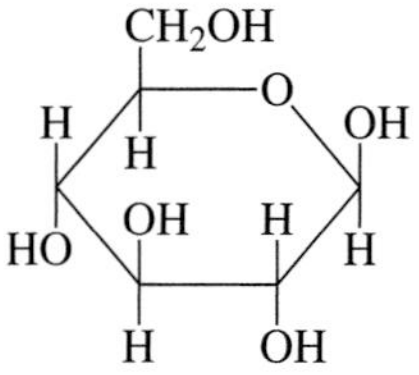

10.15 NAD, FAD, coenzyme A

10.16 Nitrogenous base < nucleoside < nucleotide < double-strand DNA < gene < genome < chromosome.

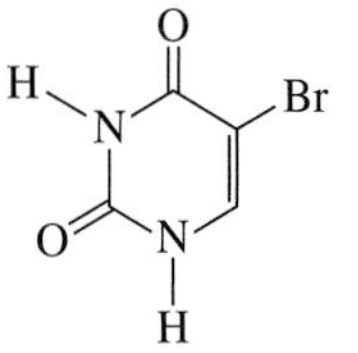

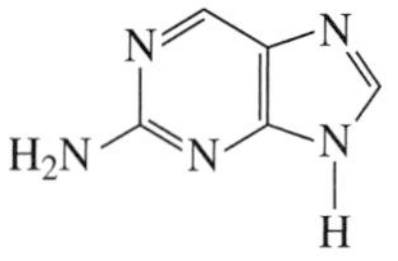

10.17 (structure)

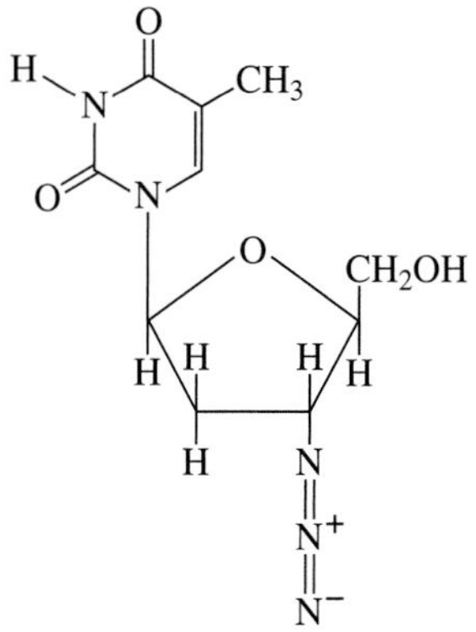

10.10 a. (structure) **b.** (structure)

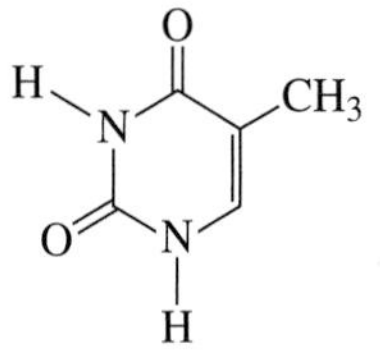

c. (structure) **d.** (structure)

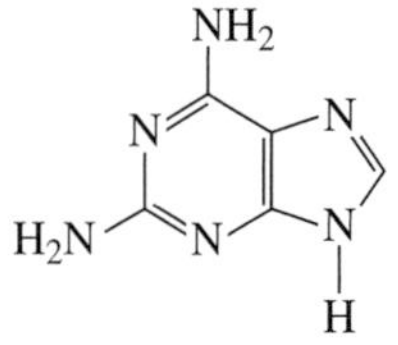

10.11 Nine

10.12 Most sites for restriction endonucleases display twofold rotational symmetry. b does not have such symmetry and is probably not a recognition site. a, c, and d are potential recognition sites.

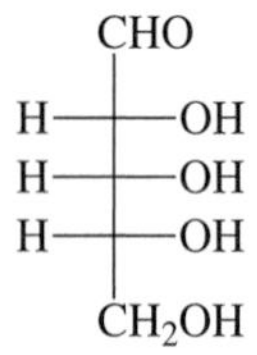

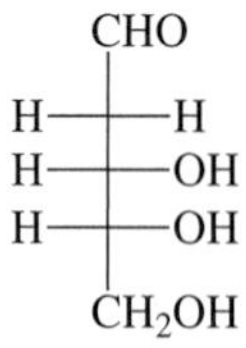

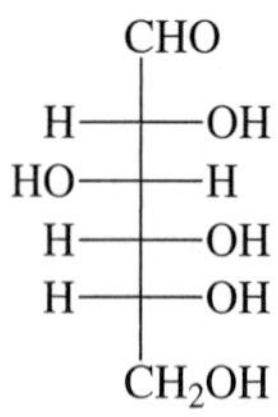

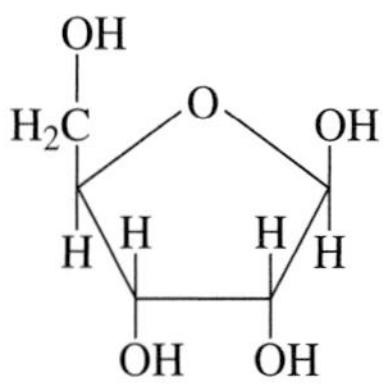

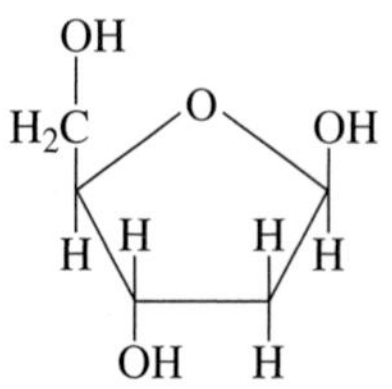

10.18 (structure)

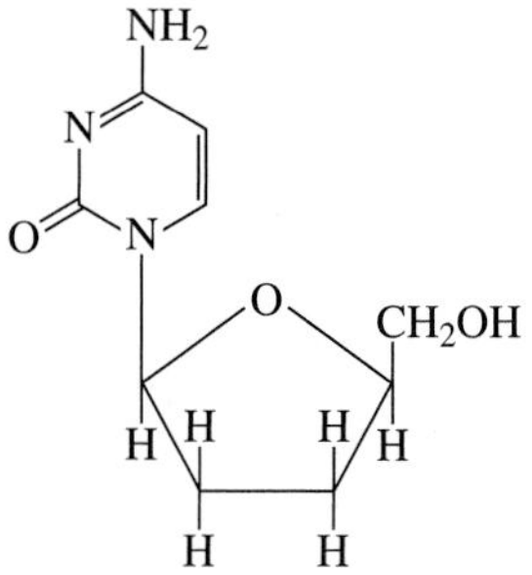

10.19 There is no 4′-hydroxyl group on cytidine for linkage of a phosphate group. The hydroxyl group is linked to C1′ via a hemiacetal bond.

10.20
$$5' \text{ AACTCGATTCTCGAACCG}$$
$$3' \text{ TTGAGCTAAGAGCTTGGC}$$

10.13 a. **b.** **c.**

```
  CHO          CHO          CHO
H—OH         H—H          H—OH
H—OH         H—OH        HO—H
H—OH         H—OH         H—OH
  CH2OH        CH2OH        H—OH
                            CH2OH
```

10.21

DNA	Histone
deoxyribose	
$O=P-O^-$	$\overset{+}{H_3}N-CH_2$ (Lys)
deoxyribose	
$O=P-O^-$	$H_2N-C-CH_2$ (Arg)

10.14 a. (structure) **b.** (structure)

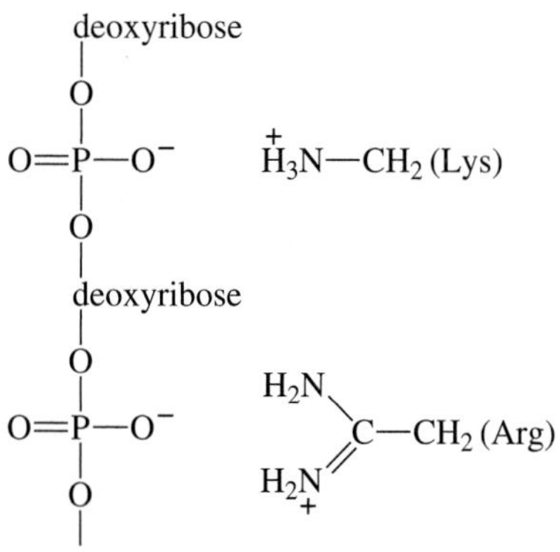

10.22

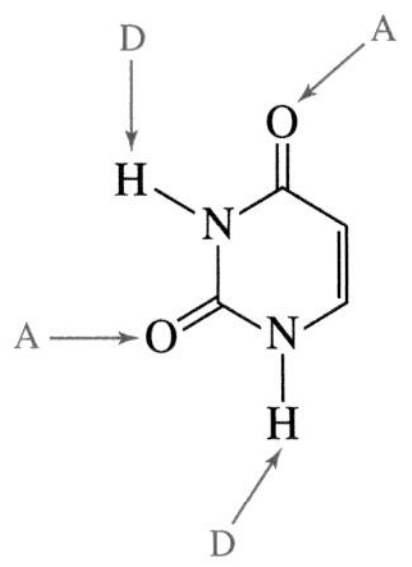

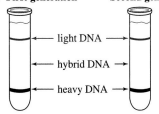

A = acceptor
D = donor

Tautomerization can change the interactions between complementary base pairs. Note that some atoms in the above structures may be donors or acceptors for hydrogen bonding.

10.23 3′ TCCCGATTGAGATTC
5′ AGGGCTAACTCTAAG

10.24 **a.** Catalyzes the interconversion of relaxed and supercoiled DNA.
b. Catalyzes the hydrolysis of phosphoester bonds in RNA.
c. A restriction endonuclease that catalyzes the hydrolysis of phosphoester bonds in DNA at the nucleotide bases,
GA̍NTC

10.25 The hydrogen bonding between GC base pairs is stronger than between AT base pairs.

10.26 a, b, c, d

10.27 This nuclease acts at the 3′ end of DNA or RNA. DNA from φX174 is closed circular and has no free 3′ end.

10.28 (a) **i.** Between N1 of thymine and C1′ of deoxyribose
ii. Between N9 of guanine and C1′ of ribose
iii. Between N1 of uracil and C1′ of ribose
iv. Between N9 of adenine and C1′ of ribose
(b) **i.** At C5′ of deoxyribose
ii. At C5′ of ribose
iii. At C3′ and C5′ of ribose
(c) Cyclic ester bonds are present only in structure d.

10.29 The sequence of bases in DNA and RNA can be "read" in order to direct the sequence of amino acids in proteins.

10.30 Neither AZT nor DDC has the 3′-hydroxyl group needed to form the next phosphoester bond in RNA or DNA synthesis. Therefore RNA and DNA elongation are blocked at this point.

Chapter 11

11.2 First generation   Second generation

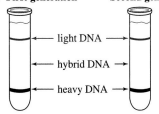

11.3 5′ TTCGAAAGGC

11.4 **a.** Catalyzes the unwinding of the DNA double helix.
b. Single-stranded DNA binding proteins stabilize and protect the unwound DNA from damage.
c. Catalyzes the ATP-dependent phosphate ester bond formation between a free hydroxyl group at the 3′ end of one fragment with a phosphate group at the 5′ end of the other.
d. Catalyzes the synthesis of an RNA primer to initiate the leading strand of DNA.
e. The major enzyme that catalyzes the formation of phosphodiester bonds in DNA.
f. Enzyme responsible for proofreading and correcting errors in newly synthesized DNA

11.5 5′ CCGUAUAGC

11.6 5′ GCUAUACGG

11.7

		DNA Polymerase III	RNA Polymerase
a.	Need for primer	Yes	None
b.	Direction of new chain elongation	5′ → 3′	5′ → 3′
c.	Proofreading functions	3′ → 5′ exonuclease	None
d.	Form of ending nucleotides	dNTPs	NTPs

11.8 **a.** 3′ AGT _ _ _ T
b. 5′ TCA _ _ _ A
c. T
d. C
e. −5, A
f. +3, T

11.9

		DNA-directed RNA synthesis	RNA-directed RNA synthesis
a.	Elongation enzyme	RNA polymerase	RNA replicase
b.	Direction for elongation	5′ → 3′	5′ → 3′
c.	Form of entering nucleotides	NTPs	NTPs
d.	Mechanism	3′-Hydroxyl nucleophilic attack	Same
e.	Proofreading	None	None
f.	Template	DNA	RNA
g.	Need for primer	None	None

11.10 **a.** A change in DNA nucleotide sequence that leads to the death of a cell and/or organism.
b. A change in DNA nucleotide sequence that does not display an observable biological effect.
c. Mutations that occur during normal genetic and metabolic functions; usually caused by errors in replication.
d. Mutations that are caused by environmental agents such as ionizing radiation, chemicals, and UV light.

11.11 a. Recognizes promoter regions and initiates synthesis of RNA.
   b. Acts as a termination factor to stop RNA synthesis.
   c. Catalyzes the formation of new phosphodiester bonds in RNA.

11.12 DNA is used primarily for storage of genetic information. RNA is used for transfer of genetic information.

11.13 Continuous replication leads to the formation of the leading strand. Discontinuous replication leads to the formation of Okazaki fragments.

11.14 Telomerase catalyzes the synthesis of specialized ends on the 3′ end of DNA.

11.15 b, c, d, e, f

11.16 a. Alkylating agent
   b. Alkylating agent
   c. Intercalating agent
   d. Base analog

11.17

Enol   Keto

11.18

It can be added to the growing RNA chain but another nucleotide cannot be added because it has no 3′-hydroxyl group.

11.19 5-Bromouracil is a thymine analog and becomes inserted into DNA at sites normally occupied by T. 5-Bromouracil can exist in keto and enol tautomeric forms. The enol form pairs with G instead of the normal A.

11.20 They have the same sequence except for T in DNA and U in RNA.

11.21 Spontaneous: a, d
   Induced: b, c, e

11.22

AZT is incorporated into the growing DNA chain at the 5′ end. Chain elongation cannot continue because the 3′-azido group blocks the formation of the next phosphodiester bond.

11.23 a.   DNA primer

   b. $2P_i$
   c. $H\cdot + \cdot OH$

11.24 The Ames test evaluates the potential of a chemical compound to promote mutations in a specialized bacterial strain. Very few chemicals actually are shown to be mutagenic by this test, but at least 90% of those positive in the Ames test are shown to be carcinogenic in animal tests. Time and money are saved in testing chemicals because of this strong correlation between mutagenesis and carcinogensis.

11.25 b, c, d, f, g, h

11.26 During transcription, the new RNA strand forms a "hybrid" with DNA. These two nucleic acids are held together by specific base pairing:

DNA ──────── RNA
G · · · · · · · · C
A · · · · · · · · U
T · · · · · · · · A
C · · · · · · · · G

The bases are held together by noncovalent hydrogen bonding.

11.27 c, d, e, f

11.28 See Section 11.2.

11.29 An RNA single-stranded molecule can fold back on itself and be held in hairpin turns by hydrogen bonding at base-paired regions.

11.30 It is not an accurate statement. There is about a 90% chance that a chemical that gives a positive Ames test (mutagenic) is a carcinogen in humans or other animals.

Chapter 12

12.2 a. Phe-Leu-Asp-Arg-Val
   b. Gly-Gly-Val-Ser-Cys

12.3 5′ C-A-U-A-A-U-C-C-U
   Yes. The genetic code is degenerate; that is, some amino acids are designated by more than one triplet of bases. See Figure 2.7.

12.4

	Helix–Turn–Helix	Zn Finger
a. Amino acid composition	Gly at turns	Cys, His
b. Number of amino acid residues	20	30
c. Sequence of amino acids	No regular sequence	See Fig. 12.20
d. Presence of α-helixes, βsheets, turns	2 α-helices, β-turn	Unknown
e. Presence of metal ions	None	Zn

12.5 Two regions on the ribosome–mRNA complex where intermediate polypeptide chains bind during protein synthesis. See Figure 12.5.

12.6 a. Met-Leu-Thr-Ser-Gly-Arg-Ser
  b. Met-Ser-Leu-Gln-Gly-Glu

12.7 c

12.8 a. Molecular complexes that are the sites for cellular protein synthesis.
  b. A sequence of three nucleotide bases on mRNA that interacts with the anticodon on tRNA; specifies the incorporation of a specific amino acid into a polypeptide.
  c. The sequence of three bases on tRNA that combines with the complementary triplet codon on mRNA.
  d. Enzymes that catalyze the linkage of an amino acid to its appropriate tRNA.
  e. Enzyme that catalyzes formation of peptide bonds in protein synthesis.

12.9 b, d

12.10 (a) ATP, (b) ATP $\longrightarrow$ AMP + PP_i, (c) amino acyl adenylate

12.11 Constitutive genes are those that code for certain protein products that must be present at all times for general cell maintenance and metabolism. Inducible genes are those that are activated by signal molecules to increase the level of a specific mRNA and protein.

12.12 b, d, e

12.13 a, b, c, d

12.14 c, e

12.15 Review Section 12.4.

12.16 An average of four phosphoanhydride bonds is cleaved for the incorporation of each amino acid into a polypeptide; 300 amino acids $\times$ 4 $\frac{\text{phosphoanhydride bonds}}{\text{amino acid}}$ =

1200 phosphoanhydride bonds.

12.17 c, d, e, f

12.18 See Figure 12.18.

12.19 a. Amino acid + tRNA + ATP $\rightleftharpoons$
      amino acyl-tRNA + AMP + PP_i
  b. An RNA enzyme that catalyzes the formation of peptide bonds in protein synthesis
  c. NTP + $(NMP)_n \xrightarrow{\text{DNA template}} (NMP)_{n+1}$ + PP_i
      RNA   lengthened RNA

12.20 No. There are only four different nucleotide bases in mRNA. Only four different amino acids could be coded.

12.21 Metals are detoxified by storage in the protein metallothionein. The protein has a great abundance of cysteine residues. Metal ions bind to the sulfur atom by coordinate covalent bonds.

12.22 a. Amide or peptide
  b. Phosphodiester
  c. Ester
  d. Phosphodiester
  e. Hydrogen bonding

12.23 This means that several copies of a polypeptide can be made from a single copy of the mRNA. This is especially important because mRNA is rather unstable and does not have a long lifetime in the cell.

12.24 Lysine and proline are incorporated at the stage of protein synthesis. Post-translational processing involves reactions adding hydroxyl groups to the amino acid side chains.

12.25 If the four code letters (A, T, G, C) arranged in groups of two, then 4^2 = 16 different combinations are possible. This is not enough to code for 20 different amino acids found in proteins.

12.26 a.

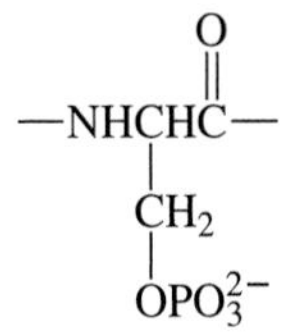

  b.

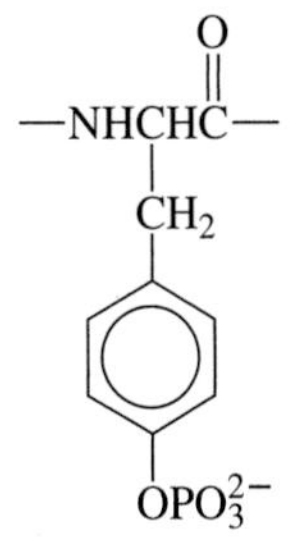

12.27 The three bases in a codon bind to three bases in the appropriate anticodon by hydrogen bonding of base pairs. For example:
Anticodon   3′ U A G 5′
Codon   5′ A U C 3′

12.28 Those amino acids biochemically altered are Tyr, Lys, Pro, Ser, and Thr.

12.29 The carboxyl terminus of ubiquitin is linked via an amide bond to an ε-amino group of a Lys residue of the target protein.

12.30 Review Section 12.4.

Chapter 13

13.2 Requires a free 3′-OH and a 5′-PO_3^{2-}. Only (a) has these requirements.

13.3 *mRNA:* 3′ ACAATAAAAGUUTTAACAGGAGCACC

13.4 Restriction enzyme cleavage sites must display twofold rotational symmetry. Only c is a possible site.

13.5 Stringent replicated plasmids are present in a bacterial cell in only a few copies, whereas several copies of relaxed replicated plasmids are present.

13.6 a, b, d, h, i

13.7 a. Cleavage of DNA by hydrolysis of phosphoester bonds at specific nucleotide sequences
   b. An enzyme that catalyzes the addition of deoxyribonucleotides to the unblocked 3′-hydroxyl ends of single- or double-stranded DNA. Used for adding homopolymer tails.
   c. Catalyzes synthesis of DNA at high temperatures (72°C). Used for the polymerase chain reaction.
   d. Links together 3′ and 5′ ends of DNA. See Exercise 13.2.

13.8 Many plasmids are replicated in a relaxed fashion, so many copies are present in the host cell. Plasmids can easily be screened for the presence of inserted DNA. The major disadvantage is that large eukaryotic DNA fragments cannot be inserted.

13.9

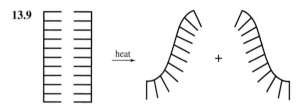

13.10 a. 3′ GAACTTTCACAACCA

   b. 3′ GAACTTCGACAACCA

13.12 Genetic diseases are not contagious because they are not caused by bacterial or viral infection. They are caused by genetic deficiencies, which cannot be spread by contact.

13.13 a, c, d

13.14 c

13.15 *Similarities:* Both catalyze hydrolysis of phosphodiester bonds.
Differences: Nucleases are of two types: exonucleases and endonucleases. The restriction enzymes are always endonucleases, whereas other nucleases may be in either group. Restriction enzymes usually show specificity by acting at specific base sequences. Nucleases vary in specificity, some acting on all phosphodiester bonds and others at specific types of bonds.

13.16 b

13.17 c

13.18 G 3′   +5′ A—A—T—T—C
   C—T—T—A—A 5′   3′ G

13.19 5′ A—A—T—T—C———————G 3′
                   | Section A |
         3′ G———————C—T—T—A—A 5′

13.20 G—A—A—T—T—C———————G A—A—T—T—C
                   | Section A |
         C—T—T—A—A G———————C—T—T—A—A G

13.21 *RNA probe:* 3′ UAACCAUGUU

13.22 AUG is the codon for the first amino acid, *N*-formylmethionine. There must be a promoter region to the left (upstream) of AUG. This is a segment of 20–200 base pairs specific for the gene to be transcribed. Within the promoter region there must be the Pribnow box, TATAAT at base −10, and at base −35, there must be the sequence TTGACA.

13.23 (a) 1, (b) 1, (c) 7, (d) 10

13.24 Increasing the temperature of the reaction mixture enhances the rate of the DNA polymerization reaction.

Chapter 14

14.2 a. Oxidation–reduction
   b. Isomerization and rearrangement
   c. Group transfer
   d. Hydrolytic cleavage
   e. Nonhydrolytic cleavage
   f. Bond formation using energy from ATP

14.3 a. Dehydrogenase
   b. Isomerase
   c. Transferase
   d. Hydrolase
   e. Lyase
   f. Ligase

14.4 *Correct:* a, d
   Incorrect: b, stage III of metabolism is spontaneous; therefore, $\Delta G^{\circ\prime}$ for the process is negative. c, the hydrolysis of storage molecules, proteins, starch, and triacylglycerols is the first step in catabolism.

14.5 *Catabolism:* energy release, convergence of reactions, oxidation reactions, production of ATP, degradation

14.6 $\Delta G^{\circ\prime} = -64.1$ kJ/mol.

14.7 a. 2   **d.** 1   **g.** 0
   b. 1   **e.** 1   **h.** 0
   c. 0   **f.** 0   **i.** 1

14.8 a. 2
   b. Phosphoester
   c. Complexed with Mg^{2+}
   d. *N*-Glycosidic

14.9 a. NAD^+, **b.** CH_3CH_2OH

14.10 a.

$$CH_3C\overset{\displaystyle O}{\underset{\displaystyle H}{<}}$$

   b. $NADH + H^+$

14.11

$$\overset{\displaystyle O}{\underset{\displaystyle O_-}{\overset{\|}{{}^-O-\underset{|}{P}-OH}}} \longleftrightarrow \overset{\displaystyle O^-}{\underset{\displaystyle O_-}{\overset{\|}{O=\underset{|}{P}-OH}}} \longleftrightarrow \overset{\displaystyle O^-}{\underset{\displaystyle O}{\overset{|}{{}^-O-\underset{\|}{P}-OH}}}$$

14.12 All of these biomolecules have an adenine moiety.

14.13

	Standard Conditions	Normal Cellular Conditions
Temperature	25°C	Variable
Pressure	1 atm	1 atm
Concentrations	1 *M* of reactants and products	Metabolites are present in many different concentrations

14.14 *Autotrophs:* Organisms that are "self-feeding"; that is, they can synthesize all of their organic chemical needs from inorganic carbon supplied as CO_2.
Heterotrophs: Organisms that "feed on others"; that is, they synthesize organic chemicals from the other organic compounds they consume.

14.15 $\Delta G^{\circ\prime}$ = standard free energy change for a process under standard conditions and at pH 7.0
ΔG° = standard free energy change for a process under standard conditions

14.16

$$^-O-\overset{\overset{O}{\|}}{\underset{\underset{O_-}{|}}{P}}-O-\overset{\overset{O}{\|}}{\underset{\underset{O_-}{|}}{P}}-O^- \longleftrightarrow O=\overset{\underset{\underset{O_-}{|}}{P}}{}-O-\overset{\overset{O}{\|}}{\underset{\underset{O_-}{|}}{P}}-O^- \longleftrightarrow$$

$$^-O-\overset{\overset{O^-}{\|}}{\underset{\underset{O}{|}}{P}}-O-\overset{\overset{O}{\|}}{\underset{\underset{O_-}{|}}{P}}-O^- \quad \text{etc.}$$

14.17 See Figure 14.13b

14.18

$CH_3CH_3 < CH_2=CH_2 < CH_3CH_2OH < CH_3C\overset{O}{\underset{H}{<}}$
(least oxidized)

$< CH_3COOH < \overset{COOH}{\underset{COOH}{|}} < CO_2$ (most oxidized)

14.19 $K'_{eq} = 19$; $\Delta G^{\circ\prime} = -7.3$ kJ/mol.
The reaction is thermodynamically favorable in the direction written.

14.20 (lowest) AMP $\approx$ 2,3-BPG < ADP $\approx$ ATP <
1,3-BPG < PEP (highest).

14.21
a. To the right
b. To the right
c. To the right
d. To the left
e. To the right

14.22 $K'_{eq} = 1$; $\Delta G^{\circ\prime} = 0$.
The reaction is at equilibrium. The rate is equal in both directions. No energy is released.

14.23 Only one set of reactions is required after acetyl CoA. This means less energy input is necessary and fewer resources are required for catabolism.

14.24
a. $K'_{eq} = 851$
b. $K'_{eq} = \frac{[\text{glucose 6-phosphate}][\text{ADP}]}{[\text{glucose}][\text{ATP}]}$
c. 85.1

14.25
a. $RCOOH + CH_3OH$
b. $RCOOH + NH_3$
c. $RCOOH + H_2N - CHR'COOH$
d. $\underset{OH}{\overset{|}{C}}H_2CH_2CH_2COOH$
e. $\underset{NH_2}{\overset{|}{C}}H_2CH_2CH_2COOH$

14.26 *Blueprint (directions):* DNA
Materials: Foodstuffs in the form of fats, proteins, and carbohydrates
Energy: The ATP cycle; oxidation of foodstuffs with energy transfer to the form of ATP; use of ATP in biosynthesis, movement, etc.

14.27
a. Phosphoryl group transfer from glucose 6-phosphate to H_2O.
b. The reaction will proceed to the right as written.
c. The enzyme will have no effect on the $\Delta G^{\circ\prime}$ for the reaction so the answer is -13.7 kJ/mol. The enzyme will allow the reaction to proceed to equilibrium faster.

14.28 The process of photosynthesis, $CO_2 \rightarrow$ glucose, is one of reduction. Note that most of the carbons and oxygens in glucose are in the alcohol oxidation stage $\underset{OH \quad OH}{\overset{-CH-CH-}{}}$. Even the aldehyde carbon $\overset{O}{\underset{}{-\overset{\|}{C}-H}}$ is reduced in glucose compared to $O=C=O$. Note that reducing agents (NADPH) are required for photosynthesis.

14.29 $\overset{COO^-}{\underset{}{|}}$
$\underset{}{CH_2}$
$\underset{}{N-CH_3}$
$\underset{}{C=NH_2^+}$
$\underset{}{NH_2}$
Creatine

14.30 The overall process of glucose catabolism is one of oxidation. Carbons in glucose ($\underset{OH}{\overset{|}{-C-}}$ and $\underset{O}{\overset{\|}{-C-H}}$) are oxidized to become $O=C=O$. The reactions are coupled to the formation of NADH from NAD^+.

14.31 $K'_{eq} = 2.2 \times 10^5$.

14.32
a. $RCOOH + CH_3-SH$; greater than one.
b. $ADP + HPO_4^{2-}$; greater than one.
c. Glucose 6-phosphate + ADP; greater than one.

Chapter 15

15.2
a. Nonhydrolytic cleavage
b. Bond formation using energy from ATP
c. Hydrolysis
d. Nonhydrolytic cleavage

15.3 Lactate $\rightarrow$ pyruvate $\rightarrow$ PEP $\rightarrow$
glyceraldehyde 3-phosphate $\rightarrow$ glucose
See Figure 15.8.

15.4 Glucose $\rightarrow$ glucose 6-phosphate $\rightarrow$ fructose 6-phosphate
Glucose $\rightarrow$ glucose 6-phosphate $\rightarrow$
glucose 1-phosphate $\rightarrow$ UDP-glucose

UDP-glucose + fructose 6-phosphate $\rightarrow$
sucrose 6-phosphate
Sucrose 6-phosphate + $H_2O \rightarrow$ sucrose

15.5 Reactions c, e, and f would be blocked.

15.6 a. Fructose 6-phosphate + ATP $\rightleftharpoons$
    fructose 1,6-bisphosphate + ADP
  b. UDP-galactose + glucose $\rightleftharpoons$ lactose + UDP
  c. Starch + H_2O $\rightleftharpoons$ glucose
  d. Glycerol + ATP $\rightleftharpoons$ glycerol 3-phosphate + ADP
  e. Glycerol 3-phosphate + NAD^+ $\rightleftharpoons$
    dihydroxyacetone phosphate + NADH + H^+
  f. UDP-galactose + glucose $\rightleftharpoons$ lactose + UDP

15.7 All three polysaccharides are composed of glucose monomers; however, there are differences in the type of bonding. See Table 8.3.

15.8 a. Glucose 6-phosphate
  b. Fructose 6-phosphate
  c. Mannose 6-phosphate

15.9 (a) ATP, (b) glucose 6-phosphate, (c) ATP

15.10 (a) Inhibition, (b) inhibition, (c) stimulation

15.11

	Glycolysis	**Gluconeogenesis**
a. Starting substrates	Glucose, other carbohydrates	pyruvate, lactate, glycerol
b. Energy requirements	2 ATP generated	6 phosphoanhy-dride bonds: 4 ATP; 2 GTP used
c. NAD^+/NADH	2 NADH generated	2 NAD^+ generated

15.12 Glucose + ATP $\rightleftharpoons$ glucose 6-phosphate + ADP

Fructose 6-phosphate + ATP $\rightleftharpoons$
    fructose 1,6-bisphosphate + ADP

PEP + ADP $\rightleftharpoons$ pyruvate + ATP

All three reactions are essentially irreversible because of a large negative $\Delta G^{o\prime}$.

15.13 a. Free and unmodified glucose
  b. Free and unmodified glucose
  c. Glucose 1-phosphate
  d. Free and unmodified glucose

15.14   acetaldehyde
In yeast: Pyruvate $\longrightarrow$ + $\longrightarrow$ ethanol
                CO_2

In human muscle: Pyruvate $\longrightarrow$ lactate

15.15 Blood sugar is glucose.

15.16 a. Glucose 1-phosphate + 3 ADP + 2 P_i + 2 NAD^+ $\longrightarrow$
    2 pyruvate + 3 ATP + 2 NADH + 2 H^+ + 2 H_2O
  b. Glyceraldehyde 3-phosphate + 2 ADP + P_i $\longrightarrow$
    lactate + 2 ATP + H_2O
  c. Glycerol + ADP + P_i + 2 NAD^+ $\longrightarrow$
    pyruvate + ATP + 2 NADH + 2 H^+ + H_2O
  d. Pyruvate + NADH + H^+ $\longrightarrow$
    ethanol + CO_2 + NAD^+

15.17 Glucose + ATP $\xrightarrow{\text{hexokinase}}$
    glucose 6-phosphate + ADP

Glucose 6-phosphate + H_2O $\xrightarrow{\text{glucose 6-phosphatase}}$
    glucose + P_i

15.18

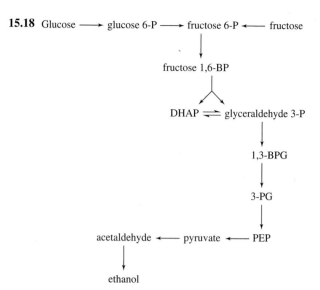

15.19 Maltose + 4 ADP + 4 P_i + 4 NAD^+ $\longrightarrow$
    4 pyruvate + 4 ATP + 4 NADH + 4 H^+ + 3 H_2O

15.20 Glucose derived from lactose hydrolysis is oxidized by glycolysis to pyruvate. The pyruvate is converted to lactate. Galactose is converted to glucose 6-phosphate (Figure 15.3), which is converted to lactate.

15.21 a. Half of the pyruvate molecules would be labeled with ^{14}C in the methyl carbon.
  b. The carboxyl carbon (C1) of pyruvate is labeled with ^{14}C.
  c. The carbonyl carbon (C2) of pyruvate is labeled with ^{14}C.

15.22 The carboxyl carbon of pyruvate is labeled with ^{14}C.

15.23 No carbon atoms in glucose would be labeled.

15.24 Glucose + 2 ATP + H_2O + glycogen$_n$ $\longrightarrow$
    glycogen$_{n+1}$ + 2 ADP + 2 P_i
Therefore, two phosphoanhydride bonds are required.

15.25 Glucose + 2 ADP + 2 P_i + 2 NAD^+ $\longrightarrow$
    2 pyruvate + 2 ATP + 2 NADH + 2 H_2O + 2 H^+
Glucose + 2 ADP + 2 P_i $\longrightarrow$ 2 lactate + 2 ATP + 2 H_2O

	Glycolysis	**Fermentation**
a. Starting substrate	Glucose	Glucose
b. Final product	Pyruvate	Lactate
c. ATP yield	+2	+2
d. NADH yield	+2	0

15.26 2 glycerol $\longrightarrow$ 2 glycerol 3-phosphate $\longrightarrow$
    DHAP $\longrightarrow$ glyceraldehyde 3-phosphate
Glyceraldehyde 3-phosphate + DHAP $\longrightarrow$
fructose 1,6-bisphosphate $\longrightarrow$ fructose 6-phosphate $\longrightarrow$
    glucose 6-phosphate $\longrightarrow$ glucose

15.27 The phosphoryl group at C1 is linked as an anhydride bond. The phosphoryl group at C3 is linked as a phosphate ester. To answer, see Table 14.6.

15.28 Lactate $\longrightarrow$ PEP
Lactate + NAD^+ + ATP + GTP + H_2O $\rightleftharpoons$
    PEP + GDP + ADP + P_i + NADH + H^+
Pyruvate $\longrightarrow$ PEP
Pyruvate + ATP + GTP + H_2O $\rightleftharpoons$
    PEP + ADP + GDP + P_i

15.29 $\Delta G^{\circ\prime} = -16.7$ kJ/mol.

15.30 a.

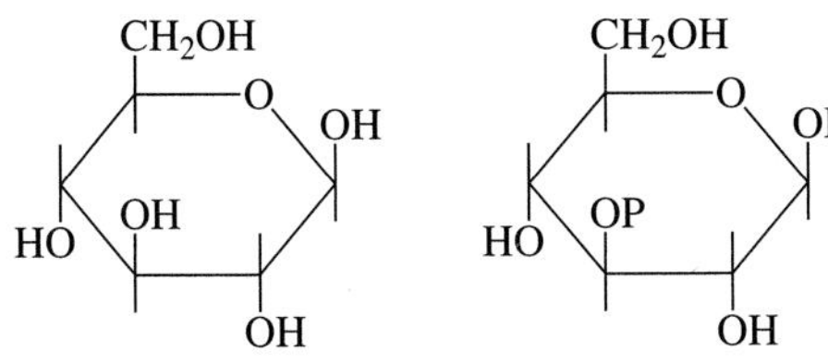

b. The phosphoryl groups are held by a phosphoester bond in both the 3 and 6 positions. The energy would be the same.

Chapter 16

16.2
1. c 4. f 7. h
2. e 5. i 8. g
3. a 6. b 9. d

16.3 Substrate-level ATP (or GTP): 4
NADH: 10
FADH$_2$: 2

16.4

$$CH_3CCOO^- + CO_2 + NADH + H^+ \rightleftharpoons$$

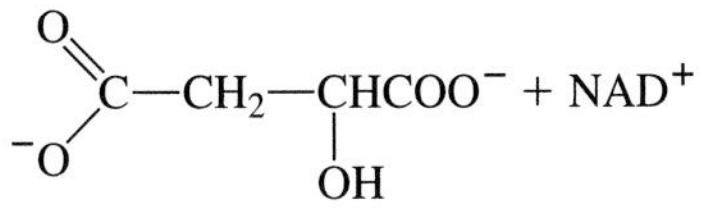

Pyruvate is carboxylated and the keto group is reduced to a hydroxyl group.

16.5 a. Oxaloacetate
b. Succinyl CoA
c. Oxaloacetate
d. α-Ketoglutarate
e. Oxaloacetate
f. α-Ketoglutarate
g. Oxaloacetate
h. Oxaloacetate

16.6
1. c
2. a
3. d
4. b

16.7

ATP (GTP)	NADH	FADH$_2$
0	+1	0

16.8 a.

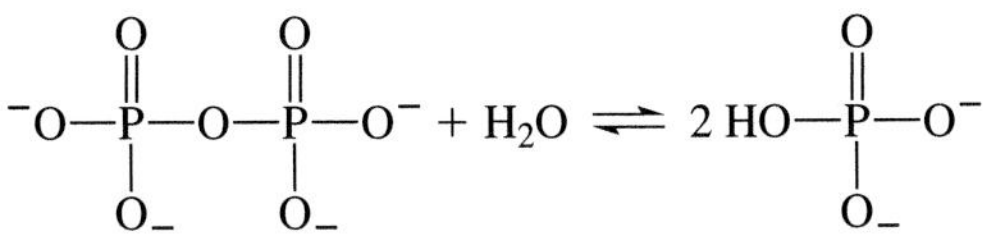

b. Phosphoanhydride bond
c. PP$_i$ is the product of some reactions of ATP:
ATP → AMP + PP$_i$. The continued hydrolysis of PP$_i$ releases more energy for a coupled reaction.

16.9 c

16.10 Both display the same general biochemistry with identical enzymes and chemical mechanisms. They differ in the structure of the entering substrate:

16.11

ATP	NADPH	FADH$_2$
−1	+2	0

16.12 Both pathways cause the oxidation of glucose. The products differ, pyruvate versus ribose 5-phosphate. NADH is produced in glycolysis, NADPH in the phosphogluconate pathway. They serve very different biological functions, glycolysis for energy metabolism, the phosphogluconate pathway to prepare intermediates for biosynthesis.

16.13 a. Isomerization–rearrangement
b. Oxidation–reduction and nonhydrolytic cleavage

16.14 NADPH has a phosphoester group not present in NADH. (See Figure 16.7.) Both cofactors function in redox reactions.

16.15 a. Cytoplasm
b. Mitochondrial matrix
c. Cytoplasm
d. Cytoplasm
e. Mitochondrial matrix

16.16 NADH is usually formed from NAD$^+$ during catabolic reactions.
NADPH is usually used as a reducing agent for anabolic processes.

16.17 Isocitrate dehydrogenase
α-Ketoglutarate dehydrogenase complex
Succinate dehydrogenase
Malate dehydrogenase

16.18 Glucose-6-phosphate dehydrogenase
6-Phosphogluconate dehydrogenase

16.19 a. If levels of ATP are high then the pyruvate dehydrogenase complex that causes the production of more ATP is inhibited.
b. High levels of ADP are indicative of low ATP levels. Stimulation of isocitrate dehydrogenase will increase ATP levels.
c. Pyruvate carboxylase catalyzes the synthesis of oxaloacetate, which reacts with acetyl CoA.
d. NADH is a product of the citric acid cycle. An abundance of NADH slows the citric acid cycle.
e. If fatty acids are abundant, there is less need for acetyl CoA from pyruvate oxidation.

16.20 a. Pyruvate + 4 NAD$^+$ + FAD + ADP + P$_i$ + 2 H$_2$O → 3 CO$_2$ + 4 NADH +4 H$^+$ + FADH$_2$ + ATP
b. Acetyl CoA + 3 NAD$^+$ + FAD + ADP + P$_i$ + 2 H$_2$O → 2 CO$_2$ + 3 NADH + 3 H$^+$ + FADH$_2$ + ATP + CoASH
c. 2 acetyl CoA + NAD$^+$ + 2 H$_2$O → succinate + 2 CoA + NADH + H$^+$
d. Glucose 6-phosphate + 2 NADP$^+$ + H$_2$O → ribose 5-phosphate + CO$_2$ + 2 NADPH + 2 H$^+$

16.21 Ethanol → acetaldehyde → acetate
↓
to glyoxylate ← acetyl CoA
cycle

16.22 Substrate-level ATP (or GTP): 1
NADH: 4
FADH$_2$: 1

16.23 All enzymes that have TPP as a prosthetic group and catalyze the decarboxylation of α-keto acids would be slowed.

16.24 $^-OOC—C≡C—COO^-$; $^-OOC—COO^-$

16.25 a.

$$CH_3C\overset{O}{\underset{H}{\diagup}} + NADH + H^+$$

b. $FAD + NADH + H^+$

c.

$$Lipoamide\overset{S}{\underset{S}{\diagdown|}} + FADH_2$$

d.

$$^-OOC\diagdown\underset{H}{C}=\underset{COO^-}{C}\diagup H + FADH_2$$

16.26 $PEP + ADP \rightleftharpoons$ pyruvate $+ ATP$
$1,3\text{-}BPG + ADP \rightleftharpoons 3\text{-}PG + ATP$

16.27 $*COO^-$
$|$
$*CHOH$
$|$
CH_2
$|$
COO^-

16.28 See Table 16.2.
Reaction 1: hydrolysis of thioester
Reaction 7: addition to double bond

16.29 d

16.30 NADPH is a product of the phosphogluconate pathway. If there is already a sufficient quantity of NADPH then there is no need to produce more.

Chapter 17

17.2 b, c, d, e, f, g
a. The electron carriers involved in electron transport are present in the mitochondria.

17.3 a. -70.4 kJ/mol
b. -43.4 kJ/mol
c. -3.9 kJ/mol
d. -39.6 kJ/mol
e. -49.2 kJ/mol
$ADP + P_i \rightarrow ATP + H_2O$ $\Delta G^{\circ\prime} = +30.6$ kJ/mol;
ATP is synthesized by a, b, d, e.

17.4 (a) NADH, (b) FADH₂, (c) cyt *c*, (d) CoQH₂, (e) NADH, (f) succinate

17.5 (a) CoQ, (b) O₂, (c) cyt *a/a₃*, (d) cyt *c*, (e) FAD, (f) O₂

17.6 Heterotrophs cannot synthesize all needed organic compounds and so must ingest complex carbon- and nitrogen-containing compounds in their diets. Phototrophs absorb energy from solar radiation and use the energy to make ATP and NADPH, which are used to make carbohydrates.

17.7 a. Electron-transfer agent
b. Catalytic unit for ATP synthase in oxidative phosphorylation
c. Electron-transfer agent
d. Transmembrane channel for proton flow; component of ATP synthase
e. Electron-transfer agent
f. Electron-transfer agents

17.8 a. Electrons flow from H₂O to NADP⁺ to make NADPH
b. True
c. True
d. True
e. Cyclic photophosphorylation results in production of ATP but no reduced product NADPH or O₂ evolution.
f. True

17.9 Secondary pigments absorb light in wavelength ranges in which chlorophyll is not as efficient; thus, they supplement the photochemical action spectrum of the chlorophylls.

17.10 *Cyclic:* In photosynthesis, formation of ATP is linked to a cyclic flow of electrons within photosystem II; no NADPH or O₂ is produced.
Noncyclic: Synthesis of ATP is linked to a one-way flow of electrons from H₂O through photosystems I and II to NADP⁺, which produces NADPH.

17.11 a. Primary acceptor of light energy.
b. Secondary pigment for absorption of light energy at wavelengths where chlorophylls are not efficient.
c. Photosystem I transfers electrons from P700 to NADP⁺.
d. An electron-transfer protein containing iron and sulfur in photosystem I.
e. A blue, copper-containing, electron-transfer protein in photosystem II.
f. A protein containing manganese that transfers electrons from H₂O to P680⁺.
g. An integral protein in the thylakoid membrane that acts as a protein channel to regulate the formation of the proton gradient.
h. A peripheral thylakoid membrane protein that catalyzes synthesis of ATP.

17.12 a. 3
b. 1, 2
c. 2, 4
d. 4, 5

17.13 a. 3
b. 4
c. 1
d. 2
e. 6
f. 5

17.14 Rubisco catalyzes the addition of inorganic carbon (CO₂) to ribulose 1,5-bisphosphate to form a six-carbon intermediate; this is followed by cleavage of the intermediate, resulting in two molecules of the three-carbon compound, 3-phosphoglycerate. Its oxygenase activity involves the addition of O₂ to ribulose 1,5-bisphosphate to produce phosphoglycolate and 3-phosphoglycerate.

17.15

NAD Linked	FAD Linked
Cytoplasm	Mitochondria
Mitochondria	
Mitochondria	
Mitochondria	
Mitochondria	

17.16 *Mitochondrial electron transport:* Electrons flow from NADH and FADH, to O₂. Energy released is coupled to ATP formation. Electrons flow downhill. This process occurs in animal and plant mitochondria.

Chloroplast electron transport: Electrons flow from H_2O to $NADP^+$. Since electrons flow uphill, energy from the sun is required. Products are NADPH and O_2.

17.17 Primarily iron and copper

17.18 Primarily iron, copper, and magnesium

17.19 2,4-Dinitrophenol is acting as an uncoupling agent.

17.20 See Section 17.3.

17.21 (a) 38, (b) 12, (c) 76, (d) 15, (e) 37

17.22 Green leaves in summer contain all pigments, but chlorophyll is the most abundant. Cold weather in the fall triggers a decline in chlorophyll synthesis and an increase in its degradation. The red and yellow carotenoids do not degrade as readily.

17.23 **Calvin Cycle**

$$6\,CO_2 + 12\,NADPH + 12\,H^+ + 18\,ATP + 12\,H_2O \longrightarrow$$
$$C_6H_{12}O_6 + 12\,NADP^+ + 18\,ADP + 18\,P_i$$

Gluconeogenesis

$$2\text{ pyruvate} + 4\,ATP + 2\,GTP +$$
$$2\,NADH + 2\,H^+ + 6\,H_2O \longrightarrow$$
$$\text{glucose} + 2\,NAD^+ + 4\,ADP + 2\,GDP + 6\,P_i$$

17.24 These cell organelles are somewhat similar in structure and also perform similar functions. They both are involved in production of ATP for energy metabolism. Mitochondria produce ATP by oxidation of nutrients and electron transfer to O_2. Chloroplasts produce ATP by absorption of light energy and electron transfer.

17.25 *Substrate level:* The synthesis of ATP from ADP using energy from the direct metabolism of a high-energy reactant.
Oxidative: The synthesis of ATP from ADP using energy from mitochondrial electron transfer from NADH and $FADH_2$ to O_2.

17.26 2 glyceraldehyde 3-phosphate $\longrightarrow$ 2 DHAP
2 DHAP + 2 glyceraldehyde 3-phosphate $\longrightarrow$
2 fructose 1,6-bisphosphate

2 fructose 1,6-bisphosphate $\longrightarrow$ 2 fructose 6-phosphate
Fructose 6-phosphate $\longrightarrow$ glucose 6-phosphate $\longrightarrow$
glucose 1-phosphate
Glucose 1-phosphate + UTP $\longrightarrow$ UDP-glucose
UDP-glucose + fructose 6-phosphate $\longrightarrow$
sucrose 6-phosphate $\longrightarrow$ sucrose

17.27 Cysteine residues are often used to bind metals in metalloproteins. The side chain sulfur acts as a metal coordinating group.

17.28 $Fe^{2+} + Cu^{2+} \longrightarrow Fe^{3+} + Cu^{1+}$

17.29 **a.** Oxidation–reduction
b. Phosphoryl group transfer
c. Isomerization
d. Carboxylation

17.30 The yellow carotenoids are present in fall leaves.

17.31 Reactants oxidized: RH, NADPH
Reactants reduced: O_2

Chapter 18

18.2 Bile salts are amphiphilic molecules. Their nonpolar regions interact with nonpolar regions of dietary fats. Their polar regions form hydrogen bonds with water, thus making them somewhat soluble; they form emulsions in water.

18.3 **a.** The reductase catalyzes a very early step in cholesterol synthesis.
b. Would increase fatty acid oxidation and produce more energy as ATP.
c. A high concentration of fatty acid, as indicated by palmitic acid, inhibits acetyl CoA carboxylase, which would eventually increase fatty acid synthesis.

18.4 **a.** Chylomicrons; high concentration of lipids, which have lower density than proteins
b. Low density
c. High density
d. Chylomicrons
e. High density
f. High density

18.5

$$CH_3(CH_2)_7CH{=}CH(CH_2)_7CO$$

18.6 They form the outer boundary of the molecules. They have regions that are polar, which face the outside, and nonpolar regions, which point toward the inside.

18.7 **a.** Acetoacetate + acetyl CoA
b.
$$CH_3CH_2CH_2CS{-}ACP + NADP^+$$
(with O double bond on carbon)
c. Mevalonate + 2 $NADP^+$ + CoASH
d. Cholesterol + linoleate
e. Squalene oxide + $NADP^+$
f. CH_3CH_2CSCoA + acetyl CoA
(with O double bond)
g. Acetyl CoA + oxaloacetate
h. Palmitic acid + ACP-SH
i. Acetoacetyl CoA + CoASH
j. Glycerol + 3 myristic acid

18.8 d, e, b, h, e, c, f, g, a

18.9

β Oxidation	Synthesis
c, e, f,	a, b, d, g

18.10 a, c, d, f, i, j

18.11 The reductase catalyzes the rate-limiting step in cholesterol synthesis. The rate of cholesterol synthesis will be lowered.

18.12 **a.** Hydrolase
b. Ligase
c. Oxidoreductase
d. Lyase
e. Ligase
f. Oxidoreductase

ANSWERS TO STUDY EXERCISES

18.13 a. Mitochondria
   b. Mitochondria
   c. Mitochondria
   d. Cytoplasm
   e. Cytoplasm

18.14 (a) FAD, (b) NAD^+, (c) coenzyme B_{12}, (d) NAD^+, (e) biotin

18.15 See Section 16.4.

18.16 None of the carbons in palmitate is labeled. The labeled carbon is lost as CO_2.

18.17 a. CH_3CH_2CH=$CHCH_2CSCoA$
          $\overset{\|}{O}$

   1. Enoyl-CoA isomerase
   2. Enoyl-CoA hydratase
   3. Hydroxyacyl-CoA dehydrogenase
   4. β-Ketothiolase
   5. One complete β-oxidation cycle
   b.
          $\overset{O}{\overset{\|}{}}$
      CH_3CH=$CHCH$=$CHCSCoA$

   1. 2,4-Dienoyl-CoA reductase
   2. Enoyl-CoA isomerase
   3. Steps 3, 4, and 5 in part a.

18.18

Stage	NADH	FADH$_2$	ATP
CoA activation	0	0	−2
β oxidation (8 cycles)	8 (mitochondria)	8	0
Citric acid cycle (9 cycles)	27 (mitochondria)	9	9
Subtotal	35	17	7
Total = 146			

18.19 See Section 20.4.

18.20 The apolipoproteins must interact with both nonpolar molecules (lipids) and with water.

18.21 See Section 18.6 and Table 18.5.

18.22 Glucose → glucose 6-phosphate $\xrightarrow{glycolysis}$ pyruvate → acetyl CoA → malonyl CoA → fatty acids

18.23 NADH may be considered a "temporary" fuel molecule. Its energy is used to make ATP during electron transport. If NADH concentrations are sufficient then there is no need to make more by β oxidation.

18.24 Activation of fatty acids by CoASH

18.25 Fatty acids are highly reduced compared to carbohydrates. Since catabolism consists of oxidation, the more steps that a fuel molecule can pass through, the greater the amount of energy released.

18.26 Carnitine assists the transport of fatty acids across the mitochondrial membrane, where they are metabolized. If fewer fatty acids are transported, then more will be stored in their usual form, as triacylglycerols.

18.27 Acetyl CoA is the major product of β oxidation. A surplus of acetyl CoA would slow the process of β oxidation.

18.28
          $^*CH_3$
   $CH_2\overset{*}{C}H_2\overset{*}{C}CH_2COO^-$
   $\overset{|}{OH}\quad\overset{|}{OH}$

18.29 a. Isomerase
   b. Isomerase and 2,4-dienoyl-CoA reductase

Chapter 19

19.2 1. d   **5.** g
     2. e   **6.** f
     3. a   **7.** c
     4. b   **8.** c

19.3 1. Branched-chain α-keto acid dehydrogenase
     2. Homogentisate oxidase
     3. Phenylalanine hydroxylase
     4. Various enzymes, including some in urea cycle
     5. Various enzymes in heme metabolism
     6. Various enzymes in heme metabolism
     7. Unknown
     8. Hypoxanthine-guanine phosphoribosyltransferase

19.4 a. Transamination
        Alanine → pyruvate
   b. Oxidation–reduction
        Branched-chain dehydrogenase
   c. Transfer of one-carbon units
        Glycine → serine
   d. Transfer of one-carbon units
        Methylation of bases in RNA
   e. Activation of acyl groups
        See Figure 19.13.

19.5 b, d, c, f, a, e

19.6 1. a
     2. c
     3. b
     4. a

19.7 a. 1, 2, 3
     b. 1, 2, 4, 5
     c. 1, 2
     d. 1, 2, 4, 6
     e. 1, 2, 4

19.8 a, b, c, e

19.9 See Figure 19.16.

19.10 a. [^{15}N]aspartate → AMP

   b. [^{14}C]glycine (C2) → AMP

19.11 a. Oxidoreductase
b. Transferase
c. Hydrolase
d. Oxidoreductase

19.12 Iron and molybdenum

19.13 Individuals with phenylketonuria must restrict the amino acid phenylalanine from their diets.

19.14 a. $HN = NH$; $H_2N—NH_2$; NH_3
b. $H_2C = CH_2$; $H_3C—CH_3$
c. $H_2C = NH$; $H_3C—NH_2$

19.15 A lack of dietary protein would cause lowered levels of the amino acids tyrosine and phenylalanine. These amino acids are the precursors for the skin pigment melanin.

19.16 a.
$$
\begin{array}{c}
COO^- \\
| \\
C=O \\
| \\
CH_2 \\
| \\
COO^-
\end{array}
$$

c.
$$
\begin{array}{c}
COO^- \\
| \\
C=O \\
| \\
CH \\
H_3C \quad CH_3
\end{array}
$$

e.
$$
\begin{array}{c}
COO^- \\
| \\
C=O \\
| \\
CH_2 \\
| \\
CH_2 \\
| \\
CH_2 \\
| \\
NH_2
\end{array}
$$

b.
$$
\begin{array}{c}
COO^- \\
| \\
C=O \\
| \\
CH_3
\end{array}
$$

d.
$$
\begin{array}{c}
COO^- \\
| \\
C=O \\
| \\
CH_2 \\
| \\
\phi
\end{array}
$$

19.17 a. Alanine
b. Valine
c. Lysine

19.18 a. Tryptophan
b. Phenylalanine or tyrosine
c. Glutamate
d. Arginine
e. Proline

19.19 a. Dehydrogenation: decarboxylation; formation of CoA ester
b. Transamination; decarboxylation
c. Two hydroxylations; decarboxylation

19.20 Alanine $\longrightarrow$ pyruvate $\longrightarrow$
acetyl CoA $\longrightarrow$ citric acid cycle $\longrightarrow$
electron transport and oxidative phosphorylation

19.21 See Figure 19.16, the urea cycle.
Flow of nitrogen: alanine $\longrightarrow$ glutamate $\longrightarrow NH_4^+$

19.22 a. α-Ketoglutarate
b. Oxaloacetate
c. Oxaloacetate
d. α-Ketoglutarate

19.23 a, b, e

19.24 It depletes cellular concentrations of α-ketoglutarate.

19.25 a, b, c, e, g, h, i

19.26 Humans have no metabolic pathway for synthesis of Phe. Tyrosine is synthesized from dietary Phe by hydroxylation.

19.27 The physiological form of ammonia is the ammonium ion, NH_4^+. Since this species is charged, it is unable to diffuse through cell membranes.

19.28 The purines and pyrimidines do not yield ATP during degradation. Most are converted to uric acid or are recycled for use as nucleotides. The purines and pyrimidines are very expensive to make (in terms of energy) and it would not be economical to use them as fuel molecules.

19.29 Birds do not excrete urea, so they do not have the enzymes of the urea cycle. Birds excrete excess nitrogen as uric acid.

19.30 See Window on Biochemistry 19.2.

Chapter 20

20.2 a. Glucose + 2 P_i + 2 ADP + 2 NAD^+ →
2 pyruvate + 2 ATP + 2 NADH + 2 H^+ + 2 H_2O
b. Glucose 1-phosphate + 3 ADP + 2 P_i + 2 NAD^+ →
2 pyruvate + 3 ATP + 2 NADH + 2 H^+ + 2 H_2O
c. Glyceraldehyde 3-phosphate + P_i + 2 ADP + NAD^+ →
pyruvate + 2 ATP + NADH + H^+ + H_2O
d. Glucose + 2 ADP + 2 P_i → 2 lactate + 2 ATP + 2 H_2O

20.3 a. 2 pyruvate + 4 ATP + 2 GTP + 2 NADH + 2 H^+ +
6 H_2O → glucose + 2 NAD^+ + 4 ADP + 2 GDP + 6 P_i
b. 2 PEP + 2 ATP + 2 NADH + 2 H^+ + 6 H_2O →
glucose + 2 ADP + 4 P_i + 2 NAD^+
c. 2 lactate + 4 ATP + 2 GTP + 6 H_2O →
glucose + 4 ADP + 2 GDP + 6 P_i

20.4 Palmitic acid + ATP + 7 FAD + 7 NAD^+ + 8 CoASH +
8 H_2O → 8 acetyl CoA + AMP + 2 P_i + 7 $FADH_2$ +
7 NADH + 7 H^+

20.5 a. Oxaloacetate
b. Pyruvate
c. 2-Keto 3-methylbutanoic acid

20.6 a. Aspartate
b. Alanine

20.7 a. Acetoacetate, acetone
b. Hydroxybutyrate, acetoacetate
c. Hydroxybutyrate

20.8 a. Insulin
b. Epinephrine
c. Glucose 6-phosphate
d. ATP

20.9 a. Acetyl CoA **b.** Insulin
c. Glucagon **d.** Glucagon
e. Epinephrine

20.10 See answer to Problem 15.11.

20.11 d, e, b, h, e, c, f, g, a

20.12 a. 22
b. 29

20.13 a. 3107 Calories
b. 4780 Calories

20.14 c, d

20.15
a. Enhances cell permeability to glucose; stimulates glycogen synthesis.
b. Increases cyclic AMP in liver; decreases glycogen synthesis.
c. Increases cyclic AMP in muscle; decreases glycogen synthesis.

20.16 b, c, d

20.17 Type I: low blood supply; low concentrations of myoglobin; low activities of enzymes for aerobic metabolism.
Type II: extensive blood supply; high levels of mitochondria; high levels of enzymes for aerobic metabolism.

20.18 b

20.19 See answer to Problem 14.23.

20.20 Lactate → pyruvate → PEP → glyceraldehyde 3-phosphate → glucose 6-phosphate

20.21 See Section 20.2 and Window on Biochemistry 20.1.

20.22
a. A, B   b. C, D
c. D   d. A
e. A   f. D
g. A   h. A
i. B, C   j. A

20.23 a, d, f

20.24 Acetoacetate → acetoacetyl CoA → 2 acetyl CoA → citric acid cycle

20.25
a. Cytoplasm
b. Cytoplasm
c. Cytoplasm
d. Mitochondria
e. Cytoplasm

20.26 $2\ ATP + 2\ GTP + 4\ H_2O \longrightarrow 2\ ADP + 2\ GDP + 4\ P_i$

20.27 Dietary glucose and sucrose immediately are absorbed into the blood, raising "blood sugar" rapidly. An increase in blood glucose will cause the pancreas to release insulin. Insulin slows energy mobilization by increasing glycogen and triacylglycerol synthesis. Starch is more slowly degraded and slowly releases glucose into blood thus lessening insulin secretion.

20.28 Untreated diabetics are not able to utilize glucose so ketone bodies are a major source of energy.

20.29 About 50 days.

20.30 a, d

20.31 The net reaction for the Cori cycle (see Problem 20.26) shows the hydrolysis of four phosphoanhydride bonds, at first indicating wasteful hydrolysis. However, it must be realized that the two parts of the cycle occur in different types of cells. The glucose that is synthesized in liver is transported to skeletal muscle where it is degraded with production of abundant ATP.

20.32 $\Delta G^{\circ\prime} = -12.5$ kJ/mol

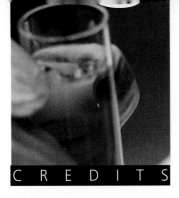

This page constitutes an extension of the copyright page. We have made every effort to trace the ownership of all copyrighted material and to secure permission from copyright holders. In the event of any question arising as to the use of any material, we will be pleased to make the necessary corrections in future printings. Thanks are due to the following authors, publishers, and agents for permission to use the material indicated.

Photographs

Chapter Part 1. 1: © Ryan McVay /PhotoDisc. **3:** ©PhotoLink /PhotoDisc. **5:** © Bruce Coleman /Bruce Coleman Limited **6:** ©NASA **7:** © George Holton/Photo Researchers **7:** Stamp courtesy of Prof. C. M. Lang. Photography by Gary J. Shaffer, University of Wisconsin–Stevens Point **9:** © A. Barrington Brown/Science Source/Photo Researchers **13:** Figure 1.7 a & b © 1997 PhotoDisc **14:** Figure 1.7 a & b © 1997 PhotoDisc **14:** Figure 1.7 c & d © 1997 PhotoDisc **14:** Figure 1.7 c & d © 1997 PhotoDisc **15:** © Jim Sternberg/Photo Researchers **19:** Figure 1.11a ©Biophoto Associates/Photo Researchers **20:** Figure 1.11b ©Alfred Pasieka/SPL/Photo Researchers. **25:** Figure 1.13 © M. Schliwa/Visuals Unlimited. **Chapter 2. 29:** © Chad Baker/PhotoDisc **30:** © Gary Meszaros/ Visuals Unlimited **31:** Stamp courtesy of Professor C. M. Lang. Photography by Gary J. Shaffer, University of Wisconsin–Stevens Point **34:** ©James Chatters/ Applied Paleoscience **39:** © Stanley Fledger/Visuals Unlimited **43:** © Jean-Loup Charmet/SPL/Photo Researchers. **Chapter 3. 47:** © Lawrence Lawry /PhotoDisc **48:** © Douglas Faulkner/Photo Researchers **49:** Stamp courtesy of Professor C. M. Lang. Photography by Gary J. Shaffer, University of Wisconsin–Stevens Point. **Chapter 4. 69:** © Ryan McVay/ PhotoDisc **76:** Stamp courtesy of Professor C. M. Lang. Photography by Gary J. Shaffer, University of Wisconsin–Stevens Point **82:** © Rich Treptow/ Visuals Unlimited **95:** © Chad Baker /PhotoDisc. **Chapter 5. 97:** © Ryan McVay /PhotoDisc. **98:** Stamp courtesy of Professor C. M. Lang. Photography by Gary J. Shaffer, University of Wisconsin–Stevens Point. **103:** © John Gerlach/Visuals Unlimited **106:** Figure 5.11b © Tripos, Inc. **107:** Figure 5.12 © SPL/Photo Researchers. **111:** Courtesy of Alexander McPherson. **116:** © Alfred Pasieka/SPL/Photo Researchers. **119:** © Meckes/Ottawa/ Photo Researchers **120:** © D. Cavagnaro/Visuals Unlimited **Chapter 6. 125:** © Don Farrall/PhotoDisc. **126:** Courtesy of Dr. Alexander McPherson. **126:** Stamp courtesy of Professor C. M. Lang. Photography by Gary J. Shaffer, University of Wisconsin–Stevens Point **127:** Figure 6.1 courtesy of Rod Boyer, photographer Dr. Harvey Blankespoor, Hope College. **136:** © American Chemical Society **148:** © Mark Tuschman. **Chapter 7. 155:** © Jim Wehtje/PhotoDisc. **156:** Stamp courtesy of Professor C. M. Lang. Photography by Gary J. Shaffer, University of Wisconsin–Stevens Point. **Chapter 8. 179:** © Don Farrall /PhotoDisc. **180:** Stamp courtesy of Professor C. M. Lang. Photography by Gary J. Shaffer, University of Wisconsin– Stevens Point. **190:** © Leonard Lessin/Peter Arnold, Inc. **193:** Figure 8.18a & b © CNRI/SPL/Photo Researchers **193:** Figure 8.18a & b © CNRI/SPL/ Photo Researchers. **197:** Figure 8.23 © Bill Beatty/Visuals Unlimited. **202:** Courtesy of Dr. Alexander McPherson. **Chapter 9. 207:** © Jason Reed/ PhotoDisc **208:** © Tripos **208:** Figure 9.1 (a &b) © Tripos. **212:** © Carl Purcell/Photo Researchers. **213:** © William J. Webber/ Visuals Unlimited. **215:** Figure 9.5 © P. Motta/Dept. of Anatomy/Univ. La Sapienza Rome/ SPL/Photo Researchers. **220:** Stamp courtesy of Professor C. M. Lang. Photography by Gary J. Shaffer, University of Wisconsin– Stevens Point. **233:** © Fred Hossler/Visuals Unlimited. **235:** © Victor Englebert/Photo Researchers **241:** ©Chad Baker /PhotoDisc **Chapter 10. 243:** © Nick Koudis /PhotoDisc. **244:** © Millard M. Sharp/Photo Researchers. **252:** Stamp courtesy of Professor C. M. Lang. Photography by Gary J. Shaffer, University of Wisconsin–Stevens Point. **254:** Figure 10.10 (left, right, and middle) © Tripos, Inc. **262:** © Alphacell Corp **263:** © Tom McHugh/Photo Researchers **266:** Figure 10.22b © Andrew Leonard/CDC/

Photo Researchers. **Chapter 11. 271:** © Randy Allbritton/PhotoDisc. **272:** Stamp courtesy of Professor C. M. Lang. Photography by Gary J. Shaffer, University of Wisconsin–Stevens Point. **275:** © J. Cairns/Cold Spring Harbor Laboratory Press. **293:** © Matt Meadows/Peter Arnold, Inc. **294:** © Hybridon, Inc. **Chapter 12. 305:** © Ryan McVay/PhotoDisc **306:** Stamp courtesy of Professor C. M. Lang. Photography by Gary J. Shaffer, University of Wisconsin–Stevens Point **317:** Reprinted with permission from ©*Science Magazine* (AAAS), Aug. 11, 2000, Volume 289, p. 923, Figure 4. American Association for the Advancement of Science. **318:** © Cabisco/ Visuals Unlimited **319:** Figure 12.9 b ©Secchi-Lecaque-Roussel-UCLAF/ CRNI/SPL/Photo Researchers, Inc. **322:** © Kjell B. Sandved/Photo Researchers. **330:** © Tom McHugh/Photo Researchers. **Chapter 13. 337:** © Lawrence Lawry /PhotoDisc. **338:** Stamp courtesy of Professor C. M. Lang. Photography by Gary J. Shaffer, University of Wisconsin–Stevens Point. **349:** © Photo courtesy of Incyte Genomics, Palo Alto, CA. **353:** © Steven Kagan/Gamma Liaison. **354:** © SIU/Visuals Unlimited. **357:** Figure 13.12 courtesy of Agricultural Research Service, USDA. **358:** Figure 13.13a © R. Benali–S. Ferry for *Life* Magazine/Gamma Liaison. **358:** Figure 13.13b © Dr. Ron Evans/The Salk Institute **363:** © Cole Group/PhotoDisc. **Chapter 14. 365:** © Ryan McVay/PhotoDisc **366:** Stamp courtesy of Professor C. M. Lang. Photography by Gary J. Shaffer, University of Wisconsin–Stevens Point. **367:** © Martin Bond/SPL/Photo Researchers **379:** © Joseph Fontemot/Visuals Unlimited **Chapter 15. 393:** © Lawrence Lawry/PhotoDisc. **395:** © Mihojac Studios/Gamma Liaison. **401:** © Steve Kaufman/Peter Arnold, Inc. **405:** © Dennis MacDonald/Photo Edit **416:** Stamp courtesy of Professor C. M. Lang. Photography by Gary J. Shaffer, University of Wisconsin–Stevens Point. **Chapter 16. 423:** ©Harnett/Hanzon /PhotoDisc **425:** Stamp courtesy of Professor C. M. Lang. Photography by Gary J. Shaffer, University of Wisconsin–Stevens Point. **433:** © Richard Megna/Fundamental Photographs. **437:** © Farrell Grehan/Photo Researchers **443:** © Bill Banazewski/Visuals Unlimited **Chapter 17. 449:** © Lawrence Lawry/PhotoDisc. **450:** Stamp courtesy of Professor C. M. Lang. Photography by Gary J. Shaffer, University of Wisconsin–Stevens Point. **463:** © David J. Cross/Peter Arnold, Inc. **468:** Figure 17.17a © Biophoto Associates/ Photo Researchers **469:** © Linde Waidhofer/Gamma Liaison **470:** ©1996, Harbor Branch Oceanographic/T. Frank. **472:** © Linde Waidhofer/Gamma Liaison. **481:** Courtesy of Agricultural Research Center, USDA. **Chapter 18. 487:** © Jim Wehtje/PhotoDisc. **488:** © Robert Vanmarter/Gamma Liaison **492:** ©Tripos, Inc. using SYBYL **497:** © K. D. McGraw/Rainbow. **507:** Figure 18.19a © SPL/Photo Researchers. **507:** Figure 18.19 b ©SPL/Photo Researchers. **509:** © Warner Lambert Export Limited. **512:** Stamp courtesy of Professor C. M. Lang. Photography by Gary J. Shaffer, University of Wisconsin–Stevens Point. **Chapter 19. 523:** ©Randy Allbritton /PhotoDisc. **524:** © Mathew Klein/Photo Researchers. **526:** Figure 19.2 © C. P. Vance/ Visuals Unlimited. **545:** Stamp courtesy of Professor C. M. Lang. Photography by Gary J. Shaffer, University of Wisconsin–Stevens Point. **548:** © Jane Thomas/Visuals Unlimited. **Chapter 20. 557:** © PhotoLink /PhotoDisc. **566:** Stamp courtesy of Professor C. M. Lang. Photography by Gary J. Shaffer, University of Wisconsin–Stevens Point.

Illustrations and Text

Chapter 1. 18: Reprinted from David S. Goodsell, "Inside a Living Cell," *Trends in Biochemical Sciences,* vol. 16, p. 203–206 (1991), with permission from Elsevier Science. **Chapter 5. 99:** © Irving Geis. Geis Archives Trust. Rights owned by Howard Hughes Medical Institute. Not to be used without permission. **100:** Figure 5.3 adapted from *Biochemistry,* 3rd Edition by Mary K. Campbell, copyright © 1999 by Saunders College Publishing, reproduced by permission of the publisher. **101:** A. L. Lehninger, D. L. Nelson, and M. M. Cox, *Principles of Biochemistry,* 2nd Edition (1993), Worth Publishers. **105:** From Garrett and Grisham, *Biochemistry* (1995), Saunders

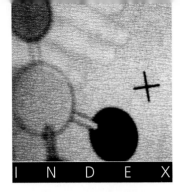

I N D E X

Page references in **boldface** refer to text passages in which a compound or concept is defined or introduced; page references followed by an "f" indicate figures and page references followed by a "t" indicate tables.